Engine Combustion Instrumentation and Diagnostics

Other SAE books on this topic

Advanced Engine Technology
By Heinz Heisler
(Order No. R-163)

Introduction to Internal Combustion Engines
By Richard Stone
(Order No. R-278)

ENGINE COMBUSTION INSTRUMENTATION AND DIAGNOSTICS

Hua Zhao
Nicos Ladommatos

SAE INTERNATIONAL®

Society of Automotive Engineers, Inc.
Warrendale, Pa.

eISBN: 978-0-7680-4034-0

Library of Congress Cataloging-in-Publication Data

Zhao, Hua, 1963-
Engine combustion instrumentation and diagnostics / Hua Zhao,
Nicos Ladommatos.
p. cm.
Includes bibliographical references and index.
ISBN 0-7680-0665-1
1. Automobiles—Motors—Combustion. I. Ladommatos, Nicos. II. Title.

TL210 .Z46 2001
621.43'028'7—dc21 00-063512

Society of Automotive Engineers, Inc.
400 Commonwealth Drive
Warrendale, PA 15096-0001 U.S.A.
Phone: (724)776-4841
Fax: (724)776-5760
E-mail: publications@sae.org
http://www.sae.org

ISBN 0-7680-0665-1

SAE Order No. R-264

Contents

Acknowledgements

H. Zhao's first experience with laser diagnostics in combustion flows began fifteen years ago, when he was granted a scholarship by the Ministry of Education of China and the British Council, for which he is much obliged.

N. Ladommatos's experience with internal combustion engine research and diagnostics extends over fifteen years, following his arrival at Brunel University. Brunel has proven to be a fertile ground for such work, and he is indebted to the University for all the support he has received over the years.

During the last ten years, various organizations have provided support to our research, particularly the Engineering and Physical Sciences Research Council (EPSRC), Ford Motor Company, and the European Community. Special thanks are due to Dr. Tom Ma and Roy Horrocks of Ford Motor Company for their technical support and sponsorship.

In writing this book, we are most grateful to Dr. John E. Dec of Sandia National Laboratories for supplying many original photographs from his extensive laser measurements inside a DI diesel engine. Further, we would like to express our thanks to Dr. Peter O. Witze and Dr. Paul Miles of Sandia National Laboratories for their useful comments on the manuscript and direction to relevant literature.

Finally, we would like to thank our wives, Liz and Katrina, for their encouragement and support during our undertaking of this time-consuming but enjoyable task. Thanks are also due to the numerous graduate students and research assistants, from whose work we have learned so much.

Hua Zhao
Nicos Ladommatos

Preface

Since the birth of internal combustion (IC) engines more than 100 years ago, engine technology has made tremendous progress in terms of efficiency and reliability. In the last two decades, the environmental implications of engine exhaust emissions have resulted in greater effort in reducing harmful pollutants to meet ever-stringent legislation. Increased knowledge and understanding of the combustion and pollutant formation processes in IC engines are clearly necessary to increase their efficiency and cleanliness, and to evolve and invent new engines with increased efficiency and lower emissions. The traditional engineering approach has been mostly to treat the engine combustion chamber as a black box and make measurements in the inlet and the exhaust systems. As the processes of improvement in efficiency and emission abatement have reached points of diminishing returns, there is an increasing need for measurements directly inside the combustion chamber, where the combustion and pollutant formation processes actually take place. Therefore, this book is particularly concerned with diagnostic techniques for in-cylinder measurements. The equipment used for exhaust gas measurements are only briefly discussed since these are well established, and they are readily available commercially.

Owing to the wide range of techniques and, particularly, the rapid development of modern diagnostics for engine combustion research in the last decade, it is difficult, even for a specialist, to acquire an adequate overview of the achievements in this field. A newcomer to engines research must wade through a number of research monographs and many widely scattered journal or conference papers in order to assimilate the current state of the art and the tools available for research. This is obviously inconvenient for both graduate students or research and development engineers who wish to pursue engines research. While supervising research students and staff in their laboratories and in working with research and development engineers in industry, we found that, in the literature, there is a lack of a systematic approach to engine instrumentation and in-cylinder measurements. The purpose of this book, therefore, is to provide, between one set of covers, a rather complete description of instrumentation and in-cylinder measurement techniques for IC engines. This book is written primarily for researchers and engineers involved in advanced research and development of internal combustion engines, as well as for postgraduate students. It provides an introduction to the instrumentation and experimental techniques that are suitable for in-cylinder measurements and, at the same time, includes sufficient details for the readers to be able to set up and apply these techniques to IC engines. Additional emphasis here is on the applications of these techniques to improve the understanding of combustion and pollutant formation processes in IC engines.

In-cylinder engine flow and combustion diagnostics may be divided into two groups: those that use physical probes, and those that sense the properties remotely. The latter is often referred to as optical diagnostics. Engine diagnostics can also be classified as classical or modern diagnostics.

"Classical" implies a diagnostic technique that does not involve the use of modern inventions, such as the laser. Some optical approaches, e.g., direct photography, absorption and emission spectroscopy, Schlieren/shadowgraphy, are considered classical diagnostics, although lasers have been introduced as more powerful sources of illumination in some of these techniques. There can be no doubt that modern techniques, such as laser diagnostics, have increased our knowledge and understanding of engine flow and combustion. The fact that classical diagnostic tools are old tools does not imply that they are obsolete. Most engine research conducted today still involves classical diagnostics, despite the increasing use of modern diagnostics, particularly laser techniques. Every tool has its difficulties and limitations. It is important to recognize the complementary nature of the classical and modern tools in the experimental studies. The same argument applies to the techniques involving physical probes.

The spectrum of diagnostic techniques included in the book covers both classical and modern physical probes and optical techniques. The thread that ties both classical and modern techniques, or physical probes and optical techniques, is that they all, in some way, provide information on the state of the cylinder charge at a given time during the engine cycle. The state of the mixture in the cylinder is determined by thermodynamic, chemical, and fluid dynamic considerations. In general, the thermodynamic state at a given spatial location is determined by any two of the following three properties: pressure, temperature, or density. The chemical state is mainly defined by knowledge of the concentrations of the chemical species in the mixture. The nature of the flow is characterized by both large and small-scale structures. In this book, the diagnostic techniques will be tied to the characterization of these properties, including the thermodynamic, chemical, and fluid dynamic states of the in-cylinder mixture before, during, and after combustion. This approach ensures that the development of diagnostic techniques is linked to an identified need, as it is the end that justifies the means but not vice versa. This approach is particularly useful for researchers or engineers who are interested in the diagnostic tools that are capable of solving a specific research problem, irrespective of the nature of each technique, whether it is classical or modern. At the same time, by presenting different techniques for the same diagnostic purpose, the limitations of each technique can be understood better for intelligent application and successful outcome.

Although the applications of laser diagnostics, especially laser spectroscopic approaches, to IC engines are relatively new, and some of the techniques are undergoing continuous improvement, laser diagnostics have received much research attention and been demonstrated in flames and some practical combustion devices in the last couple of decades. The majority of these techniques have matured to a point where a text including this subject area will not be out of date in the near future.

The book begins with an introduction to general engine instrumentation, including engine dynamometers and engine controls, exhaust gas measurements, measurements of air flow and fueling. Because most in-cylinder measurements are carried out using optical techniques, optical access to the cylinder is discussed and examples of common optical engine setups are given. Chapter 2 considers pressure measurements and their use in engine performance and heat release analysis. Chapter 3 presents detailed analyses of the fast flame ionization detector (FID) and its performance. Chapter 4 discusses the design principles and construction of high-speed intermittent sampling probes and the operation of the total cylinder dumping technique. Chapter 5 introduces the fundamentals of spectroscopy and the principles of major laser spectroscopic techniques, and discusses some general experimental considerations in the application of laser diagnostics. Techniques for in-cylinder flow measurements are described in detail in Chapter 6, including laser Doppler anemometry (LDA) and particle imaging velocimetry (PIV). Chapter 7 considers the measurements of in-cylinder fuel distribution in spark ignition (SI) engines, and this is followed by the measurements of liquid fuel in both diesel and SI engines in Chapter 8. Chapter 9 describes the visualization techniques for in-cylinder combustion. Chapters 10 discusses the two-color method for in-cylinder diesel combustion temperature and soot loading measurements, and Chapter 11 presents in-cylinder soot measurement techniques. In-cylinder gas temperature measurements are discussed in Chapter 12. Chapter 13 concludes with a description of approaches that permit measurements of in-cylinder combustion species.

Because the book is written by authors who have combined knowledge and experience in both IC engines research and combustion diagnostics, it bridges the gap between researchers in engine development and specialists in the development of these techniques. The authors' complementary skills in classical and modern diagnostics make it possible for the first time to bring together all the available tools to the engines research community, so that the appropriate approaches, irrespective of their nature, can be adopted for a specific research need. Another feature of the book is that each chapter is individually referenced and references are called out in the text by authors and year. References are in long form with titles, which is a particularly useful feature for readers to pre-screen the nature of the research work in each reference.

Hua Zhao
Nicos Ladommatos
September 1999

Nomenclature

1. Symbols in Roman

a	size parameter, $\pi d/\lambda$
A	area, cm^2 Einstein transition probability for spontaneous emission, sec^{-1}
A_o	Avogadro's number, 6.023×10^{23} molecules/mole
$b_{J,J'}$	Placzek-Teller coefficient
B	rotational constant Einstein transition probability for stimulated emission, $m^3/J\ sec^2$
B_v	rotational constant corrected for vibration
c	speed of light in vacuum, 3×10^8 m/s
c_p	specific heat at constant pressure, kJ/kg K
c_v	specific heat at constant volume, kJ/kg K
C	concentration, cm^{-3}
D	centrifugal distortion constant diameter
D_v	centrifugal distortion constant corrected for vibration
E	electric field, V/m
E_i	internal energy of a molecule at state i, cm^{-1}
E_r	rotational energy, cm^{-1}
E_v	vibrational energy, cm^{-1}
f	focal length, meter
g	degeneracy
h	Planck's constant, 6.63×10^{-34} J/s specific enthalpy, kJ/kg K
h_c	heat transfer coefficient, $kJ/cm^2\ K$

I_i	incident laser intensity, W/cm^2
I_v	laser spectral intensity, W/cm^2 Hz
J	rotational number
k	Boltzmann's constant, 1.38×10^{-23} J/K
K_{ext}	extinction coefficient, cm^{-1}
K_{abs}	absorption coefficient, cm^{-1}
L	length, cm
m	refractive index mass, kg
M	magnification
n	polytropic exponent real part of the refractive index
N	total number density, cm^{-3} engine speed, rev/sec
p	pressure, kN/cm^2
P	scattered light signal, Watts
$P_\perp$	vertically polarized scattered light signal, Watts
$P_{\parallel}$	parallel polarized scattered light signal, Watts
Q	quenching rate constant heat release heat transfer
Q_{abs}	absorption efficiency
Q_{ext}	extinction efficiency
Q_{rot}	rotational partition function
Q_v	vibrational partition function
R	specific gas constant, kJ/kg K
R_{tot}	responsivity of a detection system
S	signal
t	time, s
T	temperature, K

u	specific internal energy, kJ/kg velocity component, m/s
U	internal energy, kJ
v	vibrational number
V	volume, cm^3
V_c	collection volume, cm^3
w	$1/e^2$ beam radius

2. Symbols in Greek

α	polarizability of a molecule soot absorption parameter sound absorption coefficient
α'	mean invariant of the derived polarizability tensor
α_e	vibrational-rotation interaction constant
α_o	equilibrium polarizability
β	rate of change of polarizability
β_e	vibrational coefficient correction to centrifugal distortion constant
χ_i	mole fraction of species *i*
ε	emissivity
ϕ	angle fuel/air equivalence ratio
γ	ratio of the specific heats
γ'	anisotropy invariant of the derived polarizability tensor
γ_o	anisotropy of the equilibrium polarizability tensor
η	calibration factor
η_c	detection efficiency
λ	relative air/fuel ratio
μ	induced dipole
θ	angle
ρ	density, kg/m^3

π	pi, 3.1415926
σ	scattering cross section, cm^2
σ_i	scattering cross section of species i, cm^2
$\partial\sigma/\partial\Omega$	differential cross section, cm^2/sr
ν	frequency, sec^{-1}
$\tilde{\nu}$	wavenumber, cm^{-1}
ω_e	equilibrium vibrational frequency, cm^{-1}
Ω	solid angle, sr

3. Abbreviations

AC	alternating current
ADC	analog to digital converter
A/F	air/fuel ratio
AM	amplitude modulation
ATDC	after top dead center
BDC	bottom dead center
BTDC	before top dead center
CA	crank angle
CARS	coherent anti-Stokes Raman scattering
CCD	charge coupled device
CI	compression ignition
COV	coefficient of variation
CW	continuous wave
DC	direct current
DI	direct injection
EGR	exhaust gas recirculation
ELPI	electrical low-pressure impactor
FARLIF	fuel/air ratio laser induced fluorescence

FFT	fast Fourier transform
FID	flame ionization detector
FM	frequency modulation
FTIR	Fourier transform infrared
He-Ne	helium-neon
IC	internal combustion
ICCD	intensified charge coupled device
IDI	indirect injection
imep	indicated mean effective pressure, bar
LDA	laser Doppler anemometry
LDV	laser Doppler velocimetry
LEA	laser extinction and absorption
LIEF	laser induced exciplex fluorescence
LIF	laser induced fluorescence
LII	laser induced incandescence
LIS	laser induced scattering
LRS	laser Rayleigh scattering
MBT	minimum advance for best torque, CA
MCP	micro channel plate
Nd:YAG	neodymium YAG laser
NO_x	nitric oxides ($NO + NO_2$)
PDA	phase Doppler analyzer
PIV	particle image velocimetry
PLIF	planar laser induced fluorescence
SI	spark ignition
SMD	Sauter mean diameter
SMPS	scanning mobility particle sizer
SRS	spontaneous Raman scattering

TDC	top dead center
TLAF	two line atomic fluorescence
TLMF	two line molecular fluorescence
uHC	unburned hydrocarbon
UV	ultraviolet

Chapter 1

General Engine Instrumentation

Outline

1.1 Introduction

1.2 Dynamometers and Engine Control

- 1.2.1 Introduction
- 1.2.2 Electrical Dynamometers
- 1.2.3 Water Dynamometers
- 1.2.4 Dynamometer Operating Envelope
- 1.2.5 Dynamometer Operating Modes

1.3 Air Flow Rate Measurement

- 1.3.1 Air Box with Sharp-Edged Orifice Plate
- 1.3.2 Viscous-Flow Meter
- 1.3.3 Positive Displacement Flow Meter
- 1.3.4 Corona Discharge Flow Meter
- 1.3.5 Hot Wire Flow Meter
- 1.3.6 Air Flow Measurement from Exhaust Gas Analysis

1.4 Fuel Consumption Measurement

- 1.4.1 Average Fuel Consumption Rate
- 1.4.2 Instantaneous Fuel Consumption Rate
- 1.4.3 Fuel Spill-Back

1.5 Acquisition of Instantaneous Engine Variables

- 1.5.1 Transducers and Signal Conditioning
- 1.5.2 Multiplexing
- 1.5.3 Analog to Digital Converter
- 1.5.4 Sample-and-Hold
- 1.5.5 Clock Signal
- 1.5.6 Cycle Reference Signal
- 1.5.7 Sampling Rate
- 1.5.8 Shaft-Encoder

1.6 Diesel Fuel Line Pressure and Needle Lift

- 1.6.1 Fuel Pipe Pressure
- 1.6.2 Injector Needle Lift

1.1 Introduction

To study the performance and emission of an engine, one must control the speed of the engine and the load applied to the engine. This is realized by means of a dynamometer, as shown in Fig. 1.1. One also must instrument the engine to measure such parameters as fuel flow rate, air flow rate, various temperatures, and the spark or fuel injection timing. In addition, exhaust gas measurements are an essential part of any engine combustion study. In order to understand the complex flow and combustion processes in the cylinder, optical access is required for in-cylinder optical measurements.

This chapter begins with an introduction to general engine instrumentation required to perform engine combustion studies. This includes: dynamometers, air and fuel flow measurements, measurements of high-pressure fuel line pressure and fuel injection timing, and digital data acquisition systems. This is followed by a description of instruments used in exhaust gas analysis. Finally, provision of optical access to combustion chamber is discussed.

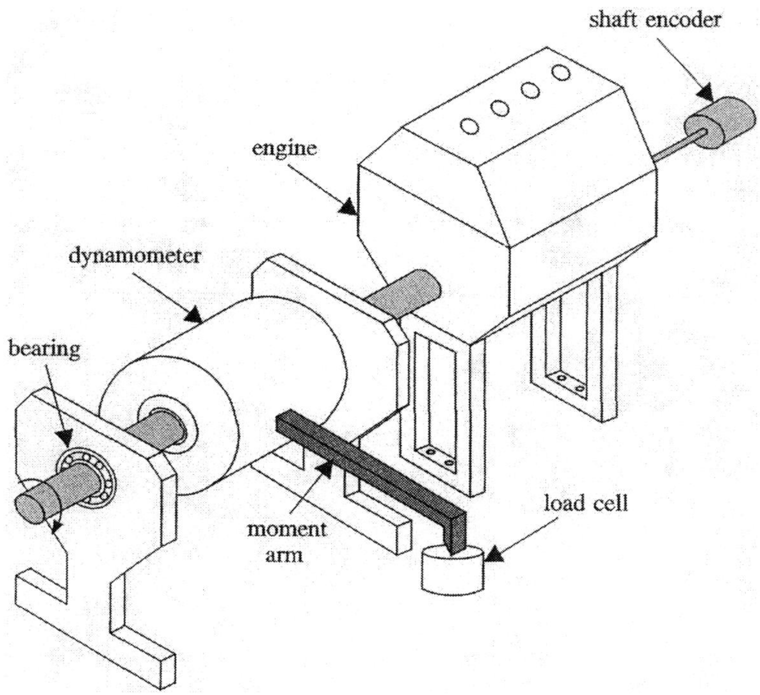

Fig. 1.1 Schematic diagram of dynamometer installation.

1.2 Dynamometers and Engine Control

1.2.1 Introduction

The main purpose of an engine dynamometer or brake is to resist the rotation of the engine shaft. The amount of resistance offered by the brake to the shaft rotation is adjustable. In steady-state testing, the use of an engine dynamometer permits engine tests to be carried out in one of three modes: constant engine speed, constant torque, and constant throttle or fueling.

Early brakes relied on resisting the rotation of the engine shaft through friction. The concept of a friction brake is illustrated schematically in Fig. 1.2. The rotation of the engine shaft is resisted by the brake ring through friction. Tightening the nut shown in the figure increases the resistance offered to the engine shaft.

The brake ring is stopped from rotating by the moment arm. The force required to stop the moment arm rotating is measured by means of a spring balance. When the force on the moment arm is measured in this way, it is usual to incorporate a dashpot (damper) in the arm as shown in

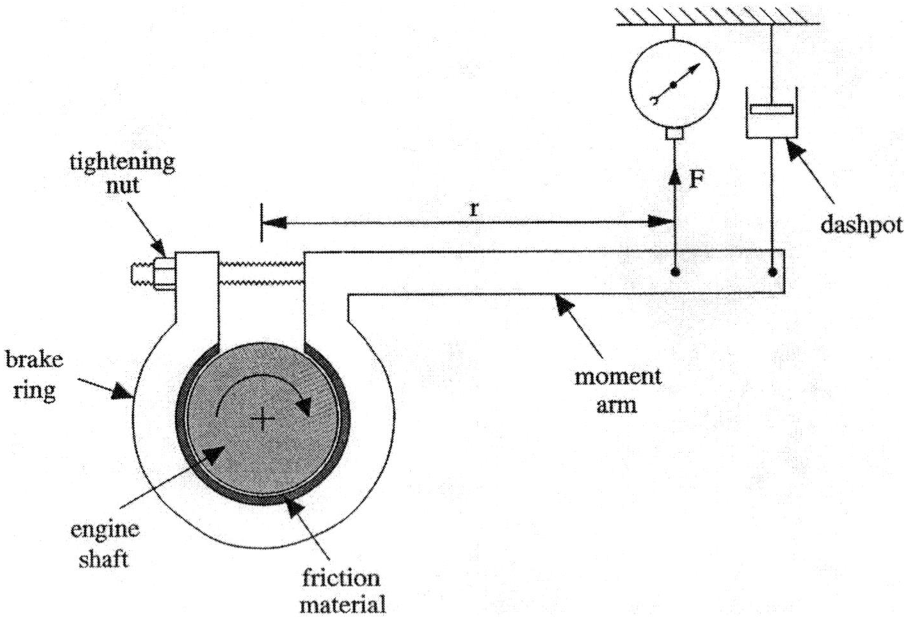

Fig. 1.2 Conceptual arrangement of an engine brake.

Fig. 1.2, in order to damp out oscillations [Stone, 1992]. The torque that the engine must develop in order to overcome the brake resistance can be calculated:

$$T = Fr \tag{1.1}$$

where T = torque
r = radius of moment arm
F = restraining force required to stop the moment arm rotating

The power developed by the engine and absorbed by the dynamometer can be calculated from:

$$P = T\omega \tag{1.2}$$

where P = power
ω = angular speed of dynamometer shaft
= $2\pi N$
N = engine rotational speed (in rev/s)

Therefore, the dynamometer allows the engine to develop torque by resisting the rotation of the engine crankshaft. For a given engine torque, the engine speed at which this occurs decides the engine power output and the power that has to be absorbed by the dynamometer. Thus, a dynamometer must be capable of allowing the engine to develop torque at different engine speeds.

As discussed above, in Fig. 1.2 the moment arm is coupled to the engine shaft through friction. Evidently, other forms of coupling could be employed. One of the commonest forms is an electromagnetic field. The strength of the field can be varied in order to adjust the resistance offered by the brake to the rotation of the engine shaft.

As shown in Fig. 1.1, the dynamometer shaft is supported in bearings. The casing of the dynamometer is stopped from rotating by means of the arm, and the force required to stop the arm rotating is measured by means of a load cell. Dynamometers that rely on friction to resist the rotation of the engine shaft are nowadays obsolete. Most modern dynamometers are electrical, although many water dynamometers are still in use.

It should be mentioned, in passing, that torque-measuring transducers are also available commercially, which can be incorporated into the transmission shaft coupling the engine to the dynamometer. Their cost is, however, significantly higher than that for a moment arm and load cell system combinations, as shown in Fig. 1.1.

1.2.2 Electrical Dynamometers

In electrical dynamometers, the shaft rotation drives some form of electrical generator. The strength of the electromagnetic field coupling the rotating and stationary parts of the dynamometer can be adjusted in order to increase or decrease the resistance offered to the engine rotation. A great advantage of electrical dynamometers is that the electromagnetic coupling can be altered almost instantly and, in this way, the resistance offered to the engine shaft rotation can be changed very rapidly (within a fraction of a second). This is useful, for example, when a sudden opening in the engine throttle must be resisted by the dynamometer so that no appreciable increase in engine speed occurs.

With electrical dynamometers the engine power is converted to electrical energy which can be transferred away from the dynamometer by cables. Alternatively, the electrical energy can be dissipated within the dynamometer as heat and carried away from the dynamometer by cooling water.

Direct-current (DC) dynamometers comprise a DC generator mounted as shown in Fig. 1.1. The strength of the electromagnetic field within the generator, coupling its rotating and stationary parts, can be adjusted to vary the resistance to the engine shaft rotation. Sometimes, in the case of small DC dynamometers having electrical power output of a few kilowatt, this power is dissipated in air-cooled electrical resistance heaters, thus eliminating the need for cooling water. A great advantage of the DC dynamometer is that the function of the generator can be reversed, and the dynamometer can be used as an electric motor. This allows the engine to be motored for starting and running without combustion. DC dynamometers are capable of very rapid changes in the resisting torque offered to the engine, and are amenable to computer control. However, they are not capable of rapid speed changes because their rotating parts have high inertia.

Alternating-current (AC) dynamometers comprise an AC generator. As in the case of DC dynamometers, the function of the generator can be reversed and the dynamometer can be used as an electric motor to motor the engine. AC dynamometers are also capable of very rapid changes in resisting torque, and are also amenable to computer control. However, fast speed changes are constrained by the inertia of their rotating parts.

Eddy-current dynamometers are found widely in research laboratories. In this type of dynamometer, the electrical power generated is dissipated as heat within the dynamometer itself, and this energy is carried away by cooling water. Eddy-current machines are of relatively simple and robust construction [Plint and Martyr, 1995]. However, unlike DC and AC dynamometers, they cannot be used to motor the engine. Eddy-current dynamometers are readily controllable by a computer, and their resisting torque can be altered almost instantly. Faster speed changes than those for DC and AC machines are possible due to lower rotor inertia.

1.2.3 Water Dynamometers

A significant number of hydro-kinetic or hydraulic dynamometers are still in common use. The rotating and stationary parts in this type of dynamometer are coupled by means of water that continuously transfers momentum from the moving to the stationary parts. Altering the resistance offered to the shaft rotation is rather slow, taking several seconds to accomplish. Also, this type of dynamometer cannot be used to motor the engine.

1.2.4 Dynamometer Operating Envelope

As explained above, the purpose of a dynamometer is to resist the rotation of the engine shaft. The engine operator can test an engine in a number of ways. For example, the operator can select the torque and the speed at which the engine is to be tested. The operator then supplies sufficient fuel to the engine to develop that required engine torque at the required engine speed. The dynamometer must be capable of providing a resistance to the engine shaft rotation commensurate to the engine torque at the required engine speed. Ideally, the dynamometer must be capable of offering a resistance to the engine shaft so as to match whatever combination of torque and speed the engine may be capable of.

As an example, Fig. 1.3 shows the torque/speed operating envelope for an eddy-current dynamometer. The shaded area in the diagram includes all the combinations of engine torque and speed that the dynamometer is capable of resisting. Figure 1.3 shows that the dynamometer envelope has five elements, labeled (i) to (v). These five elements arise due to the following operating characteristics [Plint and Martyr, 1995]:

(i) Torque line at low speeds. Along this line the torque is decided by the maximum eddy currents that can be generated within the dynamometer at low dynamometer speeds.
(ii) Maximum torque line. This line defines the maximum torque that the dynamometer shaft can withstand.
(iii) Maximum power line. This line shows the maximum power that can be dissipated by the dynamometer, which, in turn, is decided by the maximum energy that the dynamometer cooling water can carry away from the dynamometer.
(iv) Maximum speed line, defined by the maximum speed that the dynamometer can withstand.
(v) Residual torque line arising from bearing friction, aerodynamic resistance within the dynamometer, and residual magnetization of some of the dynamometer components.

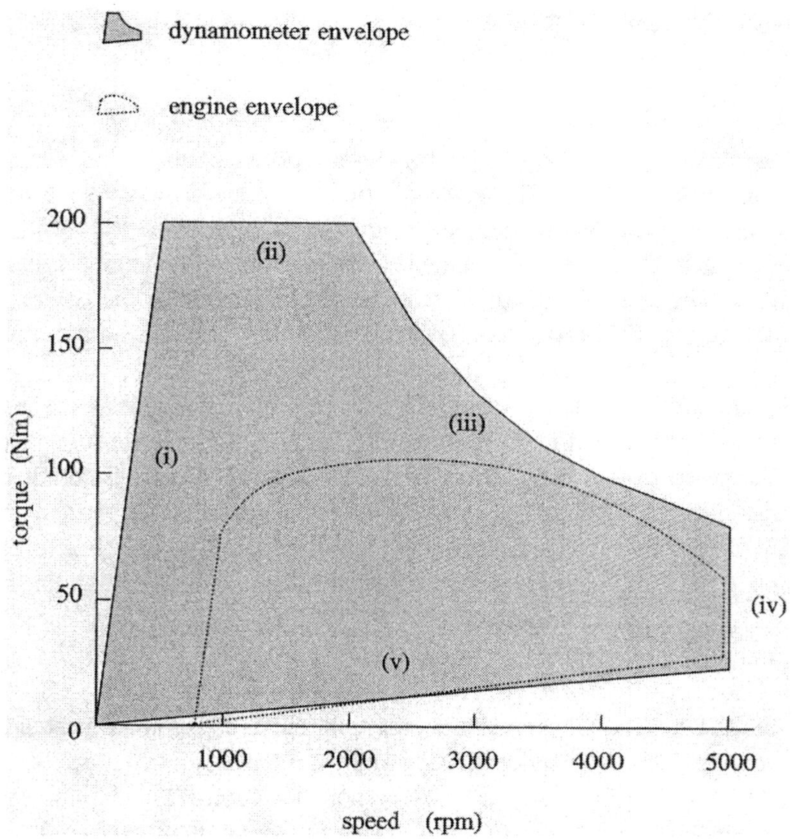

Fig. 1.3 Operating envelope for an eddy-current dynamometer and for an IC engine connected to it.

By way of an example, Fig. 1.3 also shows the operating envelope (broken line) of an internal combustion engine. It can be seen that the engine envelope lies entirely within the dynamometer envelope. Therefore, this internal combustion engine could be tested with this dynamometer throughout all the running conditions that the engine is capable of.

The maximum torque for the engine above was about 100 Nm. Consider now the case of a larger engine of similar design being coupled to this dynamometer. Suppose that this larger engine has a maximum torque of, say, 150 Nm. This engine could not be tested at high torque/high speed combinations, because the power developed by this engine would exceed the power that the dynamometer could absorb at high speeds.

1.2.5 Dynamometer Operating Modes

Most research engines are run under one of three test modes:

- Constant engine speed mode
- Constant torque mode
- Constant throttle opening or fueling mode

Constant Engine Speed Mode

In this mode, the objective is to always keep the engine speed constant while the engine is being tested at different torque output levels. For example, the operator may wish to keep the speed always at 2000 rpm, but change the engine output torque from 50 to 100 Nm by opening the engine throttle (or, in the case of diesel engines, by increasing the engine fueling). What is needed, is a dynamometer that can offer an increased resistance to the engine to match the greater torque output of 100 Nm, so that the engine speed will stay constant at 2000 rpm.

When a water dynamometer is used, it is controlled manually as follows. With the engine operating at 2000 rpm and 50 Nm, the engine operator first increases the engine throttle opening, and this causes the engine speed to rise above 2000 rpm. The dynamometer resistance offered to the engine is then increased manually by the operator, and this lowers the engine speed back to 2000 rpm. The torque output from the engine is noted and, say, it is 85 Nm. The throttle is opened slightly more and the dynamometer resistance is adjusted again to bring the engine speed down to 2000 rpm. The torque is again noted and if not 100 Nm, the procedure is repeated until 100 Nm at 2000 rpm is obtained.

Electric dynamometers are capable of maintaining engine speed constant automatically, as follows. The dynamometer electronic controller is first switched into the "constant speed" mode and the controller speed is "set" at 2000 rpm on the dynamometer control panel. The engine operator then gradually opens the engine throttle until the torque value displayed on the dynamometer control panel reaches 100 Nm. While the operator is opening the engine throttle, the dynamometer controller monitors electronically the dynamometer speed. It automatically increases the resistance offered to the engine in order to stop the engine speed rising above the set value of 2000 rpm.

Constant Torque Mode

In this mode, the objective is to keep the torque developed by the engine always constant, while the engine is being tested at different throttle openings (or, in the case of diesel engines, at different fueling levels). As in the case of the constant speed mode, this control mode can be achieved either manually or automatically.

In the case of electric dynamometers, the engine torque can be kept constant automatically, as follows. The dynamometer electronic controller is first switched into the "constant torque" mode and the controller torque set, for example, at 100 Nm. The engine operator then gradually opens the engine throttle. While the operator is opening the engine throttle, the dynamometer controller electronically monitors the torque resistance offered to the engine. The controller automatically keeps the resistance offered to the engine constant at the set value of 100 Nm. A consequence of keeping the torque resistance offered to the engine constant while the engine throttle opening is increased is that the engine speed rises. This is to be expected, since the increased throttle opening raises the fuel flow rate supplied to the engine. A relevant question is, what happens to the extra fuel, since the engine output torque is kept always constant at 100 Nm? The answer is that the power output from the engine increases. Or, in more detail, the increased fuel supply is reflected, first, in the engine cycle now being repeated more frequently (albeit at the same torque output) and, secondly, in increased mechanical losses within the engine and the dynamometer.

Constant Throttle Opening or Fueling Mode

In this mode, the objective is to keep the engine throttle opening (or, in the case of diesel engines, fueling) always constant while the engine is being tested at different combinations of torque and speed. This test mode is useful for obtaining full-throttle torque/speed curves for an engine (also called maximum power curves). Before the test starts, the engine is operating at it maximum throttle opening and at its lowest stable speed. The test is then conducted, as follows. First, the initial engine speed and torque output are noted. The dynamometer resistance is then reduced slightly, and this causes the engine speed and torque to both rise a little. Both the engine speed and torque are noted again. Again the dynamometer resistance is reduced slightly, and the speed and torque noted. This procedure is repeated, each time reducing the dynamometer resistance slightly until the engine reaches its maximum operating speed. Incidentally, the same procedure can be performed at lower constant throttle settings. Alternatively, a related procedure can be carried out whereby the engine is always operating at maximum power but relatively small changes are made to the fueling, with each speed/torque combination, in order to keep exhaust emissions or cylinder gas pressure below specified limits.

A number of other test modes are sometimes encountered in engine research, which involve various combinations of engine speed, torque, and throttle opening. Some of these are described by Plint and Martyr [1995].

1.3 Air Flow Rate Measurement

It is possible to make an approximate estimate of the air mass flow rate supplied to the engine, from the following expression (no EGR):

$$\dot{m} = \eta_v \rho_i V_s \frac{N}{n_s 60} \text{ kg/s} \tag{1.3}$$

where ρ_i = intake density measured in the inlet manifold (kg/m^3)
= $p_i/(RT_i)$
p_i = absolute pressure in inlet manifold (Pa)
T_i = air temperature in inlet manifold (K)
R = specific gas constant for air (J/kgK)
η_v = volumetric efficiency based on inlet manifold conditions
n_s = constant (n_s = 1 for two-stroke and n_s = 2 for four-stroke engines)
V_s = engine total swept volume (m^3)
N = engine speed (rpm)

More often than not, estimates of the air mass flow rate into an engine are not sufficiently accurate. Therefore, it is common practice in engine research laboratories to measure the air flow rate to the engine. The measured air mass flow rate can then be used, for example, in the calculation of: the engine volumetric efficiency; the air/fuel ratio supplied to the engine; and the amount of exhaust gas recirculation [Ladommatos et al., 1996].

An engine researcher is often interested in one of the following two air flow rate measurements: instantaneous and quasi-steady. The nature of these measurements is explained below.

The air flow rate into the engine cylinder during the induction process is, of course, highly unsteady. The air flow rate changes from zero to its maximum value and back to zero within a fraction of a second. Researchers are often interested in the *instantaneous* air flow rate into the engine cylinder during the induction process. For example, they might be interested in the maximum value of air flow rate and when this occurs during the induction stroke; or, they might be interested in the flow reversal that can occur just before the inlet valve closes, when for a brief period the air can flow out of the cylinder and back into the inlet manifold. Measurement of the instantaneous air flow rate requires specialized fast-response instruments. The output signal from these instruments can respond to large changes in air flow rate within a few milliseconds.

Often, researchers are not particularly interested in the instantaneous air flow rate into the engine. Instead, they are interested in the average air flow rate over the entire engine cycle. This is called the *quasi-steady* air flow rate. This can be measured in a number of ways. One of them involves measurement of the total amount of air supplied to the engine over a period of time lasting several minutes. For example, if 3 m^3 of air was supplied to the engine over a 5 minute period, then the average or quasi-steady air volume flow rate to this engine would be 10 liter/s. Another way involves smoothing out the instantaneous air flow rate variations, so that the flow past the air flow meter is steady.

The following sections outline a number of practical methods and instruments which are used in research laboratories for the measurement of quasi-steady and instantaneous air flow rate to internal combustion engines.

1.3.1 Air Box with Sharp-Edged Orifice Plate

The simplest of the quasi-steady air flow rate methods is the air box, shown in Fig. 1.4(a). The engine intake system is connected to a large air box. The air drawn from the box by the engine is replenished by atmospheric air which enters the box through a calibrated sharp-edged orifice plate or a venturi. The volume of the air box must be sufficiently large to damp out the air pulsation generated by the engine, so that the air flow through the orifice plate is as far as possible steady. The air mass flow rate through the orifice plate can be calculated from:

$$\dot{m} = C_d A \sqrt{2\rho\,\Delta p}\ \text{kg/s} \tag{1.4}$$

where C_d = orifice discharge coefficient
A = orifice area (m^2)
Δp = pressure drop across the orifice plate (Pa)
= $\rho_i g h$
h = height difference in the manometer liquid columns in Figure 4(a) (m)
g = acceleration due to gravity (9.81 m/s^2)
ρ_i = density of manometer liquid (kg/m^3)
ρ = air density (kg/m^3)
= $p_o/(RT_a)$
p_o = atmospheric absolute pressure (Pa)
T_a = atmospheric temperature (K)
R = specific gas constant for air (J/kgK)

Designs of suitable orifice plates can be found in BS 1042 (1964), including suitable values for the discharge coefficient, C_d. In practice, a value for $C_d = 0.6$ will be sufficiently accurate for most purposes. The above equation is applicable to incompressible flow, and for good accuracy it is best to select an orifice plate design that limits the pressure drop, Δp, to less than 1.2 kPa.

If the air mass flow rate through the orifice is pulsating rather than steady, then this will be because the pressure drop across the orifice plate is pulsating. This presents a number of problems. First, the values of C_d given in BS 1042 are strictly applicable to steady flow through the orifice plate. The second problem is more complicated. The above equation shows that the mass flow rate $\propto (\Delta p)^{0.5}$. This means that the pressure drop across the orifice plate does not

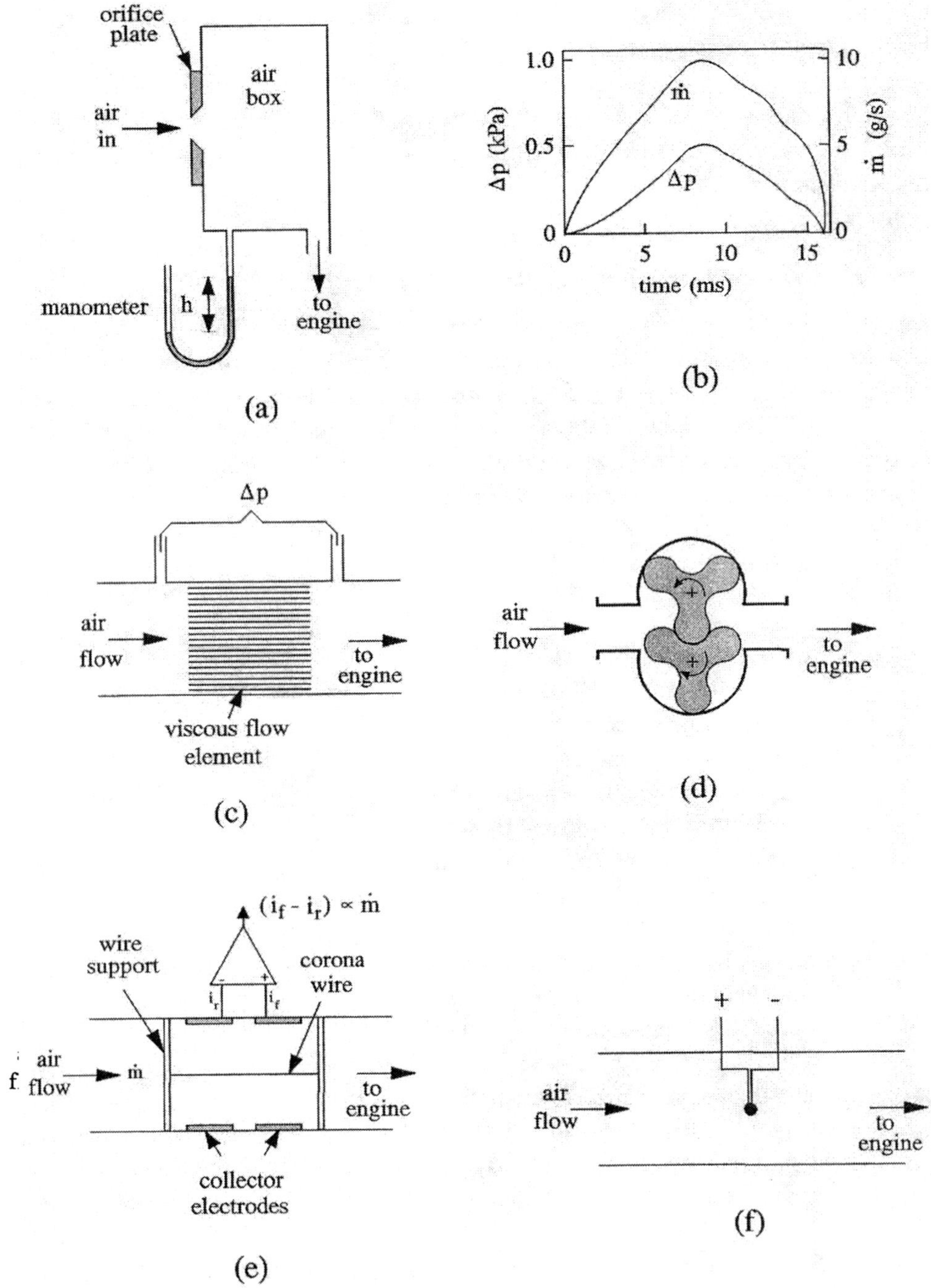

Fig. 1.4 Air flow measuring devices.

fluctuate in direct proportion to the fluctuations in mass flow rate. Therefore, the engine operator cannot estimate a simple time-average value of the pulsating pressure drop and use this average in the above expression. This is illustrated in Fig.1.4(b), where the pressure drop and mass flow rate are shown for an induction period lasting 16 ms. In all cases, the mass flow rate was calculated from the pressure drop using the above expression with: $C_d = 0.6$, $A = 0.00049\ m^2$, and $\rho = 1.2\ kg/m^3$. Using the data for Fig.1.4(b), three alternative average mass flow rates were calculated over the induction period of 16 ms.

(i) *Time-averaged mass flow rate*

$$\bar{\dot{m}}_1 = \frac{\int_{t_1}^{t_2} \dot{m}(t)\, dt}{t_2 - t_1} = 6.37\ g/s$$

where: the instantaneous air mass flow rate $\dot{m}_t = C_d A \sqrt{\rho\, \Delta p(t)}$

t_1 = start of air mass flow
t_2 = end of air mass flow

(ii) *Using time-averaged pressure drop*

$$\bar{\dot{m}}_2 = C_d A \sqrt{\rho\, \overline{\Delta p}} = 7.14\ g/s$$

where

$$\overline{\Delta p} = \frac{\int_{t_1}^{t_2} \Delta p(t)\, dt}{t_2 - t_1} = 490\ Pa$$

(iii) *Using range-averaged pressure drop*

$$\bar{\dot{m}}_3 = C_d A \sqrt{\rho\, \overline{\overline{\Delta p}}} = 7.21\ g/s$$

where

$$\overline{\overline{\Delta p}} = \frac{\Delta p_{max} + \Delta p_{min}}{2} = 500 \text{ Pa}$$

The first average value of 6.37 g/s is the true mean of the mass flow rate over the induction period $(t_2 - t_1)$. The second value is +12% in error due to the nonlinear relationship between the mass flow rate and pressure drop. The third average is what an engine operator might calculate using an average value of the pressure drop, obtained by "judging" the midpoint between the upper and lower swings of a manometer liquid column. It can be seen that this mass average is also in error by about 13%.

In view of the above errors which may be introduced due to the pulsating nature of the flow, it is important that a large air box is used in conjunction with orifice plates, in order to damp down the air flow pulsation. The required volume of air box for steady flow through the orifice can be calculated from the following expression [Kastner, 1947; Plint and Martyr, 1995]:

$$V_b = \frac{417 \times 10^6 n_s^2 d^4}{n_c V_s N_m^2} \tag{1.5}$$

where n_s = constant ($n_s = 1$ for two-stroke and $n_s = 2$ for four-stroke engines)
d = orifice plate diameter (m)
n_c = number of engine cylinders
V_s = engine total swept volume (m^3)
N_m = minimum engine speed at which measurements are required (rpm)

The above expression shows that a larger box volume will be required when the engine is four-stroke, single cylinder, having a small displacement and running at low engine speed. Consider, for example, an orifice having a diameter of 0.025 m and used to measure the air flow to a single-cylinder, four-stroke engine having a displacement of 0.5 liters and running at 1500 rpm. A box having a volume of almost 600 liters will be needed, which is 1200 times the engine displacement. It is important to realize that, according to the above expression, the box volume required is very sensitive to the orifice diameter. For example a 20% increase in orifice diameter results in the doubling of the box volume required.

Often, a significant disadvantage of the air box is a long tube that is often required to connect the box to the engine. The long tube can affect the breathing characteristics and the performance of the engine significantly [Stone, 1992]. An alternative approach is to use a large-diameter flexible tube made of "floppy" material supported by a wire spiral. This type of tube has a smaller effect on the engine performance, and it also helps to reduce the flow pulsations at the orifice plate [Lilly, 1984].

In spark ignition engines the air flow rate is at a minimum when the engine is idling (throttle almost closed) and a maximum at high load and high speed (throttle wide open). The turn-down ratio in air mass rate can be very large, on the order of 30:1 or greater. The consequence of this is that a single orifice plate size would not be sufficient for measuring the air flow rate throughout the entire engine operating range, for the following reasons. The square relationship between the air mass flow rate and pressure drop across the plate, means that if the air mass rate is reduced by ten times, then the pressure drop across the plate would be reduced by 100 times. This makes accurate measurement of the pressure drop and accurate prediction of the mass rate progressively more difficult at low air mass rates. In practice, greater accuracy can be obtained by using a number of interchangeable orifice plates, with progressively smaller hole diameters when the air mass flow rate is lower. In this way, the pressure drop across the plate can be kept higher, so that it can be measured more accurately.

1.3.2 Viscous Flow Meter

Fig. 1.4(c) shows a schematic of a viscous air flow meter, which is suitable for measurement of quasi-steady air flow. The meter comprises numerous parallel flow passages. In order to ensure that the flow through the meter is laminar, each passage has a small flow area (typically 0.1 mm^2 and 70 mm long). In principle, this makes the pressure drop across the meter proportional to the air flow velocity or volumetric flow rate through the meter:

$$\dot{m} = k\rho\Delta p \tag{1.6}$$

where Δp = pressure drop across the viscous meter
ρ = air density
k = calibration constant

Viscous flow meters require calibration at steady flow against a known standard device, such as an orifice plate.

Generally, a viscous flow meter is more suitable for use in pulsating flows than an orifice plate, because the mass rate through the viscous meter is directly proportional to the pressure drop across it. Even so, the accuracy of viscous meters does deteriorate somewhat with pulsating flow, when the flow rate is no longer exactly proportional to the pressure drop across the meter. Stone [1989] showed how the mean pressure drop and the average flow rate could be calculated accurately when pulsating flow is present. The direct proportionality between mass rate and pressure drop also allows viscous flow meters to be used over a larger turn-down ratio in air flow than is the case with orifice plates.

A significant disadvantage of viscous flow meters is their sensitivity to fouling of the small flow area passages, which causes change of calibration. Good filtering of dust and other particles upstream of the meter is essential. However, filtering cannot prevent lacquering of the flow passages, which can also cause change of calibration [Lilly, 1984].

1.3.3 Positive Displacement Flow Meter

Perhaps the most accurate method for measuring the quasi-steady air flow rate is to use a positive displacement meter. Some of these meters have a similar construction to the Roots blower, as shown in Fig.1.4(d). The air mass flow rate through a positive displacement meter can be calculated from:

$$\dot{m} = \rho_i V_d n \text{ kg/s} \tag{1.7}$$

where ρ_i = density at meter inlet (kg/m^3)
= $p_i/(RT_i)$
p_i = absolute pressure at meter inlet (Pa)
T_i = air temperature at meter inlet (K)
R = specific gas constant for air (J/kgK)
V_d = air volume displaced per revolution of the meter rotors
n = rotational speed of meter rotors (rev/s)

The accuracy of positive displacement meters is not susceptible to pressure pulsation, and this type of meter has a larger turn-down ratio in comparison to orifice plates and viscous flow meters. Positive displacement meters cause a small pressure drop in the flow (typically, around 1 or 2 kPa). Also, the inertia of the meter rotors can constrain rapid changes in engine air flow and thereby reduce the engine transient response.

1.3.4 Corona Discharge Flow Meter

Figure 1.4(e) shows the corona discharge meter manufactured by Lucas-Dawe. This meter is capable of measuring the instantaneous air mass flow rate through it. Referring to Fig. 1.4(e), the corona wire is maintained at a high voltage of around 10 kV. This causes ionization of the air within the meter and a corona discharge. As a result, an ion flow is established between the wire and the two tubular collector electrodes. With no air flowing through the meter, the ion current flow to each of the two electrodes is the same, and $i_f = i_r$. When air is flowing through the meter, the ion streams to the two electrodes are deflected. This causes an imbalance between the ion currents flowing to the two electrodes. The difference in the two current flows is proportional to the air mass flow rate through the meter [Cockshott et al, 1983]. That is, air mass rate $\propto (i_f - i_r)$.

The corona discharge meter responds very rapidly (within about 1 ms) to changes in air mass flow rate. The meter is also capable of distinguishing and measuring flow in both the forward and reverse directions. The rapid response and bidirectionality of this meter make it particularly suitable for investigating the highly pulsating air flow through internal combustion engines, where reverse flows can occur at the start and end of the induction process. In the case of multi-cylinder engines, differences in the air mass flow rate between cylinders can also be detected when the meter is installed at the entrance to the inlet manifold. The meter introduces very low pressure drop, it has a turn-down ratio of about 100:1 (using four electronically switched ranges), and its accuracy is about 1%. Disadvantages of the meter include sensitivity to air temperature and humidity, and the need for periodic recalibration at steady flow against a known standard device, such as an orifice plate.

1.3.5 Hot Wire Flow Meter

Figure 1.4(f) shows a schematic diagram of a hot wire air flow meter, which, in principle, is capable of measuring instantaneous air mass flow rate. The wire is maintained at a higher temperature than that of the air stream, by means of an electrical supply. This type of anemometer relies on the principle that variations in the air flow rate past the wire cause variations in the heat loss rate from the wire. This heat loss rate can be calculated from the electrical energy supplied to the wire. In the case of a cylindrical hot wire, the heat loss rate can be related to the air mass rate through the flow meter by King's law [Weaving, 1990]:

$$q = I^2 R \cong a + b(\rho V)^n \tag{1.8}$$

where q = heat loss rate from the hot wire
I = current flowing through the hot wire
R = electrical resistance of the hot wire
ρ = air density
V = air flow velocity past the wire
a, b, n = constants determined by calibration

Using the mass continuity equation,

$$\dot{m} = \rho VA \quad \text{then} \quad I^2 R = A + B(\dot{m})^N$$

where A, B, and N, are constants appropriate to the hot wire meter, determined by calibration of the meter against a standard air mass flow rate device. The calibration may only hold good over a limited range of pressure and temperature. It can be seen from the above expression that the air mass flow rate is related to the electrical energy dissipated in the wire by a nonlinear function. After calibration, it is possible to estimate the mass flow rate through the meter by monitoring the electrical energy dissipated in the hot wire. Alternatively, the current supply to the wire can be varied by an electronic control circuit, so as to keep the wire resistance, and hence the wire temperature, constant. The variation in the wire supply current can then be related to the variation in mass flow rate through the meter.

In principle, hot wire meters can respond almost instantaneously to changes in air flow mass rate. For a fast response, the hot wire thermal inertia must be very small, which means that the wire diameter must also be very small (<100μm). This makes the hot wire fragile. In practice, larger wire diameters are used in order to improve robustness at the expense of slower response to changes in air mass flow rate. It should be noted that the hot wire meter can measure flow rate in both the forward and reverse directions. However, unlike the corona discharge meter discussed above, the hot wire meter is incapable of discriminating the flow direction.

1.3.6 Air Flow Measurement from Exhaust Gas Analysis

The above sections outlines several methods for measuring air flow rate directly, using instrumentation installed in the intake system of the internal combustion engine. However, it is also possible to estimate the quasi-steady air mass flow rate indirectly, often with similar accuracy. This can be achieved by measuring the fuel flow rate to the engine and calculating the air/fuel ratio using exhaust gas analysis. Heywood [1988] deals with such calculations. In general, accurate laboratory measurements of fuel flow rate can usually be achieved more easily than accurate measurements of air flow rate.

1.4 Fuel Consumption Measurement

Figure 1.5 shows a number of methods for measuring fuel consumption rate. The methods illustrated in Figs. 1.5(a) to 1.5(c) are suitable for measuring average fuel consumption, while Figures 5(d) and 5(e) show two methods suitable for instantaneous flow rate.

1.4.1 Average Fuel Consumption Rate

Figure 1.5(a) shows the burette method for measuring average fuel consumption. This method relies on recording the time taken (usually several minutes) for the engine to consume a certain volume of fuel. The system configuration shown in Fig. 1.5(a) works as follows. With no measurement being made, valve B is kept open and vent A closed. When a measurement is to be made, vent A is opened briefly so that the burette fills with fuel flowing through the open valve B. When the burette has filled with fuel, valve B is closed and vent A is kept open. Fuel is now supplied to the engine only from the burette. The consumption of the fuel in one of the calibrated volumes of the burette is then timed with a stopwatch, by observing the level of the fuel meniscus descending past two of the spacers. The average fuel flow rate is then calculated using the timed period. The burette usually includes a choice of calibrated volumes, so that a larger calibrated volume could be chosen when the engine fuel consumption rate is high. This device gives the volumetric fuel consumption rate. To convert this measurement to a gravimetric rate, the fuel density must be measured separately, using a suitable technique. This method has the merit of simplicity, and with careful use it can provide measurements of the average gravimetric fuel flow rate with an accuracy of around 1%. Usually, 2 to 3 minutes are required to complete a measurement of the average fuel consumption rate. The burette method can be automated by installing optoelectronic liquid-level sensors and using an electronic control system to open and close valves A and B, and in order to start and stop an electronic timer.

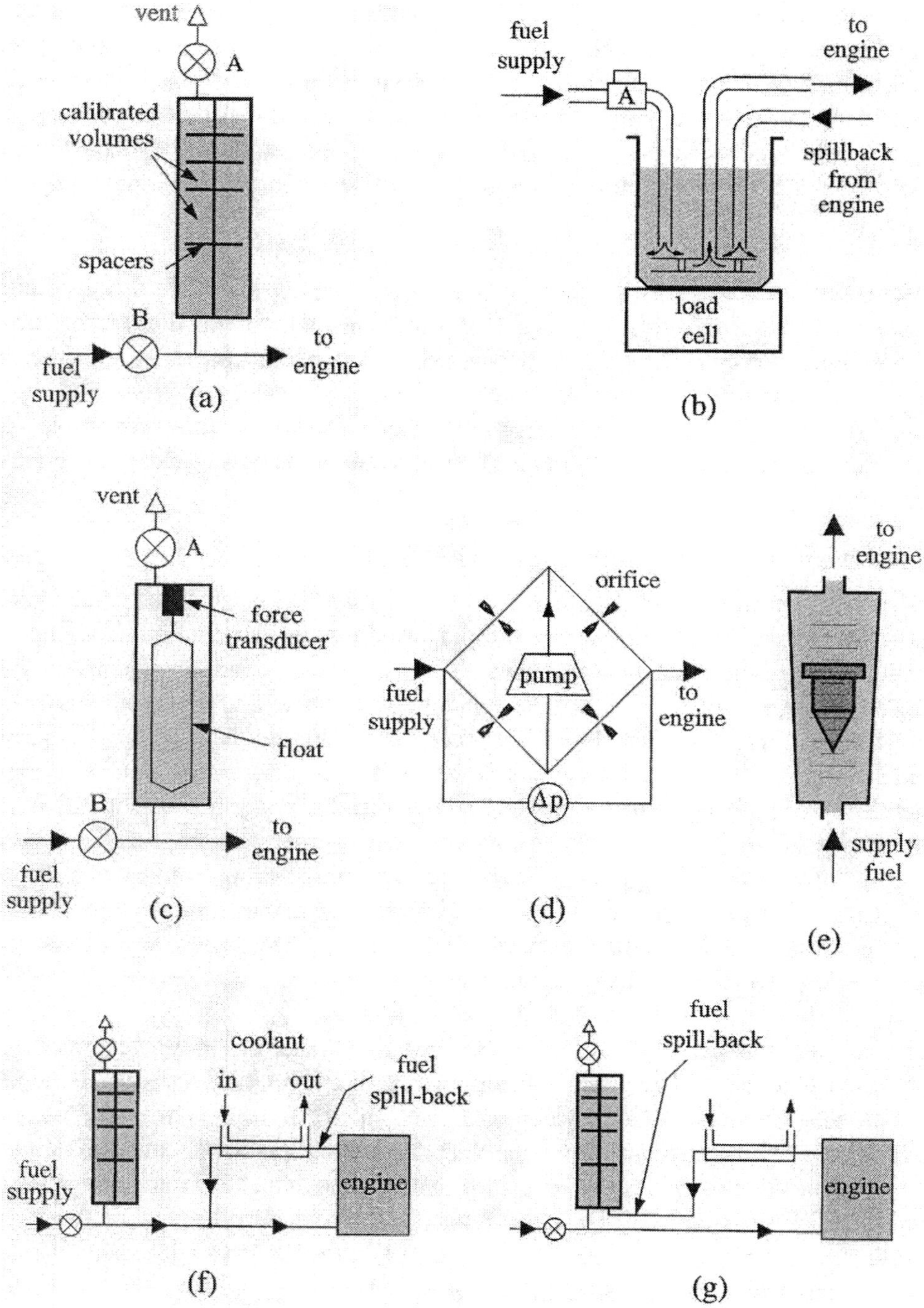

Fig. 1.5 Devices for measuring fuel consumption.

Figure1.5(b) shows a weighing method for measuring the average gravimetric fuel consumption rate. The advantage of this method is that it measures directly the fuel mass rather than the fuel volume consumption. The weight of fuel in the vessel at any instant is measured using the load cell. With no measurement being made, valve A is opened periodically to ensure that the vessel has sufficient fuel. When a measurement of fuel consumption is to be made, valve A is kept closed and fuel is drawn by the engine from the vessel. The mass of fuel consumed over a period of time is then measured using the load cell. The system can be readily automated by means of electronic control, which can monitor the weight of the vessel and control the operation of valve A and a timer.

Figure 1.5(c) shows another method for measuring the average gravimetric fuel consumption rate. Again, the advantage of this method is that it measures directly the mass rather than the volume consumption. A cylindrical float is positioned inside a cylindrical vessel, and the force on the float due to buoyancy is measured by means of a force transducer. When fuel from the vessel is consumed by the engine, the change in buoyancy force is directly proportional to the mass of fuel consumed. This method can also be readily automated using electronic control.

1.4.2 Instantaneous Fuel Consumption Rate

The Flo-Tron system in Fig. 1.5(d) is capable of measuring the instantaneous fuel mass rate supplied to the engine. This system has a rapid response and can, therefore, monitor fuel consumption under transient engine conditions. This system uses a hydraulic equivalent of the electrical Wheatstone bridge. Four orifices make up the bridge, and the pump shown in the figure establishes a reference fuel flow which recirculates around the bridge. When no fuel enters or leaves the bridge, the reference fuel flows through the bridge are balanced. However, when fuel passes through the bridge on its way to the engine, the bridge becomes unbalanced and it produces a pressure-difference signal (Δp) which is directly proportional to the instantaneous mass flow rate of the fuel passing through the bridge. The mass flow rate indicated by the meter is not affected by changes in the fuel viscosity or density. The manufacturers of the instrument quote: a turn-down ratio of over 1000:1; accuracy better than 0.5%; and a system response of less than 0.5 s to a step change in engine fuel mass flow rate.

When great accuracy is not required, a variable-area flow meter or "rotameter," as shown in Figure 1.5(e), can be used. This type of meter consists of a short conical float, often made of metallic materials, which is free to move up or down a tapered transparent tube. Fuel entering from the bottom of the tapered tube raises the float due to the action of viscous and momentum forces. The extent of the rise in the float provides an indication of the flow rate through the meter with an accuracy of a few percent. This type of meter is calibrated by the manufacturer, against a standard reference meter, for a given fluid. Changes in the density or viscosity of the fluid affect the calibration and lead to loss of accuracy. It is, however, possible to apply density and viscosity correction factors to take into account changes in the density and viscosity of the fluid so as to maintain the meter calibration. For example, a suitable correction factor for changes in density is:

$$\frac{m_1}{m_2} = \sqrt{\frac{(w - \rho_1)\rho_1}{(w - \rho_2)\rho_2}} \text{ (liquids)}$$

$$\frac{m_1}{m_2} = \sqrt{\frac{\rho_1}{\rho_2}} \text{ (gases)}$$

where m = mass flow rate
w = density of the float material
ρ = fluid density
subscript 1 refers to the fluid density for which the meter was calibrated
subscript 2 refers to the new fluid density

Closer inspection of the above expressions shows that the equation for gases is a simplification of the one for liquids, based on the fact that in the case of gases $w >> \rho$.

For changes in viscosity a suitable correction factor is:

$$\frac{m_1}{m_2} = \frac{(w - \rho_1)\, \rho_1\, \mu_2}{(w - \rho_2)\, \rho_2\, \mu_1} \text{ (liquids)}$$

$$\frac{m_1}{m_2} = \frac{\rho_1 \mu_2}{\rho_2 \mu_1} \text{ (gases)}$$

where μ is viscosity of the fluid.

Again, the expression for gases is a simplification of the one for liquids.

Turbine flow meters can provide instantaneous measurements of the fuel volumetric flow rate. Accuracy is often worse than 1% and the turn-down ratio is usually around 10:1 [Stone, 1992]. The calibration is affected to a small extent by changes in the fuel viscosity, especially with meters of small size. Turbine meters can respond rapidly to changes in the fuel flow rate, the response time being dependent mainly on the rotor inertia.

Positive displacement volumetric fuel flow meters are also available with better accuracy and larger turn-down ratios than for turbine flow meters. Positive displacement flow meters can, however, introduce substantial pressure drops across the engine fuel supply line.

1.4.3 Fuel Spill-Back

In the case of diesel engines and fuel-injected gasoline engines, the fuel supplied to the engine is not all used by the engine. A significant proportion of this fuel supply returns from the engine. In fact, in the case of fuel-injected gasoline engines installed in vehicles, most of the fuel pumped to the engine is returned to the vehicle tank.

In the laboratory, the net fuel consumption by an engine can be calculated provided both the fuel supply and the "spill-back" returned from the engine are measured accurately. An alternative approach is to return the spill-back downstream of the flow meter, after being cooled in a heat exchanger, as shown in Fig. 1.5(f). However, this approach can give rise to inaccuracies, because the spill-back can often contain small quantities of air and fuel vapor. A more satisfactory approach would be to return the spill-back to the measuring device, as shown in Fig. 1.5(g), where the entrained air and fuel vapor are vented. A similar method for dealing with spill-back is shown in Fig. 1.5(b).

1.5 Acquisition of Instantaneous Engine Variables

Research on internal combustion engines often requires the observation and recording of rapidly changing variables such as the cylinder gas pressure, the movement of the diesel injector needle, and the diesel fuel injection pressure. The magnitude of these variables is often changing greatly over a period of a few milliseconds.

Specialist instrumentation is often needed in order to monitor these physical variables. While the output voltage from such instruments can be observed on a single- or multi-channel oscilloscope, a permanent record of these voltage is often required for later observation and analysis. Nowadays, it is routine practice in engines research laboratories to record these variables in digital form. First, this requires the conversion of the physical variables into voltages. The resulting voltages are then converted into digital form using a data acquisition system. It is common practice nowadays to use a personal computer based acquisition system. This comprises a card that is plugged into one of the expansion slots of the personal computer (PC). The card is usually controlled by specialized PC software supplied by the card manufacturer. An example of such a system is shown in Fig. 1.6. The main elements of this PC-based data acquisition system are discussed below.

1.5.1 Transducers and Signal Conditioning

Figure 1.6 shows a number of transducers and their signal conditioning amplifiers. These are external to the data acquisition system. The transducers convert fluctuations in physical variables into fluctuations in electrical quantities. For example, a thermocouple converts temperature into a weak voltage (a few millivolts). The electrical output from most transducers is not suitable as it stands for input into a data acquisition system. A signal conditioning amplifier is usually needed to modify the transducer output. For example, a piezoelectric pressure transducer converts changes in cylinder gas pressure into a varying electrical charge. A signal-conditioning amplifier is used to convert this charge into a voltage and amplify it.

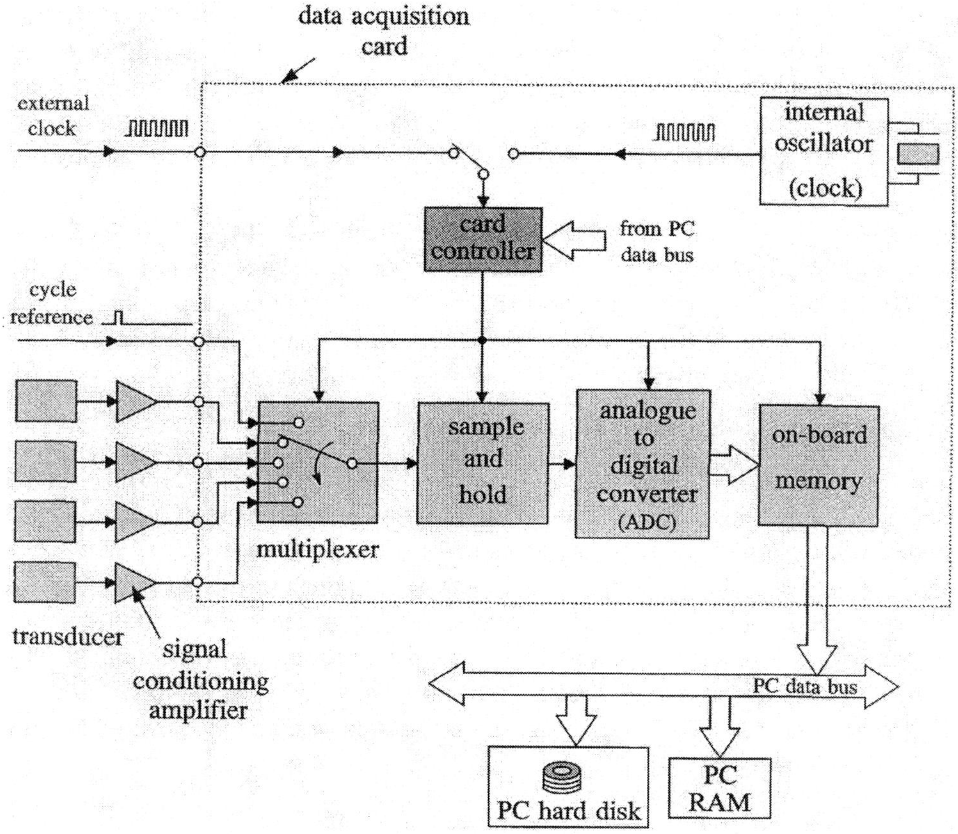

Fig. 1.6 PC-based data acquisition system.

Often, the electrical output from a transducer may not be related linearly to changes in a physical variable. One example is the thermocouple. The output voltage from a copper-constantan thermocouple increases by 0.39 mV if the temperature rises from 0 to 10°C, and by 0.47 mV when the temperature rises from 90 to 100°C (cold junction at 0°C). A thermocouple signal-conditioning amplifier will amplify and linearize the thermocouple voltage output. In this way, the amplifier output will always increases by 1 V when the temperature rises by 10°, irrespective of whether the temperature rise was from 0 to 10°C or from 90 to 100°C.

1.5.2 Multiplexing

Often, engine research requires a great number of rapidly changing variables to be recorded simultaneously by a data acquisition system. These may include cylinder gas pressure, diesel injector needle movement, diesel injection pressure, output from a spark-discharge transducer, throttle angle during engine load transient testing, instantaneous air flow into the engine, pressure pulsation in the inlet tract of a gasoline engine, output from specialized instruments such as a fast flame ionization detectors, combustion optical sensors, fast-response surface thermocouples, and so on.

In principle, it would be desirable for all the engine transducer voltages to be recorded simultaneously. For this to be achieved, each transducer voltage would require its own digitizing and data handling subsystem. Furthermore, all these subsystems will have to operate simultaneously and synchronously. In practice, this approach is rarely adopted. Instead, the alternative approach shown in Fig. 1.6 is commonly adopted, which is substantially less costly and complex.

The system shown in Fig. 1.6 processes the individual transducer voltages one after the other. A selector circuit (labeled "multiplexer" in the figure) selects the voltage from the first transducer amplifier and offers it to the digitizing subsystem. When the digitization and storage of the value has been completed, the multiplexer selects the voltage for the second transducer for digitization and storage. This sequential process continues until all the transducer voltages have been digitized and stored.

A multiplexing system, such as the one shown in Fig. 1.6, is not capable of recording all the transducer signals simultaneously. The recording of individual transducer voltages is separated by delays of several microseconds. For example, if the multiplexer takes 1 μs to switch from one input voltage to the next, and the digitization and storage require a further 4 μs, then a delay of 5 μs would separate the individual transducer voltage recordings. Furthermore, if the total number of transducer voltages being recorded is, say, 8, then an accumulated delay of 40 μs would separate the recording of the first and eighth transducer. In principle, compensation for these delays is possible after the completion of the engine tests, by processing the stored digital data using appropriate software.

1.5.3 Analog to Digital Converter

Figure 1.6 shows the voltage selected by the multiplexer passing onto a sample-and-hold circuit. It is more convenient to leave the description of this circuit until the operation of the analog to digital converter has been discussed first. The task of the analog to digital converter (ADC) shown in Fig. 1.6 is to convert each transducer voltage into a digital value which is then stored in digital memory.

In general, the input to an ADC is voltage (e.g., 2 V) and the output is an integer value (e.g., 51). An ADC can accept input voltages within a certain range, for example, 0 to 10 V. The integer values out of the ADC are also within a certain range, for example, 0 to 256. Thus, a voltage input of 10 V will generate an output integer value of 256; 5 V will generate an output integer value of 128, and so on.

The integer values generated by the ADC are stored into temporary digital memory, as shown in Fig. 1.6, and subsequently transferred to the PC memory where they can be stored permanently on the hard disk. A 10 V input into the ADC will be transferred into a PC file as [10 V × (256/256)] = 10 V, while an input of 1.25 V will generate an output integer value of 32 from the ADC which will be stored in a PC file as [10 V × (32/256)] = 2.25 V.

In the case of the above ADC, the input voltage must increase by 0.039 V (10/256 = 0.039 V) before the output integer value is incremented by 1. If the input voltage increases by less than 0.039 V, then the ADC output will not be incremented. This causes a maximum uncertainty of 0.039 V, as illustrated in Table 1.1.

The maximum absolute uncertainty in the above results (Table 1.1) is 0.038 V. This can be interpreted as follows: if we examine the output from the ADC, we will be unsure by as much as 0.039 V of what the actual value of the input voltage is. This maximum uncertainty is not so important if the input voltage is around 10 V, since the uncertainty is only 0.4% of the 10 V. However, when the input voltage is around 0.6 V, then we will be unsure by as much as 6% of what the actual input voltage is. This could be of considerable practical importance. For example, it would be unsatisfactory to measure the engine air inlet temperature as 20°C with an uncertainty of 1.2°C.

A conversion resolution of 0.039 V is often unsatisfactory for internal combustion engine research, because small but significant changes in the variable being digitized (such as cylinder gas pressure) may be missed. For example, suppose that 10 V corresponds to the maximum cylinder gas pressure of 120 bar. This means that gas pressure fluctuations smaller than 0.47 bar will be missed; this would be clearly unacceptable.

Let us return, briefly, to the results of Table 1.1. The table demonstrates that the percentage uncertainties are small when the voltage input to the ADC is as close as possible to 10 V. In general, it is important for the output range of the signal conditioning amplifiers to be the same as

Table 1.1 Uncertainty due to limited resolution of an A/D converter

ADC input (V)	ADC output (N)	Value in PC file (V)	Uncertainty (V)	Uncertainty (%)
V_i		$V_o = 10 \times (n/256)$	$e = V_o - V_i$	$e = [(V_o - V_i)/V_i] \times 100$
10	256	10	0	0
9.999	255	9.961	−0.038	−0.38
5.038	128	5	−0.038	−0.75
5	128	5	0	0
2.538	64	2.5	−0.038	−1.50
2.5	64	2.5	0	0
1.290	33	1.289	−0.001	−0.08
1.288	32	1.25	−0.038	−2.95
1.25	32	1.25	0	0
0.663	16	0.625	−0.038	−5.73
0.625	16	0.625	0	0

the input range of the data acquisition system (e.g., 0 to 10 V, 0 to 5 V, –5 to +5 V). In practice, most signal conditioning amplifiers have a variable gain, so that their output range can be adjusted so that it is approximately the same as the input range of the data acquisition system. Sometimes, signal-conditioning amplifiers may allow only limited or no adjustment of their output range. Fortunately, some commercially available data acquisition systems allow their input range to be adjusted so that it is approximately the same as the output range of the signal conditioning amplifiers. The adjustment of the acquisition system input range could be on an individual channel basis. This allows several transducer-conditioning units with different output ranges to be used without compromising digitization accuracy.

ADCs having more than 256 output integer steps are readily available. For internal combustion engine research it is usual to employ an ADC with at least 4096 output steps.

Manufacturers of data acquisition systems do not normally quote how many output integer steps the system ADC is capable of. Instead, they quote the number of ADC bits, from which the output steps can readily be calculated. For example, an 8-bit ADC is capable of $2^8 = 256$ output steps, a 10 bit one is capable of $2^{10} = 1024$ steps, and a 12 bit one is capable of $2^{12} = 4096$ steps.

1.5.4 Sample-and-Hold

Data acquisition systems use a sample-and-hold circuit at the input to the ADC, as shown in Fig. 1.6. Just before the ADC begins the conversion of the input voltage, the sample-and-hold circuit "samples" the input voltage level and holds it at a constant level during the ADC conversion process. This ensures that the ADC input voltage does not change during the conversion, which could lead to significant conversion errors.

1.5.5 Clock Signal

A data acquisition system digitizes and stores the input voltages from a set of transducers. The system carries out this task, for the entire set of transducers, every time it is commanded to do so by a "clock." The clock output is a series of pulses. Every pulse commands the acquisition system to digitize and record the entire set of transducer voltages. This task must be completed between successive clock pulses.

Two alternative clock systems can be used, as shown in Fig. 1.6:

1. A "time clock" residing within the data acquisition card; this clock could, for example, output a train of pulses every 1 ms.
2. A "crank angle clock" external to the data acquisition system; this clock could, for example, output a train of pulses every 1 degree of the crankshaft revolution.

Using an external "crank angle clock" (clock 2 above) is far preferable to using an internal "time clock" for the following reasons. The events occurring in an engine are dependent on the rotation of the crankshaft, which causes the motion of the piston, the motion of the inlet and exhaust valves, the fuel and spark delivery, and so on. It is important to realize that all these events are not time dependent, but, instead, they rely on the rotation of the crankshaft. If a "time clock" is used, then problems could arise when the acquired data are being interpreted after the engine tests have been completed. This is because the rotational speed of the engine crankshaft changes by a small, but significant, amount during the course of each engine cycle, and from one engine cycle to the next.

1.5.6 Cycle Reference Signal

The data acquisition system must always record an engine cycle reference signal, as shown in Fig. 1.6, alongside the voltage inputs from the various engine transducers. This signal allows the transducer voltages to be related to the position of the mechanical components of the engine, such as the piston or the inlet and exhaust valves. The cycle reference signal must be obtained from an engine crankshaft sensor. Usually, this sensor will provide one output signal every engine revolution, for example, at piston top dead center (TDC). In the case of four-stroke engines, two output signals will be obtained during a complete engine cycle, one for each crankshaft revolution. This presents a problem, because it will normally be necessary to distinguish whether a reference signal occurred during the power or exhaust revolution of the engine. This distinction may be needed, for example, when this reference signal is to be used for automating the analysis of digitized data using computer software.

Figure 1.7 shows several ways of distinguishing whether a reference signal occurred during the power or exhaust revolution of the engine. Here, it will be assumed that the reference signal is in suitable form for processing by digital electronic circuits. One method, shown in Fig. 1.7(a) is to skip one of the two reference signals using a simple electronic circuit such as a toggle-type flip-flop. It should be noted, however, that two alternative outputs from the flip-flop are possible, depending on which of the two reference signals arrived first at the flip-flop when the latter was switched on. A flip-flop reset switch as shown in Fig.1.7(a) can be used to select the most convenient of the two output alternatives. The reset switch is operated momentarily so as to select the desired output, while observing the cylinder gas pressure and flip-flop output on an oscilloscope.

A more sophisticated approach, shown in Fig. 1.7(b), involves the cylinder gas pressure signal being supplied to a simple threshold detecting circuit which gives a digital "high" output when the gas pressure exceeds, say, 5 bar. The reference signal (assumed here to correspond to TDC), together with the threshold detector output can then form the inputs to an AND gate. This gate then provides a "high" output at the TDC of the engine power revolution only. The TDC occurring during the exhaust revolution is ignored.

Yet another approach, shown in Fig. 1.7(c), involves using a signal from the engine camshaft (rather than the cylinder gas pressure signal). The camshaft signal could be at a digital "high" level during the engine power revolution and at "low" level during the exhaust revolution. Again, the crankshaft and camshaft signals form the inputs to an AND gate which provides a "high" output at the TDC of the engine power revolution only.

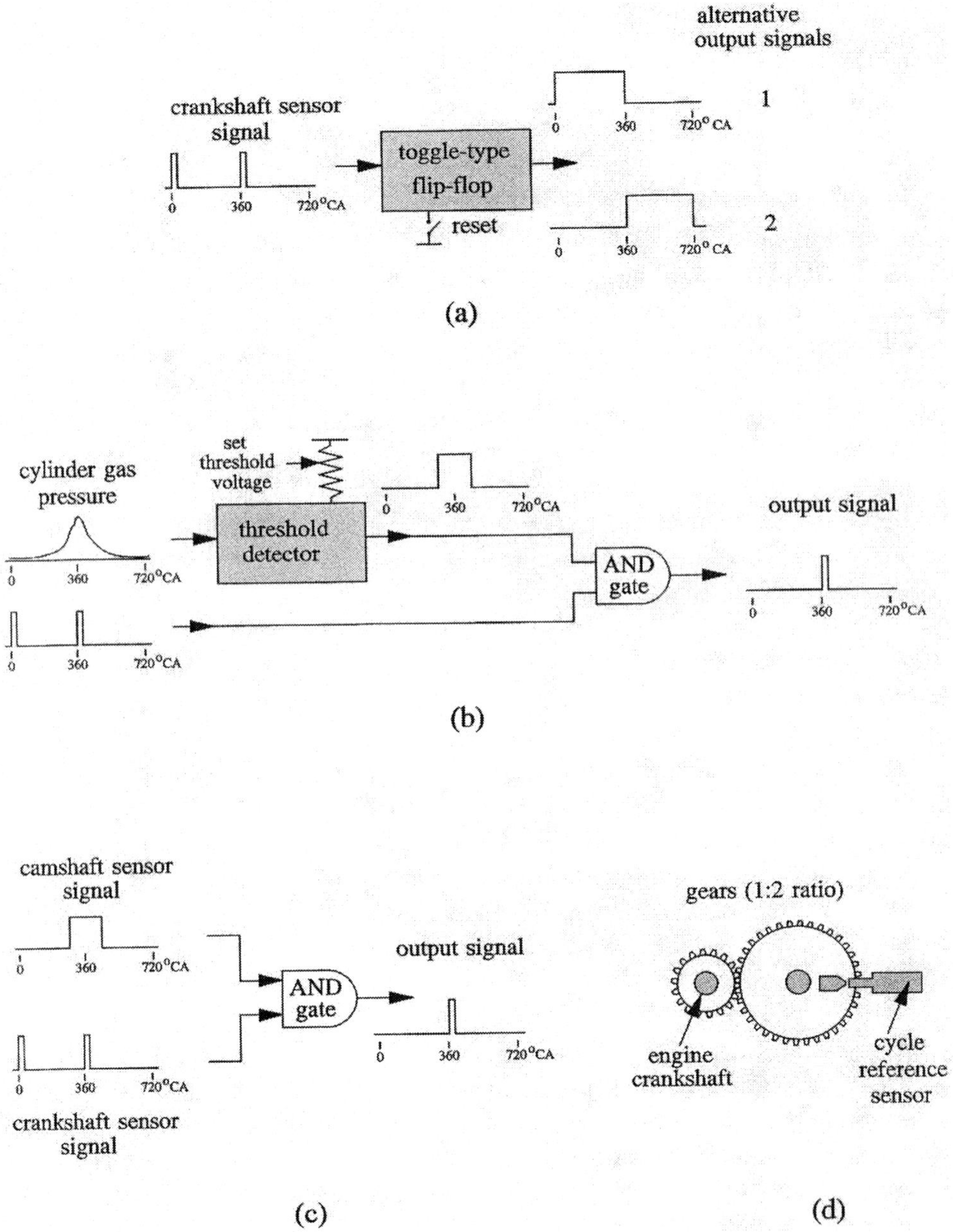

Fig. 1.7 Various methods of producing a cycle reference signal.

The last approach, shown in Fig. 1.7(d), is to mechanically halve the crankshaft speed. Since the second gear rotates at half the crankshaft speed, the signal from the second gear will occur only once per engine cycle and always on the same revolution of the crankshaft. A set of good quality gears is required in order to ensure precise phasing in their rotation without backlash.

Finally, the reader is strongly discouraged from obtaining the cycle reference signal from the camshaft. It may be tempting to do so, as the camshaft is already rotating at half crankshaft speed. However, this would be an unsatisfactory thing to do. In most automotive engines and in many research engines the camshaft is driven by the crankshaft through a flexible tooth-belt. As a result, the phasing of the crankshaft and camshaft rotations varies during the engine cycle due to the flexibility of the belt.

The cycle reference signal must normally be set accurately with respect to the position of the piston. The choice of method for setting the reference signal depends on what type of reference sensor is used. For example, if the sensor allows, this can be done "statically" by removing the spark plug or the fuel injector from the engine to gain access into the engine cylinder. The position of the piston can then be monitored physically while the engine crankshaft is turned manually and slowly. The reference signal can then be set when the piston reaches the appropriate position (e.g., TDC). Alternatively, the reference signal can be set "dynamically" by means of a proximity sensor mounted into the engine head, for example, using the spark-plug or injector hole. With the engine being motored, the proximity sensor can then sense when the piston is at TDC, and the reference signal set accordingly. Finally, it would be unsatisfactory to motor the engine and set the reference signal to coincide with the peak cylinder gas pressure. This is because the pressure peak may not coincide with the piston TDC due to asymmetries in the heat loss from the compressed gas during the compression and expansion strokes. (Note: When an engine is driven by an electrical motor or dynamometer without combustion, it is said to be motored. The corresponding engine test is called a motoring test.)

1.5.7 Sampling Rate

Suppose that the voltages from, say, eight engine transducers are to be recorded once every degree of crankshaft rotation, when the engine is running at 2000 rpm. Then, each set of eight voltages must be digitized and stored at a rate corresponding to 12,000 times/s. However, it is important to realize that 12,000 times/s is the rate at which the entire set of eight input voltages must be recorded. With a multiplexed system, the ADC must, in fact, digitize at a rate of $8 \times 12{,}000 = 96{,}000$ times/s. Similarly, the digitized data must be stored at a rate of 96,000 values/s.

The above digitization and storage rate of 96,000 values/s can be achieved fairly easily by modern data acquisition systems. However, in internal combustion engine research, digitization and storage rates far in excess of 96,000 values/s are often required, for the reasons explained below.

Often, engine combustion events such as the pre-mixed combustion period in diesel engines, or the flame kernel initiation in spark-ignition engines, last for only a few degrees crank angle (°CA). If a detailed study of these events is to be made it would be necessary to record the cylinder gas pressure at least every 0.5°CA. Furthermore, suppose that data must be recorded from the eight engine transducers at a fairly high engine speed of 4000 rpm.

The number of ADC digitizations and system data storage operations per second can now be calculated as:

$$\frac{360°\text{CA}/\text{rev}}{0.5°\text{CA}/\text{sample}} \times \frac{4000\ \text{rev}/\text{min}}{60\ \text{sec}/\text{min}} \times 8\ \text{inputs} = 384{,}000\ \text{samples}/\text{sec}$$

or one digitization every 2.6 μs.

In practice, the period of 2.6 μs per digitization is not available in its entirety to the ADC. This is because the multiplexer itself requires some time to switch over to the next channel to be digitized. This switch-over can take as much as 1 μs, leaving only 1.6 μs for the ADC to digitize the input voltage. Fortunately, there are several ADCs available commercially which can cope with such short digitization rates. However, problems can often arise with the storage of the digitized data. Fast data acquisition systems normally have on-board memory (buffer) into which the digitized voltages from the ADC can be dumped very quickly. However, the buffer is normally of limited capacity, and can easily fill up while the ADC is still digitizing and delivering data to the on-board memory. When the buffer fills up, it has to be emptied by transferring the digitized data to other storage media, such as the PC random access memory (RAM) or the PC hard disk (see Figure 1.6). This operation takes time which may not be available if, at the same time, the ADC continues to send digitized data to the buffer. Some commercially available systems may not be able to cope with this situation. There are several remedies to this problem, which compromise to varying extends the usefulness of the data collected. One of them is to reduce the number of transducers being recorded, if this is possible. Another is to lower the recording frequency from, say, once per 0.5°CA to once per 1°CA. Lowering the recording frequency eases the data acquisition task in two ways. First, it reduces the total quantity of data that is collected. In this way, the entire amount of data required may fit into the on-board buffer, obviating the need to empty the buffer while the ADC is still digitizing and dispatching data. Second, when the recording frequency is low, it may be possible to store the data dispatched by the ADC directly in PC RAM, or even directly onto the PC hard disk. Writing data from the ADC directly onto the hard disk allows vast amounts of data to be collected continuously, albeit at relatively low recording speeds.

In practice, however, lowering the recording frequency may not be an option. This is because the Shannon sampling theory requires that the recording frequency must be at least twice the highest frequency present in the signal [Doebelin, 1975]. In fact, somewhat more than twice the highest frequency is usually required in order to achieve good accuracy when the digitized signal is subsequently reconstructed. A simple example will help clarify the implications of the Shannon theory. Suppose that the cylinder gas pressure in a spark-ignition engine exhibits oscillations having frequency of 16 kHz due to pre-ignition (knock). At 4000 rpm, the data acquisition system must be capable of capturing the entire set of eight engine variables (including cylinder gas pressure) every 0.75°CA (i.e., 32,000 times/s). Therefore, a reduction in recording frequency from, say, once per 0.5°CA to once per 1°CA would not allow the knock frequency to be determined accurately from the digitized data.

1.5.8 Shaft-Encoder

Figure 1.6 shows the data acquisition system receiving an incoming (external) clock signal which is normally generated by a shaft encoder driven directly by the engine crankshaft. A shaft encoder generates an output signal consisting of a train of clock pulses, for example, one pulse per degree of crankshaft rotation. There are several types of shaft encoder available commercially. A common optoelectronic type consists of a plastic disc with a photographically produced pattern of transparent and opaque areas, through which light is directed to photosensors. The disc is driven by the engine crankshaft. There are two main signal outputs from this type of encoder: the first comprises a train of clock pulses; the second comprises just a single pulse per revolution of the crankshaft. This single pulse can be physically aligned with, say, the top dead center position of the piston, and used as the cycle reference signal discussed previously in a previous section (see also Figure 1.6). Although this type of optoelectronic shaft encoder is used often for internal combustion engine research, it can be susceptible to damage from engine-induced vibrations. A more robust type of shaft encoder uses a metal tooth-wheel. A sensor detects the passage of the teeth and generates a train of clock pulses, one per tooth. A second sensor detects the passage of an additional single tooth (axially displaced), and provides the once-per-revolution cycle reference pulse. An alternative approach for generating clock pulses is to use the starter-ring teeth on the rim of the engine flywheel, in conjunction with a suitable sensor. This approach has the disadvantage that the flywheel is unlikely to have 360 teeth or an integer fraction of 360. Finally, it is worth mentioning that electronic phase-lock-loop circuits can be used to electronically halve or double the number of pulses generated per revolution by gearwheel teeth or optical shaft encoders.

1.6 Diesel Fuel Line Pressure and Needle Lift

1.6.1 Fuel Pipe Pressure

In diesel engines, it is often necessary to record the pressure in the fuel injection system. Usually, of interest is the fuel pressure at three main locations: the inlet to and outlet from the pipe conveying the fuel from the pump to the injector, and the pressure within the injector itself. The maximum pressure at these locations can often exceed 1000 bar. Furthermore, between successive injections, the residual pressure at these locations does not fall to zero, but remains at a high level due to the presence of a non-return valve at the exit from the fuel pump.

A transducer monitoring the pressure at these locations must meet a number of requirements, including:

(i) It must be small, so that it does not introduce significant cavities in the fuel line or the injector passages, which might interfere with the fuel pressure dynamics.

(ii) It must not alter significantly the stiffness of the fuel line and injector, which might also interfere with the fuel pressure dynamics.
(iii) It must have a sufficiently high frequency response so that it can respond adequately to the rapid fuel pressure pulsation occurring during the injection process; at the same time, it must also be capable of recording the steady residual pressure between injections.
(iv) It must have a high natural frequency in order to avoid resonant output oscillations.

Strain-gauge type pressure transducers are capable of meeting the above requirements, and are widely used for injection pressure measurements. The strain gauges are used in Wheatstone bridge configuration, which often includes compensation for changes in the gauge temperature. Miniature piezoelectric pressure transducers can also be used. These have excellent dynamic response to the fast pressure pulsation, but are not capable of measuring reliably the steady residual pressure between injections. This is because the electrical charge generated by piezoelectric transducers is prone to dissipation through leakage.

Figure 1.8 shows the principle of operation of the strain-gauge pressure transducer. The figure also shows examples of the transducer installation in the high-pressure fuel line and on the injector itself. Figure 1.8(a) shows the strain-gauge pressure transducer wired in a Wheatstone bridge configuration. The bridge consists of two identical strain gauges and two identical fixed resistors (R_1 and R_2). Usually the strain gauges have the same resistance as the fixed resistors (e.g., 2000 ohm). Often just the two gauges are incorporated physically into the structure of the pressure transducer, although there are many examples of pressure transducers incorporating the two gauges and the two fixed resistors within their casing. The active strain-gauge shown in the figure measures the pressure in the fuel line, while the dummy gauge compensates for changes in the fuel temperature. This compensation eliminates drift in the instrument output due to changes in the fuel temperature. The pressure transducer operates as follows. A change in the fuel pressure causes a strain in the active gauge, and this changes its resistance. The Wheatstone bridge and associated electronic circuits detect this change in resistance and convert it to an output voltage that is linearly related to fuel pressure. However, changes in the fuel temperature also cause unwanted changes to the active gauge resistance, which must be compensated for. This is achieved by means of the dummy gauge. This gauge is also exposed to the fuel temperature, but is carefully positioned so that it is not strained by fuel pressure fluctuations. In this way, changes in fuel temperature affect the resistance of both gauges equally and cancel each other out.

Figure 1.8(b) shows a convenient way of installing the pressure transducer in the high-pressure fuel pipe without having to cut the pipe. A transducer boss is welded (brazed) onto the pipe. A small diameter communicating-hole is then drilled through the boss and the fuel pipe. An alternative boss design (not illustrated) involves clamping the boss onto the pipe and sealing the fuel pressure by means of small O-rings.

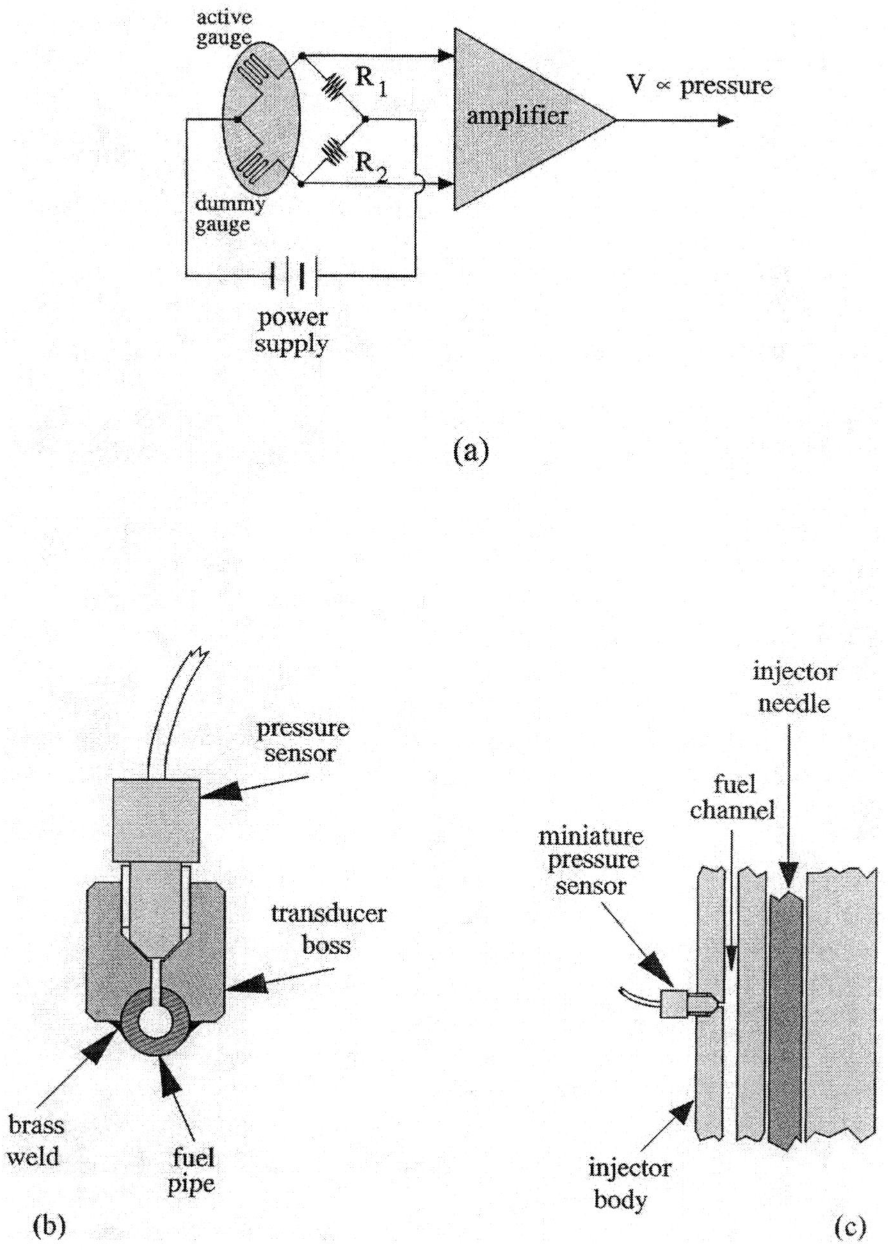

Fig. 1.8 Sensing fuel line pressure.

1.6.2 Injector Needle Lift

The movement of the injector needle is another parameter often monitored on diesel engines. The motion of the needle can help determine the start of fuel delivery to the combustion chamber and can reveal the presence of more than one injection event per engine cycle.

Often the injector needle movement is monitored by means of a variable inductance system, which operates on the frequency modulation (FM) principle illustrated in Fig. 1.9(a). The inductive sensor coil is part of an electronic oscillator circuit which operates at a frequency of approximately 2 MHz. The oscillator frequency is partly determined by the inductance of the sensor coil. When the injector needle rises, the armature rod penetrates further into the sensor coil, causing a change in the coil inductance and the oscillator frequency. The changes in oscillator frequency are detected by electronic circuits which convert them to changes in output voltage from the instrument. The electronic circuits within the instrument are usually designed so that the output voltage is related linearly to the displacement or "lift" of the injector needle. Figure 1.9(b) shows the way in which the sensor coil can be incorporated into the fuel injector.

The FM unit described above can also be used with a sensor that relies on a change in capacitance rather than inductance. This alternative arrangement is also shown in Fig. 1.9(a).

A simple alternative method of monitoring the injector needle motion is shown in Fig. 1.9(c). The movement of the needle, which is made out of a ferromagnetic material, alters the gap reluctance, which, in turn, alters the electromagnetic coupling with the coil. This induces an electromotive force (emf) in the coil. The voltage across the coil terminals is usually up to 100 mV or so, and can be monitored after amplification. This voltage output is proportional to the velocity of the injector needle, rather than its displacement. The coil required with this method is considerably bulkier than for the FM system described above. This feature tends to restrict its application to larger injectors, or to specialist applications where the injector was designed to accommodate the bulkier coil.

Other types of sensor are available which could be used to monitor the injector needle movement. These include Hall-effect semiconductors and differential transformers. The principle of the linear variable differential transformer (LVDT) is illustrated in Fig. 1.10. The primary coil, on the left-hand side of the diagram, is continually being excited by a sinusoidal voltage (V_{ex}) having a constant amplitude (say, 10 V) and a constant frequency (say, 10 kHz). Both of the two secondary coils have induced in them sinusoidal voltages of the same frequency (10 kHz). However, the amplitude of the voltage induced in each of these two coils depends on the position of the coil core, which is made of a ferromagnetic material and is joined to the injector needle. When the core is pushed up from the midpoint position by the injector needle, the electromagnetic coupling of the upper secondary coil improves. As a result, the amplitude of the induced voltage in the upper coil increases and that in the lower coil decreases. When the core is pulled down from the midpoint position, the reverse happens, and the induced voltage in the lower coil increases while that in the upper coil decreases (a 180 deg voltage phase shift also occurs). The two secondary coils are often wired in series-opposition, as shown in Fig. 1.10. The amplitude modulated (AM) output signal (V_{out}) is demodulated by a phase-sensitive AM detector. The voltage output from the detector is linearly related to the movement of the core and, hence, the displacement of the injector needle [Doebelin, 1975].

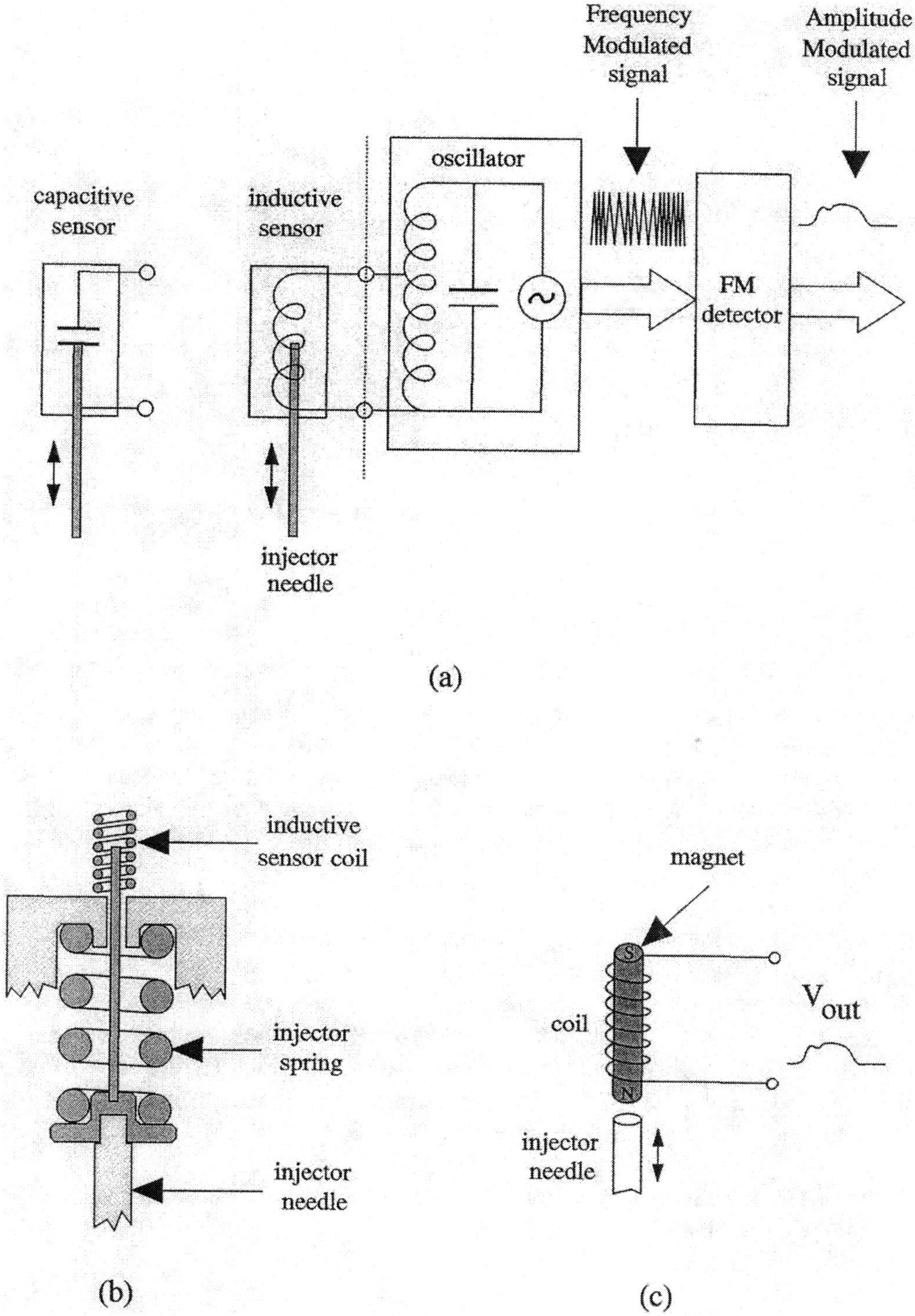

Fig. 1.9 Sensing injector needle movement.

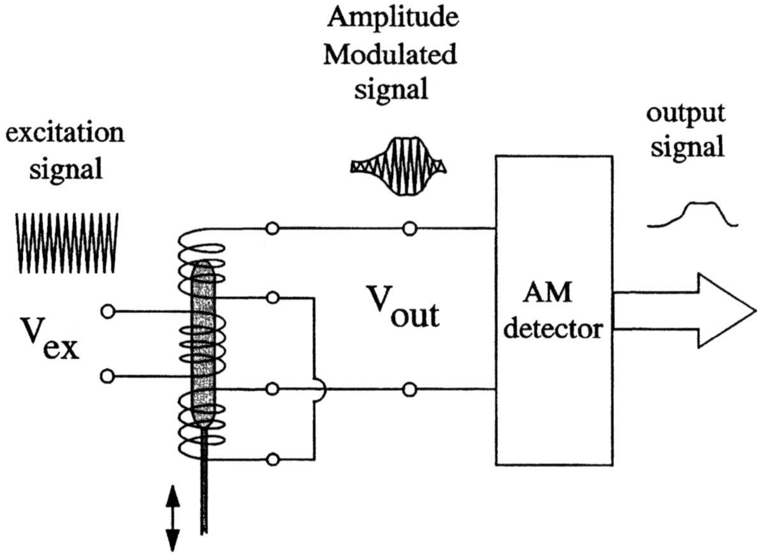

Fig. 1.10 Linear variable differential transformer (LVDT) system.

1.7 Exhaust Gas Analysis

1.7.1 Introduction

Measurements of exhaust gas emissions are needed not only because of legislation, but also because of the insight that they provide into engine combustion processes. The exhaust gases that are currently legislated include: carbon monoxide (CO), unburned hydrocarbons (μHC), nitrogen oxides (NO_x), plus particulate matters from diesel engines. In addition, oxygen and carbon dioxide (CO_2) are also measured to determine the air/fuel ratio. Measurement techniques for these emissions will be described in the following sections.

Exhaust emissions are covered by regulations in most of the industrial countries. Standard running procedures for the vehicles to be checked are specified by several types of test cycles, depending on the vehicle types and countries [Degobert, 1995]. Each cycle consists of a sequence of speeds as a function of time. To simulate these cycles, vehicles are placed on chassis dynamometers within a well-controlled environment (temperature, relative humidity). The chassis dynamometer reproduces the load and inertia of the vehicle when driven on the road. The mechanical energy transmitted to the rollers by the wheels is absorbed by a braking system.

For trucks and buses, it is difficult to use chassis dynamometers and hence, tests are performed on the engine itself, which is mounted on an engine test bench equipped to produce the standard running cycles.

For SI engine equipped vehicles, the constant volume sampling (CVS) method is used. In this method, the exhaust gases are diluted with filtered ambient air to maintain a constant total flow rate (exhaust + air) under all running conditions. This simulates the dilution conditions of the exhaust in the ambient air at the tail pipe exit. Throughout the test a fixed fraction of the diluted mixture is collected in plastic bags, giving cumulative emission results.

For diesel engines, the risk of condensation of the heavier hydrocarbons in the bag requires the use of a heated line (190°C) for the continuous analysis of μHC, an integrator calculating the cumulative emissions over the cycle. A dilution tunnel of sufficient length is also needed before the sampling of particulates takes place. Other pollutants are then collected in plastic bags downstream of this dilution tunnel as in SI engines.

In most engine research laboratories, exhaust measurements are often carried out on an engine test bench. Exhaust gas is sampled directly from the exhaust systems and analyzed without dilution.

In the following sections, the principle and operation of each instrument will be described.

1.7.2 CO and CO_2 Analyzer

Infrared analyzers are used for measuring concentrations of carbon monoxide (CO) and carbon dioxide (CO_2). Their principle of operation is based on the absorption of infrared radiation by these gases. For example, carbon monoxide absorbs at about 4.6 μm; whereas carbon dioxide absorbs at about 4.2 μm. The amount of radiation absorbed is given by Beer's law,

$$a_\lambda = 1 - \exp(-C_i Q_\lambda L) \tag{1.9}$$

where C_i = the concentration of species i
Q_λ = the absorption efficiency
L = the optical path length

Fig. 1.11 shows the key components of a nondispersive infrared gas analyzer (NDIR) employing a differential detector. Because NDIR instruments use the total absorption over a given wavelength, no dispersing element (prism or grating) is required to resolve detailed spectral lines, hence the term "nondispersive."

The infrared sources are normally heated filaments that emit blackbody radiation over a wide range of wavelengths. In Fig. 1.11(a) the detector cells are filled with the gas that is to be measured, so that they absorb the radiation in the absorption band of that gas. The two cells are identical and are separated by a diaphragm that moves between the plates of a capacitor. The energy absorbed in each cell causes the cell pressure to rise. One beam passes through an absorption cell containing the gas to be analyzed (sample cell), and the other beam passes through a cell (reference cell) filled with a non-absorbing gas (nitrogen). The detector cell will absorb less radiation on the top than on the bottom, because the attenuation in the sample cell causes the diaphragm to deflect in proportion to the difference in the rates of energy absorption. Since the

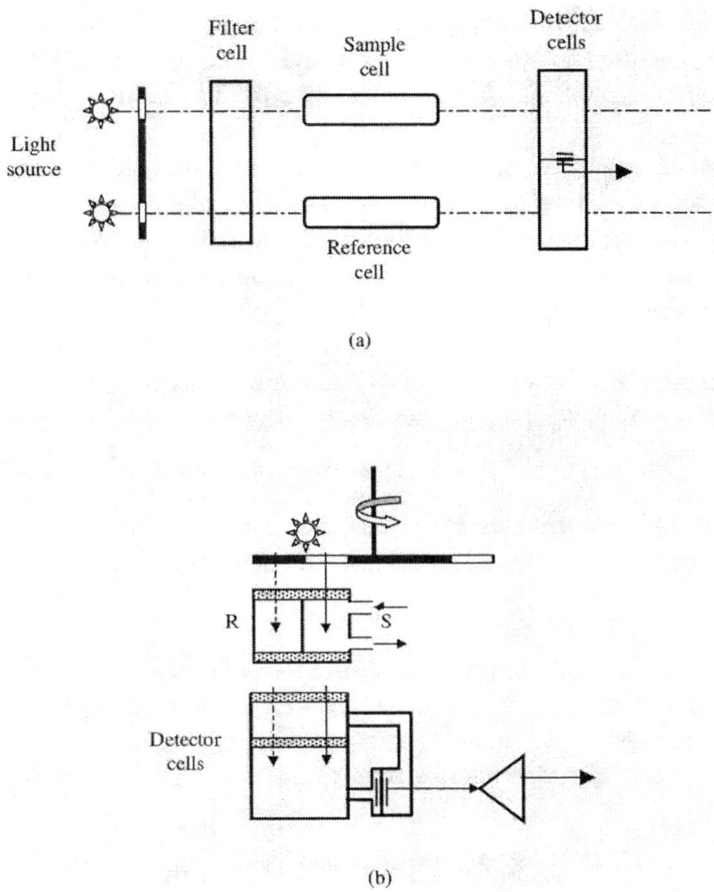

Fig. 1.11 NDIR analyzers with differential detectors.

deflection will depend on the gas concentration in the sample cell, the amount of deflection can be sensed through the capacitance change and calibrated to read in units of concentration. The infrared sources are normally modulated at a frequency of about 10 Hz by an optical chopper for zero referencing.

Since there is an overlap between the absorption bands of CO and CO_2, the presence of one will affect the reading of the other that is to be analyzed. This problem is eliminated by placing a filter cell between the infrared sources and the sample and reference cells. For example, the filter cell would be filled with CO if CO_2 is to be measured, hence any CO in the sample will not lead to further infrared absorption (and vice versa if CO is to be analyzed).

In order to minimize the drifts caused by aging of two independent infrared sources or any fouling between the emitter and the receiver, single-beam instruments have been developed, which have a modulator that causes the beam to pass alternately through the reference cell and the sample cell. This can be realized through the arrangement shown in Fig. 1.11(b). The detector cells are arranged in series with an infrared transparent window between them. The difference in the amount of radiation that can be absorbed by the two detector cells varies with the infrared radiation incident on them. In the reference mode, the infrared beam is strong enough to heat up both cells to a similar degree. When the infrared radiation passes through the sample cell, attenuated infrared results in more absorption in the top cell than in the bottom cell, which can then be calibrated to give the gas concentration.

NDIR correlation analyzers incorporating a solid-state infrared detector have also been developed, as shown in Fig. 1.12. A chopper disc has two cells in the form of filled segments, one with nitrogen and the second with the gas to be measured. The gas concentration in the sampling chamber is determined from the ratio of the detector outputs with the nitrogen and the gas cells, respectively.

To minimize window fouling, the sample should be filtered to remove particulates and condensation (of water vapour or high molecular weight hydrocarbons) by cooling and removing the condensate at low temperature before warming the sample to ambient temperature.

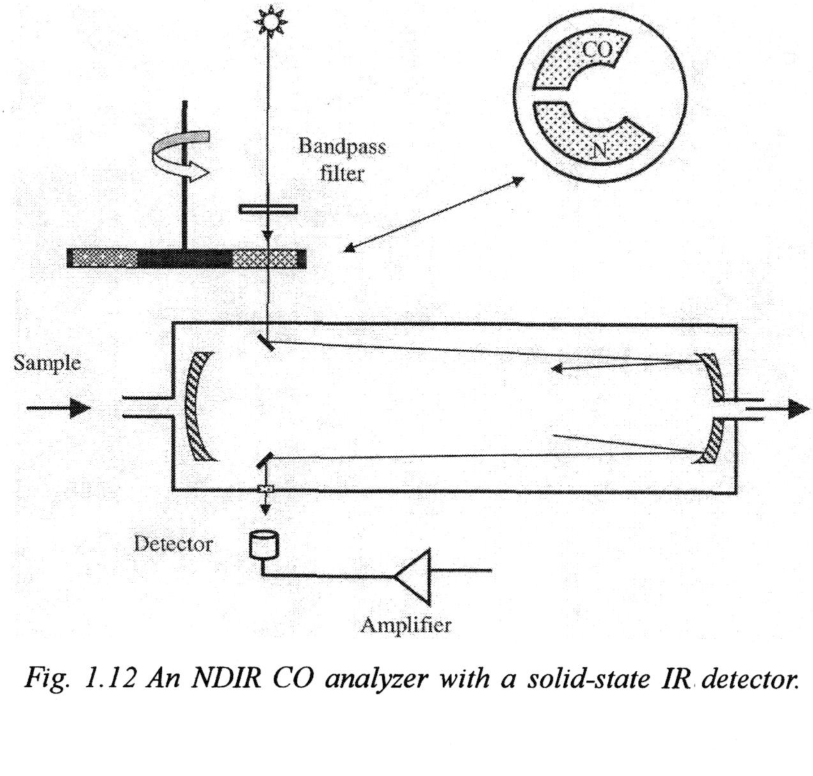

Fig. 1.12 An NDIR CO analyzer with a solid-state IR detector.

1.7.3 NO_x Analyzer

Nitric oxide (NO) is measured by the chemiluminescence technique. Chemiluminescence is the light emission from an atom or molecule that is electronically excited by a chemical reaction. The chemical reaction used in the measurement of NO is

$$NO + O_3 \Rightarrow NO_2^* + O_2 \Rightarrow NO_2 + O_2 + h\nu$$

which states that, following the reaction of NO with ozone, nitrogen dioxide (NO_2) is produced at an excited state and emits light ($h\nu$) as it returns to its ground electronic state. The amount of light emitted is proportional to the concentration of nitric oxide if the reactor is at a fixed temperature.

Fig. 1.13 shows the arrangement of a NO_x analyzer. The ozone is generated by electrical discharge in oxygen at low pressure. The light emission in the reactor is measured by a photomultiplier, and then amplified before it is converted into NO concentration readings. In view of likely variations in the photomultiplier gain or other parameters that affect the output, it is essential to calibrate the analyzer regularly during an experiment. Calibration gases are used to set the zero (with nitrogen) and check the span (with known amount of NO).

In addition to nitric oxide, the exhaust of an engine also contains nitrogen dioxide. NO_x is used to denote the sum of the nitrogen oxides. Nitrogen dioxide is measured by passing the sample through a catalyst that converts the nitrogen dioxide to nitric oxide. By switching the converter in and out of the sample line, the concentrations of NO and NO_x ($NO + NO_2$) can be determined.

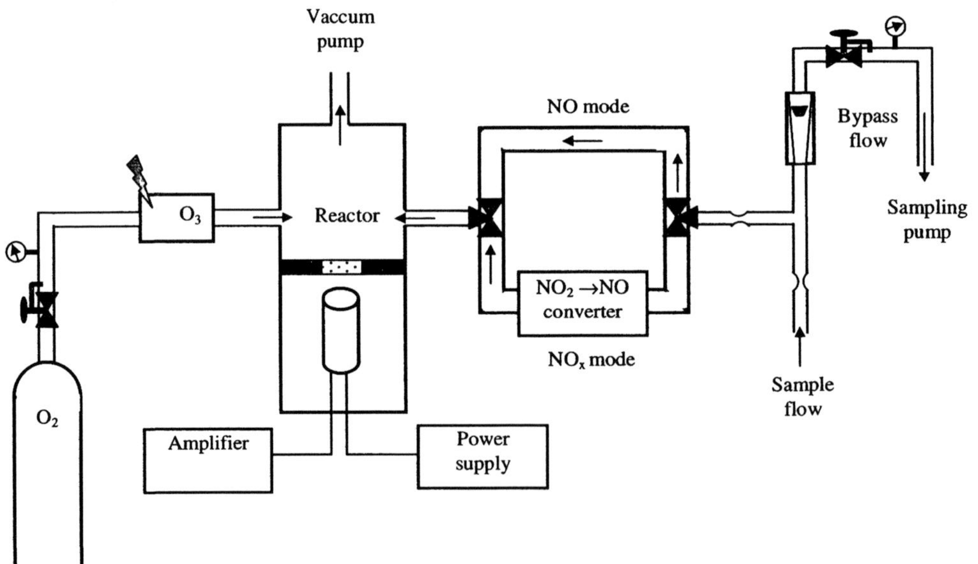

Fig. 1.13 Schematic of a chemiluminescence NO_x analyzer.

A particularly ingenious NO_X analyzer has been developed by Cambustion Ltd. The system has a very fast response (about 4ms), since the reaction chamber has been miniaturized and the chemiluminescence is detected remotely by a photomultiplier via a bundle of optical fibres. The system is capable of making cycle-resolved measurements. The principal feature is similar to a fast FID that is to be described in detail in Chapter 4.

1.7.4 Measurement of Unburned Hydrocarbons

1.7.4.1 Total Unburned Hydrocarbons

Infrared analyzers described in Section 1.7.2 can also be used for measuring the unburned hydrocarbons in the exhaust. However, their output is difficult to interpret as different hydrocarbon species have different absorption spectra. For an instrument calibrated with n-hexane, aliphatics yield nearly 100% response, while olefins and aromatics do not, as shown in Table 1.2. This type of analyzer is normally incorporated in the portable systems used for qualitative measurements.

Table 1.2 Relative response of NDIR and FID instruments: calibration with n-hexane

Hydrocarbon	Average relative responses based on carbon	
	NDIR	FID
Paraffins:		
• methane	30	110
• ethane	100	102
• propane	103	100
• i-butane	101	–
• n-butane	106	100
• i-pentane	99	100
• n-pentane	104	102
• n-hexane	100	100
• n-heptane	97	101
Olefins:		
• ethylene	9	100
• propylene	31	100
• 1-butene	53	–
• 1-pentene	57	–
• 1-hexane	61	–
Aromatics:		
• benzene	2	98
• toluene	13	98
Acetylenics:		
• acetylene	1	99
• methylacetylene	16	99
• ethylacetylene	32	99

The preferred technique for the total unburned hydrocarbon measurement is the flame ionization detector (FID). Introduction of hydrocarbons into a hydrogen flame produces, in a complex process that will be described in more detail in Chapter 3, electrons and positive ions. The ions generated can be detected in the form of an electric current between an electrode at a high voltage and a ground electrode (Fig. 1.14). Then the current flow corresponds very closely to the number of carbon atoms present. Table 1.2 shows that the response is slightly dependent on the molecular structure. The presence of some atoms in the molecule can suppress the ionization current from the carbon atoms, particularly the oxygen in primary alcohol [Degobert, 1995]. Care should be exercised when oxygenated compounds (alcohols, aldehydes, and ethers) are present in the exhaust.

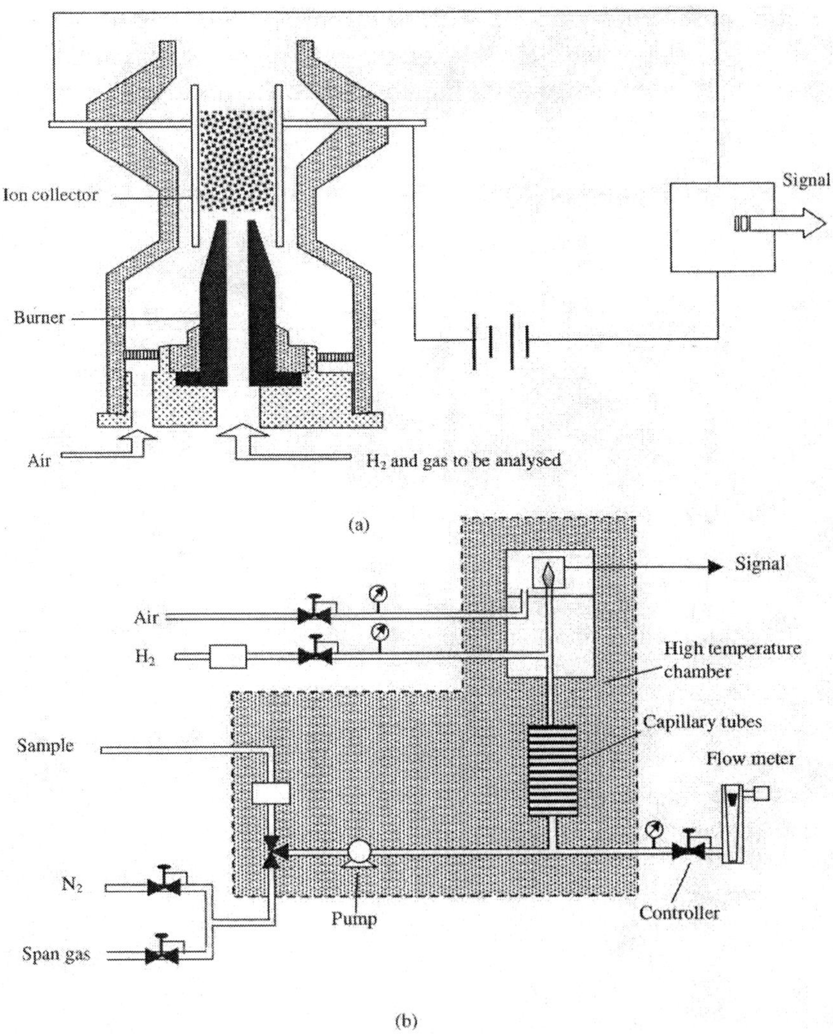

Fig. 1.14 Schematic of an FID hydrocarbon analyzer (a) and an enlarged view of its burner (b).

Fig. 1.14 shows a complete FID analyzer and an enlarged view of its burner. The sample is mixed with hydrogen and burned in air. Air of high purity is used to reduce the risk of introducing hydrocarbons or other species. The hydrogen and sample flows must be closely regulated, as the instrument response is directly related to the sample flow rate, and dependent on the fuel flow rate. The flows are regulated by maintaining fixed pressure differences across devices such as capillary tubes, as shown in Fig. 1.14(b), or critical flow orifices.

In Fig. 1.14(a), a potential of about 100 volts is applied between the electrodes. The signals are small and have to be amplified. The calibration of an FID system is achieved by zeroing the instrument with nitrogen and using calibration gases of known hydrocarbon concentration (propane in nitrogen) close to that in the sample to be analyzed.

Because the flame ionization detector gives no information about the type of hydrocarbons in the exhaust or their average hydrogen to carbon ratio, it is preferable to report measurements as ppmC (parts per million carbon) rather than as ppm CH_4 or C_3H_8 or equivalent.

To prevent heavy fractions of hydrocarbons from condensing, the sampling line is normally heated, particularly when sampling from a diesel engine exhaust, where the exhaust gas temperature is lower than that in an SI engine.

For cycle-resolved measurements, a fast response FID system may be used, which will be described in detail in Chapter 3.

1.7.4.2 Detailed Analysis of Unburned Hydrocarbons

The flame ionization detector itself cannot distinguish between different hydrocarbon species. However, chromatographic methods can be used for identifying particular hydrocarbon species in the exhaust gas.

Chromatography is not itself an analytical method, but a method for separating the components of a mixture without destroying the elements of the mixture. Besides being a nondestructive technique, another advantage of chromatography is that minimal fractions of samples are required.

Chromatographic methods are based on the differential migration of the components in the mixture to be separated as the mixture is flowing through long tubes (columns) coated with certain absorbents. There are two main categories of methods used to analyze exhaust pollutants: gas chromatography (GC), and liquid-phase chromatography (LC). In GC, the gas sample and carrier gas are passed into coated capillary columns several dozen meters in length. Since different molecules pass through the column at different rates, when the carrier gas leaves the column it contains the different molecules in discrete groups at one instant, and thus the different molecules can be identified by the residence time. The resolution of compounds attainable by high-performance liquid-phase chromatography (HPLC) is about an order of magnitude lower than can be achieved with capillary-column GC, hence it is often used as a high-resolution pre-fractionation technique for complex mixtures prior to analysis by a GC [Williams, 1990]. In addition, HPLC is superior to GC for the analysis of high-molecular weight polyaromatic compounds (PAC) that can be found in diesel exhaust.

At the exit of a chromatographic system, the isolated components must be detected, identified, and quantified by measuring a particular physical or chemical property specific to the components to be analyzed. The most widely used detector in GC is the flame ionization detector (FID). Other detectors are available for detection of a specific group of compounds. For example, a thermoionic detector (also called nitrogen phosphorus detector, NPD) is highly sensitive for measuring nitrogen and phosphorus derivatives; a flame photometry detector (FPD) is used for sulphur-specific detection.

1.7.5 Fourier Transform Infrared Emission Analyzer

Fourier transform infrared (FTIR) emission analysis is a new technique for the measurement of automotive exhaust emissions. In an FTIR analyzer, infrared light from a suitable source passes through a scanning Michelson interferometer, and subsequent Fourier transformation of the interference fringes (interferogram) then gives a plot of intensity versus frequency. When a gas sample is placed in the beam after the interferometer, it absorbs particular frequencies, so that their intensities are reduced in the interferogram and the ensuing Fourier transform is the infrared absorption spectrum of the sample. Since each component is identified by its characteristic frequency on the absorption spectrum, simultaneous detection of multicomponents can be carried out for any species with an infrared absorption band.

When the FTIR emission analyzer first appeared on the market, it received a lot of attention because FTIR technology provides simultaneous and continuous analysis of multicomponent gases including CO, CO_2, NO, N_2O, NO_2, formaldehyde, methanol, and other primitive hydrocarbons. However, the technique has not found wider applications as originally expected, due to problems of cross sensitivity of different species, slow response, and less accuracy than conventional analytical techniques. In spite of the problems, the technique is highly valuable in measuring unregulated pollutants. Several reports have been published recently for the continuous measurement of nitrous oxide (N_2O) [Ballantyne et al., 1994], nitrogen dioxide (NO_2) [Majewski et al., 1995], ammonia (NH_3) [Laurikko et al., 1995], and light hydrocarbons [Laurikko et al., 1993]. Adachi et al. [1997] carried out a detailed study on the trade-off between the sensitivity and response of an FTIR system, which was then used in the measurement of exhaust gases from SI, diesel, and gas engines.

1.7.6 Oxygen and Air/Fuel Ratio Measurements

Measurement of oxygen concentration in the exhaust is used to evaluate the air/fuel ratio of the in-cylinder charge.

1.7.6.1 Galvanic Oxygen Sensor

The oxygen sensor used in portable analyzers is of the disposable galvanic cell type. The cell comprises a PTFE membrane with a gold coating that acts as the cathode. A silver or lead anode is also immersed in the electrolyte (potassium chloride gel). When oxygen diffuses through the membrane, a current is generated that is proportional to the partial pressure of the oxygen in the sample.

1.7.6.2 Lambda Sensor

A zirconium oxide sensor can also be used to measure the oxygen content in a gas mixture. When heated above 300°C, the zirconium oxide behaves as a solid electrolyte, allowing the movement of oxygen ions through it. As shown in Fig. 1.15(a), oxygen is ionized by the platinum electrode on each side of the zirconium oxide. The movement of oxygen ions through the zirconia is due to a difference in oxygen concentration between the two sides of the zirconia. When an oxygen ion moves through the zirconia, it picks up two electrons from one platinum electrode and leaves them on the other one, where it recombines to an oxygen atom. This leads to a potential difference between the two electrodes, which is proportional to the difference in oxygen concentration. Therefore, if the oxygen concentration is known on one side, the reference side, one can calculate the oxygen concentration on the other side by measuring the potential difference between the electrodes. The reference side is normally exposed to air that has an almost constant oxygen concentration of 20.8%.

The sensor can either be mounted in the exhaust stream or to take a sample of the exhaust to the analyzer. It provides a relatively fast response compared to other types of oxygen sensors. Fig. 1.15(b) shows the sensor voltage output as a function of the oxygen content in the exhaust. It is evident from this diagram that the characteristics of the sensor voltage allow particularly accurate correlation within a very narrow λ range. The zirconium oxide sensor, configured as shown Fig. 1.15, is commonly known as the lambda sensor. It is used for the closed-loop control in a gasoline-fueled car fitted with a three-way catalyst.

In addition to the oxygen content, the temperature of the zirconia body in the lambda sensor is also important since it affects the zirconia's oxygen ion conductivity. The optimum operating temperature is approximately 600°C, and the minimum temperature is 300°C. Therefore, lambda sensors are normally heated by a built-in electrical heater for optimum performance.

1.7.6.3 Lamdascan Air/Fuel Ratio Analyzer

A lambda sensor configured as shown in Fig. 1.15 is only sensitive within a narrow λ range around $\lambda = 1.0$. To extend the range of operation of a zirconium oxide sensor, a heated platinum catalyst is incorporated in the Cussons Lamdascan air/fuel ratio analyzer. In this analyzer, a sample of the exhaust passes through a heated platinum catalyst to ensure complete combustion so that the zirconium oxide sensor operates within its most sensitive range ($\lambda = 1.0$), as shown in Fig. 1.16. When the AFR is such that the residual oxygen concentration is below the optimum measuring range, additional oxygen is added, that is, the sample is doped (usually with air) prior to

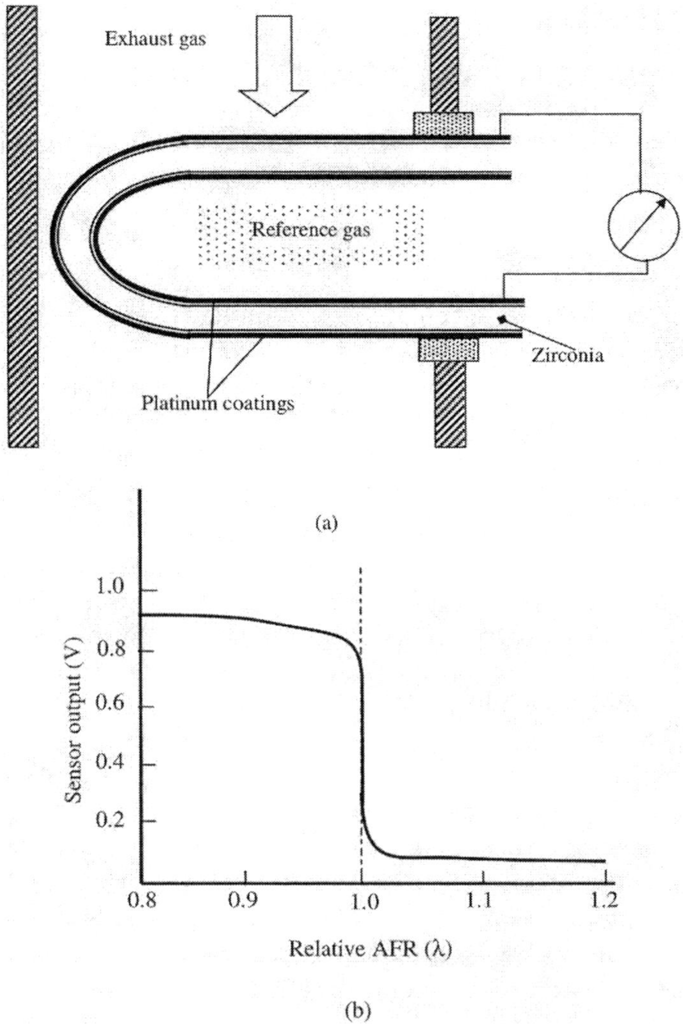

Fig. 1.15 Schematic of a zirconium oxide sensor (a) and its voltage curve at 600°C (b).

the catalyst, so that once again combustion is completed and an accurately measurable residual oxygen remains. The AFRs are then calculated by a microprocessor using the measured value and the pre-programmed hydrogen/carbon and oxygen/carbon ratios of the fuel in use.

1.7.6.4 Universal Exhaust Gas Oxygen (UEGO) Sensor

In order to meet the wide range of air/fuel ratio control in lean-burn gasoline engines, the universal air/fuel ratio heated exhaust gas oxygen (UEGO) sensor has been developed. The sensor is made of three solid zirconia substrates (Fig. 1.17). The first substrate is used as the oxygen pumping cell; the second acts as an oxygen sensing cell; and the third acts as the wall to the oxygen reference cavity. Platinum electrodes are printed on both the first and second substrate

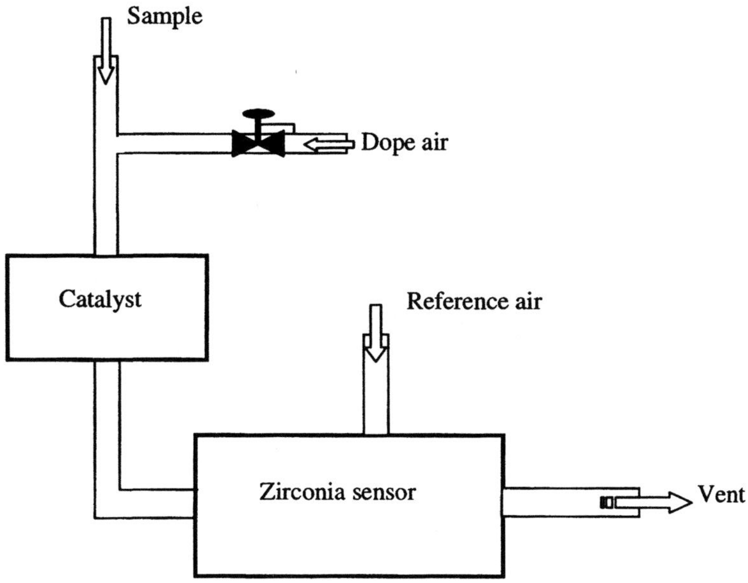

Fig. 1.16 Schematic of a Cussons Lamdascan analyzer.

elements. An insulated electrical heating element is attached to the surface of the pumping cell to heat the whole assembly. The exhaust gas is drawn into the gap between the pumping and sensing cells through microscopic holes in the porous layer.

Meanwhile, a very small constant current (I_{cp}) is supplied to the sensing cell. This current helps pumping oxygen to the reference cavity in order to keep oxygen partial pressure constant in the cavity. The pumping current I_p is the sensor output, and it is controlled to keep the oxygen concentration in the sensing gap at a constant level. In the lean region, oxygen molecules are pumped out of the gap (Fig.1 .17[b]). In the rich region, the direction of the current is reversed and oxygen molecules are brought in from outside to compensate for the loss of oxygen due to its reactions with carbon monoxide and hydrogen in the sensing gap. Therefore, a forward current appears proportional to the concentration of residual oxygen in the exhaust gas of a lean mixture, and a reverse current proportional to the sum of the hydrogen and carbon monoxide concentration due to rich mixture combustion. The air/fuel ratio is then calculated from the data on residual oxygen, hydrogen, and carbon monoxide obtained from the pumping current [Yamada et al., 1992].

A UEGO sensor has a similar dimension to a lambda sensor. It is small and compact, and can be easily installed onto the engine exhaust. It has a much faster response time than any other oxygen analyzers. Commercial UEGO-based air/fuel ratio analyzers have a dynamic range of A/F 10–30, oxygen concentration 0–25%.

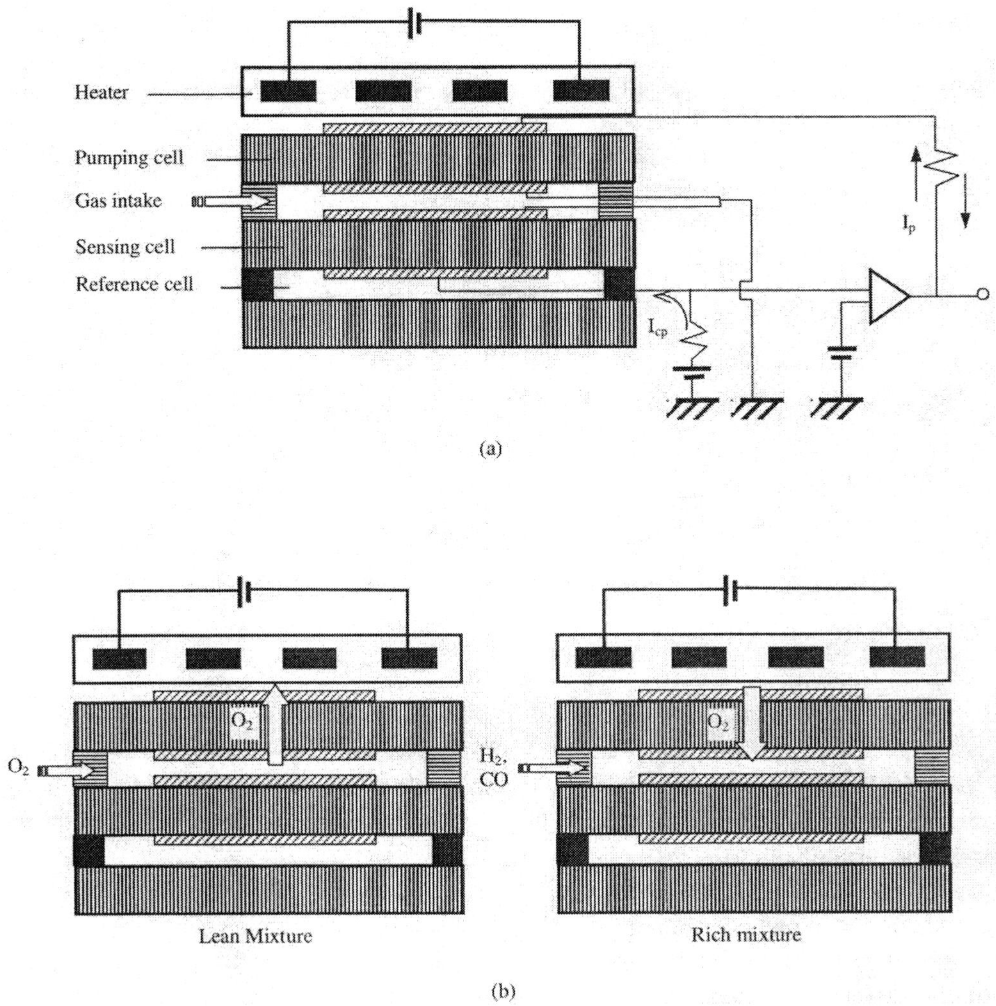

Fig. 1.17 UEGO sensor structure (a) and operational principle (b).

1.7.6.5 Paramagnetic Oxygen Analyzers

Paramagnetic oxygen analyzers are probably the most accurate. Oxygen is one of a few gases that are paramagnetic. If a gas mixture (exhaust gas) flows through a magnetic field, a force is exerted virtually exclusively on the paramagnetic molecules. The force created is then measured and related to the oxygen levels in different ways. There are two main types of oxygen analyzers based on the paramagnetic principle: thermomagnetic analyzers and magnetodynamic analyzers.

The principle of the thermomagnetic oxygen analyzer is illustrated by Fig. 1.18. The gas to be analyzed circulates in an annular chamber, of which the two branches are connected by a horizontal tube with three zones in series: a cold zone in the field of a permanent magnet, a zone heated by a filament placed inside the field of the magnet, and a third zone heated by a second filament outside the magnetic field. The heated filaments in the cross-tube form part of a Wheatstone

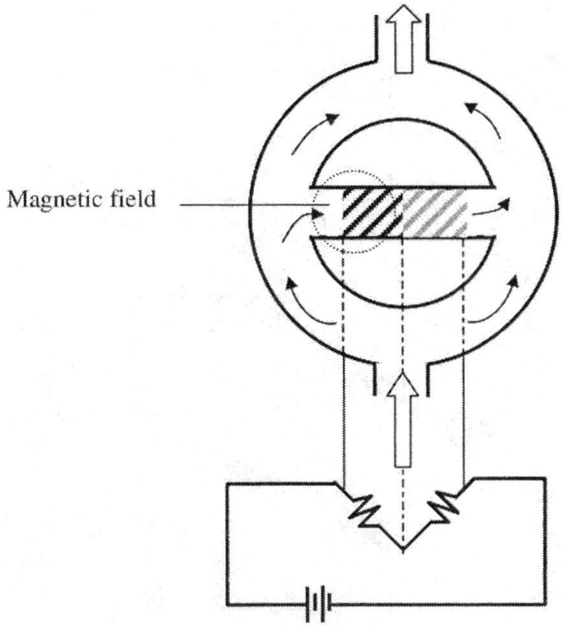

Fig. 1.18 Schematic of the thermomagnetic paramagnetic oxygen analyzer.

bridge. The oxygen molecules are attracted to the magnetic field and are heated in the first heated zone. Their paramagnetism decreases and they are expelled by the colder molecules with greater susceptibility, creating a magnetic wind inside the tube. The flow of gas cools the filament, thereby changing its resistance and leading to an imbalance in the Wheatstone bridge, which can be related to the oxygen level.

The basis of the magnetodynamic oxygen analyzer is shown in Fig. 1.19. The dumbbell-shaped spheres, made of diamagnetic material, are repelled by the magnetic field, and will reach an equilibrium position when the repulsion force is balanced by the torque from an energized coil. When the oxygen concentration changes, the magnetic field strength will also change, and the dumbbell will reach a new equilibrium position. The current required by the coil to maintain the dumbbell in its position is directly related to the oxygen partial pressure of the sample. If the analyzer is operated at a constant pressure, then it can be calibrated in terms of the oxygen concentration. Compared with thermomagnetic analyzers, magnetodynamic analyzers are not influenced by changes in the transport properties of the sample gas or the interference of hydrocarbons and other combustible gases.

1.7.7 Exhaust Smoke and Particulates

1.7.7.1 Measurements of Particulate Mass

Particulates consist principally of combustion-generated carbonaceous material (soot) on which some organic compounds have become absorbed. They can be partitioned with an extraction

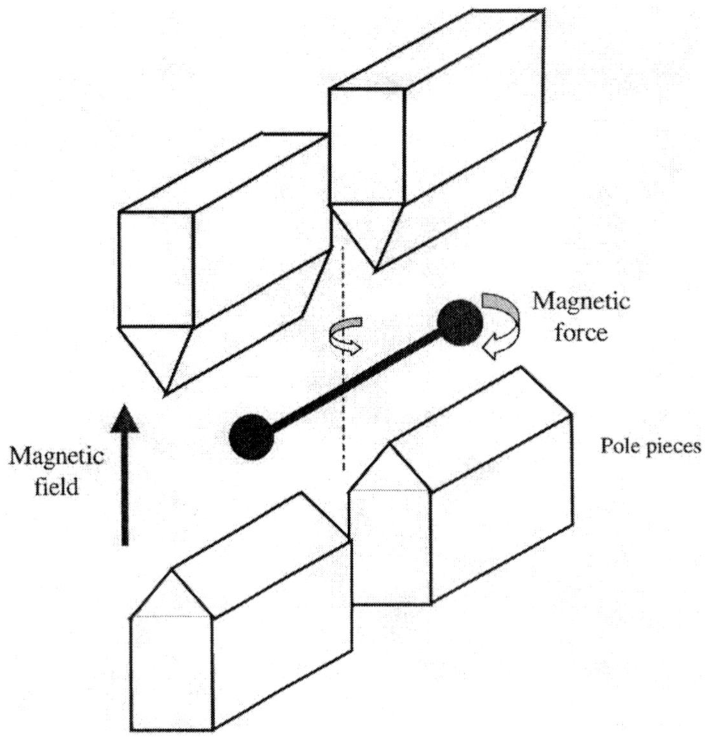

Fig. 1.19 Schematic of a magnetodynamic paramagnetic oxygen analyzer.

solvent into a dry soot fraction and a soluble organic fraction (SOF), or a volatile organic fraction (VOF), depending on the separation process involved. Typically 15 to 30 mass percent is extractable, though the range of observations is much larger (about 10–90 percent).

A variety of systems have been developed for measuring the smoke level in diesel exhaust. The most widely used system is of the Bosch smoke meter type. An automatic version of the Bosch system is the AVL smoke meter, in which a controlled volume of exhaust gas is drawn through a filter paper. The change in the reflectance of the filter paper then corresponds to the smoke level. A value of ten is assigned to a filter paper with no reflection, and a value of zero is assigned to a clean filter paper. In this instrument, it is primarily the carbonaceous material or soot contained in the exhaust gas that is measured. An estimate of the mass of particulates can be obtained from Greeves and Wang's correlation [1981],

$$\text{Particulates (g/m}^3) = 1.024 \times \text{smoke (g/m}^3) + 0.505 \times \mu\text{HC (g/m}^3) \tag{1.10}$$

where μHC is obtained by converting an FID reading of unburned hydrocarbons to g/m^3, and the smoke reading is obtained indirectly by converting the measured Bosch number or filter smoke number (FSN) to the filter load or soot density (g/m^3) based on the MIRA correlation:

$$\text{Smoke (g/m}^3) = A_0 + A_1 \times \text{FSN} + A_2 \times (\text{FSN})^2 + A_3 \times (\text{FSN})^3 \tag{1.11}$$

where $A_0 = -0.0019827868, A_1 = 0.033537412, A_2 = -0.0031243902, A_3 = 0.0031122530$.

Greeves and Wang's correlation (Eq. [1.10]) considers that particulates comprise soot and unburned hydrocarbons with fixed proportions. In practice, the percentage of soot and unburned hydrocarbons varies with operation conditions. Therefore, it should be used only as an estimate of particulates.

Another commonly used method of characterizing particulates is to determine the mass of particulates deposited from the exhaust onto Teflonized fiberglass filters. In the standard mass emission measurement procedure, dilution tunnels are used to simulate the physical and chemical processes the particulate emissions undergo in the atmosphere. In the dilution tunnel, the raw exhaust gases are diluted with ambient air to a temperature of 50°C or less, and a sample stream from the diluted exhaust is filtered to remove the particulate material. Particulates are collected on a filter paper for a fixed volume of exhaust gases.

There are three main methods used to determine soluble or organic fraction of diesel particulates: Soxhlet extraction, thermogravimetric analysis (TGA), and vacuum-oven sublimation (VOS). In the extraction method, the collected sample is subject to boiling organic solvent for a number of hours, during which time the solvent is recycled automatically by a siphoning system. The amount of condensed and adsorbed material is measured either by determining the sample weight loss after extraction or by determining the weight of extracted material after evaporating the solvent. Various solvents have been recommended for the extraction of organic fractions, among which dichloromethane (DCM) is recommended by the EPA [Perez et al., 1984]. The composition of the extracted organic fractions can be further analyzed by gas chromatography (GC), high-performance liquid chromatography (HPLC), or GC/mass spectrometry [Williams, 1990].

In the thermogravimetry and vacuum-oven sublimation methods, the hydrocarbon portion of the particulates is measured by determining weight loss on heating in either a flowing inert gas stream (TGA) [Cuthbertson et al., 1979] or under vacuum (VOS) [Shimpi and Yu, 1981; Halsall et al, 1987]. Extraction methods have the advantage of allowing the investigator to look at the chemical composition of the adsorbed hydrocarbon fraction, whereas TGA and VOS have the advantage of being much less time-consuming.

1.7.7.2 Measurements of Particulate Size

In the last few years, there has been growing interest in and concern relating to the potential health effects of particulate size and composition. There are two main types of particle size analyzers suitable for measurements of exhaust particles: cascade impactors, and electrical mobility analyzers. In an impactor type particle sizer, particles are blown through an orifice onto an impact plate, which forces the stream to suddenly change direction by 90 degrees. Particles with sufficient inertia cannot follow the streamline and impact against the plate, whereas the smaller particles can follow the flow and are then forced through a smaller orifice. Because of the higher flow speed induced by the smaller orifice, smaller particles will be captured in the second stage. Several successive stages can therefore be used to classify the particles by decreasing sizes.

Deposits at each stage are then weighed. By charging particles upstream, the weighing process can be automated by measuring the charges collected at each stage using electrometers. One such instrument is the electrical low-pressure impactor (ELPI). An ELPI is better suited to measuring the mass particle size distribution between 0.3 and 10 μm.

For measuring the size distribution of sub-micron particles, a scanning mobility particle sizer (SMPS) is used. The SMPS uses an electrical-mobility detection technique, and its main components are shown in Fig. 1.20. During operation, the particle-laden sample passes through a single-stage, inertial impactor, in order to remove large particles outside the measurement range. Next, the flow passes through a bipolar ion neutralizer, which imparts a high level of positive and negative ions. The charged and neutral particles now enter the differential mobility analyzer (DMA), where particles are separated according to their electrical mobility. Electrical mobility is inversely related to particle size. An electric field inside the DMA influences the flow trajectory of the charged particles. Only positively charge particles within a narrow range of electrical mobility have the correct trajectory to pass through an open slit near the DMA exit. By exponentially ramping the electric field in the DMA over a user-selected period of time, the entire particle size distribution is measured to a high degree of accuracy. It may take a few minutes to scan through the size range of 10 nm to 1 μm. After exiting the DMA, the classified particles enter a condensation particle counter (CPC), which provides a measurement of particle concentration. Thus, the SMPS can provide both the particle size distribution and the concentration of each particle size.

1.8 Optical Access

1.8.1 Introduction

The necessity for optical access to a combustion chamber arises from the application of optical techniques to observe fluid flow, mixture formation, and combustion and pollutant formation inside the combustion chamber. One optical access is sufficient for direct observation of combustion. For most optical techniques, two or more optical accesses are required when a light source is employed to illuminate or excite in-cylinder species. Numerous research engines have been designed to allow the optical investigation of in-cylinder flow and combustion phenomena. Their design features are usually matched to the requirements of the optical techniques, thus each design tends to have a rather limited scope of application. For example, optical engines are frequently constructed to atypically low compression ratios. Some optical engines can only operate for a very short period before the optical windows are blocked by combustion deposits. However, the amount of information gained from measurements in these engines has enabled engineers to visually "see" the influence of a particular engine modification on in-cylinder flow, mixture distribution, and combustion so that an optimized engine can be designed.

1.8.2 Materials for Optical Windows

There are a number of transparent materials that can be used as optical windows. They may be classified as plastics, optical glass, and crystals. Plastics include materials made of

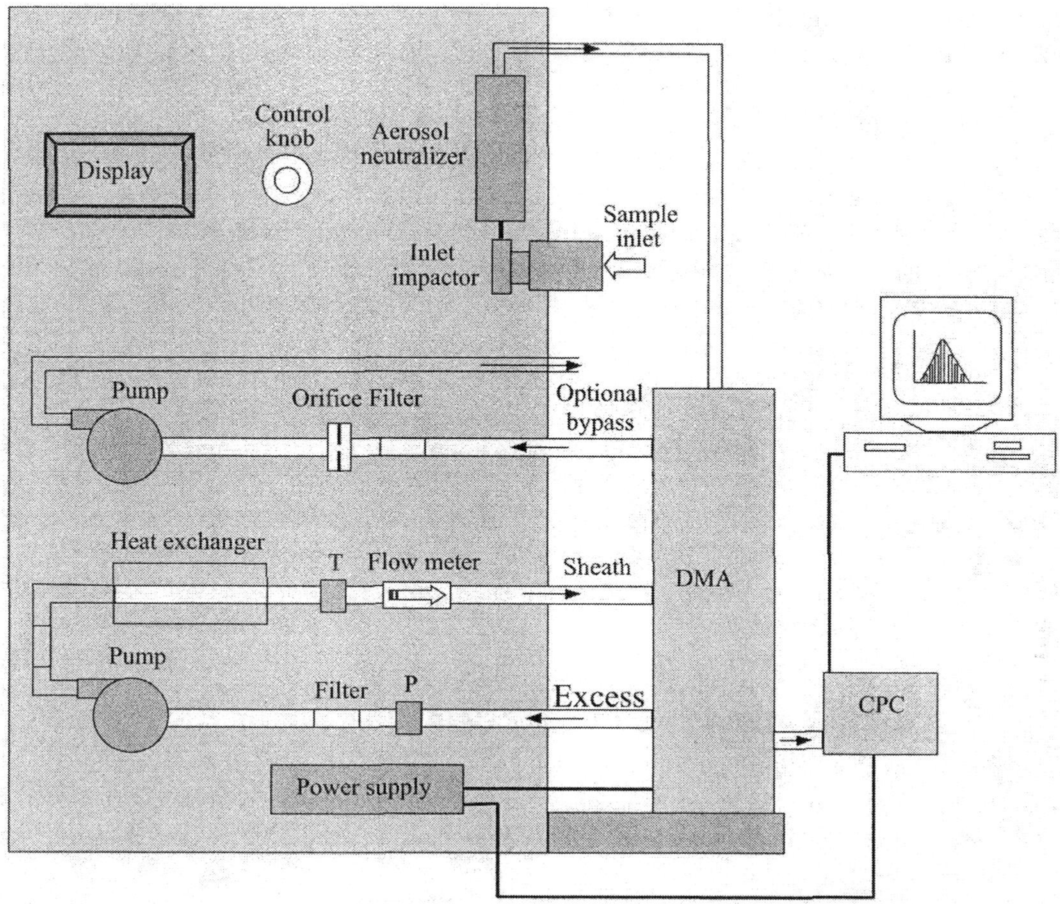

Fig. 1.20 Schematic of a scanning mobility particle sizer (SMPS).

polymethylmethacrylate (e.g., Plexiglas) or polycarbonate (Lexan). They are strong enough to be used for motoring operation. For firing operation, the high pressure and temperature do not permit the use of plastics.

For applications in the near ultraviolet (UV) and visible range, three types of optical glasses have been used as optical windows. BK7 is a low-index crown glass. It exhibits excellent transmission from 350 nm to 2000 nm. At the UV end of the spectrum, the material is not usable at wavelengths lower than 300 nm. The other two are fused silica and fused quartz (Pyrex®). Both materials exhibit transmission characteristics that bear some similarity to the optical glass with which they are closely related. However, their simpler chemical composition leads them to have improved transmission in the UV region of the spectrum. The fundamental difference between these two materials is in the method of manufacture. Most fused quartz is produced by the melting and re-fusing of silica sand and natural quartz. By comparison, fused silica is produced by the flame hydrolysis of a silica halide. The resultant material is thus much purer, and free from the sites that could cause absorption in the UV region or problems with internal absorption in the use of high-power lasers. Consequently, the transmission range of fused silica is extended to 170 nm, while that of fused quartz is limited to 250 nm.

Sapphire is single-crystal aluminium oxide. It has excellent mechanical and optical properties. It is extremely hard and has a strength that is retained even at elevated temperatures. In addition, it is resistant to virtually all forms of chemical attack. It has a transmission band that extends from the UV at around 180 nm, throughout the visible, and to the IR region at 5500 nm. Due to its higher refractive index, sapphire suffers more reflection losses than other optical materials. The high cost of this versatile material tends to limit its use to special applications.

1.8.3 Arrangement of Optical Access

A production engine is normally constructed with multiple cylinders of identical design. Each cylinder consists of a cylinder head, a cylinder block, and a crankcase. The cylinder head wall, cylinder block liner, and piston top form the combustion chamber, as shown in Fig. 1.21. There are therefore three ways to gain optical access to the combustion chamber through: the top (cylinder head), side (cylinder liner), or bottom (piston). More than one optical access is often required for most optical techniques.

1.8.3.1 Optical Access through the Cylinder Head

The construction of a cylinder head determines the size and implementation of optical access through the top. Two-stroke engines achieve the intake and exhaust flows through ports installed in the crankcase and cylinder block. The lack of overhead valves in a two-stroke engine thus permits replacement of the cylinder head by an optical window, allowing a complete view of the whole combustion chamber from top. Fig. 1.22 shows a modern production two-stroke SI engine modified for autoignition and knocking combustion studies [Konig and Sheppard, 1990]. A specially designed single-cylinder engine has been used extensively at Princeton University for the last twenty years [Felton et al., 1988]. This engine is ported and runs nominally as a two-stroke. There are six exhaust ports equally spaced through 180° and six equally spaced intake ports on the opposite side of the cylinder block. The intake ports are fitted with inserts to generate the desired initial flow field. One set of insets directs the flow 30° from the cylinder radius and 30° upward to provide a nearly solid body rotation. Otherwise, the non-swirl ports direct the flow radially inward and upward at 30° to improve scavenging efficiency. Intake air is compressed to 7 bar before it is delivered to the intake ports, in order to produce more realistic pressure and temperature in the chamber. This engine has been operated as both homogeneous charged and direct-injection engines [see, e.g., Ghandhi et al., 1994].

In order to study diesel combustion, Scott [1969] employed a loop scavenged two-stroke engine running on a four-stroke cycle. A diesel fuel injector was mounted on one side of the combustion chamber, as shown in Fig. 1.23. The top of the combustion chamber was covered with a large optical window so that a complete view of the fuel spray and combustion could be achieved. In order to achieve more realistic diesel combustion in small direct injection diesel engines, the

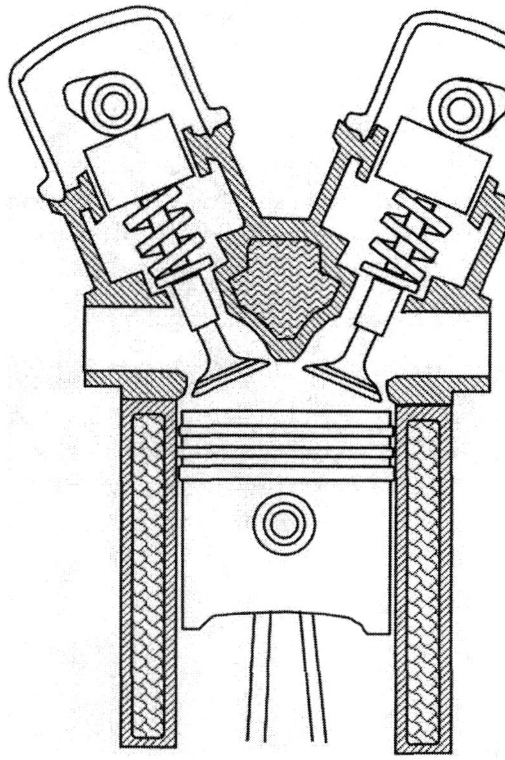

Fig. 1.21 A single-cylinder engine with overhead valves.

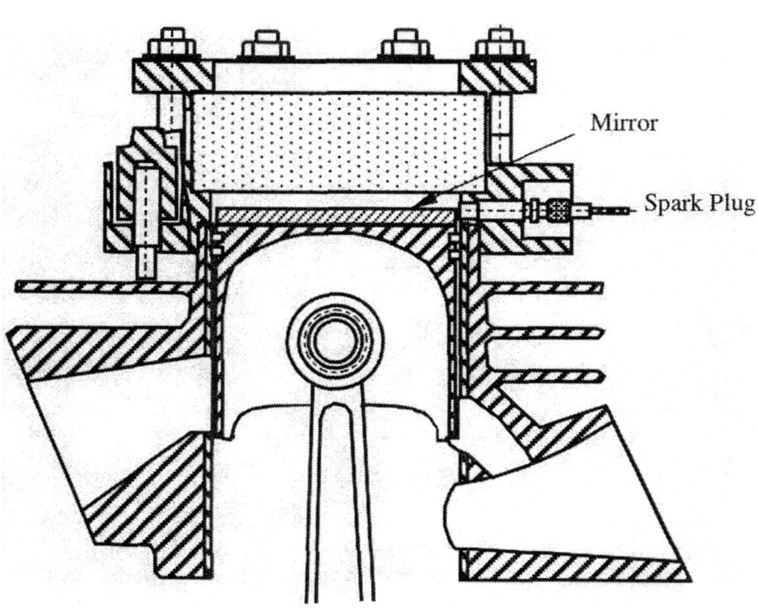

Fig. 1.22 A two-stroke engine with transparent cylinder head.

cylinder head window was modified to incorporate a centrally mounted fuel injector in the same loop scavenged two-stroke engine, as shown in Fig. 1.24.

Early four-stroke SI engines were constructed with side valves. One of the most popular side valve engines adopted for combustion visualization is the L-head engine, as shown in Fig. 1.25. The valves and camshaft are located in the cylinder block. The whole combustion chamber can be made visible by replacing a part or all of the cylinder head with a transparent window [Withrow and Rassweiler, 1938].

A four-stroke SI engine has been used extensively for in-cylinder flow and combustion studies with a transparent cylinder head at Sandia National Laboratory [Johnston, 1979]. The engine head configuration is shown schematically in Fig. 1.26. The intake and exhaust valves are actuated radially, driven by pushrods through a linkage mechanism that transforms the original axial pushrod motion to a radial one. Swirling flow can be generated inside the combustion chamber by a shroud on the intake valve. A fused-silica window above the piston provides a full field of view of the combustion chamber.

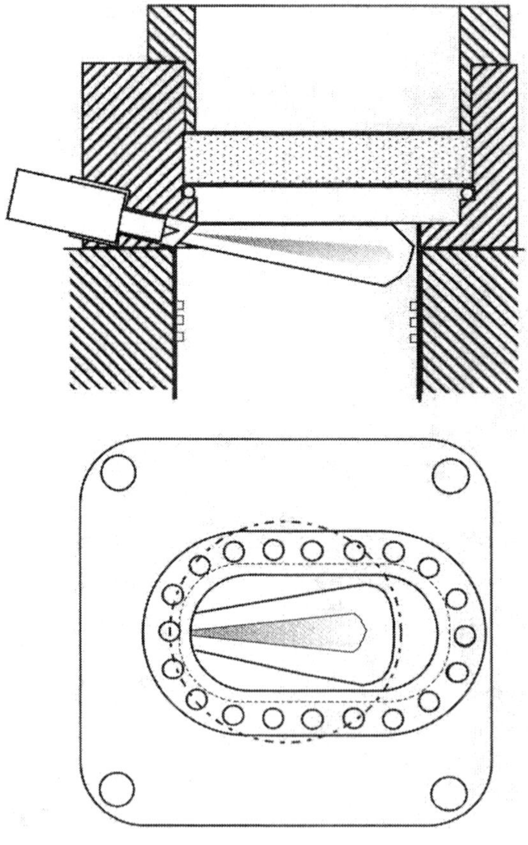

Fig. 1.23 Loop scavenged two-stroke diesel engine with a transparent cylinder head and side injection.

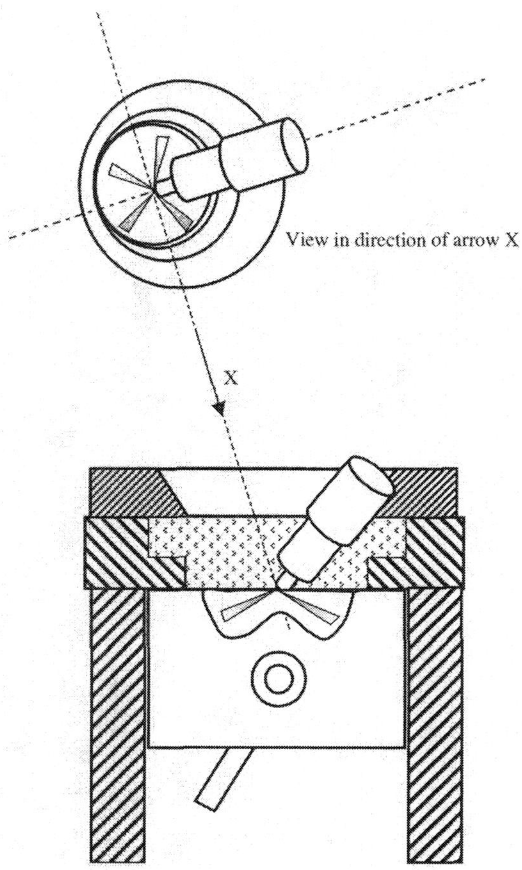

Fig. 1.24 Loop scavenged two-stroke diesel engine with a transparent cylinder head and overhead central fuel injection.

Modern engines, however, are constructed with overhead valves, which seriously limit the space for optical access from the top. Spicher et al. [1991] commissioned a specially designed two-valve cylinder head with optical access, as shown in Fig. 1.27. Inlet and exhaust valves and the spark plug are located within the tub-shaped main combustion chamber. A quartz window is mounted on the other side. The end-gas region is therefore located under the window, through which autoignition and knocking combustion can be observed by high-speed photography and the Schlieren technique. Alternatively, a similar setup can be achieved by replacing the pre-chamber in an IDI diesel cylinder head [Hands et al., 1990] with an optical window.

For relatively large DI diesel engines, one of the two exhaust valves can be replaced by an optical window, as shown in Fig. 1.28. Since the intake valves are unchanged, the in-cylinder flow field is preserved. By positioning a multi-hole injector in an orientation so that the axis of one of the sprays passes directly below the window in the cylinder head, the atomization and combustion of this spray can be studied. Using four equally spaced holes in an injector nozzle, Dec et al. [1991] were able to observe one complete spray plume and its combustion.

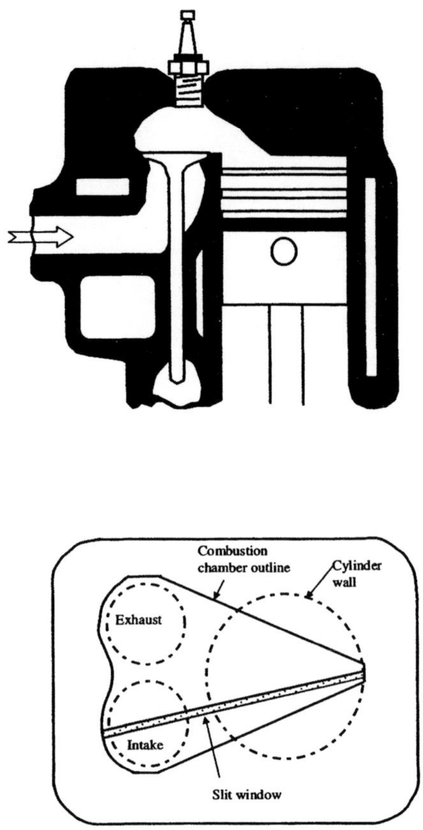

Fig. 1.25 A single-cylinder four-stroke engine with an L-head combustion chamber.

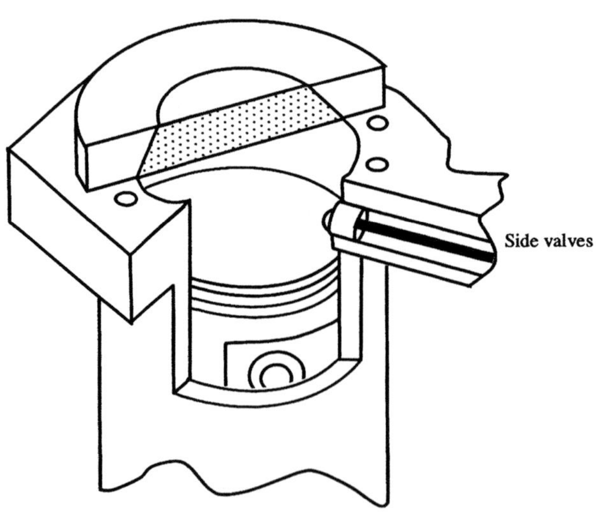

Fig. 1.26 A four-stroke engine with radially positioned side valves and a transparent cylinder head.

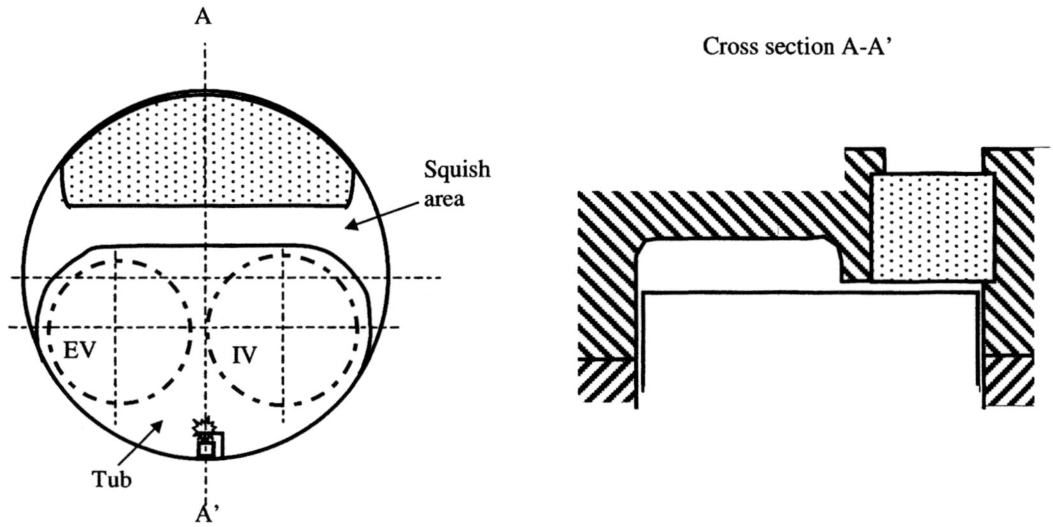

Fig. 1.27 A specially designed cylinder head with overhead valves and an optical window.

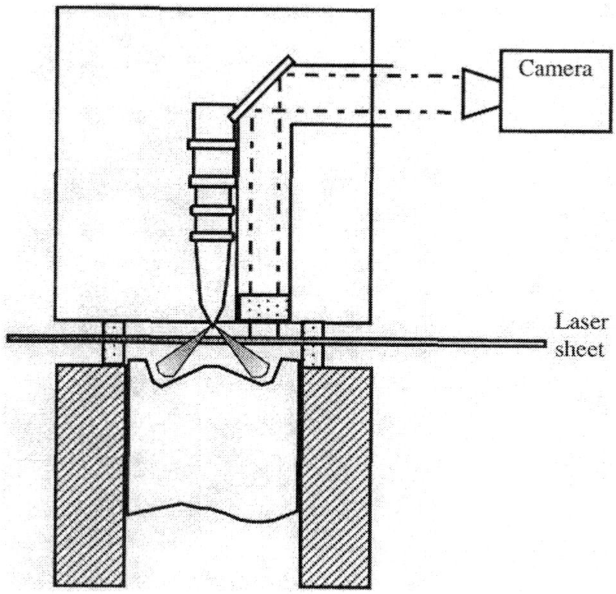

Fig. 1.28 A DI diesel engine with one of the exhaust valves replaced with an optical window.

A swirl chamber in an IDI engine can be used to study the interaction of diesel spray with air, and diesel combustion. Fig. 1.29 shows the cylinder head geometry of an IDI diesel engine in the author's laboratory. The engine is based on a single-cylinder Lister-Petter DI diesel engine. Its original cylinder head consists of two parts. The top half is cast iron and contains the valve gear, which is retained in the modified engine. The lower half is replaced with an aluminium block, in which is machined a swirl chamber and valve seats. The swirl chamber is connected to the main combustion chamber via an interchangeable throat of variable diameter. The swirl chamber is fitted with optical windows on both ends, through which diesel spray atomization, combustion, and soot formation can be studied by means of shadowgraph, light extinction, and two-color methods. The swirl ratio can be altered by either the engine speed or the throat diameter. The third window in the orthogonal direction provides the optical access for the laser beam when right-angle detection is required. Because the windows are shielded from the fuel sprays by the rotating airflow, the engine can be operated continuously for several minutes, limited only by the rise in the cylinder head temperature. It takes a couple of hours of accumulated engine running before the windows must be taken off and cleaned.

1.8.3.2 Optical Access through the Piston

The use of overhead valves in modern engines has led to alternative arrangements for optical access. A widely used and well-established concept to obtain insight into the combustion chamber of a four-stroke engine is based on mounting an optical window in the piston, as shown in Fig. 1.30. An elongated piston, with a quartz window in its crown and a slot on one or both sides, reciprocates about a slanted mirror at 45°, which is fixed to the engine block. The radiation

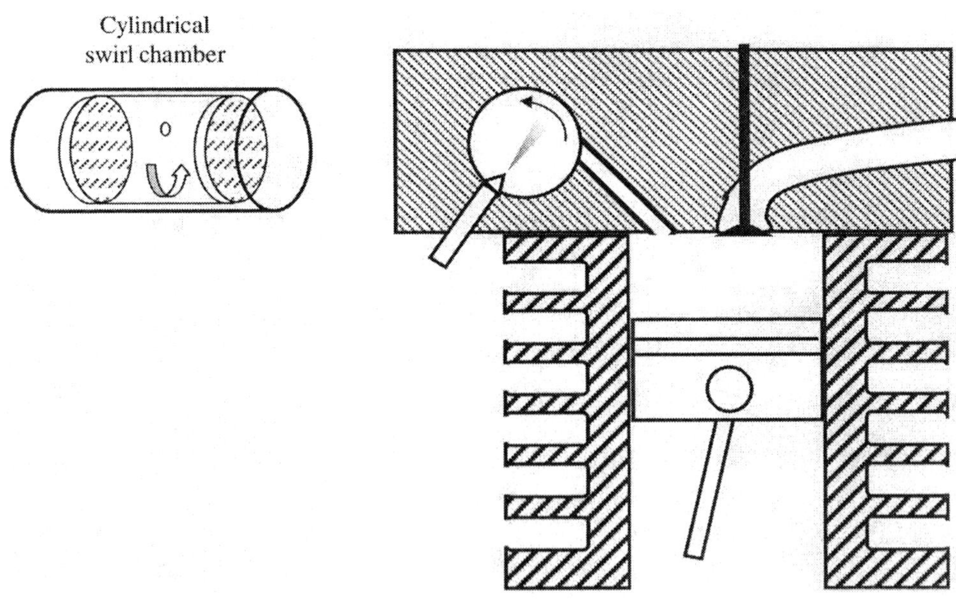

Fig. 1.29 An IDI diesel engine with a swirl chamber and optical windows.

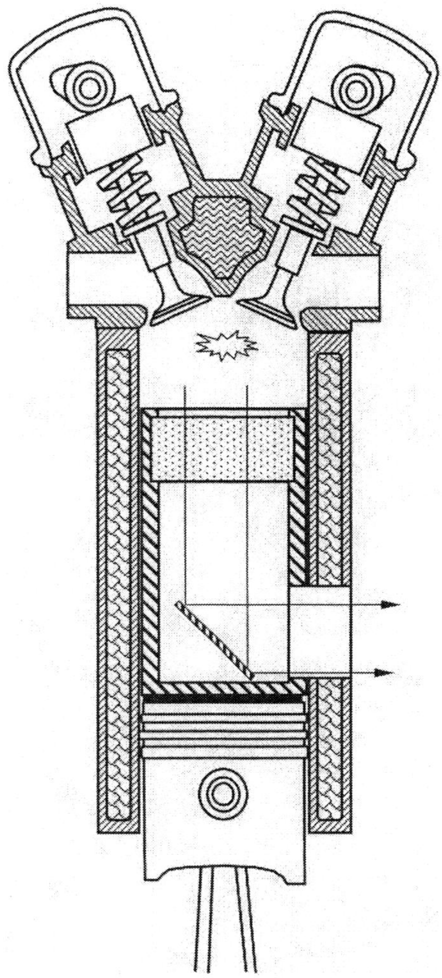

Fig. 1.30 A single-cylinder engine with a transparent piston.

received through the window by the mirror is then reflected horizontally to an imaging device. This design is often referred as the Bowditch extension, named after the engineer at General Motors, who first proposed and developed the setup [Bowditch, 1961]. The benefit of this configuration is in the use of a realistic cylinder head without the need for modifications, and the large optical access to a large portion of the combustion chamber.

For four-stroke SI engines, the top piston surface is normally flat. Direct injection diesel and gasoline engines use complicated piston shapes to create the desired fuel/air mixing and efficient combustion. It is therefore desirable to maintain the piston geometry in such engines. In addition, the piston window can be rapidly covered up with fuel and soot deposits, limiting measurements to the first few engine cycles only. In this case, the optical access through one of the two exhaust valves may be advantageous.

1.8.3.3 Optical Access through the Side

In order to apply spatially resolved laser diagnostics to IC engines, additional windows are mounted between the cylinder head and cylinder block for the admission and exit of a laser beam or sheet. Alternatively, a quartz ring can be installed to provide a 360° view from the side. Because any contact between the piston and the glass ring might result in window destruction, the piston trajectory should be well centered to avoid any piston slap onto the glass liner. It may be desirable to keep piston rings from contacting the optical window. Lubrication-free sealing between the cylinder liner and piston can be achieved by using piston rings made from polyamide-imide (Torlon®), which are supported by an inner steel expansion ring [Lee et al., 1999].

When a larger side view into the cylinder is required, the complete cylinder liner may be replaced with a glass liner. The design of the liner assembly must accommodate the extreme hardness of the quartz or sapphire liner by appropriate mounting structures and thermal properties which can lead to large and destructive thermal stresses. Thermal stresses can be decreased by skip firing at elevated temperatures. A detailed description of the design and installation of a sapphire liner was given by Bates [1988].

The curved walls of the optical liner, however, produce distorted images of in-cylinder flow and combustion, particularly near the wall. Although in most optical measurements, the distortion is negligible, it can be a problem for Schlieren imaging. In order to overcome this problem, square piston engines have been used at MIT for flow visualization [Namazian et al., 1980] and observation of flame propagation [Gatowski and Heywood, 1984], at Daimler-Benz for various laser diagnostics [e.g. Braumer et al., 1995], and at Mitsubishi Motors for the development of their gasoline direct injection (GDI) engines [Kume et al., 1996]. Fig. 1.31 shows schematically the MIT square-piston optical engine. The square-piston engine has a square cross section with two flat quartz windows located on opposite sides of the cylinder block. A square piston is mounted from a cross-head to reciprocate between flat cylinder walls, where large optical windows provide an unobstructed and undistorted view of in-cylinder flow and combustion. Any swirl generated in the intake flow, however, is bound to be affected when a square cylinder is used.

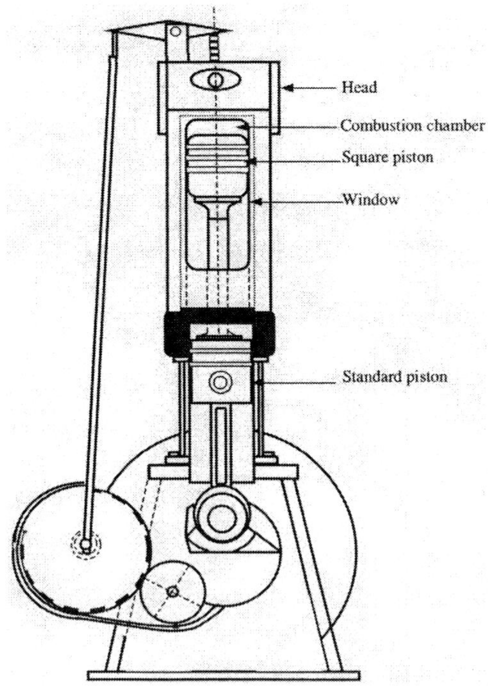

Fig. 1.31 The MIT square piston optical engine.

In order to apply the Schlieren technique to an engine with realistic cylinder geometry, Richman and Reynolds [1984] used a hologram to correct for the Schlieren image distortion due to the curved surface of a quartz cylindrical liner. However, the correction technique involves complex optical setup, and the improvement in the resulting Schlieren image was limited. A more successful approach is to machine the quartz liner so that its outer surface is shaped as a corrective lens [Fujikawa et al., 1988].

Image distortion due to the curved surface of a cylindrical liner can also introduce errors to particle image velocimetry (PIV) measurements, when particle images are recorded through the curved surface [Reeves, 1994]. This is particularly the case when PIV images are analyzed by the autocorrelation method.

In many applications, research engines are equipped with a combination of optical access described above. The side windows are normally used in laser-based techniques, together with a third window on the side, top, or piston for the observation of scattered light. For example, a single-cylinder research engine with a sapphire liner and a piston with a transparent window was used for combustion and flow visualization at General Motors [Bates, 1988]. Multiple optical accesses to the swirl and main combustion chambers have been employed to study combustion in both chambers simultaneously in IDI diesel engines [Scott, 1969; Hotta et al., 1998].

1.8.4 Operation of Optical Research Engines

When a single-cylinder engine is built, it is not mechanically balanced as a multi-cylinder engine is. This can be compensated with balance shafts or weights. To overcome the shortcomings of atypical gas dynamics in a single-cylinder engine, a four-cylinder engine with four Bowditch extensions was built at Volkswagen [Hentschel et al., 1989].

In additional to providing the optical accesses, another design requirement for an optical engine is the clean operation of optical windows. The clean operation of optical elements includes the use of unlubricated bronze-Teflon or graphite-plastic piston rings, minimizing the oil loss through the valve stems into the combustion chamber, and minimizing oil pumping from the crankcase into the mirror housing in the extended piston. At higher engine speeds, the extraction of oil mist from this area may be necessary. In addition, special fuels are sometimes used in place of normal fuels. In spark ignition engines, propane or preferably, isooctane can be used in order to minimize window fouling from combustion deposits, which permits continuous engine operation for several minutes or longer without cleaning the windows. In diesel engines, the optical windows, especially the one in the piston crown, are obscured in a matter of seconds due to soot deposits. Oxygenated fuels have been used to suppress soot formation in order to prolong engine operation, though combustion characteristics could be affected.

1.8.5 Engine Simulators or Test Rigs with Large Optical Access

There has been a wide range of test rigs used in the past to study engine combustion. Researchers with a strong desire to validate their mathematical models of engine combustion have, by and large, used constant-volume combustion chambers. These devices provide excellent optical access and allow independent control of the major influencing parameters. They have been used for fundamental studies of spray formation, ignition delay and combustion [e.g. Wang et al., 1995; Wiartalla et al., 1995; Siebers, 1998], and flame propagation [Arcoumanis et al., 1994]. A combustion bomb is also particularly suited to the development and calibration of laser diagnostics [Dyer, 1979]. However, they do not simulate the real engine conditions, because there is no piston movement or inlet and exhaust valves present in a reciprocating engine. Absence of the above elements changes the air flow characteristics drastically.

Rapid compression machines (RCM) have been widely used for knocking combustion [e.g., Hayashi et al., 1984], diesel combustion [Kamimoto et al., 1981], and auto-ignition studies [Taylor et al.,1950]. A rapid compression machine is not intended to be an engine; rather it permits a complete view of combustion processes without the complications of turbulence, swirl, etc. With more space for optical access, various optical diagnostics can be easily applied to an RCM. However, an RCM is incapable of simulating combustion in the expansion stroke. To overcome this difficulty, Kobori and Kamimoto [1995] developed a rapid compression and expansion machine (RCEM) in order to study diesel combustion under more realistic conditions.

1.9 Summary

In this chapter, general engine instrumentation is described, including engine dynamometers, air flow and fuel flow measurements, and the acquisition of engine data. The principle and operation of exhaust gas analyzers are outlined. Measurement techniques for particulate matter and their size distribution are discussed. Finally, extensive examples of optical research engines are presented, and their pros and cons are discussed so that appropriate design of optical access can be selected according to the research need.

References

Adachi, M., Inoue, K., Yamagishi, Y., and Ishida, K. (1997), "Discussion of Operating Parameters and Analysis Capability for a Fourier Transform Infrared Emission Analyzer," SAE Paper No. 971018.

Arcoumanis, C., Hull, D.R., and Whitelaw, J.H. (1994), "An Approach to Charge Stratification in Lean-Burn, Spark-Ignition Engines," SAE Paper No. 941878.

Ballantyne, V.F., Howes, P., and Stephanson, L. (1994), "Nitrous Oxide Emissions from Light Duty Vehicles," SAE Paper No. 940304.

Bates, S. (1988), "A Transparent Engine for Flow and Combustion Visualisation Studies," SAE Paper No. 880520.

Bowditch, F.W. (1961), "A New Tool for Combustion Research—A Quartz Piston Engine," *SAE Trans.*, Vol., 69, pp. 17–23.

Braumer, A., Sick, V., Wolfrum, J., Drewes, V., Zahn, M., and Maly, R. (1995), "Quantitative Two-Dimensional Measurements of Nitric Oxide and Temperature Distributions in a Transparent Square Piston SI Engine," SAE Paper No. 952462.

BS 1042 (1964), *Methods for the Measurement of Fluid Flow in Pipes, Part 1: Orifice Plates, Nozzles and Venturi Tubes,* British Standards Institution, London.

Cockshott, C.P., Vernon, J.P., and Chambers, P. (1983), "An Air Mass Flow Meter for Test Cell Instrumentation," Fourth International Conference on Automotive Electronics, Institution of Mechanical Engineers, Mechanical Engineering Publications (MEP), London.

Cuthbertson, R.D., Stinton, H.C., and Wheeler, R.W. (1979), "The Use of a Thermogravimetric Analyser for the Investigation of Particulates and Hydrocarbons in Diesel Engine Exhaust," SAE Paper No. 790814.

Dec, J., Zur Loye, A.O., and Siebers, D.S. (1991), "Soot Distribution in a DI Diesel Engine Using 2-D Laser Induced Incandescence Imaging," SAE Paper No. 910224.

Doebelin, E.O. (1975), *Measurement Systems: Application and Design*, McGraw-Hill Kogakusha, Tokyo.

Degobert, P. (1995), *Automobiles and Pollution*, SAE International.

Dyer, T.M. (1979), "Characterisation of One- and Two-Dimensional Homogeneous Combustion Phenomena in a Constant Volume Bomb," SAE Paper No. 790353.

Felton, P.G., Mantzaras, J., Bomse, D.S., and Woodin, R.L. (1988), "Initial Two-Dimensional Laser Induced Fluorescence Measurements of OH Radicals in an Internal Combustion Engine," SAE Paper No. 881633.

Fujikawa, T., Ozasa, T., and Kozuka, K. (1988), "Development of Transparent Cylinder Engines for Schlieren Observation," SAE Paper No. 881632.

Gatowski, J.A., Heywood, J.B., and Deleplace, C. (1984), "Flame Photographs in a Spark-Ignition Engine," *Combust. Flame*, Vol. 56, pp. 71–81.

Ghandhi, J.B., Felton, P.G., Gajdeczko, B.F., and Bracco, F.V. (1994), "Investigation of the Fuel Distribution in a Two-Stroke Engine with an Air-Assisted Injector," SAE Paper No. 940394.

Gatowski, J., and Heywood, J., (1984), "Flame Photographs in a Spark-Ignition Engine," *Combustion and Flame*, Vol. 50, No. 1, pp. 71–81.

Greeves, G. and Wang, C.H.T. (1981), "Origins of Diesel Particulate Mass Emission," SAE Paper No. 810260.

Halsall, R., McMillan, M.L., and Schwatz, B.J. (1987), "An Improved Method for Determining the Hydrocarbon Fraction of Diesel Particulates by Vacuum Oven Sublimation," SAE Paper No. 872136.

Hands, T., Collings, N., and Ma, T. (1990), "Investigations on Research Engines with Visible and Infrared Optical Access," *Proc. of Inst. Mech. Engrs*, Part D, Vol. 204, No. 3.

Harrington, D. (1983), "Interactions of Direct-Injection Fuel Sprays with In-Cylinder Air Motions," SAE Paper No. 831728.

Hayashi, T., Taki, M., Kojima, S., and Kondo, T. (1984), "Photographic Observation of Knock with a Rapid Compression and Expansion Machine," SAE Paper No. 841336.

Hentschel, W., Hessel, A., and Schindler, K. (1989), "Experimental Investigation of Spray Formation and Combustion in a Real Diesel Engine," Autotech. Seminar, C399/19, IMechE.

Heywood, J.B. (1988), *Internal Combustion Engine Fundamentals*, McGraw-Hill, New York.

Hotta, Y., Nakakita, K., Inayoshi, M., Ogawa, T., Sato, T., and Yamada, M. (1998), "Combustion Improvement for Reduced Exhaust Emissions in IDI Diesel Engines," SAE Paper No. 980503.

Johnston, S.C. (1979), "Precombustion Fuel/Air Distribution in a Stratified Charge Engine Using Laser Raman Spectroscopy," SAE Paper No. 790433.

Kamimoto, T., Kobayashi, H., and Matsuoka, S. (1981), "A Big Size Rapid Compression Machine for Fundamental Studies of Diesel Combustion," SAE Paper No. 811004.

Kastner, L.J. (1947), "The Air Box Method of Measuring Air Consumption," *Proc. Institution of Mechanical Engineers*, 157.

Kobori, S. and Kamimoto, T. (1995), "Development of a Rapid Compression-Expansion Machine Simulating Diesel Combustion," SAE Paper No. 952514.

Konig, G. and Sheppard, C.G.W. (1990), "End Gas Autoignition and Knock in a Spark Ignition Engine," SAE Paper No. 902135.

Kume, T., Iwamoto, Y., Iida, K., Murakami, N., Akishino, K., and Ando, H. (1996), "Combustion Control Technologies for Direct Injection SI Engines," SAE Paper No. 960600.

Ladommatos, N., Balian, R., Horrocks, R., and Cooper, L. (1996), "The Effect of Exhaust Gas Recirculation on Combustion and NO_X Emissions in a High-Speed Direct-Injection Diesel Engine," SAE Paper No. 960840.

Laurikko, J. and Nylund, N. (1993), "Regulated and Unregulated Emissions from Catalyst Vehicles at Loa Ambient Temperatures," SAE Paper No. 930946.

Laurikko, J. and Aakko, P. (1995), "The Effect of Ambient Temperature on the Emissions of Some Nitrogen Compounds: A Comparative Study on Low, Medium and High Mileage Three-Way Catalyst Vehicles," SAE Paper No. 950933.

Lee, S., McGee, J., Quay, B., and Santavicca, D., (1999), "A Comparison of Fuel Distribution and Combustion During Engine Cold Start for Direct and Port Fuel Injection Systems," SAE Paper No. 1999-01-1490.

Lilly, L.R.C. (editor) (1984), *Diesel Engine Reference Book*, Butterworths, London.

Majewski, W.A., Ambs, J.L., and Bickel, K. (1995), "Nitrogen Oxides Reactions in Diesel Oxidation Catalyst," SAE Paper No. 950374.

Namazian, M., Hansen, S.P., Lyford-Pike, E.J., Sanchez-Barsse, J., Heywood, J.B., and Rife, J. (1980), "Schlieren Visualization of the Flow and Density Fields in the Cylinder of a Spark-Ignition Engine," SAE Paper No. 800044.

Perez, J.M., Lipari, F., and Seizinger, D.E. (1984), "Cooperative Development of Analytical Methods for Diesel Emissions and Particulates, Solvent Extractables, Aldehydes, and Sulfate Methods," SAE Paper No. 840413.

Plint, M. and Martyr, A. (1995), *Engine Testing: Theory and Practice*, Butterworth-Heinemann, Oxford.

Reeves, M. (1994), PhD Thesis, Loughborough University of Technology.

Richman, R.M. and Reynolds, W.C. (1984), "The Development of a Transparent Cylinder Engine for Piston Engine Fluid Mechanics Research," SAE Paper No. 840379.

Scott, W.M. (1969), "Understanding Diesel Combustion through the Use of High Speed Moving Pictures in Colour," SAE Paper No. 690002.

Shimpi, S.A. and Yu, M.L. (1981), "Determination of a Reliable and Efficient Diesel Particulate Hydrocarbon Extraction Process," SAE Paper No. 811183.

Siebers, D.L. (1998), "Liquid-Phase Fuel Penetration in Diesel Sprays," SAE Paper No. 98080.

Sinnamon, J.F., Lancaster, D.R., and Steiner, J.C. (1980), "An Experimental and Analytical Study of Engine Fuel Spray Trajectories," SAE Paper No. 800135.

Spicher, U., Kroger, H., and Ganser, J. (1991), "Detection of Knocking Combustion Using Simultaneously High-Speed Schlieren Cinematography and Multi Optical Fibre Technique," SAE Paper No. 912312.

Stone, R. (1989), "Air Flow Measurement in Internal Combustion Engines," SAE Paper No. 890242.

Stone, R. (1992), *Introduction to Internal Combustion Engine*, 2nd Edition, MacMillan, London.

Taylor, C.F., Taylor, E., Livengood, J.C., Russell, W.A., and Leary, W.A. (1950), "Ignition of Fuels by Rapid Compression," SAE Paper No. 500178.

Wang, Y., Shu, G., Yang, C., Ju, Y., and Zhao, K. (1995), "Combustion Process of Diesel Spray in High Temperature Air," SAE Paper No. 950856.

Weaving, J.H. (editor) (1990), *Internal Combustion Engineering: Science and Technology*, Elsevier Applied Science, pp. 806–811, London.

Wiartalla, A., Backer, H., and Durnholz, M. (1995), "Influence of Injection System Parameters on Spray Development Combustion and Soot Formation by Optical Measurement Techniques in a Model Combustion Chamber," SAE Paper No. 950233.

Williams, P.T. (1990), "Sampling and Analysis of Polycyclic Aromatic Compounds from Combustion Systems: A Review," *J. Inst. Energy*, March 1990, pp. 22–30.

Withrow, L. and Rassweiler, G.M. (1938), "Motion Pictures of Engine Flames Correlated with Pressure Cards," *SAE Trans.*, Vol. 43, pp. 185–204.
Yamada, T., Hayakawa, N., Kami, Y., and Kawai, T. (1992), "Universal Air-Fuel Ratio Heated Exhaust Gas Oxygen Sensor and Further Applications," SAE Paper No. 920234.

Chapter 2

In-Cylinder Pressure Measurement and Combustion Analysis

Outline

2.1 Introduction

The purpose of an internal combustion engine is to convert chemical energy into mechanical work. The mechanical work is realized through the action of gas pressure on the reciprocating piston. The in-cylinder gas pressure therefore directly affects the power output and performance of the engine. In the quest for increasingly powerful and efficient IC engines, ever since the very early age of IC engines, attempts have been made to accurately measure cylinder pressure.

The earliest pressure measurements were carried out by an indicator mechanism, which was inherited from years of steam engine development and was completely mechanical. A series of pressure volume (p-V) diagrams were produced by a recording stylus, which was moving in proportion to the cylinder pressure, on paper wrapped around a cylindrical drum that was made to oscillate synchronously with piston motion. Hence, a p-V diagram is still sometimes referred as indicator diagram. Several other methods of in-cylinder pressure measurements were discussed by Obert [1973].

A typical system for measuring in-cylinder pressure in modern laboratories is shown in Fig. 2.1. The transducer measures instantaneous pressure, which can be displayed on an oscilloscope or recorded by a fast data acquisition system. The heart of the system is a piezoelectric transducer. Thus, this chapter begins by outlining the principle of the piezoelectric transducer. Next, we describe the preparation and mounting of the pressure transducer in an engine. Then, the procedures for recording and converting the transducer output to absolute pressure are presented. The rest of the chapter is devoted to the use of in-cylinder pressure data for engine performance and combustion diagnostics. Combustion analysis based on measured pressure data is divided into two main categories: the first one involves direct analyses of cylinder pressures; the second is heat release analysis based on the first law of thermodynamics.

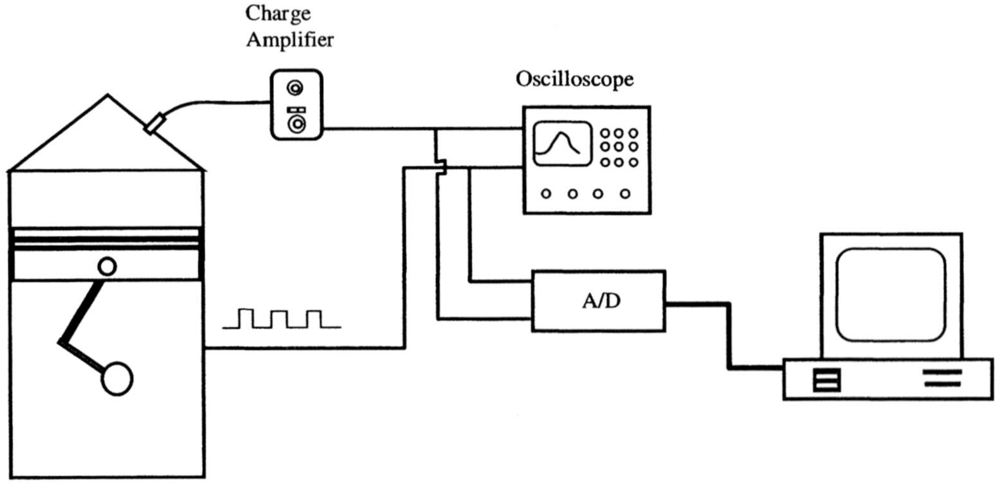

Fig. 2.1 Experimental setup for in-cylinder pressure measurements.

2.2 Principle of the Piezoelectric Pressure Transducer

Fig. 2.2(a) shows the construction of a quartz piezoelectric pressure transducer manufactured by Kistler Instrument Ltd. The quartz crystal is cut as a cylinder (with two 180 deg or three 120 deg sectors). As the cylinder pressure increases, the quartz crystals are compressed through the deflected metal diaphragm exposed to the combustion chamber. Due to compression, the inner and outer surfaces of the crystal cylinders are electrically charged. This electrical charge, measured in picocoulombs (pC), is exactly proportional to the force acting on the crystals. This unique property is called the piezoelectric effect (pressure electricity). There are two primary piezoelectric effects: (1) the transverse effect: charge on the x-planes of the crystal from force acting upon the y-plane, as in the case shown in Fig. 2.2(a); and (2) the longitudinal effect: charge on the x-planes of the crystal from force acting upon the x-plane. The longitudinal effect is utilized in the construction of the piezoelectric pressure transducer by AVL Ltd., as shown in Fig. 2.2(b). In this transducer, the quartz is cut into a number of wafers electrically connected (not shown) in parallel, and the potential difference is measured between the plane surfaces. Both types of arrangements have similar sensitivity.

Note that a rise in temperature will cause the housing to expand and thereby relieve the pre-compressed crystals from load. To compensate, a metal wafer is added below the quartz (Fig. 2.2[a]) so that it can expand with the housing. Changes in temperature affect the transducer output by varying the Young's modulus and resonant frequency of the quartz crystals as well as by causing expansion/contraction of the diaphragm and housing of the transducer and its mounting. Transducer manufacturers have attempted to achieve this goal by attaching a metal heat shield to the face of the diaphragm and/or supplying the pickup with internal coolant passages.

2.3 Pressure Transducer Preparation

Piezoelectric pressure transducers are available in different sizes and with different sensitivities. The selection of a transducer is much dependent on the installation requirements. The use of a relatively large, water-cooled transducer is preferred in order to maximize the sensitivity and minimize the temperature effect due to thermal shock. For knocking combustion detection, the natural frequency of the transducer should be much higher than the knocking frequency of the engine, which will be discussed in Section 2.5.1.4.

The pressure transducer should normally be mounted flush with the cylinder wall or as close to it as possible through a small communicating passage. This minimizes the lag in the pressure signal and should avoid introducing any resonance in the connecting passage. However, a single-passage mounted pressure transducer can provide accurate measurements if there is negligible pressure drop across the passage and acoustic resonance is absent. It was also found that remote mounting via multiple slots suffered the least from thermal shock. Remote mounting via small passages, however, is not feasible in diesel engines in which the small passages could easily be blocked. Transducers should be installed according to manufacturer's specifications. With some transducers, the output sensitivity changes during tightening. With

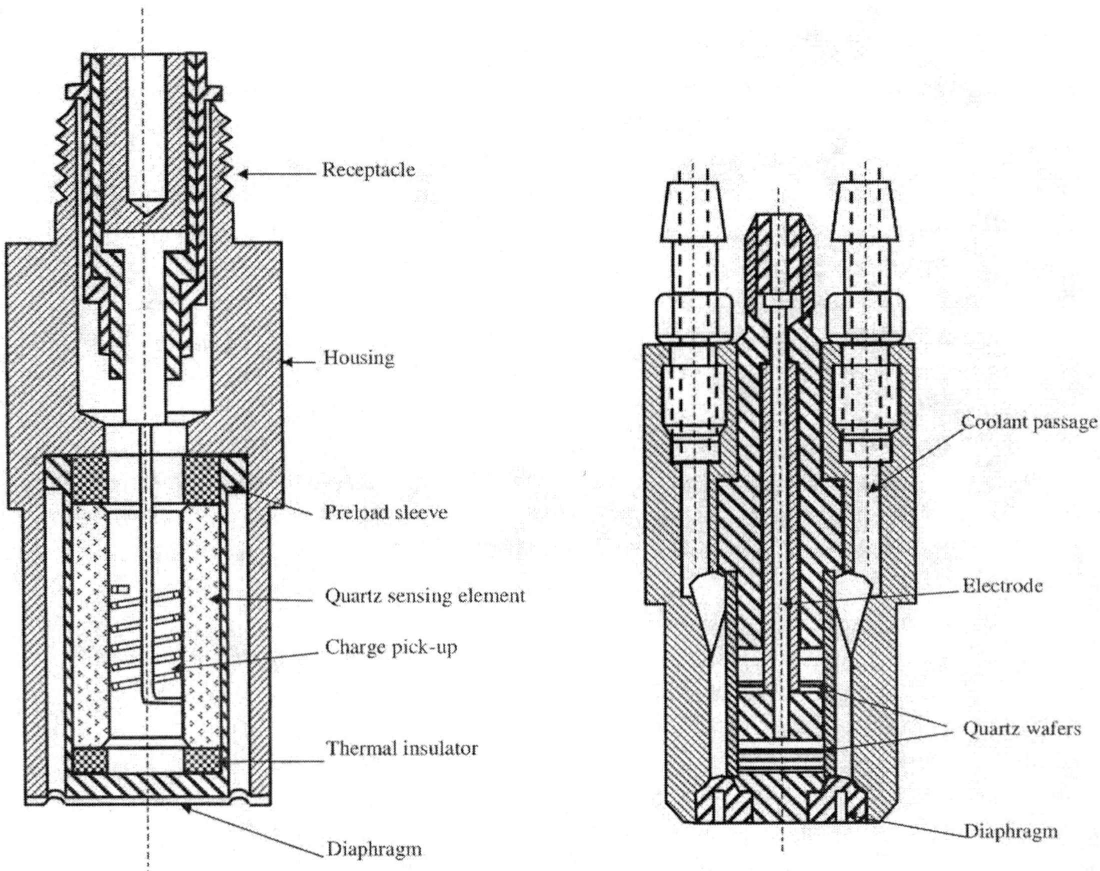

(a) Piezoelectric pressure transducer by Kistler Instrument Ltd.   (b) Piezoelectric pressure transducer by AVL Ltd.

Fig. 2.2 Schematics of piezoelectric pressure transducers.

water-cooled transducers, the variations of pressure in the water supply can cause fluctuations in the output of the transducer. Such variations should be eliminated.

As discussed in the previous section, the output of the transducer is in the form of a very small electrical charge, which mandates the use of a highly insulating and low-noise coaxial cable for connection to a charge amplifier, as shown in Fig. 2.1. The charge amplifier converts the electrical charge yielded by the piezoelectric sensor into a proportional voltage signal. Calibration is necessary to determine the relationship between the pressure input and voltage output for this system. This relationship must be linear.

Calibration can be carried out by a dead weight tester to supply calibration pressures and an oscilloscope to record the charge amplifier output. Initially, the sensitivity of the charge amplifier is set at the value supplied for the transducer by the manufacturer. The transducer is then loaded with the dead weight tester to at least the maximum pressure it will experience during engine operation. The sensitivity and range of the charge amplifier should be adjusted to provide some even value of output sensitivity or calibration factor, such as 10 bar/V. Use of standard output sensitivity simplifies data reduction. The linearity of the transducer should then be verified by checking the change in output voltage generated by using progressively smaller weights for loading and unloading the transducer. The transducer should be linear over the calibration range.

Because the charge generated from the transducer will leak with time, the transducer should be calibrated with the charge amplifier set at the long time-constant and on the range to be used. Drift of the charge amplifier output is normally caused by low insulation resistance in the transducer or cable and connections. If a rapid drift in charge amplifier output is shown on the oscilloscope during calibration, the cable connections should be separated and cleaned. If drift is still present, the transducer should be removed and vacuum baked at 370–395°C for about an hour, which will increase the resistance of the transducer.

2.4 Pressure Data Acquisition and Processing

2.4.1 Acquisition of the In-Cylinder Pressure Signal

The charge amplifier output is normally recorded on a crank angle basis. This is accomplished by using the output of a shaft encoder as the external clock of the data acquisition system. Acquisition of the in-cylinder pressure on a crank angle basis is necessary because the events occurring in an engine are dependent on the crankshaft rotation, which causes the motion of piston, fuel and spark delivery, etc. Since the rotational speed of the engine crankshaft changes by a small, but significant, amount during the course of each engine cycle, and from one cycle to the next, the use of a time clock will lead to problems of irregular timings in crank angles. The data acquisition rate depends on the resolution of the shaft encoder. For spark ignition engines, the rates of pressure rise are normally relatively slow, and acquisition at 1°CA provides sufficient resolution. If knocking combustion is being studied, pressure data should be digitized at much smaller time intervals, which sometimes necessitates the use of absolute time based data acquisition to resolve the details of the oscillating pressure trace. In diesel engines, the rates of pressure rise can be extremely high, and acquisition at ¼°CA intervals is frequently used to accurately define the pressure trace.

2.4.2 Conversion of Transducer Output to Absolute Pressure

The charge amplifier output, E, is in volts. By design, piezoelectric transducers respond to pressure differences by outputting a charge referenced to an arbitrary ground. The relationship between the output voltage and actual pressure is defined at any crank angle θ by

$$E(\theta) = \frac{p(\theta)}{C} + E_b \tag{2.1}$$

where E_b is the bias voltage with zero pressure, and C (bar/v) is the gain (calibration) factor of the system. Thus, to quantify absolute pressures, the transducer output must be directly correlated with pressure at some point in the cycle. This process is termed "pegging" or "referencing."

Randolph [1990a] summarized nine methods of pegging. The most often used method is referencing at inlet bottom dead center (IBDC), and this involves setting cylinder pressure at IBDC equal to intake manifold absolute pressure (MAP), i.e., p_{IBDC} = MAP. The absolute pressure is then determined from the output voltage at any crank angle by

$$p = C \cdot (E - E_{IBDC}) + p_{IBDC} \tag{2.2}$$

where E_{IBDC} is the charge amplifier output at IBDC where the absolute pressure is determined.

This is a very accurate procedure in untuned intake systems or at very low speeds in tuned systems. In order to minimize the interference of noise, an average of three or five points, namely the transducer output at IBDC, one and/or three CA degrees before IBDC, and one and/or three CA degrees after IBDC, should be used to reduce the potential for error associated with relying on a single point measurement. Absolute manifold pressures can be monitored by a manometer or accurately measured by a piezoresistive pressure transducer installed in the intake manifold.

If the engine-tuning effect becomes significant, then the forced polytropic coefficient method may be adopted. This method is based on the fact that compression follows the polytropic equation (PV^n = constant) in most internal combustion engines. The change in pressure between two points during compression is therefore given by

$$\Delta p = p_1 \left[\left(\frac{V_1}{V_2} \right)^n - 1 \right] \tag{2.3}$$

The cylinder volume V is calculated from the cylinder geometry at the crank angle θ by

$$V = V_c \{1 + \tfrac{1}{2}(r_v - 1)[R + 1 - \cos\theta - (R^2 - \sin^2\theta)^{1/2}]\} \tag{2.4}$$

where V_c is the clearance volume and r_v is the compression ratio. R is the ratio of connecting rod length to crank radius.

Using Eq. (2.1), the change in pressure between two points can also be written as

$$\Delta p = C \cdot \left[E(\theta_2) - E(\theta_1) \right] \tag{2.5}$$

Finally, combining Equations (2.3) and (2.5) leads to

$$E_b = E(\theta_1) - \frac{E(\theta_2) - E(\theta_1)}{\left[\left(\frac{V_1}{V_2} \right)^n - 1 \right]} \tag{2.6}$$

Equation (2.6) defines the bias voltage for any θ_1 and θ_2. To increase the accuracy, several values of θ_1 and θ_2 should be used and the resulting array of bias voltages averaged to provide a representative value. The main drawback of this procedure is the assumption of the value of the polytropic constant n, which normally varies between 1.25 and 1.35. In addition, hardware failures, wrong reference crank angle phasing, or odd engine characteristics may cause actual compression not to be polytropic.

A more accurate method is to install a piezoelectric transducer in the cylinder liner, so that it measures cylinder pressure when the piston rings are below the measurement location, and it measures crankcase pressure when the piston rings are above the measurement location. Because crankcase pressure is relatively constant and easily measured, this serves as an effective reference. Alternatively, a piezoresistive transducer that provides absolute pressure readings may be mounted low in the cylinder bore, where it is only exposed to the cylinder contents when the piston is near BDC. Because in-cylinder gases at this point are relatively cool, the effect of temperature on the absolute pressure readings is reduced. The accuracy of pressure readings is improved if the transducer is calibrated when engine is hot but not running. Referencing the output of the in-cylinder pressure transducer to the absolute pressure measurement at IBDC will then give the absolute in-cylinder pressure readings.

Pegging may be performed every cycle or only once for a series of cycles. Pegging for every cycle eliminates the problem of long-term drift inherent in piezoelectric devices, and hence is desirable whenever possible.

2.4.3 Inspection of In-Cylinder Pressure Data

Complete engine pressure data is subsequently used in engine combustion and performance analysis. However, before large amounts of data are recorded, at least one set of data should be carefully analyzed to check the accuracy and reliability of the entire test setup and recording procedures. This is best done on pressure traces obtained from a motored engine.

The first check of motored data is to verify the amplitude and phasing of the peak cylinder pressure with respect to volume when the engine is motored. The peak cylinder pressure can be estimated from the polytropic relationship:

$$p_{max} = p_{IBDC} \cdot r_v^n$$

The peak cylinder pressure should occur slightly before TDC because of heat loss and leakage. If the phasing between piston position and crank angle is done correctly, the angle by which peak pressure precedes TDC generally decreases with increasing engine speed and typically ranges from 0.7 to 0.9 deg for spark ignition engines, 1.1 to 1.3 deg for IDI diesel engines, and 0.8 to 1.0 deg for large DI diesel engines.

A logarithmic p-V diagram can then be used to further determine the scaling, phasing, and transducer performance [Randolph, 1990a]. Since the compression process, as defined from the intake valve closure (IVC) to TDC (Fig. 2.3), is polytropic, the compression line should be straight with a slope of –n. If a curvature appears during the first part of the compression line, it indicates an incorrectly assigned reference pressure being used. The wrong clearance volume, on the other hand, will introduce a curvature at the opposite end of the compression stroke near TDC. Deviation from a straight line in the central portions of the compression curve is one indication that the pressure data are faulty for properly phased data.

2.5 SI Engine Performance Analysis Using In-Cylinder Pressure

2.5.1 Direct Analysis of In-Cylinder Pressures

2.5.1.1 The Effect of Spark Timing on Engine Performance

The in-cylinder pressure data after correct scaling and referencing are readily plotted as a function of time, normally on a crank angle basis. Fig. 2.4 illustrates such a pressure time (p-t) diagram from an SI engine. The dotted line represents the p-t diagram when the engine is motored. A lot of useful information can be learned from the p-t diagram. Following spark discharge, there is a period during which the energy release from the developing flame is too small for the pressure rise due to combustion to be discerned. Because the flame becomes turbulent as it grows and propagates across the combustion chamber, the pressure rises steadily above the motoring pressure. The pressure reaches its peak value after TDC, and then decreases as the cylinder volume continues to increase due to the downward piston movement.

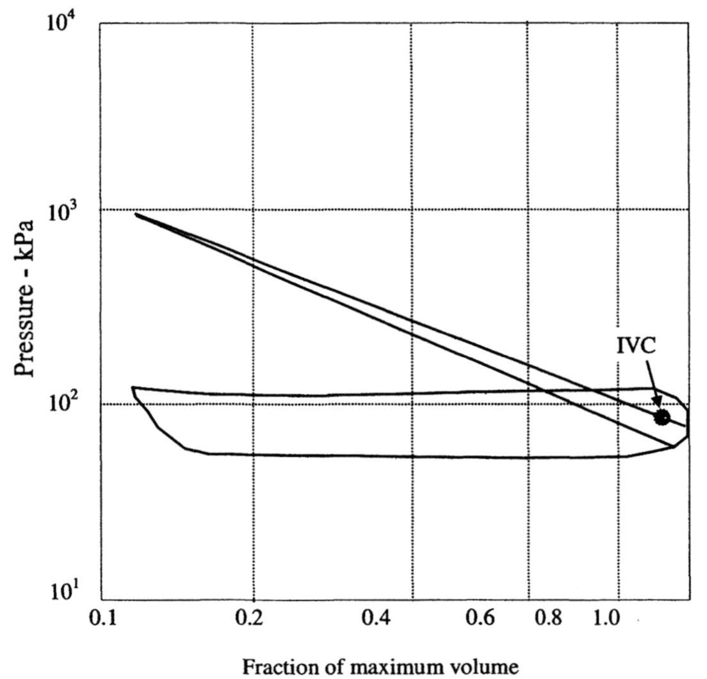

Fig. 2.3 Schematic representation of a motoring log p-V diagram.

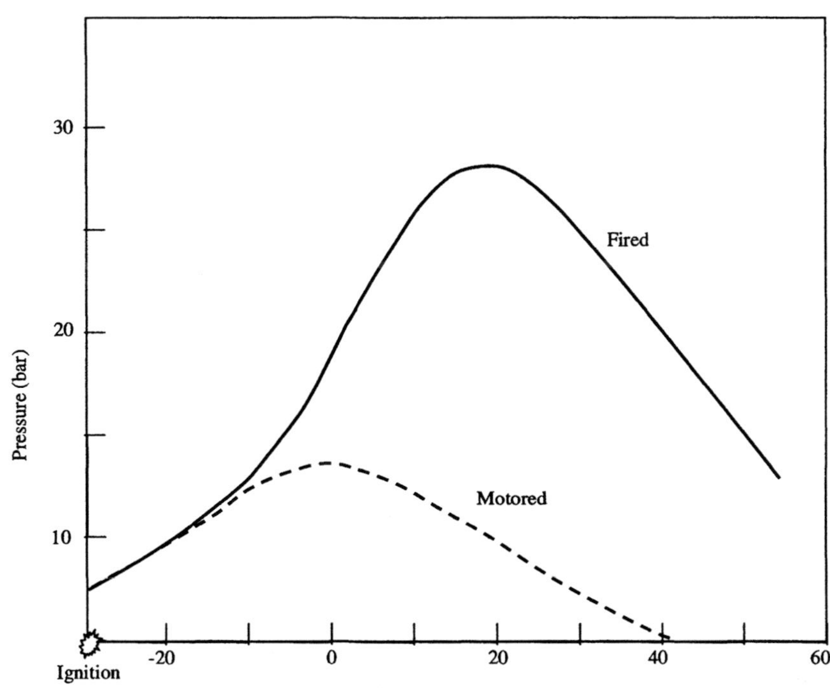

Fig. 2.4 In-cylinder pressure as a function of crank angle in an SI engine.

The shape of the p-t diagram is affected by the geometry of the combustion chamber, the position and number of spark plugs, in-cylinder mixture strength and distribution, flow field, and engine conditions, as well as the spark timing. The spark timing determines the relative phase of combustion to the cylinder volume and hence the maximum power output. Fig. 2.5 shows three pressure traces obtained with spark timings at 10, 30, and 50°CA BTDC. Fig. 2.5 reveals that if the ignition point lies close to TDC, less work is required for the compression stroke, which is a negative work. If the peak cylinder pressure lies near TDC, more work is delivered to the piston on the power stroke. If the start of the combustion process is progressively advanced before TDC, the negative compression work increases. If the combustion period is progressively delayed by retarding the spark ignition point, the peak cylinder pressure is reduced as it occurs later in the expansion stroke. Both changes reduce the net work output of the engine. At 30°CA BTDC, the maximum brake torque output is obtained, which occurs when the magnitude of these two opposing trends just offset each other. The spark timing is termed MBT, which stands for Minimum ignition advance for Best Torque. Practical experience suggests that the maximum pressure should occur at about 17°CA ATDC at MBT timing. Therefore, a quick check of the p-t diagram will reveal any anomalies in the combustion process due to engine design or malfunction of engine components.

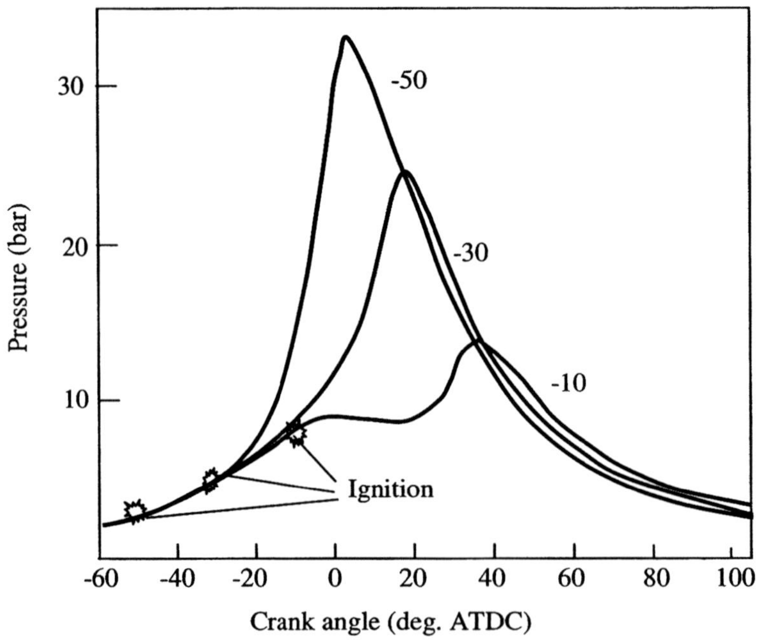

Fig. 2.5 In-cylinder pressures with different spark timings.

2.5.1.2 Calculation of Indicated Work and Pumping Work

Additional information can be gained if the in-cylinder pressure is plotted against the cylinder volume, i.e., p-V diagram or indicator diagram. Fig. 2.6 includes p-V diagrams for wide-open throttle (WOT) and throttled operations of a four-stroke SI engine.

The p-V diagram can show whether valve timings and flow areas have been correctly designed. If the exhaust valves open too early, the expansion power of the gas is somewhat wasted; if they open too late, the higher pressure during the exhaust stroke will increase the pumping work. If the intake valves open too early, exhaust gas may enter the intake manifold; if too late, the pumping work is increased.

At WOT, the intake stroke (cd) occurs essentially at atmospheric pressure being less because of fluid friction losses in the intake system; at part-load the pressure falls appreciably below atmospheric. The work transfer between the piston and the cylinder gases during the inlet and exhaust strokes is shown as area B + area C and is termed the pumping work, W_p. The work delivered to the piston over the compression and expansion strokes, shown as area A + area C, is defined as the gross indicated work per cycle, which is most commonly referred to simply as the gross indicated work. Hence, the pumping work can be obtained by integrating around the p-V diagram over exhaust and intake strokes,

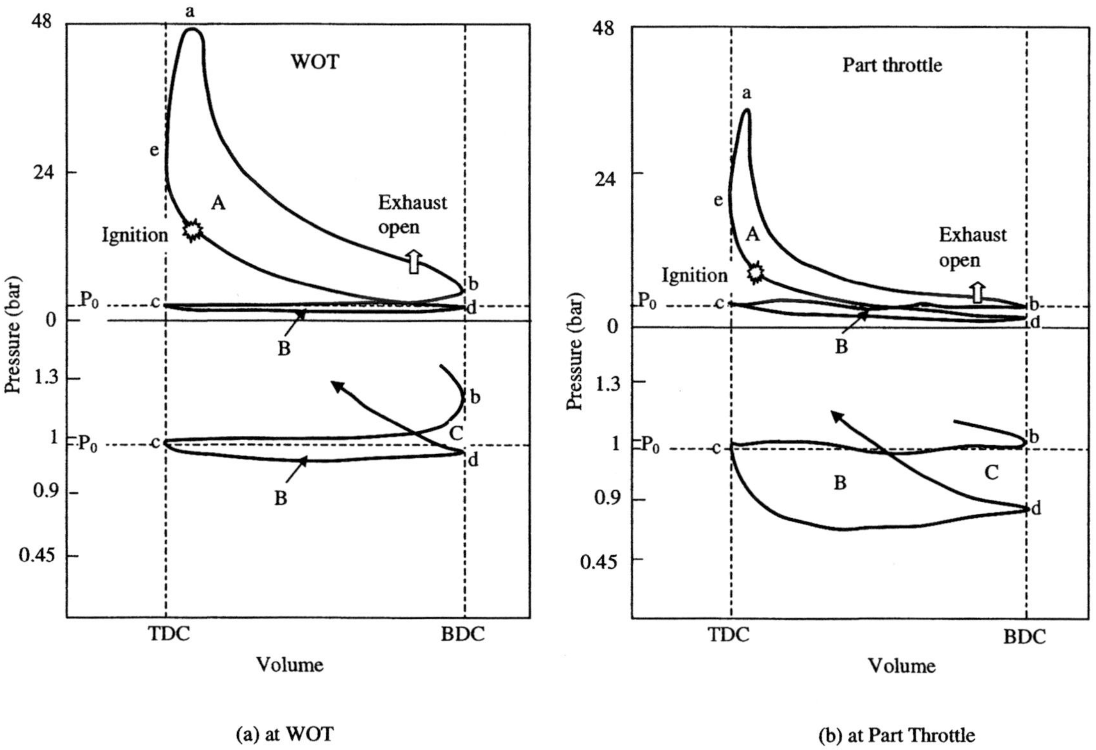

Fig. 2.6 p-V diagrams of a four-stroke SI engine.

$$W_p = \int_{bcd} pdV \tag{2.7}$$

and the gross indicated work per cycle is calculated from the compression and expansion strokes,

$$W_{i,g} = \int_{deab} pdV \tag{2.8}$$

For the four-stroke engine, the net indicated work per cycle is also used, and it is calculated from the four strokes,

$$W_{i,n} = \int_{bedea} pdV$$

and $W_{i,n}$ equals (area A – area B).

The indicated power per cylinder is then related to the indicated work per cycle by

$$P_i = \frac{W_i N}{n_R} \tag{2.9}$$

where n_R is the number of crank revolutions for each power stroke per cylinder, and it has a value of 2 for a four-stroke engine. N is the engine crank-shaft speed in rev/s. The net indicated power output represents the sum of the useful work available at the shaft and the work required to overcome all the engine losses. Measurements of brake power output by a dynamometer then allow the frictional power to be determined (frictional power = indicated power – brake power).

Another useful parameter that can be obtained from the p-V diagram is the indicated mean effective pressure (imep),

$$\text{imep} = \frac{W_i}{V_d} \tag{2.10}$$

which indicates how good the engine design is, and is independent of the engine size.

In discussing indicated quantities of the four-stroke engine, the definition used (gross or net) should always be explicitly stated.

2.5.1.3 Analysis of Cyclic Variations in an SI Engine

The cylinder pressure in an SI engine varies from cycle to cycle, because the flame development and propagation differ significantly for each cycle, due to the variations in the mixture strength and distribution and flow field. Cycle-by-cycle variability is often measured from pressure-related parameters. These include the maximum cylinder pressure P_{max}, the crank angle at which this maximum pressure occurs θ_{max}. Another important measure of cyclic variability is the coefficient of variation (COV) in individual imep values. It is defined as the ratio of the standard deviation in imep and the mean imep

$$COV = \frac{\sqrt{\sum_{1}^{n}(imep - \overline{imep})^2 / (n-1)}}{\overline{imep}} \tag{2.11}$$

where n is the number of cycles. A few hundred cycles are required to secure a reliable sample size.

In the analysis of cyclic variability, the values of P_{max} or COV are plotted against θ_{max}. Fig. 2.7 shows schematically the relationship between P_{max} and θ_{max}. In this figure, points A and A′ correspond to the MBT timings, while all other points correspond to either retarded or advanced

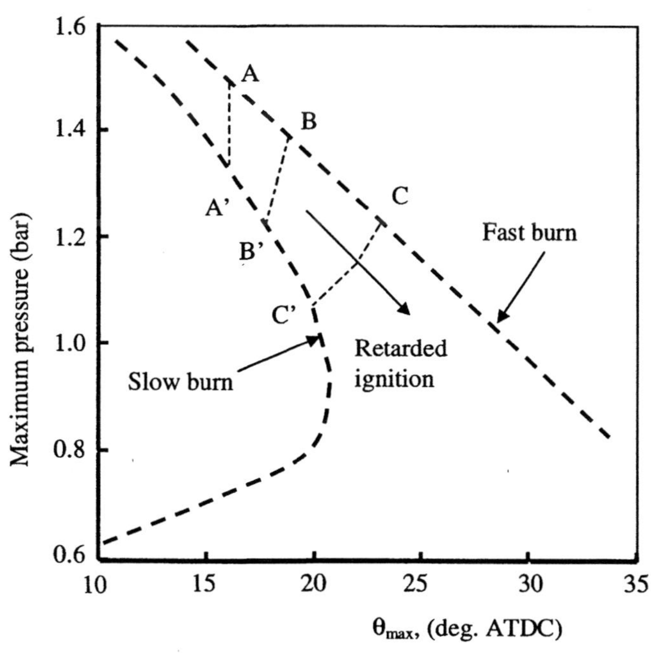

Fig. 2.7 Schematic of variation in maximum cylinder pressure and crank angle, at which it occurs, in individual cycles, adapted from Matekunas [1983].

timings. One notices that the relationship between P_{max} and θ_{max} is approximately linear for fast burn in the entire range, while for the slow burn case this linearity stops at point D′. For further retarding, there is a hook-back region which corresponds to very late phasing of combustion, when the effect of expansion due to the piston motion becomes an important factor.

2.5.1.4 Detection of Knocking Combustion in an SI Engine

Under normal combustion in an SI engine, the spark-ignited flame moves steadily across the combustion chamber until the charge is fully burned. For some conditions, abnormal combustion may take place as a result of surface ignition or knocking combustion.

Surface ignition is ignition of the fuel-air charge by a hot spot in the combustion chamber, such as overheated valves or spark plugs, glowing combustion chamber deposits, etc. It may occur before the spark ignition (pre-ignition) or after normal ignition (post-ignition). The former can be easily discerned from measured cylinder pressure traces. The latter may be more difficult to identify depending on the time and magnitude associated with the post-ignition flame.

Knock is the most important abnormal combustion phenomenon in an SI engine. Knocking combustion is caused by the autoignition of the end gas (the unburned gas ahead of flame front) due to the compression by piston and expanding flame front. It is characterized by high-frequency pressure oscillations inside the cylinder, as shown in Fig. 2.8(a), which produce the sharp metallic noise called "knock." The frequency of these oscillations depends on the combustion chamber size and in-cylinder temperature and pressure. For typical automotive engines, the pressure oscillates around 5–7 kHz. In order to resolve it, a pressure transducer should be selected with a natural frequency much higher than 10 kHz. An accurate measurement of these oscillations require sampling frequencies on the order of 100 kHz, which requires a departure of normal pressure recording procedures. In most applications, as described in previous sections, pressure is recorded on a crank angle basis by using the shaft encoder output to drive the data acquisition system clock. To log pressure at 100 kHz or higher, pressure traces have to be recorded on an absolute time basis. Two approaches can be utilized: the first approach is to sample the entire cycle at a 100 kHz sampling rate for many cycles [Zhao et al., 1992]; or the signal is sampled every degree crank angle up to the time of knock, then is switched to high sampling rate during the knock period and back to crank degree sampling [Chun and Heywood, 1989]. The first method requires large memory and restricts the number of cycles that can be recorded at one time. The second approach requires hardware capable of generating time windows and appropriate sampling pulses at a given window.

When an engine is knocking lightly, the pressure surge in the combustion chamber will affect the pressure trace only to a slight degree. The presence of knock will be greatly magnified, however, if dp/dt is plotted. As knock occurs, dp/dt goes negative rapidly, indicating the start of knocking combustion [Obert, 1973]. However, Barton et al. [1970] found that the rate of pressure change could be better isolated from the mechanical vibration of the engine with a bandpass filter. They defined the knock intensity as the root-mean-square (rms) value of the rate of pressure change between 1 and 5 kHz, averaged over a 10-second interval; and trace

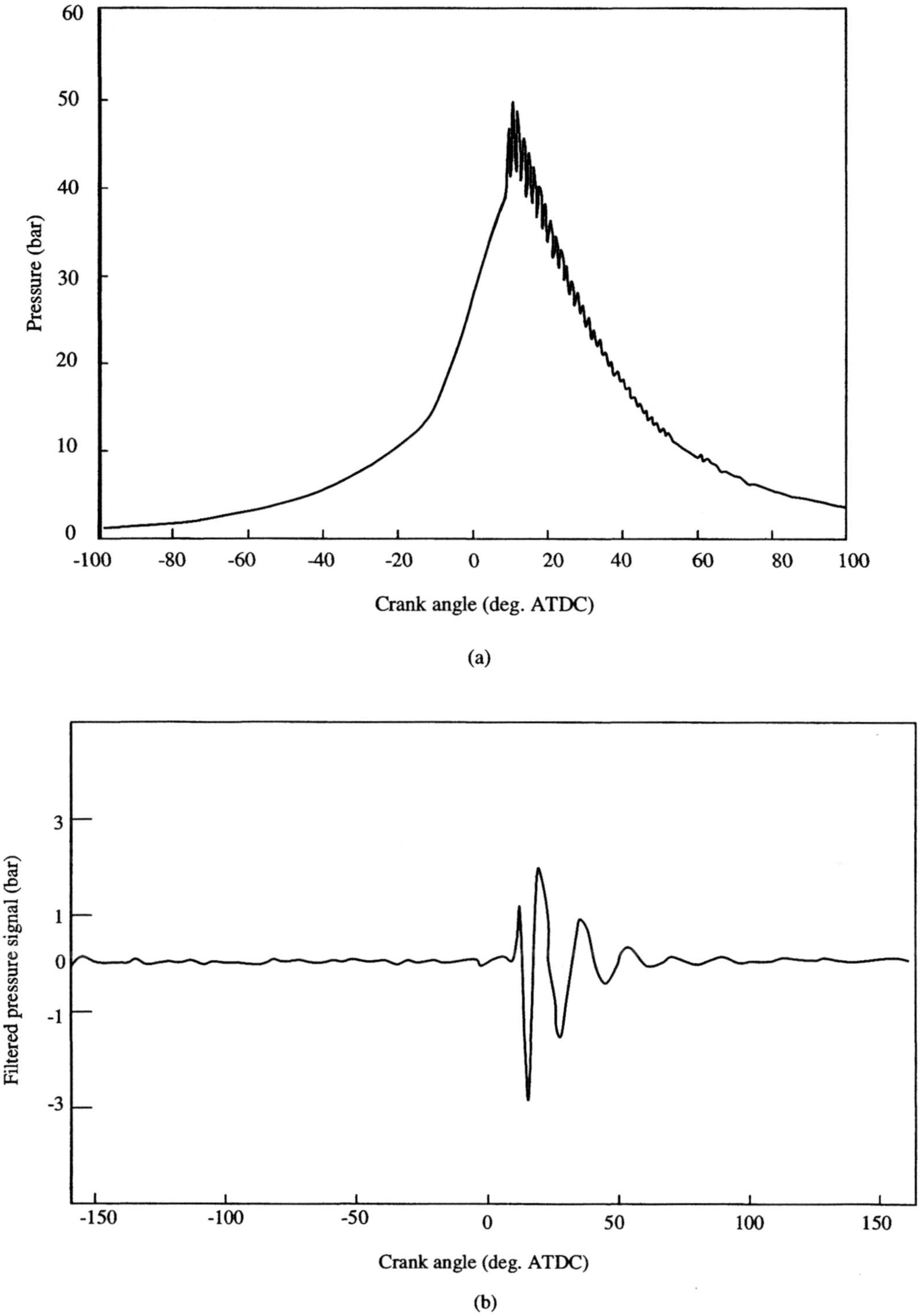

Fig. 2.8 Cylinder pressure signal of a knocking cycle (a) and its bandpass-filtered output (b).

knock as occurring when the rms rate of pressure change for 20 percent of the engine cycles exceeded a value twice that for non-knocking cycles.

In studying cycle-to-cycle variability of knock in SI engines, Leppard [1982] found that knock had a characteristic frequency of 6 kHz in the two engines. In processing the cylinder pressure signal from a pressure transducer adapted to the spark plug, he introduced a bandpass filter into the detection circuit. The resulting bandpass filtered signal is similar to that shown in Fig. 2.8(b). From the bandpass-filtered signal, he compared as knock criteria the maximum pressure amplitude, the integral of the pressure oscillations, and the integral of the square of the pressure oscillations. He found that there were no statistically significant differences in their comparative merit. Thus, he chose the maximum amplitude of the bandpass-filtered cylinder pressure signal as the measure of knock intensity. The same definition of knock intensity was also adopted by Chun and Heywood [1989] and Zhao et al. [1992] in characterizing knocking combustion. The start of knocking combustion was determined from the bandpass-filtered signal (Fig. 2.8[b]), when the peak of the filtered signal was several times greater than the averaged background noise.

Another method used in quantifying the knock level is known as power spectral density method, which is the integral of the Fourier spectrum of the knock signal for a limited range of frequencies.

The manner in which the knock pressure signals are interpreted may be expected to result in different indices of severity for knock in a given engine at a specified condition. Therefore, parameters for quantifying the knock level should be used as a relative measure of the level of knocking intensity because there is no clear evidence that one is more relevant to engine damages caused by knocking combustion. This is further complicated by the fact that the knock level picked up by a pressure transducer depends on its location relative to the end gas region in the combustion chamber.

2.5.2 Heat Release Analysis in an SI Engine

2.5.2.1 Introduction

In addition to the engine performance parameters discussed in the previous section, in-cylinder pressure is widely used to derive the heat release rate or mass fraction burned profiles from IC engines. The mass fraction burned analysis is normally used to analyze combustion in premixed charge engines. The heat release analysis is normally applied to study combustion process in engines with fuel directly injected into the cylinders. This is because not all the fuel is burned with sufficient air in a direct injection engine, as it is in a premixed charge engine.

Fig. 2.9 shows a mass fraction burned profile derived from the in-cylinder pressure measurements in a spark ignition engine. The rate at which fuel-air mixture burns increases from a low value immediately following the spark discharge to a maximum about halfway through the burning process, and then decreases to close to zero as the combustion process ends. Therefore, useful information can be obtained from the curve to characterize different stages of the combustion process. Three parameters are normally used to characterize the energy-release aspects of combustion: flame development angle, $\Delta\theta_d$, rapid-burning angle, $\Delta\theta_b$, and overall burning angle, $\Delta\theta_o$. The flame development angle is that from the point of ignition to that when a small proportion of the charge, usually 10% but sometimes 5% or less, is burned. The rapid burning period is usually characterized by the number of crank angles required for the fraction from 10% to 90% of the charge to be burned. The sum of the former two defines the overall burning angle. The flame development process is influenced primarily by the mixture state, composition, and motion in the vicinity of the spark plug. Once a flame kernel has been established, the major portion of the charge burns as the flame propagates to the chamber walls, defining the rapid burning period. This stage is obviously influenced by conditions throughout the combustion chamber. The final stage, where the remainder of the charge burns to completion, cannot be easily quantified because energy-release rates are comparable to other energy-transfer processes that are occurring, mainly heat transfer to walls.

2.5.2.2 Rassweiler and Withrow Method

One well established method was developed by Rassweiler and Withrow [1938] for estimating the mass fraction burned profile as shown in Fig. 2.9 from the cylinder pressure and volume data. In their analysis, the cylinder pressure rise during any crank angle interval, $\Delta\theta$, is considered to be

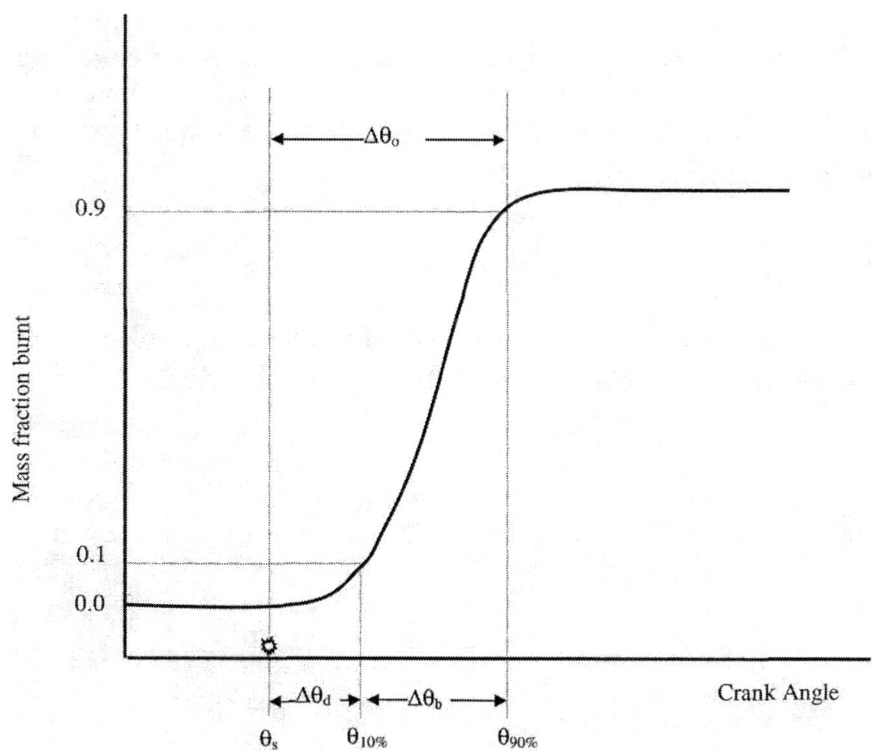

Fig. 2.9 Definition of flame-development angle, $\Delta\theta_d$, and rapid-burning angle, $\Delta\theta_b$, on a mass fraction burned curve.

caused by the pressure rise due to volume change, Δp_v, and the pressure increase due to combustion, Δp_c:

$$\Delta p = \Delta p_v + \Delta p_c \tag{2.12}$$

As the crank angle (θ_i) increments to its next value (θ_{i+1}), the volume changes from V_i to V_{i+1}, and the pressure changes from p_i to p_{i+1}. The pressure change due to the piston movement is related to the volume change by the polytropic relationship

$$\Delta p_v = p_i \left[\left(\frac{V_i}{V_{i+1}} \right)^n - 1 \right] \tag{2.13}$$

Combining Eqs. (2.12) and (2.13), Δp_c can be evaluated:

$$\Delta p_c = p_{i+1} - p_i (V_i / V_{i+1})^n \tag{2.14}$$

where V_i is the cylinder volume given by Eq. (2.4).

Because the combustion process is not occurring at constant volume, the pressure rise due to combustion is not directly proportional to the mass of fuel burned. The pressure rise due to combustion has to be referenced to a datum, e.g., the clearance volume at top dead center, V_c, to take into account of the volume effect:

$$\Delta p_c^* = \Delta p_c (V_i / V_c) \tag{2.15}$$

If it is assumed that the referenced pressure rise due to combustion is proportional to the mass fraction burned (χ), then, the mass fraction burned at the end of the interval $\Delta\theta$ is given by

$$\chi = \frac{m_{b,i}}{m_{b,total}} = \frac{\sum_1^i \Delta p_c^*}{\sum_1^N \Delta p_c^*} \tag{2.16}$$

The end of combustion occurs after N increments, and is defined by the pressure rise due to combustion becoming zero.

This method depends on using an appropriate value of the polytropic index n. Suitable values of the polytropic index can be found for before and after combustion from the motored log(p)-log(V). An averaged value during combustion can be used within the range of 1.2 to 1.3 for spark ignition engines.

This method has the advantage of being easy to use though it contains several approximations. The polytropic constant used in the calculation is often assumed to be a constant throughout the analysis. However, the polytropic constant varies from compression to expansion and changes during the combustion process. The effect of heat transfer is not accounted for in the model. It is included only to the extent that the polytropic constant is different from the ratio of specific heats. In addition, the pressure rise due to combustion is proportional to the amount of fuel chemical energy released rather than the mass of mixture burned. Therefore, the Rassweiler and Withrow method is really a scaling relationship: it produces a mass fraction burned profile that automatically goes from zero at ignition to unity at the end of combustion, irrespective of whether combustion is in fact complete or the accuracy of the cylinder pressure and mass flow rate data.

To overcome these shortcomings, more rigorous analyses based on the first law of thermodynamics have been developed. There are two approaches. The first approach yields directly the rate of heat release using cylinder pressure. The second approach generates the fuel mass fraction burned directly using cylinder pressure data. The heat release rate and fuel mass fraction burned can be readily converted into each other by the calorific value of the fuel used.

2.5.2.3 One-Zone Heat-Release Analysis

2.5.2.3.1 Net Heat Release Rate

One-zone models define the state of the cylinder contents in terms of average properties and do not distinguish between burned and unburned gases. The cylinder contents are effectively modeled as homogeneous. A one-zone model offers the benefit of including heat transfer and gas flow phenomena more simply. The combustion process is considered as a separate heat addition process, and the contents of the chamber are regarded as a single fluid. Straightforward heat transfer and crevice models can be used to complete the energy balance. For many engineering applications, a one-zone model is adequate for engine combustion analysis.

A one-zone heat-release analysis method has been developed and evaluated at MIT [Gatowski et al. 1984; Chun and Heywood, 1987; Cheung and Heywood, 1993]. The heat-release model for the contents of the engine combustion chamber begins with the first law of thermodynamics for an open system, as shown in Fig. 2.10. Using the sign convention shown in the figure, the first law is written as

$$dQ_{ch} = dU_s + dW + \sum h_i dm_i + dQ_{ht} \tag{2.17}$$

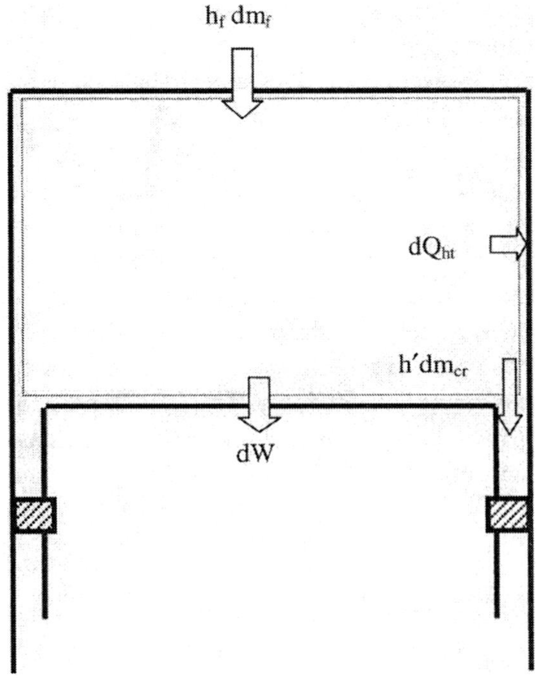

Fig. 2.10 Open system boundary of a combustion chamber for heat release analysis.

where U_s is the sensible energy of the charge, which is separated from that due to change in composition. By "sensible" we mean changes in u or h that result from changes in temperature alone, and we exclude changes due to chemical reaction or phase change. The term Q_{ch} represents the chemical energy released by combustion. dW is the piston work and equals to pdV. The last term Q_{ht} represents the heat transfer to cylinder walls. The mass flux term represents flow across the system boundary. It includes the flow into and out of crevice regions, and the fuel injected into the chamber for a DI engine, i.e.,

$$\sum h_i dm_i = \left(h' dm_{cr} - h_f dm_f\right)$$

Assuming that U_s is given by m·u(T), where T is the mean charge temperature and m is the mass within the system boundary, then

$$dU_s = mc_v(T)dT + u(T)dm \tag{2.18}$$

For SI engines operating with a premixed fuel-air charge, dm is caused by the flow into and out of the crevices regions ($dm = -dm_{cr}$). (Note: Direct injection SI engines will be considered in Section 2.6.2.2, together with DI diesel engines.) Therefore, Eq. (2.17) becomes

$$\delta Q_{ch} = mc_v(T)dT + pdV + (h' - u)dm_{cr} + \delta Q_{ht} \tag{2.19}$$

or

$$dQ_{ch}/dt = mc_v(T)dT/dt + pdV/dt + (h' - u)dm_{cr}/dt + dQ_{ht}/dt \tag{2.20}$$

where $dm_{cr} > 0$ when flow is flowing into the crevice
$dm_{cr} < 0$ when flow is flowing out of the crevice into the cylinder
h' is evaluated at cylinder conditions when $dm_{cr} > 0$
h' is evaluated at crevice conditions when $dm_{cr} < 0$

This equation can be used in a few ways. The term dQ_{ch}/dt on the left is termed the apparent gross heat release rate. When it is combined with the heat transfer term, dQ_{ht}/dt, the combination ($dQ_{ch}/dt - dQ_{ht}/dt$), is termed the net heat release rate, dQ_n/dt, which is the combustion energy release rate less heat lost rate to the walls.

Eq. (2.18) can be further simplified by the ideal gas law, $pV = mRT$, with R assumed constant. It follows that

$$\frac{dp}{p} + \frac{dV}{V} - \frac{dm}{m} = \frac{dT}{T} \tag{2.21}$$

which can be used to eliminate T from Eq. (2.18) to give

$$\frac{dQ_n}{dt} = \frac{\gamma}{\gamma - 1} p \frac{dV}{dt} + \frac{1}{\gamma - 1} V \frac{dp}{dt} + (h' - u + c_v T)\ dm_{cr} \tag{2.22}$$

where γ is the ratio of specific heats, C_p/C_v. Eq. (2.22) shows that the net heat release rate can be calculated from the changes in the cylinder volume and cylinder pressure if crevices flow is ignored. In calculating the net heat release rate, γ is the most important thermodynamic property used. This equation is often used with a constant value of γ within a range of 1.3 to 1.35. In fact, the value of γ for both burned and unburned gases decreases with increasing temperature and varies with composition. As the mean charge temperature increases during compression and combustion and then decreases during expansion, γ should vary. The right values for γ during combustion that will give most accurate heat release analysis are not clearly defined. For spark ignition engines, Chun and Heywood [1987] and Cheung and Heywood [1993] obtained γ values by matching the output of the one-zone analysis to that of a more rigorous two-zone analysis for the same cylinder pressure generated by an engine simulation code. They found that accurate heat release results were achieved by using γ values approximated by a constant during the combustion process, and by linear functions of temperature during the compression and the post combustion expansion processes. The γ values are summarized in Table 2.1.

The net heat release profile obtained by integrating the first two terms on the right-hand side of Eq. (2.22), normalized to give unity at its maximum value, is interpreted as the mass fraction burned versus crank angle, as shown in Fig. 2.9. Strictly speaking, the profile thus obtained represents the energy release fraction.

2.5.2.3.2 Gross Heat Release Rate

Additional insight into the combustion process can be obtained if the heat transfer term, dQ_{ht}/dt, can be evaluated at different crank angles. In this way, a close approximation to the gross heat release rate can be obtained. The convective heat transfer rate to the combustion chamber walls can be calculated from the relationship,

$$\frac{dQ_{ht}}{dt} = Ah_c(T - T_w) \tag{2.23}$$

where A is the chamber surface area, T is the mean gas temperature, T_w is the mean wall temperature, and h_c is the heat transfer coefficient averaged over the chamber surface area. h_c is normally estimated from either Annand's or Woschni's correlation.

Annand's correlation for the spatially averaged instantaneous heat transfer coefficient is given by:

$$\left(\frac{h_c B}{k}\right) = a\left(\frac{\rho \bar{S}_p B}{\mu}\right)^b \tag{2.24}$$

where B is the diameter of the cylinder bore, and the value of a varies with intensity of charge motion and engine design. $\bar{S}_p$ is the mean piston speed, and it is related to the stroke L and engine speed via

$$\bar{S}_p = 2LN$$

Table 2.1 Values of γ for different fuels, equivalence ratios (ϕ), and residual burned gas fractions (X_b), adapted from Cheung and Heywood [1993]

Isooctane

	Compression		Combustion		Expansion	
ϕ	$X_b = 8.5\%$	$X_b = 22.0\%$	$X_b = 8.5\%$	$X_b = 22.0\%$	$X_b = 8.5\%$	$X_b = 22.0\%$
0.7	$1.412-1.457 \times 10^{-4}T$	$1.410-1.385 \times 10^{-4}T$	1.290	1.295	$1.330-3.604 \times 10^{-5}T$	$1.322-3.098 \times 10^{-5}T$
0.9	$1.403-1.493 \times 10^{-4}T$	$1.405-1.476 \times 10^{-4}T$	1.270	1.280	$1.455-1.071 \times 10^{-4}T$	$1.360-5.783 \times 10^{-5}T$
1.0	$1.400-1.540 \times 10^{-4}T$	$1.403-1.520 \times 10^{-4}T$	1.255	1.267	$1.477-1.262 \times 10^{-4}T$	$1.430-1.020 \times 10^{-4}T$
1.1	$1.398-1.583 \times 10^{-4}T$	$1.401-1.546 \times 10^{-4}T$	1.235	1.240	$1.410-7.877 \times 10^{-5}T$	$1.307-2.621 \times 10^{-5}T$
1.2	$1.396-1.610 \times 10^{-4}T$	$1.399-1.567 \times 10^{-4}T$	1.211	1.213	$1.313-2.794 \times 10^{-5}T$	$1.289-1.443 \times 10^{-5}T$
1.3	$1.392-1.630 \times 10^{-4}T$	$1.397-1.585 \times 10^{-4}T$	1.189	1.191	$1.284-1.087 \times 10^{-5}T$	$1.285-9.701 \times 10^{-6}T$

Propane

	Compression		Combustion		Expansion	
ϕ	$X_b = 8.5\%$	$X_b = 22.0\%$	$X_b = 8.5\%$	$X_b = 22.0\%$	$X_b = 8.5\%$	$X_b = 22.0\%$
0.7	$1.423-1.518 \times 10^{-4}T$	$1.422-1.483 \times 10^{-4}T$	1.292	1.297	$1.334-3.812 \times 10^{-5}T$	$1.324-3.207 \times 10^{-5}T$
0.9	$1.420-1.626 \times 10^{-4}T$	$1.420-1.593 \times 10^{-4}T$	1.271	1.281	$1.466-1.123 \times 10^{-4}T$	$1.368-6.201 \times 10^{-5}T$
1.0	$1.414-1.598 \times 10^{-4}T$	$1.413-1.556 \times 10^{-4}T$	1.255	1.267	$1.475-1.249 \times 10^{-4}T$	$1.434-1.041 \times 10^{-4}T$
1.1	$1.417-1.729 \times 10^{-4}T$	$1.412-1.575 \times 10^{-4}T$	1.236	1.242	$1.430-8.863 \times 10^{-5}T$	$1.313-3.001 \times 10^{-5}T$
1.2	$1.411-1.677 \times 10^{-4}T$	$1.411-1.602 \times 10^{-4}T$	1.212	1.215	$1.325-3.437 \times 10^{-5}T$	$1.292-1.623 \times 10^{-5}T$
1.3	$1.410-1.717 \times 10^{-4}T$	$1.410-1.628 \times 10^{-4}T$	1.191	1.192	$1.291-1.487 \times 10^{-5}T$	$1.284-9.880 \times 10^{-6}T$

Methanol

	Compression		Combustion		Expansion	
ϕ	$X_b = 8.5\%$	$X_b = 22.0\%$	$X_b = 8.5\%$	$X_b = 22.0\%$	$X_b = 8.5\%$	$X_b = 22.0\%$
0.7	$1.427-1.680 \times 10^{-4}T$	$1.424-1.595 \times 10^{-4}T$	1.258	1.253	$1.314-2.937 \times 10^{-5}T$	$1.320-3.328 \times 10^{-5}T$
0.9	$1.426-1.829 \times 10^{-4}T$	$1.420-1.690 \times 10^{-4}T$	1.265	1.270	$1.391-7.785 \times 10^{-5}T$	$1.326-4.321 \times 10^{-5}T$
1.0	$1.420-1.809 \times 10^{-4}T$	$1.423-1.814 \times 10^{-4}T$	1.260	1.271	$1.449-1.162 \times 10^{-4}T$	$1.395-8.827 \times 10^{-5}T$
1.1	$1.420-1.863 \times 10^{-4}T$	$1.423-1.850 \times 10^{-4}T$	1.2239	1.242	$1.331-4.454 \times 10^{-5}T$	$1.291-2.312 \times 10^{-5}T$
1.2	$1.419-1.916 \times 10^{-4}T$	$1.422-1.888 \times 10^{-4}T$	1.214	1.215	$1.287-1.962 \times 10^{-5}T$	$1.285-1.793 \times 10^{-5}T$
1.3	$1.423-2.043 \times 10^{-4}T$	$1.417-1.845 \times 10^{-4}T$	1.192	1.192	$1.276-1.185 \times 10^{-5}T$	$1.282-1.406 \times 10^{-5}T$

With normal combustion, $0.3 \le a \le 0.8$ with $b = 0.7$, and a increases with increasing intensity of charge motion in the cylinder. Gas properties, e.g., density ρ, dynamic viscosity μ, and thermal conductivity k, are evaluated at the cylinder average charge temperature, $\bar{T}_g$,

$$\bar{T}_g = \frac{pVM}{m\tilde{R}} \tag{2.25}$$

Woschni's correlation for h_c is given by

$$h_c\left(W/m^2K\right) = 3.26B(m)^{-0.2}p(kPa)^{0.8}T(K)^{-0.55}w(m/s)^{0.8} \tag{2.26}$$

where the average gas velocity in the cylinder, w, for a four-stroke engine is expressed as follows:

$$w = \left[C_1\bar{S}_p + C_2\frac{V_d T_r}{P_d V_r}(p - p_m)\right] \tag{2.27}$$

where V_d is the displacement volume, p is the instantaneous cylinder pressure, p_r, V_r, T_r are the gas pressure, volume, and temperature at some reference state (inlet valve closure or start of combustion), and p_m is the motored cylinder pressure at the same crank angle as p.

For the gas exchange period: $C_1 = 6.18 + 0.417\frac{v_s}{\bar{S}_p}$

For the rest of cycle: $C_1 = 2.28 + 0.308\frac{v_s}{\bar{S}_p}$

where $v_s = B\omega_p/2$ and ω_p is the rotation speed determined from a paddle wheel used to measure the swirl velocity.

In general, the heat transfer integrated over the duration of the combustion period is 10 to 15 percent for gasoline engines and up to 25 percent of the total heat released for diesel engines.

Crevice effects account for the second most important contribution to the difference between the gross and net heat release. According to Gatowski et al. [1984], assuming $\gamma(T) = a + b \cdot T$, the factor by which dm_{cr} is multiplied in Eq. (2.22) can be expressed as

$$(h' - u + c_v T) = T' + \frac{T}{\gamma - 1} - \frac{1}{b}\ln\left(\frac{\gamma - 1}{\gamma' - 1}\right) \tag{2.28}$$

Assuming the overall effect of crevice volumes is equivalent to a single aggregate crevice volume where the gas is at the same pressure as the combustion chamber, but at the cylinder wall temperature, i.e.,

$$m_{cr} = \frac{pV_{cr}}{RT_w} \quad \text{or} \quad dm_{cr} = \left(\frac{V_{cr}}{RT_w}\right) dp \tag{2.29}$$

Inserting Eqs. (2.28) and (2.29) into Eq. (2.22), one obtains

$$\frac{dQ_n}{dt} = \frac{\gamma}{\gamma-1} p \frac{dV}{dt} + \frac{1}{\gamma-1} V \frac{dp}{dt} + V_{cr}\left[T' + \frac{T}{\gamma-1} - \frac{1}{b}\ln\left(\frac{\gamma-1}{\gamma'-1}\right)\right]\frac{dp}{dt} \tag{2.30}$$

Fig. 2.11 shows the integrated heat-release results obtained from an experimental pressure trace using Eq. (2.30). The inclusion of heat transfer and crevice models allows the chemical energy released Q_{ch} to be determined, as well as the net heat release. The difference between the final value of Q_{ch} and the total energy input (m_f·CV) should represent the combustion inefficiency, which can be determined from the exhaust gas composition. The difference may also be caused by inaccuracies in the cylinder pressure data and the heat-release analysis, which can be determined from the measured and predicted combustion inefficiencies.

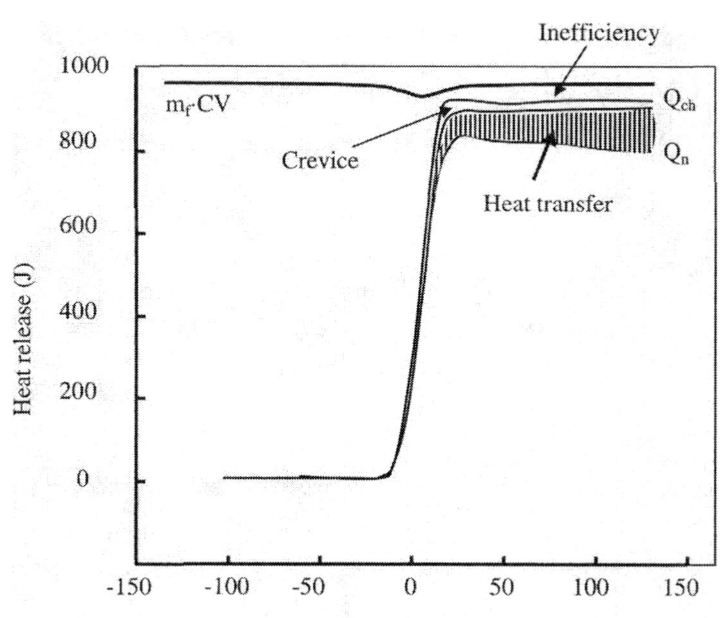

Fig. 2.11 Heat-release profile showing the effects of heat-transfer, crevices, and combustion efficiency, adapted from Heywood [1988].

2.5.2.4 Two-Zone Fuel Mass Fraction Burned Analysis

A two-zone model was first developed by Krieger and Borman [1966]. The two-zone model assumes that during combustion, the combustion chamber is divided into a burned volume and an unburned volume separated by an infinitesimally thin flame front. Each volume is assumed to be in thermodynamic equilibrium and both systems are considered to be at the same pressure. The differential equations based on the first law of thermodynamics for the unburned and burned gases, ignoring the crevice or blow-by, can be written as

$$\frac{d(m_u u_u)}{dt} = -p\frac{dV_u}{dt} + \frac{dQ_u}{dt} \tag{2.31}$$

$$\frac{d(m_b u_b)}{dt} = -p\frac{dV_b}{dt} + \frac{dQ_b}{dt} \tag{2.32}$$

Subscripts u and b represent the unburned and burned systems, respectively. Q_u and Q_b are heat losses to the walls by the unburned and burned gas, respectively, and can be calculated from Eq. (2.23). The difficulty, however, in calculating the heat transfer comes in determining the area of the chamber walls wetted by the burned gas. This is normally done by approximation methods. Krieger and Borman arbitrarily assumed that only a fraction of the piston area was exposed to the burned gas, equal to the volume fraction, V_b/V. Ferguson [1986] argued that the fraction of cylinder area contacted by burned gas should be proportional to the square root of mass fraction burned to reflect the fact that the burned gas occupies a larger volume fraction of the cylinder than the unburned gas because burned gas has a different density than unburned gas.

Assuming that R_u is constant, Eq. (2.31) can be rearranged to find the unburned gas temperature change,

$$dT_u = \frac{T_u\left(\frac{dp}{p}\right) + \frac{dQ_u}{m_u R_u}}{\left(\frac{1}{R_u}\right)\left(\frac{\partial u_u}{\partial T_u}\right) + 1} \tag{2.33}$$

Assuming R_b is independent of temperature and pressure, one may solve for the mass fraction burned rate dm_b,

$$dm_b = \frac{A - B + dQ_b}{(u_b - u_u) + \left(\frac{R_u T_u}{R_b} - T_b\right)\frac{\partial u_b}{\partial T_b}} \tag{2.34}$$

where

$$A = (m_u R_u dT_u - pdV)\left(\frac{1}{R_b}\frac{\partial u_b}{\partial T_b} + 1\right)$$

$$B = dp \cdot V \cdot \left(\frac{1}{R_u}\frac{\partial u_b}{\partial T_b} + \frac{M_b}{V}\frac{\partial u_b}{\partial p} + \frac{V_u}{V}\right)$$

The initial temperature of the burned gas at the spark ignition is assumed to be the adiabatic flame temperature calculated by solving

$$h_b(T_{b,0}) = h_u(T_u)$$

The rate of change of burned gas volume can then be calculated from

$$dV_b = V_u\left(\frac{dm_b}{m_u} - \frac{dT_u}{T_u} + \frac{dp}{p}\right) + dV \tag{2.35}$$

The volume relationship, $V = V_b + V_u$, conservation of mass, $m = m_b + m_u$, and the equation of state for both burned and unburned gases are also required in calculating the mass fraction burned from the cylinder pressure.

The thermodynamic properties of the cylinder gases can be accurately determined from several sources (e.g., JANAF Tables, 1971). Polynomial functions for various fuels (in the vapor phase) for calculating the specific heat capacity at constant pressure and the specific enthalpy as a function of temperature are given by Heywood [1988], from which the internal energy can be computed.

The most accurate approach, for burned mixture property and composition calculations, is to use a thermodynamic equilibrium program at temperatures above 1700 K and a frozen composition below 1700 K. The NASA equilibrium program is readily available for this purpose, and is well documented [Gordon and McBride, 1971; Svehla, and McBride, 1973].

For heat-release rate analysis, estimates of internal energy and gas constant of the combustion products are adequate. Several techniques for estimating the thermodynamic properties of burned gases for engine applications have been developed. One commonly used approach developed by Krieger and Borman [1966] is to describe the internal energy and gas constant as a function of temperature, pressure, and equivalence ratio in polynomials. For lean mixtures ($\phi \leq 1$) of C_nH_{2n} with air, the specific internal energy of combustion products is given by a single set of equations

$$u = K_1(T) - K_2(T)\phi \quad \text{kJ/(kg of air)} \tag{2.36}$$

where

$$K_1 = 0.692T + 39.17 \times 10^{-6}T^2 + 52.9 \times 10^{-9}T^3 - 228.62 \times 10^{-13}T^4 + 277.58 \times 10^{-17}T^5$$

$$K_2 = 3049.33 - 5.7 \times 10^{-2}T - 9.5 \times 10^{-5}T^2 + 21.53 \times 10^{-9}T^3 - 200.26 \times 10^{-14}T^4$$

with the gas constant given by

$$R = 0.287 + 0.020\phi \quad \text{kJ/[(kg of air)K]} \tag{2.37}$$

A modification is introduced to account for dissociation if the temperature is above 1450 K:

$$u_{corr} = u + 2.32584\exp(A + B + C) \quad \text{kJ/kg of original air} \tag{2.38}$$

where

$$A = 10.41066 + 7.85125\phi - 3.71257\phi^3$$

$$B = (-15.001 - 15.838\phi + 9.613\phi^3) \times 103 / T$$

$$C = [0.154226\phi^3 - 0.38656\phi - 0.10329 + (118.27\phi - 14.763)/T] \times \ln(p \times 14.503)$$

and

$$R_{corr} = R + 0.004186\exp\{[11.98 - 25442/T - 0.4354\ln(p \times 14.503)]\phi + 0.2977\ln(\phi)\} \quad \text{kJ/[(kg original air)K]} \tag{2.39}$$

For a fuel of composition C_nH_{2n} the stoichiometric gravimetric fuel/air ratio, ϕ, is 0.0676. Thus, if the internal energy of the products is wanted on a basis of per unit mass of products, it is necessary to divide equations for u and R by ($1+0.067\phi$).

For rich mixtures, sets of equations are available, each set applying to a specific value of equivalence ratio [Krieger and Borman, 1966].

The two-zone analysis provides a more accurate thermodynamic analysis of the SI engine combustion process. However, the calculation procedure is more complex than the single-zone approach, making it less suitable for analysis of hundred of cycles of pressure data. The inclusion of heat transfer requires additional geometric assumptions concerning flame location. To include crevices in the analysis, a model for the composition of the gas flowing into the crevice region must be developed. Due to this complexity, crevice models are usually omitted in most two-zone models, despite the fact that their impact can be significant.

2.6 Diesel Combustion Analysis Using In-Cylinder Pressure

2.6.1 Direct Analysis of In-Cylinder Pressures

2.6.1.1 Diesel Combustion Performance Monitoring Using Pressure-Time Diagrams

The p-t diagram for a four-stroke naturally aspirated diesel engine is shown in Fig. 2.12. An important parameter in diesel combustion is the ignition delay between the start of injection (SOI) and start of combustion (SOC). The knowledge of the start of injection together with ignition delay (ID) is essential to the optimization of the diesel combustion process because these parameters determine the phase of combustion in a diesel engine just as the spark timing does in a SI engine. The start of injection is measured from the output of a needle lift sensor, as described in Chapter 1. Because of the more rapid pressure rise associated with diesel combustion, the start of combustion can be accurately determined from the dp/dt diagram.

Diesel combustion can produce very high pressure, because of the rapid heat release associated with multiple ignitions in a diesel engine. A diesel engine is designed to withstand a certain peak cylinder pressure, above which engine components may be damaged. Excessive peak cylinder pressure can occur due to high boost pressure in a turbocharged engine, or faulty fuel injection control. It is therefore important to be able to monitor the peak cylinder pressure during engine experiments.

The rapid pressure rise following the ignition delay may create such shock forces which cause part of the engine to vibrate. The resulting engine noise is known as diesel knock. Since the pressure during combustion is the exciting force for the undesirable vibrations, the maximum rate of pressure rise measured from the maximum slope of the p-t diagram is one such index. The maximum rate that an engine can withstand depends mainly upon the stiffness of the connecting rod, the crankshaft, and the crankcase. Too rigid components lead to excessive weight and cost. The reduction of diesel combustion noise is largely dependent on the optimization of the fuel injection pattern using in-cylinder pressure measurements.

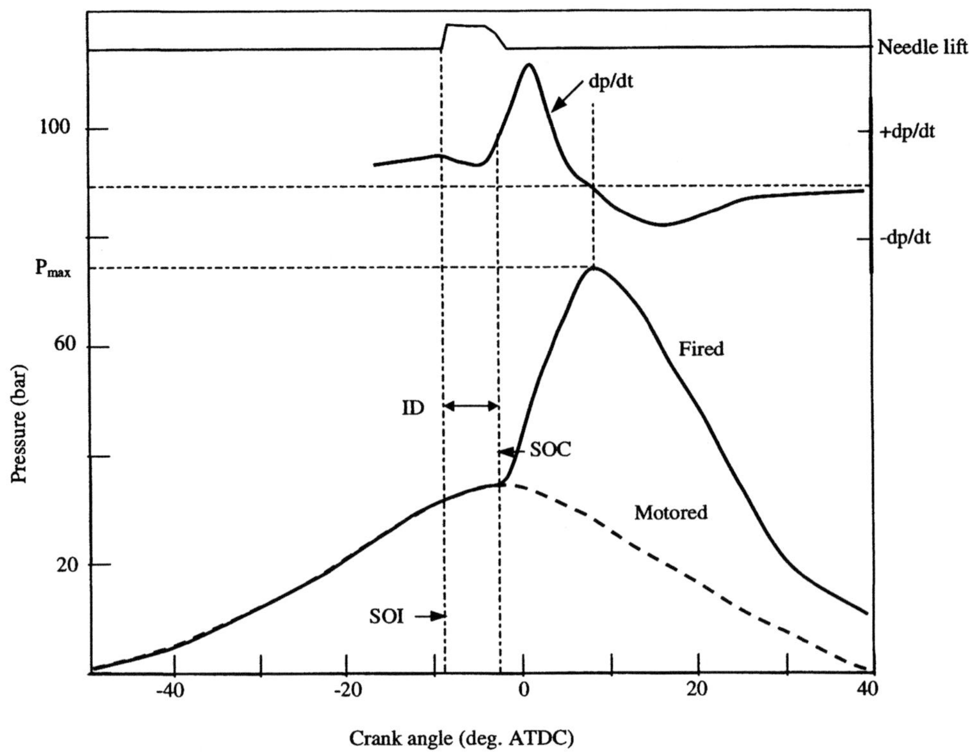

Fig. 2.12 In-cylinder pressure, injector needle lift, and dp/dt from a four-stroke DI diesel engine.

2.6.1.2 Analysis of Pressure-Volume Diagrams

Because of the absence of a throttle in the intake system, the p-V diagram for a naturally aspirated four-stroke diesel engine is similar to that of an SI engine at WOT, as shown in Fig. 2.6(a). If the engine is turbocharged, both the intake and exhaust pressures will be above atmospheric, the exact values depending on the engine operating conditions and turbocharger design. Since the air intake is not throttled at part load, the negative work area W_p is much smaller than the SI engine. Indicated work and pumping work are calculated from the p-V diagram using Eq. (2.7) and Eq. (2.8), respectively. Indicated mean effective pressure can then be calculated using Eq. (2.10).

The logarithmic p-V diagram is another useful method for diesel combustion analysis. During compression and expansion processes of an engine, the relationship between pressure and volume is closely described by the polytropic relationship pV^n = constant. Portions of the compression and expansion process on a logarithmic p-V diagram are therefore essentially straight lines, as shown in Fig. 2.13. Moreover, the beginning and end of combustion will be marked by a departure from the straight line since the effect of combustion is equivalent to heat addition with consequent change in the values of n. Due to the leaner mixture used in a diesel engine, the values of n are higher than those for the SI engine. It is also true that the values of n for expansion rise slowly due to continued burning on the expansion stroke of the diesel engine. Thus, a log p-log V diagram can be used as one way of observing the progress of combustion.

2.6.2 Heat Release Analysis in Diesel Engines

2.6.2.1 Introduction

The method of heat release analysis for diesel engines is similar to that described above for SI engines, and it starts with the first law of thermodynamics,

$$dQ_{ch} = dU_s + dW + \sum h_i dm_i + dQ_{ht}$$

There are, however, a number of problems in the application of this equation to diesel combustion:

(1.) The fuel/air mixture distribution is not uniform and varies with both time and space.
(2.) The composition of the burned gases is not uniform and not known.
(3.) The heat-transfer prediction is less well defined.

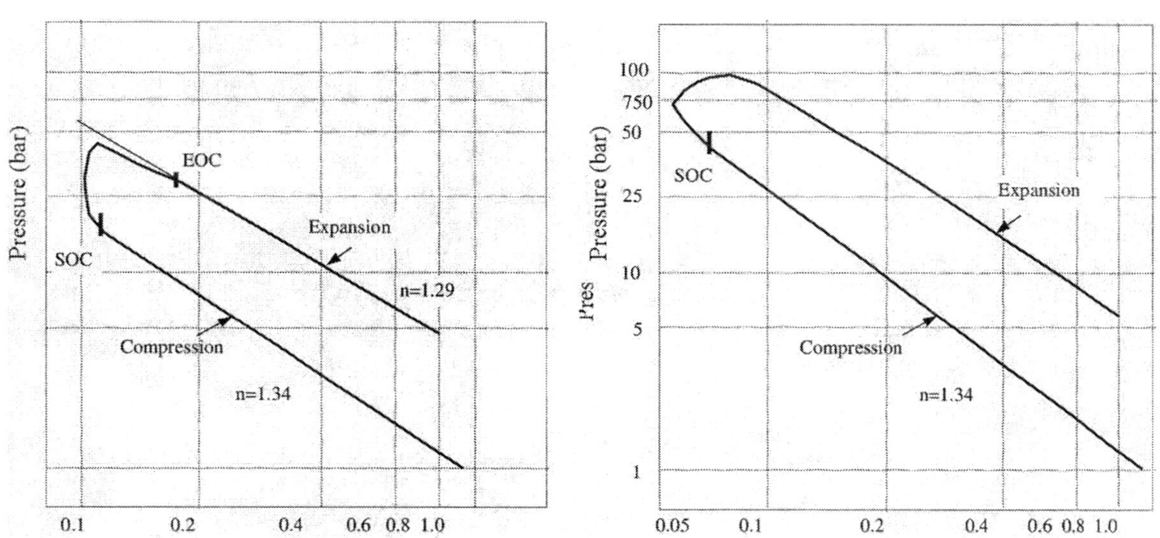

(a) Logp-log V diagram from a 4-stroke SI engine
(b) Logp-log V diagram from a 4-stroke turbocharge diesel engine

Fig. 2.13 Logarithmic p-V diagrams from four-stroke gasoline and diesel engines.

Because of the difficulties in dealing with these problems, heat-release methods, whether simple or sophisticated, give less accurate results than those for SI engines with premixed charge.

2.6.2.2 Direct-Injection Diesel Engines

2.6.2.2.1 Introduction

Two approaches have been developed to obtain combustion information from pressure data based on the application of the first law of thermodynamics to an open system, Eq. (2.17),

$$\frac{dQ}{dt} - p\frac{dV}{dt} + \dot{m}_f h_f = \frac{dU}{dt} \tag{2.40}$$

where crevice flow effects have been omitted. One method produces a fuel mass burning rate; the other generates heat-release rate directly. In both approaches, the cylinder contents are assumed to be at a uniform temperature at any moment during the combustion process. Included in the assumption is that the entire cylinder contains a homogeneous mixture of air and combustion products at each instant. The burning is assumed to take place incrementally as a homogeneous combustion and thus acts as a uniform heat source. Phenomena such as temperature gradients, non-equilibrium compositions, fuel vaporization, and mixing are not considered.

The stages of the overall compression-ignition diesel combustion process can be readily identified from a typical heat-release rate diagram for a DI diesel engine, as shown in Fig. 2.14. They are defined as the ignition delay (ab), premixed or rapid combustion phase (bc), mixing-controlled combustion phase (cd), and late combustion phase (de). Using such diagrams, the effect of injection parameters and engine design on the combustion process can be assessed.

2.6.2.2.2 Heat-Release Analysis

If U and h_f in Eq. (2.40) are taken to be the sensible internal energy of the in-cylinder charge and the sensible enthalpy of the injected fuel, respectively, then dQ/dt becomes the net heat release rate (the difference between the chemical energy released from the fuel and the heat loss from the engine). If we further assume that the contents of the cylinder can be modeled as an ideal gas, the net heat-release rate in a DI diesel engine is given by

$$\frac{dQ_n}{dt} = \frac{\gamma}{\gamma - 1} p \frac{dV}{dt} + \frac{1}{\gamma - 1} V \frac{dp}{dt} \tag{2.41}$$

in which crevice flow effects are omitted, and the sensible enthalpy of the fuel $h_{s,f} \approx 0$ and is ignored.

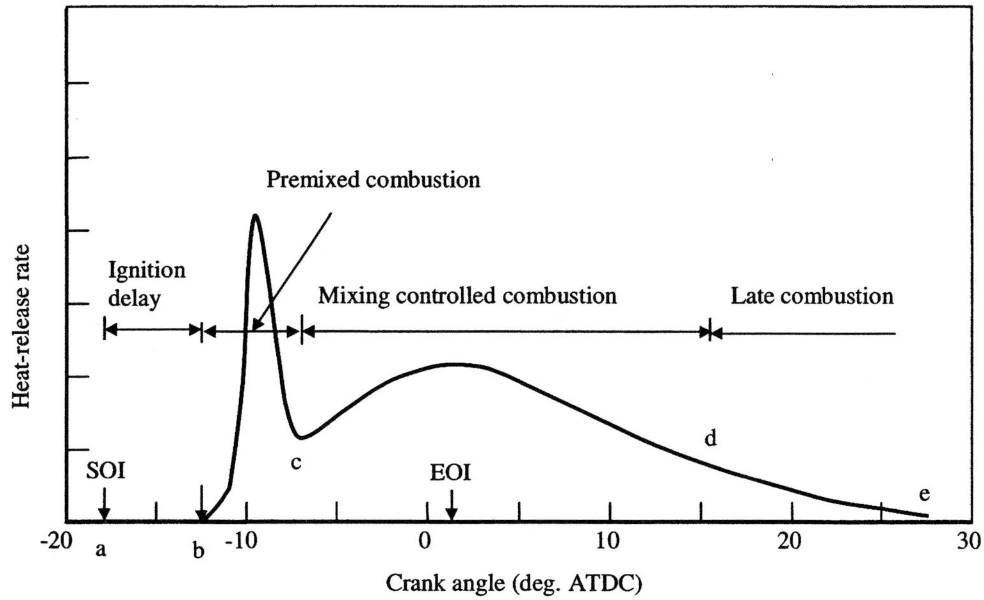

Fig. 2.14 Idealized DI diesel engine heat-release rate diagram showing different combustion phases.

Eq. (2.41) provides a simple way to evaluate the heat release rate in a DI diesel engine with the knowledge of the ratio of specific heats γ. It is expected that γ for diesel combustion should have values appropriate to air at end-of-compression stroke temperatures prior to combustion ($\approx$1.35) and to burned gases at the overall equivalence ratio following combustion ($\approx$1.26–1.3).

The gross heat-release rate dQ_{ch}/dt can be obtained if the heat transfer rate term dQ_{ht}/dt is modeled as described previously. The integral of the gross heat-release rate over the combustion process should then equal the product of the mass of fuel injected m_f and the calorific value of the fuel CV, i.e., $Q_{ch} = m_f CV$.

For most applications, Eq. (2.41) is adequate for diesel engine combustion analysis. More elaborate methods of heat-release analysis based on Eq. (2.39) have also been developed to include sophisticated models for the gas properties before, during, and after combustion, for heat-transfer and crevice effects [Gatowski, et al., 1984]. Fig. 2.15 illustrates the relative magnitude of gross and net heat release, heat transfer, crevice effects, and heat of vaporization and heating up of fuel for a turbocharged four-stroke DI diesel engine. The first dip in the heat release curve is due to the effect of fuel vaporization and fuel heating up from injection temperature to typical compression air temperature. The largest effect is the heat loss to the engine blocks, amounting to 25% of the total heat released.

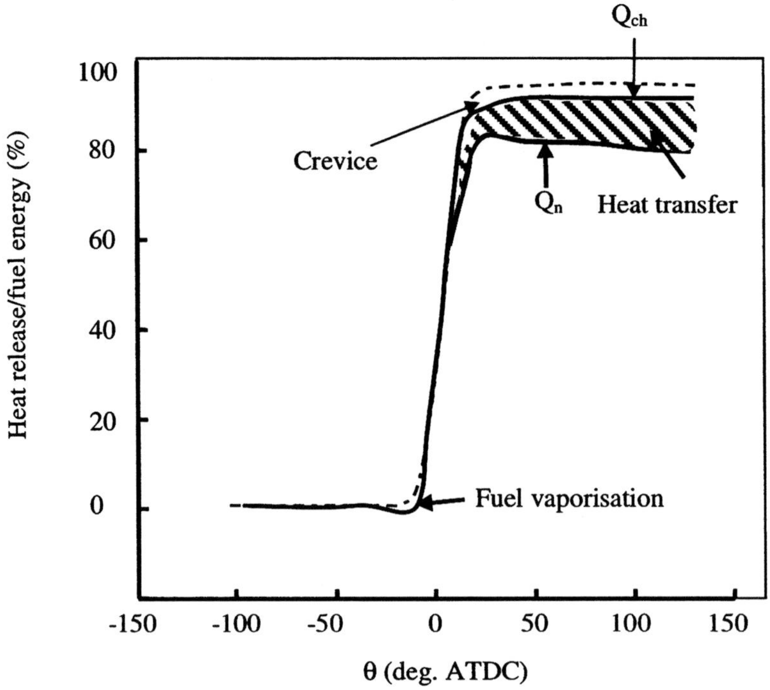

Fig. 2.15 Gross and net heat release profile for a turbocharged DI diesel engine.

2.6.2.2.3 Fuel Mass Burning Rate Analysis

The other approach yields the fuel mass burning rate directly [Krieger and Borman,1966]. If U is taken to be the internal energy of the cylinder charge including both the sensible energy and the energy of formation of the relevant species in the charge, the first law of thermodynamics can be written as

$$\frac{d(mu)}{dt} = -p\frac{dV}{dt} + \frac{dQ_{ht}}{dt} + h_f\frac{dm}{dt} \tag{2.42}$$

where Q_{ht} is the heat transfer to the gas within the combustion chamber.

The specific internal energy, u, and the gas constant, R, of the gases in the cylinder are in general a function of p, T, and the equivalence ratio, ϕ,

$$u = u(T, p, \phi) \text{ and } R = R(T, p, \phi)$$

Therefore

$$\frac{du}{dt}=\frac{\partial u}{\partial T}\frac{dT}{dt}+\frac{\partial u}{\partial p}\frac{dp}{dt}+\frac{\partial u}{\partial \varphi}\frac{d\varphi}{dt} \tag{2.43}$$

$$\frac{dR}{dt}=\frac{\partial R}{\partial T}\frac{dT}{dt}+\frac{\partial R}{\partial p}\frac{dp}{dt}+\frac{\partial R}{\partial \varphi}\frac{d\varphi}{dt} \tag{2.44}$$

The equivalence ratio is given by

$$\varphi=\varphi_0+\left(\frac{m}{m_0}-1\right)\frac{1+\left(\frac{F}{A}\right)_o}{\left(\frac{F}{A}\right)_s} \tag{2.45}$$

and

$$\frac{d\varphi}{dt}=\frac{1+(F/A)_o}{(F/A)_s\, m_o}\frac{dm}{dt} \tag{2.46}$$

The subscript "o" denotes the initial value prior to fuel injection, and the subscript "s" denotes the stoichiometric value. (F/A) is the fuel/air ratio. It then follows that

$$\frac{1}{m}\frac{dm}{dt}=\frac{-(RT/V)(dV/dt)-(\partial u/\partial p)(dp/dt)+(1/m)(dQ/dt)-CB}{u-h_f+D(\partial u/\partial \varphi)-C[1+(D/R)(\partial R/\partial \varphi)} \tag{2.47}$$

where

$$B=\frac{1}{p}\frac{dp}{dt}-\frac{1}{R}\frac{\partial R}{\partial p}\frac{dp}{dt}+\frac{1}{V}\frac{dV}{dt}$$

$$C=\frac{T(\partial u/\partial T)}{1+(T/R)(\partial R/\partial T)}$$

$$D=\frac{[1+(F/A)_o]m}{(F/A)_s\, m_o}$$

Krieger and Borman found that the effect of dissociation of the combustion products was negligible in a DI diesel engine. Without dissociation, u = u(T, ϕ), and R can be taken as constant. Then, Eq. (2.47) is simplified to

$$\frac{dm}{dt}=\frac{\left[1+(c_v/R)\right]p(dV/dt)+(c_v/R)V(dp/dt)-(dQ/dt)}{h_f+(c_v/R)(pV/m)-u-D(\partial u/\partial \varphi)} \tag{2.48}$$

Eq. (2.47) or Eq. (2.48) can be solved numerically for m(t) given m_o, ϕ_o, p(t) and appropriate properties of the working fluid (Eqs. (2.36) and (2.37)) and for the heat transfer term dQ_{ht}/dt. Terms dp/dt and dV/dt are obtained from the cylinder pressure data and engine geometry. The result obtained is an apparent fuel mass burning rate. It is best interpreted, after multiplying by the calorific value of the fuel, as the fuel chemical energy or heat release rate. The actual fuel-burning rate cannot be known because not all of the fuel will burn with sufficient air available locally to produce products of complete combustion. This is why heat release rate analysis is preferred in diesel engine combustion analysis.

Fig. 2.16 shows a fuel mass burning rate profile in a DI diesel engine based on the measured cylinder pressure using the Krieger and Borman's approach. Annand's correlation was used in the heat-transfer model. The first major peak representing the premixed phase of combustion is obvious. It should also be noted that chemical energy continues to be released well into the expansion process.

2.6.2.3 Indirect-Injection Diesel Engines

An indirect-injection (IDI) diesel engine contains a small combustion chamber and a main combustion chamber, which are connected via a narrow passage. The pressures in each of the two chambers are different, particularly during the combustion period. As the combustion starts in the pre-chamber or swirl chamber, the heat released in this chamber causes its gas pressure to rise faster than that in the main chamber. Due to the pressure difference between the two chambers, a flow of air, fuel, and burned gases goes into the main chamber where additional heat is released by combustion. To analyze the heat release in IDI engines, the effect of the pressure difference between the chambers should be included. Sometimes the practical difficulties in installing two separate pressure transducers on each chamber may prevent the use of cylinder pressures from each chamber. In this case, either main chamber or auxiliary chamber pressure data alone may be used. This, however, will introduce errors in the heat release analysis. The magnitude of the error will depend on the design of the combustion chamber and on engine speed and load. With narrower passages and higher loads and speeds, the pressure difference between the two chambers becomes larger, hence greater error in heat release rates.

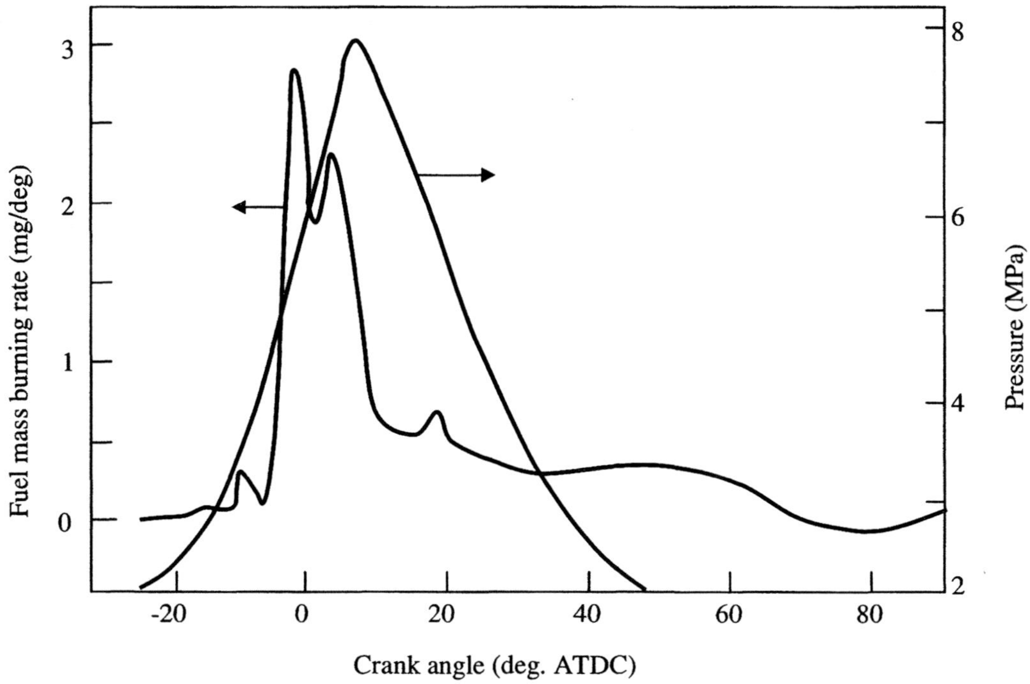

Fig. 2.16 Fuel mass burning rate calculated from pressure using the Krieger and Borman method, adapted from Heywood [1988].

The heat release analysis of an IDI diesel engine follows the application of the first law to both chambers respectively, as shown in Fig. 2.17. In the main chamber, the first law becomes

$$\frac{dQ_1}{dt} - p_1 \frac{dV_1}{dt} + h_{2,1} \frac{dm}{dt} = \frac{dU_1}{dt} \tag{2.49}$$

and in the pre-chamber, it is given by

$$\frac{dQ_2}{dt} + h_f \frac{dm_f}{dt} - h_{2,1} \frac{dm}{dt} = \frac{dU_2}{dt} \tag{2.50}$$

where dm/dt is the mass flow rate between the chambers with positive flow from the pre-chamber to the main chamber. If $dm/dt < 0$, $h_{2,1} = h_1$; if $dm/dt > 0$, $h_{2,1} = h_2$. As in the previous section, U_1 and U_2 are defined as sensible internal energies, and h_f as the sensible enthalpy of the fuel. dQ_1/dt and dQ_2/dt are the net heat release rates, i.e., the difference between the gross heat release rates and the rates of heat transfer to the walls.

From the ideal gas model for the working fluid in each chamber,

$$dU_1 = m_1C_vdT_1 + U_{2,1}dm \quad \text{and} \quad dU_2 = m_2C_vdT_2 - U_{2,1}dm$$

together with the relations

$$\frac{dp_1}{p} + \frac{dV_1}{V} = \frac{dT_1}{T} \quad \text{and} \quad \frac{dp_2}{p} + \frac{dV_2}{V} = \frac{dT_2}{T}$$

Eqs. (2.49) and (2.50) can be written as

$$\frac{dQ_1}{dt} = \frac{\gamma}{\gamma - 1}p_1\frac{dV_1}{dt} + \frac{1}{\gamma - 1}V_1\frac{dp_1}{dt} - (c_p + c_v)T_{2,1}\frac{dm}{dt} \tag{2.51}$$

$$\frac{dQ_2}{dt} = \frac{1}{\gamma - 1}V_2\frac{dp_2}{dt} + (c_p + c_v)T_{2,1}\frac{dm}{dt} \tag{2.52}$$

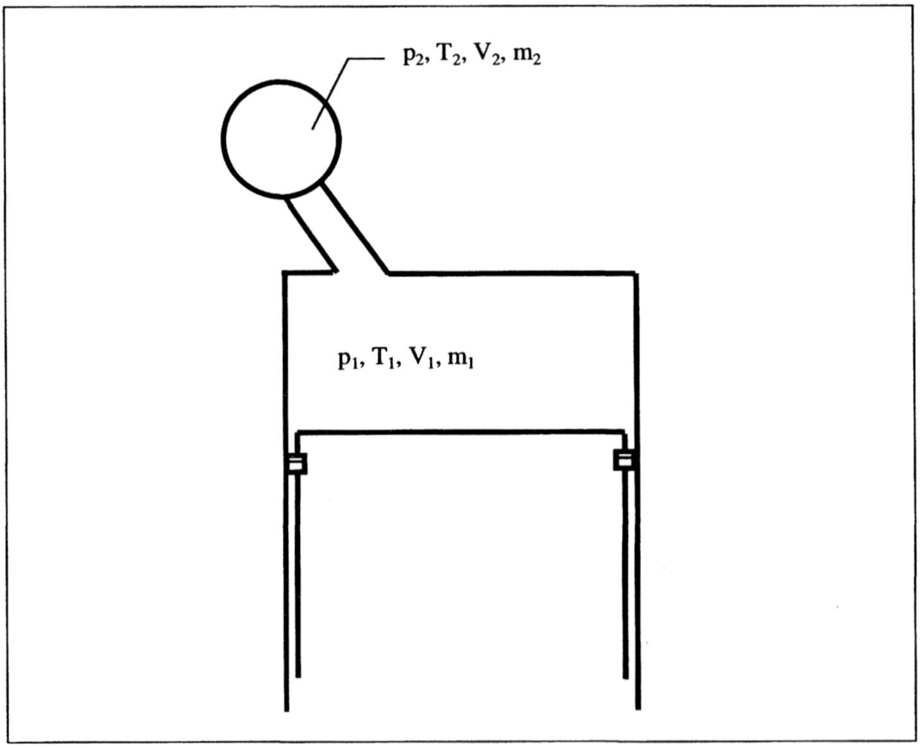

Fig. 2.17 Schematic of an IDI diesel engine with definition for heat release analysis.

When the above two equations are added together, the following equation for the total net heat-release rate is obtained,

$$\frac{dQ_n}{dt} = \frac{dQ_1}{dt} + \frac{dQ_2}{dt} = \frac{\gamma}{\gamma - 1} p_1 \frac{dV_1}{dt} + \frac{1}{\gamma - 1}\left(V_1 \frac{dp_1}{dt} + V_2 \frac{dp_2}{dt} \right) \tag{2.53}$$

The heat-release rate can be determined with either individual or ensemble-averaged pressure data because dQ_n/dt is linearly dependent on p_1 and p_2. Averaging the pressure traces over many engine cycles minimizes the noise problem. As for the DI diesel engine, the value of γ is not known accurately. Kort et al. [1982] tested the sensitivity of the heat-release rate to the value of γ and found that only modest changes in the heat release profile resulted for γ of 1.30 and 1.35. A value of $\gamma = 1.35$ was then used because it was closely matched to the expected value for air at the end of the compression stroke.

The net heat release rate calculated from this equation is the gross heat release rate due to combustion less the heat transfer to the walls, energy transfers due to crevices, fuel vaporization and heat-up, i.e.,

$$\frac{dQ_n}{dt} = \frac{dQ_{ch}}{dt} - \frac{dQ_{ht}}{dt} - \frac{dQ_f}{dt} \tag{2.54}$$

Fig. 2.18 shows an idealized rate of heat-release profile as a function of crank angle calculated from Eq. (2.53) [Kort et al., 1982]. Included in the diagram are the approximate contribution from each of the terms in Eq. (2.54). The negative value of dQ_n/dt before the start of fuel injection is due to heat loss to the combustion chamber walls. The start of fuel injection reinforces slightly the negative value of dQ_n/dt due to fuel evaporation and the heating term dQ_f/dt. As soon as combustion starts, the release of chemical energy (dQ_{ch}/dt) becomes the dominant term and results in a positive net heat-release rate around TDC. The shaded area represents the total heat-release from combustion. Although the baseline for the heat release calculation (the bottom dashed line) is not known precisely, it is found that a straight line, between the start-of-combustion and the crank angle where dQ_n/dt approaches zero, gives the integrated heat-release, which is in good agreement with that calculated from $Q_{ch} = \int Q_{ch}/dt \cdot dt = m_f \cdot CV$.

Other approaches to analyze heat releases in an IDI diesel engine have also been proposed and developed. Following the approach of Krieger and Borman for DI diesel engines described in the previous section, Watson and Kamel [1979] calculated fuel mass burning rate, and heat release rate in an IDI engine. The energy conservation equation for an open system was applied to the main chamber and the pre-chamber, respectively. Gas flow through the connecting passages was calculated from the one-dimensional, quasi-steady, compressible flow equation through an orifice. With appropriate models for convective and radiative heat

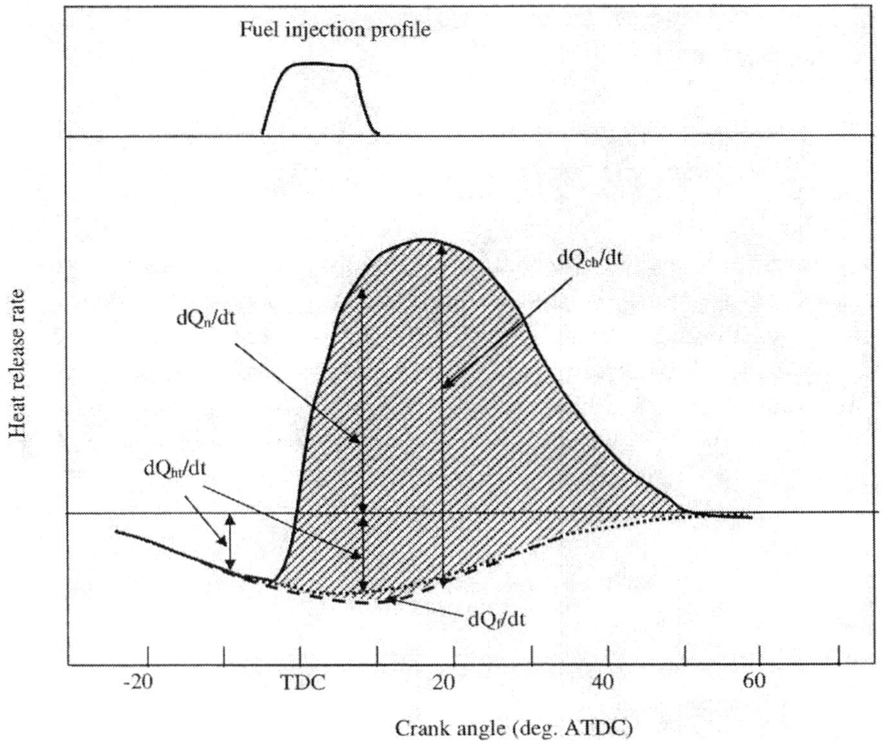

Fig. 2.18 Heat-release rate versus crank angle for an IDI diesel engine, adapted from Kort et al. [1982].

transfer and for gas properties, the model calculated directly the fuel burning rates and determined heat release rates in both the main and pre-chamber by multiplying the fuel burning rates by the fuel heating value.

2.7 Summary

Although in-cylinder pressure measurement is the most widely used technique for engine combustion diagnostics, there has been a lack of a systematic approach to this subject. On many occasions, the use of a pressure transducer and the interpretation of its output have been taken for granted without awareness of the precautions necessary in the process. In this chapter, measurements and analysis of in-cylinder pressure for combustion diagnostics have been discussed in detail, including the selection, installation, and calibration of the pressure transducer and methods of heat release analysis from in-cylinder pressure measurements. It is hoped that the information presented in this chapter will help engine researchers to optimize the use of pressure transducers for engine design and combustion analysis.

References

Annand, W.J.D. (1963), "Heat Transfer in the Cylinders of Reciprocating Internal Combustion Engines," *Proc. Instn. Mech. Engrs.*, Vol. 177, No. 36, pp. 973–990.

Barton, R.K., Lestz, S.S., and Duke, L.C. (1970), "Knock Intensity as a Function of Engine Rate of Pressure Change," SAE Paper No. 700061.

Cheung, H. and Heywood, J.B. (1993), "Evaluation of a One-Zone Burn-Rate Analysis Procedure Using Production SI Engine Pressure Data," SAE Paper No. 932749.

Chun, K.M. and Heywood, J.B. (1987), "Estimating Heat-Release and Mass-of-Mixture Burned from Spark-Ignition Engine Pressure Data," *Combustion Science and Technology*, Vol. 54, pp.134–144.

Chun, K.M. and Heywood, J.B. (1989), "Characterisation of Knock in a SI Engine," SAE Paper No. 890156.

Ferguson, C.R. (1986), *Internal Combustion Engines*, J. Wiley & Sons.

Gatowski, J.A., Balles, E.N., Chun, K.M., Nelson, F.E., Ekchian, J.A., and Heywood, J.B. (1984), "Heat Release Analysis of Engine Pressure Data," SAE Paper No. 841359.

Gordon, S. and McBride, B.J. (1971), "Computer Program for the Calculation of Complex Chemical Equilibrium Composition, Rocket Performance, Incident and Reflected Shocks, and Chapman-Jouguet Detonations," *NASA Publication SP-273.*

Heywood, J.B., (1988), *Internal Combustion Engine Fundamentals*, McGraw-Hill Book Company, New York.

JANAF Thermochemical Tables, 2nd Ed., NSRDS-NB537, NIST, June 1971.

Kort, R.T., Mansouri, H., Heywood, J.B., and Ekchian, A., (1982), "Divided-Chamber Diesel Engine, Part II: Experimental Validation of a Predictive Cycle-Simulation and Heat Release Analysis," SAE Paper No. 820274.

Krieger, R.B. and Borman, G.L. (1966), "The Computation of Apparent Heat Release for Internal Combustion Engines," ASME Paper 66-WA/DGP-4.

Lancaster, D. (1975), "Measurement and Analysis of Engine Pressure Data," SAE Paper No. 750026.

Leppard, W.R. (1982), "Individual-Cylinder Knock Occurrence and Intensity in Multicylinder Engines," SAE Paper No. 820074.

Matekunas, F.A. (1983), "Modes and Measures of Cyclic Combustion Variability," SAE Paper No. 830337.

Obert, E.F. (1973), *Internal Combustion Engines and Air Pollution*, Harper & Row, New York.

Randolph, A. (1990a), "Methods of Processing Cylinder Pressure Transducer Signals to Maximise Data Accuracy," SAE Paper No. 900170.

Randolph, A.(1990b), "Cylinder Pressure Transducer Mounting Techniques to Maximise Data Accuracy," SAE Paper No. 900171.

Rassweiler, G.M. and Withrow, L. (1938), "Motion Pictures of Engine Flames Correlated with Pressure Cards," *SAE Trans.*, Vol. 83, pp.185–204. (Reprinted as SAE Paper No. 800131, 1980).

Stone, R. and Green-Armytage, D.I. (1987), "Comparison of Methods for the Calculation of Mass Fraction Burnt from Engine Pressure-Time Diagrams," *Proc. I. MechE.*, Part D, Vol. 201, No. 1, pp. 61–67.

Svehla, R. A. and McBride, B. J., (1973), "Fortran IV Computer Program for Calculation of Thermodynamic and Transport Properties of Complex Chemical Systems," *NASA Technical Note TND-7056.*

Watson, N., and Kamel, M. (1979), "Thermodynamic Efficiency Evaluation of an Indirect Injection Diesel Engine," SAE Paper No. 790039.

Woschni, G. (1967), "Universally Applicable Equation for the Instantaneous Heat Transfer Coefficient in the Internal Combustion Engines," SAE Paper No. 670931.

Zhao, H., Collings, N., and Ma, T. (1992), "Characterisation of Knock and Its Effect on Surface Temperature," SAE Paper No. 920514.

Chapter 3

Fast-Response Flame Ionization Detector

Outline

3.1 Introduction

The exhaust gas emitted from internal combustion engines usually contains unburned hydrocarbons (HC) which can be harmful to the environment. These unburned hydrocarbons are a collection of many different compounds, each with its own molecular structure. In internal combustion engine research, it is sometimes useful to know which hydrocarbon compounds are present in the exhaust and the quantity of each compound. Such an analysis of the exhaust gas, however, is time consuming and requires fairly complex and specialized equipment. More often, a simpler and quicker analysis is performed. This simpler analysis gives the *total* amount of hydrocarbon compounds in the exhaust gas, but does not provide any information about which hydrocarbon compounds are present. The instrument used most often for this simpler analysis is the Flame Ionization Detector (FID).

As discussed in Chapter 1, a conventional FID system continuously extracts a small fraction of the exhaust gas stream and measures the total amount of hydrocarbons present in the exhaust stream. The measurement is usually provided to the user as the concentration of hydrocarbons (HC) in the exhaust gas on a volumetric basis; that is, the fraction of a given volume of exhaust gas occupied by the various hydrocarbon compounds.

A conventional FID system provides a continuous measurement of the HC concentration in the exhaust. Its output continuously adjusts to changes in HC concentration of the exhaust gas. However, the speed with which a conventional FID system adjusts is rather slow. For example, a sudden increase in HC concentration can take several seconds before it is fully registered by the instrument output.

In internal combustion engine research, it is sometimes useful to monitor the instantaneous concentration of hydrocarbons in various parts of the engine. For example, of interest might be the instantaneous HC concentration in the exhaust gas flowing past the exhaust valve just as the exhaust valve opens. Also of interest might be the HC concentration within the engine cylinder, close to the spark plug, just before ignition. It is known that the mixture strength at the spark plug can influence flame development and the mean effective pressure of an engine cycle.

Such instantaneous measurements of HC concentration cannot be performed with a conventional FID system. However, fast FID systems are available which allow such instantaneous measurements to be made, both in the exhaust system and in the cylinders of internal combustion engines. The output from a fast FID system is a continuous signal corresponding to the instantaneous concentration of HC in the exhaust or cylinder gas. The fast response of this type of instrument allows many of the instantaneous features of the exhaust event, and of the compression and power strokes to be resolved [Summers and Collings 1995b; Collings, 1988; Summers, 1996]. A fast FID instrument that has found wide use in internal combustion engine research and development is manufactured by Cambustion Ltd. in the U.K. [Cambustion].

This chapter will review the theoretical principles on which the design and operation of a fast FID system is based. It will then discuss practical aspects of the design and operation of such a system, drawing from the experience of a number of researchers who have used this type of instrument for internal combustion engine research. Applications of the fast FID system for in-cylinder fuel sampling will be discussed in Chapter 7.

3.2 Flame Ionization

As mentioned above, the unburned hydrocarbons in the engine exhaust gas include numerous compounds, each with its own molecular structure. Similarly, internal combustion engine fuels include numerous hydrocarbon compounds. The molecules of these hydrocarbons are made up of only carbon and hydrogen atoms, bonded together in a great variety of configurations.

Consider what happens when just one molecule of these hydrocarbon compounds is introduced into a hydrogen-air flame. The molecule is pyrolyzed and the carbon and hydrogen are oxidized. When this takes place, ions are produced through complex chemi-ionization processes, which are not yet fully understood. In contrast, the hydrogen flame itself is virtually free of ions when no hydrocarbon molecules are involved [Cambustion].

The generation of ions takes place in the reaction zone of the hydrogen flame, where the hydrocarbon molecules are burned; the concentration of ions reaches its maximum level very close to the reaction zone. The number of ions produced per molecule of the hydrocarbon compound burned depends primarily on the number of carbon atoms in the molecule. The structure of the molecule is a much weaker influence on the number of ions produced. Thus, burning a methane molecule will produce a certain number of ions corresponding to the single carbon atom in the methane molecule, while burning a molecule of propane, which has three carbon atoms, will produce approximately three times as many ions as the methane molecule.

Consider now what happens when a certain volume of a hydrocarbon compound is burned. Then, the total number of ions produced will depend, first, on how many carbon atoms there are in each molecule and, second, on the number of molecules in this volume. In principle, an instrument could be constructed which would be capable of counting the number of molecules provided:

(a) A way could be found of measuring the number of ions produced.
(b) The number of carbon atoms per molecule is known.

These are the basic principles underlying the operation of an FID instrument.

In internal combustion engine research and development, the concentration of hydrocarbons in the exhaust or cylinder gas is usually of interest. A "sample" to be analyzed by the FID system is a volume of exhaust or cylinder gas containing a certain amount of hydrocarbons. The function of the FID instrument is to measure the concentration of hydrocarbons in this sample; in other words, it must measure the fraction of the sample volume occupied by the hydrocarbons.

3.2.1 Flame Ionization Detector

Figure 3.1 shows a simplified representation of a flame ionization detector. A potential difference is applied between the anode and cathode. The fuel is mixed with the hydrocarbon-bearing sample, and the resulting mixture is burned in a diffusion flame formed by the combustion of hydrogen in air. As the products of combustion pass between the electrodes, the positively charged ions in the products migrate to the cathode while the negatively charged electrons migrate to the anode. As a result, a current is established between the electrodes. Hydrogen is used as the fuel because the burning of hydrogen molecules does not produce ions by means of chemi-ionization as the burning of hydrocarbon molecules does. The ions collected by the electrodes of Fig. 3.1 are therefore due to the presence of the hydrocarbons in the sample alone.

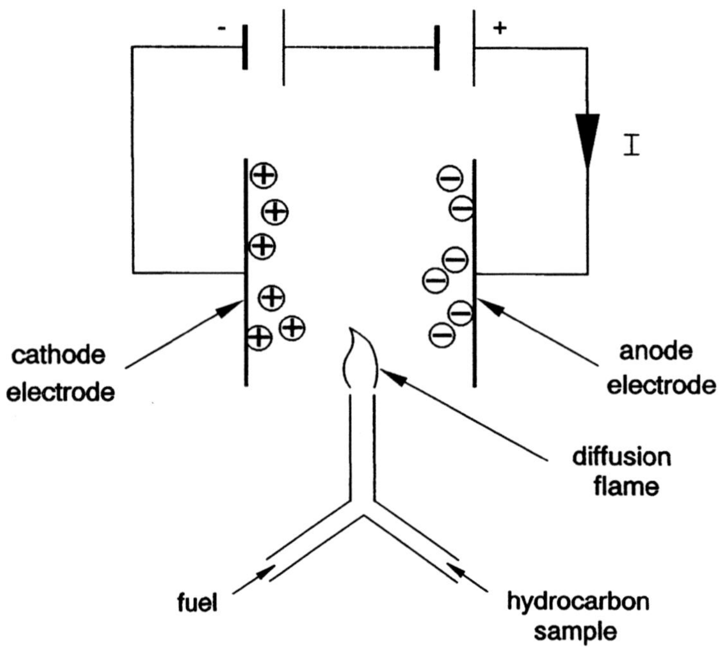

Fig. 3.1 Simplified representation of a flame ionization detector.

The magnitude of the ionization current (I) depends on the number of ions collected by the cathode per unit time. In turn, the number of ions collected per unit time is dependent on three variables: the volume flow rate of the hydrocarbon-bearing sample, Q; the number of hydrocarbon molecules per unit volume of the sample, C; and the number of carbon atoms per molecule of the hydrocarbon, N (i.e., for methane, N = 1; for propane, N = 3; for a mixture of hydrocarbons, the value of N is a weighted average).

Therefore,

$$I \propto QCN \tag{3.1}$$

A theoretical estimate of the current produced by the detector can be made as follows. (For simplicity, the sample is assumed to contain a single hydrocarbon compound only.):

Let I = current from FID (A)
Q = volume flow rate of sample gas (m^3/s) (sample comprises the hydrocarbon compound plus the gas in which the hydrocarbon is diluted)
$[HC]$ = volumetric (molar) concentration of hydrocarbon in sample in (ppm)
= (volume or moles of hydrocarbon compound)/(total volume or moles of sample)
C = number of hydrocarbon molecules per unit volume of sample (m^{-3})
N = number of carbon atoms per molecule of hydrocarbon compound
p_{HC} = absolute pressure of the sample gas containing HC (N/m^2)
T = sample absolute temperature (K)
R_o = universal gas constant (8314.4 J/(mol K))
n = Avogadro's number (6.023×10^{23} molecules/mol)
j = number of ion pairs (one +ve and one −ve ion per pair) produced per carbon atom burned ($j \approx 10^{-6}$ pairs for alkanes—depends somewhat on molecular structure)
e = electronic charge (1.6×10^{-19} coulombs)

Now,

$$C = n\,[HC]\frac{p}{R_o T}10^{-6} \tag{3.2}$$

From Eq. (3.1) above,

$$I \propto Qn[HC]\frac{p}{R_o T}N\ 10^{-6} \tag{3.3}$$

Assuming 100% collection efficiency of ions by the FID electrodes

$$I = Qn\,[HC]\frac{p}{R_o T}Nje\ 10^{-6} \tag{3.4}$$

Substituting numerical values for n, j, e, R_o, and taking $p = 0.001$ bar and $T = 293°K$:

$$I = 0.4Q[HC]10^{-10} \tag{3.5}$$

where Q is estimated at the sample pressure and temperature, e.g., $p = 1$ bar and $T = 293°K$.

In practice, usually, we would like to know the volumetric concentration of the hydrocarbon in the sample, i.e., [HC]. However, a practical FID instrument would be measuring the current, I. Therefore, we would like the current, I, to be directly proportional to [HC]. Then, as the hydrocarbon concentration [HC] varies, the output current, I, from the instrument will also vary as a linear relationship. It is clear from Eq. (3.5) above that for a linear relationship between [HC] and I, both N and Q must have constant values. If a single-component hydrocarbon is contained in the sample, the number of carbon atoms per molecule, N, will be fixed by the molecular structure; however, practical steps will have to be taken to keep the volume flow rate of the sample, Q, supplied to the FID flame constant. Thus, if N and Q are constant, then I = constant × [HC]. Hence, the current output can then be used to measure the volumetric concentration of the hydrocarbon in the sample.

Example:

Sample: Nitrogen containing 0.1% by volume propane (C_3H_8)
Sample flow: 3.33×10^{-7} m^3/s (20 cc/min)

From Eq. (3.5), substituting, [HC] = 1000 ppm, $Q = 3.33 \times 10^{-7}$ m^3/s, and N = 3,

Estimated current flow: 4×10^{-14}

Practical FID systems do not rely on an analytical expression, such as Eq. (3.5) above, in order to estimate the [HC] concentration from the output current. Instead, first, they convert the current from the detector to voltage and amplify the resulting signal so that the output from the system has a magnitude of several volts. Second, the system is calibrated, using several calibration samples with known hydrocarbon concentration (e.g., 0.1%, 1%, etc., propane in nitrogen) so that a known relationship between sample concentration and output voltage is established. This relationship is usually a linear one over a wide range of hydrocarbon concentration (5×10^3 to 5×10^4 ppm propane in air), but can depart from linearity for very rich mixtures. (A stoichiometric mixture of propane in air is approximately 42,000 ppm by volume or 4.2 percent.) [Peckham, 1993].

3.3 Calibration of a Flame Ionization Detector

3.3.1 Hydrocarbon Concentration

As explained above, in practical FID systems, the sample volume flow rate, Q, is maintained constant. Therefore, it is possible to calibrate a FID instrument so that a linear relationship between the output voltage and the volumetric concentration of a given hydrocarbon (e.g., propane) in the sample is established. However, it should be remembered that even with Q maintained constant, the voltage output will depend not just on the concentration. As Eq. (3.5) above shows, the number of carbon atoms per molecule of the sample hydrocarbon also affects the output voltage.

For example, with Q constant:

- A sample of 0.1% propane (C_3H_8) in nitrogen will produce a voltage output from an FID instrument which is three times greater than that obtained from a sample of 0.1% meth ane (CH_4) in nitrogen.
- Similarly, a sample of 0.1% hexane (C_6H_{14}) will produce a voltage output which is six times greater than that from a 0.1% methane sample.

It is clear that the FID system output, for a given hydrocarbon concentration, is dependent on the number of carbon atoms per molecule. Therefore, a calibration curve relating the voltage output from the instrument to the concentration of, say, propane will not be applicable to, say, methane. Thus, the calibration curve produced with propane is not a "universal" one, and is not applicable to other hydrocarbon compounds.

As a solution to this problem, the calibration curve (obtained using methane or propane) is often expressed as a relationship or a graph of FID voltage output against concentration in units of carbon, e.g., ppmC or ppmC_1 (i.e., parts per million carbon; sometimes ppmC_1 is used as an alternative to ppmC). In this way, the calibration curve becomes, in effect, a relationship between the voltage output of the instrument and the concentration of carbon atoms (rather than molecules of a specific compound) in the sample. A calibration curve of voltage output against concentration of, say, propane (C_3H_8) can be converted to a more generally applicable calibration curve of voltage output against ppmC_1, by simply scaling up the voltage axis by a factor of 3. Similarly, on a calibration curve for the same FID instrument obtained with hexane (C_6H_{14}) the voltage axis will have to be scaled up by a factor of 6, while a calibration curve obtained with methane (CH_4) will not need any scaling.

This solution has an additional advantage. The calibration curve in terms of ppmC_1 is valid even when the sample to be investigated (e.g., exhaust gas) comprises a mixture of two or more hydrocarbon compounds, each compound having a different number of carbon atoms per molecule. The FID system will then measure the concentration in terms of C_1; that is, in effect, it will measure the concentration of carbon atoms in the sample, rather than the concentration of hydrocarbon molecules.

3.3.2 Sensitivity of Calibration to Molecular Structure

The previous section discussed the dependence of the FID system output on the number of carbon atoms in a hydrocarbon molecule. For example, for a given concentration, it was said that the voltage output for propane (C_3H_8) will be three times that for methane (CH_4). In other words, the output is simply scaled up by the number of carbon atoms in the molecule. This is true in the case of propane and methane because they both belong to the same family of hydrocarbon compounds (alkane family).

Unfortunately, the FID output for different HC compounds is not always scaled up simply by the number of carbon atoms in a molecule. In some cases, two molecules having the same number of carbon atoms will produce somewhat different levels of output from an FID system, because of differences in the way the molecules are constructed. This is illustrated in the example below:

HC Compound	HC Family	Number of carbon atoms per molecule	Number of bonds between carbon atoms	FID system output per
ethane (C_2H_6)	alkane	2	1	1.0 V
ethene (C_2H_4)	alkene	2	2	0.95 V
ethyne (C_2H_2)	alkyne	2	3	1.30 V

In the above example all three compounds have two carbon atoms per molecule. Yet, they generate different FID system outputs per molecule, due to differences in the bonding between the carbon atoms in each molecule. Table 3.1 gives an indication of the sensitivity of the FID current output to molecular structure for common hydrocarbon families.

A significant practical problem would arise if an FID system that was previously calibrated with an alkane compound (e.g., methane or propane) was then used to measure the concentration of sample containing a mixture of compounds of the alkene and alkyne families. The concentration measurement obtained from the FID would be in error, because the calibration curve would be based on an alkane compound. As Table 3.1 shows, the detector output for alkanes is different from that for alkenes and alkynes. Particular care is therefore needed when an FID is used to measure the concentration of a mixture that might contain one or more compounds from a family that is different from the one for which it was calibrated. This situation will arise when sampling from the cylinder of a spark-ignition engine operating on gasoline. This fuel contains a wide range of compounds belonging to many different hydrocarbon families. Similarly, the exhaust gas contains a range of compounds from different hydrocarbon families.

Table 3.1 Sensitivity of FID detector current output to hydrocarbon type

Hydrocarbon family	Current output Relative to methane
Methane	1.00
Alkanes	1.00 × N
Alkenes	0.95 × N
Alkynes	1.30 × N
Aromatics	1.00 × N

Note: Relative current response for a hydrocarbon compound having N atoms per molecule, is defined as:

$$\frac{\text{Current output for the hydrocarbon at concentration X}}{\text{Current output for methane at same concentration X}}$$

3.4 Fast-Response FID System

3.4.1 Overview of a Fast FID System

Figure 3.2 shows a schematic diagram of a fast-response FID system for in-cylinder measurement of hydrocarbon concentration. The sample, consisting of the engine cylinder contents, is carried to the ionization flame via three capillaries, labeled in Fig. 3.2 as sample transfer tube, expansion tube, and FID tube. The sample from the engine cylinder first enters the relatively long sample transfer tube and travels to the expansion tube. The sample emerges from the expansion tube into the constant pressure (CP) chamber. Prior to emerging into the CP chamber, a small proportion of the sample is diverted into the FID tube, reaching the base of the hydrogen flame. The hydrocarbon molecules in the sample are burned in the flame, thereby generating an ionization current. This current is subsequently electronically processed and amplified into an output voltage, usually in the range 0 to 5 or 0 to 10 volts.

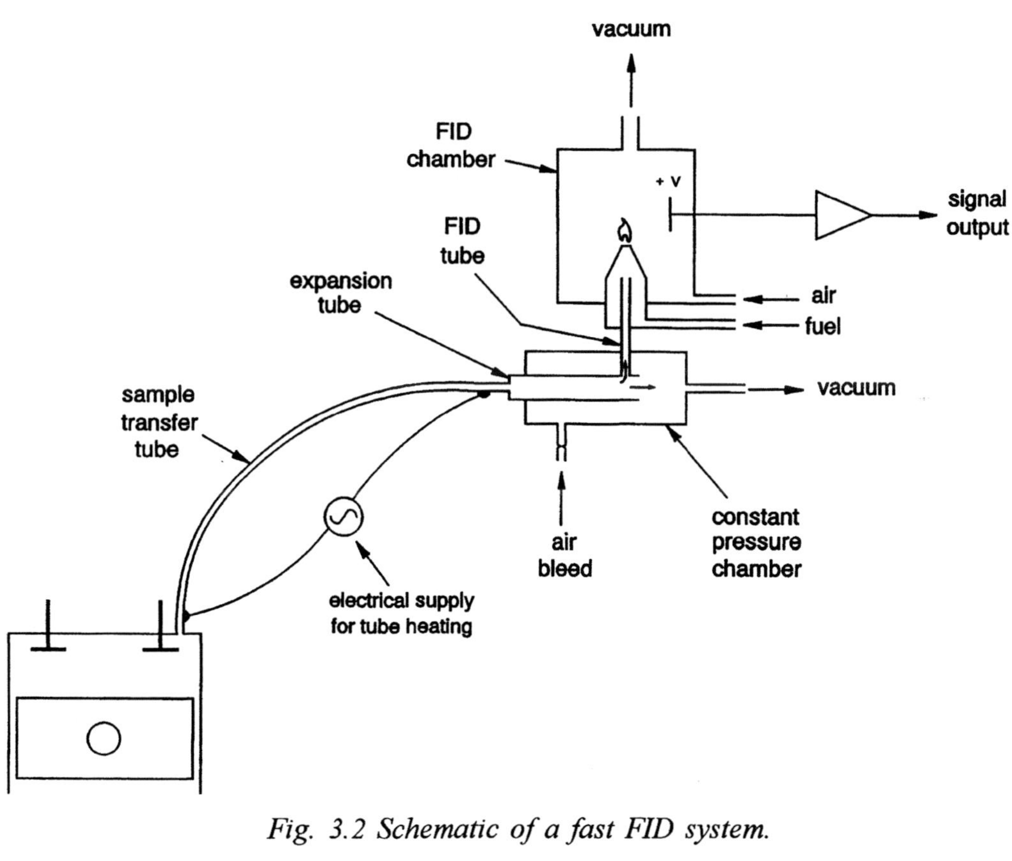

Fig. 3.2 Schematic of a fast FID system.

3.4.2 Fluid Mechanics of a Fast Flame Ionization Detector System

3.4.2.1 Flow in the FID Tube

Figure 3.3 shows the fast FID system in greater detail. The pressure level in the CP chamber is often maintained below atmospheric level, so that sample can be sucked into the CP chamber. In turn, the FID chamber pressure is maintained below that of the CP chamber. This allows sample to flow into the FID chamber, via the FID tube. The flow rate of the sample through the FID tube will depend on a number of factors, including the pressure difference between the CP and FID chambers, and the dimensions of the tube. It is necessary to maintain the flow rate of the sample through the FID tube constant, for the following reason: As discussed previously, the magnitude of the FID output depends on both the sample flow rate into the FID flame, as well as the HC concentration in the sample (see Eq. [3.5] above). Since only the HC concentration is of interest, the sample flow rate via the FID tube to the flame must be kept constant. In order to achieve this, the static pressure drop across the FID tube must be kept constant, since this pressure drop is the major factor affecting the flow rate through an FID tube of given dimensions. In this way, the signal output from the FID system will be responding only to changes in the HC concentration in the sample.

In order to maintain the pressure drop across the FID tube constant, the absolute static pressure at points A and B in Fig. 3.3 must be kept constant. Consider, first, the absolute static pressure at point A. The static pressure at this point will be equal to the bulk pressure in the CP chamber, even if the flow velocity in the expansion tube varies, provided the flow emerges with parallel streamlines at point C (exit of the expansion tube). Also, point B must be close to point C so that fluid friction between the two points is negligibly small. In addition, the exit of the expansion tube at point C must not be choked, because this will introduce a substantial pressure discontinuity at point C. Thus, the flow velocity at C (exit of the expansion tube) must be well below the sonic value. Provided these precautions are observed, the expansion tube should still operate satisfactorily even if the exit to the sample transfer tube (i.e., entry to the expansion tube) becomes choked. Finally, the static pressure at point B will normally be equal to the bulk pressure in the FID chamber.

As Figs. 3.2 and 3.3 show, the bulk of the sample flowing through the expansion tube is dumped into the CP chamber at point C; only a small fraction is diverted via the FID tube into the base of the FID flame. The flow rate of the sample bulk discharging at point C varies widely due to the very large variation in the engine cylinder gas pressure. The CP chamber must, therefore, be able to cope with this large variation in flow discharge while still maintaining a constant pressure. In order to minimize consequent pressure fluctuations in the CP chamber, a number of measures can be taken: for example, the volume of the CP chamber can be increased; or the diameter of the sample transfer tube can be decreased so that the sample flow rate is reduced by greater frictional resistance. Summers [1996] encountered the problem of CP pressure fluctuations when using a fast FID system to measure residual hydrocarbon concentration within the cylinder of a single-cylinder gasoline engine fitted with a mass-produced cylinder head. The CP chamber pressure fluctuations persisted even when the CP chamber volume was roughly doubled by the

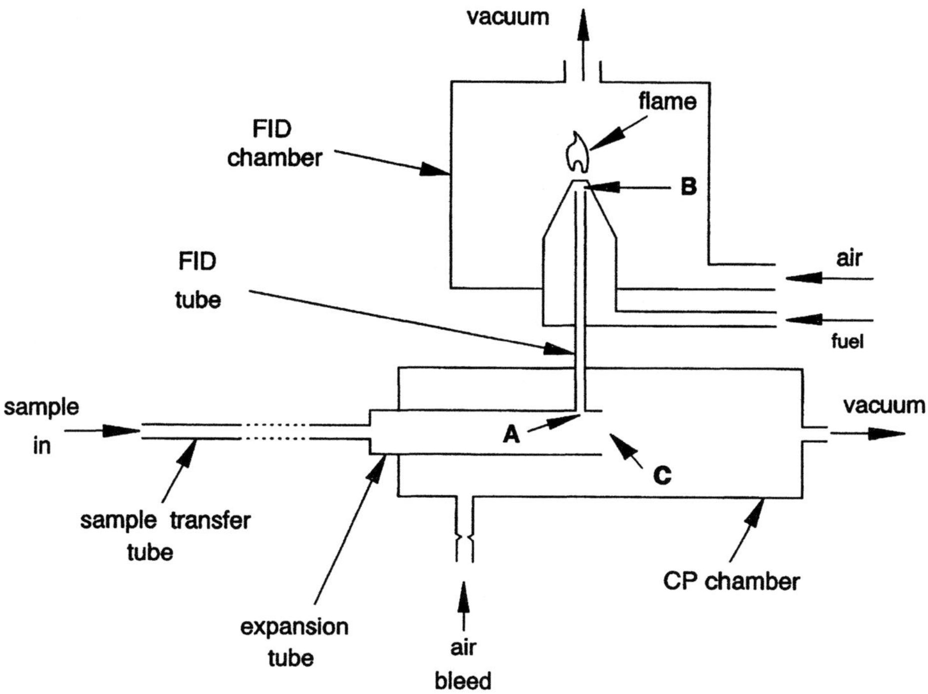

Fig. 3.3 Details of a fast FID system sampling arrangement.

addition of a further 20 cm^3. The CP pressure fluctuations still caused an unacceptable swing of about 40% in the pressure difference across the FID tube. When the CP chamber was extended by a massive 3700 cm^3, the CP pressure fluctuations were very much reduced. More importantly, the time constant of the fluctuations became much longer, with no detectable CP pressure fluctuation during the short period of interest which occupied only a very small fraction of the engine cycle. Tests with enlarged CP chamber volume were also conducted by Cambustion Ltd. on an engine running at high load and low speed. This running condition gives high sample flow discharge into the CP chamber, exacerbating the problem of CP pressure fluctuations. A sample transfer tube having a length of 300 mm, a diameter of 0.4 mm, and heated to 150°C was used; CP pressure was 0.47 bar and pressure drop across the FID was 0.13 bar. When the CP chamber volume was extended by attaching an additional volume of 2 liters, the CP chamber pressure fluctuations were reduced to about ±0.0013 bar or ±1 percent of the pressure drop across the FID tube [Cambustion].

An option open to the user is to open the CP chamber to the atmosphere or remove it altogether. If this is done, the instrument output will be useful only when the cylinder pressure is substantially above the atmospheric level. Thus, the FID system signal would not be useful during the following periods: early during the compression stroke, late during the expansion stroke, and during the exhaust and induction strokes.

Finally, it is worth pointing out that in the case of the FID chamber it is considerably easier to maintain the bulk pressure constant. This is because the various flows into this chamber are small and very close to steady state.

From the above discussion it should be clear that:

(i) The sample flow rate to the FID flame must be kept constant.
(ii) The static pressure difference between points A and B is the major factor controlling the sample flow rate to the FID flame.
(iii) The CP chamber system requires careful design and operation in order to maintain the flow rate to the FID flame constant.

3.4.2.2 Flow in the Sample Transfer Tube

The sample from the source (engine cylinder or exhaust system) flows along the sample transfer tube toward its exit. All along the transfer tube length, the sample pressure and the density fall, causing the flow to accelerate. On reaching the entry to the expansion tube, the sample velocity falls to a much lower level, because the diameter of the expansion tube is usually several times larger than that of the transfer tube. The function of the expansion tube is, indeed, to reduce the sample velocity in the expansion tube to well below the sonic level, so as to avoid compressibility effects within the tube and choking at the exit of the expansion tube (point C in Fig. 3.3). Should the expansion tube exit become choked, the FID instrument would cease to operate correctly. Choking of the expansion tube exit causes the static pressure within the tube to rise above the CP chamber pressure. The expansion tube then becomes pressurized, and this destroys the pressure isolation of the FID tube entry (point A in Fig. 3.3).

Consider now, in more detail, the flow in the sample transfer tube. Assume that the sample behaves as a perfect gas. Starting with the full equations for continuity, momentum, and energy applicable to compressible flow of a perfect gas in a constant-area duct, it can be shown [Cambustion] that for capillary flow (i.e., tube diameter small) the sample temperature is very nearly constant (isothermal flow) along the tube, except for: a short settling length near the entrance; and a short length close to the exit of the tube when the exit is either choked or close to choking. Here it will be assumed that the flow is isothermal over the entire length of the tube. The errors arising from this assumption have been shown to be insignificant [Summers and Collings, 1995a; 1995b], mainly because the non-isothermal lengths of the tube at both the entry and exit are very short [Cheng et al., 1989] (less that 10 mm of tube length when tube diameter < 1 mm).

Assuming that the flow through the transfer tube is isothermal, it can be shown [Cambustion] that the mean flow velocity, U, at a distance, x, along the tube length is given by:

$$\frac{1}{U} = \sqrt{\frac{4fL}{RTd}\left(\frac{p_1^2}{p_1^2 - p_2^2} - \frac{x}{L}\right)} \tag{3.6}$$

where U = mean flow velocity
x = distance along tube from its entry
f = friction factor
L = total length of tube
R = specific gas constant
T = gas temperature
d = tube diameter
p_1 = inlet gas pressure
p_2 = outlet gas pressure

Assumptions:

- Isothermal gas flow along entire tube length. (Assumption becomes inappropriate at tube exit if exit is choked.)
- One-dimensional flow.
- Compressible unchoked flow. (In practice, sometimes the flow will be close to or choking at exit when sampling from the engine cylinder.)
- Tube has constant flow area.
- Perfect gas behavior.
- The Reynolds number, Re, is constant along the tube. (This is approximately true for isothermal flow.)
- The friction factor, f, is constant along the tube (see discussion below).

The value of the Reynolds number will have to be estimated from the tube geometry, flow velocity, and fluid properties. The estimated value for Re can then be assumed to be constant along the tube length for isothermal flow. The value of Re can also be used to estimate the value of the friction factor, f, using the following expressions: for laminar flow (i.e., Re $\leq$ 2500), f = 16/Re; and for turbulent flow (Re values up to 50,000), $f = 0.079\ Re^{-1/4}$. Note that since the value of Re was assumed to be constant along the tube length, it follows that the value of f must also be constant along the tube length. An engine cylinder gas pressure of 12.5 bar has been estimated to result in Re $\approx$ 10,000 when the transfer tube length is about 300 mm, the tube diameter is 0.25 mm, and the CP chamber pressure is 0.7 bar [Summers, 1996]. It is interesting to note from these figures that in-cylinder sampling with these transfer tube dimensions will result in turbulent flow inside this tube for most of the compression stroke.

An assumption made in the derivation of Eq. (3.6) above is that the flow through the sample transfer tube will be isothermal. In practice, although the sample gas temperature from the engine cylinder can be high, cooling of the sample by heat transfer to the tube walls is very rapid due to the small tube diameter. In the case of the transfer tube used as an example in the preceding paragraph, the cylinder gas temperature will be reduced close to the wall temperature within the first 10 mm or so from the tube entry (turbulent flow conditions), or within a few percent or so of the tube length [Summers, 1996]. This is not a surprising result, in view of the high length/diameter ratio of this capillary tube (1200).

In practice, cooling of the sample gas in the transfer tube may reduce the sample temperature sufficiently for liquid fuel fractions or water vapor to condense within the tube. This problem can be overcome by electrically heating the transfer tube walls to around 150°C. In practical sampling systems the transfer tube is usually made of stainless steel, which has quite high electrical resistivity. It is therefore feasible to heat the sample transfer tube directly by conducting a modest amount of electrical current through the stainless tube walls, as shown in Fig. 3.2. The current supply could be controlled by means of a thermocouple sensing the tube wall temperature, so that the desired wall temperature could be maintained. Thermal insulation could also be applied to the external surfaces of the tube to help maintain the desired wall temperature.

When sampling from the engine cylinder, the flow velocity through the transfer tube increases during the compression and combustion processes. Depending on tube geometry and peak cylinder gas pressure, it is likely that the flow velocity at the outlet of the tube will reach the sonic value. The flow at the tube outlet then becomes choked; then, the flow through the transfer tube no longer depends on the pressure beyond its outlet (i.e., the CP pressure). It is possible to use Eq. (3.6) above in order to estimate the pressure ratio across the transfer tube at which the tube exit becomes choked.

When the tube exit becomes choked, the flow velocity at the tube exit (i.e., at $x/L = 1$), becomes equal to the sonic value, that is, $U = (\gamma RT)^{0.5}$. Substitution of these values for x/L and U into Eq. (3.6) above gives:

$$\frac{p_1}{p_c} = \left(\frac{4fL\gamma}{d} + 1\right)^{1/2} \tag{3.7}$$

where γ = ratio of specific heat capacities for sample gas
p_c = pressure at the plane of the tube exit ($p_c > p_2$)

After the flow at the tube exit plane has become choked, the pressure at the tube exit plane, p_c, is greater than the pressure p_2 in the expansion tube. Therefore, the pressure of the sample leaving the tube falls rapidly from p_c to p_2 in the expansion tube by dissipation through shocks. It is important that the expansion tube is sufficiently long to ensure that shocks die down and the flow becomes settled before the entry to the FID tube is reached (point A in Fig. 3.3).

Equation (3.7) above can provide an approximate estimate for the choking pressure ratio across the tube, in the case of typical sample transfer tube geometries. Some examples are shown below when: $\gamma = 1.4$ and $f = 0.0064$ (f is assumed to be constant since it is only a weak function of Re for turbulent flow conditions):

L (m)	d (m)	p_1/p_c
0.2	0.00020	6.1
0.3	0.00020	7.4
0.2	0.00025	5.4

Therefore, if in a practical system the ratio of (sample pressure/CP chamber pressure) is greater than the value of p_1/p_c, the tube exit will be choked. Consider, for example, the case of sampling from the engine cylinder. For convenience, take the ratio $p_1/p_c \approx 6$. Suppose that the CP chamber pressure is 1 bar. Then, during the engine compression stroke the exit of the sample transfer tube will become choked when the cylinder gas pressure has risen to 6 bar (ignoring variations in f, and assuming $f = 0.0064$). As the cylinder gas pressure continues to rise above 6 bar, the transfer tube exit remains choked, and the exit plane pressure, p_c, rises above p_2. For instance, when the pressure in the cylinder has risen to $p_1 = 50$ bar, then $p_c = 8.3$ bar compared with $p_2 = 1$ bar in the CP chamber. Thus, a pressure drop of 7.3 bar must occur within the expansion tube through the action of shocks.

Finally, it is worth noting that after choking has occurred at the exit to the transfer tube, the mass flow rate through the tube is not constant. As the cylinder pressure, p_1, continues to rise so will the density and mass flow rate of the sample continue to increase.

3.4.2.3 Flow in the Expansion Tube

The function of the expansion tube is to help maintain the pressure at the entry to the FID tube (point A in Fig. 3.3) constant, so that the sample flow rate through the FID tube remains fixed at all times. Taking a closer look at the construction of the fast FID sampling system, shown in Fig.3.3, it can be seen that the junction between the sample transfer and the expansion tubes forms a sudden expansion with a diameter ratio usually lying in the range of 5 to 15.

Sudden flow expansion is a topic that has been researched extensively over the years. In general, the flow arriving at the entry to the expansion tube forms an expanding jet. The behavior of the jet flow, as it goes through a sudden expansion, very much depends on the flow rate. At low flow rates, such as would be encountered when the fast FID system is used for sampling from the engine exhaust system, the flow emerging from the transfer tube quickly expands to occupy the whole of the expansion tube cross section. Then, the flow velocity in the expansion tube is small and can be considered to be incompressible.

However, as the sample pressure at the entrance to the transfer tube rises, the flow rate at the exit to this tube also rises, and eventually reaches choking conditions. Choking occurs when the engine cylinder gas pressure rises to around 4 to 6 bar. As the cylinder pressure rises beyond

6 bar or so, the flow emerging into the expansion tube undergoes significant changes, believed to be as follows [Summers, 1996]. When the cylinder pressure is a little over 6 bar or so, the flow emerges into the expansion tube as a supersonic jet. There, the jet rapidly loses energy, becoming subsonic and spreading to fill the entire cross section of the expansion tube. As the cylinder pressures rises further, the flow velocity becomes high enough for normal shocks to start forming in the expansion tube itself. Energy dissipation across the shocks causes the flow to become subsonic. At still higher engine cylinder pressures, when the piston is approaching top dead center, a series of oblique shock waves are established in the expansion tube. These persist for some distance along the tube, until they are dissipated by fluid friction. Finally, the cylinder pressure can become so high during combustion that flow at the exit of the expansion tube itself reaches the sonic level, and flow at the tube exit becomes choked.

It is important to consider briefly how the above flow regime can interfere with the correct operation of the FID instrument [Summers, 1996]. The reader is reminded that the pressure close to the expansion tube exit (i.e., at point A in Fig. 3.3) must remain constant at all times. For this reason, if possible, the tube design must ensure that compressibility effects are dissipated before the flow reaches the tube exit; also, the expansion tube exit must not become choked. Consider now the first two flow regimes outlined in the preceding paragraph (supersonic jet and normal shocks). With these regimes, compressibility effects are likely to extend along the expansion tube for a few diameters [Summers, 1996]. Since the fast FID expansion tube is usually constructed with a length/diameter ratio of about 20, compressibility effects are not likely to reach the FID tube exit. Consider now the case of the oblique shock waves. These waves may persist right through the expansion tube and interfere with the pressure isolation at the entry to the FID tube (location A in Fig. 3.3). In practice, it may be possible to solve this problem by extending the length of the expansion tube so as to allow the shocks to dissipate. A better solution might be to reduce the diameter of the sample transfer tube, so as to lower the sample flow rate through both the transfer and the expansion tubes. This solution would also deal with the problem of choking at the expansion tube exit.

3.4.3 System Response

In considering the response of a fast FID system it is necessary to distinguish between the transit time and the time constant. The transit time is a time delay between a fresh sample entering the transfer tube and the instrument signal starting to respond. The time constant decides the frequency response of the FID system. A short time constant is usually highly desirable, because it will allow the FID system to respond adequately to rapid variations in the sample concentration.

3.4.3.1 Transit Time

It is more convenient to explain and analyze the concept of transit time in a simple case of a tube having uniform cross sectional area, as illustrated at the top of Fig. 3.4. Consider the case when the concentration at the entry to the transfer tube is changing, as shown in Fig. 3.4(a). Obviously, a time delay will elapse before any change in entry concentration appears at the tube exit. This time delay is the transit time, as indicated in Fig. 3.4(a). In Fig. 3.4(a) the transit time is shown to be the same for all points on the concentration profile. Thus, the phenomenon of transit time simply causes the entire outlet concentration curve to be delayed in time. The shape of the outlet concentration profile is not affected by this phenomenon (although other phenomena cause the outlet profile to change).

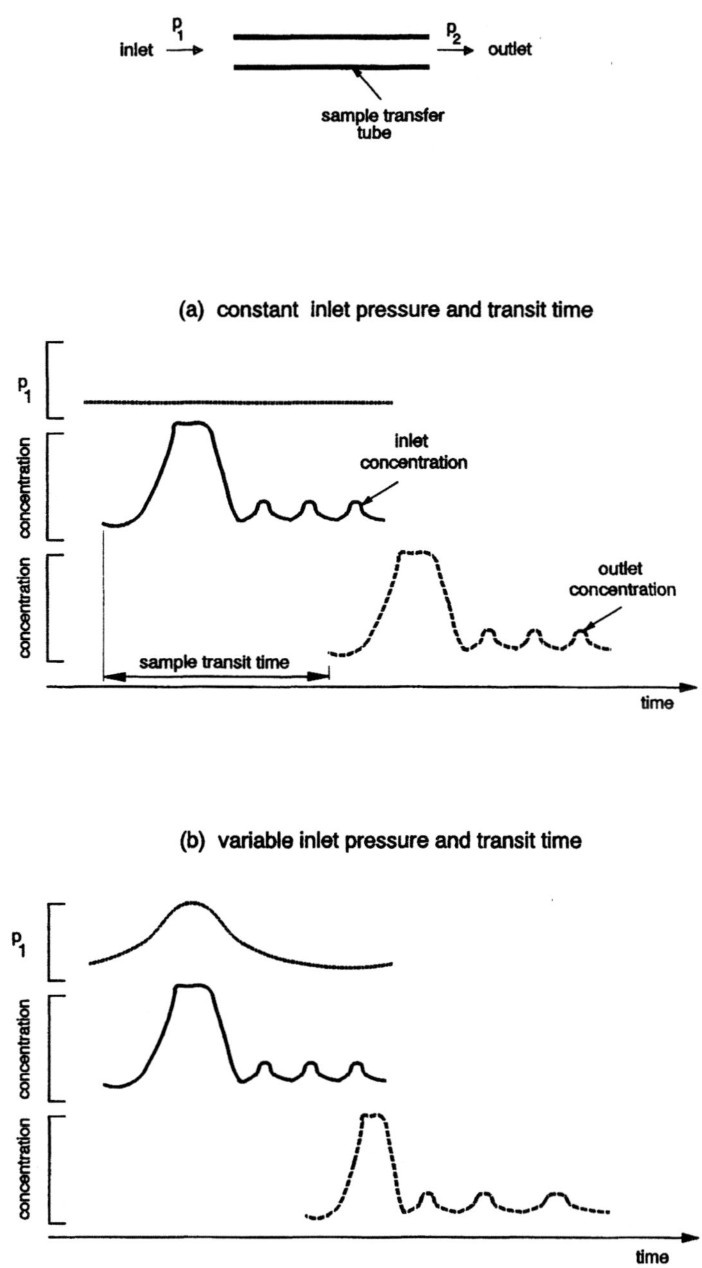

Fig. 3.4 Distortion of outlet concentration profile due to variable transit time.

As pointed out in the preceding paragraph, Fig. 3.4(a) shows the transit time as being constant throughout the concentration profile shown. Unfortunately, this may not be the case in many practical applications. A hypothetical example of variable transit time is shown in Fig. 3.4(b). This figure shows both the hydrocarbon concentration and the pressure of the sample at the tube inlet changing with time. To begin with, the inlet pressure increases, causing the sample to travel faster through the tube. The transit time decreases, and this manifests itself as shrinking of the outlet concentration profile in the horizontal direction. Subsequently, after reaching a peak, the pressure starts to decrease, and the transit time now increases. This manifests itself as stretching of the outlet concentration profile. As a result, the outlet concentration profile becomes distorted in the horizontal direction, as illustrated in Fig. 3.4(b). Sampling from the engine cylinder involves very large pressure fluctuations. For example, the cylinder pressure during the exhaust stroke is around 1 bar, but during the combustion process it can rise beyond 50 bar. Consequently, although the transit time during the exhaust stroke may be around 6 ms, at peak pressure it may drop to one-third of this value.

In the case of a fast FID system, the total sample transit time is composed of the individual transit times for the transfer, expansion, and FID tubes.

The transit time depends on several parameters: the tube length and diameter, pressure difference across the tube, sample temperature, Reynolds number, and fluid properties. The transit time for isothermal flow through a tube having length L can be obtained by simply integrating Eq. (3.6) above with respect to distance along the tube (x) for flow velocity (provided that the pressures p_1 and p_2 are constant with respect to time):

$$\tau_t = \sqrt{\frac{16fL^3}{9RTd} \cdot \frac{\left(p_1^3 - p_2^3\right)^2}{\left(p_1^2 - p_2^2\right)^3}} \tag{3.8}$$

The assumptions for this expression are the same as the ones listed for Eq. (3.6). Equation (3.8) can be used to estimate the transit time in each of the three main flow elements of a fast FID sampling system, namely: the sample transfer tube, the expansion tube, and the FID tube (see Fig. 3.3). The estimated sample transit times for each of these three elements can then be combined into the total transit time between the sample transfer tube entry and the base of the FID flame. A further time delay may be involved before the FID electrical output responds, and this is associated with the sample burning in the flame, collection of ions, and any delays in electronic processing of the ionization current.

In practice, the calculation of transit time from Eq. (3.8) above may not be straightforward in the case of the expansion tube. This is because this equation requires the pressure at the entry and exit to each tube (p_1 and p_2). In the case of the FID tube this is not a problem, because pressures p_1 and p_2 will, nominally, be the CP and the FID chamber pressures. In the case of the transfer tube, there is also no major difficulty. The sample pressure p_1 will usually be known. The outlet pressure p_2 can be assumed to be approximately equal to the CP chamber pressure provided the exit of the expansion tube is not choked. The main difficulty arises with the expansion tube. The pressure p_2 at the exit to the expansion tube will be the CP chamber pressure for correct operation of the system. However, the inlet pressure p_1 for the expansion tube may not be known without performing a series of complex calculations. A solution to this problem is to assume that the flow in the expansion tube is incompressible and calculate the transit time from the flow velocity and length of the expansion tube. The assumption of incompressibility is a very good one, except at the highest engine cylinder pressures occurring near top dead center during combustion. The flow velocity in the expansion tube can be obtained from mass continuity using the mass rate through the sample transfer tube. The calculation steps needed to obtain an approximate estimate of the transit time for the expansion tube are outlined below:

(i) Calculate the flow velocity, $U_{x=0}$, at the transfer tube entry ($x = 0$) from Eq. (3.6) above, taking: p_1 to be the sample pressure, p_2 the CP chamber pressure (or critical pressure at transfer tube exit if transfer tube is choking), and a tentative value for the friction factor, f (say, 0.01).

(ii) Calculate the sample density at the transfer tube entry assuming ideal gas behavior, $\rho_{x=0} = p_1 / (RT_{x=0})$.

(iii) Calculate the Reynolds number at entry to the transfer tube, $Re_{x=0} = \rho_{x=0} U_{x=0} d / \mu$.

(iv) Use the value of $Re_{x=0}$ to improve the estimate of the friction factor using:
$f = 16/Re_{x=0}$ for laminar flow, (i.e., when $Re_{x=0} \leq 2500$)
or $f = 0.079\ Re_{x=0}^{-1/4}$ for turbulent flow up to $Re_{x=0} < 50{,}000$

(v) Recalculate, the transfer tube entry flow velocity from Step 1 and iterate until there is no further improvement in the value of the entry velocity.

(vi) Calculate the mass flow rate, m, through the transfer tube using the entry density and entry velocity and the cross-sectional area of the transfer tube. The mass flow rate through the expansion tube will be the same as that through the transfer tube, since the expansion tube has to swallow whatever mass rate emerges from the transfer tube (provided the expansion tube is not choking).

(vii) Calculate the density, ρ_{ex}, in the expansion tube using the CP chamber pressure, p_{cp}, and temperature, T_{cp}; that is, $\rho_{ex} = p_{cp} / (RT_{cp})$

(viii) Now calculate the velocity and transit time in the expansion tube from, respectively, $U_{ex} = m / (\rho_{ex} A_{ex})$ and $\tau_{ex} = L_{ex} / U_{ex}$ where A_{ex} and L_{ex} are the expansion tube flow area and length, respectively.

Examples:

Eq. (3.8) for transit time and the iterative procedure outlined above for the expansion tube were used to obtain estimates of the transit time for the three main flow elements of a fast FID system. Results of this analysis are shown in Table 3.2. The dimensions of the various tubes were chosen merely for illustrative purposes; a more extensive discussion of the choice of tube dimensions appears in a later section (Section 3.4.3.5). It should also be pointed out that these results are calculated for the case of a step change in concentration at the entry to each tube. In the case of a practical FID sampling system, a step input is, in principle, possible only for the sample transfer tube. Thus, the transit times shown in Table 3.2 allow useful comparisons between the different tubes, but are only rough estimates of the transit times that would be obtained in the various sampling elements of a practical fast FID system.

Table 3.2 Analysis of the transit time and time constant for the fast FID sampling system

FID TUBE (L = 20 mm)						
d mm	p_1 bar	p_2 bar	Re	Exit Mach No.	Transit time ms	Time constant ms
0.15	0.30	0.25	12	0.02	2.5	0.17
0.15	0.50	0.45	21	0.02	2.5	0.22
0.15	1.00	0.95	44	0.02	2.5	0.31
0.20	0.50	0.45	51	0.04	1.4	0.22

Note: p_1 = sample pressure; p_2 = CP chamber pressure; FID tube length = 20 mm

SAMPLE TRANSFER TUBE (L = 200mm)							EXPANSION TUBE (L = 25mm)				
d mm	p_1 bar	p_2 bar	Re	Exit Mach No.	Transit time ms	Time const. ms	d mm	Re	Mach No.	Transit time ms	Time const. ms
0.2	0.75	0.5	33	0.02	28.5	1.15	1.2	6	0.001	101.2	22.30
0.2	1.5	0.5	213	0.14	7.6	0.80	1.2	36	0.004	15.8	8.82
0.2	3.0	0.5	928	0.60	3.3	0.73	1.2	155	0.017	3.6	4.23
0.2	6.0	ch	2300	ch	2.6	0.20	1.2	383	0.041	1.5	2.68
0.2	12.0	ch	5068	ch	2.4	0.17	1.2	845	0.091	0.7	0.35**
0.2	24.0	ch	11167	ch	2.2	0.15	1.2	1861	0.201	0.3	0.15**
0.2	48.0	ch	24583	ch	2.0	0.13	1.2	4097	0.440	0.14	0.07**
0.4	0.75	0.5	266	0.09	7.1	1.15	1.2	89	0.010	6.3	5.58
0.4	1.5	0.5	1702	0.55	1.9	0.80	1.2	567	0.061	1.0	0.53**
0.4	3.0	ch	3382	ch	1.8	0.19	1.2	1128	0.122	0.5	0.26**
0.4	6.0	ch	7439	ch	1.7	0.16	1.2	2480	0.268	0.23	0.11**
0.4	12.0	ch	16346	ch	1.5	0.14	1.2	5449	0.588	0.10	0.05**
0.4	24.0	ch	–	ch	–	–	1.2	–	ch	–	–
0.4	48.0	ch	–	ch	–	–	1.2	–	ch	–	–

Notes: p_1 = sample pressure; p_2 = 0.5 bar (i.e., CP chamber pressure) unless transfer tube exit is choking when p_2 is set equal to critical pressure; "ch" indicates that the tube exit is choking; ** indicates that when Re > 400 in expansion tube, the flow in this tube is assumed to be turbulent (see text); sample transfer tube length = 200 mm and expansion tube length = 25 mm

The results in Table 3.2 permit a number of practically useful observations. Consider, first, the FID tube. The flow through the tube is well within the laminar regime with both tube diameters of 0.15 and 0.20 mm. The transit time through the tube was estimated at 2.5 ms with a tube diameter of 0.15 mm, but it could be reduced to 1.4 ms by increasing the tube diameter to 0.20 mm. This increase in diameter, however, increases the volume flow rate of the sample at the FID tube exit, from 0.15 to 0.68 cm^3/s. Clearly, an increase in the diameter of the FID tube is beneficial in lowering the transit time through the tube. However, in practice, care would be needed because the increased sample flow rate may make the flame unstable; if the flow rate becomes too high, the flame may be extinguished altogether. It is also interesting to note from Table 3.2 that in the case of the 0.15 mm diameter FID tube the transit time is controlled entirely by the pressure difference across the tube (not by the individual pressure levels of the FID or CP chambers).

Consider now the flow through the sample transfer tube. The results for a sample tube diameter of 0.2 mm allow a very important observation to be made: when the sample pressure is low (around 0.75 to 3 bar) the transit time is very sensitive to the sample pressure, p_1. For example, the transit time for the transfer tube is reduced from 28.5 to 7.6 ms when the sample pressure is doubled from 0.75 to 1.5 bar. As the sample pressure rises to 3 bar the transit time falls to 3.3 ms and becomes quite insensitive to further increases in sample pressure, maintaining a value of around 2 ms up to a sample pressure of 48 bar. The insensitivity of the transit time to sample pressure above about 3 bar coincides with the flow in the tube becoming turbulent and the exit from the sample tube becoming choked. The great variability of the transit time at low sample pressure has important practical implications. It means that when sampling from the engine exhaust, for example, when the sample pressure varies close to the atmospheric level, great care is needed in interpreting instantaneous measurements of hydrocarbon concentration (the consequences of variable transit time were illustrated in Fig. 3.4 and discussed in Section 3.4.3.1). A similar comment applies to measurements made within the engine cylinder during the exhaust and intake strokes. In contrast, measurements made during most of the compression and expansion strokes will be subject to nearly a constant transit time through the sample transfer tube. The results in Table 3.2 for the larger sample tube diameter of 0.4 mm are also of considerable interest. They show that the transit becomes stable at under 2 ms, but now at the much lower sample pressure of 1.5 bar. Clearly, the 0.4 mm diameter will be a better choice for a sample transfer tube intended for measurements in the engine exhaust because of its shorter transit time and greater insensitivity to sample pressure variations. The larger tube diameter of 0.4 mm does, however, increase greatly the sample flow rate dumped into the CP chamber, and this can increase the problem of avoiding CP chamber pressure fluctuations. The 0.4 mm diameter tube may also be more suitable for in-cylinder measurements during the exhaust and induction strokes, although the problem of CP pressure fluctuations may become much more severe because during the compression and expansion strokes very high flow rates of sample would be dumped into the CP chamber. Table 3.2 illustrates an additional severe problem that can arise with the use of the 0.4 mm diameter transfer tube. It can be seen that at sample pressures of 24 bar or higher, the expansion tube exit becomes choked. This condition is highly undesirable as it can introduce a severe pressure rise at the entrance to the FID tube (point A in Fig.3.3), rendering the fast FID system temporarily inoperative. As this undesirable condition is of no practical value, no calculated results are shown for it in Table 3.2.

Before discussing the results for the expansion tube, it is useful to recall that the flow rate through the expansion tube is decided by the operation of the sample transfer tube. The expansion tube merely "swallows" the quantities of sample delivered to it by the transfer tube. Under most flow conditions, the transfer tube has a negligible influence on the sample flow rate, largely due to the small fluid friction resistance offered by the expansion tube. Consider now the results in Table 3.2 for the expansion tube, when it is supplied with flow from the 0.2 mm diameter transfer tube. An important observation from Table 3.2 is that at low sample pressures the transit time through the expansion tube can be very long. For example, with a transfer tube diameter of 0.2 mm and a sample pressure of 0.75 bar, the transit time through the expansion tube is 101.2 ms. However, the transit time for the expansion tube falls very rapidly to 3.6 ms as the sample pressure rises to 3 bar; at sample pressures beyond 6 bar the transit time tends to settle below 1 ms. As discussed in the preceding paragraph, a variable transit time is undesirable, as it causes output signal distortion. Again, Table 3.2 shows that the larger transfer tube diameter of 0.4 mm would be more suitable for exhaust sampling applications because, first, it reduces the transit time and, second, it makes the transit time of the expansion tube less sensitive to sample pressure variations when the sample pressure is low.

Overall, the results of Table 3.2 indicate that with careful choice of sampling tube dimensions, a fast FID sampling system can be designed to have reasonably short transit times which are not greatly sensitive to sample pressure swings. However, the results also demonstrate that careful choice of tube dimensions is essential.

Eq. (3.8) for transit time is applicable only when the sample inlet pressure is constant and the sample flow through a tube is steady. When sampling from the engine cylinder or the exhaust, the sample pressure will vary substantially. Summers [1996] and Summers and Collings [1995a] constructed a dynamic model which predicted the varying transit time during the expansion and exhaust strokes of a gasoline engine running at 900 rpm and about half load. The dynamic model was partly based on the above steady-state analysis, but with provision for varying sample pressure. The length and diameter, respectively, of the various tubes were: transfer tube 306 and 0.25 mm, expansion tube 23 and 1.15 mm, FID tube 20 and 0.2 mm; the CP and FID chamber pressures were, respectively, 0.48 and 0.4 bar. The results show that when the cylinder pressure was above 5 or 6 bar, the transit times predicted by the dynamic model for the transfer, expansion, and FID tubes were close to the values given by the steady-state equations outlined above. However, below cylinder pressures of about 2 bar, the agreement between the dynamic and steady-state models was rather poor. Thus, greater care is required when using the steady-state Eq. (3.8) above for exhaust sampling or in-cylinder applications during the gas exchange processes.

Summers went on to validate experimentally the results predicted by the dynamic and steady-state models, through tests in a Ricardo E6 gasoline engine operating at 900 rpm and a range of loads between near-idling and half full-load. The fast FID sampling system dimensions and chamber pressures were as given in the preceding paragraph. The results for all the engine loads tested show good agreement (within 6%) between the measured transfer time and that predicted by the dynamic model for the FID instrument as a whole. The agreement between the measured transfer time and that predicted by the steady-state expressions was good at half full-load (within 1%) and poorer at near-idling (within 43%). The poorer agreement at near-idling was ascribed to low cylinder gas pressures, resulting in poor predictive accuracy with the steady-state expressions.

In summary, when considering the transit time through an FID system caution is required, for the following reasons:

- When the sample pressure is low (on the order of 1 to 3 bar), variations in the transit time are sensitive to sample pressure fluctuations.
- When the sample pressure is low (on the order of 1 to 3 bar) and the sample pressure is changing rapidly, the transit time predicted by the steady flow equations outlined above may be in considerable error.

3.4.3.2 Time Constant

The time constant for a fast FID system can be defined as the 10–90% response time of the FID electrical output to a step change in concentration at the transfer tube entry. The time constant for the fast FID system will depend on the design of the sampling system and on the operating conditions.

Figure 3.5 shows how the time constant occurs in the case of a capillary sampling tube. When a step change in concentration occurs at the entry to the tube, the concentration at the entry has no axial or radial variation; this is illustrated in Fig. 3.5(a). After the new sample has entered the tube, wall friction causes the material that was initially near the wall to travel through the tube slower than the material at the center of the tube; this is shown in Fig. 3.5(b). As a result, a concentration profile develops, with the concentration varying in the radial direction, as shown in Fig. 3.5(b). However, this concentration profile does not survive intact the journey of the sample through the tube. Molecular diffusion and turbulent transport cause the sample to migrate radially outward, as illustrated in Fig. 3.5(c). This tends to destroy the radial concentration variations. Provided the journey of the new sample through the tube is sufficiently long, the radial variations in concentration will disappear altogether, as shown in Fig. 3.5(d). Although the radial variations in concentration eventually may disappear, an axial variation in concentration (i.e., along the tube) will still persist, as shown in Fig. 3.5(d). It is worth commenting that for capillary tubes, the time necessary for the radial migration to be completed is usually very short in comparison to the total journey time of the sample through the tube [Summers, 1996]. This is a consequence of the small diameters used for FID transfer tubes (about 0.2 to 0.5 mm).

When the new sample with this axial variation in concentration arrives at the exit of the tube, the sample concentration is initially lower, increasing to the true level over a period of time, as shown in Fig. 3.5(d). The time constant is the time needed for the concentration to build up from 10% to 90% of the true level.

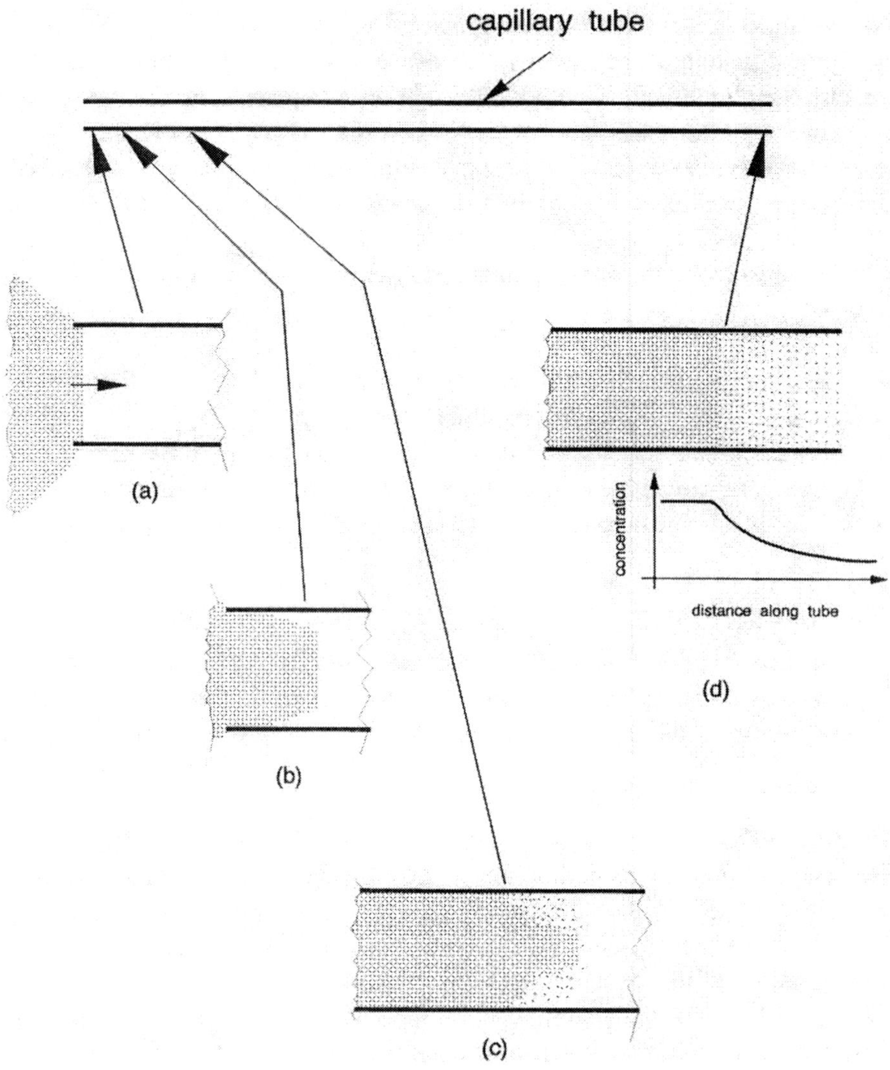

Fig. 3.5 Physical processes occurring during the passage of fresh sample through a fast FID capillary sampling tube.

With turbulent flow, the axial concentration variations at the tube exit can be expected to be less pronounced than in the case of laminar flow. This is because laminar flow has an extended concentration profile while the profile for turbulent flow is much flatter. The flatter profile leads to much less variation in concentration along the tube length. Therefore, the time constant with turbulent flow can be expected to be shorter than for laminar conditions.

When the fast FID system is used in internal combustion engines, it is useful to be able to predict the value of the time constant. This permits the design of the FID sampling system to be optimized so as to minimize the instrument time constant, and can also be helpful in interpreting subsequent data collected from engine tests.

It is possible to estimate the time constant for both laminar and turbulent incompressible flow conditions using an analysis first performed by Taylor [1953] and later extended by Smith [1989] for laminar compressible flow [Summers, 1996]. This analysis applies only to the simple cases of tubes having uniform cross-sectional area.

Laminar flow:

For laminar, isothermal, incompressible flow:

$$\tau_c = 1.24\,L\sqrt{\frac{1}{\Delta p}\cdot\frac{p}{RT}} \tag{3.9}$$

where L = total length of tube
R = specific gas constant
T = gas temperature
p = gas pressure in tube
Δp = pressure difference across tube

Or, more appropriately, for laminar, isothermal, compressible flow:

$$\tau_c = 1.24L\sqrt{\frac{1}{RT}\cdot\frac{\left(p_1^2 + p_2^2\right)}{\left(p_1^2 + p_2^2\right)}} \tag{3.10}$$

where p_1 = inlet gas pressure
p_2 = outlet gas pressure

Note that the time constant, τ_c, is independent of the tube diameter.

Assumptions:

- Isothermal flow.
- Axial variation in concentration assumed to be small (i.e., concentration varies within the tube radially and with time only).
- Cross-stream (i.e., radial) diffusion is dominant (i.e., axial diffusion is negligible).
- Tube has constant flow area.

- Perfect gas behavior.
- The Schmidt number $Sc = \mu/(D\rho) \approx 0.7$ (for most gases, the value of $\mu/(D\rho)$ lies in the range 0.6 to 0.8; i.e., the value of D has roughly the same magnitude as the kinematic viscosity, ν).
- The time required for radial molecular or turbulent transport is much shorter than the sample transfer period through the tube (note that this assumption may not be satisfied in the case of short tubes which have small transfer times, because there may not be sufficient time available for radial transport to be completed).

As explained previously in Section 3.4.3.1, a difficulty arises with the application of Eqs. (3.9) and (3.10) for the calculation of the time constant in the case of the expansion tube. This is because an appropriate value for either the expansion tube inlet pressure or pressure drop across the tube is required, and these quantities may not be known.

This difficulty is overcome by estimating the time constant for the expansion tube from:

$$\tau_c = 3.62\sqrt{\frac{kL}{U^3}} \tag{3.11}$$

where k is the virtual (or effective) diffusion coefficient; for laminar flow, k is given by [Summers and Collings 1995b],

$$k = \frac{d^2U^2}{192D}$$

A suitable rough value for the coefficient of molecular diffusion, D, would be 20×10^{-6} m^2/s

Turbulent flow:

Analogous expressions for turbulent flow conditions are given below. Assuming that the friction factor is related to the Reynolds number by the expression, $f = 0.079\ Re^{-0.25}$ (for values of the Reynolds number up to around 50,000).

For turbulent, isothermal, incompressible flow:

$$\tau_c = 1.44 L\, Re^{-\frac{3}{16}} \sqrt{\frac{1}{\Delta p}\cdot\frac{p}{RT}} \tag{3.12}$$

Or, more appropriately, for turbulent, isothermal, compressible flow:

$$\tau_c = 1.44 L \, Re^{-\frac{3}{16}} \sqrt{\frac{1}{RT} \frac{\left(p_1^2 + p_2^2\right)}{\left(p_1^2 - p_2^2\right)}} \tag{3.13}$$

The above expressions (3.12) and (3.13) for turbulent flow can be fairly inaccurate, but can provide useful predictions of trends. Both expressions are for smooth pipes.

As mentioned above, Eqs (3.12) and (3.13) containing p_1 and p_2 cannot be used readily in the case of the expansion tube. Instead, a rough estimate of the time constant for the expansion tube can be obtained from Eq. (3.11), but with the virtual diffusion coefficient now evaluated for turbulent flow conditions using the following expression [Summers and Collings 1995b],

$$k = 5.05 du \sqrt{\frac{f}{2}}$$

For Re values up to 50,000, the expression for the friction factor, $f = 0.079\ Re^{-1/4}$, could be used to simplify the above equation for k, giving:

$$k = dU \, Re^{-\frac{1}{8}}$$

Of course, the overall time constant for a fast FID instrument will not simply be the sum of the individual time constants for the transfer, expansion, and FID tubes, respectively. In general, the overall time constant for the instrument, τ_{tot}, comprising n component parts, can be obtained from the following expression:

$$\tau_{tot} = \left[\sum_{i=1}^{n} \tau_i^2 \right]^{1/2} \tag{3.14}$$

The above equation (3.14) applies regardless of the mechanisms involved in each time constant.

Examples:

Eqs. (3.10) and (3.13) for laminar and turbulent compressible flow, respectively, together with Equation (3.11) for the expansion tube (with the value of the virtual diffusion coefficient calculated, as appropriately, for laminar or turbulent conditions) have been used to estimate the time constants for the three main sampling elements of a fast FID system. The resulting estimates of the time constants are shown in Table 3.2. The reader is reminded that these results were calculated for a step change in concentration at the entry to each tube. In the case of practical fast FID systems, however, a step input can occur only at the entry to the sample transfer tube.

The results in Table 3.2 allow a number of practically useful observations. First, it can be seen that the value of the time constant for the FID tube is short, on the order of 0.2 ms. Also, for a given pressure drop across the tube, the value of the time constant is quite insensitive to the pressure levels in the FID and CP chambers. Consider now the results in Table 3.2 for the sample transfer tube with 0.2 mm diameter. The first important observation is that the time constant for laminar conditions is substantially greater (0.7 to 1.2 ms) than for turbulent conditions (<0.2 ms). A second observation of practical significance is that when the sample pressure rises, the value of the time constant falls. For example, when sample pressure rises from 0.75 to 3 bar, the time constant falls from 1.15 down to 0.73 ms. A further increase in sample pressure from 6 to 48 bar lowers the time constant even more, from 0.2 to 0.13 ms. It is interesting to note, in passing, that with sample pressures above 6 bar, the transfer tube exit becomes choked and, as it happens, the flow within the tube changes to the turbulent regime. Consider now the results in Table 3.2 for the transfer tube when its diameter is doubled from 0.2 to 0.4 mm. One interesting and practically significant observation is that the mass flow rate through the tube is now higher, and this brings about transition to turbulence at lower sample pressures (e.g., at 3 bar with the 0.4 mm diameter tube compared with 6 bar with the 0.2 mm diameter tube). As can be seen from Table 3.2, a transition to the turbulent regime is beneficial because it sharply lowers the time constant.

Finally, consider the results in Table 3.2 for the expansion tube. First, consider the results for the expansion tube when it is supplied with sample by the 0.2 mm diameter transfer tube. At low sample pressures (<1.5 bar) the time constant for the expansion tube can be very long (>8.2 ms) when compared with the corresponding time constants for either the sample transfer tube (>0.8 ms) or the FID tube (~0.2 ms). Apparently, the expansion tube time constant is long because the flow regime in the expansion tube is laminar. As the sample pressure rises further to 6 bar, the time constant for the expansion tube falls, but even at 6 bar its value is significant (2.68 ms). However, at 12 bar its value tumbles to 0.35 ms and continues to decline with rising sample pressure, reaching a value of 0.07 ms at 48 bar. This sharp fall in the time constant with sample pressure of 12 bar is associated with transition of the flow regime in the expansion tube from laminar to turbulent. It should be noted here that when the Reynolds number rises beyond 400, the flow in the expansion tube is assumed to have switched to turbulent conditions. This assumption is based on Summers's work, which showed that turbulent conditions occur in the expansion tube at much lower Reynolds numbers than for flow in pipes [Summers, 1996]. This early transition to turbulence is related to the fact that the flow emerges into the expansion tube as a fast jet. For example, it is known [White, 1974] that in the case of free jets, transition to turbulence occurs at very low values of Re of around 12. (See also discussion on flow regimes in the transfer tube in Section 3.4.2.3 above.) The choice of Reynolds number value of 400 for a switch from laminar to turbulent conditions is somewhat arbitrary; it reflects a compromise between unbound jets, which become turbulent at very low Re values, and enclosed fully-developed flows, which become turbulent at much higher Re values (greater than 2500).

Consider now the results in Table 3.2 for the expansion tube when it is supplied with sample by the sample transfer tube with the larger diameter of 0.4 mm. The larger diameter has two main effects on the time constant. First, at a given sample pressure, it lowers the value of the time constant. Second, the transition to turbulent flow occurs at the much lower sample pressure of 1.5 bar, compared with 12 bar with the 0.2 mm diameter transfer tube. Therefore, an increase in the transfer tube diameter from 0.2 to 0.4 would be beneficial when sampling from low sample pressure sources such as the engine exhaust; this is because the larger diameter will reduce the time constant substantially. However, as Table 3.2 shows, the larger transfer tube diameter will also cause the exit of the expansion tube to choke at high sample pressures (24 bar or more); this will make the fast FID system inoperative above this pressure. This point has been discussed in Section 3.4.2.1 and at the start of Section 3.4.2.2 above; Section 3.4.2.3 discusses various flow regimes in the expansion tube.

The equations presented in this and the previous section for transit time and time constant are based on Taylor's work [1953; 1954], and their derivation assumes fully developed steady-state flow. This assumption may not be entirely realistic in the case of high flow rates and rapidly changing sample pressures.

Summers [1996] validated experimentally predictions of the time constant for a fast FID system as a whole under the following two conditions: first, when the sample pressure was constant and flow through the FID system was steady; and, second, when the sample pressure was changing rapidly and the flow through the system was unsteady.

The steady-flow experiments were carried out with the following FID system configurations [Summers, 1996]: sample pressure constant at the atmospheric level; CP chamber pressures ranging from 0.35 to 0.75 bar (absolute); expansion tube lengths from 10.5 to 22.5 mm; and transfer tube length and diameter, 323 and 0.5 mm, respectively. The corresponding predictions were made using the steady flow equations outlined earlier in this section. The experimental results agreed very well with the predictions (within a few percent) when the flow in the expansion tube was assumed to be turbulent (despite Reynolds numbers much lower than 2500).

The unsteady-flow experiments were conducted in a Ricardo E6 engine running under a variety of speed/load combinations. The corresponding predictions were made with a quasi-dynamic model, which was based partly on the steady-state analysis outlined earlier in this section. When the flow in the expansion tube was modeled as being turbulent, the agreement between the predicted and measured values for the time constant of the instrument as a whole was only fair. Closer agreement was achieved when the flow in the expansion tube was modeled as being entirely disorganized; the expansion tube was considered to be merely a mixing box. The equations outlined above do not apply to such a regime, because they are based on well-defined laminar or turbulent velocity profiles. Estimation of the expansion tube time constant using the "mixing box" approach is described in Summers [1996] and Summers and Collings [1995b].

3.4.3.3 Hydrogen Flame

Referring to Fig. 3.3, it can be seen that the sample containing HC emerges from the FID tube into the base of the FID flame. The sample emerging from this tube forms a slowly expanding jet which entrains and mixes with the surrounding hydrogen fuel. The mixture is then carried to the outer fringe of the laminar diffusion flame where it mixes with air and burns.

A transit time will be involved between the sample exiting the FID tube and burning in the flame. This transit time arises because the sample first has to mix with the hydrogen, then the mixture has to be transported to the flame fringe, where it is burned to produce ions. Both, the sample jet from the FID tube and the surrounding hydrogen flow are laminar. Mixing between the jet and the fuel is by molecular diffusion, and involves a characteristic mixing time, τ_f, given by [Summers, 1996]:

$$\tau_f = \frac{r_j^2}{D_{H_2}} \tag{3.15}$$

Taking the sample jet radius, r_j, as 0.15 mm and the diffusion coefficient for hydrogen as 40×10^{-6} m^2/s, gives $\tau_f = 0.6$ ms.

The transport of the sample/hydrogen mixture to the outer fringe of the flame envelope incurs an additional transit period. A minimum value of this period can be estimated by assuming that the sample jet emerging from the FID tube entrains hydrogen and thereby carries the mixture to the flame front. If the flame diameter is assumed to be around 3 mm and the sample/hydrogen mixture velocity is 30 m/s [Summers, 1996], this transport delay period would be 0.1 ms. In practice, this transport period is substantially longer because the sample jet momentum is dissipated significantly. The actual value for this transport period would probably be several time this value.

3.4.3.4 Electronic Signal Conditioning

The ion current collected in the FID chamber has to be conditioned before a usable and relatively noise-free output is available from the fast FID system. Time constants are likely to be associated with some of the signal conditioning units. For example, an appropriate low-pass filter designed to reduce output noise may involve a time constant of around 0.2 ms [Summers, 1996].

3.4.3.5 Selection of Sampling System Dimensions, Chamber Pressures, and Fuel

This section gives some guidance on how to select appropriate dimensions for the various tubes of a fast FID system. It also discusses the factors affecting the selection of CP and FID chamber pressures, and the type of fuel for the FID flame. The tube dimensions and chamber pressures suggested below are only a rough guide and would require considerable optimization to suit a particular application. Optimization can be carried out using the various expressions given in the preceding sections. A certain amount of experimental trial and error may also be involved before finalizing the choice of tube dimensions and chamber pressures [Cambustion; Summers and Collings, 1995a and 1995b]. Various applications for the fast FID system are considered in turn:

(a) Sampling from the engine cylinder during the compression or expansion strokes

(i) **Transfer tube**: Diameter should be small (e.g., 0.20 to 0.4 mm) to limit sample flow.

Advantages: Limited sample flow into the CP chamber, helps to reduce pressuriza tion of the chamber and, consequently, fluctuations in the pressure drop across the FID tube (particularly at high cylinder pressures produced at high engine loads); limited sample flow reduces risk of choking at the expansion tube exit, which could destroy the pressure isolation of the FID tube inlet; reduced amount of sample extracted from the cylinder, minimizes the impact on the phenomenon being observed (especially at low engine speed and high load).

Disadvantages: The transit times during the exhaust and intake strokes are likely to be longer and more sensitive to sample pressure; the time constant can be very long, particularly for the expansion tube.

Note: With transfer tube diameters in the range 0.2 to 0.4 mm, the transit time tends to a fixed value as pressure rises beyond a few bar, making data interpretation during the compression and expansion processes easier. The time constant for the sample transfer tube is not affected by reducing the transfer tube diameter.

(ii) **CP chamber**: Pressure in the region of 0.5 bar (absolute); lower CP chamber pressure would be needed at low engine load and near idling if flow in the transfer tube is not to reverse during the intake stroke; large CP volume may be necessary for sampling during compression and expansion in order to dampen CP chamber pressure fluctuations; alternatively, the CP chamber could be opened to the atmosphere or removed altogether, in which case the instrument output will be useful only when the cylinder pressure is substantially above the atmospheric level.

(ii) **FID chamber**: Pressure about 0.05 to 0.1 bar below the CP chamber pressure; when the FID chamber pressure is much below the CP chamber pressure, the sample flow to the flame can be high and the linearity of the output signal may be worsened when the HC concentration in the sample is high (sampling from the cylinder usually involves high HC concentration in the sample); conversely, when the CP chamber and FID chamber pressure are very close to each other, the flow through the FID tube becomes more sensitive to pressure fluctuations in the CP chamber (i.e., the instrument pressure independence is worsened).

(b) Sampling from the cylinder during the exhaust stroke

(i) **Transfer tube**: Diameter should be larger (e.g., 0.3 to 0.6 mm) than for case (i) above.

Advantages: During the exhaust stroke the transit time is shorter, which could be helpful in the interpretation of results; also, the transit time is less sensitive to sample pressure (cylinder pressure) during the exhaust stroke (also helpful in interpretation of results); flow in the expansion tube is more likely to be turbulent, which reduces the time constant for this tube; flow in the expansion tube becomes turbulent at lower sample pressures, causing a sharp drop in the time constant for this tube

Disadvantages: Excessive sample flow into the CP chamber during the compression and expansion strokes could make the pressure fluctuations in the chamber unaccept-able; also, excessive sample flow through the expansion tube during the compression and expansion strokes may cause the exit from the expansion tube to choke, destroying the pressure isolation of the FID tube and rendering the fast FID instrument inoperative at high pressures; the amount of sample extracted from the cylinder may be excessive and influence the phenomenon being observed (e.g., when sampling close to the spark-plug electrode gap.

(ii) **CP chamber**: Pressure in the region of 0.5 bar (absolute).

(iii) **FID chamber**: Pressure around 0.05 to 0.1 bar below the CP chamber pressure.

(c) Sampling from the cylinder during the intake stroke

(i) **Transfer tube**: Diameter in the region of 0.3 to 0.6 mm.

Advantages: As for (b) above.
Disadvantages: As for (b) above.

(ii) **CP chamber**: Pressure should be low, for example, in the region of 0.3 to 0.4 bar (absolute) or less. The CP chamber pressure must be sufficiently low to enable sample to be drawn through the transfer tube continuously even at low engine load when the cylinder pressure during intake can fall well below the atmospheric level.

(iii) **FID chamber**: Pressure about 0.05 to 0.1 bar below the CP chamber pressure; however, too low an FID chamber absolute pressure may result in the flame becoming unstable.

(d) Sampling from the exhaust system during blow-down

(i) **Transfer tube**: Diameter in the region 0.3 to 0.6 mm.

Advantages: As for (b) above.
Disadvantage: As for (b) above.

(ii) **CP chamber**: Pressure in the region of 0.5 bar (absolute).

(iii) **FID chamber**: Pressure around 0.05 to 0.1 bar below the CP chamber pressure.

In all the above applications, an expansion tube having a length of 20 to 25 mm and a diameter of around 1.0 to 1.2 mm could be used. The FID tube diameter could be in the region of 0.15 to 0.2 mm, and the tube length would typically be about 20 to 30 mm. The length of the transfer tube should be made as short as the installation of the system will allow, in order to reduce the transit time and time constant of the instrument.

The fuel for the FID flame can be either pure hydrogen or a mixture of 40 percent hydrogen/60 percent helium. The use of H_2/He mixtures significantly reduces the linearity of the instrument output when the HC concentration in the sample is high. Therefore, using pure hydrogen as the fuel maximizes the range of sample HC concentrations over which the instrument output is linear [Cambustion].

3.4.3.6 Output Signal Linearity

A calibration curve for the fast FID instrument comprises a curve of output signal voltage plotted against HC concentration in the sample (e.g., ppm C_3, or C_1). This curve should normally be a linear one over a wide range of sample HC concentrations. For example, in the case of samples made up of propane-in-air, the linear range could extend from 5×10^3 to 5×10^4 ppm C_3H_8 (the stoichiometric mixture for propane-in-air is approximately 4.2×10^4 ppm C_3H_8 by volume or 4.2 percent). However, the curve may depart from linearity for very rich mixtures [Peckam, 1993; Cambustion].

The linearity of the output signal is affected by the following factors:

(i) Fuel type for FID flame: Pure H_2 gives significantly better linearity at high sample HC concentrations than a mixture of H_2/He does, possibly due to diffeences in the flame geometry. Therefore, if a mixture of H_2/He is used, the maximum HC concentration in the sample at which the system output is still linear is significantly reduced [Cambustion].

(ii) Pressure drop across FID tube: High sample flow rate to the flame can reduce linearity at high HC concentrations. (Note that sampling from the engine cylinder usually involves samples having high HC concentrations.) Cambustion found the that instrument output was fairly linear (within 2% of ideal) for sample concentrations up to 50,000 ppm propane with the following sampling arrangement: FID tube diameter, 0.2 mm; CP chamber pressure, 0.5 bar; pressure drop across the FID tube up to 0.13 bar; and pure hydrogen as the fuel. With H_2/He as the fuel, good linearity at high HC concentrations was obtained with FID tube pressure drops up to about 0.04 bar (i.e., with reduced sample flow to the flame) [Cambustion].

(iii) Sample flow supplied to the flame: A high sample flow to the flame can also reduce the maximum HC concentration in the sample at which the instrument output is still linear. The sample flow to the flame is controlled by the pressure drop across the FID tube, and this has been discussed in (ii) above. The sample mass flow rate is also controlled by the density of the sample gas through the FID tube, which is influenced by the CP chamber pressure. Dispensing with the CP chamber or operating with a CP chamber pressure at the atmospheric level increases the density and mass flow rate of the sample through the FID tube. To counteract this, the mass rate can be reduced by lowering the pressure drop across the FID tube.

A practical problem relating to the FID system calibration is that commercial suppliers of calibration gases do not normally supply hydrocarbon/air mixtures that approach the lower flammability limit. Thus, mixtures of hydrocarbon gases such as propane and air for calibration purposes often have to be obtained by mixing separate streams of air and propane. Please note that great caution is required in handling flammable mixtures, especially if they are in confined spaces. Alternatively, supplies can be obtained of propane/nitrogen mixtures having high propane concentration.

3.5 Response of the Fast FID when Installed in the Engine Cylinder

This section will consider the response of a fast FID system when used for in-cylinder sampling in internal combustion engines.

Figure 3.6 shows sketches of various fast FID output signals that might be obtained under various engine running conditions. Figure 3.6(a) shows the salient features of a typical fast FID output signal. The signal can be seen to rise rapidly when fresh sample from the cylinder arrives at the ionization flame. The signal ultimately reaches a "plateau" corresponding to the HC concentration in the engine cylinder gas before the ignition spark. When the flame arrives at the sampling point, the HC in the sample is consumed, causing the output signal to fall sharply. The duration of this sharp fall is indicative of the time constant for the instrument. Figure 3.6(b) shows the way in which the instrument output signal responds to changes in the air/fuel ratio supplied to the engine cylinder. The signal plateau level can be seen to rise when the mixture supplied is made richer and vice versa when it is made leaner. Figure 3.6(c) shows a "spike" introduced onto the signal when the outlet from the expansion tube chokes. This phenomenon may occur when cylinder pressure is around its maximum level, and is likely to be associated with a choice of a large sample transfer tube diameter. It is evident from this figure that choking of the expansion tube exit renders the upper part of the output signal unusable.

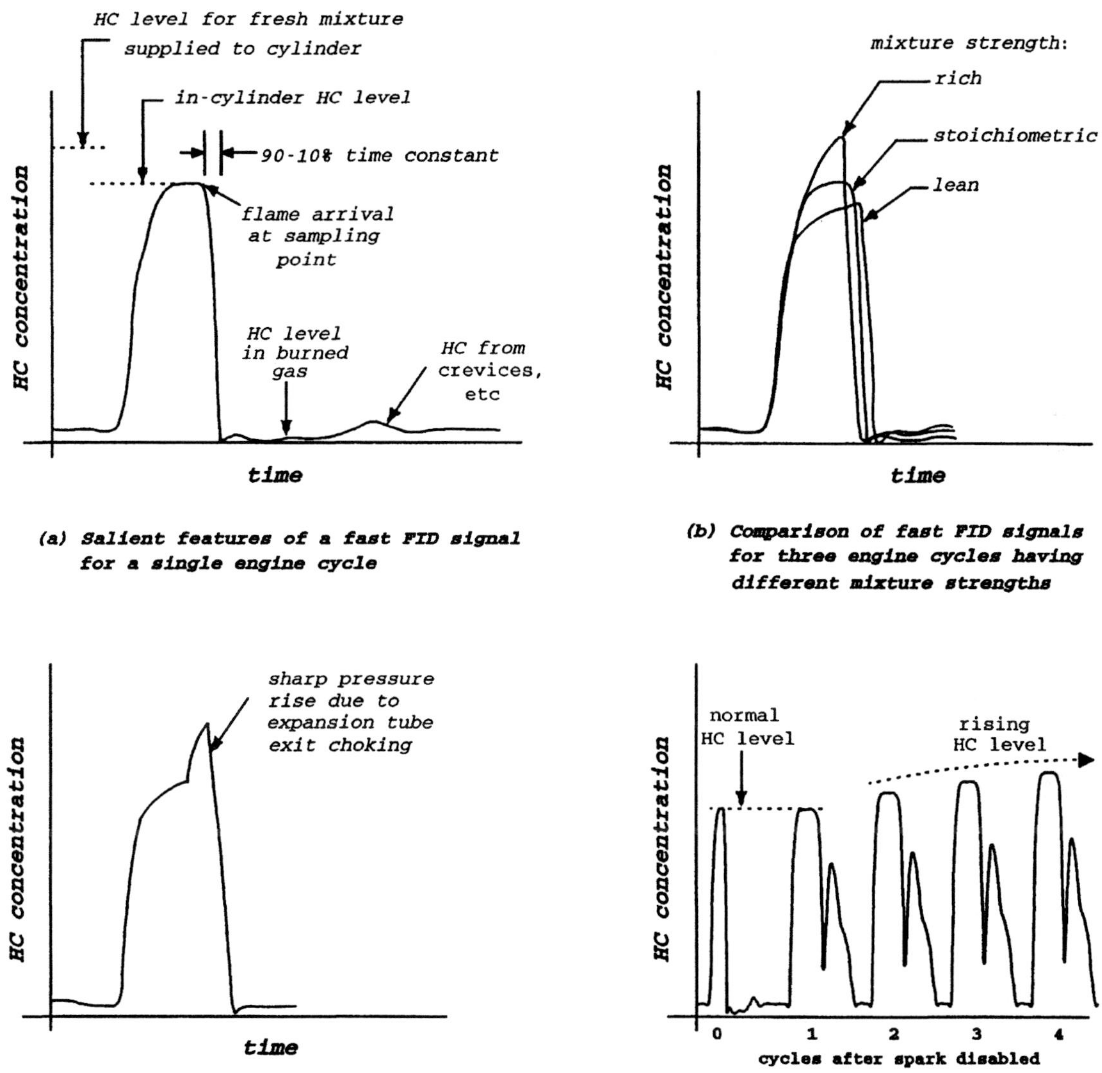

(a) Salient features of a fast FID signal for a single engine cycle

(b) Comparison of fast FID signals for three engine cycles having different mixture strengths

(c) Choking at expansion tube exit

(d) increase in HC level due to expulsion of residuals

Fig. 3.6 Fast FID output signals obtained under various running conditions.

3.5.1 Adequacy of Sampling Period Available

Figure 3.7 shows a sample entering the capillary tube for a finite period only. This results in the sample within the tube having a finite length. During the passage of the sample through the tube, the leading end of the sample suffers from dispersion, as discussed previously in Section 3.4.3.2. The trailing end of the sample also suffers from dispersion, as illustrated in Fig. 3.7(a). When the sample arrives at the edge of the tube, the concentration profile has the shape shown in Fig. 3.7(b). Only the middle section of the sample has remained unaffected by dispersion, and only here is the HC concentration at the true level for the sample entering the tube. Clearly, it is vital that an unadulterated middle section of the sample exists (i.e., $\Delta l > 0$). Otherwise, the true HC level cannot be discerned at the tube outlet.

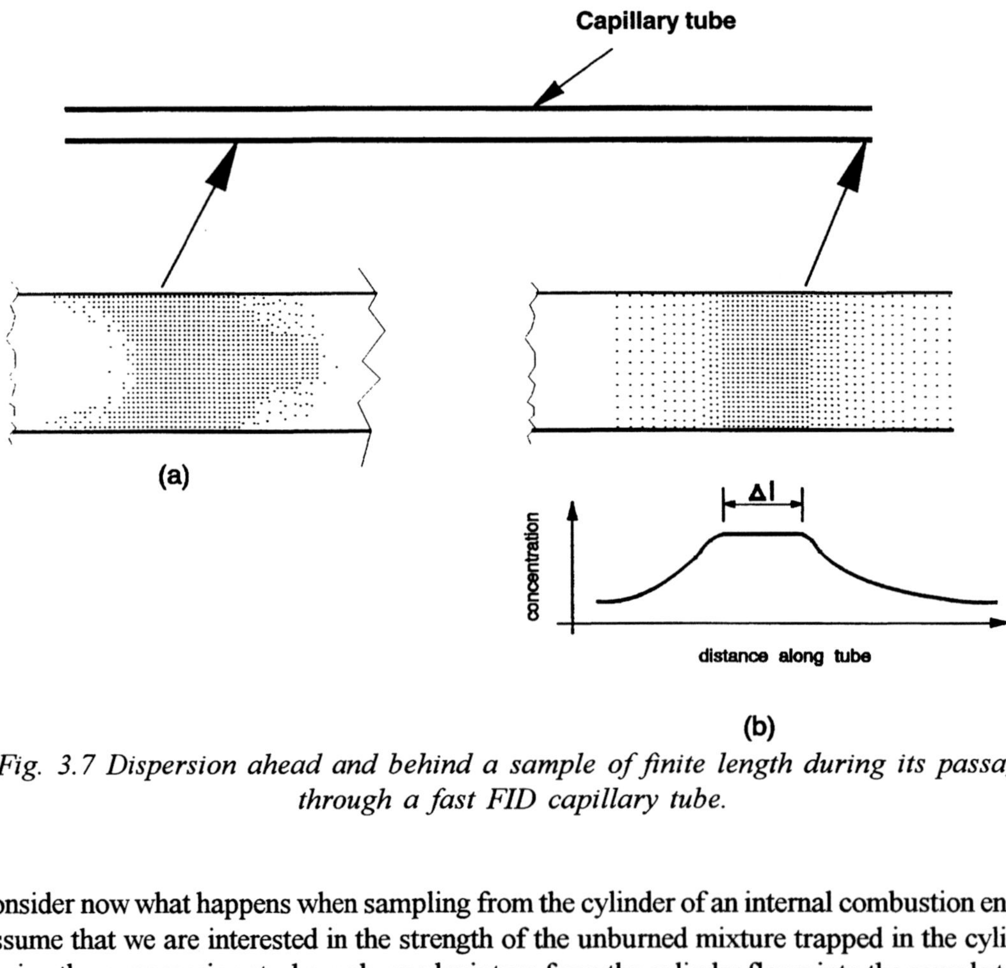

(b)

Fig. 3.7 Dispersion ahead and behind a sample of finite length during its passage through a fast FID capillary tube.

Consider now what happens when sampling from the cylinder of an internal combustion engine. Assume that we are interested in the strength of the unburned mixture trapped in the cylinder. During the compression stroke, unburned mixture from the cylinder flows into the sample transfer tube of the fast FID system. Ahead of the sample is burned gas from the preceding engine cycle. At some point after top dead center, the combustion flame arrives at the sampling point. Burned gas now enters the sample transfer tube. Therefore, the period during which unburned air fuel mixture enters the sampling system is finite. The situation is analogous to that shown in

Fig. 3.7, discussed above. This finite sampling period must be long enough so as to accommodate dispersion effects ahead and after the sample, while still providing a sufficient quantity of unadulterated sample. The following paragraph will consider this point further.

Figure 3.8 shows, hypothetically, what happens when sampling from the engine cylinder at three sampling points, labeled A, B, and C. Consider first the output signal from the fast FID instrument when sampling at point C. Fresh mixture forced into the sample transfer tube during compression suffers from dispersion of its leading edge, causing the output signal to rise with a time constant τ_c ms. The signal reaches a plateau and stays there for a period of t_c ms, until the arrival of the flame at sampling point C causes the signal to decline to a near-zero level. It is this period t_c which must have a finite value, if the mixture strength in the engine cylinder is to be measured correctly. Figure 3.8 shows that a plateau is also reached successfully in the case of sampling from point B. However, when sampling is from point A, a plateau is not reached by the output signal, for the following reasons. When sampling from this point, the leading edge of the sample is being dispersed and diluted as it passes through the sampling tubes of the FID system. However, due to the close proximity of sampling point A to the spark plug, burned gas enters the transfer tube before dispersion of the leading edge of the sample is completed. As a result, there is no unadulterated sample sandwiched between the leading and trailing edges of the gas traveling through the transfer tube. The consequence is the output signal cannot rise fully to a plateau level corresponding to the HC concentration in the engine cylinder mixture.

A useful technique for checking whether a stable plateau has been reached involves disabling abruptly the ignition spark. This is done as follows. The engine is operated at steady state and at a convenient instant the spark is abruptly disabled. This eliminates ignition of the mixture and flame arrival at the sampling point during subsequent engine cycles. If the system is operating correctly, the plateau level reached by the *first* unfired cycle should be the same as that for the last fired cycle; this is illustrated in Fig. 3.6(d). If the height reached by the first unfired cycle is greater, then the time available for the instrument to respond is not sufficient. The sampling system should then be redesigned with different tube dimensions, or the engine speed should be reduced if this is possible.

The elimination of the flame through spark disablement allows unburned mixture from the engine cylinder to flow through the sampling system of the fast FID for a longer period [Galliot et al, 1990; Ladommatos, 1992a]. Incidentally, the plateau level should not change for the first unfired cycle, but should increase progressively with each successive unfired cycle, as illustrated in Fig. 3.6(d). This progressive increase should occur because the combustion residuals remaining in the engine cylinder during the last fired cycle are gradually being expelled [Galliot et al, 1990; Ladommatos, 1992a]. In contrast the last fired and first unfired cycles should have the same plateau levels. This is because the first unfired cycle still contains the residuals of the last fired one.

A technique similar to the abrupt spark disablement can be achieved by skip-firing the engine. This involves the engine running at steady state, but with the spark plug disabled for, say, one in ten engine cycles. It is important to allow several fired cycles between the unfired ones, so that the combustion residuals can return to their normal level just before the next skipped cycle.

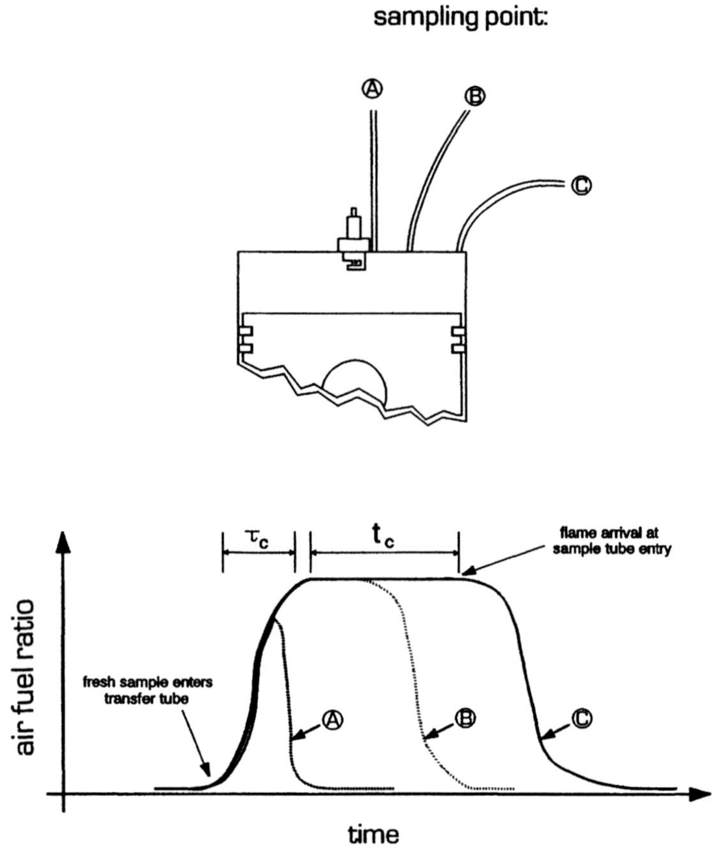

Fig. 3.8 Sketch of fast FID output that might be obtained at three sampling points.

In general, the interpretation of the HC concentration measurements in the cylinder for both abrupt spark disabling and skip-firing require considerable care so as to take full account of the presence of the combustion residuals trapped in the engine cylinder.

Crawford and Wallace [1996] carried out engine tests in conjunction with a fast FID, which addressed the problem discussed above: that is, whether a sampling point close to the spark plug would allow a stable plateau in the fast FID output signal to be reached. Sampling was carried at two locations simultaneously: close to the spark plug (about 7 mm), and far away from the spark plug (close to the opposite cylinder wall). A CFR engine running on propane at speeds between 900 and 1650 rpm was used. The sampling system geometry is shown in Table 3.3., Set 3. The principal parameter measured was the duration of the plateau in the output signal (marked t_c in Fig. 3.8). The results show that when sampling *far* from the spark plug, a stable plateau was reached; the plateau duration reduced from about 7 to 6 ms as the engine speed rose from 900 to 1650 rpm. Similarly, when sampling *close* to the spark plug, a stable plateau was also reached; its duration reduced from around 2 to 1 ms as the speed rose from 900 to 1650 rpm. Crawford and Wallace [1996] concluded that even at 1650 rpm, there was nothing in the fast FID system behavior to indicate that the limit of its dynamic response was encountered.

Table 3.3 Salient features of fast FID sampling systems used by various investigations

Set	Sample Transfer Tube		Expansion Tube		FID Tube		CP Chamber Pres.	FID Tube Pres. Drop	Fuel	Transfer Tube Temper.	Sampling Point Location and Engine	Reference
	L mm	d mm	L mm	d mm	L mm	d mm	bar	bar		°C		
1	306	0.25	23	1.2	20	0.2	0.49	0.08[1]	propane		cylinder wall, Ricardo E6, 900 rpm	Summers, 1996
2	150	0.33	23	1.2	20	0.2	0.41	0.07[1]	gasoline	160	cylinder wall, Ricardo Hydra, 1040 rpm, full load	Summers, 1996
3	260	0.30		1.2		0.15	1.01	0.53	natural gas	~25	near and far from spark plug, CFR, 900-1650 rpm	Crawford and Wallace, 1996
4	100[2]	0.2[2]	25[2]	0.8[2]	30[2]	0.15[2]	1.01		propane		spark plug base, Ricardo Hydra, 1200 rpm	Cheng et al, 1989
5	100[2]	0.2[2]	25[2]	0.8[2]	30[2]	0.15[2]	0.4-0.67	0.13	propane		spark plug base, Ricardo Hydra, 1200 rpm	Galliot et al, 1990
6	250	0.2	25	0.8	30	0.15	0.21		natural gas		pre-chamber wall of gas engine, 1500 rpm, high load	Charlton et al, 1992
7	210	0.2	25	0.8	30	0.15	1.01	0.54	gasoline	150	spark plug base, port-injected six-cylinder engine, 1200 rpm	Ladommatos, 1992a
8	210	0.2	25	1.2	30	0.15	0.95	0.31	gasoline	150	cylinder wall, fast-burn production engine, 1200 and 1700 rpm	Ladommatos, 1992b
9	450	0.5							propane		on moving piston crown, Ricardo E6, 500 rpm, no load	Peckham and Collings, 1994
10	250	0.25							propane		Engine head flame face, Ricardo E6, 1300 rpm, medium load	Peckham and Collings, 1994
11	150	0.69	20	1.2	25	0.3			propane		In exhaust, 150mm from exhaust valve, Ricado Hydra, 900, 1600, and 2500 rpm, low to high load	Min et al, 1994
12	300	0.4					0.47	0.13		150	In-cylinder, high load and low speed	Cambustion
13	100	0.3	20	1.0	30	0.3	1.01	0.05	indolene[3]	200	Away from spark plug, 600-900 rpm low and high loads	Fox et al, 1992

Notes: [1] nominal pressure drop; [2] inferred; [3] gasoline with tightly specified composition

Crawford and Wallace also investigated the effect of engine load on the duration of the signal plateau. When sampling *far* from the plug, the duration reduced from around 6 to 3 ms as the load was increased from zero to high load. When sampling *close* to the plug, the duration of the plateau reduced from around 1.5 to 1 ms as the load was increased.

Crawford and Wallace also compared the HC concentration measured simultaneously *far* and *close* to the spark plug for successive engine cycles. Their results show that the HC concentration levels measured at the two locations differed by only about 3 percent.

Cheng et al. [1989], also sampled close to the spark plug (7 mm from the electrode gap) using a sampling system summarized in Table 3.3., Set 4. Their results at 1200 rpm, when running on propane, show an extensive plateau in the output signal, the duration of which varied between about 2.4 and 4.4 ms due to cycle-to-cycle variability.

3.5.2 Measurements of the Time Constant

Cheng et al [1989] made a rough estimate of the time constant as being around 1 ms, for the system summarized in Table 3.3., Set 4. This estimate was made by measuring the fall-off time (90–10%) in the fast FID signal when the combustion flame arrived at the sampling point, as illustrated in Fig. 3.6(a). Cheng et al. assumed that the flame traverse time across the sampling point inlet was on the order of 0.05 ms and therefore negligible. Using a the same technique, Summers [1996] also provided measurements of the time constant for the sampling geometry summarized in Table 3.3, Set 1, installed on a Ricardo E6 variable compression ratio engine running on propane at 900 rpm. Sampling was far away from the spark plug. Summers found that the time constant varied between about 1 ms at high load and high compression ratio to 2.25 ms at low load and low compression ratio (the corresponding transit time ranged from 5 to 8 ms). In a separate series of tests on a different engine (single cylinder Ricardo Hydra), Summers estimated the time constant as 2 ms (configuration FID as in Table 3.3, Set 2). The technique used by Cheng et al. [1989] and Summers [1996] also allows a rough estimate of the time constant to be made for the system used by Crawford and Wallace (Table 3, Set 3); their FID signal output suggests a time constant value of around 3 ms at low engine load and high speed. Likewise, the same technique allows an estimate of about 1 ms for the fast FID system used by Charlton et al. [1992] for tests on a pre-chamber gas engine (system summarized in Table 3.3, Set 6).

Collings [1988] measured the time constant of a fast FID system using sample gas at atmospheric pressure and a sample tube length of 40 mm. The measurements were made by traversing rapidly a jet of sample gas across the sample tube end, and noting the time constant for the sudden rise in the output signal. The traverse of the jet was made progressively faster and faster, until no further reduction in the time constant could be observed. In this way, the time constant was estimated at about 1 ms.

3.6 Calibration and Pressure Independence of a Fast FID System

A calibration curve for a fast FID system comprises output voltage from the instrument plotted against HC concentration. A calibration curve can be produced using, for example, propane-in-air mixtures having a range of known propane concentrations. As discussed above, this curve should normally be a linear one over a wide range of HC concentrations (5×10^3 to 5×10^4 ppm propane in air) [Peckam, 1993], but may depart from linearity for very rich mixtures (stoichiometric mixture ~42,000 ppm by volume or 4.2 percent).

Calibration of the fast FID system can be carried out before or after it has been installed on the engine. Calibration before installation onto the engine involves offering a calibration gas of known HC concentration at the sampling point (entrance to sample transfer tube) and noting the instrument output voltage. The calibration can be at atmospheric or high pressure. Alternatively, the sampling system can be installed onto the stationary engine and then the engine cylinder can be flooded with calibration gas.

It is also possible to calibrate the output signal in-situ while the system is installed in a running engine. For example, Cheng et al. [1989] and Galliot et al. [1990], both calibrated the instrument before installing it into the engine (fueled by propane) and subsequently verified the calibration curve by motoring the engine while it was being supplied with air fuel mixture as normal. Although the maximum cylinder pressure of a motored engine is several times lower than that when the engine is fired, calibration in a motored engine corresponds more closely to conditions in a fired engine than does bench calibration. Galliot et al. [1990] found that the HC concentration in the cylinder gas of a motored engine (plateau level of the FID output signal) corresponded well with the HC concentration measured in the fresh propane/air mixture being supplied to the engine.

A useful in-situ calibration technique developed by Fox et al. [1992] (see also Crawford and Wallace, [1996]; and Cambustion) allows the fast FID to be calibrated while it is installed in a running engine. Calibration gas is supplied close to the outlet of the fast FID expansion tube, as shown in Fig. 3.9. During the engine intake stroke, the calibration gas is sucked into the expansion tube, and from there into the FID flame via the FID tube. This generates a calibration signal at the fast FID output during the intake stroke, as shown in Fig. 3.9. During the subsequent compression stroke, the calibration gas is expelled from the expansion tube into the CP chamber by sample gas flowing from the engine cylinder, and the normal FID signal output is reestablished. Care is required in designing and setting up this calibration arrangement so as to ensure that the calibration gas sucked into the expansion tube at each cycle is not diluted by ambient air. It is also possible to use this in-situ calibration method when a CP chamber is fitted to the sampling system [Cambustion]. However, the CP chamber must be at a higher pressure than the pressure in the engine cylinder during the intake stroke. Furthermore, care must be taken to ensure that the calibration gas sucked into the expansion tube is not diluted by the contents of the CP chamber (a mixture of unburned fuel and air and combustion products). Therefore, it is essential that the CP chamber is well purged with calibration gas and, also, vented effectively.

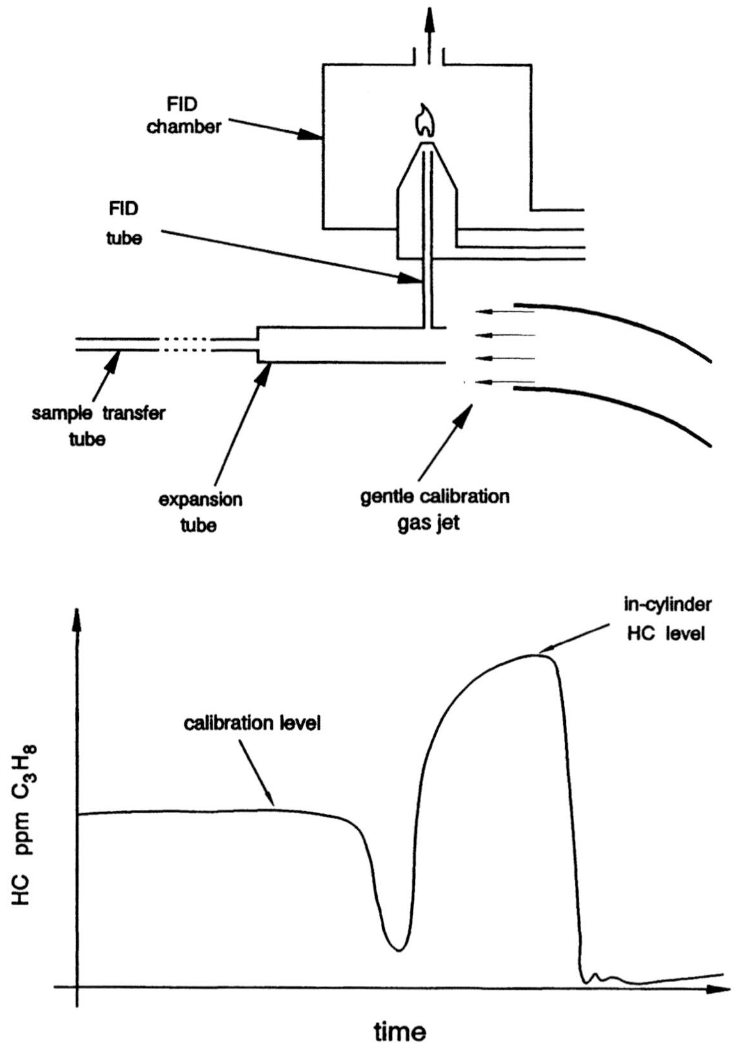

Fig. 3.9 In-situ calibration of the fast FID system.

The HC concentration measured in the engine cylinder gas during the compression stroke will generally be lower than the HC concentration of fuel in the fresh air fuel mixture being supplied to the engine, due to the fact that the fresh mixture entering the cylinder is diluted by the combustion residuals remaining in the engine cylinder at the end of the exhaust stroke. It is therefore not possible to use the HC concentration measured in the engine cylinder gas to estimate the air/fuel ratio supplied to the engine, unless the amount of trapped residuals is known. Also, a great deal of caution is required in interpreting cycle-to-cycle fluctuations in HC concentration. These could be caused by fluctuations in one or more of the following: the amount of combustion residuals, the amount of fresh mixture admitted to the cylinder, the mixture strength, and the extent of mixture stratification within the cylinder.

As discussed in Section 3.4.3, it is of great importance that the fast FID output signal for a given HC concentration is not affected by sample pressure variations. This is particularly important for in-cylinder sampling applications, where the cylinder gas pressure varies widely. In order to check for pressure independence, Peckham and Collings [1992] motored the engine while it was being supplied with air propane mixture and ensured that the signal plateau levels during the compression and exhaust strokes were equal. This method relies on the air fuel mixture being homogeneous during the compression stroke, a requirement met more readily with gaseous fuels than gasoline. An alternative to motoring the engine is to skip-fire or interrupt the spark, as discussed previously. The signal for the unfired cycles should show an extensive plateau covering the period between early in the compression stroke to late in the expansion stroke; the plateau should be flat for pressure independence.

Summers [1996] carried out experiments in a single-cylinder engine aimed at measuring the in-cylinder concentration of unburned HC in the products of combustion. A problem encountered by Summers was substantial pressure fluctuations in the CP chamber during engine operation. The pressure across the FID tube fluctuated between 0.07 and 0.1 bar during the engine cycle, causing the sample flow rate to vary significantly, even when the CP chamber volume was increased. Fortuitously, it was possible to make the required measurements during a period in the engine cycle when the pressure drop across the FID tube was stable for a sufficiently long period (see also Section 3.4.2.1).

3.7 Summary

The fast FID system has been developed for instantaneous measurements of hydrocarbon concentrations in the cylinder as well as in the exhaust. In this chapter, the principle and design of a fast FID system are described in detail. In the successful operation of the fast FID system, the pressure independence of the system must be checked and maintained. The transient time must be considered in the interpretation of the fast FID output from a rapidly varying measurement source such as in-cylinder measurements.

The fast FID system has found fairy wide application, ranging from intake fuel transport, in-cylinder fuel concentration, and exhaust unburned hydrocarbons. The application of fast FID systems to in-cylinder fuel concentration measurement will be described in Chapter 7. The use of fast FID systems to measure in-cylinder residual gas concentration will be described in Chapter 13.

By combining a specially designed chemiluminscence detector (CLD), described in detail in Chapter 1, with the heated sampling system of the fast FID described in this chapter, Cambustion Ltd. has produced the first fast CLD system that is capable of transient NO concentration measurements with a response time of less than 4 milliseconds. The envisioned applications include injection optimization for DI gasoline research, transient fuel and EGR calibration, and combustion analysis. The system is currently being evaluated in a number of research laboratories and results have yet to be published. Much of the discussion presented in this chapter, however, applies directly to the fast CLD system with respect to its pressure independence and transient time problem.

References

Cambustion Ltd., *User Manuals for HFR 400 Fast FID and HFR 400 Sampling Spark Plug*, Portugal Place, Cambridge, U.K., 1998.

Charlton, S.J., Jager, D.J., and Tawfig, M.E. (1992), " In-Cylinder Measurement of Mixture Strength in a Turbocharged Natural Gas Engine," *Proc. Instn. Mech. Engrs.*, Vol. 206, Part A, pp. 167–175.

Cheng, W.K., Galliot, F., and Collings, N. (1989), "On the Time Delay in Continuous In-Cylinder Sampling from IC Engines," SAE Paper No. 890579.

Collings, N. (1988), "A New Technique for Measuring HC Concentration in Real Time in a Running Engine," SAE Paper No. 880517.

Crawford, J.G and Wallace, J.S. (1996), "Validation Tests for a Fast Response Flame Ionisation Detector for In-Cylinder Sampling Near the Spark Plug," SAE Paper No. 961201.

Fox, J.W., Min, K.D., and Heywood, J.B. (1992), "Mixture Preparation in a SI Engine with Port Fuel Injection during Starting and Warm-Up," SAE Paper No. 922170.

Galliot, F., Cheng, W.K., Cheng, C.O., Sztenderowicz, C., and Heywood, J.B. (1990), "In-Cylinder Measurements of Residual Gas Concentration in a Spark Ignition Engine," SAE Paper No. 900485.

Ladommatos, N. (1992a), "Cyclically Resolved Measurements of Hydrocarbons in the Cylinders of Internal-Combustion Engines by Means of a Fast Flame-Ionisation Detector," *J. Inst. Energy*, Vol. LXV, pp. 94–101.

Ladommatos, N. (1992b), "Investigation of Gasoline Films Accumulating in the Inlet Ports of Spark-Ignition Engines," *J. Inst. Energy*, Vol. LXV, pp. 102–104.

Ladommatos, N. and Rose, D. (1996), "On the Causes of In-Cylinder Air Fuel Ratio Excursions during Load and Fuelling Transients in Port-Injected Spark-Ignition Engines," SAE Paper No. 960466.

Ladommatos, N. and Rose, D. (1995), " Monitoring of the Mixture Strength in the Cylinders of a Port-Injected Gasoline Engine during Transient Operation," *Proc. Instn. Mech. Engrs.*, Vol. 209, Part D, pp. 127–134.

Li, C.F., Dual, Q.C., Maroteaux, F., and Murat, M. (1994a), "Time-Resolved Measurement of Fuel Transient Behaviour and Cycle to Cycle Variation of Local Fuel/Air Equivalence Ratio in Gasoline Engines," SAE Paper No. 940989.

Li, C.F., Murat, M., Jullien, J., Zhu and M.J. (1994b), "A Fast-Response Catalytic Hot-Wire Sonic Nozzle Probe to Measure Fuel Concentration in a Fuel/Air Mixture," *Measurement Science and Technology*, 5, pp. 151–156.

Min, K., Cheng, W.K., and Heywood, J.B. (1994), "The Effects of Crevices on the Engine-Out Hydrocarbon Emissions in SI Engines," SAE Paper No. 940306.

Meyer, R.C. and Thring, R.H. (1995), "Mixture Preparation Measurements," SAE Paper No. 950069.

Peckham, M. and Collings, N. (1992), "Study of Engine Wall Layer Hydrocarbons with a Fast-Response FID," SAE Paper No. 922237.

Peckham, M. and Collings, N. (1994), "'Static' and 'Flying' FID Sampling of In-Cylinder Crevice Hydrocarbons," I. Mech. E. Seminar "Measurement and Observation Analysis of Combustion in Engines," London.

Peckham, M.S. (1993), Study of Engine Wall Layer Hydrocarbons with a Fast Response FID, PhD Thesis, Cambridge.

Rose, D., Ladommatos, N. and Stone, R. (1994), " In-Cylinder Mixture Excursions in a Port-Injected Engine Using Fast Throttle Opening," SAE Paper No. 940382.

Shin, Y., Min, K. and Cheng, W.K. (1995), "Visualisation of Mixture Preparation in a Port-Fuel Injection Engine during Warm-Up," SAE Paper No. 952481.

Sleightholme, G.R. (1990), "In-Cylinder Measurements of Charge Inhomogeneity in a Spark-Ignition Engine," SAE Paper No. 900484.

Smith R. (1989), "Loss of Frequency Response along Sampling Tubes for the Measurements of Gaseous Composition at High Temperature and Pressures," *J. Fluid Mechanics*, Vol. 208, pp. 25–43.

Summers, T. and Collings, N. (1995a), "Modelling the Transit Time of a Fast Response Flame Ionisation Detector during In-Cylinder Sampling," SAE Paper No. 950160.

Summers, T. and Collings, N. (1995b), "Signal Reconstruction Applied to a Fast Response Flame Ionisation Detector," SAE Paper No. 952541.

Summers, T. (1996), *Fast-Response FID Measurement of SI Engine Residual Gas Hydrocarbon Concentration*, DPhil Thesis, University of Cambridge.

Taylor, G.I. (1953), "Dispersion of Soluble Matter in Solvent Flowing Slowly through a Tube," *Proceedings of the Royal Society of London*, A, 219, 186–203.

Taylor, G.I. (1954), "The Dispersion of Soluble Matter in Turbulent Flow through a Pipe," *Proceedings of the Royal Society of London*, A, 223, 446–468.

White, F.M. (1974), *Viscous Fluid Flow*, McGraw-Hill, 1st Ed.

Chapter 4

Fast-Acting In-Cylinder Sampling Valves and the Total Cylinder Dumping Method

Outline

4.1 Introduction

Over the years, fast-acting sampling valves have been developed which can obtain small gas samples from the combustion chamber of an operating engine at any desired crank angle, once every engine cycle. The development of these sampling valves has been prompted mostly by the desire to gain greater understanding of the chemical and physical nature of combustion and pollutant formation processes in internal combustion engines.

The requirements of an in-cylinder sampling valve can be summarized as follows:

(i) Sample at specific times of the engine cycle, which should be capable of being selected.
(ii) Sample from different locations in the combustion chamber.
(iii) Gas leakage through valve seat should be minimized to reduce sample contamination by leakage gases during the periods when sample valve is closed. For example, a small leakage of unburned mixture during the compression stroke can have an overwhelming effect on the study of unburned hydrocarbons left after combustion.
(iv) The sampling duration should be small (if possible <1 ms), especially during parts of the engine cycle when the cylinder gas composition is changing very rapidly.
(v) The fraction of the gas sample coming from crevices and wall quench layers should be as small as possible relative to the sample volume.
(vi) The sample entering the valve should be quenched as rapidly as possible so as to arrest chemical reactions in the sample.
(vii) Valve surfaces in contact with the high-temperature gas sample should be made from materials that are non-catalytic to the sample gas.
(viii) The sampling valve should operate with high repeatability over a long life span, and should be easily constructed, assembled, adjusted, and serviced.
(ix) The timing of the valve opening and the sampling duration should be easy to control.

As early as 1927, Lovell et al. used a poppet sampling valve, actuated mechanically by the camshaft through a pushrod and a rocker to sample combustion chamber end gases from a gasoline engine operated under knocking and non-knocking conditions. Mechanically operated valves were also used by other investigators in both gasoline and diesel engines until 1938, when Bartholomew et al. reported the use of an electromagnetically operated sampling valve to obtain unburned mixture from a gasoline engine installed in a vehicle. Thereafter, electromagnetic actuation prevailed, although hydraulic and electro-hydraulic actuation has been used in a significant number of sampling valves. Tables 4.1 to 4.4 summarize the salient features of sampling valves used for research on diesel and gasoline engines since 1927. The earlier uses of sampling valves were mainly concerned with understanding the phenomenon of knock and mixture preparation in gasoline engines, as well as the process of compression ignition in diesel engines. Subsequently, sampling valves were used widely in both gasoline and diesel engines to improve understanding of the combustion process. In recent years, sampling valves have been used mainly in research aimed at reducing pollutant emissions from both gasoline and diesel engines.

4.2 Brief Summary of Sampling Valves

A number of commonly used sampling valve concepts are shown in Fig. 4.1. Fig. 4.1(a) shows a poppet-type valve (outward-opening). With no current supply to the electromagnet, the poppet is seated firmly against its seat due to the force provided by the compressed return spring. The gas pressure in the engine cylinder acting on the poppet surface provides an additional closing force, which helps keep the poppet valve shut. When the current supply to the electromagnet is switched on, the armature is attracted by the electromagnet. The movement of the armature

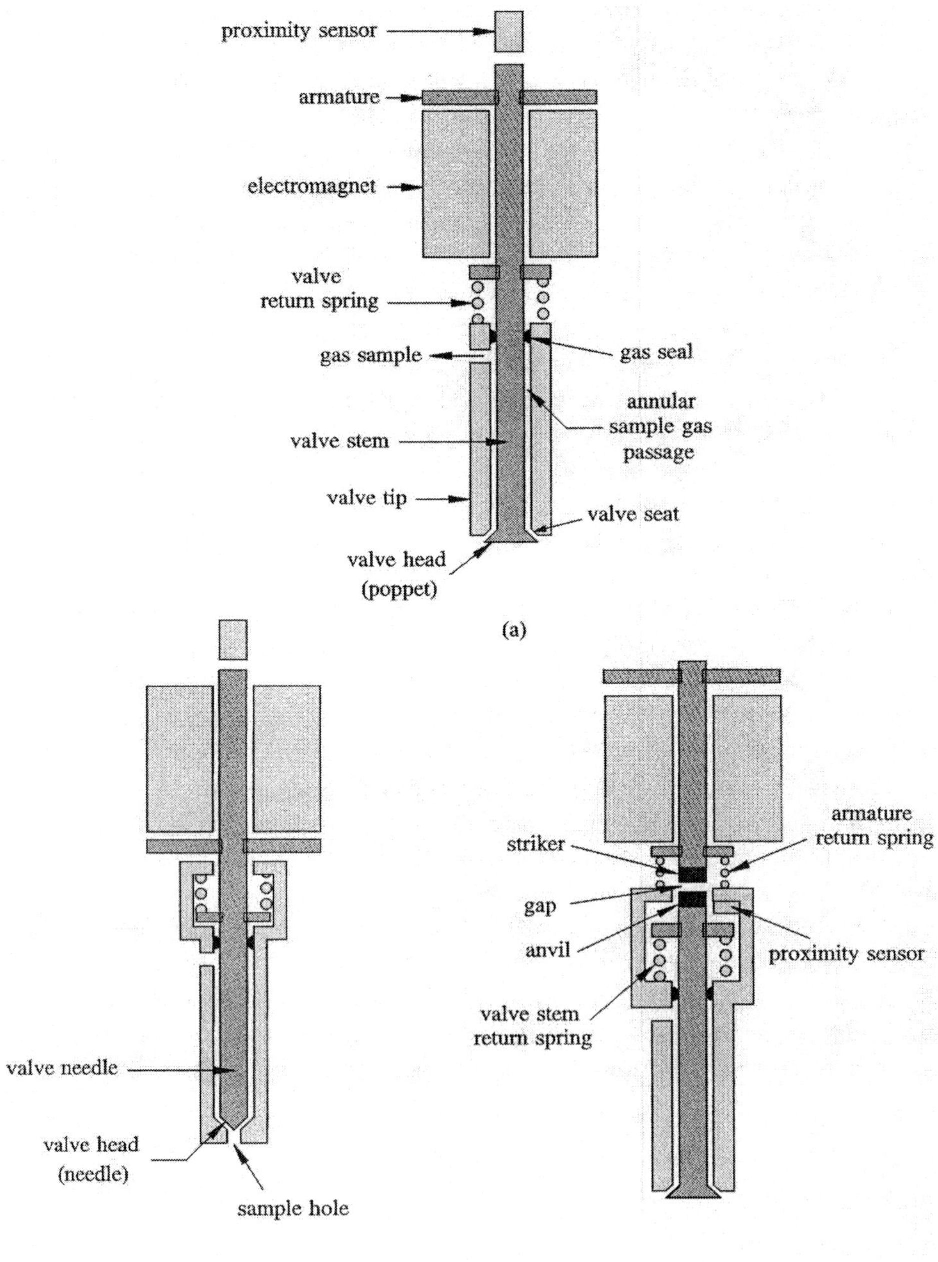

Fig. 4.1 Design concepts for sampling valves.

Table 4.1 Operating parameters for various sampling valves used in spark ignition engines

Sampling period ms	V_{sample}/ cycle at STP cm^3	Valve lift mm	Application	Reference
1.65	~1		4-stroke, 1200 rpm, fuel gasoline, sampled end gas CO, CO_2, O_2, H_2, HC	Lovell et al., 1927
0.65	~1		4-stroke, 500–1200rpm, fuel gasoline, sampled O_2 and CO_2	Withrow et al., 1930
0.7–2.5	0.27	0.25	4-stroke, 600 rpm, fuel gasoline and benzene, sampled end gas CO, CO_2, O_2, aldehydes, peroxides	Egerton et al., 1935
			4-stroke engine installed in car, distribution of gasoline/air mixture distribution between the 8 cylinders	Bartholomew et al., 1938
18.6	7.5		4-stroke, 1000rpm, sampling during compression stroke fuel/air mixture distribution between 6 cylinders	Donahue et al., 1950
>0.33	0.27		4-stroke, 1500 rpm, fuel mainly isooctanes and benzenes, sampled end gas aldehydes and peroxides	Downs et al., 1951
2–3	25–60		2-stroke, up to 9000 rpm, sampling CO, CO_2, and O_2 once per 100 or 200 cycles	Asanuma and Yanagihara, 1962
0.5–1.5		0.25	4-stroke, 900 rpm, isooctane/DIB, sampled end gas with detailed analysis	Alperstein and Bradow, 1965
1.4–3	5–6		1275 rpm, isooctane, sampled CO, CO_2, NO_x, HC	Zvonow et al., 1968
1.3–2.5			2000 rpm, fuel gasoline, valve built into piston, sampling top land crevice	Furuhama and Tateishi, 1972
			Multiple end-gas sampling in one cycle at every 5 deg. at 800 rpm	Luck et al., 1973
1.5–2	1–1.2		2- and 4-stroke, 1400–2500 rpm, high load, fuel gasoline, sampling AFR once per 10 or 20 cycles, He carrier gas	Ayusawa et al., 1978
4	0.4	0.5	1200–1300 rpm, fuels propane and gasoline, valve built into spark plug	Matsui et al., 1979
<1		<0.1	700–1900 rpm, 3.4–6.6 bar bmep, fuel propane, sampling quench layer HC	LoRusso et al., 1984
~2			1000 rpm, low-full load, fuel isooctane, sampling quench layer HC	Weiss and Keck, 1981
>1			500 rpm, fuel propane, measured AFR, HC	Johnston and Green, 1982
2.5	~5.5		4-stroke, 1800 rpm, fuel gasoline, sampling air fuel mixture 1 per 10 cycles	Ozasa et al., 1991

Table 4.2 Construction details for various sampling valves used in spark ignition engines

Actuator	Valve type	Armature And valve stem	Tip dia.	Seat		Sample hole dia.	Materials			Reference
			mm	dia. mm	angle deg.	mm	seat	valve head	body	
Mechanical	poppet	linked	22 CL	~8						Lovell et al., 1927
Mechanical	poppet	linked	22 CL							Withrow et al., 1930
Mechanical	poppet	not linked	22 CL	~10						Egerton et al., 1935
Electro-magnetic	needle	linked								Bartholomew et al., 1938
Mechanical	poppet	linked								Donahue et al., 1950
Electro-mechanical	needle	linked	18	~3.2	45	0.64	CI	SS		Downs et al., 1951
Electro-magnetic	needle poppet		14			>5				Asanuma and Yanagihara, 1962
Electro-magnetic	needle	Linked	14		0		Q	G	SS	Alperstein and Bradow, 1965
Hydraulic (diesel pump)	poppet		18	~10	45		T	I	SS	Zvonow et al., 1968
Electro-magnetic	needle		18			0.3				Furuhama and Tateishi, 1972
Mechanical	slide					0.23				Luck et al., 973
Electro-magnetic	needle	linked	6			2-2.5				Ayusawa et al., 1978
Electro-magnetic	needle	linked				0.8				Matsui et al., 1979
Electro-hydraulic	poppet	linked		6.4	45		HS	HS	HS	LoRusso et al., 1980
Electro-magnetic	needle	linked				2	C			Weiss and Keck, 1981
Electro-magnetic	needle		6.4			1.4	C	SS		Johnston and Green, 1982
Electro-magnetic	needle		HT					SS		Ozasa et al., 1991

C = copper; CI = cast iron; G = 24K gold; HS = high-strength steel; I = inconel; SS = stainless steel;
T = tungsten; Q = quartz; CL = cooled tip; HT = Heated tip; seat diameter given at seat top;
tip diameters of 22, 18, and 14 mm correspond to standard spark plug threads

Table 4.3 Operating parameters for various sampling valves used in diesel engines

Sampling period ms	Sampled volume/cycle at atmos. press. & temp. cm^3	Valve lift mm	Application	Reference
0.3–1.1	0.13	0.05	IDI, 600 rpm, low- to high-load, aldehydes, CO, CO_2, HC, O_2	Drinkwater and Egerton, 1938
0.3	0.05	0.01	IDI, 1200 rpm, pre- and post-ignition aldehydes, peroxides, CO, CO_2, NO_2, HC, O_2	Garner et al., 1952
~2		0.1–0.25	IDI and DI, 1600 rpm, AFR, CO, CO_2, NO, HC_3	Bennethum et al., 1975
1–3	>4		DI, 300 mm bore, 500 rpm, 3.5–7 bar bmep, up to 115 mm tip penetration, CO, CO_2, NO, HC	Whitehouse et al., 1977
0.8–1.6			IDI, 1000 rpm, 5.5 bar bmep, up to 18.5 mm tip penetration, CO, CO_2, NO, HC, O_2	Duggal et al., 1978
1			DI, 1250 rpm, mid-load, CO, CO_2, NO_x, HC, O_2, soot	Aoyagi et al., 1980
1.4			1200 rpm, 4.8 bar bmep, CO, CO_2, O_2, soot density, particle size	Fujiwara et al., 1984
1.2–1.8			DI, 1440 rpm, 8.25bar bmep, particulates, carrier gas	Norris-Jones et al., 1984
1			DI, 1200 rpm, mid-load, CO, CO_2, NO_x, HC, soot	Barbella et al., 1989
>0.8		0–0.4	DI and IDI, 2000-2500 rpm, 1–10 bmep, CO, CO_2, NO_x, HC, soot, N_2 carrier gas	Lepperhoff and Houben, 1990; Lepperhoff, 1990; Wolfe et al., 1993; Backer et al., 1997
>1	2–20		DI, 2000 rpm, <50% load, NO_x, O_2, soot density, particle size	Ning et al., 1991
0.8–3	>1	0–0.4	DI, 2000 rpm, 1–8 bar bmep, CO, CO_2, NO_x, HC, O_2, soot density, particle size	Ladommatos et al., 1996; Zhao et al., 1996
1			2-stroke, 1000 rpm, CO, CO_2, NO_x, O_2, H_2, HC, soot, He and N_2 carrier gas	Shimazaki et al., 1996
1.7	5		DI, 1000 rpm, 0.9 bar bmep, particulates, air dilution gas	Kato et al., 1997

Table 4.4 Construction details for various sampling valves used in diesel engines

Actuator	Valve type	Armature and valve stem	Tip dia.	Seat		Sample hole dia.	Materials			Reference
			mm	dia. mm	angle deg.	mm	seat	valve head	body	
Mechanical	poppet	not linked	~13	~9						Drinkwater and Egerton, 1938
Hydraulic DP	poppet	linked	CL				SSS	SSS		Gamer et al., 1952
Electro-magnetic	needle	linked	5.7		45	1.25	TI			Bennethum et al., 1975
Electro-magnetic	poppet	not linked	~12 CL	~6					SS	Whitehouse et al., 1977
Hydraulic DP	poppet	not linked						SS	SS	Duggal et al., 1978
Electro-magnetic	needle	linked	6			1 and 1.5		SS		Aoyagi, 1980
Electro-magnetic	needle					1.5				Fujiwara et al. 1984
Electro-magnetic	needle	linked								Norris-Jones et al., 1984
Electro-magnetic	needle	linked								Barbella et al., 1989
Electro-hydraulic	poppet		6 HT							Lepperhoff and Houben, 1990; Lepperhoff, 1990; Wolfe et al., 1993; Bäcker et al., 1997
	needle									Ning et al., 1991
Electro-magnetic	poppet	not linked	8	6	45		C	HS	HS	Ladommatos et al., 1996; Zhao et al., 1996
Electro-magnetic	needle		CL			0.5			SS	Shimazaki et al., 1996
Electro-magnetic	needle	linked								Kato et al., 1997

C = copper; HS = high-strength steel; SS = stainless steel; SSS = stainless steel faced with stellite; TI = titanium; CL = tip cooled; HT = tip heated to 350°C; DP = driven by diesel pump opens the poppet valve, and this allows the gas sample to flow past the valve seat and out of the valve as shown. When the current supply to the electromagnet is then interrupted, the return spring, assisted by the gas pressure in the cylinder, close the valve.

Fig. 4.1(b) shows a needle-type valve (inward-opening). The operation of this valve is similar to that of the poppet-type valve described above, except for the following significant difference. The gas pressure in the engine cylinder acts to open the valve. Thus, the spring return force must be sufficiently high to counteract this gas pressure force, and ensure that a good seal is obtained when the valve is shut.

Fig. 4.1(c) shows a "percussion" type of sampling valve. Note that in the percussion valve, the armature is not rigidly joined to the stem of the poppet valve. Instead, a small gap of a few millimeters exists between the spindle of the armature and the stem of the poppet. When the current is switched on, the armature together with the striker are accelerated. After traveling across the gap, the striker impacts the anvil at the end of the poppet stem, and the poppet opens for a brief period. The poppet is shut by a strong compression spring, and the armature together with the striker are returned by a second spring. An important feature of the percussion valve operation is that the electromagnet can be switched on for much longer than the sampling period. Although the percussion principle is applied to a poppet-type valve in Fig. 4.1(c), it could be applied equally successfully to a needle valve. In fact, Drinkwater and Egerton used the percussion principle in a mechanically operated sampling valve as far back as 1927 (see Tables 4.3 and 4.4).

In the case of all valve concepts shown in Fig. 4.1, a proximity sensor will normally be required in order to monitor the movement of the valve stem and head.

Sometimes liquid or gaseous coolant is channelled through internal passages (not shown in Fig. 4.1) to the valve tip in order to ensure reliable and durable operation; also, in order to ensure that any chemical reactions taking place in the sample gas entering the valve are "frozen" (stopped).

Fig. 4.2(a) shows the practical realization of an electromagnetic needle valve used by Bennethum et al. [1975] for sampling combustion gases from indirect injection (IDI) and direct injection (DI) diesel engines. Fig. 4.2(b) shows an electro-hydraulically operated poppet valve used by LoRusso et al. [1980; 1984] to investigate unburned hydrocarbons (HC) in the quench layer close to the combustion chamber wall. In the case of the electro-hydraulically actuated valve, the force needed to open the valve is provided by hydraulic pressure acting on a piston manufactured on the stem of the valve. The supply of high-pressure hydraulic oil to this piston is controlled by an electromagnetically driven spool valve. When the high-pressure oil supply to the valve stem piston is interrupted by the spool valve, the valve poppet is shut by the combined action of the return spring and the gas pressure in the engine cylinder.

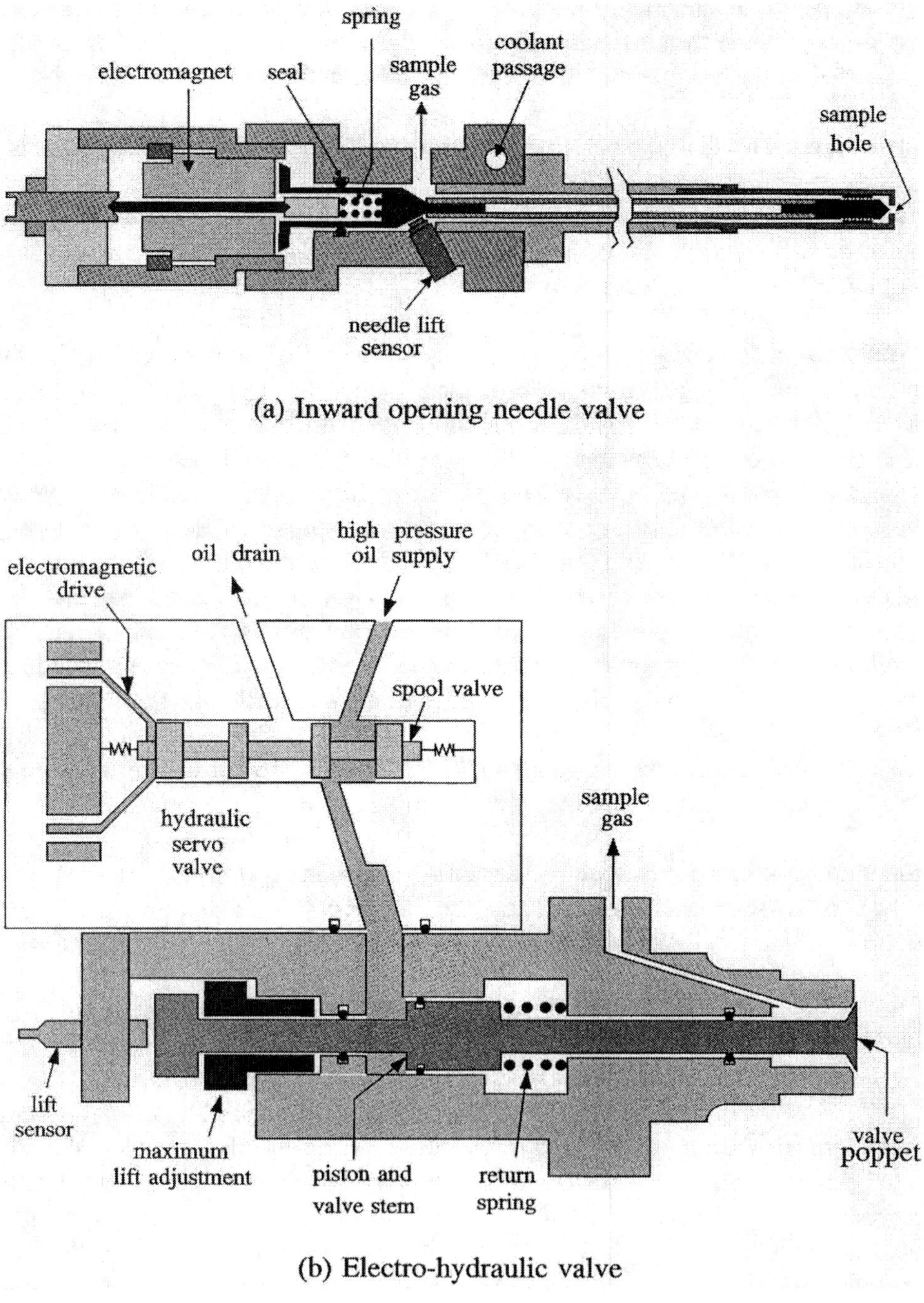

Fig. 4.2 Examples of practical sampling valve designs.

4.3 General Design Considerations

4.3.1 Materials for Valve Components

The most common material for sampling valve construction has been stainless steel, particularly for valve parts in contact with combustion gases. Stainless steel has good corrosion and thermal resistance. Mild steel and high-strength steel have also been used successfully [Ladommatos et al., 1996], although the service life of the valve may be reduced under severe operating conditions involving high-temperature corrosive gases. Whitehouse et al. [1977] experienced erosion problems in the water cooling channels around the tip of a sampling valve used in a diesel engine. The channels, which were made of valve steel (BSS EN52), were severely eroded, possibly due to either cavitation induced by vibration or electrochemical corrosion. The valve tip was redesigned using stainless steel (AISI 316).

Brazed joints exposed to the hot sample gases may have a limited life span due to corrosion attack by oxides of nitrogen.

4.3.2 Choice of Valve Type

4.3.2.1 Inward and Outward-Opening Valves

For sealing the sampling valve, the choice lies mainly between outward-opening poppet and inward-opening needle valves.

Minimization of crevices around the valve inlet is necessary, particularly when the amount of gas sample per engine cycle is small. Provided the poppet can be machined flush with the combustion chamber surface, no crevice will be left. Needle valves will always have a small crevice, although this can be minimized.

In the case of the poppet valve, sealing is assisted by gas pressure acting on the poppet, while the opposite is true in the case of the needle valve.

Needle valves can have smaller tip diameters, and this is a significant advantage in many applications where access to the combustion chamber can be difficult. The smaller tip diameter also reduces the disturbance to the gas flows and the combustion process in the engine cylinder.

Bennethum et al. [1975] found that poppet valves are more prone to leakage than needle valves, perhaps due to greater susceptibility to contamination.

The sample shape within the engine cylinder with needle valves tends to be roughly hemispherical while that for poppet valves tends to be roughly toroidal. For a given sample volume, the maximum reach of the sample will tend to be somewhat greater with the needle valve.

4.3.2.2 Rotary Valve

Rhee et al. [1978] employed a radically different valve opening technique. The tip of the sampling valve is shown in Fig. 4.3. The tip consists of a tubular outer casing of about 6.4 mm in diameter, blanked off from the engine cylinder, except for an eccentrically located 0.4 mm diameter hole. Inside the outer casing is located, concentrically, an oscillating tube, which is able to rotate back and forth though part of one complete revolution. The motion of this oscillating tube is synchronized with the engine crankshaft by means of a hydraulic actuator and control system. This oscillating tube has attached on its inner rim an eccentrically located sample tube of 2.5 mm diameter. During each engine cycle, the open end of the sample tube is moved circumferentially by the oscillating tube, so that the open end of the sample tube aligns briefly (>0.5 ms) with the sample hole. Thus, a small amount of gas sample flows from the engine cylinder, via the sample hole, into the sample tube and out of the valve for collection and analysis. When the sample tube is not aligned with the sample hole, the cylinder gas passes through the sample hole into the oscillating tube, from where it flows to waste. Grooves cut on the outside rim of the oscillating tube carry helium gas which helps cool the valve tip.

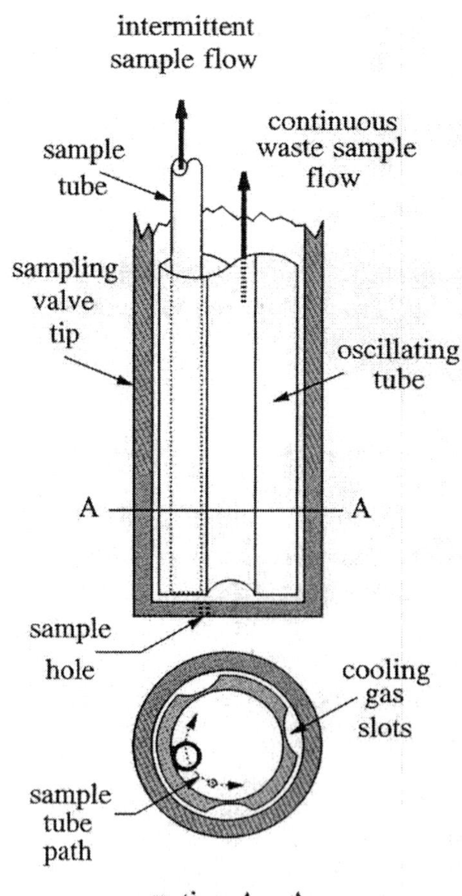

Fig. 4.3 Oscillating-tube sampling valve tip.

Motion and synchronization of the oscillating tube is significantly more complicated than that for electromagnetically actuated inward- and outward-opening valves. The oscillating tube has to rotate freely within the outer casing of the valve tip. Sealing of the sample tube, when not taking a sample, is necessary, but relatively easy since downstream of the sample hole the system is not pressurized. A major advantage of this system is that there is a constant sample flow through the sample hole, even when sample is not being taken by the sampling tube. Thus, the influence of the boundary layer on the sample taken may be smaller than that for poppet and inward-opening valves [Rhee et al., 1978].

4.3.3 Actuation

The moving parts of the valve have to be accelerated within a very short period of time during each valve opening. The weight of the moving parts must therefore be kept to a minimum. This reduces the force required to open and close the valve and improves control over the opening of the valve. For example, Nightingale [1975] constructed the valve stem out of thin tubing in order to reduce weight.

4.3.3.1 Electromagnetic Actuation

The main requirements of an electromagnet for sampling valve applications are, first, a high traction force and, second, high armature acceleration.

The electromagnet traction force must be high in order to overcome the force of the return spring keeping the valve closed. Usually, the return spring will be a strong one, in order to ensure good sealing of the valve when closed.

When the electromagnet is energized, the acceleration of the armature is required to be high in many valve designs, so that the valve can be opened as rapidly as possible.

It might be thought that an electromagnet with a high traction force would also have a higher armature acceleration. In practice, this may not always be the case, because a larger traction force can be accompanied by a disproportionately heavier armature [Seilly, 1981].

4.3.3.1.1 Increasing the Electromagnet Traction Force

Fig. 4.4 shows a conventional electromagnet. The traction force is generated between the traction faces and the armature shown in the figure. For a given traction face area, the traction force can be improved by increasing the magnetic flux. However, in practice, the maximum flux is limited by magnetic saturation of the electromagnet material [Seilly, 1981]. Therefore, if a greater traction force is required, usually then the traction face area of the electromagnet and armature must be increased, as demonstrated below.

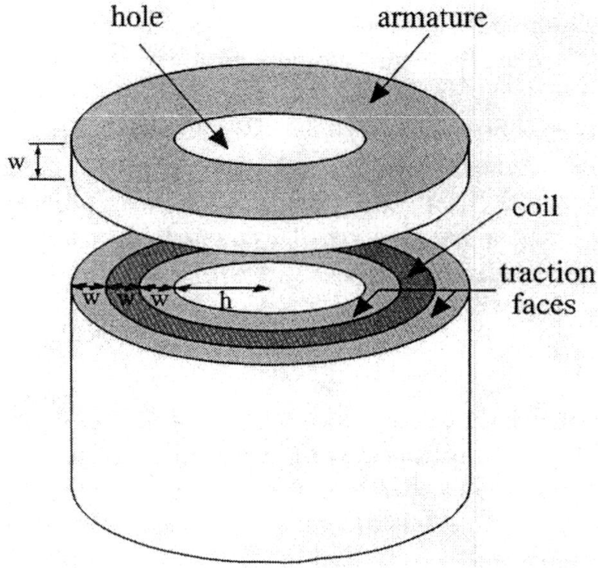

Fig. 4.4 General structure of conventional electromagnet.

Let f_a = armature acceleration
F_a = electromagnetic force on the armature
w_a = width of magnetic flux path (equal to thickness of the armature)
h_a = radius of hole at center of electromagnet and armature
ρ_a = density of armature material
k_e = constant
m_a = armature mass

From the discussion above, for a given electromagnetic flux, the electromagnet traction force on the armature,

$$F_a \propto \text{total traction face area} \propto \pi(4h_a w_a + 6w_a^2)$$

and,

$$F_a = k_e \pi\left(4h_a w_a + 6w_a^2\right)$$

Therefore, the traction force can be increased by either increasing the width of the flux path, w_a, while keeping the hole radius, h_a, constant, or vice versa by increasing the hole radius, h_a, while keeping the width of the flux path, w_a, constant.

The mass of the armature, m_a, is given by:

$$m_a = \rho_a \pi \left(6h_a w_a^2 + 9w_a^3\right)$$

Therefore, the armature acceleration, f_a, is given by:

$$f_a = \frac{F_a}{m_a} = \frac{2k_e}{3\rho_a w_a} \qquad (4.1)$$

Consider, again, ways in which the electromagnet force, F_a, can be increased:

(i) Increasing the width of the flux path will increase the electromagnet force, but this will be at the expense of lower armature acceleration.
(ii) Increasing the radius of the central hole will increase electromagnet force, F_a, and this will not affect acceleration of the armature; this is so, because the above expression for acceleration, f_a, is independent of hole radius, h_a.

It can therefore be seen that the latter is a more favorable option when more traction force is required. However, in practice, this option may also result in a less compact design for the electromagnet. Often, practical electromagnets are designed with $h_a = 0$, and as their size increases the motion of their armature becomes slower.

4.3.3.1.2 Increasing the Armature Acceleration

In the above example, it is assumed that thickness of the armature will be equal to the width of the flux path, w_a, in order to provide an unrestricted flux path. In practice, as the electromagnet is operated transiently, the flux may not penetrate deeply, and the thickness of the armature could be reduced without significant loss of performance [Kajima and Kawamura, 1995]. This would reduce the mass of the armature and help improve its acceleration. Also, construction of the armature from a light material with good magnetic properties would also help reduce mass.

Another way of improving armature acceleration is to improve the build-up of the magnetic flux. In order to assist the rapid build-up of magnetic flux, eddy currents in the armature need to be minimized. Laminating the armature may be impractical, but the alternative technique suggested by Ohdaci et al. [1991] of slitting the armature may be possible.

Kajima and Kawamura [1995] have established that reducing the number of winding turns on the electromagnet coil will provide a faster current response due to reduced self-inductance. The rate at which the magnetic field can be established is, however, still limited by the supply voltage. Voltage control has been used by Ohdaci et al. [1991] to provide rapid initial flux growth, and they report a corresponding improvement in device response. The voltage used, however, is ultimately limited by the dielectric strength of the winding insulation. Since the duty cycle for the electromagnet coil is short, it is possible to operate the coil with currents considerably above the continuous rating, and thereby increase the acceleration of the armature.

Finally, Seilly [1981] describes the development of helically and conically shaped electromagnets which enable large traction forces to be obtained without significant compromise in armature acceleration. Pull-in forces as high as 275 N have been achieved by such specialized helenoid and colenoid electromagnets [Yates, 1988; Seilly, 1979, 1981].

4.3.3.2 Hydraulic Actuation

An important characteristic of hydraulic actuators is that they can provide high forces while taking up a relatively small space. Furthermore, electro-hydraulically operated actuators can also provide good controllability and short response time [LoRusso et al., 1980, 1983, 1984; Adamczyk et al., 1983]. The ability of hydraulic systems to provide large opening forces allows high spring forces to be used in sampling valves. The high spring forces help to minimize leakage rates through the valve seat when the sampling valve is closed [LoRusso et al., 1980, 1983]. However, the control system for electro-hydraulically actuated sampling valves generally tends to be more complicated than that for electromagnetically actuated valves.

An example of an electro-hydraulically operated sampling valve is shown in Fig. 4.2(b). This valve was used by LoRusso et al. [1980, 1984] to sample the cylinder gases very close to the cylinder wall of a gasoline engine. The valve comprises two main components: the hydraulic servo system which controls the supply of high-pressure oil to the sampling valve; and the sampling valve itself. To open the sampling valve, current is applied to the electromagnetic drive which, in turn, shifts the hydraulic spool valve to the open position. This allows high-pressure oil to flow into the sampling valve. The oil, acting on the valve stem piston, forces the poppet of the sampling valve to open. In order to close the sampling valve poppet, current of reverse polarity is applied to the electromagnetic drive which now shifts the hydraulic spool valve in the opposite direction. The high-pressure oil in the sampling valve exhausts itself out to the oil drain, and this removes the oil pressure force holding the valve poppet open. The return spring forces the sampling valve poppet shut.

A number of investigators used hydraulic actuation in a somewhat unconventional but effective way to operate their sampling valves. For example, Duggal et al. [1978] used a diesel injector to open the poppet of their sampling valve. The diesel injector was positioned upside down and back-to-back with the sampling valve. The injector was operated by means of an auxiliary diesel fuel pump driven by the engine at half the crankshaft speed. When the injector opened, its needle pushed the sample valve stem so that the sampling valve poppet opened. The sampling timing was selected by adjusting the phasing between the diesel pump shaft and the engine drive shaft. The sampling period was varied by adjusting the injection period of the diesel injector, and periods of 0.8 to 1.7 ms could be achieved. A similar system was used by Zvonow et al. [1968] with broadly similar results. Lepperhoff and Houben [1990] reported results from a sampling valve also with hydraulic actuation by means of a diesel distribution pump, achieving a minimum sampling period of 0.8 ms and a valve lift up to 0.3 mm [see also Bäcker et al., 1997]. Garner et al. [1952] also used a diesel pump to actuate hydraulically a poppet sampling valve, but with their arrangement the valve lift and, hence, the gas flow rate was severely restricted. A mechanical phasing system similar to one used by Egerton et al. [1935] was interposed between the diesel pump and the engine drive shaft to allow wide variation of the sampling timing.

4.3.3.3 Mechanical Actuation

The first reports describing the use of in-cylinder sampling valves were published in the literature during the 1920s and 1930s. These early sampling valves were mechanically actuated by linkages driven by the engine crankshaft or camshaft. For example, Lovell et al. [1927] used a poppet sampling valve, actuated mechanically by the camshaft through a pushrod and rocker to sample combustion chamber end-gas from a gasoline engine operated under knocking and non-knocking conditions. Withrow et al. [1930] also used a poppet sampling valve, actuated mechanically by a cam driven at half engine speed to sample in-cylinder gases from a gasoline engine. Egerton et al. [1935] used a mechanical system driven by the crank shaft to open and close a poppet sampling valve, with the valve sampling timing, sampling duration, and valve lift all being adjustable. In order to obtain this degree of adjustment, the mechanical actuation system had to be quite complicated. The valve-opening cam was driven through a differential gear system which allowed the valve sampling timing to be altered. The cam acted on the poppet valve stem via a rocker arm.

Donahue and Kent [1950] also used a mechanically actuated sampling system to investigate mixture maldistribution between the six cylinders of an automotive spark ignition engine. Six poppet-type sampling valves were installed, one in each cylinder, and actuated from an auxiliary cam shaft driven at engine half speed. The six valves extracted cylinder mixture during the compression stroke, between 120 and 18°CA before top dead center. Daniel and Wentworth [1962] used a mechanically-actuated poppet valve to sample hydrocarbons within the quench layer close to the combustion chamber wall.

Taylor and Rogowski [1954] used a simple automatic mechanical sampling valve. This was installed in the exhaust port of a two-stroke engine, and was used to investigate scavenging effectiveness. The valve opened automatically by the action of the stagnation pressure of the cylinder gases leaving the exhaust port. Fig. 4.5 shows a diagram of the device and the way it was used. The principles of its operation are as follows. When the exhaust port is uncovered, the stagnation pressure from the escaping cylinder gases opens the sampling valve automatically. The sample flows through the valve and via the sample tube into the sample vessel. This vessel is continuously bled to atmosphere via the bleed valve. The pressure in the sample vessel can be adjusted by varying the opening of the bleed valve. As the blow-down process continues through the exhaust port, the stagnation pressure acting on the sampling valve head to keep it open decreases. Eventually, the balance of the spring and pressure forces on the valve head is such that the valve head closes automatically. The stagnation pressure at which the sample valve closes is chosen by adjusting the sample vessel pressure. In this way, it is arranged for the sampling valve to close automatically before the scavenging process starts (i.e., before fresh air fuel mixture enters the cylinder to help expel the burned gas residuals). Closure of the valve before the scavenging process starts ensures that no fresh air fuel mixture enters the sampling valve. Thus, the sampling valve supplies the sample vessel only with the burned gases residing in the cylinder late in the expansion stroke. The face area A, shown in Fig. 4.5, must be large enough to ensure that sufficient pressure force is developed so as to open the sample valve quickly. The return-spring for the valve head is a weak one, so that the main force tending to close the valve is provided by the sample vessel back pressure. The spring helps to start the build-up of the back pressure in the sample vessel when the engine is first started, and once this has been accomplished, the role of the spring becomes quite unimportant.

4.3.4 Percussion and Linked Valve Systems

4.3.4.1 Overview

Most sampling valves reported in the literature employed electromagnetic actuation. In most cases, the armature was linked rigidly to the valve stem, as illustrated in Figs. 4.1(a) and (b), and in Fig. 4.2(a). In this type of design, the electromagnet is normally switched on prior to the valve opening and the armature must undergo high acceleration, so that the valve can open very rapidly. Likewise, when the electromagnet is switched off, the armature must undergo high acceleration in the reverse direction, under the action of the valve return spring, so that the valve closes very rapidly. High armature acceleration is therefore essential when the armature is linked to the valve stem.

A number of successful valve designs have been based on the "percussion" principle, whereby the armature is not linked with the valve stem [Whitehouse et al., 1977; Yates, 1988; Ladommatos et al., 1996; Zhao et al., 1996]. This valve design is illustrated in Fig. 4.1(c). In this design, the striker is accelerated, while moving over the gap shown. The striker then strikes the anvil causing the valve to open. Because the sampling valve is open for only a tiny fraction of the engine cycle, there is ample time for the electromagnet to be switched on well before the valve is required to open. The armature can gather speed gradually, while the striker moves across the gap. This type of valve design allows a substantial amount of kinetic energy to be built up in the moving armature, before the striker strikes the anvil. Thus, a strong valve spring force can be overcome with a relatively small electromagnet size.

Egerton et al. [1935] used the percussion principle in conjunction with a mechanical actuation system. A heavy rocker/pushrod combination was accelerated, causing a striker to travel over a gap of about 1.25 mm and strike a blow to the rear of the sampling valve stem. This system operated with high repeatability and reliability, allowing a sampling period as short as 0.7 ms to be achieved on a spark ignition engine operating at 600 rpm.

Drinkwater et al. [1938] also used a percussion system in conjunction with mechanical actuation, and this is illustrated in Fig. 4.6. The cam disc having a spirally-shaped perimeter displaces the striker away from the anvil. Once each cam revolution, the rocker is released and accelerated by the main spring. The striker strikes the anvil and causes the valve to open for a brief period.

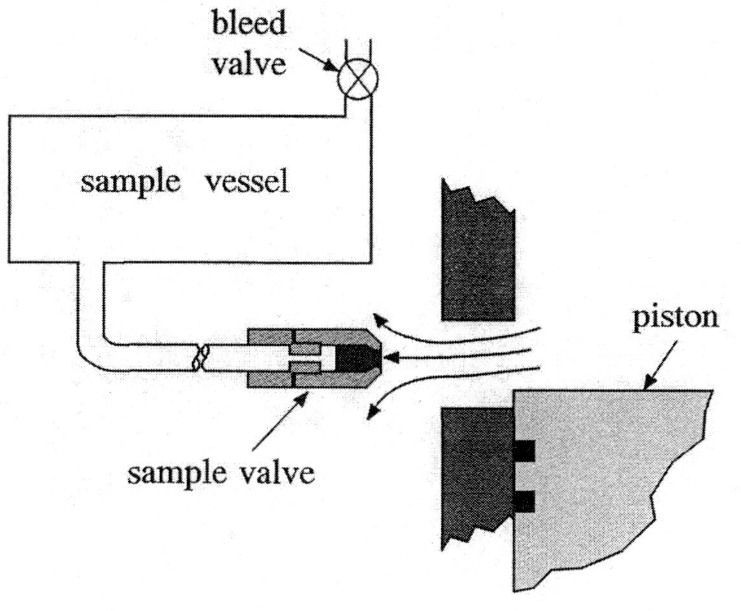

(a) sampling arrangements

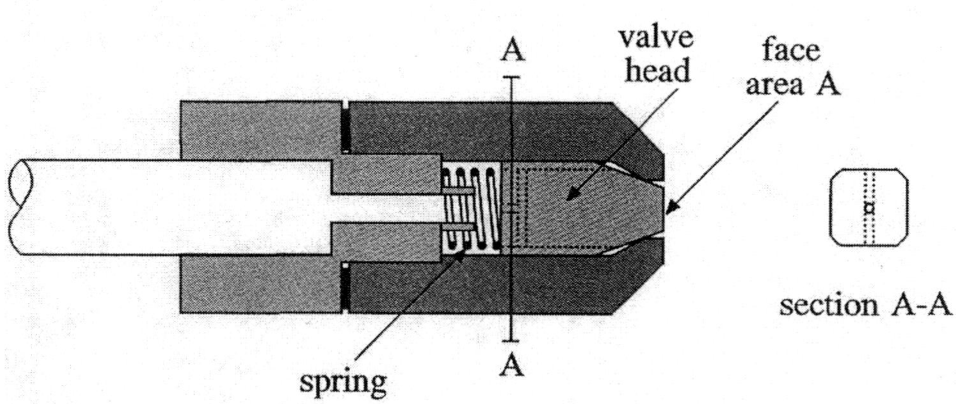

(b) details of the automatic sampling valve

Fig. 4.5 Sampling valve for two-stroke engine scavenging studies.

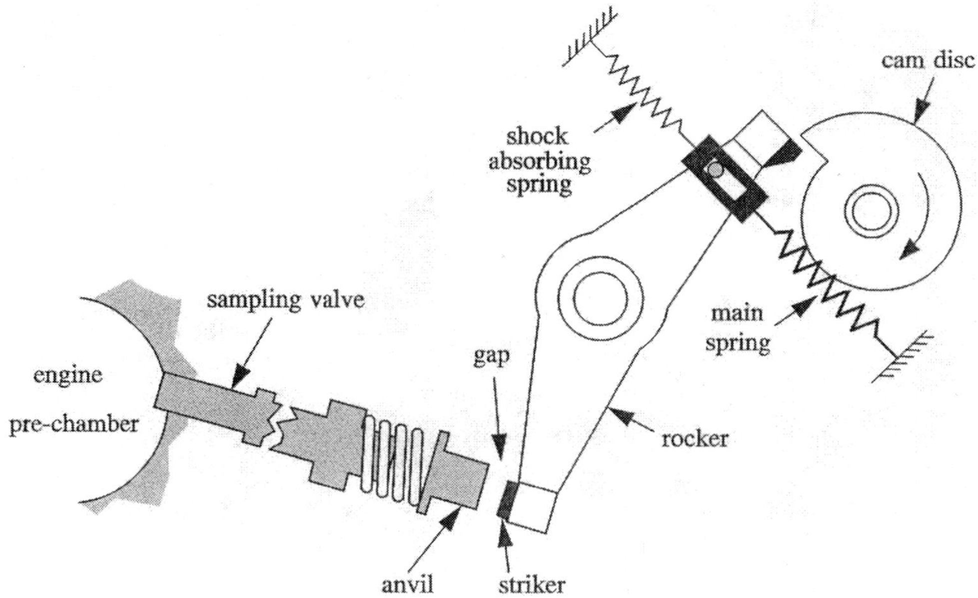

Fig. 4.6 Mechanically actuated sampling valve.

4.3.4.2 Dynamic Response of a Percussion System

Fig. 4.1(c) illustrates, schematically, a sampling valve employing the percussion principle. Ideally, when the striker strikes the anvil, it should deliver only sufficient momentum to the valve stem to achieve momentary opening of the valve, but not interfere further with the motion of the valve stem. In practice this would require that the striker vacates the volume swept by the anvil before the anvil returns, thereby avoiding a second collision between the striker and the anvil, which would only delay the valve closing. This aspect of the valve operation was considered in more detail by Zhao et al. [1996]. It was shown that a second collision between the striker and the returning anvil is very difficult to achieve in practice for the following reasons:

(i) First, the natural frequency of the valve stem/return spring mechanism is usually greater than that of the armature/return spring mechanism. Values of 61 Hz and 11 Hz, respectively, were calculated by Zhao et al. [1996] for their sampling valve. The former mechanism requires a high spring constant in order to ensure adequate sealing of the valve in the closed position. In contrast, the spring constant for the armature/return spring mechanism is kept low in order not to create excessive load for the electromagnet. In an attempt to raise the natural frequency of the latter mechanism, the latter spring constant could be increased and the rating of the electromagnet could also be increased to provide the required force. However, the benefit would be small, since the resulting mass of the armature will become relatively large to provide the volume of ferromagnetic material for the electromagnetic operation.

(ii) Second, the valve closure is accelerated by the high gas pressure in the combustion chamber. This is demonstrated by the shorter sampling times achievable when the engine piston is near top dead center, when cylinder gas pressures are high.

4.4 Electronic Control and Monitoring of Valve Operation

A sampling valve user will normally be interested in controlling the sampling valve as follows:

(i) Choose the instant during the engine cycle when the valve should open.
(ii) Choose the period for which the valve remains open.

The sampling valve control unit must therefore perform the following functions:

(i) Provide a trigger signal at some reference engine angle so that the valve opening can be synchronized with the engine cycle.
(ii) Provide a way of delaying the opening of the valve, after the trigger signal, so that the crank angle at which the valve opens can be chosen.
(iii) Provide a way of controlling the duration for which the valve remains open.
(iv) Monitor the timing of the valve opening and the period the valve remains open.

Often a sampling valve can be controlled using an engine shaft encoder. A convenient trigger signal could be the once-per-cycle signal given out by the shaft encoder at, say, bottom dead center (start of compression). Pulse counting of the individual crank angle pulses will provide control of the time of opening, this timing being independent of engine speed.

Fig. 4.7 shows a relatively simple digital control system for controlling and monitoring the valve operation. The system is shown only in outline and can be adapted to suit particular user requirements. The bottom part of Fig. 4.7 (enclosed in the bottom broken-line rectangle) controls the valve electromagnet. The top part of the diagram (enclosed in the top broken-line rectangle) monitors the valve operation.

4.4.1 Control of Valve Electromagnet

Fig. 4.7 includes a shaft encoder driven by the engine crankshaft. This encoder provides two output signals. The first, is a pulse once every engine cycle; that is, once every two engine revolutions (720°CA) in the case of a four-stroke engine. This once-per-cycle pulse could be arranged to occur, for example, at bottom dead center (start of the compression stroke). The second output signal from the shaft encoder is a pulse every degree crank angle of crankshaft rotation.

Consider now the bottom part of Fig. 4.7, which controls the valve electromagnet. The first counter, labeled "valve on" in Fig. 4.7 is pre-loaded with the crank angle at which the electromagnet should be switched on. This crank angle can be pre-loaded using, for example, thumbwheel

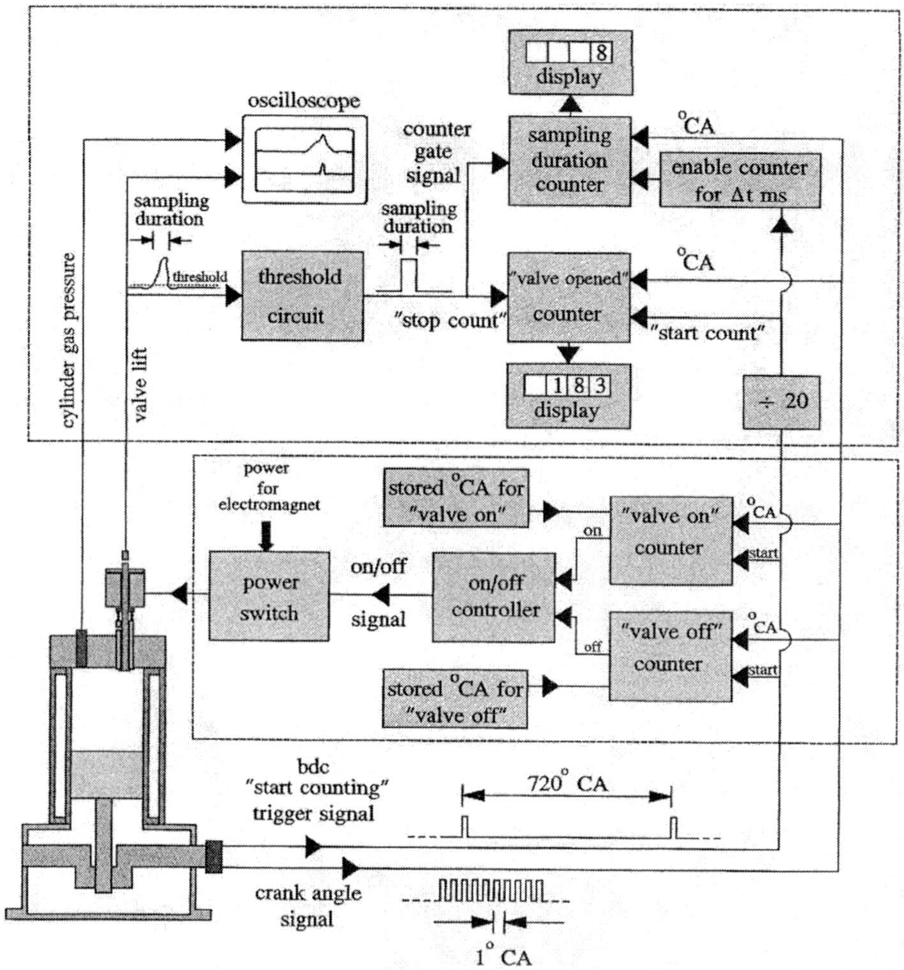

Fig. 4.7 Control and monitoring system for electromagnetic sampling valve.

switches (labeled "stored °CA for valve on" in Fig. 4.7). The once-per-cycle bottom dead center signal from the shaft encoder starts this "valve on" counter. Once started, this counter then counts the shaft encoder crank angle degrees and continuously compares the instantaneous count with the pre-loaded value of "stored °CA for valve on." When the pre-loaded value is reached, a signal is sent to the on/off controller, which, in turn, switches the power supply to the electromagnet on. The second counter shown in Fig. 4.7 and labeled "valve off," is operated in precisely the same way as the first counter, except that it is pre-loaded with the crank angle at which the electromagnet should be switched off. Both counters are started simultaneously by the once-per-cycle pulse.

It can thus be seen that the "valve on" counter controls the timing (i.e., the °CA) at which the valve electromagnet is switched on. On the other hand, the "valve off" counter controls the timing (i.e., the °CA) at which the valve electromagnet is switched off and, thereby, the period for which current is supplied to the electromagnet. For example, if a great number of °CA elapses before the "valve off" timer switches off the current to the valve electromagnet, then the valve will remain open for a longer period.

It should be noted that when the electromagnet current is switched on by a signal from the "valve on" counter, the sampling valve poppet does not open instantaneously. A period of several °CA elapses before the poppet valve opens. This is because a finite period of time is required for the electromagnetic flux to be established in the valve electromagnet coil. In addition, due to mechanical inertia, time is required before the electromagnet armature and the valve poppet connected to it starts to open. Therefore, it is necessary for the electromagnet current to be switched on ahead of the time (°CA) at which the poppet valve is actually required to physically start opening. The dead time or time (°CA) delay between the electromagnet current being switched on and the poppet valve starting to open depends on a number of variables (e.g., engine cylinder gas pressure) and is difficult to predict. For this reason, it is necessary to have a sensing system which detects the timing (°CA) at which the poppet actually starts to open and to display this timing to the valve operator; such a system, able to monitor the opening and closing of the poppet valve, will be outlined in Section 4.4.3.

Similar comments apply to the closing of the valve poppet. Once the current to the electromagnet is switched off by means of a signal from the "valve off" counter, a period of several °CA elapses before the poppet starts to actually close. Again, this is because a finite period of time is required for the electromagnetic flux to collapse in the electromagnet coil. In addition, due to mechanical inertia, time is required before the electromagnet armature and the poppet valve connected to it start to move. It is therefore necessary to switch off the electromagnet current ahead of the timing (°CA) at which the poppet valve is physically required to start closing. As this time delay is difficult to predict, it is necessary to have a system sensing the motion of movement of the poppet valve and display this to the operator (see Section 4.4.3).

If the two counters used are not able to reset themselves automatically, then some means of resetting the two counters once per engine cycle will be needed, perhaps through utilizing somehow the once-per-cycle bottom dead center signal.

4.4.2 Power Supply to the Electromagnet

The amount of power that can be supplied to the electromagnet is limited by a number of considerations, including magnetic saturation and overheating of the coil. Because the electromagnet is switched on for a very small proportion of the engine cycle, it is possible to supply it with a heavy current that is greater than its continuous rating. In some designs, this can increase the force provided by the electromagnet. However, if a circuit malfunction occurs and the power supply to the electromagnet stays on continuously, then the electromagnet may be damaged due to overheating. Some means of avoiding this situation may be necessary, such as a timing circuit that switches power off should power to the electromagnet stay on for longer than a fraction of a second. Alternatively, current to the electromagnet could be supplied by discharging a bank of capacitors through the coil. This has the inherent advantage of allowing current to flow for a limited period only. Capacitor banks have been used by a number of investigators [Bartholomew et al., 1938; Downs et al., 1951; Rifkin and Walcutt, 1956]. For example, Nightingale [1975] used a trigger signal from a crankshaft sensor to operate a silicon-controlled rectifier discharging a number of capacitors through the electromagnet of the sampling valve. The charging voltage across the capacitor bank and the number of capacitors were adjustable so as to allow variation of the magnitude and duration of the current pulse.

4.4.3 Monitoring of Valve Operation

The top part of Fig. 4.7, shows a system for monitoring the timing of valve openings and closings. Monitoring of valve operation is carried out by observing the valve opening timing and sampling duration, say, once every 20 or so engine cycles. This means that at an engine speed of 2000 rpm, the valve operation is observed (sampled) once every second or so. In order to achieve this, the once-per-cycle signal is divided by 20 by means of a digital electronic divider. The resulting once-per-twenty cycles signal is used to start the counter labeled "valve-opened" in Fig. 4.7, which determines the crank angle at which the sample valve has opened. This counter counts the crank angle degrees from, say, bottom dead center, until counting is stopped by the valve lift signal arriving from the sample valve proximity sensor. This count is then displayed, and it corresponds to the crank angle at which the valve opened.

A second counter, labeled "sampling duration counter" in Fig. 4.7, determines the sampling period in degrees crank angle. The once-per-twenty cycles signal is also used to enable the second counter for a period lasting around two revolutions, so that it is ready and waiting to start counting. This counter is started and stopped (gated) by the "squared" valve-lift signal supplied by the threshold circuit. During this brief gate period of 1 or 2 ms, the counter counts the crank angle degrees, and this count is then displayed.

An oscilloscope may be used to monitor the cylinder gas pressure and valve lift signals for correct operation.

4.4.4 Sensing of Flame Arrival

LoRusso et al. [1980] mounted a thermocouple and an ionization probe close to the valve tip in order to allow measurement of the wall temperature and flame arrival timing at the sampling point. A flame ionization gap, used to control the valve sampling timing, was also installed close to the sampling valve tip by Downs et al. [1951].

4.4.5 Sensing and Choice of Valve Lift

The valve lift is chosen so that the required gas sample volume is obtained with as small a sampling duration as possible. Typical valve lifts quoted in the literature are 0.1 to 0.3 mm, although examples of valve lifts as low as 0.01 mm and as high as 0.5 mm can be found in the literature (see Tables 4.1 and 4.3). It is important to have accurate information on the motion of the valve head, so that the valve opening timing and the sampling duration can be determined accurately.

Usually, direct monitoring of the motion of the valve head is difficult. This is because valve heads of both the poppet and the needle-type are quite inaccessible. To overcome this problem, the motion of the valve stem some distance away from the valve head is monitored instead. The actual motion of the valve head is usually somewhat different than that monitored at, say, the rear end of the valve stem due to the elasticity of the stem. For example, LoRusso et al. [1980] conducted a bench test during which the lift of both the poppet and the rear of the valve stem were measured simultaneously. It was found that, for a poppet lift of 0.01 mm, the corresponding lift at the rear of the valve stem was about 0.02 mm. The example worked out in Section 2 of Appendix D gives an elongation of the valve stem of 0.04 mm when the valve head impacts on its seat at closure.

Sometimes, the valve motion sensed by a proximity sensor installed at the end of the valve stem shows oscillations when the valve closes. These oscillations can be due to bounce of the valve head on its seat, following the closing impact. Another possibility is that the valve stem tends to oscillate due to the closing impact, and this is picked up by the proximity sensor. The valve stem may be oscillating from side to side (radially) or longitudinally due to elastic elongation and contraction of the stem. Rifkin and Walcutt employed an interesting method for suppressing the oscillations of the valve stem, which they ascribed to bounce of the valve head on its seat. At the rear end of the stem they attached a metal cup having a volume of about 2 cm^3, which they filled with tantalum powder. This provided sufficient damping for the valve stem to stop oscillating.

4.4.6 Choice of Sampling Period

The in-cylinder gas composition can change very rapidly during the combustion process. Fig.4.8(a) shows, for example, that the entire premixed-combustion phase in diesel engines can be as short as 12 degrees crank angle. Therefore, a sampling valve must be capable of taking a gas sample over a very short period of a few degrees crank angle. If the sampling period is longer than a few degrees, then the sample composition would be an average of the wide range of gas compositions that occur while the sampling valve is taking the gas sample. The short duration, on the other hand, will tend to limit the volume of the sample obtained per engine cycle, so a larger valve lift would be required to ensure adequate sample volume. In practice, the combination of a large valve lift and very short valve sampling period (<1 ms) is very difficult to achieve. Most successful valve designs reported in the literature achieve a minimum sampling period of around 0.5 to 1 ms, which corresponds to 4 to 8°CA at 1500 rpm, which may be somewhat longer than desirable for certain stages of the gasoline and diesel engine combustion process.

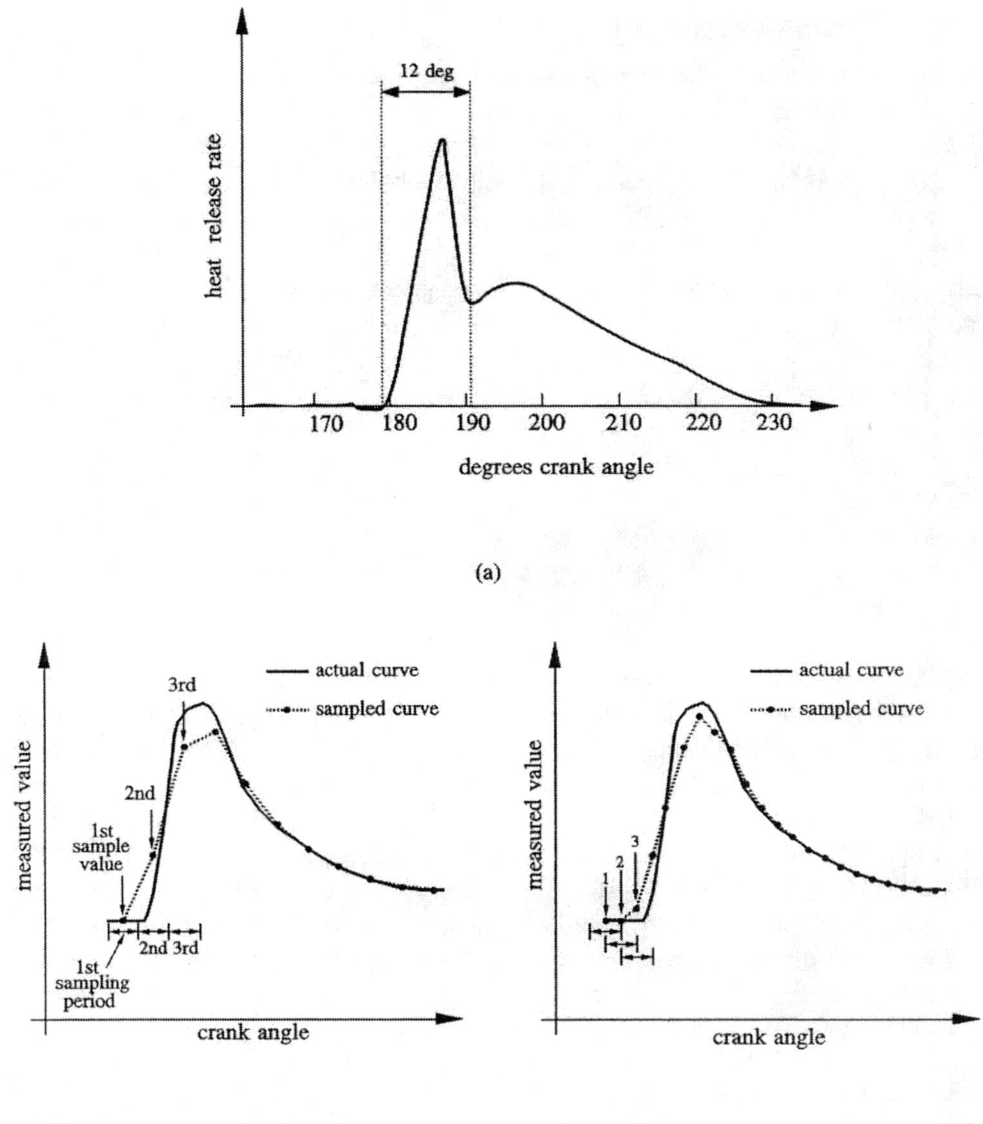

Fig. 4.8 Effects of sampling duration.

A detrimental effect of a long sampling period is illustrated in Fig. 4.8(b). It can be seen that rapid changes in the actual curve cannot be reproduced accurately by the sampled curve. Fig. 4.8(c) shows how this problem can be alleviated somewhat by staggering the sampling periods. The sampling period itself remains the same and it is still long; however, the sampled curve is a somewhat more accurate representation of the actual curve.

Several investigators varied the sampling period in order to maintained a constant sample volume (at atmospheric conditions) per valve opening. This requires short sampling periods when sampling around top dead center, when cylinder pressure is high, and longer periods during the compression and expansion period.

4.4.7 Skip-Cycle Sampling

In most applications of sampling valves in both spark ignition and diesel engines, the valve is opened once every engine cycle. At high engine speeds, the valve may not be capable of responding sufficiently rapidly, and as a result one opening per cycle cannot be achieved. This problem may be overcome by opening the valve once every two, or more, engine cycles. However, it should be noted that skip firing does not affect the valve sampling period, which will become longer as the engine speed rises.

The technique of skipping several cycles between sampling was employed by Isonomy and Yanagihara [1962] to measure the scavenging efficiency of small two-stroke motorcycle engines running up to 9000 rpm.

4.4.8 Single-Cycle and Continuous Sampling

In practice, it is possible to arrange for a sampling valve to open just once and take a small gas sample during an individual engine cycle. The volume of this gas sample will usually be, at most, a few cm^3. Such relatively small gas volumes can be analyzed using chromatographic techniques.

Often, an alternative technique is employed, in which the sampling valve is opened continuously during every consecutive engine cycle. This provides a quasi-steady stream of gas which is made up of the samples from individual successive engine cycles. Unless prevented from doing so, these individual gas samples usually become mixed within the sampling valve and in the subsequent sample handling systems. The composition of the mixed gas stream then represents the average in-cylinder gas composition, at a particular part of the engine cycle (averaged over several engine cycles).

The mixed gas stream can either be supplied directly to an analytical instrument or stored in a suitable container for analysis later. In many cases, the analytical instruments used are ordinary exhaust gas analyzers. If the flow rate of the mixed gas stream is sufficiently high, then the stream can be supplied directly to one or more gas analyzers. If the stream flow rate is not sufficient, it can be diluted by means of an inert gas such as helium in order to increase the flow rate, and then supplied to one or more analyzers. An alternative method is to store the gas stream in a suitable container until a sufficiently large volume is accumulated, and then supply the analyzer(s) with the stored sample.

A number of investigators have developed techniques for single-cycle sampling, whereby a sample is extracted and analyzed from a single engine cycle [Ayusawa et al., 1978; Matsui et al., 1979]. The techniques will be discussed in Chapter 7, where applications of sampling valves to in-cylinder hydrocarbon concentration measurements are discussed.

4.5 Sample Volume and Sample Shape in the Engine Cylinder

4.5.1 Minimum Sample Volume

In Appendix B, two important conclusions are reached regarding the volume occupied by the sample in the engine cylinder, before the sample is expelled when the sampling valve opens. For a given sampling period and valve lift, it is shown that:

(i) The volume occupied by the sample is independent of cylinder gas pressure.
(ii) The volume occupied by the sample is not greatly affected by cylinder gas temperature.

The sample volume at cylinder conditions is an important parameter, because it determines the maximum distance from the valve inlet from which the sampled material is drawn. The greater the volume of the sample, the greater will be the distance from the wall from which the gas sample is drawn into the valve.

In most in-cylinder sampling applications, the volume of sample extracted during each valve opening is within the range 0.1 to 1 cm^3 (at cylinder gas conditions). This corresponds, at most, to between 0.4 and 4% of the cylinder gas volume in the case of automotive diesel and gasoline engines.

In practice, it would be desirable to keep the sample size small so as to reduce the impact on the combustion process. However, a small sample size can also have significant disadvantages. It is usually inevitable that a fraction of the sample will be drawn from the boundary layer or crevices adjacent to the sampling point. Unless interested in sampling from the boundary layer itself, the influence of the layer on a small sample can be very substantial. For instance, the concentration of NO close to the cylinder wall can be up to an order of magnitude lower than the corresponding level in the bulk gas.

It is useful to consider briefly the proportion of the sample that may originate from the boundary layer. For example, if the sample volume at cylinder conditions had a hemispherical shape with a 5 mm radius, then roughly 15% of the sample volume or 25% of the sample mass would be within the boundary layer. The mass fraction is considerably greater than the volume fraction, due to the higher density of the cooler gas in the boundary layer. Shimazaki et al. [1996] estimated that 8% of a 0.5 cm^3 sample volume (at cylinder conditions) originated from within the boundary layer. This included the gas trapped in the dead-volume of the needle-valve sample hole, which had a 0.5 mm diameter and 10 mm length.

Several investigators determined experimentally the minimum acceptable sample volume per engine cycle. The technique used by these investigators involved progressively increasing the sample volume extracted per engine cycle (e.g., by increasing the sampling period or the valve lift), while monitoring the sample composition. When no further change in sample composition was detected, the sample volume was deemed to be the minimum that was acceptable.

Ayusawa et al. [1978] used this technique on a gasoline engine in conjunction with a needle-type sampling valve. Their results show that when sampling occurred before spark discharge, a minimum sample volume of 1 cm^3 (at cylinder conditions) was required before HC concentration in the sample stopped increasing. When sampling occurred after spark discharge, the corresponding minimum sample volume was 4 cm^3 (at cylinder conditions).

Similar tests were carried out by Zvonow et al. [1968] in a spark ignition engine. They found that a minimum volume about 5 cm^3 (at atmospheric conditions) was required before the CO_2, CO, and NO_x concentrations in the sample became independent of sample size. The results of Zvonow et al. are also interesting because they show a common minimum sample volume of about 5 cm^3 for the three species CO_2, CO, and NO_x. This implies that the quench layer might be about the same for all three species. One might have expected the NO_x quench layer to be somewhat greater in view of the great sensitivity of NO_x formation rate to gas temperature.

Whitehouse et al. [1977] also used the same technique in a 300 mm bore diesel engine operating at 500 rpm, with the sample valve tip penetrating 63 mm into the cylinder. They found that a minimum sample volume of 4 cm^3/cycle (at atmospheric conditions) was required before the NO level became independent of the sample size.

Daniel and Wentworth [1962] found that when a poppet-type sample valve was installed almost flush with the combustion chamber upper surface, no appreciable change in HC concentration was observed with sample volumes greater than around 0.4 cm^3 (cylinder gas conditions). However, when the valve was installed with a small recess, the minimum sample volume was considerably larger at, roughly, 1.5 cm^3.

4.5.2 Boundary Layer Thickness

The preceding section emphasized the influence of the boundary layer on the sample composition, and outlined a commonly used experimental technique for determining the minimum acceptable sample volume. This section considers in a little more detail the thickness of the boundary layer.

Theoretical flow analysis for a line sink was carried out by LoRusso et al. [1983] and by Weiss and Keck [1981]. Fig. 4.9(d), shows the gas temperature profile close to the combustion 0chamber wall of a spark ignition engine at 40°CA ATDC, during the expansion stroke. The profile shown was estimated by LoRusso et al. [1983] using compressible turbulent boundary layer analysis, for the following gasoline engine running condition: 1250 rpm, 3.8 bar bmep, and compression ratio of 7. The profile shows a thermal sub-layer reaching out to a distance of about 0.08 mm from the wall, where the gas temperature is about 1200 K. Beyond the sub-layer, the profile is logarithmic, with the thermal boundary layer growing gradually up to a distance of about 5 mm, where the temperature reaches the bulk gas temperature of about 2200 K.

Over the years, there have been a number of measurements of thermal boundary layer thickness in running spark ignition engines, often accomplished by means of Schlieren photography, which tends to emphasize temperature-induced density gradients in the cylinder gas. For example, Lyford-Pike and Heywood [1984] made such measurements in an optically accessible experimental engine with a disc-shaped combustion chamber, running at 1380 rpm and 0.71 bar inlet manifold pressure. Their

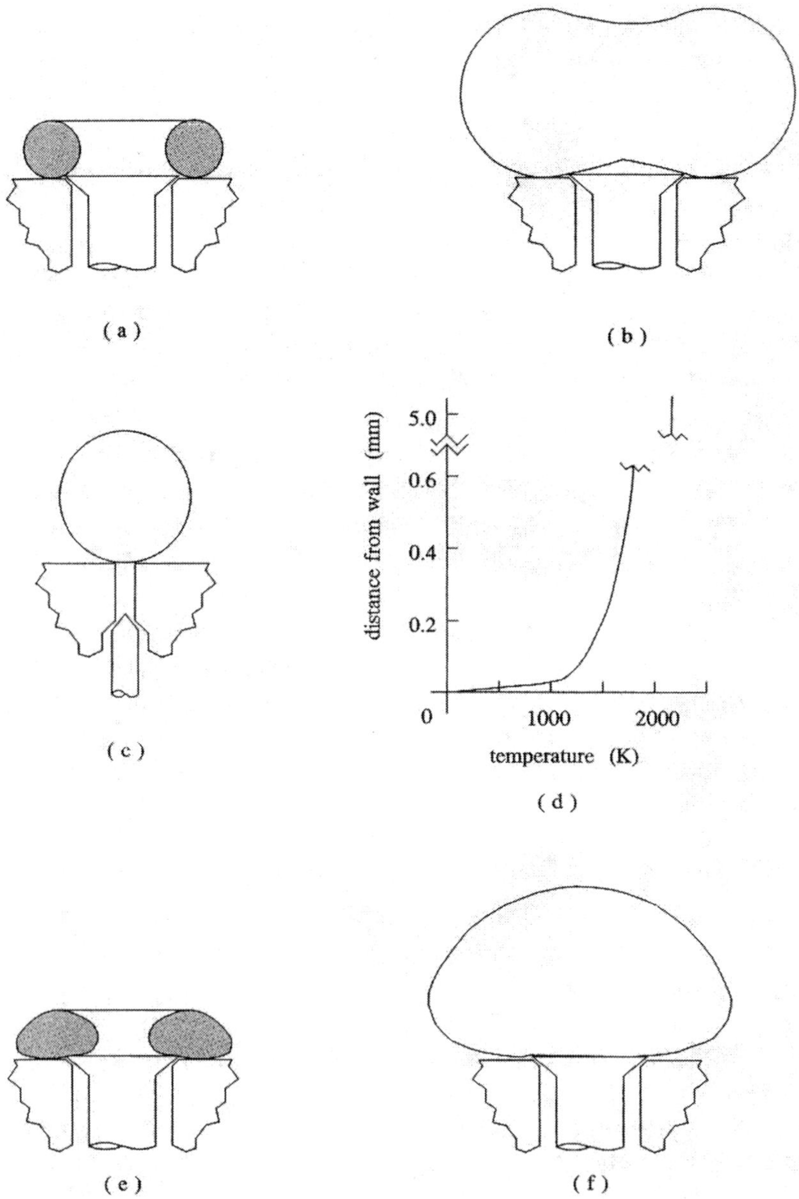

Fig. 4.9 Sample volume shapes with needle and poppet valves.

data show that the thermal boundary layer on the cylinder wall surface increased steadily during the compression stroke, from about 0.1 mm at 45° CA BTDC to just over 1 mm at 10° CA BTDC. During the expansion stroke the layer grew further, from about 1 mm at 30° ATDC to a maximum of about 2 mm at 120° CAATDC. At a given engine running condition and crank angle, the thickness of the thermal boundary layer varied considerable at different locations around the combustion chamber. For example, the layer on the head and piston surfaces was up to 2 to 3 times thicker during compression and expansion than on the cylinder wall surfaces.

Daniel [1957] also estimated the thickness of the thermal boundary layer as 0.05 to 0.4 mm using Schlieren photography, the thickness being dependent of the engine running conditions.

4.5.3 Shape of Sample Volume Inside the Engine Cylinder

4.5.3.1 Preliminary Consideration of Idealized Sample Shapes

The shape of the sample volume at cylinder conditions is another important parameter that must be considered carefully when interpreting results obtained from in-cylinder sampling. Figures 4.9(a), (b), and (c) show highly idealized sample volumes corresponding to different sampling conditions. Consider, first, the case of the poppet sampling valve in Figs 4.9(a) and (b). When the sample volume is very small, then the idealized shape of the sample, as shown in Fig. 4.9(a), is, roughly, of toroidal or doughnut shape. When the sample volume is large, its idealized shape tends toward a spherical shape, as shown in Fig. 4.9(b). In the case of inward-opening valves (needle valves), the sample volume also tends towards a spherical shape, as shown in Fig. 4.9(c).

In reality, the shape of the sample volume will be far from circular or spherical. This is because the flow of the sample close to the wall will be impeded by wall friction and influenced by the presence of the boundary layer (Fig. 4.9[d]). The presence of a boundary layer distorts the shape of the sample volume close to the wall, roughly as shown in Figs. 4.9(e) and (f). In Fig. 4.9(e), the sample volume is still toroidal, but the cross section of the toroid is closer to semicircular than circular. Similarly, in Fig. 4.9(f), the shape of the sample is shown to be closer to hemispherical than spherical.

As a rough approximation, the sample shape in Fig. 4.9(e) can be idealized into a toroidal shape with semicircular cross section, having radius, r_c, which can be calculated from:

$$r_c = \frac{1}{\pi}\sqrt{\frac{2\Delta V_c}{a}} \tag{4.2}$$

where r_c = idealized sample radius at cylinder conditions
ΔV_c = sample volume at engine cylinder pressure and temperature
a = poppet seat diameter

Similarly, as a rough approximation the sample shape in Fig. 4.9(f) can be idealized into a hemispherical shape, having a radius, r_c, which can be calculated from:

$$r_c = \left(\frac{3\Delta V_c}{2\pi}\right)^{1/3} \tag{4.3}$$

These toroidal and hemispherical sample shapes in the engine cylinder are very approximate, but may suffice for many applications. More accurate estimates of the sample shapes will be dealt with in the following sections, using analyses presented by Weiss and Keck [1981] and by LoRusso et al. [1983].

4.5.3.2 Detailed Consideration of Hemispherical Sample Shape

The analysis by Weiss and Keck [1981] is two-dimensional, and it considers the sample volume to be flowing into a point sink, as shown in Fig. 4.10(a). The sink represents the opening of the sampling valve into which the sample flows. In the case of an inward-opening needle-type valve, such as the one shown in Fig. 4.10(b), the approximation of a point sink is a reasonable one provided that the valve sample hole is of small diameter. It is also a satisfactory two-dimensional approximation for a poppet sampling valve, when the sample volume is small and forms a toroid.

The analysis of Weiss and Keck [1981] is two-dimensional and considers the sample volume to be in two parts, as shown in Fig. 4.10(a). The upper part of the sample lies outside the influence of the velocity boundary layer which has a thickness δ. The lower part of the sample volume lies within the velocity boundary layer, and its shape is influenced by the presence of the boundary layer, resulting in the highly flattened shape shown in Fig. 4.10(a).

Referring to Fig. 4.10(a), for the upper part of the sample, which lies outside the velocity boundary layer, Weiss and Keck [1981] assumed a circular shape (two-dimensional analysis). For the lower part of the sample, which lies within the velocity boundary layer, Weiss and Keck used a similar solution of the boundary layer equations for two-dimensional incompressible flow into a point sink on a plane surface.

The two-dimensional analysis of Weiss and Keck [1981] is extended to the three-dimensional hemispherical and toroidal shapes, by assuming axial symmetry about the axes of the hemisphere and the toroid.

Next, the two-dimensional sample shape shown in Fig. 4.10(a) will be evaluated. Consider, first, the circular part of the sample volume, which lies outside the boundary layer (i.e., $z > \delta$). The coordinates x, $z(x, r_c)$ of the point shown can be evaluated as follows:

In general,

$$x^2 + z^2 = r_c^2$$

Hence,

$$z = \left(r_c^2 - x^2\right)^{1/2}$$

where r_c is the constant value of the radius of the circular part of the sample volume (lying outside the boundary layer) as shown in Fig. 4.10(a)

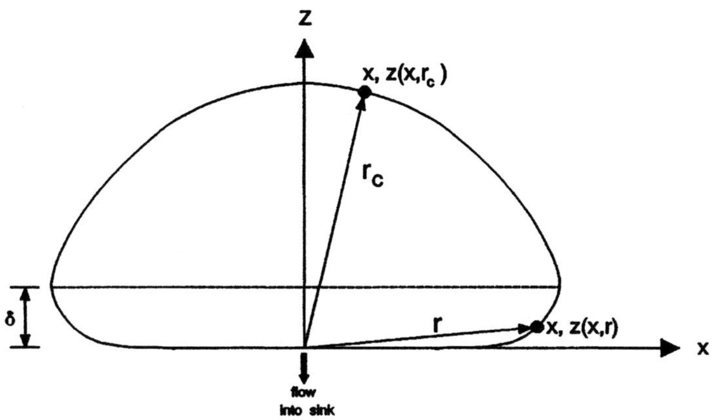

(a) Analysis of shape

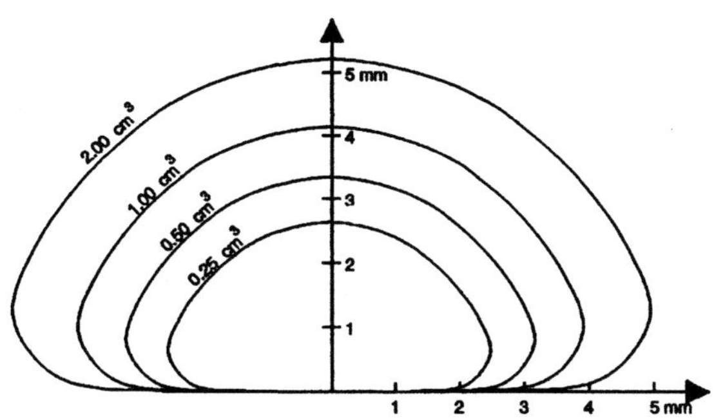

(b) Dimensions of four sample volumes

(dimensions in mm are at cylinder gas p and T)
(volumes in cm^3 are at atmospheric p and T)

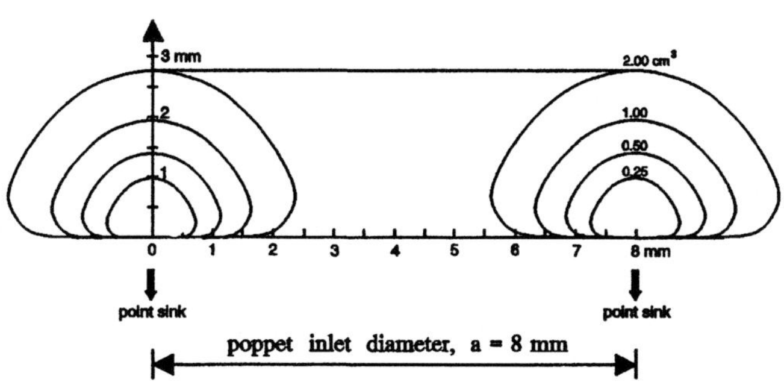

(c) Calculated flow into a poppet valve

(dimensions in mm are at cylinder gas p and T)
(volumes in cm^3 are at atmospheric p and T)

Fig. 4.10 Analysis of sample volume shape.

If the portion of the sample volume lying outside the boundary layer is assumed to have a hemispherical shape, the constant value of the radius, r_c, can be evaluated from Eq. (4.3). Therefore,

$$z = \left[\left(\frac{3\Delta V_c}{2\pi} \right)^{2/3} - x^2 \right]^{1/2} \tag{4.4}$$

This expression allows the value of z to be evaluated for a given value of x. Thus, the circular (hemispherical) portion of the sample shape lying outside the boundary layer can now be plotted.

Often, the sample volume, ΔV_c, may not be known at the engine cylinder conditions. For example, more often than not, the volume flow rate of the sample, V_o, is measured at atmospheric temperature, T_o, and pressure, p_o, after the sample leaves the valve, by means a suitable flow device. The sample volume per engine cycle, ΔV_c, can then be determined from the following expression:

$$\Delta V_c = \frac{P_o}{P_c} \frac{T_c}{T_c} \Delta V_o = \frac{P_o}{P_c} \frac{T_c}{T_c} V_o \frac{60n}{N} \tag{4.5}$$

where V_o = volume flow rate from sample valve at atmospheric p_o and T_o
ΔV_o = volume of sample per engine cycle at atmospheric p_o and T_o
N = engine revolutions per minute
n = number of engine revolutions per cycle
T_c = engine cylinder gas temperature
p_c = engine cylinder gas pressure
T_o = atmospheric temperature
p_o = atmospheric pressure

Of course, if the sample flow rate is measured at a temperature and pressure other than the atmospheric values, the values of temperature and pressure of the sample at the flow measuring device should be used instead of p_o and T_o.

The engine cylinder gas temperature, T_c, will need to be estimated. Assuming that the cylinder gas pressure, P_c, is being monitored by means of a pressure transducer, then T_c can be calculated from:

$$T_c = T_i \frac{P_c}{P_i} \frac{V_c}{V_i} \frac{M_c}{M_i} \tag{4.6}$$

where P_c = gas pressure in the cylinder
V_c = cylinder volume during the sampling period
P_I = cylinder gas pressure at the end of the engine induction process (i.e., when the engine inlet valve closes)
V_I = cylinder volume at end of the engine induction process
T_I = cylinder gas temperature at the end of the engine induction process (i.e., when the engine inlet valve closes)
M_s = molar mass (molecular weight) of the gas sample
M_i = molar mass of the fresh cylinder charge admitted into the engine cylinder during the induction process

Consider now the non-circular part of the sample volume, which lies inside the velocity boundary layer δ. The coordinates x, z(x, r) of the point shown in Fig. 4.10(a) can be evaluated from the following expression [Weiss and Keck, 1981]:

$$z = -\delta\left(\frac{x}{r_c}\right)^{3/2} \ln\left[1-\left(\frac{x}{r_c}\right)^3\right] \qquad (4.7)$$

From this equation, the value of z can be evaluated for different values of x. The velocity boundary layer thickness, δ, can be evaluated, approximately, from:

$$\delta = \left(v \cdot t_v\right)^{1/2} \qquad (4.8)$$

where v = kinematic viscosity of the cylinder gas within the boundary layer
t_v = time interval during which mass flows into the point sink (i.e., sampling period)

An indicative value of the boundary layer thickness can be obtained from:

$$v = 3.5 \times 10^{-4}\ m^2/s$$

$$t_v = 0.001\ s$$

hence,

$$\delta = 5.9 \times 10^{-4}\ m \text{ or } 0.59\ mm$$

The derivation of the above expression for z assumes that $x \approx r$ within the velocity boundary layer (i.e., when $z < \delta$). This is a reasonable approximation provided the velocity boundary layer thickness is small compared with r. This approximation is at its worst when z approaches δ.

Finally, it should be pointed out that the above analysis assumes quiescent gas conditions within the engine cylinder. In an operating engine, strong organized gas motion can exist, for example, in the form of swirl, which will tend to distort the shapes evaluated by means of the above analysis.

4.5.3.3 Numerical Example of Hemispherical Sample Shape

The above expressions were used to evaluate the sample shape at cylinder conditions for four sample volumes. Fig. 4.10(b) shows the hemispherical shapes that these four sample are predicted to have at engine cylinder conditions. The following data were used for the evaluation of the four shapes:

Sample volumes per engine cycle at atmospheric pressure and temperature

0.25, 0.5, 1.00, and 2.00 cm^3

Sample volumes per engine cycle at cylinder gas pressure (40 bar) and temperature (1800 K),

0.0375, 0.075, 0.15, and 0.3 cm^3

Kinematic viscosity, v

3.5×10^{-4} m^2/s

Sampling period, t_v

0.001 s

On Fig. 4.10(b), consider, for example, the shape corresponding to 2 cm^3 of free sample gas (at atmospheric conditions). Inside the engine cylinder, this sample would occupy a volume of only 0.15 cm^3. Its shape is described by the curve for 2 cm^3 in Fig. 4.10(b). It can be seen that inside the engine cylinder the sample volume extends to a maximum distance of a little over 5 mm, which is well beyond the boundary layer thickness.

Finally, it should be noted that the shape of the curves shown in Fig. 4.10(b) at the interface between the boundary layer and the bulk gas (i.e., at $z = \delta$) is only approximate, for the reasons given at the end of the preceding section.

An interesting observation from Fig. 4.10(b) is that at large sample volumes the sample shape can be represented with considerable accuracy, simply as a hemisphere.

4.5.3.4 Detailed Considerations of Toroidal Sample Shape

At low sampling volumes, the shape of the sample in the case of poppet valves is toroidal, roughly as shown in Fig. 4.9(e). The two-dimensional expressions for z obtained above for both the semicircular portion of the sample and for the portion lying within the boundary layer can be used for the toroidal shape as well. However, the radius, r_c, must now be obtained for a sample volume having a toroidal rather than hemispherical section. Denoting the volume of a toroidal sample volume by ΔV_c,

$$2 \cdot \Delta V_c = \left(\pi r_c^2\right)(\pi a)$$

hence,

$$r_c = \frac{1}{\pi}\left(\frac{2\Delta V_c}{a}\right)^{1/2} \tag{4.9}$$

where a is the poppet valve seat diameter

Clearly, as the sample volume, ΔV_c, is increased, the value of r_c will increase. The toroidal sample shape will be retained as long as $r_c < (a/2)$. Thus, the above expression for r_c is not appropriate when $r_c > (a/2)$. If the above expression is used to evaluate r_c and the calculated value of r_c is several times greater than a/2, this would indicate that the sample shape is no longer toroidal and that, instead, it is tending toward a hemispherical shape.

4.5.3.5 Numerical Example of Toroidal Sample Shape

Fig. 4.10(c) shows toroidal shapes of the sample within the engine cylinder. The four shapes shown were calculated using the following data:

Sample volumes per engine cycle at atmospheric pressure and temperature

$$0.25,\ 0.5,\ 1.00,\ \text{and}\ 2.00\ \text{cm}^3$$

Sample volumes per engine cycle at cylinder gas pressure (40 bar) and temperature (1800 K),

$$0.0375,\ 0.075,\ 0.15,\ \text{and}\ 0.3\ \text{cm}^3$$

Kinematic viscosity, v

$$3.5 \times 10^{-4}\ \text{m}^2/\text{s}$$

Sampling period, t_v

0.001 s

Poppet valve seat diameter, a

8 mm

On Fig. 4.10(c), consider for example the shape corresponding to 2 cm^3 of free sample gas (at atmospheric conditions). Inside the engine cylinder, this sample would occupy a volume of only 0.15 cm^3. Its shape is described by the outer curve in Fig. 4.10(c). It can be seen that the sample volume extends to a maximum distance of about 2.7 mm, which is substantially beyond the boundary layer thickness. However, it is interesting to note that even at the highest sample volume of 2 cm^3 the sample retains its toroidal shape. In fact, a substantially larger sample volume of around 5 cm^3 (at atmospheric conditions) will be needed for the radius of the toroid to become equal to the valve seat diameter of 4 mm.

Finally, the closing comments made at the end of Sections 4.5.3.2 and 4.5.3.3 above, are also applicable here.

4.5.4 Valve Location and Sampling Timing

Care is required in the positioning the sampling valve tip. In most sampling investigations, the valve tip is likely to be close to the flame face of the engine cylinder head. Therefore, if the valve opens when the piston has moved well into the expansion stroke, sampling will be concentrated close to the very top of the available cylinder gas volume. Unless the cylinder contents are very well mixed, the gas sample emerging from the sampling valve is unlikely to be representative of the bulk gas in the cylinder.

Once the sampling valve has been installed on the engine, the sampling point is fixed in space and gas sweeps past this sampling point. Therefore, samples extracted at different crank angles will be from different pockets of cylinder gas traveling past the valve tip. For example, a plot of nitric oxide concentration (NO) against crank angle will not necessarily show how NO formation develops in a single pocket of cylinder gas adjacent to the sampling point. Instead, it will show the NO concentration in different pockets of gas as they sweep past the sampling point. The extent of gas motion in the cylinder is, therefore, important.

4.6 Gas Leakage Past the Valve Seat

The gas leakage through the closed valve is normally required to be very low, particularly when measuring hydrocarbon concentration in the burned gases. In gasoline engines during the compression stroke, the leakage gases have a high concentration of unburned hydrocarbons. Even the slightest leakage can introduce a significant overestimation of the HC concentration in the burned gases within the engine cylinder.

4.6.1 Factors Controlling Leakage

The seal formed between the seat and a poppet or needle valve is normally of the metal-to-metal type. For this type of seal, the mass leakage rate is generally controlled by the following variables [Rathbun, 1965],

$$m_l = f\left(C_s, c, \sigma, \Delta p, \varepsilon, \sigma_y\right) \tag{4.10}$$

where m_l = leakage rate
C_s = circumference of the seal
c = width of the seal
σ = applied sealing stress (sealing force/seat area)
Δp = pressure difference across seal
ε = characteristic dimension representing the surface finish of the seal surfaces
σ_y = 0.2% proof strength of the seat material

If leakage is to be reduced, only ε could be altered to a significant extent without consideration of other aspects of the valve design. The values of the other five parameters are influenced significantly by additional considerations, and reduction of leakage can only be a partial influence on the values of these five parameters.

4.6.1.1 Seat Circumference

Reducing the valve seat circumference would reduce the leakage rate. Fortuitously, in practice, it is also desirable to keep the size of the valve tip small in order to ease access problems to the combustion chamber and in order to reduce the disturbance to the combustion and flow processes within the engine cylinder. On the other hand, too small a tip size would reduce the sample flow rate and may also make the task of cooling the valve tip more difficult.

4.6.1.2 Seat Width

Increasing the seat width (denoted as c in Fig. 4.11) would tend to reduce the leakage rate. For a given valve tip overall diameter, the seat width could not be increased excessively at the expense of the annular flow channel around the valve stem, because the channel must be wide enough to accommodate the sample flow. Also, an increase in seat width could not be obtained by reducing the valve stem diameter excessively, since this may reduce the stiffness and weaken the stem. Nevertheless, the seat width in most valve designs is likely to be fairly small, on the order of 0.5 mm, and it should be possible to increase its size somewhat without compromising greatly the annular channel flow area, or the strength of the valve stem, or the overall size of the valve tip.

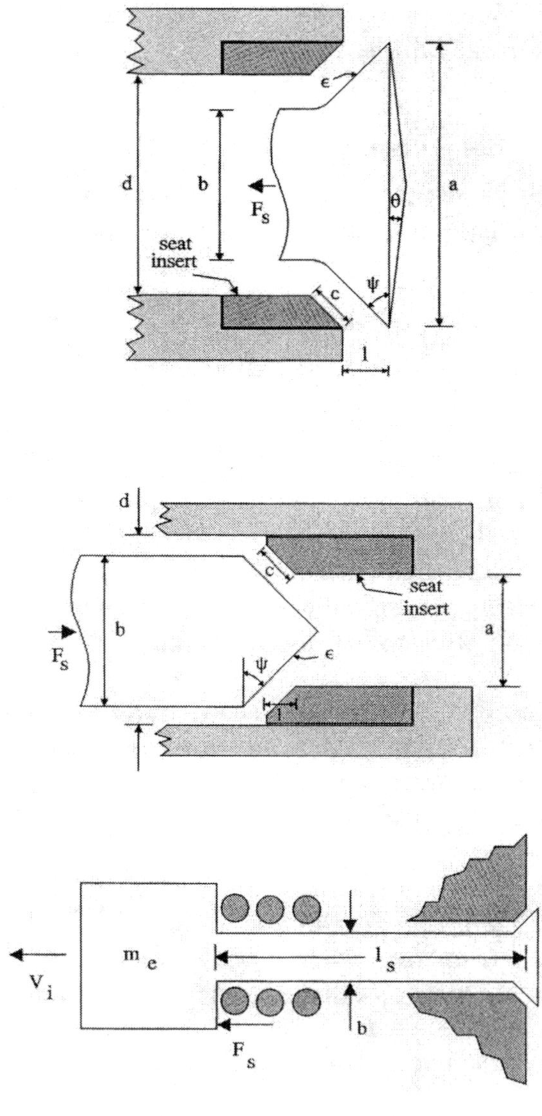

Fig. 4.11 Details of valve geometries and valve dynamic model.

4.6.1.3 Applied Sealing Stress

The sealing stress is equal to the ratio of sealing force and seat projected area. The seat area is projected normal to the direction of the sealing force. Increasing the sealing stress tends to reduce the leakage rate.

Consider, first, the sealing force. The sealing force is equal to the algebraic sum of spring closing force and the additional pressure force provided by the cylinder gases. Increasing the spring closing force may adversely affect the dynamic response of the valve. For example, greater spring force may cause the minimum achievable sampling period to become excessively large.

It is important to recognize that when the sampling valve is installed in an engine cylinder, the sealing force on the valve can be affected significantly by the magnitude of the cylinder gas pressure. In the case of inward-opening needle valve designs, the gas pressure opposes and tends to reduce the closing force provided by the valve spring or hydraulic force. With poppet valve designs, the gas pressure will assist and tend to increase the valve closing force. The gas pressure force varies throughout the engine cycle, and reaches a maximum close to top dead center.

In the case of poppet valve designs, the closing force due to the cylinder gas pressure reaches its maximum value when it is most needed. That is, when the gas pressure and, therefore, the potential for leakage are at their highest levels. The converse is true in the case of inward-opening needle valve designs.

The magnitude of the gas force can be very substantial. For example, if the peak cylinder gas pressure is, say, 60 bar, and it is acting on a poppet valve with diameter of 6 mm, then the closing force due to the cylinder gas pressure will be 170 N. In the case of an inward-opening needle valve having a 2 mm diameter sampling hole, the cylinder gas force tending to open the valve will be 19 N. Thus, it should be possible to keep the valve closed against this gas force of 19 N by means of a stronger spring. What is rather disadvantageous, is the non-availability of the additional sealing force that is present in poppet valve designs (170 N in the above example).

Consider, now, the seat projected area. A small projected area increases the sealing stress, and this ought to reduce leakage. However, the situation is not as clear-cut as this. Referring to Fig. 4.11, the projected area is given approximately by ($\pi ac\cos\psi$) (assuming $c << a$). Therefore, the projected area depends on the seat length, seat circumference, and seat angle. As discussed above, reducing the seat circumference should reduce leakage in its own right, but also reduce leakage due to a reduction in the seat projected area. A reduction in the seat length would tend to increase leakage in its own right, but also reduce leakage due to a reduction in the seat projected area. As for seat angle, large values of this angle make the projected seat area small, and this tends to increase the sealing stress (for a given sealing force) and reduces gas leakage.

The discussion in the preceding paragraph suggests that the seat projected area should be reduced by means of a smaller seat circumference and a larger seat angle.

A seat having a large seat angle has the additional advantage of allowing the design of the valve tip to be more compact, for a given seat length, c. Referring to Tables 4.2 and 4.4, the most common choice of seat angle for practical sampling valve designs has been 45°. This value was used in five out of the six valve designs for which data on seat angle is available. It should, however, be noted that too large a seat angle may restrict too much the sample flow rate past the valve seat. This is because the flow area at the seat is approximately equal to ($\pi al\cos\psi$).

4.6.1.4 Surface Finish

A finer surface finish of the seat contact surfaces tends to reduce leakage, particularly when one or both of the contact surfaces are made of hard material.

4.6.1.5 Yield Stress of the Material

Consider next the yield stress of the seat material (σ_y). This parameter could be controlled by careful choice of the seat material. It will be seen later on that the use of "soft" materials having relatively low values of σ_y could reduce leakage. The softer material could be accommodated, possibly, in the form of a separate "seat insert" as shown in Fig. 4.11. However, other considerations must be taken into account when selecting the seat material. For example, the seat material must be able to resist erosion and corrosion from the high-temperature cylinder gases flowing past the seat when the valve is open.

Rathbun [1965] carried out experiments on metal-to-metal seals which all had the same values of C_s, c, and Δp. Rathbun's results show that the leakage was largely controlled by, first, the surface finish of the mating metal surfaces and, second, the ratio σ/σ_y for the softer of the two mating surfaces. The finer the surface finish, the lower the leakage rate was likely to be. With a given surface finish, the leakage decreased strongly when the ratio of σ/σ_y was increased. This was true irrespective of the way in which the ratio σ/σ_y was increased. For example, the ratio σ/σ_y could be increased by means of a higher sealing stress obtained by applying a greater sealing force; or, alternatively, the ratio could be raised by using a seat material with lower yield stress, σ_y. In terms of sampling valve designs, a valve seat made of, say, copper would have a higher value of this ratio σ/σ_y than one made of, say, steel (for a given value of σ).

Values of yield stress at room temperature for some materials that might be considered for use in sample valve seats are listed in Table 4.5 [Perry and Chilton, 1974]. As an illustrative example, the ratio σ/σ_y is shown below for a poppet valve design having the following geometry:

Seat diameter, 8 mm

Seat width, 0.5 mm

Seat angle 45°

Combined spring and cylinder gas sealing force, 300 N

	Aluminium	Copper	Nickel	Silver	Steel
Ratio σ/σ_y	1.0	0.5	0.6	0.4	0.1

Finally, it should be noted that the above ratios of σ/σ_y discussed in the preceding section have been calculated with values of σ_y at room temperature. In practice, when the sampling valve

Table 4.5 Physical properties of various metallic materials

Material	Yield stress (0.2% proof) (at 20°C) σ_y, MPa	Ultimate stress (at 20°C) σ_u, MPa	Melting temp. t_m, °C	Thermal conductivity (0–100°C) k, W/(mK)	Corrosion resistance[a]
Aluminum (>99%)	34	89	650	222	4
Aluminum alloy 5052 (annealed)	90	193	620	139	4
Copper (>99.9%) (annealed)	69	221	1082	390	2
Gold (annealed)		121	1063	297	
Nickel (99.99%) (annealed)	59	317	1454	78	2
Nickel alloy (Inconel) (Ni77%, Cr16%, Fe7%) (annealed)	241	620	1410	15	4
Silver (>99.9%) (annealed)	83	158	960	418	4
Sterling silver (Cu 7.5%) (annealed)	138	283	891	362	
Coin silver (Cu 10%) (annealed)	159	290	879	359	
Titanium (>99.3) (annealed)	517	586	1704	16.5	5
Carbon steel AISI 1020 (0.20% C)					
(annealed)	262	447	1515	52	
(hardened)	427	620	1515	52	1–2
Carbon steel AISI 1335 (0.36% C)					1–2
(hardened)	724	869	~1500	47	
Stainless steel Type 347 (0.8% Cb)					
(annealed)	207	586	1415	16	
(cold-rolled)	827 max	1034 max	1415	16	3–4

[a] Corrosion resistance to furnace gases with incidental sulphur content:

1 poor to fair
2 fair (for mild conditions or where periodic replacement is possible)
3 fair to good
4 good (suitable when superior alternatives are uneconomic)
5 good to excellent
6 excellent

Reference: Perry and Chilton [1974]

is operating in the cylinder of an internal combustion engine the seat temperature is likely to be substantially higher. This will lower the value of σ_y in the case of all materials, and most probably improve sealing.

4.6.2 Choice of Valve Seat and Head Materials

Over the years, a considerable number of materials have been used for the construction of sampling valve seats and valve heads. The primary considerations have been good sealing against leakage of cylinder gases and high durability of the seat and head in the aggressive environment of hot corrosive cylinder gases. Broadly speaking, one of two approaches have been adopted by various investigators. The first approach has been to construct the valve seat with a relatively "soft" material that deforms to form a good seal in conformity with the valve head. The second approach has been to construct both the valve seat and head from relatively "hard" materials, and then lap the surfaces of the valve seat and head to a fine finish. Examples of these two approaches are reviewed below.

4.6.2.1 Use of "Soft" Seat Materials

The above results for the ratio σ/σ_y suggest that a valve seat made of aluminium is likely to provide the best seal. A glance at Table 4.5 shows that use of aluminium might provide two further important advantages. Firstly, higher corrosion resistance to combustion gases [Perry and Chilton, 1974]. Secondly, heat transfer away from the seat surface will be more rapid due to the high thermal conductivity of aluminum, helping to keep the seat surface at a somewhat lower temperature. However, in practice, aluminum would be unsuitable for use in a running engine because of its relatively low melting temperature. The valve seat surface is likely to disintegrate or suffer from excessive erosion caused by the high-velocity sample gases flowing past the seat surface when the valve is open. On the whole, copper, silver, and gold have more suitable properties as seat materials, save for the relatively low corrosion resistance of copper and the expense of high-purity gold. Otherwise, all three materials have fairly low values of yield stress, σ_y, high values thermal conductivity, k, and reasonably high melting temperatures. Copper has been used successfully in a number of practical sampling valve designs [Weiss and Keck, 1981, Zhao et al., 1996, Johnston and Green, 1982], often in combination with needle valve heads made of stainless steel [e.g., Johnston and Green, 1982].

An example of how a "softer" seat can reduce leakage is provided by the experience of Weiss and Keck [1981] who used a needle valve to sample quench-layer hydrocarbons in a spark ignition engine. They found that the combination of stainless steel needle head and titanium alloy seat gave unacceptable leakage rates. Switching to a copper seat reduced the leakage rates to negligible levels for a wide range of engine operating conditions.

Yates [1988] found that both a nimonic alloy (nimonic 75) and stainless steel (Type 303) gave satisfactory results for the valve head. Also, 24-karat gold gave very satisfactory results as seat material. The valve was allowed to hammer its own seat during operation in a running engine, and this resulted in very low leakage rates.

Alperstein and Bradow [1965] employed an unusual seat construction for a needle sampling valve used in a spark ignition engine. The seat angle was 0° (i.e., flat) and was formed by one end of a quartz tube with internal and external diameters of 0.5 mm and 1.5, mm respectively. The sealing end of the needle head was made of a gold disc bonded onto the stainless steel needle head. The gold disc softened somewhat when the valve was initially used in a running engine, and this helped deform the disc surface until a good seal was obtained. Alperstein and Bradow evaluated a range of materials for this disc, including lead, lead-tin-antimony alloys, silver chloride, and 22- and 24-karat gold. The 24-karat gold proved most satisfactory, showing little tendency to extrude when deformed by impact with the seat; good heat conduction; and sufficiently high melting point. This gold seal proved satisfactory over many months of continuous operation, with leakage rates on the order of 0.005% of the sample volume.

Nightingale [1975] and Zhao et al. [1996] used valves seated against square-edged seats. In the valve described by Zhao et al., repeated impact by the silver-steel poppet onto the square-edged copper seat created a conforming seat. This process was aided by the copper seat softening somewhat when installed in the running engine. Subsequent work-hardening of the copper seat resulted in a durable surface. In the case of the valve described by Nightingale, the sharp titanium edge also deformed initially to form a seat, which subsequently also work-hardened to form a durable surface.

4.6.2.2 Use of "Hard" Seat Materials

There are several examples in the literature of valves in which both the valve head and seat were made of "hard" materials. Such a combination may, for example, be selected in applications where the valve tip penetrates substantially into the combustion chamber. In such cases, the tip temperature can reach levels that may be too high for materials such as copper, silver, or gold.

Bennethum et al. [1975] used a needle-type valve in a diesel engine. With moderate tip penetrations of around 9 mm from the head flame-face and part load engine operation, a seat made of titanium proved satisfactory. However, Bennethum et al. [1975] believed that a titanium seat would have been unsuitable for combinations of high engine loads and high tip penetrations, due to high tip temperatures.

LoRusso et al. [1980] chose a high-strength, hardened, corrosion-resistant alloy for both the valve seat and poppet, in order to improve durability. However, very high sealing forces up to 1000 N were found to be necessary in order to reduce leakage rates to very small and acceptable levels when the pressure difference across the valve seat was 70 bar. Because of the high sealing forces and short sampling periods required, electro-hydraulic actuation was employed. Using information given by LoRusso et al. [1980], it is possible to calculate the value of the ratio σ/σ_y for their valve as <0.2. In addition to high sealing forces, LoRusso et al. found that a very fine surface finish (0.025 to 0.1 μm) was required for the valve head and seat surfaces, in order to reduce leakage to the very low levels required for sampling quench-layer hydrocarbons from a spark ignition engine.

Matsui et al. [1979] used a sampling valve constructed of stainless steel, in order to prevent thermal failure and corrosion attack when used in a spark ignition engine. Good sealing was achieved through careful lapping of the seating surfaces. After about 10,000 valve openings, the leakage became too high because of mechanical wear and deposits, and the seating surfaces had to be relapped.

Duggal et al. [1978] used stainless steel for the valve body and high-strength stainless steel for the poppet and the spindle of an outward-opening poppet-type sampling valve. In a somewhat different arrangement, Garner et al. used a poppet and seat both made of stainless steel, but faced with stellite. The poppet and seat were ground and lapped to ensure good seating at engine operating temperatures.

Zvonow et al. used a seat insert made of a tungsten, which is a very hard material. However, the valve head was made of a relatively softer nickel alloy (inconel), which also has good corrosion-resistance properties. Other parts of the valve in contact with combustion gases were made of stainless steel, in order to avoid corrosion.

Downs et al. [1951] tried many material combinations for the sampling needle valve head and seat materials for use in a spark ignition engine. They found that a combination of stainless steel valve head and a cast iron seat gave the best results. As for cast iron, Yates [1988] found it unsuitable for a valve poppet, due to a combination of problems: It was difficult to braze, and suffered from overheating and corrosion attack.

4.6.3 Detecting Leakage and Measuring Leakage Rate

In almost all valve designs, some leakage past the valve seat is almost certainly inevitable, although the leakage rate can usually be reduced to very low levels. During use, leakage past the sampling valve seat must be detected and measured, in order to ensure that it stays at acceptably low levels.

A simple and effective way of detecting leakage is to immerse the outlet tubing from the sample valve into a beaker of water with the valve continuously closed, while it is installed in a running engine. Bubbling through the water would indicate that leakage is present.

A convenient method of measuring leakage rate is to use a water displacement meter, as shown in Fig. 4.12(a) [Ayusawa et al., 1978]. Prior to beginning the measurement, the vent is opened so that the level of water in the two concentric cylindrical vessels is equalized. The vent is then closed and sample is supplied to the inner vessel. The volume of sample collected over a timed period is read off the graduations on the inner vessel. The pressure of the sample in the inner vessel will be equal to the atmospheric pressure. The hydrostatic pressure due to the difference in the water levels of the inner and outer vessels is negligible.

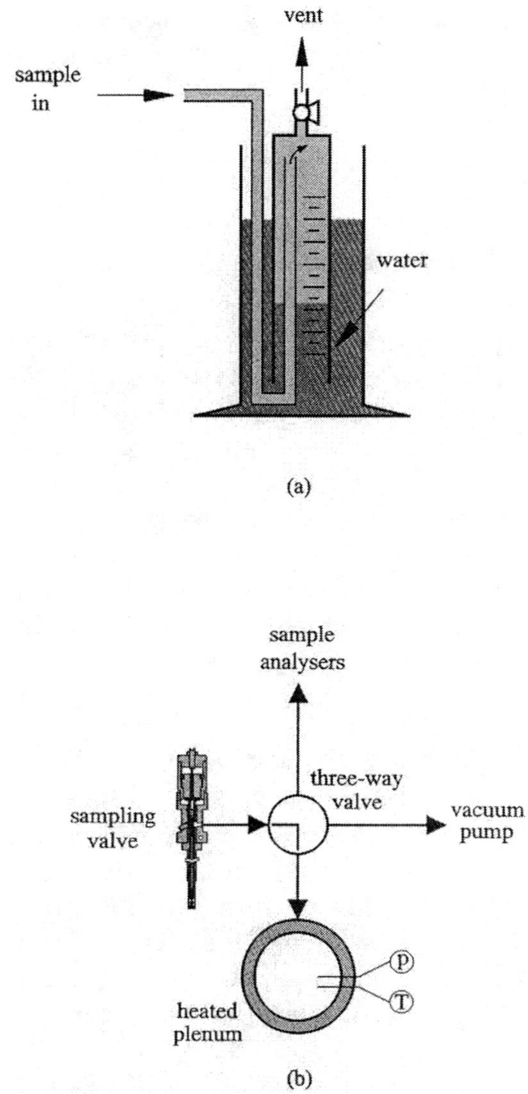

Fig. 4.12 Detection and measurement of leakage rate.

An alternative method of measuring small leakage flows is shown in Fig. 4.12(b). The sampling valve is installed on the running engine and kept continuously closed. The cylinder gases are directed into an evacuated heated manifold, maintained at constant temperature. The leaked mass, m_l, over a given period of time can be calculated from the plenum volume, V_l, pressure, p_l, and temperature, T_l, of the gas in the plenum, by assuming ideal gas behavior: $m_l = p_l V_l/(RT_l)$. Alternatively, the mass flow rate of the leaking gas into the plenum can be obtained from the rate of pressure rise:

$$\frac{dm_l}{dt} = \dot{m}_l = \frac{V_l}{RT_l}\frac{dp_l}{dt} \tag{4.11}$$

where V_l and T_l are held constant. For example, the plenum pressure can be plotted against time, and the leakage rate at a given instant can be obtained from the slope of the resulting curve at the given instant.

4.6.4 Magnitude of Leakage

4.6.4.1 Leakage Rate Measured with Valve Continuously Closed

LoRusso et al. [1983] found that the leakage rate ranged between 2×10^{-9} and 8×10^{-9} g/valve opening when the cylinder gas pressure was 10 bar and the valve was closed. A leakage rate of 5×10^{-9} g/valve opening corresponded to between 0.07% and 1% of the sample mass, depending on whether sampling was carried out at low or high sampling mass flow rates. Leakage increased to 12×10^{-9} g/valve opening after 20 hours of use in the engine, when it was deemed necessary to relap the valve seat. After relapping the valve seat five more times, leakage could not be reduced sufficiently by means of relapping, and the valve seat had to be reground.

Using a flat-ended needle tipped with a gold disk, and closing onto one end of a 1.5 mm outside diameter quartz tube, Alperstein and Bradow [1965] obtained a very small leakage rate on the order of 0.005% of the sample volume.

4.6.4.2 Leakage Rate Measured with Sampling Valve Operating

Using a "skip-cycle" method, LoRusso et al. [1983] measured the leakage rate of unburned hydrocarbons while the valve was in operation on the engine. The sample mass per cycle, Δm_s, was always kept constant at 1.5×10^{-5} g, and the valve was set to sample after combustion was completed. To start with, the valve was triggered every other engine cycle. The hydrocarbon concentration, HC_1, in the sample was measured. The valve was then operated once every three engine cycles (i.e., two cycles were skipped) and the hydrocarbon concentration, HC_2, in the sample was again measured. This procedure was repeated with the valve triggered once every 4, 5, 6, ... etc., cycles. In this way, the valve was allowed to remain closed for progressively increasing number of cycles. The values HC_1, HC_2, HC_3, ..., etc, were then plotted against the number of engine cycles skipped. When the leakage rate was very small, the resulting curve was approximately a horizontal line, indicating that the HC concentration in sample was simply that for the products of combustion, irrespective of the number of cycles skipped. However, when leakage was significant (e.g., due to a worn valve seat) the resulting curve was a straight line with positive slope, indicating that the hydrocarbon concentration in the sample increased progressively with a greater number of skipped cycles. This was due to increasing accumulation in the sample of very high HC concentration air fuel mixture leaked during the skipped cycles. LoRusso et al. [1983] estimated the dynamic leakage mass per cycle, Δm_l, from the following expression:

$$\Delta m_l = \left[\frac{d(HC_n)}{dn}\right]\left(\frac{\Delta m_s}{HC_l}\right) \tag{4.12}$$

where Δm_l = mass of cylinder gas leaked per engine cycle
Δm_s = mass of sample per engine cycle
$d(HC_n)/(dn)$ = slope of HC volumetric concentration versus n curve
n = number of engine cycles skipped
Δm_s = sample mass per engine cycle
HC_l = HC volumetric concentration in leaked gas measured *statically* (i.e., with valve continuously closed)

Note that LoRusso et al. [1983] applied the above expression when the valve was set to sample after combustion was completed.

4.6.5 Correcting for Leakage of Unburned Hydrocarbons

Weiss and Keck [1981] obtained for the case of a gasoline engine estimates of the proportion of unburned hydrocarbons in the sample that could be ascribed to leakage. These estimates were obtained using the following expression:

$$F = \frac{\varepsilon \Delta m_l}{\Delta m_s} \frac{f_{HCu}}{f_{HCs}} \tag{4.13}$$

where F = (volume of hydrocarbons in the sample due to leakage)/(total volume of hydrocarbons in the sample)
f_{HCu} = volumetric concentration of hydrocarbons in unburned mixture
= (volume of hydrocarbons in a sample of unburned air fuel mixture)/(volume of the sample of unburned air fuel mixture)
f_{HCs} = concentration of hydrocarbons in a sample of combustion products
= (volume of unburned hydrocarbons in a sample of combustion products)/(volume of the sample of combustion products)
Δm_l = mass of cylinder gas leaked per engine cycle
Δm_s = mass of sample per engine cycle
ε = fraction of leaked gas per engine cycle that is unburned air fuel mixture

The calculation of the fraction ε was based on the observation that the leakage rate when the sample valve was closed was proportional to the cylinder gas pressure, P_c. Accordingly, ε was estimated by numerical integration of the following expression:

$$\varepsilon = \frac{\int_{\theta_{IVC}}^{\theta_{FAS}} (P_c - P_o) d\theta}{\int P_c - P_o d\theta} \tag{4.14}$$

where θ = engine crank angle
θ_{IVC} = crank angle at engine inlet valve closure
θ_{FAS} = crank angle for flame arrival at the sampling valve tip
P_c = cylinder gas pressure
P_o = gas pressure at sampling valve outlet

Weiss and Keck found that the value of ε was approximately 0.3 for all of the engine conditions tested.

It is interesting to consider what might be a tolerable leakage. Rearranging the above expression,

$$\frac{\Delta m_l}{\Delta m_s} = \frac{F\, f_{HCu}}{\varepsilon\, f_{HCs}} \qquad (4.15)$$

Consider now the case of a spark ignition engine running on propane. Suppose that it is desirable that the leakage contributes less than 10% of the hydrocarbons in the sample. Thus, the maximum value of F is 0.1. A typical concentration of unburned hydrocarbons in a sample of combustion products might be 1000 ppmv, that is, $f_{HCs} = 1000$ ppmv. For an engine running on an approximately stoichiometric mixture of propane, $f_{HCu} = 40{,}000$ ppmv. Substituting in the above expression, $\varepsilon = 0.3$, $F = 0.1$, $f_{HCs} = 1000$ ppmv, and $f_{HCu} = 40{,}000$ ppmv gives,

$$\frac{\Delta m_l}{\Delta m_s} = \frac{1}{120}$$

Thus, the mass leaked per engine cycle must be less that 1% of the total sample mass per cycle.

Alternatively, the above expressions can be used to can be used to calculate the percentage of hydrocarbons in the sample, F, ascribable to leakage. This can be done by measuring the leakage rate as a fraction of the sample rate from the valve (i.e., $\Delta m_l/\Delta m_s$).

4.7 Calculation of Local Air /Fuel Ratio from Sampled Burned Gases

Once the sample from the engine cylinder has been analyzed into its constituents, it is then often possible to calculate the local air/fuel ratio at the sampling point, as outlined in the following sections.

4.7.1 Wet and Dry Analysis of the Sample

Calculation of the local air/fuel ratio requires information on the composition of the sample extracted by means of the sampling valve. The sample composition can be measured using a number of analytical techniques. It is important that the composition of the sample is measured on the same basis, that is, on a wet or a dry basis. The dry basis, for example, requires that the moisture in the gas sample be removed by cooling the sample below the dew point temperature of the water vapor in the sample. This must be done before the sample is supplied to the analytical equipment that will be used to determine its composition. The wet basis requires that all of the moisture in the sample be retained in the vapor state. This means that the sample temperature must be kept sufficiently high to avoid water vapor condensation, and certainly above the dew point of the water vapor in the sample. The elevated temperature must be retained within the sample valve, the tubes conveying the sample to the analysis equipment, and even within the analytical equipment itself. The wet basis has the advantage that the unburned hydrocarbon fractions in the sample remain in the vapor state and can all be measured, for example, using a flame ionization detector system.

4.7.2 Local Air Fuel Ratio Based on Oxygen Balance

In the case of fuels that contain only carbon and hydrogen atoms, the gravimetric air/fuel ratio can be calculated from the oxygen balance using the expression given below [Heywood, 1988)]. This calculation requires that the volumetric fractions (in %) of [CO_2], [CO], [0_2], [NO], [NO_2], and [HC] in the sample are known, either all on a dry basis or all on a wet basis. Thus, the gravimetric air fuel ratio can be obtained from:

$$\left(\frac{A}{F}\right) = 4.773\left(\frac{M_a}{M_f}\right)\left(\frac{[CO_2]+[CO]/2+[H_2O]/2+[NO]/2+[NO_2]+[O_2]}{[HC]+[CO]+[CO_2]}\right) \tag{4.16}$$

where (A/F) = gravimetric air fuel ratio [CO_2], [CO], etc
= molar (volumetric) concentrations of CO_2, CO, etc, in the sample (all in %) M_a and M_f
= molar masses of air and fuel, respectively (numerically equal to their molecular weights)

The molar mass of air, $M_a = 28.96$. The molar mass of fuel can be obtained from the expression, $M_f = 12.01 + 1.008\ y$, where y is the ratio of hydrogen to carbon atoms in each fuel molecule (H/C ratio). The molar concentration of hydrocarbons, [HC], must be expressed as C_1. The molar fraction of water (in %), [H_2O], can be estimated using the water constant, k_p, as follows,

$$[H_2O] = 0.5y\left(\frac{[CO_2]+[CO]}{\dfrac{[CO]}{K_p[CO_2]}+1}\right) \tag{4.17}$$

Bennethum et al. [1975] recommend a value of $k_p = 3.8$ for samples extracted from the engine cylinder and quenched at high temperature. The effect on the calculated air/fuel ratio of errors in the value of k_p and in the values of the various sample species $[CO_2]$, $[O_2]$, etc., is examined below.

4.7.3 Local Air Fuel Ratio Based on Carbon Balance

When the volumetric fraction of oxygen, $[O_2]$, in the sample is not known, the gravimetric air fuel ratio can be obtained from the carbon balance, using the expression given below [Heywood, 1988]. Two water concentrations are shown in the expression below:

$$\left(\frac{A}{F}\right) = \left(\frac{M_a}{M_f}\right)\left(\frac{100+[HC]-[CO]/2+3[H_2O]/2-[H_2O]_a}{[HC]+[CO]+[CO_2]} - \frac{y}{2}\right) \qquad (4.18)$$

One, $[H_2O]$, is the molar (volumetric) concentration (in %) of water vapor in the sample, which can be calculated as indicated in the preceding section using k_p. The other, $[H_2O]_a$, is the molar (volumetric) concentration (in %) of water vapor, but at the analytical equipment. The two values for water vapor concentration may be different due to water vapor condensation in the sampling valve, sample transfer tubes, and analytical equipment. If the sampling valve, sample transfer tubes, and analytical equipment are all heated so that the water vapor does not condense, then the analysis will be on a wet basis and $[H_2O]_a = [H_2O]$. If, on the other hand, the sample is chilled before being placed in the analytical equipment so that the water vapor is condensed and removed, then $[H_2O]_a$ will be very small (<1%) and can be omitted with little loss of accuracy. Unfortunately, the calculated value of (A/F) is affected by the value of $[H_2O]_a$ and some care must be exercised in deciding this value.

The effect on the calculated air fuel ratio of errors in $[H_2O]_a$ and in the measured values of the various sample species, $[CO_2]$, $[O_2]$, etc., is examined below in Section 4.10.3.1.

4.7.4 Examples of Air/Fuel Ratio Calculations Based on Oxygen and Carbon Balances

The actual compositions of three in-cylinder gas samples were used in order to calculate the gravimetric air/fuel ratio using both the oxygen and the carbon balance. The three gas samples were obtained by Aoyagi et al. [1980] from the cylinder of a DI diesel engine at three different sampling locations in the engine cylinder, where the air fuel ratio happened to be, respectively, weak, stoichiometric, and rich.

The values of the gravimetric air/fuel ratio, calculated from the compositions of these three gas samples, are shown in rows 1, 2, and 3 of Table 4.6. It can be seen from this table that the values of A/F calculated from the oxygen balance are quite close to the values calculated from the carbon balance. For example, when the in-cylinder mixture sampled is weak, row 1 shows the oxygen balance predicting an A/F value of 19.6, while the oxygen balance predicts a 2% lower value of 19.2 for the same mixture. When the sampled in-cylinder mixture is stoichiometric, row 2 shows that the carbon balance predicts a 3% lower value. Finally, when the sampled in-cylinder mixture is rich, both the oxygen and carbon balances predict the same value of A/F.

The A/F ratios in rows 1, 2, and 3 were calculated using a value of $k_p = 3.8$ and $y = 2$.

4.7.5 Error Analysis

Error analysis was carried out in order to ascertain the effect of errors in the various parameters used for the calculation of the air/fuel ratio. The error analysis was carried out first for the weak mixture and then for the rich mixture used as examples in Table 4.8 above.

4.7.5.1 Error Analysis for Weak Mixture

The results of the error analysis in the case of the weak mixture are shown in Table 4.7, and discussed below. For convenience, rows 1, 5, 8, and 10 of Table 4.7 all show the same value of the correct A/F calculated for the weak mixture. The correct value of A/F is shown repeatedly in order to aid the comparison with the erroneous A/F values.

Error: Neglecting NO, NO_2, HC_1, and CO:

Table 4.7 (rows 1 to 4) shows that when the mixture is weak, an error of about 6% is incurred when the concentrations of [NO], [NO_2], [HC_1], and [CO] in the sample are all ignored, irrespective of whether the A/F is calculated from the oxygen or the carbon balance. Careful inspection of rows 1 to 4 shows that this 6% error is almost entirely due to neglecting [HC_1].

Errors in the measured concentrations of CO_2, O_2:

Although results are not shown in Table 4.7, the effects of −5% errors in the measurement of the concentrations of [CO_2], [O_2] are as follows:

- 5% error in [CO_2]: The resulting error in A/F is +1.2% when the oxygen balance is used and +5.2% when the carbon balance is used.

- 5% error in [O_2]: the resulting error in A/F is −1.4% when the oxygen balance is used. The carbon balance does not require knowledge of [O_2].

Errors in the value of k_p:

Rows 6 and 7 of Table 4.7 show the effect on A/F of varying the value of k_p between 3 and 5 from its base value of 3.8. It can be seen that the effect on the estimate of A/F is negligible, irrespective of whether the oxygen or carbon balance is used to calculate the value of A/F.

Errors in the value of H_2O:

Row 9 shows that reducing the value of [H_2O] by −30% increases the A/F value calculated from the carbon balance by 3%. The value of A/F calculated from the oxygen balance is not affected since knowledge of [H_2O] is not required for this calculation.

Table 4.6 Calulation of gravimetric air/fuel ratio from in-cylinder sample composition at three cylinder locations

Row	[CO_2] %	[CO] %	[NO] %	[NO_2] %	[O_2] %	[HC_1] %	[H_2O] %	O bal. A/F	C bal. A/F
1	9.6	0.1	0.07	0.007	5.8	0.5	9.7	19.6 (weak)	19.2 (weak)
2	10.1	2.7	0.09	0.009	2.3	0.6	12.0	14.5 (stoich.)	14.1 (stoich.)
3	5.5	11.4	0.05	0.005	0.8	1.5	10.9	9.3 (rich)	9.3 (rich)

(i) "O bal. A/F" and "C bal. A/F" are values of the gravimetric air/fuel ratio calculated from the oxygen and the carbon balances, respectively.
(ii) CO_2%, CO%, etc. are the molar (volumetric) fractions percent.
(iii) Base values used: y = 2 and k_p = 3.8.

Errors in the value of y:

Row 11 shows that reducing the value of the atomic H/C ratio from 2 to 1.8 causes small changes in the estimated value of A/F. These changes are −1% when A/F is calculated from the oxygen balance and +2% in the case of the carbon balance.

4.7.5.2 Error Analysis for Rich Mixture

The corresponding results of the error analysis for the rich mixture are shown in Table 4.8. Again, for convenience, rows 1, 4, 7, and 9 of the Table all show the same value of the correct A/F in order to facilitate the comparisons with the erroneous A/F values.

Error: Neglecting NO, NO_2, O_2:

Row 3 of Table 4.8 shows that when the mixture is rich, an error of about −4.7% is incurred when the concentrations of [NO], [NO_2], and [O_2], in the sample are ignored. Comparing rows

Table 4.7 Effects of various errors on the calculated value of *weak* gravimetric air fuel ratios

Row	$[CO_2]$ %	[CO] %	[NO] %	$[NO_2]$ %	$[O_2]$ %	$[HC_1]$ %	$[H_2O]$ %	O bal A/F	O bal Δ(A/F) %	C bal A/F	C bal Δ(A/F) %
					Effect on A/F of neglecting, progressively, NO, NO_2, HC_1, CO						
1	9.6	0.1	0.07	0.007	5.8	0.5	9.7	19.6	–	19.2	–
2	9.6	0.1	neg	neg	5.8	0.5	9.7	19.6	–0.2	19.2	–
3	9.6	0.1	neg	neg	5.8	neg	9.7	20.6	+4.9	20.2	+5.2
4	9.6	neg	neg	neg	5.8	neg	9.6	20.7	+5.6	20.5	+6.4
					Effect on A/F of value of k_p						
5	9.6	0.1	0.07	0.007	5.8	0.5	9.7	19.6	–	19.2	–
								$k_p = 3.8$		$k_p = 3.8$	
6	9.6	0.1	0.07	0.007	5.8	0.5	9.7	19.6	–0.02	19.2	0
								$k_p = 3$		$k_p = 3$	
7	9.6	0.1	0.07	0.007	5.8	0.5	9.7	19.6	+0.02	19.2	0
								$k_p = 5$		$k_p = 5$	
					Effect on A/F of value of $[H_2O]$						
8	9.6	0.1	0.07	0.007	5.8	0.5	9.7	19.6	–	19.2	–
9	9.6	0.1	0.07	0.007	5.8	0.5	9.7	19.6	–	19.8	+3.0
					Effect on A/F of value of y						
10	9.6	0.1	0.07	0.007	5.8	0.5	9.7	19.6	–	19.2	–
11	9.6	0.1	0.07	0.007	5.8	0.5	8.7	19.4	–1.0	19.6	+2.0
								y = 1.8		y = 1.8	

(i) **neg** = value neglected.

(ii) Δ(A/F)% = deviation from base estimates of A/F (base estimates repeated in rows 1, 5, 8, and 10).

(iii) A/F values are on a gravimetric basis; CO_2%, CO%, etc are the molar (volumetric) fractions percent.

(iv) Base values used: y = 2 and $k_p = 3.8$, except where otherwise shown.

3 and 2 it can be seen that this error is almost entirely due to neglecting $[O_2]$. Note in Table 4.8 that no errors are shown for the carbon balance in rows 1 to 3. This is because knowledge of [NO], $[NO_2]$, and $[O_2]$ is not required in order to calculate A/F from the carbon balance.

Errors in the measured concentrations of CO_2, CO, and HC_1:

Results are not shown in Table 4.8. However, the effects of –5% errors in the measurement of the concentrations of $[CO_2]$, [CO], and $[HCT_1]$ are as follows:

- 5% error in $[CO_2]$: The resulting error in A/F is −1.2% when the oxygen balance is used and +1.2% when the carbon balance is used.

- 5% error in [CO]: The resulting error in A/F is +1% when the oxygen balance is used and +4.1% when the carbon balance is used.

- 5% error in $[HC_1]$: The resulting error in A/F is the same +0.4% in the case of both the oxygen and carbon balances.

Errors in the value of k_p:

Rows 5 and 6 of Table 4.8 show, again, the effect on A/F of varying the value of k_p between 3 and 5 from its base value of 3.8. The effect on the estimate of A/F is fairly small. In the case of the oxygen balance, the calculated A/F changes by −2.7 to +2.9%, while in the case of the carbon balance the change in the calculated value of A/F is 0 to −1.2%.

Errors in the value of $[H_2O]$:

Row 8 shows that reducing the value of $[H_2O]$ by −30% increases the A/F value calculated from the carbon balance by 4%. As mentioned above, the value of A/F calculated from the oxygen balance is not affected since knowledge of $[H_2O]$ is not required for this calculation.

Errors in the value of y:

Row 10 shows that reducing the value of the atomic H/C ratio from 2 to 1.8 causes small changes in the estimated value of A/F. These changes are −1.7% when A/F is calculated from the oxygen balance and +3% in the case of the carbon balance.

4.8 Total Cylinder Dumping

An alternative method to the fast-acting in-cylinder sampling valve is the total cylinder dumping valve. This valve allows rapid emptying of the entire cylinder contents into a collection bag or vessel, for subsequent analysis for CO_2, NO_x, soot particle mass, etc. Dumping of the cylinder contents is initiated at a predetermined crank angle of the engine cycle, and the dumping process can be completed in around 1 ms or so. The rapid blowdown of the cylinder contents into the collection system quenches the cylinder contents, helping to arrest the chemical reactions occurring in the cylinder contents.

The total cylinder dumping method allows selection of the crank angle at which dumping of the cylinder contents begins. Thus, this aspect of its function is similar to that of the fast-acting in-cylinder sampling valve, which also allows the crank angle at which sampling begins to be selected. The total cylinder dumping method allows the cylinder contents of a single engine cycle to be collected and analyzed. The fast-acting in-cylinder sampling valve also allows a sample from a single engine cycle to be collected, but the much smaller sample volume available required

Table 4.8 Effects of various errors on the calculated value of *rich* gravimetric air fuel ratios

Row	[CO_2] %	[CO] %	[NO] %	[NO_2] %	[O_2] %	[HC_1] %	[H_2O] %	O bal. A/F	O bal. Δ(A/F) %	C bal. A/F	C bal. Δ(A/F) %
				Effect on A/F of neglecting, progressively, NO, NO_2, O_2							
1	5.5	11.4	0.05	0.005	0.8	1.5	10.9	9.3	–	9.3	–
2	5.5	11.4	neg	neg	0.8	1.5	10.9	9.3	–0.2	9.3	–
3	5.5	11.4	neg	neg	neg	1.5	10.9	8.9	–4.7	9.3	–
				Effect on A/F of value of k_p							
4	5.5	11.4	0.05	0.005	0.8	1.5	10.9	9.3 $k_p = 3.8$	–	9.3 $k_p =$ 3.8	–
5	5.5	11.4	0.05	0.005	0.8	1.5	10.0	9.1 $k_p = 3$	–2.7	9.2 $k_p = 3$	0
6	5.5	11.4	0.05	0.005	0.8	1.5	11.9	9.6 $k_p = 5$	+2.9	9.3 $k_p = 5$	–1.2
				Effect on A/F of value of [H_2O]							
7	5.5	11.4	0.05	0.005	0.8	1.5	10.9	9.3	–	9.3	–
8	5.5	11.4	0.05	0.005	0.8	1.5	10.9	9.3	–	9.6	+4.0
				Effect on A/F of value of y							
9	5.5	11.4	0.05	0.005	0.8	1.5	10.9	9.3 y = 2	–	9.3 y = 2	–
10	5.5	11.4	0.05	0.005	0.8	1.5	9.8	9.2 y = 1.8	–1.7	9.6 y = 1.8	+3.0

(i) **neg** = value neglected.
(ii) Δ(A/F) % = deviation from base estimates of A/F (base estimates repeated in rows 1, 4, 7, and 9).
(iii) A/F values are on gravimetric basis; CO_2%, CO%, etc are the molar (volumetric) fractions percent.
(iv) Base values used: y = 2 and k_p = 3.8, except where otherwise shown.

more specialized analysis techniques. There is one fundamental difference between the total cylinder dumping method and the fast-acting in-cylinder sampling valve technique. In the latter technique a sample is collected from a single location in the combustion chamber, while in the former the entire cylinder contents are collected. Thus, the fast-acting in-cylinder sampling valve is capable of a degree of spatial resolution, while the total cylinder damping method is capable of providing space-averaged results.

4.8.1 Cylinder Dumping Valves

Fig. 4.13(a) shows a schematic diagram of a total cylinder dumping valve developed by Kittelson et al. [1986]. T he diaphragm seals the combustion chamber of the engine, allowing normal combustion to take place. In one version of the valve described by Pipho et al. [1992], the diaphragm was made of 0.018 mm thick brass shim, which was softened by heating and quenching. A cutter tube made of stainless steel was used to cut the diaphragm at a selected crank angle during the engine cycle. The cutter tube had an internal diameter of approximately 13 mm, and its bottom end was shaped and sharpened in such a way as to provide reliable cutting of the diaphragm. The cutter tube was driven downward by high-pressure helium acting on a piston which was welded onto the upper end of the cutter tube. An electronic control system activated a solenoid valve at a suitable instant during the engine cycle, and this allowed the high-pressure helium to act upon the cutter tube piston. When the diaphragm was cut by the cutter tube, cylinder gas pressure immediately caused it be folded back against the support, much like a pair

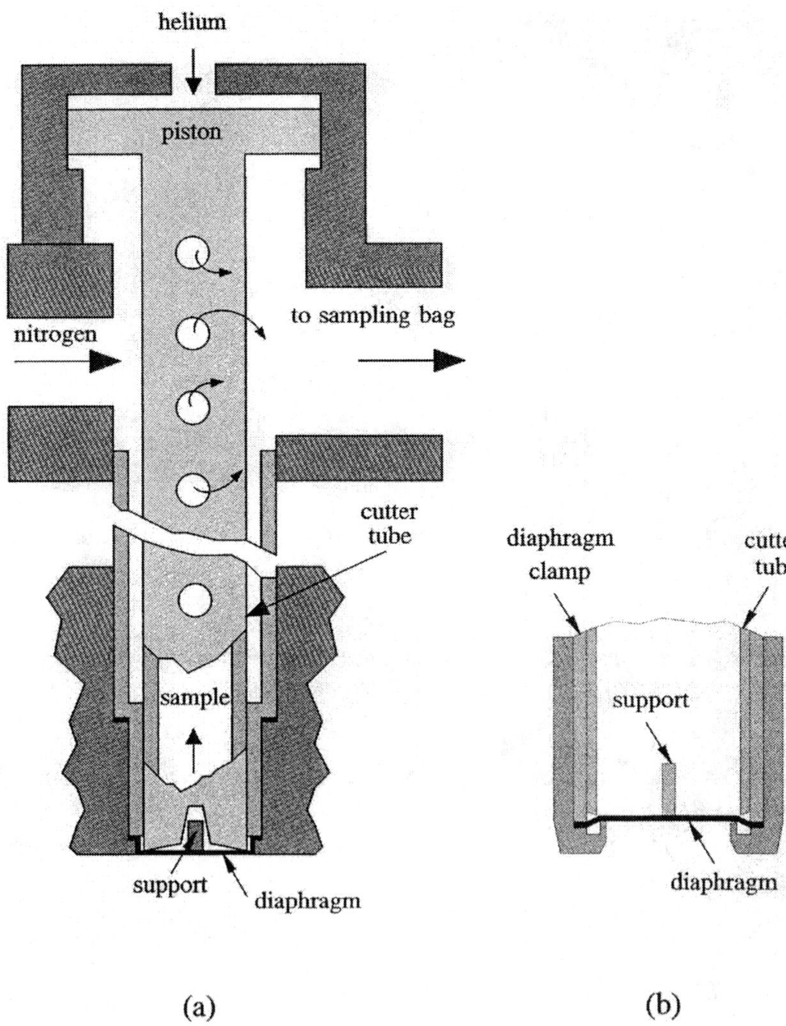

Fig. 4.13 Schematic diagram of (a) a total cylinder dumping valve and (b) an alternative tip design.

of butterfly wings. The cylinder gas was then able to flow inside the cutter tube and out of a series of small holes drilled onto the walls of this tube. Nitrogen was used to cool and dilute the cylinder gas, and the combined stream was collected into a sampling bag.

Voiculescu and Borman [1978] also described the construction and operation of a total cylinder dumping valve. The tip of their valve is shown in Fig. 4.10(b). The diaphragm was clamped by the clamping tube and further supported by a cross member along its diameter, as shown in this figure. The diaphragm was made of 0.51 mm thick deep-forming steel, and reached temperatures of 490°C when installed in a running engine. The cutter tube was made of stainless steel, except for its bottom cutting end, which was made of hardened steel and sharpened. A spring-loaded lever arm, triggered by a solenoid, drove the cutter tube downward at a selected instant during the engine cycle.

4.8.2 Cylinder Dumping Procedure

The experimental procedure used by Kittelson et al. [1988] enabled two types of cylinder dumping experiments to be performed: fired and non-fired. The fired experiments involved the engine firing only once before the cylinder contents were dumped, in order to ensure that a representative quantity of combustion residuals was present in the cylinder during the second of the two fired engine cycles. In the cylinder dumping experiment, the engine was motored at constant speed. At a convenient instant, the injection system was activated for two engine cycles only, which caused the engine to fire for only two consecutive engine cycles. During the compression stroke of the second fired cycle, the engine inlet and exhaust valves were deactivated and the cylinder contents dumped at the desired crank angle. The engine inlet and exhaust valves were deactivated in order to ensure that no extraneous contaminant gases entered the cylinder and contaminated the sample before the sample collection system was isolated from the cylinder. The non-fired experiments were almost identical, except that only one fired cycle occurred. The cylinder contents were dumped during the unfired cycle immediately following the fired one [Kittelson et al., 1988].

Elevated coolant temperatures of 127°C were used in an attempt to match, as far as possible, the combustion wall temperatures to those prevailing when the engine was fully warmed up and firing continuously at steady state. Before each engine test, the engine combustion chamber walls were thoroughly cleaned to ensure that no wall deposits were entrained in the sample. Eliminating the entrainment of these deposits was particularly important during studies of soot particle growth and oxidation in the engine cylinder. Early experiments had shown that cylinder dumping after continuous firing resulted in severe reentrainment problems involving particles that had collected on the combustion chamber walls [Pipho et al., 1992]. Cleaning of the cylinder walls prior to each test, combined with firing the engine for only two cycles, largely solved the problem of entrainment of wall deposits.

4.8.3 Cylinder Dumping Valve Performance

Kittelson et al. [1988] were able to set the start-of-cylinder-dumping crank angle accurately and with good reproducibility. This was possible because a time delay of about 5 or 6 ms between the electronic command signal and the start of cylinder dumping was amenable to calibration and, therefore, prediction. Once the start of cylinder dumping was initiated, the cylinder pressure decayed very rapidly. Typically, it reached one-half of its initial value within 0.6 to 1.3 ms (5.4 to 11.7° CA at 1500 rpm) [Pipho et al., 1991].

Extensive modeling of the NO and NO_2 reaction chemistry reported by Pipho et al. [1991] showed that measurements of the NO_2/NO ratio using the cylinder dumping method are inaccurate due to conversion of NO to NO_2 during the dumping process. However, the NO_x measurements were shown to be representative of the cylinder concentrations.

In the case of the system described by Voiculescu and Borman [1978], the diaphragm became completely open within a period of 0.09 to 0.15 ms after being cut, depending on cylinder gas pressure. The subsequent expansion of the cylinder gas lasted between 0.16 and 0.5 ms. Extensive modeling of the NO chemistry during dumping and subsequent experimental tests led Voiculescu and Borman to conclude that this method provides a reasonably accurate way of obtaining time-resolved values of the average concentration of NO in the cylinder.

4.9 Summary

In this chapter, a survey of fast-acting sampling valves and their applications to IC engines is presented. The design and implementation of various types of fast-acting sampling valves are discussed in detail. More on the application of fast-acting sampling valves will be presented in some of the later chapters.

While a fast-acting sampling valve provides localized measurements, the total cylinder dumping technique has been developed to obtain the total concentration of combustion products in the cylinder from a single cycle. The application of the total cylinder dumping technique to study in-cylinder soot and NO_x production will be described in Chapters 11 and 13.

References

Adamczyk, A.A., Kaiser, E.W., and Lavoie, G.A. (1983), "Single-Pulse Sampling Valve Measurements of Wall Layer Hydrocarbons in a Combustion Bomb," *Combustion and Flame*, 52, pp. 1–9.

Alperstein, M. and Bradow, R.L. (1965), "Combustion Gas Sampling Valve," *The Review of Scientific Instruments*, Vol. 36, No. 7, pp. 1028–1031.

Aoyagi, Y., Kamimoto, T., Matsui, Y., and Matsuoka, S. (1980), "A Gas Sampling Study on the Formation Processes of Soot and NO in a DI Diesel Engine," SAE Paper No. 800254.

Asanuma, T. and Yanagihara, S. (1962), "Gas Sampling Valve for Measuring Scavenging Efficiency in High-Speed Two-Stroke Engines," *SAE Transactions*, Vol. 70, SAE Paper No. T-47.

Ayusawa, T., Nemoto, T., Koo, Y., and Jo, S.H. (1978), "Relationship Between Local Air-Fuel Ratio and Combustion Character in Spark Ignition Engines," SAE Paper No. 780147.

Backer, H., Wiartalla, A., and Durnholz, M. (1995), "Investigation of Spray Propagation, Combustion and Soot Formation by Optical Measurement Techniques for Diesel Engine Fuels in a Model Combustion Chamber," 21st International Congress on Combustion Engines, CIMAC, Interlaken.

Backer, H., Hild, O., and Lepperhoff, G. (1997), "Investigation on Particle Size Distribution in the Cylinder of a Passenger Car DI Diesel Engine Using a Fast Gas Sampling Valve," SAE Paper No. 970875.

Barbella, R., Betoli, C., Ciajolo, A., D'Anna, A., and Masi, S. (1989), "In-Cylinder Sampling of a High Molecular Weight Hydrocarbons from a DI Light-Duty Diesel Engine," SAE Paper No. 890437.

Bartholomew, E., Chalk, H., and Brewster, B. (1938), "Carburetion, Manifolding and Fuel Antiknock Value," *SAE Transactions*, Vol. 33, pp. 141–156, 1938.

Bennethum, J.E., Mattavi, J.N., and Toepel, R.R. (1975), "Diesel Combustion Chamber Sampling—Hardware, Procedures, and Data Interpretation," SAE Paper No. 750849.

Cooper, D.E., Courtney, R.L., and Hall, C.A. (1959), "Radioactive Tracers Cast New Light on Fuel Distribution," *SAE Trans*., Vol. 67, pp. 619–639.

Daniel, W.A. (1957), "Flame Quenching at the Walls of an Internal Combustion Engine," *Proc. Sixth Symposium (International) on Combustion*, The Combustion Institute.

Daniel, W.A. and Wentworth, J.T. (1962), "Exhaust Gas Hydrocarbons—Genesis and Exodus," SAE Paper No. 48613.

Donahue, R.W. and Kent, R.H. (1950), "A Study of Mixture Distribution," *SAE Quart. Trans*., Vol. 4, No. 4, pp. 546–558.

Downs, A.D., Walsh, A.D., and Wheeler, R.W. (1951), "A Study of the Reactions that Lead to Knock in the Spark-Ignition Engine," *Phil. Trans. Roy. Soc. London*, Vol. 243, No. A870, pp. 463–524.

Drinkwater, J.W. and Egerton, A.C., (1938), "The Combustion Process in the Compression-Ignition Engine," *Proc. IMechE*, Vol. 38, pp. 415–495.

Duggal, V.K., Priede, T., and Khan, I.M. (1978), "A Study of Pollutant Formation within the Combustion Space of a Diesel Engine," SAE Paper No. 780227.

Egerton, A., Smith, F.L., and Ubbelohde, A.R. (1935), " Estimation of the Combustion Products from the Cylinder of the Petrol Engine and its Relation to Knock," *Phil. Trans. Roy. Soc. London,* Vol. 234, No, 744, pp. 433–521.

Fujiwara, Y., Fukazawa, S., Tosaka, S., and Murayama, T. (1984), "Formation of Soot Particulates in the Combustion Chamber of a Pre-Combustion Chamber Type Diesel Engine," SAE Paper No. 840417.

Furuhama, S. and Tateishi, Y. (1972), "Gases in Piston Top-Land Space of a Gasoline Engine," *Transactions of the Society of Automotive Engineers of Japan*, No. 4, pp.30–39.

Garner, F.H., Malpas, W.E., Morton, F., Reid, W.D., and Wright, E.P. (1952), "Pre-Flame Reactions in Diesel Engines (Part II)," *Inst. Petroleum*, 38, No. 341, pp. 312–343.

Gere, J.M. and Timoshenko, S.P. (1985), *Mechanics of Materials*, PWS Engineering (Boston, Massachusetts) or Wadsworth International (Belmont, California), 2nd SI Edition.

Heywood, J.B. (1988), *Internal Combustion Engine Fundamentals*, McGraw-Hill, New York.

Isonomy, T. and Yanagihara, S. (1962), "Gas Sampling Valve for Measuring Scavenging Efficiency in High-Speed Two-Stroke Engines," *SAE Transactions* T-47, pp. 420–433.

Johnston, S.C. and Green, R.M. (1982), "Engine Hydrocarbon Concentration Measurements: A Comparison Between a Sampling Probe and Raman Spectroscopy," SAE Paper No. 820499.

Kajima, T. and Kawamura Y. (1995), "Development of a High-Speed Solenoid Valve: Investigation of Solenoids," *IEEE Transactions on Industrial Electronics*, Vol. 42, No. 1, pp.1–8.

Kano, M., Saito, K., Basaki, M., Matsushita, S., and Gohno, T. (1998), "Analysis of Mixture Formation of Direct Injection Gasoline Engine," SAE Paper No. 980157.

Kato, S., Takayama, Y., Sato, G. T., Tanabe, H., and Sato, T. G. (1997), "Investigation of Particulate Formation of DI Diesel Engine with Direct Sampling from Combustion Chamber," SAE Paper No. 972969.

Kittelson, D.B., Pipho, M.J., Ambs, J.L., and Siegla, D.C. (1986), "Particle Concentrations in a Diesel Cylinder: Comparison of Theory and Experiment," SAE Paper No. 861569.

Kittelson, D.B., Pipho, M.J., Ambs, J.L., and Luo, L. (1988), "In-Cylinder Measurements of Soot Production in a Direct-Injection Diesel Engine," SAE Paper No. 880344.

Ladommatos, N., Balian, R., Horrocks, R., and Cooper, L. (1996), "The Effect of Exhaust Gas Recirculation on Combustion and NO_x Emissions in a High-Speed Direct-Injection Diesel Engine," SAE Paper No. 960840.

Ladommatos, N., Balian, R., Horrocks, R., and Cooper, L.(1996), "The Effect of Exhaust Gas Recirculation on Soot Formation in a High-Speed Direct-Injection Diesel Engine," SAE Paper No. 960841.

Lepperhoff, G. and Houben, M. (1990), "Particulate Emissions and Soot Formation Processes by Diesel Engines," IMechE Conference Paper C394/032.

Lepperhoff, G. and Houben, M. (1990), "Soot Formation and Oxidation in IDI and DI Diesel Engine," SIA Congress, SIA No. 90097.

LoRusso, J.A., Lavoie, G.A., and Kaiser, E.W. (1980), "An Electro-Hydraulic Gas Sampling Valve with Application to Hydrocarbon Emissions Studies," SAE Paper No. 800045.

LoRusso, J.A., Kaiser, E.W., and Lavoie, G.A. (1983), " In-Cylinder Measurements of Wall Layer Hydrocarbons in a Spark-Ignition Engine," *Combustion Science and Technology*, Vol. 33, pp. 75–112.

LoRusso, J.A., Lavoie, G.A., Forgacs, R.L., Jeryan, R.A., and Errante, J. (1984), "Electrohydraulic Gas Sampling Valve," *American Institute of Physics, Rev. Sci. Instrum.*, Vol. 55, No. 5.

Lovell, W.G., Coleman, J.D., and Boyed, T.A. (1927), " Studies of Combustion in the Gasoline Engine," *Industr. and Eng. Chemistry*, Vol. 19, No. 3, pp.373–76.

Luck, C.J., Burgess, A.R., Desty, D.H., Whitehead, D.M., and Pratley, G. (1973), "A Study of the Combustion of n-Heptane in an Engine using a Novel High-Speed Sampling Technique," Proc. of 14th Symposium (International) on Combustion, pp. 501–512.

Lyford-Pike, E.J. and Heywood, J.B.(1984), "Thermal Boundary Layer Thickness in the Cylinder of a Spark Ignition Engine," *Int. J. Heat Mass Transfer*, Vol. 27, No. 10, pp. 1873–1878.

Matsui, K., Tanaka, T., and Ohigashi, S. (1979), "Measurement of Local Mixture Strength at Spark Gap of SI Engines," SAE Paper No. 790483.

Nightingale, D.R. (1975), "A Fundamental Investigation into the Problem of NO Formation in Diesel Engines," SAE Paper No. 750848.

Ning, M., Yuan-Xian, Z., Zhen-Huan, S., and Hu-Guo-Dong (1991), "Soot Formation, Oxidation and its Mechanism in Different Combustion Systems and Smoke Emission Patterns in DI Diesel Engines," SAE Paper No. 910230.

Norris-Jones, S.R., Hollis, T., and Waterhouse, C.N.F. (1984), "A Study of the Formation of Particulates in the Cylinder of a Direct Injection Diesel Engine," SAE Paper No. 840419.

Ohdaci Y, Kawase Y, Murakami Y, and Inaguma Y. (1991), "Optimum Design of Dynamic Response in Automotive Solenoid Valve," *IEEE Transactions on Magnetics*, Vol. 27, No. 6, Pt. 2, pp. 5226–5228.

Ozasa, T., Suzuki, S., Kondo, H., Takahashi, H., Kataoka, M., and Oshima, Y. (1991), " Effect of Combustion Chamber Shape on a Lean-Burn SI Engine—Measurement of In-Cylinder Mixture Strength Using a Gas Sampling Technique," *JSAE Review*, Vol. 12, No. 2, pp. 4–11.

Pipho, M.J., Kittelson, D.B., and Zarling, D.D. (1991), "NO_2 Formation in a Diesel Engine," SAE Paper No. 910231.

Pipho, M.J., Kittelson, D.B., Luo, L. and Zarling, D.D. (1992), "Injection Timing and Bowl Configuration Effects on In-Cylinder Particle Mass," SAE Paper No. 921646.

Perry, R.H. and Chilton, C.H. (1974), *Chemical Engineers' Handbook*, McGraw Hill, 5th Ed. London.

Quader, A.A. (1982), "The Axially-Stratified-Charge Engine," SAE Paper No. 820131.

Rathbun, F.O. (1965), "Metal to Metal and Metal Gasket Seals," Aerospace Fluid Power Systems and Equipment Conference, Los Angeles, p. 156, SAE Paper No. 650312.

Rhee, K.T., Myers, P.S., and Uyehara, O.A. (1978), "Time- and Space-Resolved Species Determination in Diesel Combustion Using Continuous Flow Gas Sampling," SAE Paper No. 780226.

Rifkin, E.B. and Walcutt, C. (1956), "Decomposition of Tetra-Ethyl Lead in an Engine," *Industr. and Eng. Chemistry*, Vol. 48, No. 9, pp. 1532–1539.

Seilly, A.H. (1979), "Helenoid Actuators: A New Concept in Extremely Fast Acting Solenoids," SAE Paper No. 790119.

Seilly, A.H. (1981), "Colenoid Actuators—Further Developments in Extremely Fast Acting Solenoids," SAE Paper No. 810462.

Shimazaki, N., Hatanaka, H., Yokota, K., and Nakahira, T. (1996), "A Study of Diesel Combustion Process Under the Condition of EGR and High-Pressure Fuel Injection with Gas Sampling Method," SAE Paper No. 960030.

Taylor, C.F. and Rogowski, A.R. (1954), "Scavenging the 2-Stroke Engine," *SAE Trans.*, Vol. 62, pp. 486–502.

Voiculescu, I.A. and Borman, G.L. (1978), "An Experimental Study of Diesel Engine Cylinder-Averaged NOx Histories," SAE Paper No. 780228.

Weiss, P. and Keck, J.C. (1981), "Fast Sampling Valve Measurements of Hydrocarbons in the Cylinder of a CFR Engine," SAE Paper No. 810149.

Whitehouse, N.D., Clough, E., and Roberts, P.S. (1977), "Investigating Diesel Engine Combustion by Means of a Timed Sampling Valve," SAE Paper No. 770409.

Withrow, L., Lovell, W.G., and Boyed, T.A. (1930), "Following Combustion in the Gasoline Engine by Chemical Means," *Industr. and Eng. Chemistry*, Vol. 22, No. 9, pp. 945–957.

Wolfe, M., Stromberg, S., Houben, M., and Lepperhoff, G. (1993), "Influence of Engine Operating Parameters on Pollutant Formation during Diesel Engine Combustion," IMechE Conference Paper C465/040/93.

Yates, D.A. (1988), "The Compact High-Speed Gas Sampling Valve for an Internal Combustion Engine," IMechE Conference "Experimental Methods in Engine Research and Development," London.

Zhao, H., Lowry, G., and Ladommatos, N. (1996), "Time-Resolved Measurements and Analysis of In-Cylinder Gases and Particulates in Compression-Ignition Engines," SAE Paper No. 961168.

Zvonow, V.A., Srewart, H.E., and Starkman, E.S. (1968), "Hydraulically Actuated Combustion Gas Sampling Valve," *The Review of Scientific Instruments*, Vol. 39, No. 12, pp. 1820–1824.

Chapter 5

Spectroscopic Techniques and Experimental Considerations

Outline

5.1 Introduction

In the last decade, significant advances have been made in the areas of laser-based spectroscopic diagnostics, particularly the application of two-dimensional imaging of scattered laser light in internal combustion engines. Most of these laser diagnostic techniques are based on three main light scattering processes: Rayleigh scattering, Raman scattering, and laser induced fluorescence. Coherent anti-Stokes Raman scattering (CARS), a wave-mixing technique based on the Raman effect, will be discussed in Chapter 11 because of its particular use for in-cylinder gas temperature measurements.

In this chapter, readers are first introduced to the salient features of molecular spectroscopy. This is then followed by the description of the principles of laser-based spectroscopic techniques. The last section aims to acquaint the readers with the experimental equipment used in performing laser diagnostics and with the general considerations involved in their use. It is hoped that the background given here will provide the reader with a good understanding of the technical terms used in the laser spectroscopy and equipment required to perform laser-based diagnostics.

5.2 Fundamentals of Molecular Spectroscopy

5.2.1 Introduction

The objective of this section is to provide a descriptive and non-mathematical introduction to the general theory and nomenclature of molecular spectra to serve as a background for the reader who is unacquainted with spectroscopic theory. For a complete treatment of the theory of molecular spectra, the reader is referred to the books by Herzberg [1950; 1971] and Eckbreth [1996].

Laser spectroscopic techniques involve the interaction and exchange of energy between a photon and a molecule. (Mie scattering originates from particulate matter and does not involve direct interaction between a photon and a molecule, and hence is excluded from spectroscopic techniques). The nature of the interaction is specifically governed by the internal energy system of the molecule. Therefore, a good understanding of the internal energy system of a molecule is essential for the application of laser spectroscopic techniques.

The internal energy of a molecule is shared between its electronic energy, vibrational energy, and rotational energy. When the average distribution between these types of internal energy, for an ensemble of molecules, is equal, the system is said to be in thermal equilibrium. In all three modes of excitation, the energy levels are quantized. The molecules can only contain a precise quantum of energy, which is determined by the energy levels associated with each energy mode. The internal energy of a single molecule is defined by its occupation of specific electronic, vibrational, and rotational levels, i.e.,

$$E_{total} = E_{electronic} + E_{vibrational} + E_{rotational}$$

Fig. 5.1 shows the energy levels for the ground ($X^2\Pi$) and first excited electronic states ($A^2\Sigma$) of OH. In each electronic energy state, a number of vibrational levels (v' and v'') exist, in which a number of rotational energy levels (J and K) reside. Note that the energy levels are discrete as dictated by quantum mechanics. Thus, the radiation associated with changes in energy states is not continuously distributed but is discrete or quantized. The electron configurations and their vibrational and rotational states determine the energy levels possessed by a given configuration. Each configuration possesses a number of energy levels. Change from one energy level to another results from absorption (low to high) or emission (high to low) of radiation. Spectroscopy basically consists of a set of rules governing the allowable configurations, configurational changes, and energies associated with these changes.

Radiative transitions between energies occur at an optical frequency determined by the Planck's law, namely

$$E_1 - E_2 = h\nu = hc\tilde{\nu}$$

Where h is a universal constant known as Planck's constant, ν is the frequency of the light wave, and c is the speed of light. $\tilde{\nu}$ is termed the wave number (in cm^{-1}); it is typically used by spectroscopists, and is a particularly convenient energy measure, as used in Fig. 5.1. For a given wavelength of light, the number of wave numbers, $\tilde{\nu}$, is just the reciprocal of the wavelength, i.e., the number of wavelengths per unit distance. E_1 and E_2 refer to the energy levels possessed by the molecule involved in the transition.

The electronic states of diatomic molecules, as well as their vibrational and rotational energies, are discussed here due to their simplicity and relevance to most combustion diagnostics.

5.2.2 Rotational Energy of Diatomic Molecules

Molecules can rotate only at precise frequencies, successively higher frequencies forming a ladder of energies as shown on the right-hand side of Fig. 5.1,

$$E_r = hcBJ(J+1) \tag{5.1}$$

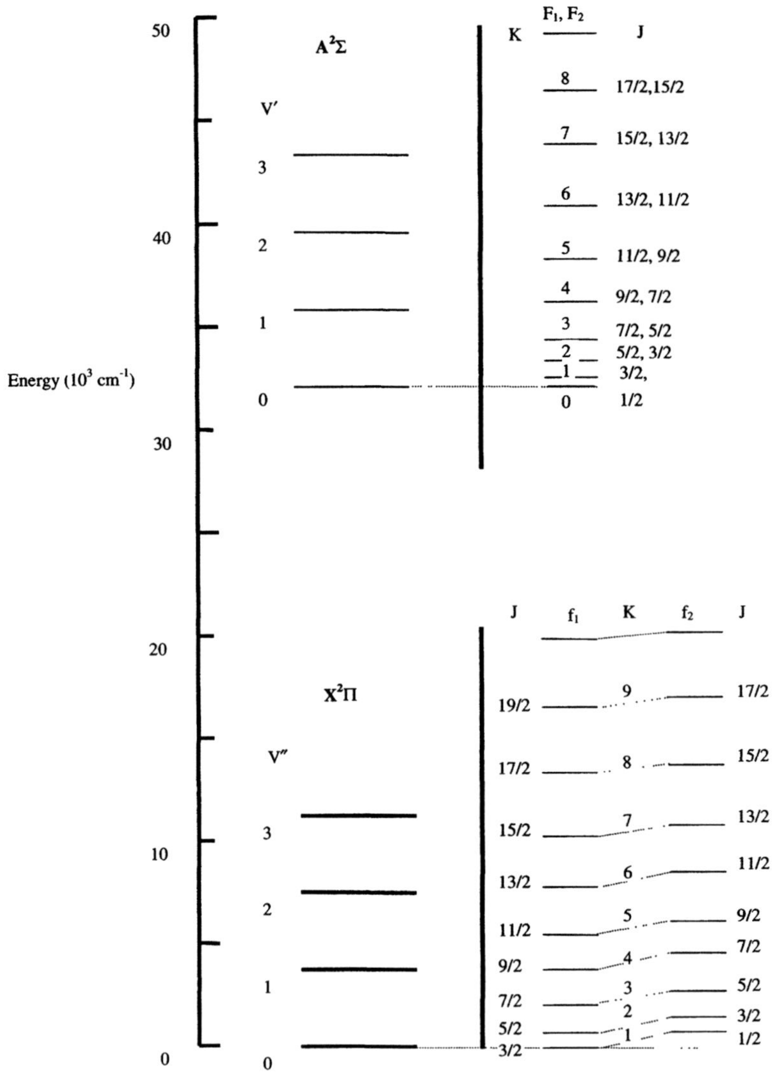

Fig. 5.1 Energy level diagram for the ground and first electronically excited states of the OH radical.

where J is the rotational quantum number, and B is termed the rotational constant. If centrifugal effects are taking into account, a more accurate description of rotational energy is given by

$$E_r = hc\left[BJ(J+1) - DJ^2(J+1)^2\right] \tag{5.2}$$

where D is introduced as the centrifugal distortion constant. Photon energies corresponding to pure rotational energy changes are such that the wavelength is in the microwave region.

5.2.3 Vibrational Energy of Diatomic Molecules

Motion of an atom relative to another in a molecule occurs through specific vibrational motions. The associated vibrational energies are modeled as harmonic oscillators and are also quantized:

$$E_v = h\nu(v + 1/2) \tag{5.3}$$

The symbol v is the vibrational quantum number, which takes integer values, 0, 1, 2, etc. For a more accurate calculation of vibrational energy, vibrating diatomic molecules should be modeled as anharmonic oscillators, which will result in higher-order terms appearing in the allowed energy:

$$E_v = hc\left[\omega_e(v + 1/2) - \omega_e x_e(v + 1/2)^2 + \omega_e y_e(v + 1/2)^3 + \ldots\right] \tag{5.4}$$

where $\omega_e = \nu / c$ is the equilibrium energy in cm^{-1} and $\omega_e x_e, \omega_e y_e \ll \omega_e$.

5.2.4 Coupling Between Vibrational and Rotational Modes

Normally, a molecule simultaneously rotates and vibrates. The vibration of the molecule will lead to a change in the rotational moment of inertia, which will cause a change in the rotational energy storage due to vibration. This effect is modeled by the following equations:

$$B_v = B_e - \alpha_e(v + 1/2) \tag{5.5}$$

$$D_v = D_e - \beta_e(v + 1/2) \tag{5.6}$$

where the subscript e refers to the equilibrium positions of the nuclei and α_e, β_e are small correction factors. Thus, the rotational constant B and its centrifugal correction D in Eq. (5.2) will depend on the magnitude of vibration, and therefore on v. The constant α_e is called the vibration-rotation interaction constant, and is important for laser Raman diagnostics.

5.2.5 Electronic States of Diatomic Molecules

The motion of electrons is also quantized; however, unlike rotations and vibrations, the energy levels do not form a simple ladder. Fig. 5.2 shows the low-lying electronic states of OH.

The symbols for the electronic energy levels indicate the character of the energy state, for instance, spin and symmetry. A brief description of the physical implications of these symbols and electronic structure is warranted here in order for readers to appreciate the salient features of molecular spectroscopy.

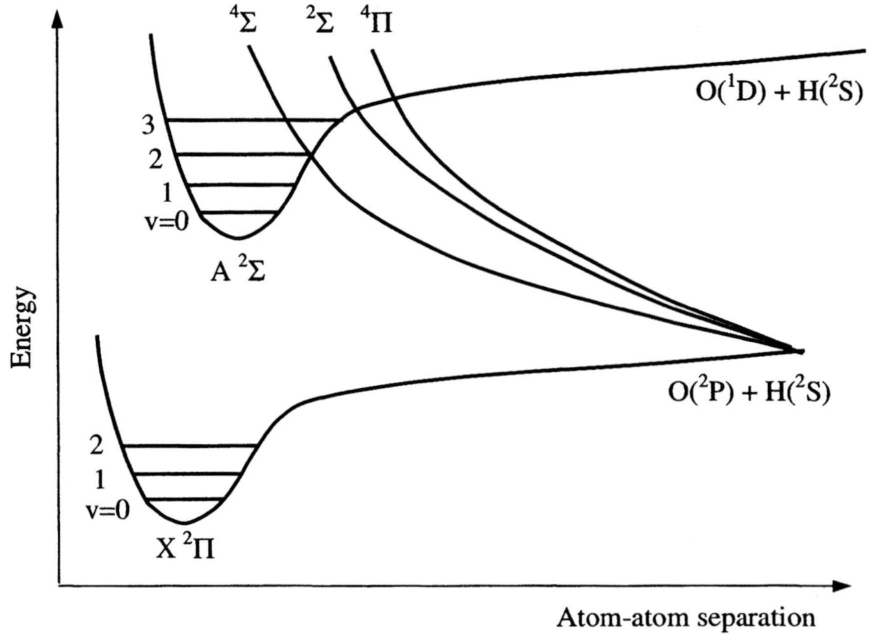

Fig. 5.2 Energy level diagram for the OH radical with electronic and vibrational levels.

Three quantum numbers, n, l, and λ, are employed to describe the state of individual electrons. n represents the approximate radial distance of an electron from the nucleus, known as the principle quantum number. l is associated with the value of the angular momentum of the electron in its orbit, which is quantized. The number l takes the value 0, 1, 2, 3, etc., up to a maximum value of n − 1, and is known as the azimuthal or orbital angular momentum quantum number. According to the value of l, the electrons are named as s ($l = 0$), p ($l = 1$), d ($l = 2$), f ($l = 3$), and so on alphabetically. The third quantum number, λ, is given by $\lambda \equiv |m_l|$, where m_l is related to the quantized component of the orbital angular momentum along the internuclear axis assuming the integer values $l, l - 1 \ldots -l$. The significance of λ is that the axial component of orbital angular momentum is $h\lambda/2\pi$. Electrons with $\lambda = 0, 1, 2, 3 \ldots$ etc. are termed σ, π, δ, ϕ electrons. Note that in the case $\lambda \neq 0$, there are two values m_l, either positive or negative. Thus, all such states are doubly degenerate. As a result of the degeneration, these levels will be split, a situation termed lambda-doubling, when the coupling of the electronic and rotational modes is considered.

In summary, individual electrons are identified by writing $nl\lambda$, thus 1sσ, 3dπ, etc. in the united atom approach [Herzberg, 1950; 1971]. In addition, a spin quantum number, m_s, is also used to describe the spin momentum of an electron, which takes half integral values $\pm 1/2$ depending on whether the spin is with or opposed to the orbital motion.

When several electrons are involved in a molecule, each electron can be considered separately and described by the quantum numbers n_i, l_i, λ_i. In general, several electronic states emerge from a given electronic configuration. Each electronic state of a molecule can usually be described by two quantum numbers, Λ and S. Λ is related to the total orbital angular momentum of the electrons along the internuclear axis, given by $\Lambda = |m_l|$. It takes the integral values 0, 1, 2, 3, etc. With $\Lambda = 0, 1, 2, 3$, etc., the electronic state is referred to by Greek capital letters Σ, Π, Δ, etc. The other quantum number of importance is the resultant spin of the electrons, S, which may be integral or half integral according to whether there is an even or odd number of outer electrons. The resultant spin of the electrons, $S = |\Sigma m_s|$, determines the multiplicity $M = 2S + 1$ of the state. Thus, any electronic state is represented by the multiplicity, written as a superscript on the left, and its Λ value. Therefore, a $^2\Sigma$ state (read as doublet sigma) has $M = 2$ (i.e., $S = 1/2$) and $\Lambda = 0$.

In denoting a transition between two electronic states of a molecule, it is customary to write the symbol for the state of higher energy first. To distinguish between electronic states of similar character, it is also common to name them by writing a letter, either small or capital, before the symbol. The ground electronic state is denoted by X, except in the case of C_2, in which the true lowest lying ground state was discovered only after a slightly higher energy level was assigned as the X state. A, B, C, etc. designate the excited electronic states, which have the same multiplicity. States of multiplicity different from the ground state, whose transitions to the ground state are normally forbidden, are denoted by small letters, a, b, c, etc.

Now we can see how the electronic states of OH, as shown in Fig.5.1, are decided. The configuration of OH in the ground electronic state is $(1s\sigma)^2\ (2s\sigma)^2\ (2p\sigma)^2\ (2p\pi)^3$, where superscripts indicate the number of electrons for a particular combination of $n\ l\ \lambda$. The first two shells, $(1s\sigma)^2\ (2s\sigma)^2$, are closed shells, i.e., shells that contain the maximum number of electrons that they can contain. Closed shells are very stable. It is generally the outer electrons that become excited, i.e., raised to higher lying energy states. This is because these electrons are less firmly bound to the nuclei due to shielding of the nuclear charge by the inner electrons. The three $2p\pi$ electrons in the OH ground electronic state have values of $n = 2$, $l = 1$, and $\lambda = 1$. m_l can take values of +1 or −1. According to the Pauli Exclusion Principle, no two electrons can possess the same n, l, m_l, and m_s. Given the two possible values of m_s, the possible values of m_l are 1, 1, −1 or −1, −1, 1. Therefore, Λ is 1 ($|\Sigma m_l|$) leading to a Π state. Similarly, m_s possibilities are −1/2, +1/2, −1/2 or +1/2, −1/2, +1/2 so the state multiplicity is 2. The ground state is therefore denoted by $^2\Pi$. The first excited state forms as one of the $2p\sigma$ electrons is excited into the $2p\pi$ level forming a closed shell. The single $2p\sigma$ electron then determines the first excited state term as $^2\Sigma$.

5.2.6 Coupling Between Rotational and Electronic Modes

Up to this point, the rotational and electronic modes have been treated independently of each other. In general, the coupling between the rotational and electronic motions must be considered in order to determine the detailed rotational energy levels. Because rotational lines are rarely resolved and used in engine combustion measurements, the detailed structure of rotational energy levels will not be discussed here. Again, interested readers are referred to the books by Herzberg [1950; 1971].

5.2.7 Population of Energy Levels

The previous discussion outlined the possible energy states that a molecule can attain by rotation, vibration, and electronic excitation. Of primary importance in laser diagnostics is how the molecules in any ensemble are distributed over these states. The distribution is prescribed by the Boltzmann equation:

$$N_j = N \cdot \frac{g_j e^{-E_j/kT}}{\sum_j g_j e^{-E_j/kT}} \tag{5.7}$$

where N_j is the number density of molecules in the jth state of energy, E_j. N is the total number density, k is Boltzmann's constant. g_j is the degeneracy of the state, which indicates the number of molecules that can occupy any given energy state. The denominator is termed the partition function and is denoted by Q.

The population of individual energy levels, within a rotational manifold of J, is given by:

$$N_J = \frac{N}{Q_{rot}} g_J (2J+1) \exp[-BJ(J+1)hc/kT] \tag{5.8}$$

where g_J is the nuclear spin degeneracy factor, which is important only for homonuclear diatomics, such as N_2, and has a value of 6 for even J levels and 3 for odd J levels for N_2. The rotational partition function, Q_{rot}, is described as:

$$Q_{rot} = \sum_{J=0}^{\infty} (2J+1) \exp\left[-hcBJ(J+1)/kT\right] \tag{5.9}$$

In the case of small hcB/kT, the summation can be replaced by integration, and one obtains

$$Q_{rot} = \frac{kT}{hcB} \tag{5.10}$$

The fractional population of an individual vibrational level is given by:

$$N_v = N\exp(-v \cdot hc \cdot \omega_e / kT)\left[1 - \exp(-hc \cdot \omega_e / kT)\right] \qquad (5.11)$$

where the last term is the vibrational partition function, Q_v, i.e.,

$$Q_v = \left[1 - \exp(-hc \cdot \omega_e / kT)\right] \qquad (5.12)$$

The partition of population among the rotational levels within a given vibrational state is therefore given by Eq. (5.8) with N replaced by N_v, i.e.,

$$N_J = \frac{N_v}{Q_{rot}} g_J (2J+1)\exp(-B_v J(J+1)hc/kT) \qquad (5.13)$$

The essential feature of population factors is that they are a function of temperature. In fact, it is precisely this feature that enables the measurement of temperature by inelastic scattering, such as Raman scattering and laser induced fluorescence. The form of the temperature dependencies for rotations and vibrations are very different, and they will be discussed in detail in Chapter 12.

5.2.8 Spectral Structure and Selection Rules

Spectral lines arise from transitions between differing electronic, vibrational, and rotational levels. Not all transitions are possible, and transitions are only allowed between electronic states for certain electron spins and certain electronic symmetry. Λ may change by 0 or ±1 only. The multiplicity, i.e., M or S, does not change at all for strong band systems. For example, the following are permitted transitions:

$$^1\Sigma \rightarrow {}^1\Sigma, \quad ^2\Sigma \rightarrow {}^2\Sigma, \quad ^1\Sigma \rightarrow {}^1\Pi, \quad ^1\Pi \rightarrow {}^1\Sigma$$

and the following would not be allowed:

$$^1\Sigma \rightarrow {}^1\Delta, \quad ^3\Sigma \rightarrow {}^3\Delta, \quad ^1\Sigma \rightarrow {}^3\Sigma$$

For a diatomic molecule, the change of vibrational quantum number is unrestricted during the electronic transition, i.e.,

$$\Delta v = 0, \pm 1, \pm 2 \ldots \text{etc.}$$

where the most important transitions for diagnostics are $\Delta v = \pm 1$.

Transitions between rotational states have the following selection rules, which are independent of Δv:

$$\Delta J = 0, \pm 1, \pm 2 \text{ only}$$

Transitions in rotational energy levels are grouped together according to the change in J values. Transitions for which $\Delta J = 0$ are termed Q-branch transitions and are only observed if $\Delta v = \pm 1$. When the change ΔJ of the rotational quantum number is ± 1, the two line series formed are referred to respectively as the R branch and P branch. Thus, the P branch lies in the shorter frequency (longer wavelength) side of the Q branch, whereas the R branch resides in the higher frequency (shorter wavelength) side of the Q branch. Similarly, if the rotational quantum number changes by ± 2, the two line series formed are named as the O branch and S branch, respectively. Therefore, from the low to high frequency, the five line series are the O branch ($\Delta J = +2$), P branch ($\Delta J = +1$), Q branch ($\Delta J = 0$), R branch ($\Delta J = -1$), and S branch ($\Delta J = -2$).

5.2.9 Spectroscopy of the OH Radical

Here we shall use the spectroscopy of the OH radical as an example to illustrate the salient features of molecular spectroscopy discussed above, and to introduce some additional features of diatomic molecules.

Two electronic states of OH are normally involved in its spectroscopy: the ground electronic state, $X^2\Pi$, and the first excited electronic state, $A^2\Sigma$, as shown in Fig. 5.1. In addition to the vibrational quantum number, v, and the rotational quantum number, J, which have been introduced in the previous sections, another quantum number, K, is used in OH spectroscopy and in spectroscopy of other similar diatomic molecules, e.g., NO spectroscopy. The quantum number K is related to the total angular momentum except electronic spin. Since the electronic spin takes values of $\pm \frac{1}{2}$, the total angular momentum quantum number that defines the rotational energy state is given by $J = K \pm \frac{1}{2}$. (This is know as Hund's case (b) in molecular spectroscopy.) As a result of the interaction between the electronic spin and the angular momentum, the rotational energy levels of the OH radical are split into doublets for each K value. The two components of the doublets are characterized by subscripts 1 and 2 for the same angular momentum number K in the rotational energies so that

$$F_1(K);\ f_1(K):\ J = K + \tfrac{1}{2}$$

$$F_2(K);\ f_2(K):\ J = K - \tfrac{1}{2}$$

F_1 and F_2 are used for the rotational levels of the upper electronic state, $^2\Sigma$; f_1 and f_2 for those of the $^2\Pi$ state. In addition, both f_1 and f_2 are further split into two slightly different energy levels because of the interaction between the orbital motion of the electrons and the rotation of the molecule, a phenomenon termed Λ-type doubling, for $\Lambda \neq 0$ states. The two sub-levels are referred to as f_i and f_i^*

Therefore, the twelve possible branches in OH spectroscopy are:

$O_{12}(K) = F_1(K-2) - f_2{}^*(K) \quad J-1 \Rightarrow J$

$P_1(K) = F_1(K-1) - f_1(K) \quad J-1 \Rightarrow J$

$P_2(K) = F_2(K-1) - f_2(K) \quad J-1 \Rightarrow J$

$P_{12}(K) = F_1(K-1) - f_2(K) \quad J \Rightarrow J$

$Q_1(K) = F_1(K) - f_1{}^*(K) \quad J \Rightarrow J$

$Q_{21}(K) = F_2(K) - f_1{}^*(K) \quad J-1 \Rightarrow J$

$Q_2(K) = F_2(K) - f_2{}^*(K) \quad J \Rightarrow J$

$Q_{12}(K) = F_1(K) - f_2{}^*(K) \quad J+1 \Rightarrow J$

$R_1(K) = F_1(K+1) - f_1(K) \quad J+1 \Rightarrow J$

$R_{21}(K) = F_2(K+1) - f_1(K) \quad J \Rightarrow J$

$R_2(K) = F_2(K+1) - f_2(K) \quad J+1 \Rightarrow J$

$S_{21}(K) = F_2(K+2) - f_1{}^*(K) \quad J+1 \Rightarrow J$

The subscripts indicate the component of the doublet of the initial and final energy levels. If the same components of the doublets are involved, only one subscript is used. The branches with only one index form strong branches, which are normally preferred in spectroscopic measurements. These branches are named depending on the change of the quantum number K as

O branch for $K-2 \Rightarrow K$

P branch for $K-1 \Rightarrow K$

Q branch for $K \Rightarrow K$

R branch for $K+1 \Rightarrow K$

S branch for $K+2 \Rightarrow K$

When a spectral band used in the spectroscopic measurement is specified, it is customary to write the electronic states involved first, followed by the vibrational numbers involved at each electronic state. For example, $A^2\Sigma \leftrightarrow X^2\Pi$ (0, 1), as used in the excitation of OH laser induced fluorescence, indicates that the OH is excited from the vibrational level 1 of the $X^2\Pi$ state to the ground vibrational level of the $A^2\Sigma$ state. In addition, specific spectral lines involved can be specified by the change of the quantum number K, e.g., $P_1(2)$ of $A^2\Sigma \leftrightarrow X^2\Pi$ (0, 1).

5.3 Raman and Rayleigh Scattering

5.3.1 Introduction

When a beam of light is passed through a transparent gas mixture, a small amount of the radiation energy is scattered. If monochromatic radiation or radiation of a very narrow frequency band, such as a laser beam, is used, the scattered energy will consist mostly of radiation of the incident wavelength, so-called *Rayleigh scattering*, and a very small amount of light at certain discrete wavelengths above and below that of the incident beam, which is referred to as *Raman scattering.*

The occurrence of Rayleigh and Raman scattering may be most easily understood in terms of the quantum theory of radiation. This treats radiation of frequency ν as consisting of a stream of particles, the so-called photons, having energy $h\nu$. Photons can be imagined to undergo collisions with molecules and, if the collisions are perfectly elastic, the photons will be deflected unchanged, which accounts for the process of Rayleigh scattering. In this case, a detector placed to collect energy at an angle to the incident beam will thus receive photons of energy $h\nu$, that is, radiation of frequency ν.

However, it may also happen that the energy is exchanged between the photon and molecule during the collision, the so-called inelastic process. The molecule may gain or lose some energy in accordance with quanta laws; i.e., the energy exchange, ΔE, must be the difference in energy between two of its allowed states. The photon will be scattered with energy $h\nu \pm \Delta E$, and the equivalent radiation will have frequency $\nu \pm \Delta E/h$.

Radiation scattered with a frequency lower than that of the incident beam is referred to as Stokes radiation, while that at higher frequency than that of the incident beam is called anti-Stokes. Since the former is accompanied by an increase in molecular energy, which can always occur subject to certain selection rules, while the latter involves a decrease in molecular energy, which can only occur if the molecule is originally in an excited state, Stokes radiation is generally more intense than anti-Stokes. Overall, however, the total radiation scattered at any but the incident frequency is extremely small, i.e., Raman scattering is much weaker than Rayleigh scattering.

5.3.2 Origin of Raman and Rayleigh Scattering

When a molecule is subject to a static electric field, it suffers certain distortion, the positively charged nuclei being attracted toward the negative pole of the field, and the electrons to the positive pole. This separation of charge centers causes an induced electric dipole moment to be set up in the molecule and the molecule is said to be polarized. The size of the induced dipole, μ, depends both on the magnitude of the applied field, E, and on the ease with which the molecule can be distorted,

$$\mu = \alpha E \tag{5.14}$$

where α is the polarizability of the molecule.

When a beam of light of frequency ν_o is passed through an ensemble of such molecules, the electric field experienced by each molecule varies according to the following equation:

$$E = E_o \sin 2\pi\nu_o t \tag{5.15}$$

Thus, the induced dipole also undergoes oscillations of frequency ν_o:

$$\mu = \alpha E = \alpha E_o \sin 2\pi\nu_o t \tag{5.16}$$

Such an oscillating dipole emits radiation of its own oscillation frequency, and we have immediately the classical explanation of Rayleigh scattering.

If, in addition, the molecule undergoes some internal motion as a result of the presence of radiation, such as vibration or rotation, which changes the polarizability, α, periodically, then the oscillating dipole will have superimposed upon it the vibrational or rotational oscillation. Supposing that the molecule has a vibration of frequency ν_{vib}, the polarizability is affected as follows

$$\alpha = \alpha_o + \beta \sin 2\pi\nu_{vib} t \tag{5.17}$$

where α_o is the equilibrium polarizability and β represents the rate of change of polarizability with the vibration, $\left(\frac{\partial \alpha}{\partial x}\right)_o$. Then we have:

$$\mu = \alpha E = \left(\alpha_o + \beta \sin 2\pi\nu_{vib} t\right) E_o \sin 2\pi\nu_o t \tag{5.18}$$

which can be rearranged to

$$\mu = \alpha_o E_o \sin 2\pi\nu_o t + 1/2\beta E_o \left\{\cos 2\pi\left(\nu_o - \nu_{vib}\right)t - \cos 2\pi\left(\nu_o + \nu_{vib}\right)t\right\} \tag{5.19}$$

The first term leads to Rayleigh scattering at the incident frequency, while the second term leads to scattered radiation shifted from the incident light by a characteristic frequency of the molecule, namely the Raman scattering. Note that it is the polarizability derivative, β, which is the source of the Raman scattering.

5.3.3 Fundamentals of Rayleigh Scattering from Molecules

Rayleigh scattering is elastic, i.e., unshifted, scattering of light from molecules or small particles. Only Rayleigh scattering from molecules will be discussed here, which forms the basis for concentration, density, and temperature measurements in a gas mixture. Rayleigh scattering from small particles will be dealt with in Chapter 11, where it is used for soot diagnostics in diesel engines

The fundamental principle governing Rayleigh scattering is the elastic collisions between gas molecules and incident laser light. The scattered light has the same wavelength as the incident light. The scattered light signal, in general, is directly proportional to the laser power, the gas density, and a differential cross section which depends on the gas that is being measured. If the scattered light from a gas mixture is detected, the above relationship about the scattered light signal can be written as,

$$P = \eta_c I_i \Omega V_c N \left(\frac{\partial \sigma}{\partial \Omega} \right) \tag{5.20}$$

where η_c = the detection efficiency of the collection system
I_i = the incident laser intensity, (W/cm^2)
N = the total number density of the gas molecules, (cm^{-3})
Ω = the solid angle of collection optics, (sr)
V_c = the collection volume, (cm^3)
$(\partial\sigma/\partial\Omega)$ = the differential cross section, (cm^2/sr), given by

$$\left(\frac{\partial \sigma}{\partial \Omega} \right) = \frac{3}{8\pi} \sigma (\cos^2 \varphi \cos^2 \theta + \sin^2 \varphi) \tag{5.21}$$

where φ is the polarization angle of the incident light relative to the scattering plane, defined by the direction of propagation of the incident light and the scattering direction, and θ is the scattering angle, which are illustrated in Fig. 5.3.

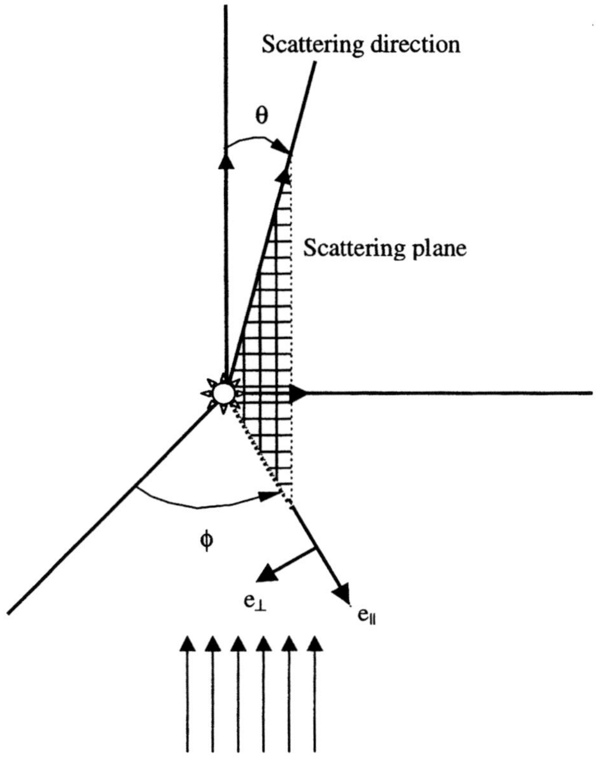

$e_\perp$: vertical polarisation; $e_\parallel$: horizontal polarisation

Fig. 5.3 Definition of the scattering plane and directions of polarization.

The term σ is the Rayleigh scattering cross section, given by

$$\sigma = \frac{8\pi}{3}\left[\frac{\pi^2(n^2-1)^2}{N_{STP}^2\lambda^4}\right] \tag{5.22}$$

where n is the real part of the refractive index of a gas at standard temperature and pressure (STP), N_{STP} is the molecular number density for an ideal gas at STP, and has a value of 2.687 × 10^{19} cm^{-3}, and λ is the wavelength in cm.

For the usual experiment, Rayleigh scattering is detected at 90° where the scattering volume is at a minimum. The scattering volume is defined by the intersection of the laser beam and the image of the aperture placed in front of the detector. From Eq. (5.21) it can be seen that the incident laser beam should be vertically polarized (polarization normal to the scattering plane) for Rayleigh scattering to be detected at a right angle. In this case, the Rayleigh scattered light signal for a mixture of gases is given by

$$P_{\perp} = \eta_c I_i \Omega V_c N \sum_{i=1}^{j} \chi_i \sigma_i \tag{5.23}$$

where $\Sigma\chi_i\sigma_i$ represents the effective Rayleigh cross section for the mixture of gases and j is the number of species. χ_i is the mole fraction of species i in the mixture and is expressed as follows

$$\sum_{i=1}^{j} \chi_i = 1$$

The total molecular number density, N, is related to the pressure, temperature, and Avogadro's number, $A_o = 6.023 \times 10^{23}$ molecules/mol, by

$$N = \frac{PA_o}{RT} \tag{5.24}$$

For a given optical detection system, η_c, Ω, and V_c are constants and the variations in the Rayleigh scattered light signal are a result of the mixture number density (i.e., the variation of temperature or pressure or both) or the species variations, or even both. Hence, an unambiguous interpretation of the Rayleigh scattered light signal requires that experiments be designed for certain special situations to supply useful information on the variables of interest. This has limited the wider application of laser Rayleigh scattering to practical combustion systems, such as internal combustion engines.

5.3.3.1 Density and Concentration Measurement by Rayleigh Scattering

For a binary mixture of fuel vapor and air, Eq. (5.23) becomes

$$P_{fa} = \eta I_i N[\chi_f \sigma_f + \chi_a \sigma_a] \tag{5.25}$$

Since $\chi_f + \chi_a = 1$, it can be written as

$$P_{fa} = \eta I_i N[\chi_f \sigma_f + (1 - \chi_f)\sigma_a] \tag{5.26}$$

where η is a constant that account for the values of η_c, Ω, and V_c.

If a calibration experiment can be performed at a standard condition of pressure, p_o, and temperature, T_o, and the Rayleigh scattering signal from a known fuel mole fraction χ_{fo} in the binary mixture is P_{fo}, and the Rayleigh scattering signal from air only is P_{ao}, i.e.,

$$P_{fo} = \eta I_i N_o \left[(1 - \chi_{fo}) \sigma_a + \chi_{fo} \sigma_f \right] \tag{5.27}$$

$$P_{ao} = \eta I_i N_o \sigma_a \tag{5.28}$$

From Eqs. (5.26), (5.27), and (5.28), the mole fraction of fuel vapor in a binary mixture can be derived as follows:

$$\chi_f = \chi_{fo} \frac{P_{fa} \dfrac{N_o}{N} - P_{ao}}{P_{fo} - P_{ao}} \tag{5.29}$$

or

$$\chi_f = \chi_{fo} \frac{P_{fa} \dfrac{p_o}{p} \dfrac{T}{T_o} - P_{ao}}{P_{fo} - P_{ao}} \tag{5.30}$$

It is clear from Eq. (5.30) that the mole fraction of fuel, χ_f, can be obtained from the Rayleigh scattering signal, P_{fa}, if calibration signals P_{fo}, P_{ao} are measured in advance. It is also necessary to know the temperature and pressure in order to derive the mole fraction of fuel from the Rayleigh signal. Generally, the pressure can be obtained with the use of a pressure transducer without any difficulty. However, measuring the temperature itself is a daunting task, which is one of the limitations preventing further application of Rayleigh scattering to internal combustion engines. In some cases, however, with a judicious design of the diagnostic system and a suitable assumption of the flow under study, Rayleigh scattering can be used to obtain the concentration of a binary mixture system without the requirement of information on temperature.

5.3.3.2 In-Cylinder Gas Temperature Measurement by Rayleigh Scattering

From Eqs. (5.23) and (5.24), the Rayleigh scattering signal, at temperature T and pressure p can be expressed as follows:

$$P = I_i \frac{pA_o}{RT} \sum_{i=1}^{j} \chi_i \sigma_i \tag{5.31}$$

and if the Rayleigh signal from the same gas mixture at a standard temperature T_o and pressure p_o is P_{cal}, then the temperature T can be obtained from the following equation:

$$\frac{T}{T_o} = \frac{\sum_{i=1}^{j} \chi_i \sigma_i}{\sum_{i=1}^{j} \chi_{io} \sigma_{io}} \frac{p}{p_o} \frac{P_{cal}}{P} \tag{5.32}$$

Equation (5.32) shows that the temperature measurement may be obtained by interpreting the total Rayleigh scattered light intensity directly, if the gas pressure, mixture composition, or Rayleigh scattering cross section of the mixture is known. Since the gas pressure can be measured precisely by a pressure transducer, the main difficulty for the accurate interpretation of the temperature lies in the precise determination of the scattering cross section of the mixture. In a homogeneous non-reacting flow, the scattering cross section of the mixture remains constant, and the gas temperature can be determined accurately using this equation. However, in a reacting flow field, since both the mixture number density and the gas composition vary simultaneously, the application of the Rayleigh scattering technique to temperature measurement becomes difficult. One approach to surmount this difficulty is to select a fuel mixture that will result in little change between the Rayleigh scattering cross sections of reactants and products [Dibble and Hollenbach, 1981], but this will be difficult to implement in any practical systems.

5.3.4 Raman Scattering and Its Measurement

A Raman scattering experiment involves the measurement of Raman spectra. Thus, before techniques based on Raman scattering are introduced, Raman spectra will be described first.

Raman spectra depend on the molecular structure of the molecules involved. For simplicity, only diatomic molecules will be considered here to illustrate the principal features of Raman spectra. Since we shall be dealing with rotational and vibrational changes, it is evident that expressions for the energy levels will be identical with those already discussed in Section 5.2. For clarity, we shall repeat any such expressions with limited explanation.

5.3.4.1 Raman Spectra of Diatomic Molecules

Raman spectrum of a diatomic molecule may involve changes in either or both the rotational and vibrational energy levels. A pure rotational Raman spectrum exists when only the rotational energy of a molecule is affected. However, when the molecule changes state from one vibrational level to another, both vibrational and rotational energies are involved.

Pure Rotational Raman Spectrum—The rotational energy of linear molecules is described by Eq. (5.2):

$$E_r = hc\left[BJ(J+1) - DJ^2(J+1)^2\right]$$

In Raman spectroscopy, the precision of measurements does not normally warrant the retention of the term involving D, the centrifugal distortion constant. Thus, we take the simpler expression:

$$E_r = hcBJ(J+1) \tag{5.33}$$

to represent the rotational energy.

Following the usual practice, we define ΔJ as ($J_{final\ state} - J_{initial\ state}$). Transitions between different rotational energy levels follow the formal selection rule for a diatomic molecule:

$$\Delta J = 0, \text{ or } \pm 2 \text{ only.}$$

The transition ΔJ = 0 is trivial since this represents no change in the molecular energy, and hence Rayleigh scattering if no vibrational energy change is involved.

For ΔJ = +2 (the Stokes transition), the change of energy level is given by:

$$\Delta \bar{\nu}_r = \Delta E_r / hc = B(4J+6)\,\text{cm}^{-1}$$

which happens when a molecule gains rotational energy from the photon during collision. J is the rotational quantum number in the lower state and takes values of 0, 1, 2, ... etc. This forms a series of lines (S branch) to the lower-frequency side of the exciting line, as shown in Fig. 5.4,

S branch: $\bar{\nu}_r = \bar{\nu}_0 - B(4J+6) \quad J = 0, 1, 2, \ldots$

Similarly, a series of lines (O branch) will appear to the higher-frequency side of the exciting line, the anti-Stokes transition, when ΔJ = –2 or the molecule losses energy to the photon,

O branch: $\bar{\nu}_r = \bar{\nu}_0 + B(4J-2) \quad J = 2, 3, 4 \ldots$

Note that the first Stokes and anti-Stokes lines are separated from the incident light frequency by 6B and that successive lines are separated from each other by 4B as shown in Fig. 5.4.

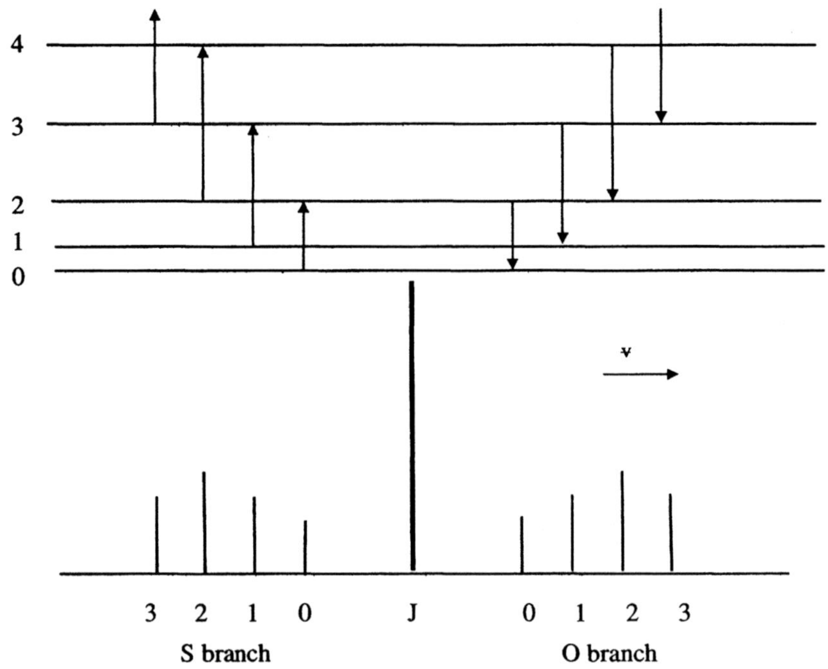

Fig. 5.4 The rotational energy levels of a diatomic molecule and the rotational Raman spectrum arising from transitions between them.

The intensity of pure rotational Raman scattering is given by

$$I \propto b_{J\pm2,J}\gamma_0^2 \tag{5.34}$$

where γ_0 is the anisotropy of the equilibrium polarizability tensor [Long, 1977]. The $b_{J',J''}$ are the so-called Placzek-Teller coefficients given by

$$b_{J+2,J} = \frac{3(J+1)(J+2)}{2(2J+1)(2J+3)} \tag{5.35}$$

$$b_{J-2,J} = \frac{3J(J-1)}{2(2J+1)(2J-1)} \tag{5.36}$$

$$b_{J,J} = \frac{J(J+1)}{(2J-1)(2J+3)} \tag{5.37}$$

The last coefficient is included for the vibrational-rational Raman scattering.

Vibrational Raman Spectra—The structure of vibrational Raman spectra is easily discussed. For every vibrational mode we can write an expression of the form Eq. (5.4):

$$E_v = hc\left[\omega_e(v+1/2) - \omega_e x_e(v+1/2)^2 + \omega_e y_e(v+1/2)^3 + \ldots\right] \quad (v = 0, 1, 2, \ldots) \tag{5.38}$$

where ω_e is the equilibrium vibrational frequency expressed in wave-numbers in cm^{-1} and x_e is the anharmonicity constant. Such an expression is very general, whatever the shape of the molecule or the nature of the vibration. For a diatomic molecule, the third and higher terms can be ignored.

Quite general, too, is the selection rule:

$$\Delta v = 0, \pm 1, \pm 2, \ldots \text{etc.}$$

Among the possible transitions between the various vibrational energy levels, the ones from $v_{upper} = 0$ to $v_{lower} = 1$ or vice versa (fundamentals) are the strongest. The probability of $\Delta v = \pm 2, \pm 3 \ldots$ decreases rapidly with increasing values of Δv.

The spectral positions of the fundamentals of vibrational Raman scattering ($\Delta v = \pm 1$), are thus given as

$$\nu = E_v / hc = \nu_0 \pm \left[\omega_e - 2\omega_e x_e(v+1)\right] \tag{5.39}$$

where the minus sign represents the Stokes line and the plus sign refers to the anti-Stokes lines. The vibrational number v in Eq. (5.39) is that of the initial vibrational energy state. The anti-Stokes lines are normally much weaker than the Stokes at low temperatures at which very few molecules exist in the excited state v = 1.

When a photon collides with a molecule, both the rotational and vibrational energies of the molecule will be affected. Thus, vibrational-rotational Raman scattering will take place, where the change in vibrational energy is accompanied by changes in rotational energy. For a diatomic molecule, Q branch transitions ($\Delta J = 0$) are much stronger than the O and S branch transitions. Furthermore, the O and S branches are spectrally quite separated from the Q branch lines, and can be neglected in the limited spectral regions about the Q branch. Ignoring centrifugal distortion but accounting for vibration-rotation interaction, the fundamental ($\Delta v = \pm 1$) Q branch ($\Delta J = 0$) vibrational-rotational spectral lines can be detected at

$$\nu = E_v / hc = \nu_0 \pm \left[\omega_e - 2\omega_e x_e(v+1) - \alpha_e J(J+1)\right] \tag{5.40}$$

Note that v and J are the vibrational and rotational quantum numbers of the initial energy states.

For diagnostic purposes, Eq. (5.40) can be simplified to

$$\nu = \nu_0 \pm \nu_{vib} \tag{5.41}$$

The second term in Eq. (5.40) is replaced by the vibrational frequency, ν_{vib}, in cm^{-1}, which is species specific. Table 5.1 lists the vibrational frequencies (cm^{-1}) for the molecules of interest in IC engines. The excitation wave number or frequency, ν_0, represents the incident light frequency in cm^{-1}. Equation (5.40) defines the spectral positions of the Q-branched vibrational Raman spectrum of a diatomic molecule. The plus and minus signs represent the anti-Stokes lines (frequency up-shifted) and Stokes lines (frequency down-shifted) respectively.

In summary, for a diatomic molecule at the ground electronic state (Σ state), the selection rules for Raman scattering are given by $\Delta v = 0, \pm 1$ and $\Delta J = 0, \pm 2$. Fig. 5.5 illustrates the rotational-vibrational Raman spectrum of a diatomic molecule. The Raman spectrum consists of a Stokes ($\Delta v = +1$) and an anti-Stokes ($\Delta v = -1$) component, each with three branches: S($\Delta J = +2$), Q ($\Delta J = 0$), and O ($\Delta J = -2$). Since the rotational lines in the Q-branch are stronger and not normally resolved, this branch is usually two orders of magnitude more intense than the S and O branches, and is thus the most characteristic feature of the Stokes and anti-Stokes components of the vibrational-rotational Raman spectrum.

5.3.4.2 Measurement of Raman Scattering

The Intensity of Raman Scattering—The Raman signal, P, from molecular species i, scattered into a solid angle Ω, is usually expressed in terms of a differential scattering cross section $(\partial\sigma/\partial\Omega)$

$$P = \eta_c I_i \Omega V_c N \left(\frac{\partial \sigma}{\partial \Omega} \right) \tag{5.42}$$

where η_c = the detection efficiency of the collection system
I_i = the incident laser intensity, (W/cm^2)
N = the number density of the Raman scattering gas molecule, (cm^{-3})
Ω = the solid angle of collection optics, (sr)
V_c = the collection volume, (cm^3)

The Stokes and anti-Stokes fundamental ($\Delta v = \pm 1$) vibrational Raman scattering cross sections for a diatomic molecule are given by [Long, 1977],

Table 5.1 Characteristic vibrational frequencies and Raman cross sections of some species of interest in IC engines

Species	Vibrational frequency (cm^{-1})	Spectral position of Q-branch Stokes line (nm)				Vibrational cross sections ($\times 10^{-30}$ cm^2/sr)			
(Excitation wavelength)		248 nm	308 nm	355 nm	532 nm	248 nm	308 nm	355 nm	532 nm
N_2	2331	263.2	331.8	387.0	607.3	13.0	5.2	2.79	0.46
O_2	1556	257.9	323.5	375.8	580.0	16.6	6.7	3.69	0.65
CO_2	1388	256.8	321.7	373.4	574.4	15.1	6.1	3.38	0.60
	1285	256.1	320.7	371.9	571.0	11.1	4.5	2.49	0.45
CO	2145	261.9	329.8	384.3	600.5	13.3	5.3	2.88	0.48
H_2O	3657	272.7	347.1	408.0	660.5	30.8	11.7	6.15	0.89
CH_4	2915	267.3	338.4	396.0	629.6	80.2	31.2	16.6	2.61
	3017	268.0	339.5	397.6	633.7	53.7	20.9	11.1	1.72
C_3H_8	2890	267.1	338.1	395.6	628.6	88.1	34.3	18.3	2.87
	1465	257.3	322.5	374.5	577.0	35.3	14.3	7.9	1.4
	857	253.5	316.3	366.1	557.4	20.8	8.5	4.8	0.89
C_8H_{18}	2939	267.5	338.6	396.3	630.6				
	1452	257.5	322.4	374.2	576.5				
	750	252.7	315.3	364.7	554.1				

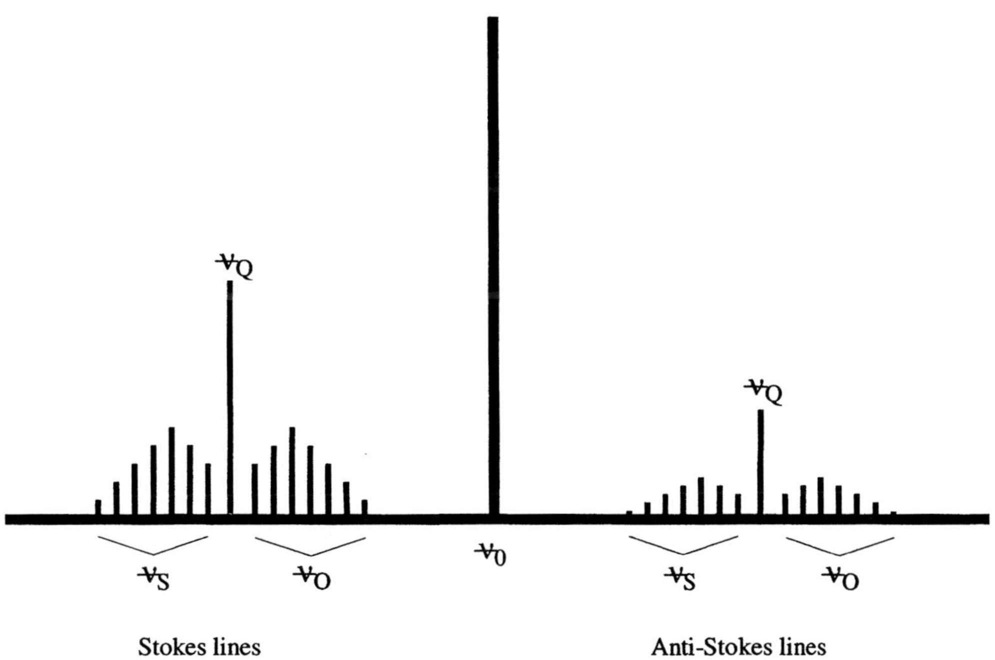

Fig. 5.5 The rotational-vibrational Raman spectrum of a diatomic molecule.

$$\left(\frac{\partial\sigma}{\partial\Omega}\right)_s = \frac{h(\nu_0 - \nu_{vib})^4 \; f(\alpha'^2, \gamma'^2, \rho)}{8mc^4 \nu_{vib}\left[1 - \exp(-hc\nu_{vib}/kT)\right]} \tag{5.43}$$

$$\left(\frac{\partial\sigma}{\partial\Omega}\right)_{as} = \frac{h(\nu_0 + \nu_{vib})^4 f(\alpha'^2, \gamma'^2, \rho)}{8mc^4 \nu_{vib}\left[1 - \exp(-hc\nu_{vib}/kT)\right]} \tag{5.44}$$

where h = Planck's constant (6.6256×10^{-34} J s),
c = the speed of light in vacuum (2.998×10^8 m/s)
k = Boltzmann's constant (1.38×10^{-23} J/K)
T = temperature (K)

$f(\alpha'^2, \gamma'^2, \rho)$ is a function of the mean (α') and anisotropy invariants (γ') of the derived polarizability tensor [Long, 1977], and the polarization state of the incident laser and scattered light ρ. For vertically polarized incident light, the values of $f(\alpha'^2, \gamma'^2, \rho)$ for the vertical, parallel and total scattered light are given as

$$f\left(\alpha'^2, \gamma'^2, {}^{\perp}\rho_{\perp}\right) = \frac{45(\alpha')^2 + 4(\gamma')^2}{45} \tag{5.45}$$

$$f\left(\alpha'^2, \gamma'^2, {}^{\perp}\rho_{\parallel}\right) = \frac{(\gamma')^2}{15} \tag{5.46}$$

$$f\left(\alpha'^2, \gamma'^2, {}^{\perp}\rho_{\perp,\parallel}\right) = \frac{45(\alpha')^2 + 7(\gamma')^2}{45} \tag{5.47}$$

The exponential terms in Eqs. (5.43) and (5.44) indicate that, the higher the gas temperature is, the larger the Raman scattering cross section becomes as the higher-energy vibrational states become more populated. For polyatomic molecules, more complex expressions for the scattering cross section arise, but the conclusions drawn from this equation regarding the temperature dependence of the cross section are still valid. It is also worth noting from the above $f(\alpha'^2, \gamma'^2, \rho)$ values that the Raman scattering intensity is independent of the scattering angle.

Table 5.1 lists room temperature vibrational Raman cross sections for species of interest in IC engines. The cross sections are taken from the tabulation by Inaba and Kobayasi [1972] for an N_2 laser at 337 nm, except those of propane (C_3H_8), which are obtained from Schrotter and Klockner [1979]. The values at 248 nm and 308 nm (excimer lasers), 355 nm and 532 nm (Nd:YAG lasers) are calculated from the original values by scaling according to $(\nu_0 - \nu_{vib})^4$. The scattering cross sections listed in Table 5.1 are obtained for the linear polarized incident and scattered light perpendicular to the scattering plane as defined in Fig. 5.3. These cross sections are valid and unchanged for any scattering angle between the incident beam and scattering observation axes.

When a photon interacts with a molecule, energy exchange may occur between both the rotational and vibrational energy levels and the photon, as discussed previously. This leads to rotational sidebands about the vibrational frequencies in the spectrum of scattered light, as shown in Fig. 5.5. The shape of the sidebands is affected by changes in temperature. Since Raman intensity is normally determined by integrating over only a portion of the sidebands, then this change in shape by temperature can influence the signal levels. In addition, the Raman scattering cross section is affected by temperature as shown by Eq. (5.43) or Eq. (5.44). These temperature effects on the Raman signal can be modeled by a bandwidth factor f(T), accounting for the fraction of the total Raman scattering actually detected. In general, Eq. (5.42) can be written as

$$P = \eta I_i N f(T) \tag{5.48}$$

where η is a factor dependent on the Raman cross section, wavelength, geometry, and optical collection efficiency, and is usually determined by an experimental calibration. Equation (5.48) shows that the Raman signal is directly related to the concentration of Raman active species N when Raman scattering is detected at the species Raman frequency defined by Eq. (5.40).

The bandwidth factor f(T) depends on the spectral location, shape, and bandwidth of the detection system, and laser line spectral profile. For example, Fig. 5.6 [Eckbreth, 1996] shows that for N_2 Raman scattering measurement, the temperature effect is small at low temperatures with a reasonable detection bandwidth (>2 nm). With a detection bandwidth of about 5 nm the requirement of knowing temperature can be eliminated in order to obtain quantitative Raman scattering measurements.

In certain cases, the temperature effect may be ignored if the gas temperature is much lower than the characteristic vibrational temperature of the species to be measured, given by

$$T_v = \left(hc\omega_e / k\right)$$

The grouping hc/k has a value of 1.44 K·cm^{-1}.

Measurement of Major Species Concentration by Raman Scattering—The forgoing analysis of Raman spectra and Raman scattering signal provides the fundamental knowledge and formulation upon which Raman species concentration and temperature measurements are based.

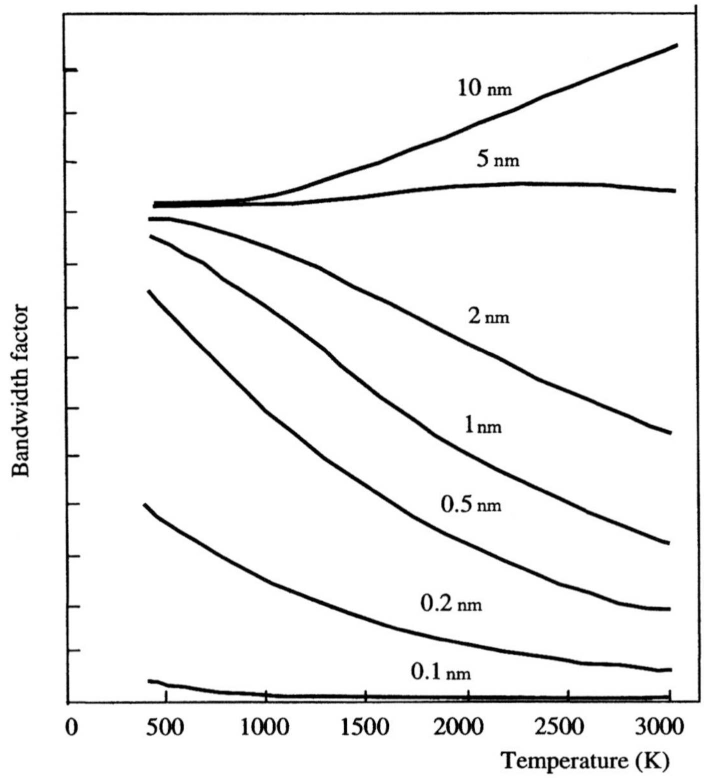

Fig. 5.6 Temperature variation of the bandwidth factor for the N_2 stokes Raman spectra for various bandpass filter bandwidths. (Laser line 532 nm; N_2 Raman signal detected at 607.3 nm)

In general, the species concentration is measured from the Raman scattered signal in some spectrally integrated portion of the Raman spectrum as defined in Eq. (5.40). After determining η by calibration from Raman scattering in a known sample by independent measurements, the species concentration N can be calculated from the measured Raman scattering from Eq. (5.42). A more complete description of species concentration measurement will be discussed in Chapter 7.

Gas Temperature Measurement by Raman Scattering—Temperature measurements by Raman scattering can be achieved via two main approaches. The first involves least-squares fitting a library of theoretical spectra to the measured spectrum. The theoretical spectrum producing the best fit identifies the temperature at the measurement point. Alternative approaches to temperature measurements employ spectral band ratio strategies, where various portions of the Raman spectrum can be monitored and the temperature determined from the resultant intensity ratio. This can be performed on the ratio of the first hot band (Δv: 1 – 2) to ground state band (Δv: 0 – 1) in the Stokes spectrum. A common technique is to ratio the Stokes to anti-Stokes vibrational scattering

$$\frac{P_{rs}}{P_{ras}} = \frac{(\nu_0 - \nu_{vib})^4}{(\nu_0 + \nu_{vib})^4} \exp(-hc\Delta\nu / kT) \frac{f^s(T)}{f^{as}(T)} \tag{5.49}$$

where $\Delta\nu$ is the energy difference between the ground and first excited vibrational levels ($\Delta v = 1$) for the fundamental Q-branch ($\Delta J = 0$). The Stokes/anti-Stokes method provides the maximum collected signal because of a wider spectral bandpass. However, the ratio becomes moderately temperature sensitive only above 700 K due to the exponential dependence of excited state population of the anti-Stokes signal with temperature. A detailed discussion on Raman thermometry will be presented in Chapter 11.

5.4 Principle of Laser Induced Fluorescence (LIF)

Laser Induced Fluorescence (LIF) is the emission of light from an atom or molecule following the excitation by a laser beam. In a laser induced fluorescence experiment, the molecule is initially at a lower electronic state before it is excited to an upper electronic energy level by a laser source. The wavelength of the laser light needs to be chosen so that it coincides with the absorption wavelength. Typically, the lower state is in the ground electronic level, because the other electronic levels are normally negligibly populated at combustion temperature.

There are five important processes following the excitation of a molecule by a laser source, as shown in Fig. 5.7. First, the molecule can be returned to its original quantum state by laser induced stimulated emission, denoted as $B_{21}I_v$. Second, absorption of an additional photon can excite the molecule to higher states, even to ionized levels, shown as Q_{ion} in Fig. 5.7. Third, the internal energy of the system can be altered in inelastic collisions with other molecules, producing rotational and vibrational energy transfers, represented by $Q_{rot,vib}$ in Fig. 5.7. In many cases, the inelastic collisions with other molecules also result in electronic energy transfer (Q_{elec}), which is often referred to as quenching. Fourth, interactions between the individual atoms of the molecule produce internal energy transfer and dissociation of molecule. The dissociation in this case is termed pre-dissociation because it is produced by a change from a stable to a repulsive electronic (unstable) state. Finally, the originally populated state, and nearby states indirectly populated through collisions, return to lower states through the emission of light, producing the laser induced fluorescence.

The LIF signal can be related to specific properties, e.g., concentration, of the absorbing species through modeling of these state-to-state transfer processes. Because the fluorescence is a function of the population of the upper state, it is necessary to solve the state-dependent population dynamics. Most treatments of LIF are based on a semi-classical rate equation analysis developed by Piepmeier [1972]. The rate equations are conceptually and mathematically more tractable than the quantum mechanical approach. In addition, it is adequate for most LIF measurements that are applicable to engine combustion.

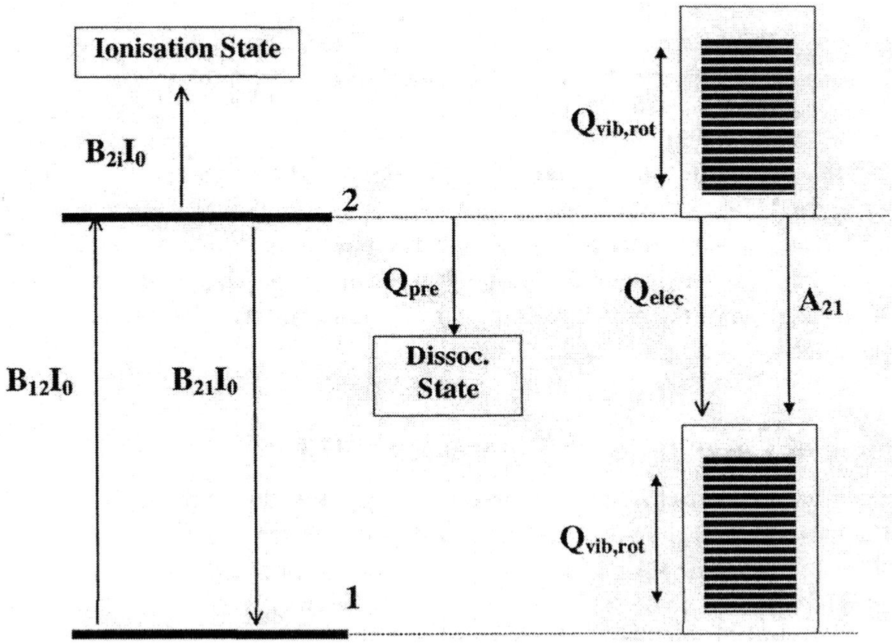

Fig. 5.7 Main energy transfer processes in LIF.

Two-energy-level models are often used as a first approximation in actual data analysis. Such a model is appropriate to LIF detection of some molecular systems with fully equilibrated or fully frozen rotational level manifolds. It also serves to illustrate the essential properties of LIF. Fig. 5.8 depicts such a two-energy-level model. For this two-energy-level model, the rate equations for the lower and upper energy level become

$$dN_1/dt = -N_1(Q_{12} + B_{12}I_v) + N_2(Q_{21} + A_{21} + B_{21}I_v) \tag{5.50}$$

$$dN_2/dt = N_1(Q_{12} + B_{12}I_v) - N_2(Q_{21} + A_{21} + B_{21}I_v + Q_{ion} + Q_{pre}) \tag{5.51}$$

where B_{12}, B_{21} are the Einstein coefficients for stimulated absorption and emission, A_{21} is the Einstein coefficient for spontaneous emission, and I_v ($Wcm^{-2}Hz^{-1}$) is the laser spectral intensity. The two energy levels are normally separated by a few electron-volts, thus collisional excitation, Q_{12}, can be omitted in most combustion measurements. Most excited states are not pre-dissociative (Q_{pre}), unless specifically chosen, and photo-ionization (Q_{ion}) can often be ignored. Eqs. (5.50) and (5.51) become

$$dN_1/dt = -N_1B_{12}I_v + N_2(Q_{21} + A_{21} + B_{21}I_v) \tag{5.52}$$

$$dN_2/dt = N_1B_{12}I_v - N_2(Q_{21} + A_{21} + B_{21}I_v) \tag{5.53}$$

Because the upper level typically has a negligible population prior to the laser excitation, the initial condition, $N_2|_{t=0} = 0$, is applied to Eq. (5.53). In addition, the total population of the molecule should be conserved, i.e., no chemical reactions occur during the measurement,

$$N_1 + N_2 = \text{constant} = N_1^0 \tag{5.54}$$

where N_1^0 is the total population prior to laser excitation. The solution to the two-level system is given by

$$N_2 = N_1^0 B_{12} I_v \tau \left(1 - e^{-t/\tau}\right) \tag{5.55}$$

where the time constant τ is $(Q_{21} + A_{21} + B_{21}I_v + B_{12}I_v)^{-1}$. For laser pulses long compared to τ, the system reaches the steady state value of

$$N_2 = N_1^0 B_{12} I_v \tau \tag{5.56}$$

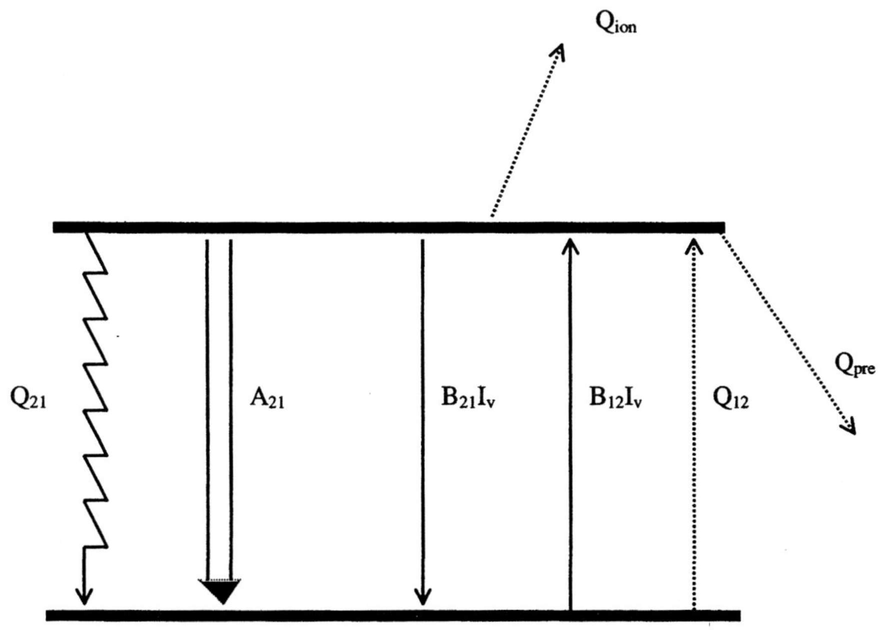

Fig. 5.8 Simple two-energy-level diagram for LIF modeling.

which can be more conveniently written as:

$$N_2 = B_{12} I_v \frac{A_{21}}{A_{21} + Q_{21}} \frac{1}{1 + I_v / I_v^{sat}} \tag{5.57}$$

where the saturation intensity I_v^{sat} is defined as:

$$I_v^{sat} = \frac{A_{21} + Q_{21}}{B_{21} + B_{12}} \tag{5.58}$$

Assuming the fluorescence is emitted equally into 4π steradians, the total number of photons N_p striking a photodetector from a collection volume V_c is given by

$$N_p = \eta \frac{\Omega}{4\pi} N_1^0 V_c B_{12} E_v \frac{A_{21}}{A_{21} + Q_{21}} \frac{1}{1 + I_v / I_v^{sat}} \tag{5.59}$$

where ε = transmission efficiency of the collection optics,
Ω = the collection solid angle, (sr)
E = the spectral fluence of the laser, (J cm^{-2} Hz^{-1}),
V_c = the sampling volume, defined by the cross-sectional area of the laser beam, A, and the axial extent along the beam, l, from which the fluorescence is detected.

For a particular experimental set up, the fluorescence signal can be expressed as

$$P_{flu} = \eta_c \Omega V_c f_1(T) \chi_m N I_v \frac{A_{21}}{A_{21} + Q_{21}} \frac{B_{12}}{1 + I_v / I_v^{sat}} \tag{5.60}$$

where $f_1(T)$ = the fractional population of the lower electronic state,
χ_m = the mole fraction of the absorbing species,
N = the total gas number density, (cm^{-3})
I_v = the spectral power of incident laser light, (W cm^{-2} Hz^{-1}).

$f_1(T)$ represents the fraction of molecules of the absorbing species that are in the specific electronic-vibrational-rotational energy level excited by the laser, and is given by the Boltzmann distribution. Thus, the grouping ($f_1(T)$ $\chi_m N$) represents the number density of the absorbing species at ground state, N_1^0.

Equation (5.60) can be further simplified for two limiting cases, corresponding to either low or high laser intensity. At low laser intensity, i.e., $I_v \ll I_v^{sat}$, it simplifies to

$$P_{flu} = \eta_c \Omega V_c f_1(T) \chi_m N I_v B_{12} \frac{A_{21}}{A_{21} + Q_{21}} \tag{5.61}$$

This is the so-called **linear fluorescence** equation, because the fluorescence is linearly proportional to laser intensity. The term $A_{21}/(A_{21}+Q_{21})$ is termed the **fluorescence efficiency** and, since $A_{21} << Q_{21}$ near or above atmospheric pressure, the fluorescence efficiency is generally much smaller than unity, significantly reducing species detectivity.

In the linear regime, one normally must to evaluate the quenching rate constant in order to make quantitative measurements of the species of interest. Because the quenching rate is sensitive to pressure, temperature, and composition, it is very difficult to make quantitative LIF measurements in atmospheric and higher-pressure combustion environments.

If the laser intensity is increased until $I_v \gg I_v^{sat}$, Eq. (5.60) then becomes

$$P_{flu} = \eta_c V_c f_1(T) \chi_m N B_{12} \frac{A_{21}}{B_{12} + B_{21}} \tag{5.62}$$

and the fluorescence is said to be in the saturation limit. In the saturation regime, the fluorescence signal is independent of both the laser irradiance and the quenching, obviating the need to measure or evaluate either one. However, complete saturation is not easily achieved due either to the specific wavelength region of the absorption or the magnitude of the saturation intensity. Saturation is not achieved at the outer edges of a laser beam. In addition, it is not possible to saturate during the entire duration of the laser pulse due to the temporal variation of the laser pulse. It is much more difficult to reach the saturation regime when planar laser induced fluorescence (PLIF) is used, as is the case for the majority of LIF applications in IC engines.

To perform quantitative fluorescence measurements on a given molecule, there are several requirements that must be satisfied. The first of these is that the molecule must have a known absorption and emission spectrum. Second, the molecule must have an absorption wavelength accessible to a laser source. A third requirement is that the rate of radiative decay of the excited state, which affects the fluorescence power directly, must be known. Fourth, excited state losses due to collisions, photo-ionization and pre-dissociation have to be taken into account. The excited state loss due to collisions, known as quenching, may be increased considerably in the presence of other molecules, especially at high pressures as encountered in IC engine applications.

5.5 Comparison of LRS, SRS, and LIF

In the previous sections, the fundamental principles of major laser-based diagnostics are described, including Rayleigh scattering (LRS), spontaneous Raman scattering (SRS), and laser induced fluorescence (LIF). These processes are compared on an energy level diagram as shown in Fig. 5.9. In Fig.5.9, ground and first excited electronic states are shown together with their vibrational and rotational energy levels. Letters v and J represent the vibrational and rotational quantum numbers, respectively. The Raman processes shown are for a diatomic molecule, to which the transition rules $\Delta v = \pm 1$ and $\Delta J = 0, \pm 2$ apply.

As shown in Fig. 5.9, Rayleigh or Raman scattering can occur upon excitation at any visible or ultraviolet wavelength. Fluorescence, however, requires a precise excitation frequency, since molecules in a particular rovibronic energy level of the ground electronic state must be excited to a specific rovibronic energy level in an excited electronic state.

The signal for the three scattering processes depicted in Fig. 5.9 can be collated from

$$P = \eta \Omega V_c \left(d\sigma / d\Omega \right) N_{vJ} I_i \tag{5.63}$$

where
- P = the power of the measured signal (W),
- η = the detection efficiency,
- Ω = the solid angle of the collection optics (sr),
- V_c = the collection volume (cm^3),
- $(d\sigma/d\Omega)$ = the differential cross section (cm^2/sr),
- N_{vJ} = the number density in the initial rovibronic level (V″, J″) of the ground electronic state (cm^{-3}),
- I_i = the irradiance of the laser beam (W/cm^2).

Eq. (5.63) indicates that for the same experimental conditions, the relative signal for each scattering method depends mainly on the differential cross-section. Approximate values for each scattering method are as follows:

Vibrational Raman	10^{-30}–10^{-28}
Rayleigh Scattering	10^{-28}–10^{-25}
Fluorescence (1 bar)	10^{-25}–10^{-20}
(Mie Scattering	10^{-27}–10^{-8})

For Rayleigh scattering, Mie scattering represents the most significant source of interference. Because of their low differential cross sections, Raman and Rayleigh scattering measurements normally require high species concentrations and clean laboratory conditions.

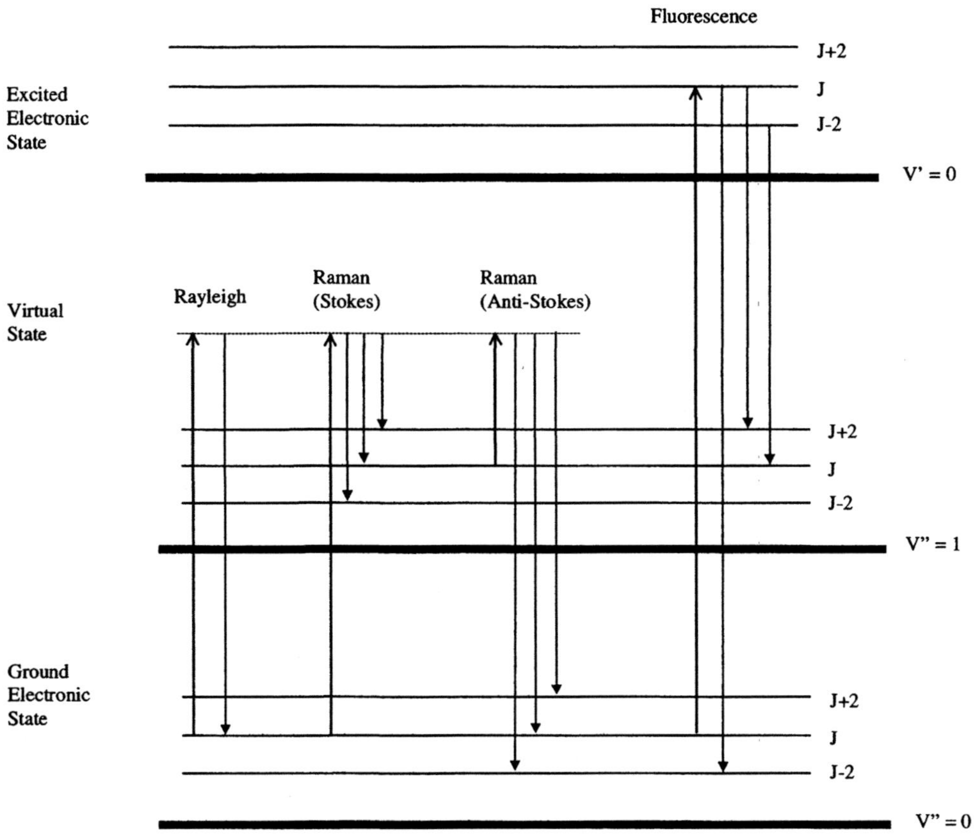

Fig. 5.9 Schematic of an energy level diagram showing Rayleigh scattering, vibrational Raman scattering, and laser induced fluorescence.

The nature of these three different processes determines the characteristics of each technique:

Laser Rayleigh scattering (LRS) is the elastic interaction between incident laser light and gas molecules. Because the intensity of the scattered light is proportional to the number density of the gas molecules, Rayleigh scattering signals can be used to deduce the gas density. Rayleigh measurements can be combined with pressure measurements to calculate gas temperature through the equation of state. The principle advantage of Rayleigh scattering is that it is the strongest of the molecular light scattering techniques, and is well suited as a density measurement technique at low gas density. When compared with Raman scattering, the scattering cross sections for Rayleigh scattering are typically three orders of magnitude larger than the corresponding vibrational Raman cross sections. The simplicity and ease of implementing a Rayleigh scattering system make it very convenient to set up the system and interpret the data. Because Rayleigh scattering is an elastic scattering process, it is not species specific. And because a Rayleigh signal has the same wavelength as the incident laser, it suffers seriously from the Mie scattering of particles and spurious scattered laser light from the surrounding surfaces. Because Mie scattering is ten to twenty orders of magnitude stronger than Rayleigh scattering, the environment must be virtually free of particles, e.g., dust in air, liquid droplet, in order for the Rayleigh signal to be detected.

Spontaneous Raman scattering (SRS) is the inelastic interaction between incident light and the electronic-mediated vibrational-rotational modes of molecules, i.e., there is an energy exchange between the photons and scattering molecules. Hence, the scattered photons will gain or lose energy to the scattering molecules, so that the Raman signal has a different wavelength from the incident light. Spectral measurement of Raman scattered light serves to identify the molecular species. The intensity of Raman scattered light is dependent on the number of molecules that are thermally populated at particular rovibronic energy levels. Since Raman peaks are spectrally well separated from the incident light source, spurious scattered light can be discriminated. Raman scattering has the advantage that the interaction time is very short (on the order of femtoseconds), so that quenching effects are not encountered. Raman scattering occurs with any laser excitation wavelength. It is capable of providing simultaneous multiple-speciesconcentration measurements. Because spontaneous Raman scattering does not involveelectronic state transitions, it can measure the concentration of stable species, such as N_2, O_2, CO_2, etc.. This is not possible with the normal laser induced fluorescence technique because these species require ultra-UV light to produce fluorescence. However, spontaneous Raman scattering is a very weak process and requires careful signal-to-noise consideration; background noise from the scattering of the unshifted laser wavelength can be a problem.

Laser induced fluorescence is an electronic absorption and emission process that produces relatively strong signal intensity with high spatial resolution. The fluorescence signal is normally red-shifted from the excitation wavelength, and hence easier to discriminate than Rayleigh scattering from stray scattered light. Fluorescence signals are directly proportional to the molecular density and can therefore be used to measure species concentration. The strong fluorescence signals allow two-dimensional measurements by using a sheet of incident light. However, quenching effects can reduce fluorescence intensities, especially by oxygen at elevated pressures, in a way that depends on the pressure and temperature, thereby making quantitative measurements difficult in the cylinder.

5.6 Experimental Considerations

The general experimental setups required to implement LRS, SRS, and LIF are similar. Fig. 5.10 shows such a schematic of a typical experimental setup. There are basically three subsystems: the laser delivery, light collection and detection, and data acquisition and processing systems. Synchronization between the engine, laser, and detection system is controlled by a PC computer and a digital delay generator, with the trigger signal coming from the engine shaft encoder. Normally, only one image is obtained at a particular crank angle in a given cycle due to speed limitations of the laser and imaging system.

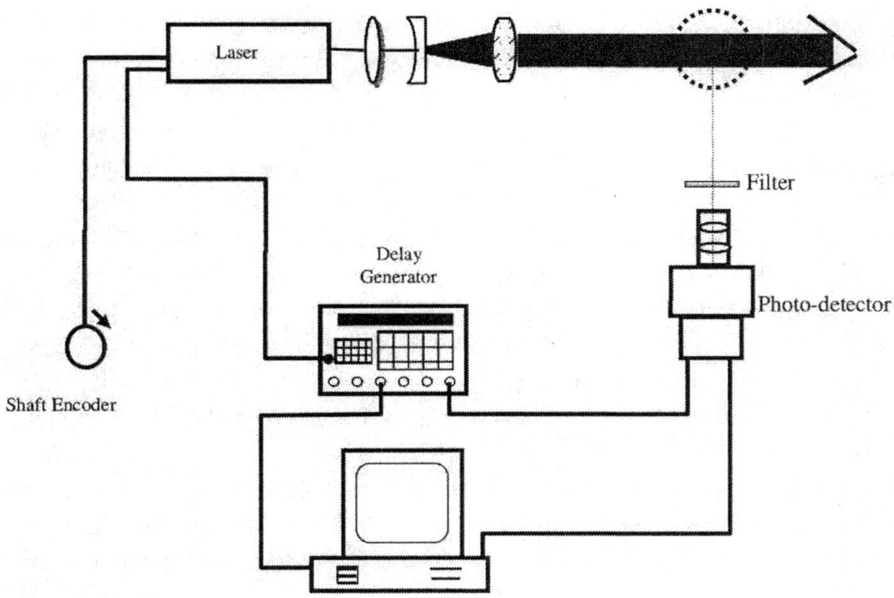

Fig. 5.10 Schematic of a typical laser scattering setup.

5.6.1 Lasers

Pulsed lasers are commonly used in light scattering and LIF measurements. Their high intensities can significantly increase the signal-to-noise ratios when gated detection is employed. Nd:YAG lasers and excimer lasers are the two most common pulsed lasers used in laser diagnostics.

Nd:YAG lasers can produce outputs at three wavelengths in the visible and UV region by frequency-mixing in optical crystals. The three wavelengths are 532 nm (2nd harmonic), 355 nm (3rd harmonic), and 266 nm (4th harmonic), respectively. For a Q-switched Nd:YAG laser, more than one joule can be generated at 532 nm. Typical pulsed Nd:YAG lasers have a repetition rate at 10 or 20 Hz. Relatively unknown to the flow and combustion research community, certain Nd:YAG lasers developed specifically for medical applications can produce power output of 20 watts at 20 kHz, providing an economical option to the copper vapor laser as the light source for high-speed visualization experiments.

For Nd:YAG lasers, direct access triggering (DAT) or external mode should be selected when the application requires laser output to be synchronized with other events, such as the operation of a camera. In the single shot mode, the laser is triggered externally by a single pulse. Lead time between the input fire command and laser output will be around a couple of hundred microseconds. Alternatively, the input of two triggering pulses can be used for more accurate control between input commands and laser output. The first trigger signal, called fire command, will cause the laser head lamps to flash. This signal precedes lasing by about 180 μs. The second trigger, called Q-switch command, fires the Pockels cell and precedes lasing by about 170 ns. When running the system in the external mode, one must flash the laser head lamps at approximately the same frequency at which the system is optimized during its manufacturing stage (typically, 10 Hz or 20 Hz). Changing lamp frequency will change thermal lensing of the YAG rod, which can result in reduced system performance. For example, to use a 10 Hz Nd:YAG laser in an engine experiment, the engine should be operated at 1200 rpm so that a single laser pulse is generated at a specified crank angle in every cycle.

Different wavelengths can be achieved separately by selecting an appropriate rare gas as the lasing medium in an excimer laser. The most frequently used excimer lasers are: XeCl, 308 nm; KrF, 248 nm; and ArF, 193 nm. They are available with pulse energies typically around a few hundred mJ and repetition rates up to several hundred hertz. Excimer lasers generally have a broad spectral output. By spectrally condensing the laser output by injection locking, excimer lasers can be tuned over a limited wavelength range (1 nm). Tunable excimer lasers have been used in UV Raman and LIF applications. Excimer lasers tend to be large, heavy, and expensive, and require a fair level of maintenance. However, they are becoming more widely used for two-dimensional LIF imaging and UV Raman applications due to their high pulse energy in UV wavelengths. An eximer laser can be triggered directly by an external device, and its frequency of operation is only limited by the maximum repetition rate.

In laser induced fluorescence experiments, fluorescence emission requires a precise excitation frequency in order to excite the molecule from a particular rovibronic level of the ground electronic state to a specific rovibronic level in another electronic state. The laser frequency must closely overlap a resonant transition in the species of interest. In some cases, fixed-frequency lasers exhibit such overlaps, such as the XeCl laser (OH) and ArF laser (NO). In many applications, tunable dye lasers must be used. Dye lasers are generally radiation pumped either by a flashlamp or by another laser. Most pulsed dye lasers intended for spectroscopic applications are laser pumped by a harmonic of Nd:YAG or excimer laser, because they possess better beam quality and spectral characteristics, e.g., narrower linewidth, better frequency stability, than flashlamp-pumped varieties. The output wavelength of a dye laser covers the whole visible spectrum, and can be altered by proper dye selection. Wavelengths in the UV range are obtained by frequency-doubling and wavemixing.

Copper vapor lasers are the most important neutral metal vapor lasers for flow and combustion diagnostics. The wavelengths of Cu vapor lasers are within the yellow (578 nm) and green (510 nm) spectrum. The beam diameter is typically 40 mm. These lasers are characterized by their high average power output (typically 1–30 watts) and an efficiency of up to one percent. They are particularly favored in high-speed visualization experiments because of their high repetition rate (up to 50 kHz) and short pulse width (~30 ns). When synchronized with a high-speed film drum camera, images can be recorded at a speed of 50,000 frames per second. Whereas most lasers are cooled, metal vapor lasers need thermal insulation so that the operating temperature for vaporizing the metal can typically reach 1500°C.

Continuous wave (CW) lasers are also used in flow and combustion diagnostics. The two important CW gas lasers are argon-ion and He-Ne lasers. Argon-ion lasers are gas lasers, and can supply over 10W in the blue-green range (mainly 488 and 514 nm). Emission is produced at several wavelengths through the use of broadband laser mirrors. Individual wavelength can be selected by means of Brewster prisms or an etalon in the laser resonator. Argon-ion lasers are frequently used for multi-component LDA (laser Droppler anemometry) measurements. He-Ne lasers are the most common and the most efficient lasers in the visible range (633 nm). The power of commercial models ranges from less than 1 mW to more than 10 mW. Their application is limited to some LDA systems and light extinction experiments.

5.6.2 Laser Beam Delivery

In most laser applications to be discussed in the subsequent chapters, it is necessary to modify or shape the laser beam using lenses and other optical elements. For point measurements, the laser beam must be focused inside the test section. Two-dimensional imaging techniques require illumination by a thin laser sheet.

5.6.2.1 Gaussian Beams

The output of most lasers (of stable laser resonators) has a Gaussian transverse intensity distribution as a function of radius r:

$$I(r) = I_0 e^{-2r^2/w^2} = \frac{2P}{\pi w^2} e^{-2r^2/w^2} \tag{5.64}$$

where P is the total beam power, and w is the radius where the intensity falls to $1/e^2$ (or 0.135) of its center line value. Diffraction causes light waves to spread transversely as they propagate (Fig. 5.11). The beam spreading is described by the beam radius as a function of the distance z:

$$w(z) = w_0 \left[1 + \left(\frac{\lambda z}{\pi w_0} \right)^2 \right]^{1/2} \tag{5.65}$$

where w_0 is the radius of the $1/e^2$ irradiance contour at the plane where the wave front is flat. This position is commonly known as the beam waist, and w_0 is called the waist radius. A waist occurs naturally at the mid-plane of a symmetric confocal cavity in a laser.

At a distance sufficiently away from the beam waist ($z >> \pi w_0/\lambda$), w(z) can be approximated by

$$w(z) = \left(\frac{\lambda z}{\pi w_0} \right) \tag{5.66}$$

Since z is much larger than $\pi w_0/\lambda$ so that the $1/e^2$ intensity contours asymptotically approach a cone of half angle θ, (Fig. 5.11)

$$\theta = \frac{w(z)}{z} = \frac{\lambda}{\pi w_0} \tag{5.67}$$

This value determines the far field divergence of the Gaussian beam. Far field is an optical term for the region at a greater distance from an aperture, as well as that which occurs at the focus of a lens.

A Gaussian beam will remain collimated for some finite distance and diverge with the spreading half angle, θ. The Rayleigh range is the distance over which a laser beam can be maintained collimated before diffraction spreading takes over, i.e., the distance on either side of the waist before the beam area doubles, and is given by

$$z_R = \frac{\pi w_0}{\lambda} \tag{5.68}$$

5.6.2.2 Focusing of Gaussian Beams

Siegman [1971] treated the general problem of Gaussian beam propagation by lenses and mirrors using matrix transformations. A less rigorous, but in many ways more insightful, approach to this problem has been developed by Self [1983]. He shows a way of modeling the transformations of

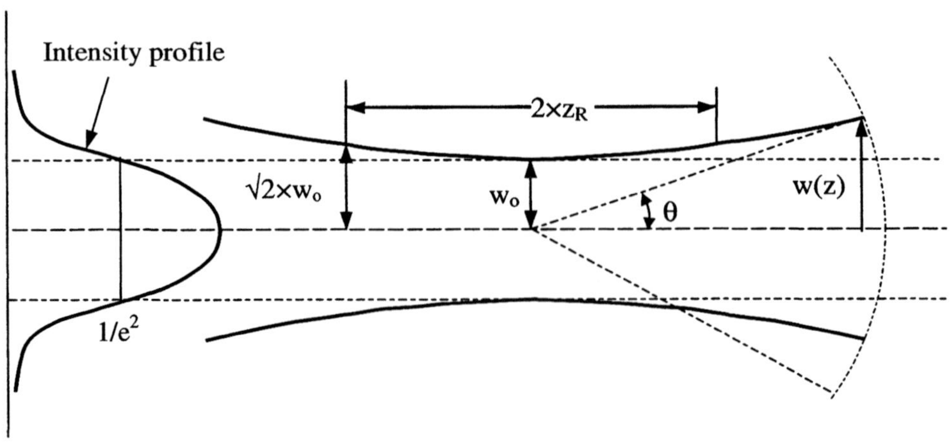

Fig. 5.11 Gaussian beam intensity distribution and geometry.

a laser beam through simple optics, under paraxial conditions, by calculating the Rayleigh range and beam waist location following each individual optical element. These parameters are calculated using a formula analogous to the well-known standard lens formula. His method of calculation is outlined below.

The standard lens equation is given by:

$$\frac{1}{s}+\frac{1}{s'}=\frac{1}{f} \tag{5.69}$$

where s and s′ are the distances from object and image to the lens, respectively; f is the focal length of the optical lens. For Gaussian beams, Self has derived an analogous formula by assuming that the waist of the input beam represents the object, and the waist of the output beam represents the image. The formula is expressed in terms of the Rayleigh range of the input beam:

$$\frac{1}{s+z_R^2/(s-f)}+\frac{1}{s'}=\frac{1}{f} \tag{5.70}$$

The magnification is given by

$$m=\frac{w_0'}{w_0}=\frac{1}{\sqrt{\left\{\left[1-\left(\frac{s}{f}\right)\right]^2+\left(\frac{z_R}{f}\right)^2\right\}}} \tag{5.71}$$

The Rayleigh range of the output beam depends on m^2 and is given by

$$z_R'=m^2 z_R \tag{5.72}$$

Therefore, the spot size and focal position of a Gaussian beam can be determined from Eqs. (5.70) and (5.71). Because the beam divergence angle θ is normally quoted by laser manufacturers, the direct relationship between the beam waist and divergence (Eq. [5.67]) can be used to substitute for the beam waist w_0 in Eqs. (5.70) and (5.71).

Two cases of particular interest are when s = 0 (the input waist coincides with the surface of the lens) and s = f (the input waist is at the front focal point of the optical system). For the case of s = 0, the equations for image distance and waist size reduce to:

$$s' = \frac{f}{1+\left(\lambda f/\pi w_0^2\right)^2} \tag{5.73}$$

$$w_0' = \frac{\lambda f/\pi w_0}{\left[1+\left(\lambda f/\pi w_0^2\right)^2\right]^{1/2}} \tag{5.74}$$

For the case of s = f, the results are:

$$s' = f \tag{5.75}$$

$$w_0' = \lambda f/\pi w_0 \tag{5.76}$$

Substituting typical values into these equations yields nearly identical results. Therefore, for most applications, the simpler, second set of equations can be used. Equation (5.76) teaches us that it is necessary to expand a laser beam before focusing it in order to minimize the spot size.

5.6.2.3 Laser Sheet Optics

Two-dimensional imaging techniques require the illumination of species by a thin laser light sheet. The essential element for the generation of a light sheet is a cylindrical lens. When using lasers with a sufficiently small beam diameter and divergence, such as argon-ion lasers, one cylindrical lens can be sufficient to generate a light sheet of appropriate shape. For other lasers, a combination of different lenses is usually required in order to generate thin light sheets of high intensity. At least one additional lens has to be used for focusing the laser beam to an appropriate thickness, as shown in Fig. 5.12(a). Frequently, another cylindrical lens is added in order to generate a light sheet of constant height (Fig. 12[b]).

The chosen optical system should be aimed at producing a thin sheet of approximately uniform illumination. In the configurations shown in Fig. 5.12, a long focal length spherical lens loosely focuses the beam, ensuring that the beam thickness varies little over the imaged section. The position and thickness of the focused beam waist in the test section is determined by the laser beam profile as well as the sheet-forming optics. For example, the typical beam profile of a Nd:YAG laser is characterized by a Gaussian distribution and the divergence of the beam is very small (<0.5 mrad). Thus, the laser beam can be focused to a very thin sheet (<200 μm). Due to the nature of electrical discharge in an excimer laser, its beam profile is normally

(a) Laser sheet forming optics with a spherical and cylindrical lens

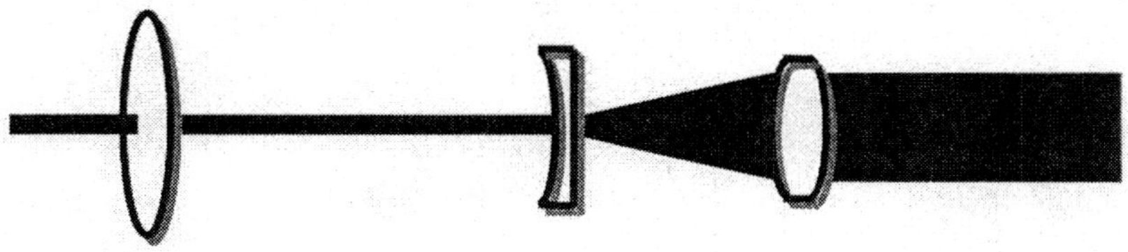

(b) Laser sheet forming optics with Galilean type beam expander

Fig. 5.12 Examples of laser sheet optics.

characterized in one direction as 'top hat' and the other direction as nearly Gaussian. The divergence in these two orthogonal directions is relatively larger (less than 3 and 1 mrad, respectively). Because of the larger divergence, the thickness of the laser sheet is restricted to more than 1 mm. In order to obtain a smaller focused beam diameter (thinner sheet), unstable resonator optics are sometimes used in excimer lasers. However, this is accompanied by a decrease in the laser output and further reductions in the focused section. The total loss can be more than 50%.

5.6.3 Light Collection and Detection System

5.6.3.1 General System

The detection system includes collection optics (e.g., lenses and spectral filters), a light detection system, and a computerized data acquisition system.

In general, the scattered light or fluorescence emission is obtained perpendicular to the direction of laser beam. This arrangement enhances spatial resolution. An aberration-corrected lens system should be used, particularly for Raman scattering where wide ranges of light frequencies are detected. For Raman scattering, the required spectral resolution is high, and hence a spectrometer is normally used rather than spectral filters.

In selection of a lens system, the major trade-off is between spatial response and light-gathering ability. The lens spatial response in the image plane of a lens with finite aperture and focusing error, (the absolute value of the difference between the "in-focus" object distance and the actual object distance), is given by [Paul et al., 1990]

$$f_L = 1.22\frac{f\#(1+M)}{M^2\delta} \tag{5.77}$$

where M is the magnification (image size/object size) of the system, and f# is the f-number of the lens given by

$$f\# = \frac{f}{D}$$

Therefore, the larger the lens is, the smaller the f# becomes.

Eq. (5.77) shows that a lens spatial response is enhanced under large f-number, low magnification conditions. The reverse is true for optimizing photon collection efficiency and, if the measurement is shot noise limited, the signal-to-noise ratio. The large collection solid angle arises from low f-numbers and large magnifications. In practice, the detector more often limits the spatial resolution of a detection system, especially when a solid state camera is used. Therefore, a compromise is to use an f-number that produces a spatial response equal to that of the detector, as this gives the maximum collection efficiency without seriously degrading the spatial resolution.

Depending on whether a single point measurement or two-dimensional imaging is required, a photodetector or an imaging device is employed. Because of its high sensitivity, a photomultiplier tube is normally used for single point light scattering experiments. For two-dimensional applications, the scattered light is recorded by photographic films, or solid state cameras. The sensitivity of solid state cameras can be increased by more than 20,000 times using an image intensifier. In an image intensifier, the photoelectrons produced by the incident photons are accelerated away from the photocathode by an applied voltage, which can be gated down to nanoseconds. These electrons undergo cascading amplification and finally strike onto a phosphor coated surface, thereby leading to an output image brighter than the original image. Such an image intensifier, referred to as MCP (micro-channel-plate), comprises millions of small diameter (10 μm) channels, where electrons are accelerated by a few hundred volt potential before striking the phosphor plate. The phosphor image produced is then optically coupled to an array detector. The coupling can be achieved via an optical lens, or coherent fiber-optic bundles. Although lens coupling can alter the magnification easily and hence different array detectors can be used, this method has poor collection efficiencies and poor spatial resolution compared to fiber-optically coupling.

Most two-dimensional images of scattered or fluorescence light are recorded with solid state array photodetectors as opposed to photographic films. Although photographic films have high spatial resolution and high data acquisition rates, they have a nonlinear response curve and require lengthy processing time. When an image intensifier is used, the spatial resolution of the detection system is limited by the image intensifier. Hence, the spatial resolution of the film is no longer an advantage. Among the types of imaging devices, the charge-coupled-device (CCD) camera is most often used. CCDs have superior noise and dynamic range performance and are available in many formats.

With charge-coupled devices (CCD), the charges created by the incoming photons are read out by transferring the charge from pixel to pixel to an on-chip register. The charge is pre-amplified and digitized before it is transferred to the computer for image processing via an interface card. The latest gated intensified CCD systems feature a very compact design, which has a built-in gate pulse generator and camera controller as opposed to the separate pulse generator and camera controller of the earlier systems. For low light detection, CCDs are generally cooled thermoelectrically by a built-in Peltier cooler to reduce the noise level, and hence increase the dynamic range of the system.

5.6.3.2 Optical Multi-Channel Detectors

For spontaneous Raman scattering and CARS measurements, it is often necessary to record the entire spectrum in order to determine the species concentration and/or temperature from the detailed spectral features. This may be achieved by scanning a monochromator so that one wavelength region at any one time is detected by a photomultiplier attached at its exit. This approach has been superseded by the simultaneous detection of the entire spectrum using optical multi-channel detectors.

Early optical multi-channel detector systems employ multiple exit slits, appropriate optics (including fibers), and separate photomultipliers. Such a system has limited spectral resolution due to geometric constraints and the number of channels is generally restricted for reasons of practicality to less than ten. In a modern optical multi-channel detector system, a linear array of detectors is attached to a spectrograph to record the entire spectrum with much higher spectral resolutions. The linear arrays used often include self-scanned photodiode arrays or charge-coupled-devices (CCDs). CCDs are very sensitive and generally possess large dynamic range and very low noise characteristics. With CCDs, MCPs can be used to increase sensitivity and signal-to-noise ratio due to synchronized shutteringoperation. When a 1-D spatial imaging mode is used, a 2-D array can be used with an imaging spectrograph. This will be discussed in the application of spontaneous Raman scattering in Chapter 7.

5.6.4 Image Processing and Calibration

If qualitative information is expected from two-dimensional measurements, the experiment and data analysis will be straightforward. To obtain quantitative results, the raw images need to be post-processed.

The first process involves the removal of systematic errors present in the imaging system. According to Seitzman and Hanson [1993], for laser induced fluorescence measurements, the digitally recorded detector signal for a single pixel of a CCD camera, S_{pix} (in number of digitization levels), can be expressed as

$$S_{pix} = \left[N_p \left(\frac{P_c}{P_0} \right) R_{tot} + S_b + S_d \right] \tag{5.78}$$

where N_p = the number of fluorescence photons leaving the collection volume within the collection solid angle,
= $P_{flu}/h\nu$
P_c = the laser energy fluence within the collection volume,
P_o = a nominal laser fluence,
R_{tot} = the total responsivity of the detection system,
S_b = the background signal associated with laser induced processes, e.g, Rayleigh and Mie scatterings and window fluorescence, and the ambient light, e.g., flame emission,
S_d = the dark current signal.

Solving Eq. (5.78) for the fluorescence signal N_p produces

$$N_p = \left[S_{pix} - (S_b + S_d) \right] (1/R_{tot}) (P_0/P_c) \tag{5.79}$$

or for the fluorescence signal received by the detector, P_{flu},

$$P_{flu} = h\nu \left[S_{pix} - (S_b + S_d) \right] (1/R_{tot}) (P_0/P_c) \tag{5.80}$$

Equations (5.79) and (5.80) show an appropriate procedure for correcting systematic errors in the recorded image. First, a combined image of the background and dark current signals (S_b+S_d) is subtracted from the recorded image (S_{pix}). Then, the new image is multiplied by a correction image accounting for the spatial variations in the responsivity of the detector. Finally, nonuniformities in the laser sheet energy are corrected by another multiplicative factor.

A combined dark and background image (S_b+S_d) can be recorded by removing the fluorescing molecules in the gas mixture when the engine is motored. Since background may vary from image to image, an average of the background can be used to minimize the effect. The responsivities are typically measured by uniformly illuminating the detector system with a light source having a spectral distribution similar to the fluorescence, as the spatial variation is often a nonuniform function of wavelength. When two-dimensional measurements are carried out inside an engine, it is more convenient to record the combined effect of responsivity ($1/R_{tot}$)

and non-uniformities in the laser sheet (P_c/P_o). The combined effect is obtained from an image recorded from a homogeneous mixture containing fluorescing molecules (S_{uni}) minus the background image ($S_b + S_d$). Hence, Eq. (5.67) can be rewritten as

$$P_{flu} = h\nu\left[S_{pix} - (S_b + S_d)\right] / \left[S_{uni} - (S_b + S_d)\right] \tag{5.81}$$

In some of the lasers used for imaging, shot-to-shot variations in the spatial profile can be significant. Thus, records of the laser sheet profile, recorded simultaneously with the fluorescence images, are preferred. The profile can be recorded directly by imaging the sheet onto another two-dimensional detector.

Other systematic errors such as steering of the laser beam by density gradients, attenuation of the laser sheet across the imaged region, laser-produced perturbations, partial saturation and radiative trapping (absorption of the fluorescence within the combustion gases) are often avoidable or negligible, and these effects must sometimes be considered.

Another source of errors is the random noise in a digital imaging system, which includes the shot noise in the photodetector and electronic noises in the detection system. In most cases, this error is negligible.

5.7 Summary

Molecular spectroscopy basically consists of a set of rules governing the allowable configurations of each energy state, their changes, and energies associated with these changes. Laser spectroscopic techniques involve the interaction and exchange of energy between a photon and a molecule, through which the concentration and temperature of a species can be obtained from the photons scattered by or emitted from molecules. The three laser spectroscopic techniques described in this chapter are laser Rayleigh scattering, spontaneous Raman scattering, and laser induced fluorescence. The nature and characteristics of these techniques are compared with respect to their applications to species concentration and temperature measurements.

Experimental systems for laser diagnostics are introduced in the second half of this chapter, and their salient features are discussed. It is hoped that the knowledge of spectroscopy and understanding of the experimental systems learned from this chapter will help the readers to perform laser based diagnostics.

References

Dibble, R.W. and Hollenbach, R.E. (1981), "Laser Rayleigh Thermometry in Turbulent Flames," *Proc. of 18th Symposium (International) on Combustion*, The Combustion Institute, Pittsburgh, Pa., pp. 1489–1499.

Eckbreth, A.C. (1996), *Laser Diagnostics for Combustion Temperature and Species,* 2nd Edition, Gordon and Breach Publishers, Amsterdam.

Herzberg, G. (1950), *Molecular Spectra and Molecular Structure, Vol. 1: Spectra of Diatomic Molecules*, Van Nostrand Reinhold Co., Cincinnati, Ohio.

Herzberg, G. (1971), *The Spectra and Structures of Simple Free Radicals*, Cornell University Press, Ithaca, N.Y.

Inaba, H. and Kobayasi, T. (1972), "Laser Raman Radar—Laser Raman Scattering Methods for Remote Detection and Analysis of Atmospheric Pollution," *Opto-Electronics,* Vol. 4, pp. 101–123.

Laurendeau, N.M. (1988), "Temperature Measurements by Light-Scattering Methods," *Prog. Energy Combust. Sci.*, Vol. 3, pp. 1–34.

Lederman, S. (1977), "The Use of Laser Raman Diagnostics in Flow Fields and Combustion," *Prog. Energy Combust. Sec.,* Vol. 3, pp. 1–34.

Long, D.A. (1977), *Raman Spectroscopy*, McGraw-Hill, New York.

Paul, P.H., Van Cruynighen, I., Hanson, R.K., and Kychakoff, G. (1990), "Applications of Planar Laser-Induced Fluorescence Imaging Diagnostics to Reacting Flows," AIAA/SAE/ASME/ASEE 26th Joint Propulsion Conference, AIAA-90-1844.

Piepmeier, E.H. (1972), "Theory of Laser-Saturated Atomic Resonance Fluorescence," *Spectrochimica Acta,* Vol. 27B, pp. 431–443.

Seitzman, J.M. and Hanson, R.K. (1993), "Planar Fluorescence Imaging in Gases," Chapter 6 in *Instrumentation for Flows with Combustion*, ed. by A.M.K.P. Taylor, Academic Press.

Self, S.A. (1983), "Focusing of Spherical Gaussian Beams," *Appl. Optics*, Vol. 22, No.5, p. 658.

Siegman, A.E. (1971), *An Introduction to Lasers and Masers*, McGraw-Hill, New York.

Schrotter, H.W. and Klockner, H.W. (1979), in *Raman Spectroscopy of Gases and Liquids*, Ed. by A. Weber, pp. 123–166, Springer-Verlag, Berlin.

Chapter 6

In-Cylinder Flow Field Measurement

Outline

6.1 Introduction

The details of fluid flow within the cylinders of IC engines are known to profoundly affect engine performance and emissions. Increasing demands for lower emissions and better fuel economy have lead to the development of combustion systems in which the control of in-cylinder fluid flow plays a vital role, such as gasoline direct injection engines. The control of in-cylinder flow is achieved primarily through the design of the inlet ports, the cylinder head, and piston crown geometry. Both large-scale fluid motion, such as swirl and tumble flows, and small-scale turbulent flow fields are employed in the optimization of combustion and emissions. On the other hand, the development of three-dimensional computational fluid dynamics (CFD) models has offered an attractive tool for engine designers to reduce engine testing and reductions in engine design cycle time. Due to its current limitations, CFD predictions always require experimental validation in new applications. Therefore, in-cylinder flow field measurements are important for the validation of CFD models, as well as for the verification of design intent and for the detailed study of fundamental in-cylinder flow processes and their interaction with combustion.

The measurement of the flow inside the cylinder may be achieved by single point measurement techniques or whole field measurement techniques. Hot-wire anemometry (HWA) or laser Doppler anemometry/velocimetry (LDA/LDV) have been used to measure the flow velocity and turbulence with high spatial resolution. HWA has the advantage of giving continuous records of velocity, but has the disadvantages that it does not readily give the direction of flow and needs careful calibration. Because the wires used in HWA are delicate, they are prone to breakage and will not stand combustion. LDA has the advantage over HWA in that it provides a direct measurement of velocity, is non-intrusive, and is independent of fluid properties. Therefore, LDA has been widely used in recent years for engine flow measurements. Witze [1980] has presented a critical comparison of HWA and LDA.

Although pointwise LDA velocity data have provided much useful insight into in-cylinder flow process and will undoubtedly continue to do so, the reality of flow within IC engine cylinders is complex, unsteady, and transient. A limited number of point measurements, while providing valuable time-averaged data and turbulence estimates, are not sufficient to fully describe the details of the unsteady flow processes taking place inside the cylinder. Therefore, whole field measurement techniques have been developed in the study of in-cylinder flows. Flow visualization, the simplest qualitative whole field technique, has been employed since the early days of engine research. For example, in-cylinder flow was studied by Clerk as early as 1921 in a low speed optical engine. The flow was visualized by the introduction of smoke, and Clerk's qualitative observation was sufficient to confirm the turbulent nature of in-cylinder flow. Another early flow visualization study in motored engines was that of Lee [1939], who used feathers in the intake air system as tracer particles, recording their motion using a high-speed cine film camera and an intense light source. More recently, microballoons (10–40 mm) have been used to mark the flow, which scatter light adequately due to their size, but being hollow, are light enough to follow the flow faithfully at low to medium engine speeds [Marko et al., 1986].

In order to characterize the complex unsteady in-cylinder flows quantitatively, particle tracking velocimetry (PTV) and particle image velocimetry (PIV) have been developed. These quantitative whole field techniques provide the ability to record instantaneous, cycle resolved, two dimensional maps of velocity within extended measurement planes inside the cylinder. In PTV and PIV measurements, small particles are introduced into the flow and subjected to intense stroboscopic illumination so that their positions may be recorded, photographically or electronically, at two or more known instants in time. In PTV, individual particles or groups of particles are identified and matched between exposures so that their displacement, and hence velocity, may be measured. PTV has been used for the characterization of induction-like flows in water analog engines [Khalighi and Huebler, 1988], the study of cyclic variability in a motored engine [Shack and Reynolds, 1991], and the study of intake port design on in-cylinder flows [Neußer et al., 1995]. More recently, high-speed consecutive PTV measurements were achieved by recording particle images at 9000 frames per second using a Kodak 4540 high-speed video camera and a 15 W copper vapor laser [Church and Farrell, 1998].

A variation on PTV is streak photography, which involves illumination of the flow with an intense light sheet that is modulated in pulses of finite duration, so that the movement of particles during the exposure period forms streaked images of individual particles. The length and direction of an individual particle image streak yields the local velocity vector. Modulating the illumination by intensity or color allows the beginning and end of the streaks, and hence the flow direction, to be identified. The streak photography technique has been employed to study large-scale flows in a water analog rig [Khalighi, 1990] and in a motored engine [Kuwahara et al., 1994].

However, the need for long trains of rapid pulses, or long illumination pulses in PTV or streak photography experiments demands the use of multi-pulsed white light sources (or high-repetition pulsed lasers) or the use of modulated CW lasers. These illumination sources, in general, provide light sheets of relatively low power density. Therefore, large seeding particles must be used to obtain sufficient light scattering for particle imaging, limiting the frequency response of velocity data. In addition, relatively low densities of trace particles are necessary to avoid the crossover or joining of particle pairs or streaks and to avoid excessive time in identifying images of individual particles. The velocity fields obtained from PTV measurements are often seen to be sparse and require significant interpretation or ensemble-averaging.

The advent of PIV has provided a means of achieving high spatial resolution two-dimensional velocity measurements. In PIV measurements, the particle movement within a thin plane in the flow is frozen by two consecutive very short high-energy laser pulses. The mean velocity within each small area of the recorded image is determined by performing a spatial correlation of the region. The spatial correlation process used in PIV does not involve the identification of individual particle images, and hence a sufficiently high density of seeding particles may be used to give more complete and detailed velocity maps than does PTV. The use of high-energy, short pulse duration laser illumination allows micron-sized particles to be used, permitting measurements of high-frequency flows. The PIV technique thus appears well suited to the study of both the formation of large-scale motion and its subsequent breakdown into small scale motions.

As noted above, laser Doppler anemometry has been widely used for pointwise mean velocity and turbulence measurements in internal combustion engines, whereas particle image velocimetry is becoming increasingly applied to the study of in-cylinder flows due to its ability to record instantaneous two-dimensional velocity maps with high spatial resolution. Thus, the major thrust of this chapter is basic LDA and PIV instrumentation and its application to internal combustion engines.

6.2 Laser Doppler Anemometry

6.2.1 The Principle of Laser Doppler Anemometry

Laser Doppler anemometry (LDA) is based on the measurement of the Doppler shift of laser light scattered from small particles carried along with the moving fluid. Because of the motion of the particles, the frequency of the scattered light will be shifted by an amount

$$\nu = \vec{K} \bullet \vec{V} \tag{6.1}$$

where $K = K_s - K_i$ is the difference between the wave vector of the scattered light K_s and that of the incident K_i, respectively, and V is the velocity vector of the particle, as shown in Fig. 6.1.

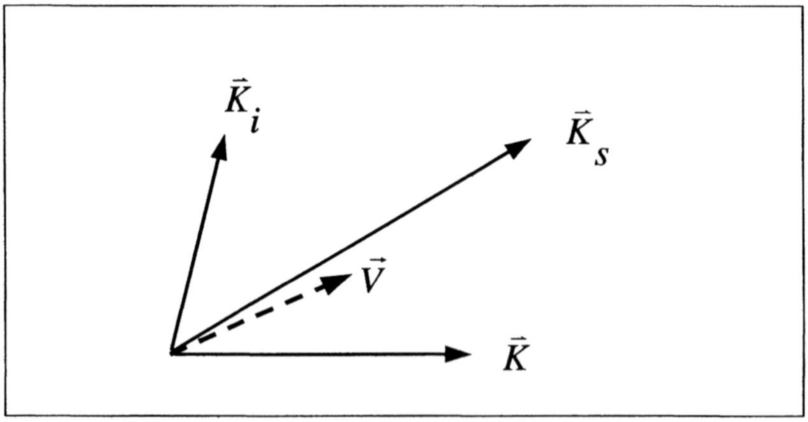

Fig. 6.1 Wave vector diagram showing directions of incident and scattered light and the velocity vector.

Since the flow velocities commonly encountered are much slower than the velocity of light, the corresponding Doppler shifts are small. For example, scattering from a particle moving at a velocity of 20 m/s will produce a shift equivalent to 0.14 parts per million in the frequency of an He-Ne laser beam. The resolution of such a small shift is beyond the capabilities of the highest-resolution optical instruments, and hence direct determination of the shift is not a practical method of measurement.

However, the optical heterodyning or light-beating effect of a square law photodetector can be utilized to measure very small Doppler shifts. Fig. 6.2 shows a simple illustration of the optical setup. Light from a laser is divided by the beam splitter, most of the light being focused to a point in the medium where the velocity measurement is to be carried out. The weaker beam, also known as the reference beam, is directed to a photomultiplier (PMT) by the mirror. The scattered light from particles in the medium will be detected by the same photomultiplier as well. It can be shown that the output from the photomultiplier contains a signal of the difference-frequency between the reference beam and the scattered light, which is of course the Doppler shift. For the experimental arrangement shown in Fig. 6.2, this given by

$$\nu_D = \frac{2v}{\lambda}\sin\left(\frac{\theta}{2}\right) \tag{6.2}$$

where v is the velocity of the moving particle, λ the wavelength of the laser beam, and θ the angle between the main and the reference beams. This arrangement is known as the reference beam technique.

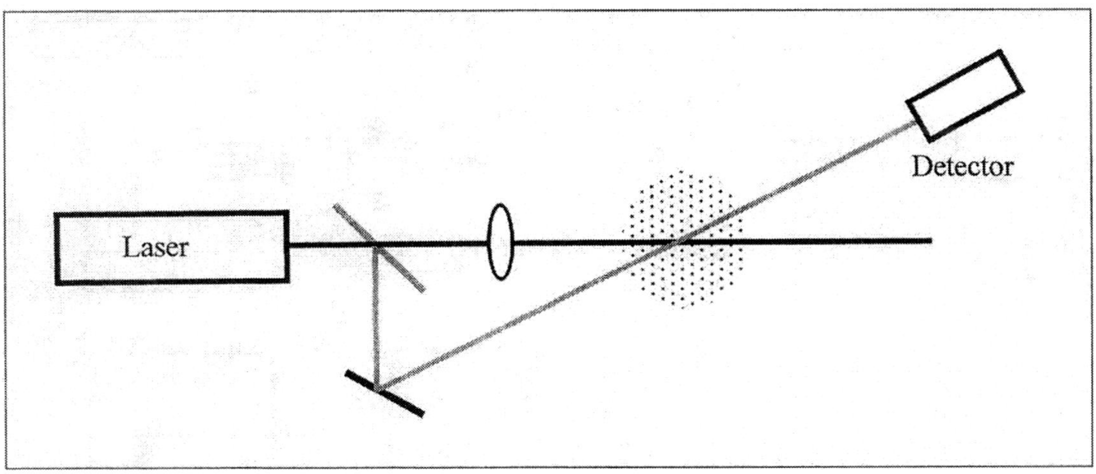

Fig. 6.2 A simple reference beam LDA system.

In order to resolve the beating frequency component containing the Doppler shift, the scattered and reference beams should be matched in intensity at the detector, which can be accomplished by providing an adjustable attenuator in the reference path.

A more commonly used technique in gas flows is the differential Doppler or dual beam technique. A simple experimental setup is shown in Fig. 6.3. The two illuminating laser beams of equal intensity are focused into a small region by a single lens. Scattered light from this region is focused onto a photomultiplier. The photomultiplier responds to the square of the total light field incident on each point of the detector surface; its output is the sum of the contributions from each point of the detector surface given by

$$I \propto \int_{A_d} |E_1 + E_2|^2 dA = \int_{A_d} |E_1|^2 dA + \int_{A_2} |E_2|^2 dA + \int_{A_d} \left(E_1 \cdot E_2^* + E_2 \cdot E_1^*\right) dA$$

or (6.3)

$$I \propto I_1 + I_2 + 2\sqrt{I_1 \cdot I_2} \cos\left[2\pi \nu_D t + \varphi\right]$$

where I_1 and I_2 are the outputs that would result from the scattered light from each beam alone, and the Doppler shift ν_D is given by Eq. (6.2).

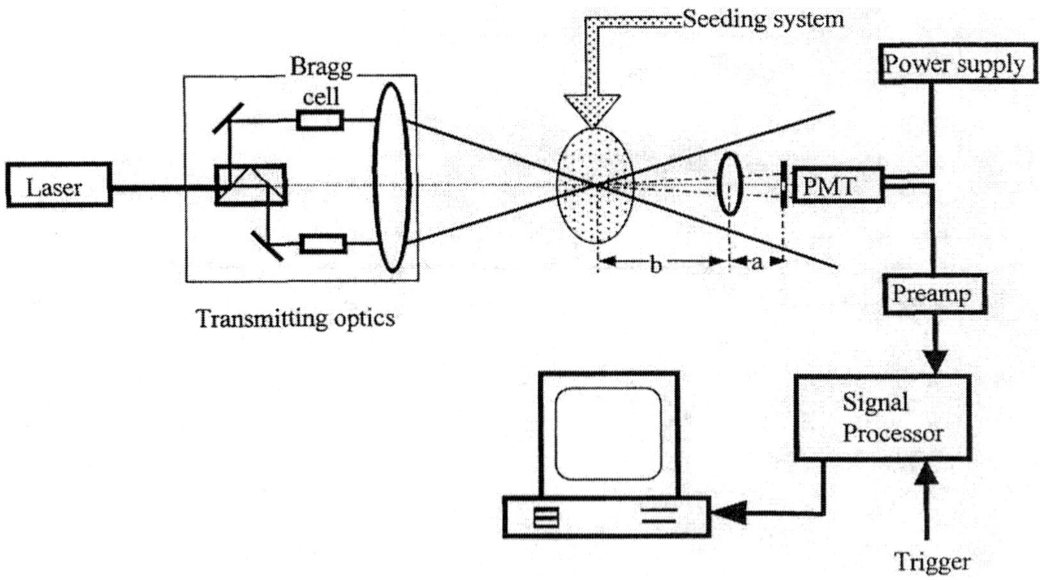

Fig. 6.3 A simple differential LDA system.

Thus, the photomultiplier output is characterized by a DC component (the first two terms in Eq. (6.3) and an AC component (the third term in Eq. [6.3]) which contains the Doppler shift.

The operation of the differential technique can also be understood in terms of the fringe pattern that exists in the crossover region due to interference between the two beams as shown in Fig. 6.4. The spacing of the interference fringe is given by

$$\Delta d = \frac{\lambda}{2\sin(\theta/2)} \tag{6.4}$$

A particle moving at a velocity v perpendicular to the fringes will thus experience a modulation of light intensity of frequency

$$\nu = \frac{2v\sin(\theta/2)}{\lambda} \tag{6.5}$$

which is the differential Doppler frequency. For this reason, the technique is sometimes also called the real fringe method.

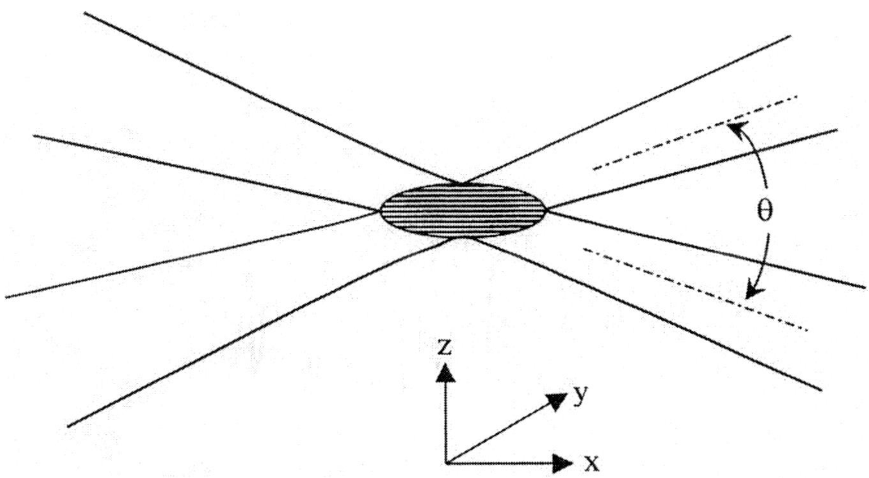

Fig. 6.4 Geometry of the probe volume.

The time variation of the scattered light from a particle moving through a fringe pattern, i.e., the AC component of Eq. (6.3), is shown in Fig. 6.5(a). It can be seen that there is a low-frequency component of intensity variation (shown by the dashed line), as well as the high-frequency component (Doppler frequency). The low-frequency component is often called the pedestal, and is caused by the Gaussian intensity distribution of a laser beam.

The differential Doppler arrangement is preferred in a majority of applications because it gives a considerably greater signal-to-noise ratio than is obtained with the reference beam technique. It is particularly suitable for situations of low particle concentration, as are frequently encountered in gas flows. Therefore, only the differential Doppler technique is discussed in the following sections.

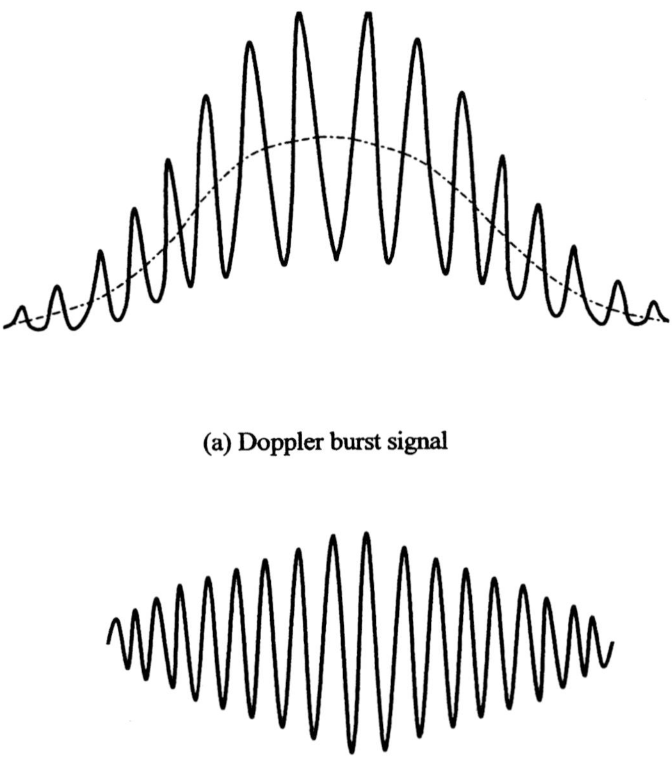

(a) Doppler burst signal

(a) Bass-pass filtered signal

Fig. 6.5 The Doppler burst signal of a differential LDA system.

6.2.2 Experimental Considerations

A laser Doppler anemometer may be designed to meet differing requirements depending on the number of velocity components to be measured, the spatial resolution, and accuracy sought, the overriding consideration always being the necessity of getting an adequate signal-to-noise ratio for signal processing.

A typical LDA system comprises four main parts (Fig. 6.3): the laser and transmitting optics, light collection and detection system, the particle generation and seeding system, and signal processing equipment.

6.2.2.1 Laser and Transmitting Optics

In general, two types of lasers have been incorporated in conventional LDA systems. An He-Ne laser is characterized by its portability and low-cost but low-power output. An argon-ion laser is preferred, particularly in engine applications where the seeding density is low. In addition, its multi-wavelength output is ideally suited to a multicomponent LDA system. Recent advances in laser diodes have resulted in the development of commercial LDA systems with extremely portable configuration.

The purpose of the transmitting optics is to split the laser output into two parallel beams of equal intensity and then focus them to form the probe volume. Beam splitting can be achieved by the use of a partial reflection beam splitter or polarizing beam splitters (Fig. 6.3). The geometry of the transmitting optics determines the dimensions of the probe volume that is defined by the intersection of the two beams. With Gaussian laser beams, the probe volume is an ellipsoid, and its dimension can be defined in terms of the half-axes (a, b, and c) of the ellipsoid in the coordinate system shown in Fig. 6.4 as follows

$$2a = 4\delta_x = \frac{d_f}{\cos(\theta/2)} \tag{6.6}$$

$$2b = 4\delta_y = d_f \tag{6.7}$$

$$2c = 4\delta_z = \frac{d_f}{\sin(\theta/2)} \tag{6.8}$$

where d_f is the beam waist of the focused laser beam, which is given by

$$d_f = \frac{4\lambda}{\pi\Delta\theta} \tag{6.9}$$

where $\Delta\theta$ is the convergence angle of the focused laser beam.

Eq. (6.2) states that, for a given wavelength of radiation, the Doppler shifted frequency is proportional to the sine of the intersection angle. Thus, the intersection angle should be chosen so that the Doppler shift in frequency over the expected velocity range is within the measuring range of the photomultiplier and signal processors, while providing an adequate number of fringes for signal processing.

6.2.2.2 Light Collection System

Light collection in the forward direction, as shown in Fig. 6.3, is preferred whenever possible, since light scattering from particles of microns is greatest in this direction. With a backward scattering configuration (Fig. 6.6), a reduction of two or three orders of magnitude in the received scattered light is expected. In this case, an adequate signal can only be achieved by the use of higher laser power. Sometimes a back scattering arrangement is unavoidable because of the available optical access to the engine.

The function of the light collection system is to ensure that the scattered light originating from the probe volume is collected and imaged onto a photomultiplier. At the same time, the system must reject any scattered light from particles other than those in the probe volume and any reflections

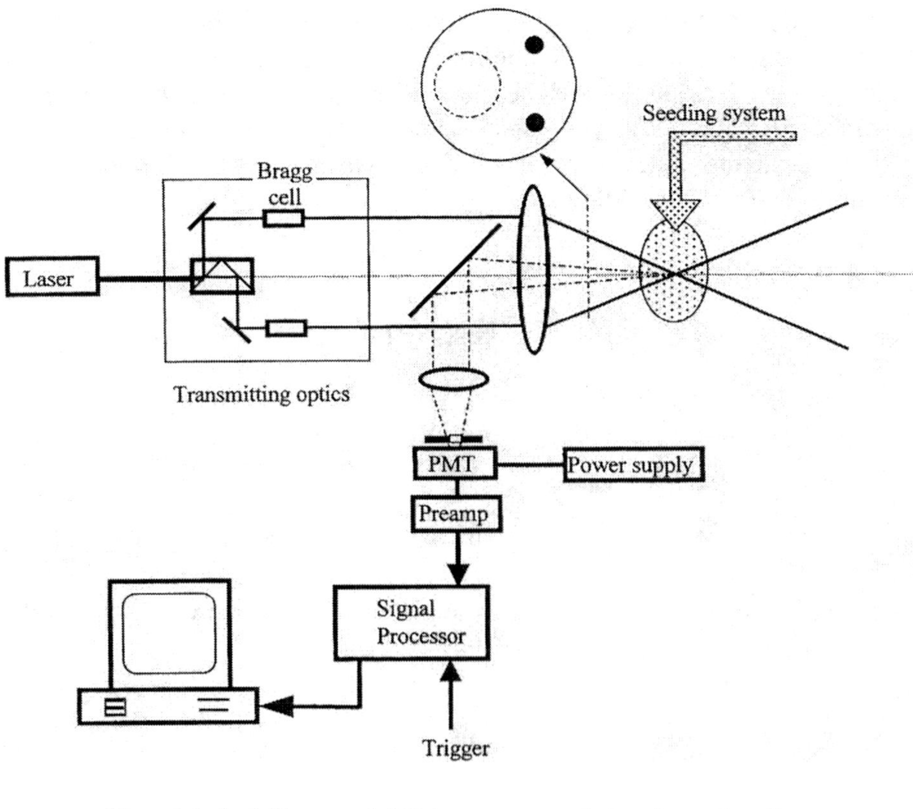

Fig. 6.6 A differential LDA system with backward collection.

from windows or smooth external surfaces. The principal discrimination against unwanted light is usually achieved by focusing light scattered from the probe volume onto an accurately positioned aperture in front of the photomultiplier.

Since the light intensity distribution across the fringe pattern is normally Gaussian, the fringes formed in the outer region of the probe volume are of a poorer quality than the remainder. It is therefore desirable to limit the detection of scattered light to the central part of the probe volume, which is defined as the measuring volume. In the forward detection configuration shown in Fig. 6.3, the measuring volume is determined by the diameter of the aperture and the position of the focusing lens

$$d_x = \frac{d_a}{M};\ d_y = \frac{d_a}{M};\ d_z = \frac{d_f}{\sin(\theta/2)} \tag{6.10}$$

where d_a is the diameter of the aperture, and M = b/a, the magnification of the optical system. The number of fringes in the measuring volume is given as

$$N_f = \frac{d_a}{\delta_f M} \tag{6.11}$$

where

$$\delta_f = \frac{1}{2}\frac{\lambda}{\sin(\theta/2)} \tag{6.12}$$

is the fringe separation.

If LDA measurements are carried out in the presence of combustion, flame luminosity can be suppressed by placing an interference filter in front of the detector, so that only the scattered laser light is detected.

6.2.2.3 Signal Processing

The primary output of an LDA measurement is a current from the photomultiplier tube. As discussed previously, the photomultiplier output consists of a DC and an AC component. The Doppler shifted frequency is found in the high-frequency part of the AC component as shown in Fig. 6.5.

Depending on the particle density in the fluid, the output can be quasi-continuous or signal bursts interspersed with periods of low-power shot noise. Burst signals occur when the concentration of seeding particles is low, as is often the case in engine flow measurements.

The Doppler frequency is commonly measured by dual-counter zero-crossing counters, which are based on two counters recording two Doppler frequencies from the same signal over a different number of cycles. The Doppler frequency is validated only if the two Doppler frequencies agree within some preset tolerance. The operation of a counter type LDA processor can be understood by referring to Fig .6.7. The timer module may be operated either in N-cycle or total-burst mode.

In principle, the LDA counter is simply a timing device that is programmed to measure the time between certain events. The burst signal from the pre-amplifier (shown in Fig. 6.5[a]) is input to the signal conditioner of the processor. The conditioner contains low and high pass filters for removing noise and the Gaussian pedestal from the Doppler signal. It also contains an amplifier that varies the amplitude of the filtered signal. The Doppler signal after the conditioner is shown in Fig. 6.5(b). Following the conditioner is a timer module. As shown in Fig. 6.7, the timing is done by counting cycles of a reference clock frequency much higher than that of the maximum Doppler signal to be measured. In the measurement of a typical Doppler burst, the timing would be initiated by a signal exceeding predetermined threshold, and stopped after a given number of cycles had occurred. In addition, the frequency of the first 10 or so cycles is compared with the frequency of the first 16 cycles, and only if they are equal to within a preset limit is the measurement considered valid. Clearly, the more cycles are included, the greater the accuracy attainable, but the number selected must be substantially less than the number of fringes in the probe volume. The frequencies (velocities) are paired with the corresponding crank angle information derived from a shaft encoder connected to the crankshaft. Either all of the data are stored for the subsequent analysis, or the signals are gated so that only those occurring within a preset crank angle window are used to build ensemble-averages from many engine cycles.

In engine applications, optical access is often limited, and backward scattering detection is sometimes the only choice available. The concentration of seeding particles is limited as well. Consequently, signals tend to be of poor signal-to-noise ratio and small amplitude. Counter-type

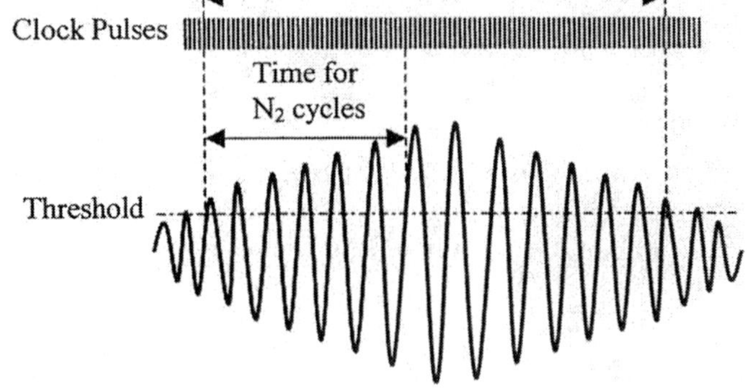

Fig. 6.7 Timing cycles of a Doppler burst for frequency measurement and validation in a counter processor.

processors cannot provide reliable estimates of Doppler frequencies. The superior ability of spectrum analysis, as compared with counters, to process noisy signals has led to the development of more powerful signal processors based on spectrum analysis in the late 80s. The procedure is to obtain the frequency spectrum by Fourier transformation, from which the Doppler frequency is determined from a well-defined peak in the spectrum [Lading, 1987]. In order to improve the resolution of the spectrum, interpolation techniques can be employed. If no peak is distinguishable or there is more than one, the measurement is rejected. This technique overcomes many of the difficulties in the use of other signal processors and is capable of extracting Doppler frequencies from bursts with very poor signal-to-noise ratios.

Both theoretical and experimental studies have shown that a spectrum analyzer can process signals at signal-to-noise ratios of about 15 dB below that at which a counter ceases to work [Lading, 1987; Tropea et al. 1988]. The spectrum analyzer is particularly useful for processing Doppler signals in the in-cylinder flow of an IC engine, where signal quality varies greatly during a single measurement. Pitcher et al. (1988) found that the wider dynamic range of a spectrum analyzer relative to a counter allowed more complete in-cycle measurements to be made, and that typically the data rate from the spectrum analyzer was three times higher than that from a counter. Because of the superior quality of a spectrum analyzer, the frequency domain signal processor using fast Fourier transform (FFT) has replaced the counter as the commercially adopted signal processor in LDA experiments.

6.2.2.4 Frequency Shifting

A fundamental problem of the LDA system shown in Fig. 6.3 is that of discriminating the direction of flow. The Doppler frequency given by Eq. (6.2) provides no information as to whether the particle is traveling up or down the fringes in the probe volume. The most common technique of discriminating flow direction involves frequency shifting to produce a frequency difference, $\Delta\nu$, between the two beams. This may be done by frequency shifting one beam, or more usually both beams (one up and the other down), so that a moving system of fringes is obtained. The speed at which the fringes move is given by

$$v_f = \frac{\Delta\nu \cdot \lambda}{2 \cdot \sin(\theta/2)} \qquad (6.13)$$

This is illustrated in Fig. 6.8. A stationary particle within the probe volume produces a signal of frequency equal to the shift frequency. Particles moving in the same direction as the fringes lower the Doppler signal frequency; particles moving in the opposite direction raise it, i.e.

$$v_D = \frac{\Delta\nu \cdot \lambda}{2 \cdot \sin(\theta/2)} \pm \frac{2v\sin(\theta/2)}{\lambda} \qquad (6.14)$$

Thus, to provide direction discrimination, the frequency shift applied must be at least equal to the Doppler shift frequency corresponding to the maximum velocity occurring in a series of measurements. A larger frequency shift is recommended to minimize the fringe biasing. The fringe biasing arises from the fact that signals from particles, which cross the fringe region at large angles, cross a smaller number of fringes so as to have less probability of being validated by the signal processors.

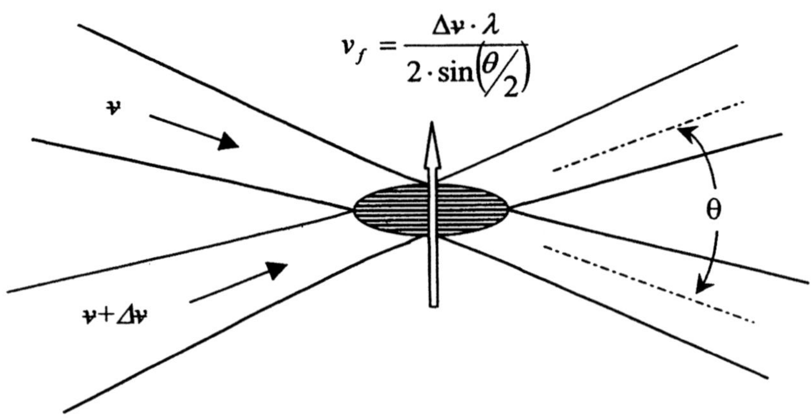

Fig. 6.8 The formation of a moving fringe system by the crossing of two beams of different frequency.

A simple and inexpensive frequency shifter is a rotating diffraction grating. Fig. 6.9 shows the usual arrangement for an LDA system incorporating a rotating diffraction grating. The laser is focused onto the grating and the grating acts as the beam splitter. The two crossing beams are produced from the ±1 orders of diffraction which are shifted by $\pm(N \cdot f_g)$, the product of the total number of lines, N, on the grating and the rotational frequency of the grating, f_g. Hence, the frequency shift imposed on the Doppler signal is $2N \cdot f_g$. The other orders of diffraction are stopped off by a beam stop..

Focusing the laser beam onto the grating is necessary to minimize the effect of grating spacing variation across the laser beam, which leads to a degradation of the diffracted beams. But focusing limits the laser power that can be used. In addition, about 30% of laser power is lost in the other orders of diffraction beams. The maximum frequency shift is limited to a few MHz by the speed of the mechanical rotation possible.

The most popular frequency shifting technique is obtained with diffraction from an acousto-optics cell (Bragg cell). The principal feature for frequency shifting with a Bragg cell is illustrated in Fig. 6.10. The laser beam is passed through a cell in which acoustic waves, excited by a piezoelectric transducer, are traveling. If the angle between the laser beam and acoustic

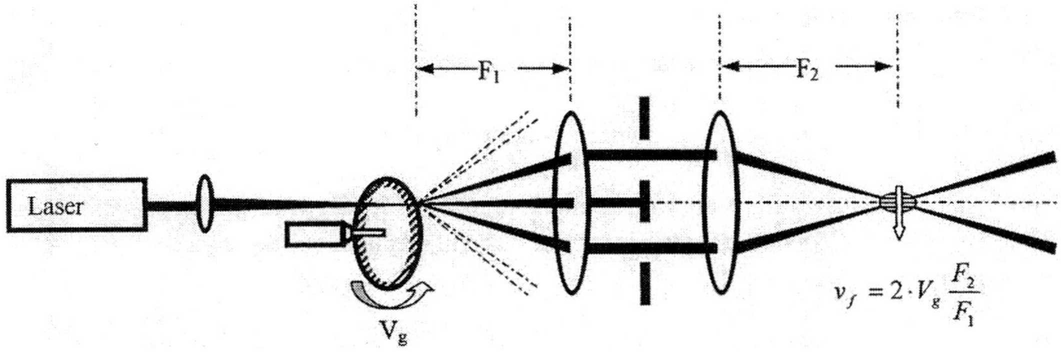

Fig. 6.9 The experimental setup of an LDA system incorporating a rotating diffraction grating

waves satisfy a certain condition (Bragg condition), diffracted beams are produced as from a moving diffraction grating. The frequency shift of the diffracted beam is equal to the frequency of the acoustic waves, which is controlled by a radio frequency generator. The deviated diffraction (shift) beam is brought back to its original direction by a prism incorporated into the acoustic cell unit. The angle of the prism required is a function of the wavelength and shift frequency.

Frequencies around 40 MHz generally give best results. This shift frequency is, however, rather large for many LDA applications, and it is above the range of some processing instruments. Hence, it is necessary to shift the signal frequency downward again electronically by mixing it with an oscillator of around 30–35 MHz before the signal processing stage. Many photomultipliers exhibit reduced response at such high frequency and hence the use of a lower shift frequency may be important. This difficulty may be overcome by placing one Bragg cell in each beam, with the two cells being excited at frequencies differing by the frequency shift required. Fig. 6.3 shows an LDA system with two Bragg cells.

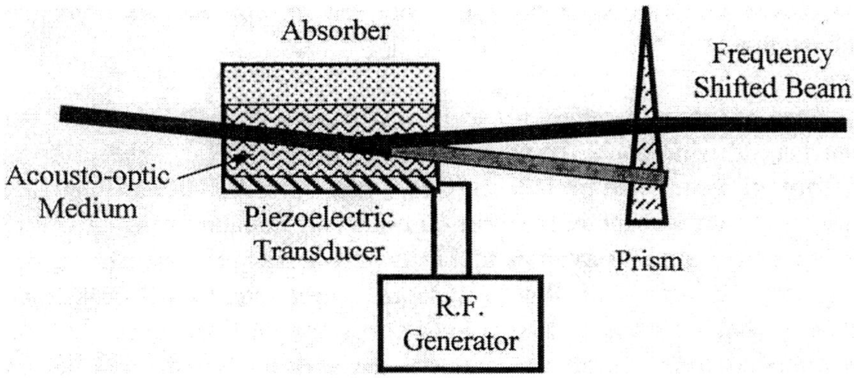

Fig. 6.10 Frequency shifting by means of an acousto-optic Bragg cell.

6.2.2.5 Seeding and Particles

LDA measurements depend on scattering by particles present in the fluid. Some suitable particles are usually present naturally in liquid flows. However, in the study of gaseous flows, the artificial addition of particles or seeding is normally required.

There are two basic requirements with regard to particles. The first is that they should follow the flow closely. The second is that they must scatter sufficient light to give an acceptable signal-to-noise ratio.

The first requirement places limitations on particle size and the density ratio between particles and fluid. The limits are specified by considering the ability of the particles to follow the turbulent fluctuations up to an upper frequency within a certain accuracy. It is estimated that for a particle-to-fluid density ratio of 1.8×10^4, particles must have diameters less than 2.7 μm to follow fluctuations up to 1 kHz and less than 0.8 μm for frequencies up to 10 kHz. A lower limit of particle size may be defined for very low velocities and density ratios in respect of Brownian motion. Thus, considering particle dynamics alone, 1 μm seems to be the optimum size, provided that the density ratio is sufficiently small.

The second requirement concerns the scattering properties of the particles as a function of size distribution, refractive index, particle shape, and concentration, as well as the incident laser power. In the size range of concern, Mie theory defines the scattering intensity as a function of refractive index, particle diameter (D), and scattering angle. The scattering cross section, which is the ratio of the total scattered light to the intensity of the incident light, has a maximum value for the size parameter $x = \pi D/\lambda$, corresponding to a diameter of 0.8 μm for $\lambda = 0.5$ μm or 1 μm for $\lambda =$ 0.633 μm. And the directional scattering intensity is strongest in the forward direction.

Optimum signal to noise ratio can only be obtained if the particle size is less than the fringe spacing. It is recommended that the ratio should be about 0.25 [Drain, 1980].

Several techniques have been employed for generating dispersions of fine particles. These include atomization of liquids, fluidization of powders, chemical reactions, and combustion-based techniques, the essential requirement being that they should supply a uniform concentration of particles into the flow.

Liquid drops are the most satisfactory forms of seeding, because they are nonabrasive, noncontaminating and, with suitable liquids, noncorrosive and nontoxic. The liquids most used are silicone oil, corn oil, water, and propylene glycol-water solution. Commercial atomizers are available. Fig. 6.11 shows an air jet atomizer for producing liquid droplets. An air jet leaving a nozzle draws up liquid from a reservoir by the venturi effect and breaks it up into a fine spray that is then carried forward with the air flow. Any large droplets present will strike the wall of the chamber and drain back into the reservoir. Droplets of 1 μm mean diameter can be produced, and the rate of production can be conveniently adjusted by controlling the flow rate. An optional cyclone can be installed downstream the atomizer so that the upper limit of particle diameter can be reduced to about 1.5 μm.

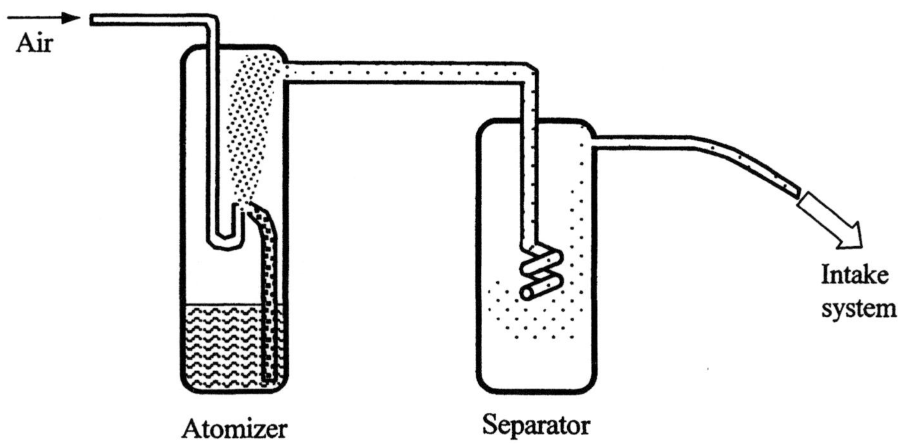

Fig. 6.11 One example of a liquid aerosol generator.

When LDA measurements are to be carried out in a high-temperature environment, the use of inert refractory solid particles is necessary. The particles can be made of MgO, Al_2O_3, or TiO_2, with Al_2O_3 being the most stable under flame conditions. The most commonly used technique of seeding with solid particles involves the use of a fluidized bed. Fig. 6.12 shows one such device [Zhao, 1989]. The device consists of a glass cylinder 4 cm in diameter and 35 cm long with a porous glass disk fixed at a position 7 cm above the bottom of the cylinder to act as a distributor for the fluidized bed. The flow passing through the device is divided into two flows, one of which is used to fluidize the bed of alumina particles. The second flow is passed to a coiled 5 mm diameter copper helix and forms a swirling flow along the axis of the cylinder. The adjustment of both flows allows the seeding levels to be controlled.

LDA measurements in firing engines require special attention due to the additional difficulties imposed by the combustion process. The reduction in data rate is one of the most serious problems under firing conditions. In addition to the optimization of optics and electronics, Lorenz and Prescher [1990] found that a particle-seeding system was critical to obtaining high data rates. Completely dry air and preparation of the seeding particles with a hydrophobic material improved significantly the particle generation process.

In LDA measurements, a small quantity of seeded air should be introduced into the main gas flow well before it enters the engine where the flow is being investigated, so that the seeding can be dispersed uniformly across the flow.

The level of seeding must be controlled as well. Too low a concentration of seeding reduces the data rate. Too much seeding results in more than one particle present in the measuring volume at a time, which causes frequency measurement to be subject to an error due to random phase fluctuation. In addition, flow characteristics may be affected by too much seeding.

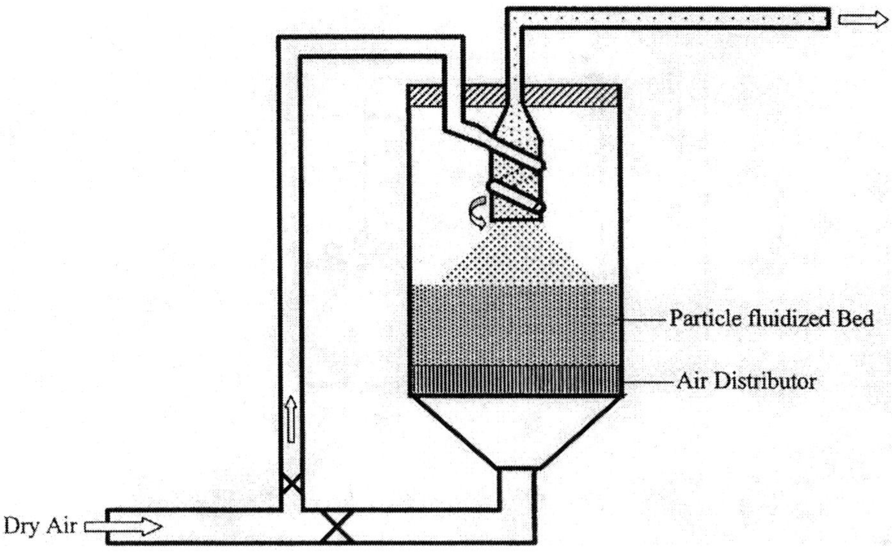

Fig. 6.12 A fluidized bed generator for seeding flows with solid particles.

6.2.2.6 Accuracy

LDA measurements in gas flows are characterized by burst signals. There is higher probability of signals from high-velocity particles than from low-velocity particles, because the particles distribute randomly in space, not in time. The probability depends on the instantaneous velocity. Consequently, the ensemble-averaged velocity becomes larger than the real value. This is called velocity bias, and the higher the turbulence intensity, the larger the velocity bias becomes.

Various schemes were proposed to correct the velocity bias. McLaughlin and Tiederman [1973] showed that it could be eliminated by using the inverse of the instantaneous velocity vector as the weighting function ($1/|u_i|$). Considering that the inverse of the instantaneous velocity is proportional to the residence time of the particle in the measuring volume, Hoesel and Rodi [1977] suggested using this time as the weighting function. The interarrival time was also considered as the weighting function for velocity bias correction [Durao, 1980]. The general bias correction formulae can therefore be written as

$$\bar{u} = \sum_1^n u_i \cdot G_i / \sum_1^n G_i \tag{6.15}$$

$$u' = \left[\sum_1^n (\bar{u} - u_i)^2 \cdot G_i / \sum_1^n G_i \right]^{1/2} \tag{6.16}$$

where the weighting function G_i of the ith measurement is given by

$G_i = 1/|u_i|$, inverse of velocity method;

$G_i = T_i$, residence time weighting;

$G_i = \Delta T_i$, interarrival time weighting.

During the signal processing stage, the register of the residence time corresponding to each measurement is thus valuable to obtain more reliable values of mean velocity and the velocity distribution function in highly turbulent flows.

It should be noted that velocity biasing occurs only if the interval between velocity measurements is large compared with the time scale of the velocity fluctuations.

If a counter processor is used, inaccuracy in velocity measurements can also be present due to fringe bias. Fringe bias arises from the fact that particles crossing the fringes at any angles other than 90 degrees will cross a smaller number of fringes in the probe volume. Because a certain number of fringe-crossings is required for a signal to be validated, these particles will experience less probability of being counted. This may be corrected for by decreasing the number of cycles required for validation by the counter, or preferably by introducing frequency shifting (Section 6.2.2.4).

6.2.3 Advanced LDA Systems

In the simple LDA system shown in Fig. 6.3, any component of velocity perpendicular to the fringes may be measured by suitable rotation of the beam splitter, but only one component can be recorded at a time. Measurements made successively in this way provide considerable information, but there are circumstances where it is necessary to make simultaneous measurements of more than one component of velocity. These may be needed not only to speed up the acquisition of data, but also enable correlation between the components of velocity to be studied, e.g., the Reynolds stress term, $\overline{u' \cdot v'}$.

Commercial LDA systems capable of simultaneous two- or three-component velocity measurements are available. Here, we use a two-component LDA system to illustrate the operating principle of a multicomponent LDA system.

To measure two components of velocity in the same probe volume, it is necessary to form two superimposed sets of fringes, which may be discriminated by polarization or more satisfactorily by color. Blue (488 nm) and green (514.5 nm) beams can be conveniently obtained simultaneously from an argon-ion laser operating in its multi-line mode. In order to measure the two components of velocity transverse to the direction of illumination, orthogonal blue and green fringes are formed in a single probe volume. A cross section of the probe volume is illustrated in Fig. 6.13(a). This is achieved by bringing to a common focus an array of four parallel beams (Fig. 6.13[b]). Fig. 6.14 shows an alternative arrangement where three parallel beams are formed and one contains both colors.

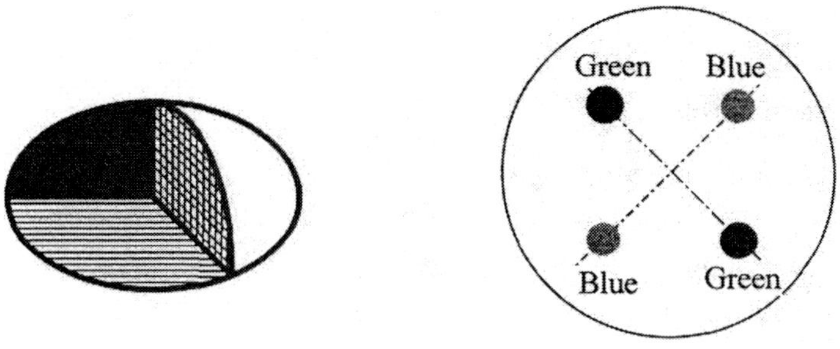

(a) Cross-section of the probe volume (b) Beam arrangement at the focus lens

Fig. 6.13 A two-color scheme for a two-component LDA.

The scattered light can be detected by two photomultipliers directly or via a beam splitter as shown in Fig. 6.14. Narrow bandpass filters are placed in front of the detectors to discriminate any light other than the laser line to be detected. It is important that the fringes of the two colors are accurately aligned. To ensure that both components of velocity are measured from the same seeding particle, the coincidence of data from the two channels needs to be checked by means of a short time window. Unless the two components are recorded within the time window, the pair of data will be rejected. Frequency shifting may be incorporated into one or both of the components by the use of a diffraction grating or Bragg cells.

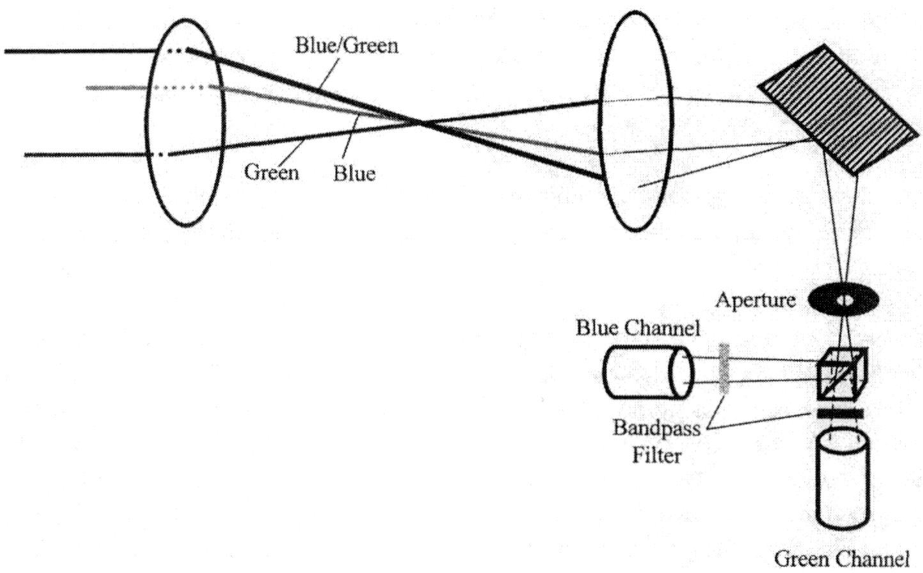

Fig. 6.14 A three-beam two-component LDA system.

Fiber optics are providing a new degree of flexibility that is of particular interest in IC engines. Most of the optical system can be located away from the engine. A commercial LDA system employing a remote fiber probe has been used to measure the in-cylinder velocity and turbulence in a reentrant diesel combustion chamber [Karimi and Jane, 1993]. The development and application of a mobile fiber optic LDA system was reported by Bopp et al. [1990]. Both systems measured the in-cylinder velocity through an optical window as a conventional modular LDA system, but with a compact probe located remotely from the rest of the LDA system. The compact probe held the receiving optics as well as the transmitting optics. Durst and Krebs [1984] and Obokata et al. [1987] described the development and application of miniaturized fiber-optic probes which could be fitted through an injector or spark plug hole in an engine. However, the problems associated with misalignment of the optics due to engine vibration were severe. This problem has been overcome in a recent study of in-cylinder flows using a Dantec fiber flow probe [Faure et al., 1998]. The probe was mounted within a windowed probe holder designed to allow both angular and axial movement of the measurement volume within the cylinder. The drawbacks are the transmission loss of laser power and received signal, which necessitates a more powerful laser and a frequency domain signal processor in order to compensate for the reduced signal-to-noise ratio.

The modular approach to LDA of the earlier days is now replaced by an integrated systems approach, where the entire system runs under a PC computer. This minimizes the setup time and ensures accuracy of the measurement. Windows-based acquisition packages with online help are available to further simplify the measurement process. In many commercial LDA systems, receiving optics are incorporated into the transmitting optics so that the whole system is very compact and self-aligned.

Automatic data analyses, using the output from the signal processor of an LDA system, are provided by commecially available software packages. Raw data are processed to provide individual velocity components and their projection in three orthogonal planes, with statistical properties including mean, turbulence intensity, skewness, and flatness of the velocity components. For multicomponent LDA systems, correlations, Reynolds stresses, velocity and flow angle histograms, and vectors, expressed in terms of magnitude and angle, can be readily displayed or printed. Velocity biasing is corrected by the software package using the particle residence time in the measuring volume. Software packages are also available to control the automatic tranversing of the table with three axes of translation. These features are greatly increasing the ease and convenience of making accurate LDA measurements.

6.2.4 Analysis of Engine Data

6.2.4.1 Ensemble-Averaging Analysis

In stationary turbulent flows, the conventional approach is to resolve the instantaneous flow velocity into (time-independent) mean and (time dependent) randomly fluctuating components.

For the non-stationary in-cylinder flows, the analogous decomposition is into a (time-varying) ensemble-mean velocity and random fluctuations about the ensemble-mean:

$$U(\theta,i) = \bar{U}(\theta) + u(\theta,i) \tag{6.17}$$

where $U(\theta,i)$ is the velocity at a certain crank angle θ in engine cycle i, and the ensemble-averaged velocity is given explicitly by

$$\bar{U}(\theta) = \frac{1}{N(\theta)} \sum_{i}^{N} U(\theta,i) \tag{6.18}$$

where $N(\theta)$ is the total number of valid velocity estimates available at crank angle θ over all cycles in the ensemble, and it is assumed that the Doppler signal is sampled uniformly to produce one velocity estimate per crank angle.

A measure of the intensity of the velocity fluctuations is provided by the rms (root mean square) or standard deviation of the instantaneous velocity about the ensemble-mean velocity:

$$u'(\theta) = \sqrt{\frac{1}{N(\theta)} \sum_{i=1}^{N} \left[U(\theta,i) - \bar{U}(\theta)\right]^2} \tag{6.19}$$

which is also known as the ensemble-rms velocity fluctuation intensity, because it is ensemble-averaged.

Since an LDA signal is only produced when a particle moves through the measuring volume, it is necessary to ensemble-average over a finite crank angle window $\Delta\theta$ around the specific crank angle of interest, θ, in order to obtain data as velocity crank angle pairs. The ensemble-averaged velocity equation thus becomes

$$\bar{U}(\theta) = \frac{1}{N_t} \sum_{i}^{N_c} \sum_{j=1}^{N_i} U(\theta \pm \frac{\Delta\theta}{2}, i, j) \tag{6.20}$$

where N_i is the number of velocity measurements recorded in the window during the ith cycle, N_c is the number of cycles, and N_t the total number of measurements. The corresponding equation for the ensemble-averaged rms velocity fluctuation is

$$u'(\theta) = \sqrt{\frac{1}{N_t} \sum_{i}^{N_c} \sum_{j=1}^{N_i} \left[u(\theta \pm \frac{\Delta\theta}{2}, i, j)\right]^2} \tag{6.21}$$

This need to ensemble-average over a finite crank angle window introduces an error called crank angle broadening, due to change in the mean velocity across the window. This error depends on the velocity gradient. The major influence of crank angle broadening is on the rms velocity fluctuation rather than on the mean value [Rask, 1981].

Alternatively, the smoothed ensemble data-reduction technique can be used [Rask, 1979]. Instead of a single value for the mean velocity, a cubic-spline smoothing routine is used to put a smooth curve through data that has been averaged into 1-degree CA windows. As a result, the ensemble-averaged mean velocity is evaluated at the exact crank angles of all data points, eliminating the crank angle broadening of window-based data reduction.

In practice, if an LDA system samples the Doppler signal uniformly with contiguous windows and yields one velocity estimate per crank angle ($\Delta\theta = 1$ deg), $\Delta\theta$ can be omitted from the calculations, that is, Eq. (6.18) and Eq. (6.19) can be applied directly to the measured data. This will be assumed to be the case in the subsequent analysis in the following sections.

6.2.4.2 Cycle-Resolved-Velocity Analysis

An important implication of Eq. (6.18) is that $\overline{U}(\theta)$ becomes a function of crank angle alone, which implies that the mean flow at a particular crank angle does not vary from one cycle to the next. Flow visualization experiments, however, have shown that this may not always be true. The cyclic variation may be due to the dynamically changing boundary conditions imposed by the piston and valve motions, not to mention fuel-injection and combustion in fired engines. If there are cycle-to-cycle variations, then in the conventional ensemble-averaged analysis, any variations will be included in $u'(\theta)$, resulting in an overestimation of the true in-cycle turbulence intensity.

Two basic approaches have been used for extracting quantitative information from cycle-resolved velocity data. Both are aimed at distinguishing between "true" turbulent velocity fluctuations and cycle-to-cycle variations in the mean velocity.

In the first approach, some form of low-pass filtering is used to estimate the mean velocity in each engine cycle. The rms deviations of the instantaneous velocities from the in-cycle mean velocities are then computed over all the cycles, and they are usually interpreted as a measure of the intensity of the "true" turbulent velocity fluctuations. We call this velocity-filtering-based analysis.

A second approach assumes that turbulent fluctuations are stochastic or random in nature, whereas large-scale fluctuations or cyclic variations in the mean velocity exhibit an appreciable degree of phase stability or coherence within individual engine cycles. The rigorous approach to assess the relative contributions of random and coherent processes is the classical statistical tool based on the autocorrelation function. The analysis attempts to identify, parameterize and extract any phase-coherent contribution to the ensemble-rms fluctuation intensity, leaving only the contribution of the stochastic or turbulent fluctuations.

Velocity-filtering analysis—the filtering approach splits the fluctuations into low- and high-frequency components relative to a filter cutoff frequency,

$$u(\theta,i) = u_{low}(\theta,i) + u_{high}(\theta,i) \tag{6.22}$$

This is equivalent to the decomposition of the instantaneous velocity into three components:

$$U(\theta,i) = \bar{U}(\theta) + u_{low}(\theta,i) + u_{high}(\theta,i) \tag{6.23}$$

The in-cycle mean velocity is therefore the sum of the ensemble-mean velocity and the low frequency fluctuation term:

$$\hat{U}(\theta,i) = \bar{U}(\theta) + u_{low}(\theta,i) \tag{6.24}$$

over which the ensemble-averaging is expected to be equivalent to the conventional ensemble-averaged mean velocity:

$$\overline{\hat{U}(\theta,i)} = \bar{U}(\theta)$$

The intensity of the low-frequency fluctuation is characterized by the rms deviations of the in-cycle mean velocity about the ensemble-mean velocity:

$$u'_{low}(\theta) = \sqrt{\overline{u^2_{low}(\theta,i)}} = \sqrt{\frac{1}{N(\theta)}\sum_{i=1}^{N}\left[\hat{U}(\theta,i) - \bar{U}(\theta)\right]^2} \tag{6.25}$$

Similarly, the high-frequency velocity fluctuation intensity is evaluated as the rms departure of the instantaneous velocity from the in-cycle mean velocity:

$$u'_{high}(\theta) = \sqrt{\overline{u^2_{high}(\theta,i)}} = \sqrt{\frac{1}{N(\theta)}\sum_{i=1}^{N}\left[U(\theta,i) - \hat{U}(\theta,i)\right]^2} \tag{6.26}$$

The conventional ensemble-rms fluctuation intensity is related to the low- and high-frequency intensities by

$$u'(\theta)=\sqrt{\overline{\left[u_{low}(\theta,i)+u_{high}(\theta,i)\right]^2}}=\sqrt{u'^2_{low}(\theta)+u'^2_{high}(\theta)+2\cdot\overline{u_{low}(\theta,i)\cdot u_{high}(\theta,i)}} \quad (6.27)$$

The low-frequency and high-frequency fluctuations are usually assumed to be statistically independent, so that the last term in Eq. 6.27 becomes zero, and the conventional ensemble-rms fluctuation intensity is given by

$$u'(\theta)=\sqrt{u'^2_{low}(\theta)+u'^2_{high}(\theta)} \quad (6.28)$$

The low- and high-frequency fluctuation intensities u'_{low} and u'_{high} are sometimes referred to as the cyclic variation in the mean velocity and the turbulence intensity, respectively.

In order to determine the low- and high-frequency fluctuation intensities, the in-cycle mean velocity $\hat{U}(\theta,i)$ has to be calculated. The first approach is a moving window average in the time (crank angle) domain [Ball et al., 1983]. In this method, a crank angle window is selected on the basis of being larger than the turbulent micro-scale, but smaller than the characteristic period of piston motion (e.g., 10°CA). By using heavy seeding and fast sampling, high data rates can be obtained to resolve most of the turbulence frequencies and scales. Because of the burst Doppler signals, the evaluation of the time-varying mean velocity within the selected window requires some sort of data interpolation between individual burst signals, particularly when there are missing data points within the selected window. The interpolation or smoothing can be done by a sample-and-hold scheme, in which a missing point is replaced by the intermediately preceding valid measurement.

The in-cycle mean velocity can also be obtained by polynomial curve fitting, which involves putting a smooth curve through data that has been averaged into 1-degree CA windows by using cubic-spline curve fitting routine [Rask, 1981]. This is equivalent to a low-pass filtering process, but its cutoff frequency is not well defined.

The third approach involves the use of a frequency-domain filter [Liou and Santavicca, 1985]. Fourier transforming the instantaneous velocity-crank-angle record U(θ, i) produces a spectrum of various frequencies. The low-pass filter is then applied in the frequency domain by setting all of the Fourier coefficients above the selected cutoff frequency f_{low} to zero and retaining the coefficients below f_{low} unaltered. Finally, the inverse Fourier transform of this truncated series back to time (crank angle) domain generates the desired estimate of the in-cycle mean velocity $\hat{U}(\theta,i)$. Since the method is based on selection of a cutoff frequency that is crank-angle dependent, it involves some arbitrariness and requires physical justification. Daneshyar and Fuller [1986] proposed the use of the inverse stroke duration as the cutoff frequency. Fansler and French [1988] assigned the value of f_{low} to the highest frequency present in the conventional ensemble-averaged velocity $\overline{U}(\theta)$.

Autocorrelation based analysis—The filtering-based technique considers the frequency content of the in-cylinder velocity fluctuations. Nothing has been said about the degree to which the fluctuations are random (chaotic, incoherent) or deterministic (organized, coherent). The classic statistical tool for assessing the relative contributions of random and deterministic processes is the autocorrelation function.

The non-stationary velocity autocorrelation function of the random, intermittent LDA data can be evaluated by [Glover, 1986; Fansler and French, 1988]:

$$R(\theta,\tau)=\frac{\frac{1}{N_p(\theta,\tau)}\sum_{i=1}^{N}u(\theta,i)\cdot u(\theta+\tau,i)}{u'(\theta)\cdot u'(\theta+\tau)} \tag{6.29}$$

where, $u(\theta,i)=U(\theta,i)-\overline{U}(\theta)$, is the fluctuation about the ensemble-averaged velocity, $u'(\theta)$ is the corresponding ensemble-rms fluctuation intensity, and θ is referred to as the reference crank angle and τ as the delay or lag. The quantity $N_p(\theta, \tau)$ is the total number of valid cross-product pairs for each (θ, τ) slot over the entire ensemble of N engine cycles. A series of discrete windows of width $\Delta\tau$ (e.g., 1°CA) in the delay τ is used to estimate $R(\theta, \tau)$.

In order to improve the statistical quality of the autocorrelation estimate, the result of Eq. (6.29) can be averaged over a reference range [Glover, 1986] that is somewhat wider than the slot width $\Delta\tau$, assuming that the velocity is locally stationary over the duration of the reference range. The final expression is therefore given by

$$R(\theta,\tau)=\frac{1}{N_{ref}}\sum_{j=1}^{N_{ref}}\left[\frac{\sum_{i=1}^{N}u(\theta_j,i)\cdot u(\theta_j+\tau,i)}{N_p(\theta_j,\tau)\cdot u'(\theta_j)\cdot u'(\theta_j+\tau)}\right] \tag{6.30}$$

Fansler and French [1988] compared three reference ranges (1, 6, 12) and found that the 12 deg. reference range used did not affect $R(\theta, \tau)$ profiles when 1200 cycles were analyzed.

The value of $R(\theta, \tau)$ expresses the extent to which the velocity fluctuations at one time are coupled to those at another time. At zero delay, the correlation is perfect, and $R(\theta, \tau)$ has its maximum value: $R(\theta, 0) = 1$. For a completely deterministic process (e.g., a periodic oscillation), the fluctuations about the mean value always remain correlated, the envelope of $R(\theta, \tau)$ persists undiminished to arbitrarily large τ. For a purely random process, the fluctuations, however, become uncorrelated or independent after a certain time delay, e.g., the integral time scale of turbulent flow. For example, for an autocorrelation curve as shown in Fig. 6.15, by virtue of its

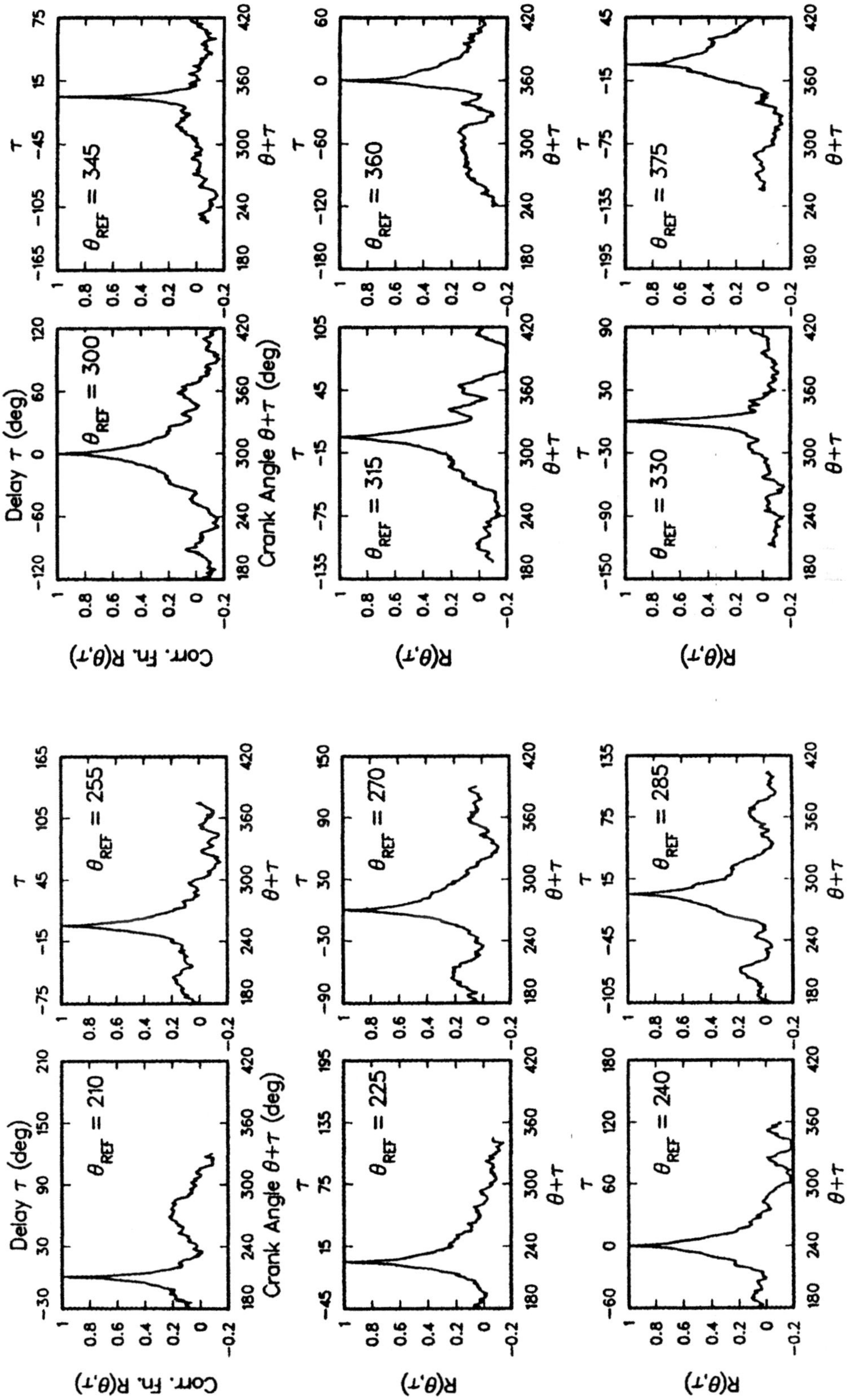

Fig. 6.15 Non-stationary velocity autocorrelation functions from an LDA system.

very rapid decay, the first peak should correspond to random velocity fluctuations, u(θ, i), about the ensemble-averaged velocity, $\overline{U}(\theta)$. The decay rate of this peak may be considered proportional to the time scale of turbulent flow. The faster the decay is, the more turbulent the flow is. In contrast, the low-amplitude structured component after the first peak persists to the maximum delays examined, which is associated with low-frequency velocity fluctuations that are deterministic within individual engine cycles [Fansler and French, 1988].

Summary of LDA data analysis—Three methods of analyzing in-cylinder velocity data have been described: (i) conventional ensemble-averaging; (ii) a low-pass/high-pass filtering scheme in either time or frequency domain; (iii) evaluation of non-stationary autocorrelation functions of the fluctuations about the ensemble-averaged velocity. The first method is the most widely used LDA data reduction scheme, but its output is averaged over many cycles, while the latter two approaches yield cycle-resolved velocity measurements.

The argument in favor of conventional ensemble-averaging is that it is unambiguous in its implementation, and that it allows comparison of data from different engines and direct compatibility between measurements and calculations. In addition, multidimensional CFD models solve the ensemble-averaged equations of motion, so that they offer no parallel to the attempt to wring additional information out of cycle-resolved velocity data. The argument against ensemble-averaging is that since the flow characteristics in individual cycles influence combustion in that same cycle, the cycle-resolved mean velocity and turbulence intensity have to be determined.

An alternative approach, which has not been included in the previous discussion, is based on conditional sampling, and it is used to identify groups of engine cycles of similar behavior, which can then be further analyzed using ensemble-averaging. This is particularly useful to resolve the relationship between flow characteristics before ignition near the spark plug, or in the unburned region ahead of the propagating flame, with some combustion parameters during individual engine cycles [Swords et al., 1982; Witze and Martin, 1986]. The combustion parameters investigated include peak cylinder pressure, flame speed, and flame arrival time.

6.2.5 Applications of LDA to IC Engines

There have been numerous applications of LDA to IC engines since 1970s. Both forward scatting and backward scattering have been employed in the measurement of in-cylinder velocities. Forward scattering is characterized by its strong signal-to-noise ratio, but it requires at least two optical accesses to the engine. Though it has a lower signal-to-noise ratio, backward scattering is preferred when only one optical access is available. Since LDA is a well-established technique, description of the LDA system in each application is kept to minimum. Because of the shear number of LDA studies reported in the literature, applications of LDA to IC engines are discussed according to the research areas, and important conclusions drawn from these studies are given.

Because of the complexity of engine flow processes, model engines with simplified geometry and running at slow speed, were used in the early studies of the effect of individual engine parameters, such as compression ratio, valve shape, etc. on the in-cylinder flow field using LDA [Morse et al., 1980; Arcoumanis et al., 1983].

Because in-cylinder flow is largely affected by the flow processes in the intake manifold and through the valves, a number of LDA studies have been carried out with steady flows. Tindal et al. [1988] investigated the steady flows through both axisymmetric and helical ports using a liquid-flow rig and a transparent port and cylinder made of acrylic plastics. The refractive index of the working fluid was matched to that of the acrylic test section so that optical access to the entire flow space was obtained without beam refraction at the walls of the port and cylinder. Measurements of the steady flows through helical and directed ports by Wigley and Hawkins [1978] and Coghe et al. [1988] showed that the flows through the two types of port were substantially different. Bicen et al. [1985] conducted a comprehensive comparison of steady and unsteady flows through axisymmetric inlet ports and cylinders based on the LDA measurements. They found that the flow in the inlet port and up to the valve exit is insensitive to confinement, the position of the piston, or flow unsteadiness.

The flow field before injection in diesel engines or ignition in SI engines is important because of its effect on mixing in diesel engines and turbulence-combustion interaction in SI engines. A comparison of intake and compression velocities between a motored and an equivalent firing cycle showed close agreement [Rask, 1979]. Thus, measurements in a motored engine should be representative of the conditions existing in a firing engine prior to combustion, and they can be obtained more easily. A number of researchers have reported LDA measurements in motored engines [e.g., Liou et al., 1984; Bopp et al., 1986; Suen et al., 1987]. A consistent conclusion emerging from these studies is that the turbulence intensity at TDC, with open combustion chambers in the absence of swirl, has a maximum value equal to about half the mean piston speed:

$$u'(\mathrm{TDC}) \approx 0.5 \cdot \overline{S}_p$$

Liou et al. [1984] showed that turbulence intensity at TDC with swirl was higher than that without swirl, but Bopp et al. [1986] found that the rms fluctuation intensity with swirl could be lower.

In recent years, multi-valve SI engines are designed with pentroof combustion chambers. For these engines, tumble has been found to accelerate the combustion process. Tumble is defined as a rotational air motion around an axis perpendicular to the cylinder axis, while swirl is the rotation around an axis parallel to the cylinder axis. A number of studies on the development of tumble and its effect on combustion have been reported. When tumble was generated by appropriate intake port design, it was found that the peak turbulence intensity occurred much closer to TDC and was significantly higher than that in the absence of tumble motion, due to the collapse of the tumble flow in the early part of the compression stroke [Arcoumanis et al.,1990; Hadded and Denbratt,1991]. The effect of the intake port design [Omori et al., 1991] and the piston shape [Baby and Floch, 1997] on the formation and breakdown of tumble were also reported.

Cycle-resolved LDA measurements have been used to investigate the cycle-to-cycle variations. Fansler and French [1988], for example, obtained cycle-resolved measurements in a motored engine, and the velocity fluctuations were analyzed both by the frequency-domain filtering method and by evaluation of non-stationary velocity autocorrelation functions. Typical results of the instantaneous, ensemble, and low-pass filtered velocities from a single engine cycle are shown in Fig. 6.16. Note that both the velocity and power spectrum curves indicate the existence of a low-frequency and a high-frequency fluctuation.

Evidence of random and organized fluctuations was obtained from the autocorrelation results shown in Fig. 6.15. In Fig. 6.15, the autocorrelation function is plotted with its reference crank angle shown at its proper position in the cycle. The lower horizontal axis represents the quantity, $\theta + \tau$, while the upper axis represents the delay, τ, relative to the reference crank angle, θ. The latter is immediately recognizable as the time when the autocorrelation function takes its maximum value, $R(\theta, 0)$. Two principal features can be readily identified from these graphs: the

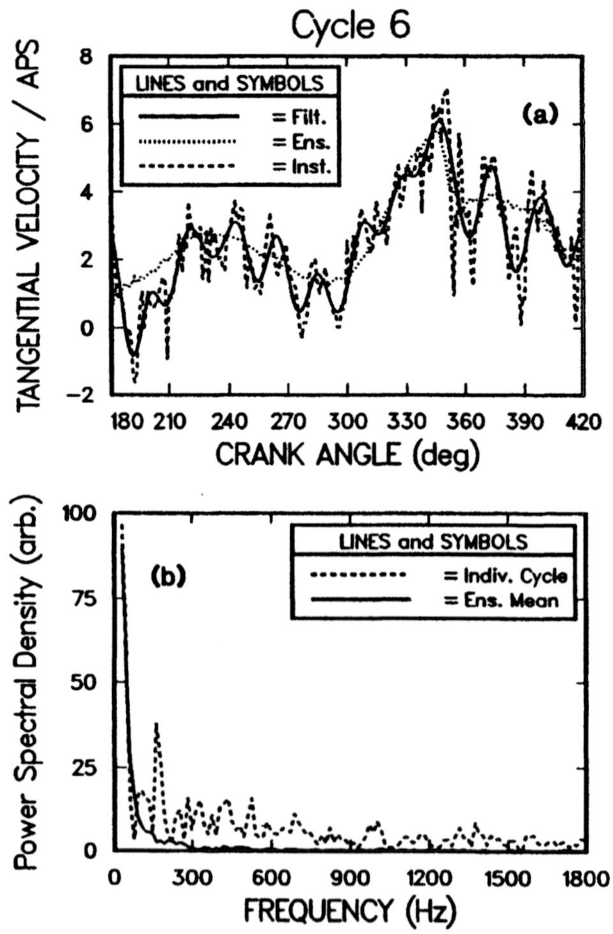

Fig. 6.16 Typical individual-cycle velocity traces and power spectra in an engine.

relatively sharp central peak, which indicates the presence of a rapidly decaying component; and a lower-amplitude structured component, which becomes evident after the first component decays and persists to the maximum delays examined. The former component associates the random fluctuation in the velocity data, while the latter corresponds to an organized structure. Fansler and French considered that these two components should correspond to the high-frequency and low-frequency fluctuations determined by the filtering technique.

Turbulence length scales have been measured using primarily two kinds of LDA techniques. The integral length scale is defined as the integral of the autocorrelation coefficient, $R(\theta, x)$, of the fluctuating velocity at two adjacent points, x and $x + \Delta x$, in the flow with respect to the variable distance between the points:

$$l_I(\theta) = \int_0^\infty R(\theta, \Delta x)dx \tag{6.31}$$

and

$$R(\theta, \Delta x) = \frac{1}{N-1} \sum_{i=1}^{N} \frac{u(\theta, x, i) \cdot u(\theta, x + \Delta x, i)}{u'(\theta, x) \cdot u'(\theta, x + \Delta x)} \tag{6.32}$$

where $u(\theta, x, i) = U(\theta, x, i) - \bar{U}(\theta, x)$ is the fluctuation velocity, $u'(\theta, x) = \sqrt{\overline{[u(\theta, x, i)]^2}}$ the rms or standard deviation of the instantaneous velocity about the ensemble-mean velocity.

Fraser and Bracco [1988] determined lateral integral length scales using a two-point, single-probe-volume LDA system in a motored single-cylinder engine. In this approach, a single elongated probe volume was generated. The measurement locations within the elongated probe volume were determined by imaging the entire probe volume onto a slit aperture, on the other side of which were the input ends of two optical fibers. Photomultipliers mounted on the other ends of the two optical fibers were connected to two counter processors, where Doppler signals were processed.

Turbulence length scale measurements by a two-probe-volume LDA technique was carried out in a motored DI diesel engine by Corcione and Valentino [1990]. The technique was realized by forming two probe volumes using separate transmitting optics for the green and blue beams of an argon-ion laser. The probe volume of the two green beams was fixed and that of the blue beam could be moved by means of rotating optics. The distance between the two probe volumes could be altered between 0 and 10 mm.

In another approach, the scanning LDA was used for the measurement of turbulence length scale [Glover et al., 1988]. The fixed mirror in the barrel extension of a single optical cylinder engine was replaced with a rotating mirror driven from the crankshaft at twice the engine speed by a toothed belt and gear arrangement. This resulted in a scan duration across a window of about 6 °CA, and its measuring volume velocity was much higher than the turbulence intensity. Thus, the probe volume was traversed rapidly through the flow, freezing the turbulence.

A number of investigations into the correlation between the flow field and combustion in fired engines have been reported [Hall and Bracco, 1986; Lorenz and Prescher, 1990; Hadded and Denbratt, 1991]. A consensus emerging from these studies is that there is a strong correlation between the turbulence intensity at the spark timing and the rate of early combustion. In order to resolve the relationship between flow characteristics before ignition near the spark plug, or in the unburned region ahead of the propagating flame, with some combustion parameter during individual engine cycles, conditional sampling of groups of engine cycles of similar behavior were used [Swords et al., 1982; Witze and Martin, 1986]. The conditional sampling parameters include peak cylinder pressure, flame speed and flame arrival time, etc.

6.3 Particle Image Velocimetry

6.3.1 Introduction

Particle image velocimetry (PIV) has its origin in laser speckle photography (LSP), which has been used for measuring surface displacement fields. In LSP, two speckle images of a solid surface, illuminated with a laser beam, are exposed onto a single photographic film before and after in-plane displacement of the surface. Given that the speckle pattern does not significantly change between the two exposures, the displacement of individual speckles in a small region of the double exposed recording gives a measure of the local surface displacement. Passing a narrow low-power laser beam through such a local area, on the recorded speckle image, produces Young's fringes, in which the orientation and spacing of the fringes are simply related to the direction and magnitude of the surface displacement of the local area [Burch and Tokarski, 1968].

Based on the same principle, laser speckle velocimetry (LSV) was adapted from LSP for the measurement of velocity field in low-speed laminar flows [e.g., Barker and Fourney, 1977; Dudderar and Simpkins, 1977]. In this case, light from a laser sheet scatters from particles added to the fluid, and the resulting speckle field generated by random interference of the scattered laser light is recorded onto a film of two successive exposures. The formation of coherent speckles in LSV requires that extremely high seeding densities be used, such that the light scattered from individual particle images can interfere at the image plane of an imaging system. In many flows, the LSV technique is limited in its application, since the seeding density needed to produce speckle is impractical with such bulk loading potentially causing significant changes in the flow behavior. Speckle at the image plane is due to random interference of overlapping particle images, and if the flow is not a simple in-plane displacement, then between exposures, the speckle pattern will be altered and the similarity of subsequent recorded speckle pattern will decrease. This results in a decrease in the visibility of Young's fringes until no fringe modulation is visible and no displacement/velocity measurable. In practice, almost all flows have out-of-plane motion, turbulence, or both, and hence the ability to record and measure complex flow fields with speckle pattern is limited.

Initial application of LSV to unsteady gas flows, however, revealed that the recordings actually consisted of well-defined images of individual particle pairs instead of the expected speckle pattern [Meynart, 1980], due to too few particle images for the optical arrangement to produce overlapping particle images and hence speckle at the film plane. In the particle imaging regime where speckle is absent, the LSV technique became known as particle image velocimetry [Pickering and Halliwell, 1984; Adrian, 1984]. The relaxed seeding requirement in PIV means that adequate seeding density can be achieved in a variety of high-speed gas flows of practical interest, while the absence of speckle permits greater tolerance to decorrelation by out-of-plane fluid motion.

Since these initial experiments over fifteen years ago, there has been considerable work on developing the PIV technique. A number of sources describing the basic principles of PIV in the context of its historical development are readily available. For example, the SPIE Milestone Series 99 edited by Grant [1994], and the review paper by Adrian [1991] and Grant [1997]. The book *Particle Image Velocimetry* by Raffel, Willert, and Kompenhans [1998] provides a comprehensive treatment of PIV techniques and theories.

The purpose of this text is to provide, in a condensed form, all of the information relevant for the understanding, planning, and use of PIV in flow velocity field measurement inside an internal combustion engine. With recent developments in digital imaging technology, an up-to-date discussion will be presented on the fully digital PIV systems.

6.3.2 Principle of Particle Image Velocimetry

The principle of the PIV technique is simple: The velocity vector is calculated from the displacement of an element of fluid in a known time interval. In its simplest form, the PIV technique requires a thin slice of the flow field to be illuminated by a laser light sheet. Particles in the flow within the light sheet scatter the light. The scattered light is detected by a camera placed at a right angle to the light sheet (Fig. 6.17). The laser light sheet is double-pulsed (switched on and off very quickly twice) at a known time interval, Δt. The first pulse of the laser freezes images of the initial positions of these particles onto the first frame of the camera. The camera frame is advanced and the second frame of the camera is exposed to the light scattered by the particles from the second laser pulse. There are thus two images, the first showing the initial positions of the particles and the second their final positions due to the movement of the flow field. Alternatively, the double exposed particle images can be recorded on the same photographic film.

The intersection volume between the region illuminated by the laser light sheet and the field of view of the imaging optics determines the region of the flow to be measured, which is denoted as the measurement region (see Fig. 6.18). The smallest resolvable area of the film or detector, the so-called interrogation area (IA), projected back onto the measurement region, determines the interrogation volume (IV) that constitutes a single velocity vector.

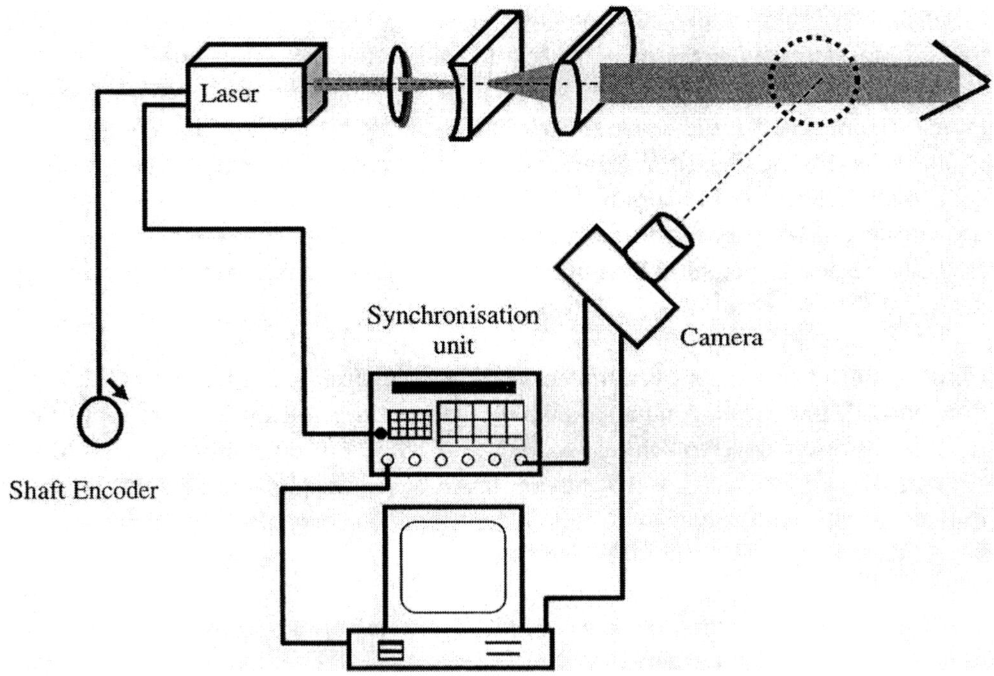

Fig. 6.17 A typical PIV experimental setup.

The particle displacement in the object plane, Δx and Δy, is found from the displacement in the image plane, ΔX and ΔY, on the recording medium,

$$\Delta x = \frac{1}{M}\Delta X \text{ and } \Delta y = \frac{1}{M}\Delta Y \tag{6.33}$$

Knowing the magnification factor of the imaging optics, M, and the time separation, Δt, between the two laser pulses, the velocity projections on the object plane, u and v, may be found from

$$u = \frac{\Delta x}{\Delta t} \text{ and } v = \frac{\Delta y}{\Delta t} \tag{6.34}$$

The results are usually presented in a velocity vector map, as shown in Fig. 6.19, which is obtained by repeating the process for each interrogation region.

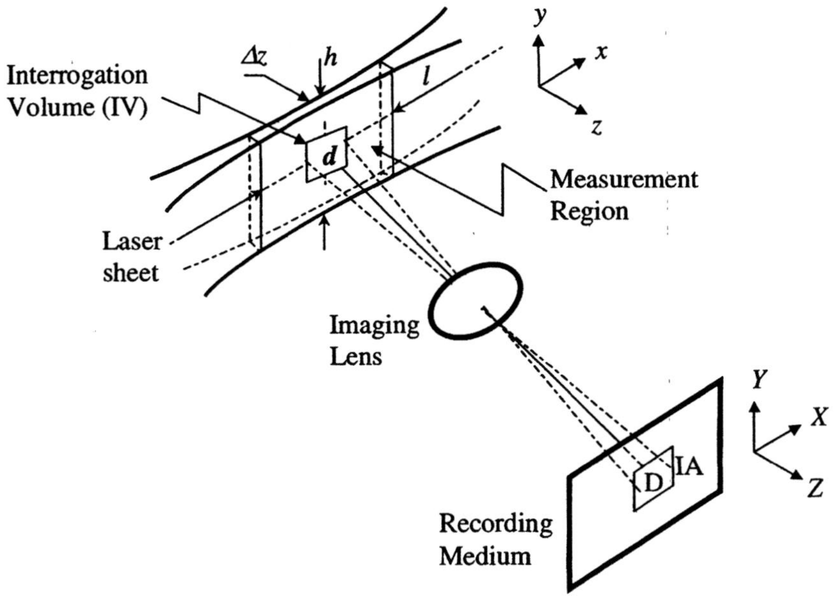

Fig. 6.18 Definition of measurement region and measurement volume.

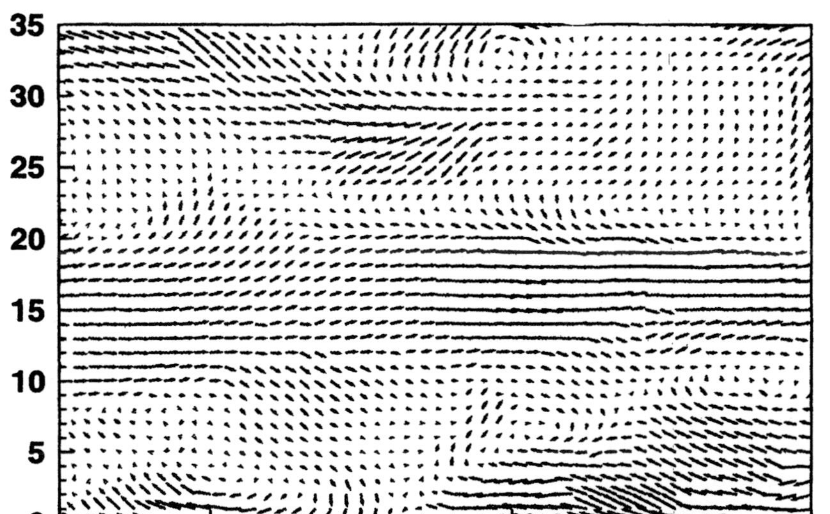

Fig. 6.19 Velocity vector map by PIV.

6.3.3 Experimental Considerations

6.3.3.1 Illumination Source and Light Sheet

Lasers provide light in a conveniently collimated and intense form, and are ideally suited as the light source for PIV measurements. The choice of a laser is controlled by a number of criteria. Central to the choice of an illumination source is its high power, because PIV measurements require illumination over an extended volume of flow, and the amount of light scattered from particles must be detectable by the camera. Pulse width must be short enough to freeze the flow.

The laser wavelength also figures strongly in the choice of lasers. Shorter wavelengths increase the average intensity of a particle image compared to longer wavelengths. Photographic emulsions are also more sensitive to blue-green wavelengths compared to red light. In some experiments, water-analog rigs are used to characterize the in-cylinder flow. Water absorbs light, and its absorption is wavelength dependent. At 0.5 μm, the absorption of light in water is significantly reduced compared to at longer wavelengths.

In PIV measurements, the most commonly used laser is the frequency doubled Nd:YAG pulsed laser. There are two types of Nd:YAG lasers that have been used in PIV experiments. The simple one is a single-cavity Nd:YAG laser with the double pulse option. Two Marx banks are used to open the Pockels cell (Q-switching device) of the laser twice during one flash-lamp discharge. This divides the available energy into two pulses. The first laser pulse is triggered externally, and a potentiometer on the double pulse option panel adjusts the timing of the second pulse. However, this reduces the energy of each pulse to much less than that of the pulses in single-pulse mode. Pulse separations are limited to typically 20 μs to 200 μs. The pulse energy and pulse width vary with the pulse separation, and this leads to particle images of uneven intensity, which must be accommodated for by the dynamic range of the recording media.

The preferred laser is a twin-oscillator, twin-amplifier frequency doubled Nd:YAG laser. Each of the two oscillators and amplifiers can be triggered separately, thus allowing infinite control of the laser pulse separations and simple control of the individual pulse energy. Alternatively, two separate lasers may be employed. These systems, however, require accurate co-alignment of the two separate outputs.

An argon-ion laser is sometimes used in slow liquid flows where large scattering signals are present. The use of a CW (continuous wave) argon-ion laser requires the addition of a mechanical or electro-optical beam modulator for double pulsed operation.

A copper vapor laser or a high-repetition Nd:YAG laser provides a burst of laser pulses of high energy and short pulse width at up to 50 kHz. By recording the particle images onto a high-speed drum film camera, time resolved PIV can be obtained by performing a cross-correlation analysis of the adjacent frames.

The illumination of flow in PIV measurements requires the formation of a thin laser light sheet within the flow. The laser light sheet is typically achieved with an arrangement of cylindrical and spherical lenses, as discussed in Chapter 5. Spherical lenses are used to focus the laser beam to the required beam waist diameter giving the light sheet thickness. Cylindrical lenses are used to expand and collimate the beam into a sheet. The width of the laser light sheet is dependent on the area of interest and the recording optics. The thickness of the laser light sheet should be selected in conjunction with knowledge about the magnitude of the out-of-plane velocities present (the velocity component normal to the light sheet) and the pulse separation for recording the flow. Large out-of-plane velocity components require a thick light sheet, reducing the intensity of the illumination and the amount of light scattered. According to Keane and Adrian [1990], the minimization of data loss due to movement normal to the light sheet is achieved when

$$w \cdot \Delta t \leq \Delta z / 4 \tag{6.35}$$

where w is the out-of-plane velocity component, Δt is the time between exposures, and Δz is the light sheet thickness.

A further consideration with regard to illumination source is beam delivery. Many PIV measurements in engines can be accomplished using a set of lenses on an optical bench. In some applications, the limited space and harsh environment require some means for the light energy to be delivered to the measurement volume. A fiber optic delivery system is difficult to apply to PIV because the requirement for very short pulses and high output beam quality limits the high-energy transmission. Though less elegant or flexible than a fiber optic system, flexible mirror arm systems may be used for beam delivery, in which a set of mirrors and optical lenses are incorporated into tubes mechanically linked together.

6.3.3.2 Seeding Particles

In engine experiments with gas flows, the supply of seeding particles is often critical for the quality and feasibility of PIV measurements. Particle seeding requirements for PIV measurements can be summarized as follows:

- Particle flow-following ability
- Particle light scattering characteristics
- Particle number density in the flow
- Tendency of particles to foul optical windows
- Tendency of particles to abrade cylinder surfaces
- Ability of particles to withstand high-temperature environment in a fired engine

The first two requirements have been discussed in detail in Section 6.2.2.5. Window fouling and abrasion damage to windows by solid particles may be severe in engine measurements. For many PIV measurements, liquid aerosol seeding is normally used. A commonly used aerosol generator is that of Laskin, as shown in Fig. 6.20. It consists of a closed cylindrical container with two air inlets and one aerosol outlet. Compressed air with 0.5 to 1.5 bar gauge pressure with respect to the outlet is fed down one or more tubes, which are sealed at their lower ends and have a series of 1 mm diameter holes in the wall (Fig. 6.20). Atomization takes place as the air jets emerge into the oil and as bubbles burst on the oil surface. Large droplets are removed by the circular impactor plate mounted inside the container. The small droplets pass through a small gap of about 2 mm between the impactor plate and the inner wall of the container. The seeding density can be adjusted by the valves at the atomization pipes and/or the additional air supply through the second air inlet. Vegetable oils are often used since they do not pose any environmental problem. However, Reeves et al. [1994] found that evaporation loss of olive oil droplets near TDC was problematic at partly throttled motoring conditions. They found that Dow Corning 550 high-temperature silicone fluid remained stable throughout the complete engine cycle.

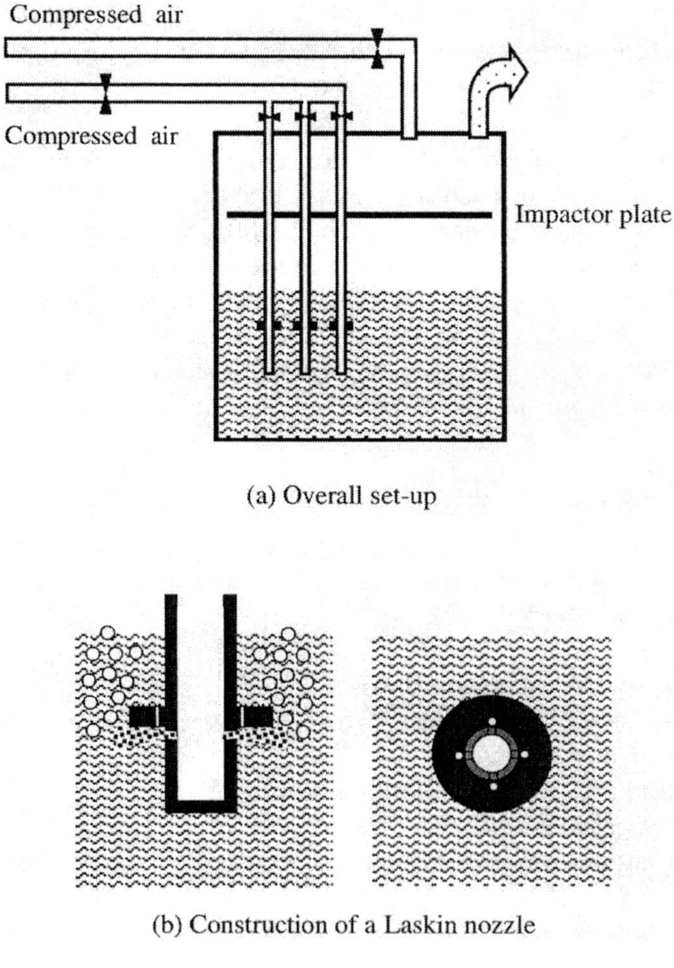

(a) Overall set-up

(b) Construction of a Laskin nozzle

Fig. 6.20 Schematic of a Laskin aerosol generator.

Many seeding systems have been designed for LDA measurements where accurate flow-following to turbulence frequencies on the order of 10 kHz are required, generating sub-micron or micron sized particles. The requirements for extended illumination in PIV measurement and sufficient light energy of scattered particle images necessitate the use of high-power lasers. However, at higher laser energies, flare or spurious scattering from surrounding surfaces increases, causing significant image noise on the recording medium. In order to reduce the spurious scattering, the use of slightly larger seeding particles may therefore be advantageous.

The seeding density must be sufficiently high. As more particle image pairs enter into the correlation calculation, the probability of a valid displacement detection increases. The actual number of image pairs captured in an interrogation area depends on three factors: the overall seeding density, the amount of in-plane displacement, and the amount of out-of-plane displacement. Keane and Adrian [1990] recommended a minimum number of 8 particle pairs in double-exposure/single-frame PIV, and more than 5 particle pairs in the single-exposure/double-frame PIV per interrogation region. The second effect that the particle image density has on the evaluation of PIV images is that it can reduce the measurement uncertainty substantially. Therefore, if a flow can be densely seeded, then both a high valid detection rate, as well as a low measurement uncertainty, can be achieved using small interrogation areas, which in turn allows for a high spatial resolution. An upper limit of seed density is reached, above which decorrelation occurs due to speckle formation. In practical gas flows, however, it is unlikely to achieve such high particle concentrations.

6.3.3.3 Recording of Particle Images

6.3.3.3.1 Introduction

The PIV recording modes can be separated into two main categories: (1) methods that capture the illuminated flow onto a single frame and (2) methods that provide a single illuminated particle image for each illumination pulse. They are referred to as single-frame/double-exposure PIV and double-frame/single-exposure PIV, respectively. There are two types of recording media: photographic films and electronic recording devices (CCD cameras). The single-frame/double-exposure approach is normally used in photographic PIV recordings, while the double-frame/single-exposure is the preferred mode of operation with digital recordings.

6.3.3.3.2 Photographic Recording and Digitization

Photographic films have been widely used in PIV systems to detect and store scattered particle images. In a typical photographic PIV setup, a film camera is placed normal to the light sheet plane (Fig. 6.17). In order to save time in the alignment of the system, a high quality and reliable focusing device is necessary. For this purpose, a photographic camera can be equipped with a device for fast focusing. A small area in the film plane can, for example, be observed by means of a CCD camera looking through an orifice in the back wall of the camera in order to control focusing, as shown in Fig. 6.21 [Reeves et al., 1994]. Alternatively, a CCD camera may be mounted in the view finder of a 35 mm camera, and aligned in such a way that the distance between the lens and the CCD sensor via the mirror in the camera body is exactly the same as that from the lens to the film plane [Raffel et al., 1998].

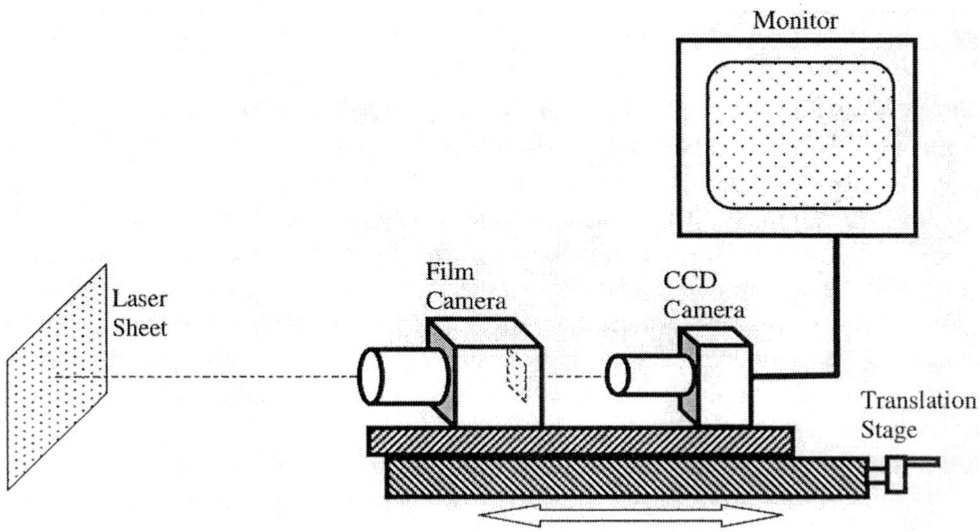

Fig. 6.21 Arrangement for focusing photographic PIV recordings.

During alignment, the laser is fired repetively at the same point in every engine cycle by means of an engine shaft encoder and an electronic timing unit producing a trigger at the specified crank angle. In the actual PIV measurement, the laser is fired at a specfic crank angle in a one-shot mode manually in conjunction with manual operation of the camera shutter on the "B" setting. During the opening period of the camera shutter, the photographic film is exposed twice to the scattered light of two laser pulses. After being developed, the photographic film containing the double exposed particle images is mounted on a two-axis translation stage. The film is divided into a large number of square interrogation regions, whose centers are located on a regular grid. The mean particle displacement vector within each interrogation region of the film or transparancy is then measured by means of appropriate interrogation schemes.

In early PIV experiments, this interrogation was carried out manually on particle images with relatively sparse seeding, which allowed the tracking of individual particles (which is sometimes referred to as particle tracking velocimetry). However, the particle tracking method is only practical in the low image density case. For quantitative and accurate measurements, a high data density is required on the PIV vector maps. Due to the number of particle image pairs present, matching pairs of particle images cannot be achieved by visual inspection of the PIV recording.

Young's fringe processing techniques were prevalent in early PIV work. In this method, each interrogation region of the double exposure image on a film is illuminated by a low-power He-Ne laser. The pairs of particle images act as interfering point sources. When the film and the observation plane (ground glass) are arranged to be in the principal focal planes of a converging lens (see Fig. 6.22), the so-called Young's fringes appear on the observation plane. The orientation and spacing of the fringes are simply related to the direction and magnitude of the image displacement, subject to a 180-degree directional ambiguity. A variety of data reduction techniques have been used to determine the fringe orientation and spacing. Typical approaches included conventional fringe counting or autocorrelation, either through Fourier transform of the digitized fringe pattern, or by presenting the fringe pattern to an optical Fourier transforming system.

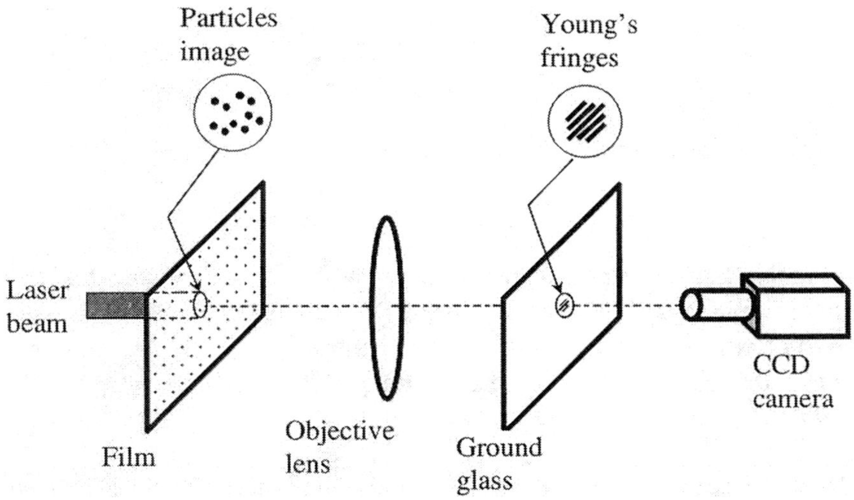

Fig. 6.22 Evaluation of PIV recordings using the Young's fringes technique.

The most common approach adopted now is to digitize the interrogation region on the double exposed film using a CCD camera equipped with a microscope objective, and to perform a fully digital computation of the two-dimensional autocorrelation function of the original particle image field. A white light source through a diffuser illuminates the film negative or transparency (Fig. 6.23). The film is divided into a large number of square regions whose centers are located on a regular grid. Each region is imaged onto a CCD (typically 768 × 512 pixels) camera with a microscope objective and extension tubes. The imaged region is then digitized with an image frame grabber. The digitized image is subdivided into a square or regtanular grid of $2^n \times 2^n$ pixel regions (e.g., 4 × 3 grid of 256 × 256 pixels) with a 50% overlap between neighboring regions as shown in Fig. 6.23. Each $2^n \times 2^n$ pixel area represents one interrogation region within which the mean particle image displacement vector is calculated using an autocorrelation method. This will be discussed later.

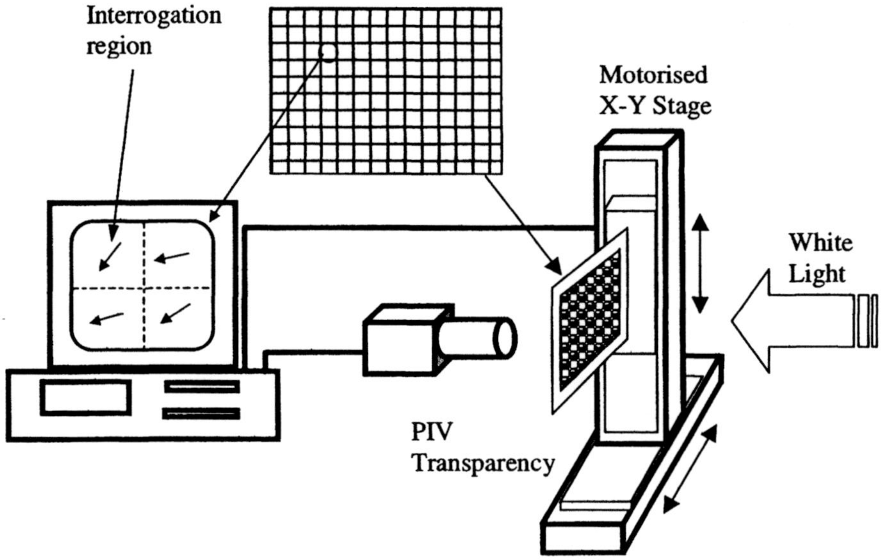

Fig. 6.23 Typical digital interrogation arrangement for photographic PIV recordings.

A fundamental problem in measuring complex flows with the single-frame/double-exposure approach concerns measurement of reversing flows. For a double-exposure PIV image, there is no information to determine the order in which the exposures are made. Thus, there is a 180 degree ambiguity in the velocity measurements. Techniques designed to resolve this problem are necessary, and will be discussed in a later section.

Photographic recording has been extented to high-speed continuous PIV measurements [Stolz et al., 1992]. The high-speed PIV system consisted of a high repetition Cu vapor laser as the light source and a rotating drum camera as the recording device. Two images shifted by 0.25 mm in space were recorded in each frame of the film. The image shift was determined by the speed of the drum camera and the pulse separation (50 μs at 20 kHz repetition rate). The presence of the image shift removed directional ambiguity. With this setup, Stolz et al. were able to obtain 200 consecutive PIV images per second. The minimum pulse separation of 50 μs at a 20 kHz repetition rate set a limit on the maximum velocity that could be measured. However, it is possible to remove the high velocity limitation by using two synchronized Cu vapor lasers [Stolz et al., 1992].

6.3.3.3.3 Digital Recording

Recent advances in electronic recording technology (particularly CCD arrays) have provided an attractive alternative to the photographic methods of PIV recording. Instant image availability and thus feedback during recording, as well as a complete absence of photochemical processing, are the apparent advantages gained with electronic imaging. CCD technology also offers improved noise and dynamic range qualities compared to photographic emulsions. Furthermore, the cross-correlation method is used in the PIV evaluation of specially designed CCD cameras (cross-correlation CCD cameras), offering significant advantages over autocorrelation analysis in terms of velocity gradient tolerance, dynamic range and signal-to-noise ratio. More importantly these cross-correlation CCD cameras remove the directional ambiguity of photographic PIV. The downside of electronic imaging is the lower spatial resolution. At present, CCD imaging arrays of up to 4000 × 4000 pixels are commercially available although very expensive. With a 4k × 4k array, the spatial resolution is a few times lower than a film of 200 line pairs/mm resolution. The present trends suggest that electronic recordings are of increased importance to PIV. But for the time being, the photographic technique is still the method of choice for PIV applications requiring high resolution over extended areas of flow field.

Two approaches are employed in the digital recording of PIV images. Single frame recording involves two exposures of the particle images on a single frame of a CCD camera analogous to photographic recording. The double exposed particle image is then digitized and transferred to the computer by means of an image acquisition card. The digitized image is subdivided into small interrogation regions and the displacement vector within each interrogation region is determined by calculating the digital autocorrelation of the double exposed particle image field. Like photographic recording, autocorrelation digital recordings suffer from directional ambiguity. Image shifting is normally required for the removal of directional ambiguity.

Double frame digital PIV separates the particle images of the two consecutive laser pulses onto two frames. The main advantages of the double frame technique are the removal of the directional ambiguity, and higher signal-to-noise ratios due to cross-correlation analysis. Standard video cameras can be used in digital PIV recordings. The standard video camera comprises an interline transfer CCD sensor which contains vertical transfer registers (storage sites) located between the active pixels, as shown in Fig. 6.24. Since two vertical pixels share a common storage site in the vertical shift registers, an image can only be stored at half the vertical resolution of the CCD sensor either by shifting the odd or even rows of pixels. This is an artifact of standard video or television transmission where each frame is split into fields containing either even or odd lines.

In order to apply the cross-correlation technique, the first of the two laser pulses is timed just before the transfer event and the second right after (see Fig. 6.25[a]). Consequently, two single exposed images with half the vertical resolution are recorded with a pulse delay as short as 1–2 μs. The maximum PIV image frame rate is limited by the camera's frame rate (25 Hz or 30 Hz) if the laser can be operated at this frequency. Because images are recorded on each field, the standard video camera can offer only half the vertical resolution of the CCD sensor.

The most versatile and commercially adopted solution to achieve double frame digital PIV is to use a full-frame, progressive scan, interline transfer CCD camera. First introduced in early 1990s, CCD technology was originally designed for machine vision applications, because it removes the artifacts associated with the interlace video imaging of standard video cameras.

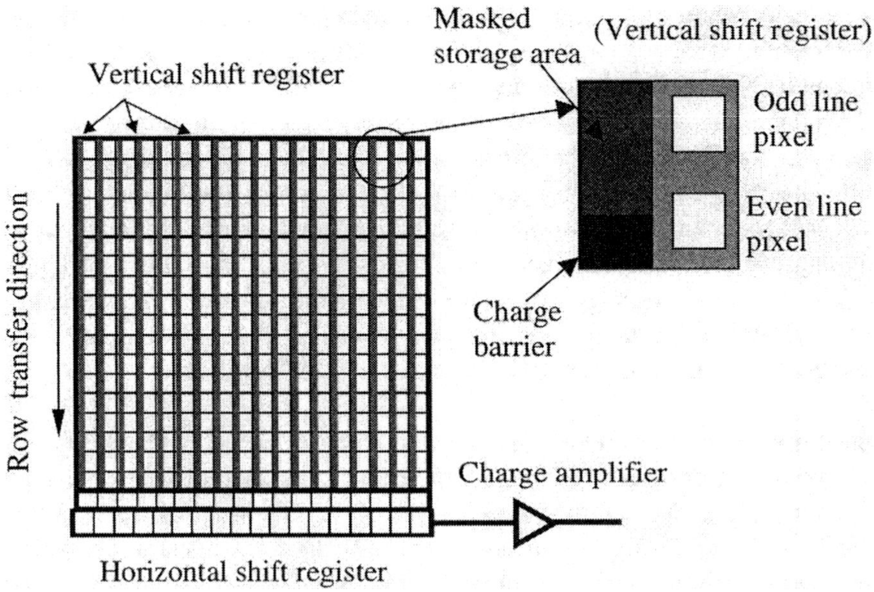

Fig. 6.24 The CCD sensor layout in a standard video camera.

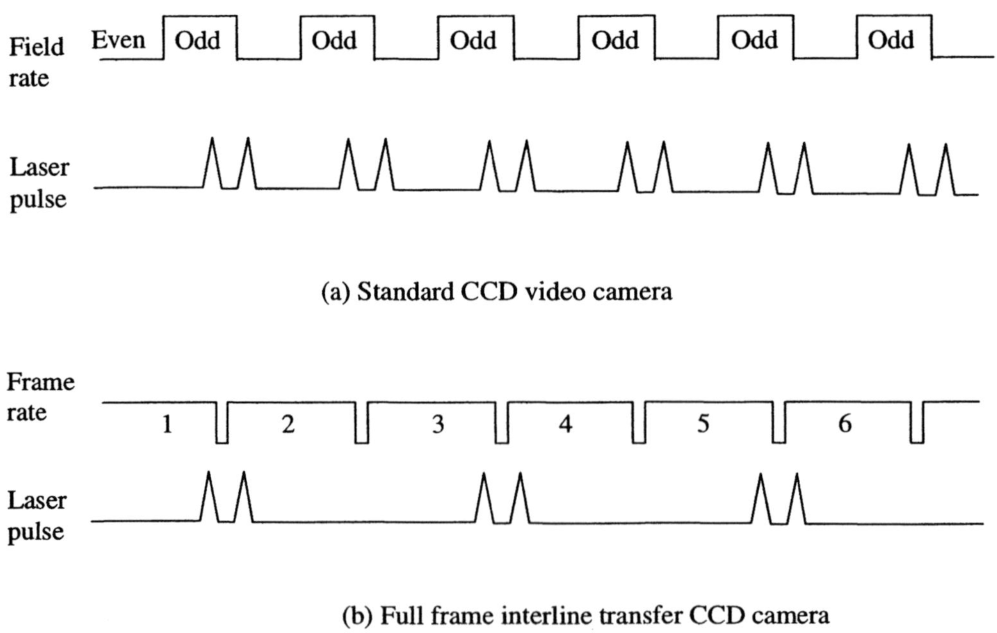

Fig. 6.25 Timing diagrams for digital PIV recordings using a video and a full frame CCD camera.

This type of CCD camera takes its name from additional vertical transfer registers located between the active pixels, as shown in Fig. 6.26. (The conventional full frame CCD sensor employed in low light imaging has only a horizontal shift register). Thus, each active pixel has its own storage site and its charge is read out sequentially (progressive scan). This permits very fast transfer of the entire exposed image into the adjoining storage sites within a few microseconds or less with high resolution. The fast transfer, then, allows two single exposed PIV images to be captured at a time delay slightly longer than the transfer time. To achieve this the laser pulses are timed so that the first pulse occurs immediately before the frame transfer takes place (e.g., on frame n), while the second pulse occurs immediately thereafter (i.e., on frame n + 1). A timing diagram for double frame PIV recording based on this sensor is given in Fig. 6.25(b). This arrangement of the laser pulses with respect to the periodic exposure cycles of the CCD sensor is known as frame straddling. The maximum PIV image frame rate is half of the camera's frame rate (normally 30 Hz) with pulse delays as short as 0.5 μs. Because these cameras also often have asynchronous reset options, they can be triggered externally and sychronized with the laser pulses.

The digitized particle images can then be transferred to a PC computer for subsequent analysis [TSI Inc., 1998] or by a dedicated digital correlator for real-time measurements [Dantec Measurement Technogy, 1998]. The interrogation is then carried out by means of the cross-correlation method. This correlation technique enables velocity vector field computation without directional ambiguity, so that no external image shifting hardware is needed when using the double frame (cross-correlation) camera.

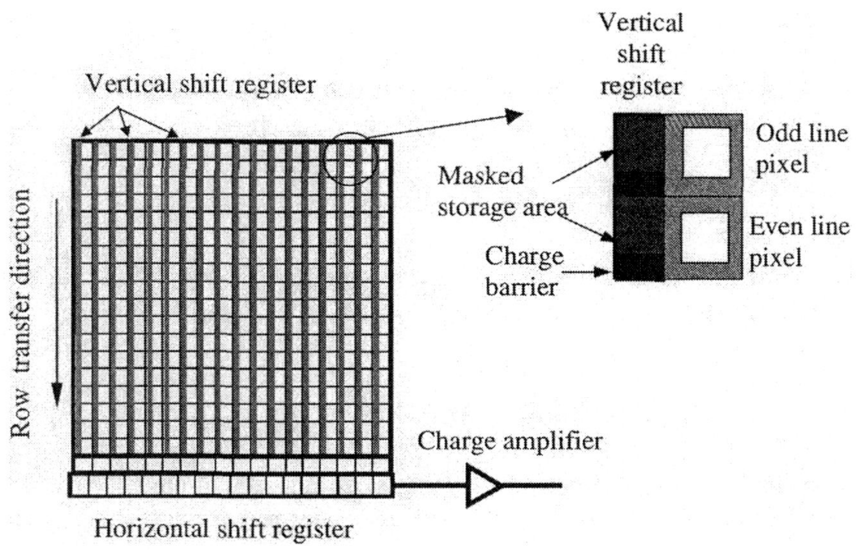

Fig. 6.26 Full frame, progressive scan, interline transfer CCD sensor layout.

6.3.3.3.4 Imaging Optics

A number of important optical imaging parameters must be considered when estimating the dynamic range of a PIV system. Sharp and small particle images minimize the uncertainty in locating the image centroid or correlation peak centroid, and hence the error in velocity measurement. Smaller particle images permit closer-spaced particle pairs of higher velocity to be resolved. Sharp and small particle images lead to high particle image intensity, since for the same amount of scattered light by the seeding particle, the light energy per unit area increases quadratically with decreasing particle images.

However, the particle image diameter is often diffraction limited by the focusing lens. The diffraction limited imaging occurs because of the finite aperture of the focusing lens. The diffraction limited minimum image diameter is given by

$$d_{diff} = 2.44 f\#(M+1)\lambda \tag{6.36}$$

where, f# is the f-number of the focusing lens, and is defined as the ratio between the focal length, f, and the aperture of the lens, A. M is the magnification factor given by

$$M = \frac{S}{s} \tag{6.37}$$

where S is the distance between the image plane and lens, and s is the distance between the lens and the object plane.

If lens aberrations are absent, the true particle image d_i can be estimated from

$$d_i = \sqrt{M^2 d_p^2 + d_{diff}^2} \tag{6.38}$$

Table 6.1 shows the calculated values of the particle images of a small particle of 1 μm in conjunction with the depth of field δ_z calculated from

$$\delta_z = 2f\# d_{diff}(M+1)/M^2 \tag{6.39}$$

It can be seen from Table 6.1 that a large aperture leads to sharp particle images, and collects more light, as well. However, in order to obtain a sharp image, the seeding particles must fall within the depth of field of the camera lens. For maximum use of the available scattered light, the depth of field should be approximately equal to the light sheet thickness. Thus, the minimum f-number, or maximum aperture for a fixed focal length, is determined by the required depth of field.

Table 6.1 Calculated values of the particle images of a 1 μm particle

$f_{\#}$	D_i (μm)	δ_z (mm)
1.4	2.3	0.25
2.8	4.6	0.5
4.0	6.5	1.1
5.6	9.1	2.0
8.0	13.0	4.2
11	17.9	7.8
16	26.0	16.6

6.3.3.3.5 Perspective Errors

When the principle of PIV was first introduced in this chapter, it was assumed that the displacement vector was in a two-dimensional plane. The two-dimensional displacement vector in the film plane, $D = X_i' - X_i$ (see Fig. 6.27) and the particle displacement vector, **d**, are related by

$$\mathbf{D} = -\mathrm{M}\cdot\mathbf{d} = -\mathrm{M}\cdot(\mathrm{u}\boldsymbol{i} + \mathrm{v}\mathbf{j})\cdot\Delta t \tag{6.40}$$

However, most practical flows are three-dimensional and the laser light sheet is of finite thickness (see Fig.6.27). In such cases, Eq. (6.40) becomes

$$\boldsymbol{D} = -\mathrm{M}\cdot\left[\left(\mathrm{d_x}\boldsymbol{i} + \mathrm{d_y}\boldsymbol{j}\right) + \mathrm{d_z}\frac{(\mathrm{x}\boldsymbol{i} + \mathrm{y}\boldsymbol{j})}{\mathrm{s}}\right] = -\mathrm{M}\cdot\Delta t\cdot\left[\left(\mathrm{u}\boldsymbol{i} + \mathrm{v}\boldsymbol{j}\right) + \mathrm{w}\cdot\frac{(\mathrm{x}\boldsymbol{i} + \mathrm{y}\boldsymbol{j})}{\mathrm{s}}\right] \tag{6.41}$$

The additional third term due to perspective effects means that a particle displacement normal to the light sheet affects the particle image displacement. This effect introduces an uncertainty in measuring the in-plane velocity components (u, v), especially at the edges of the measurement volume where (x, y) values are larger. Eq. (6.41) shows that this so-called perspective error can be reduced by a small magnification factor, shorter pulse separation, and placing the imaging system farther away from the measurement volume. If it is not properly addressed in the design of PIV measurements, the perspective error could easily go up to more than 10% of the mean flow velocity.

6.3.4 Evaluation of the Particle Image Displacement Vector

6.3.4.1 Introduction

In order to obtain a velocity vector map, particle images must be evaluated through an interrogation process, in which PIV recordings are subdivided into a number of square interrogation regions and then the velocity vector in each interrogation region is evaluated by correlation methods.

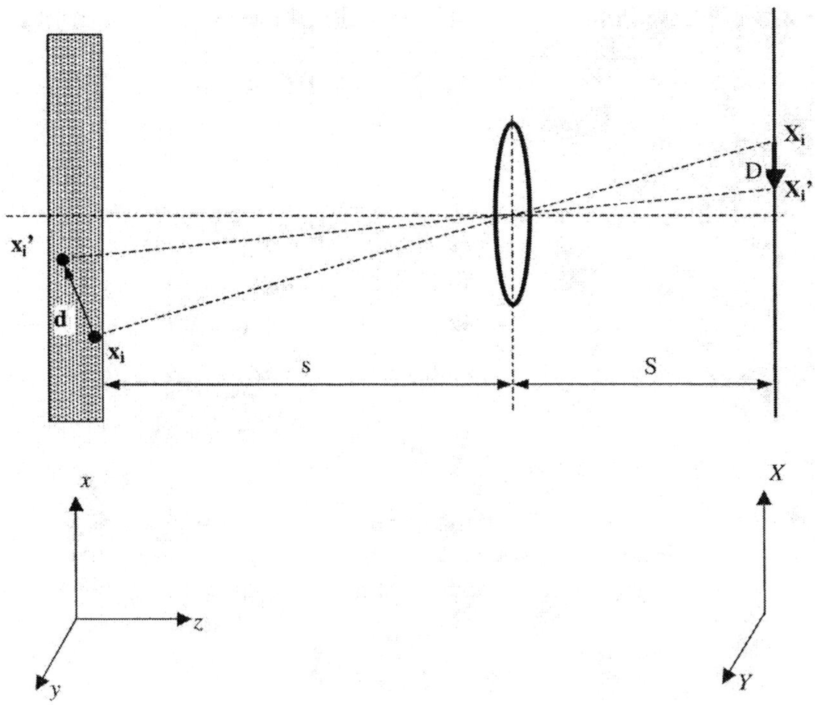

Fig. 6.27 Image formation of particle movement on the recording plane.

For single-frame/double-exposure PIV images, the particle image displacement vector is determined by calculating the digital autocorrelation of the double-exposed particle image field. For double-frame/single-exposure PIV images, cross-correlation of the two separate images is used to work out the displacement vector. In the following sections, we describe the necessary steps in the fully digital analysis of PIV recordings by means of digital correlation techniques.

The mathematical background of correlation techniques will now be described.

6.3.4.2 Particle Image Intensity Field Distribution

Before we proceed to the correlation analysis of two paired particle images, we need to obtain the particle image intensity distribution of an interrogation region in a single exposure.
Each interrogation region consists of a random distribution of particle images, which correspond to a certain pattern of N particles in the flow:

$$\Psi = \begin{pmatrix} x_1 \\ x_2 \\ \bullet \\ \bullet \\ x_N \end{pmatrix} \qquad (6.42)$$

with

$$\boldsymbol{x}_i = \begin{pmatrix} x_i \\ y_i \\ z_i \end{pmatrix}$$

Ψ represents the state of the particle ensemble at a given time t. $\mathbf{x}_i$ is the position vector of the particle i at time t, in the interrogation volume of the laser sheet.

In the following analysis, capital case letters refer to the coordinates in the image plane (see Fig. 6.27) so that

$$\boldsymbol{X} = \begin{pmatrix} X \\ Y \end{pmatrix}$$

Lower case letters are used to represent the coordinates in the object plane. Assuming that the magnification factor is M, thus:

$$x = X/M \quad \text{and} \quad y = Y/M$$

The image intensity field of a single exposure can be written as [Raffel et al., 1998]

$$I(\boldsymbol{X}, \Psi) = \sum_{i=1}^{N} P(\boldsymbol{x}_i) \cdot \tau(\boldsymbol{X} - \boldsymbol{X}_i) \tag{6.43}$$

where $\tau(\mathbf{X} - \mathbf{X}_i)$ is the point spread function of the imaging lens, and it describes the impulse response of the imaging lens. $P(\mathbf{x}_i)$ is the system transfer function, giving the light energy of the image of an individual particle i inside the interrogation volume IV and its conversion into an electronic signal.

6.3.4.3 Autocorrelation Analysis of Double-Exposed PIV Recordings

As discussed previously, a double-exposed PIV recording is evaluated by locally correlating the two images of particles on the same frame. Fig. 6.28 presents a flow chart of the digital autocorrelation method. The PIV recording is subdivided into the interrogation windows. For each window, the autocorrelation function is calculated and the position of the displacement peak is determined. The autocorrelation function is calculated over the frequency plane, employing fast Fourier transform (FFT) algorithms (to be discussed later).

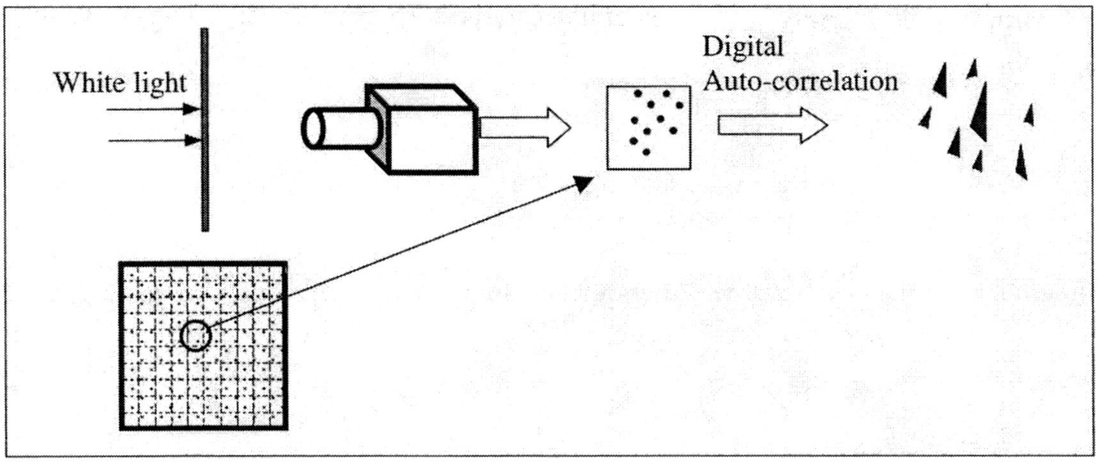

Fig. 6.28 Schematic diagram of autocorrelation analysis of PIV transparency recordings.

In the following analysis, we assume that there is a constant displacement, **d**, of all particles inside the interrogation volume (in the measurement region). Thus, the particle locations at the second exposure are given by

$$\boldsymbol{x}_i' = \boldsymbol{x}_i + \boldsymbol{d} = \begin{pmatrix} x_i + d_x \\ y_i + d_y \\ z_i + d_z \end{pmatrix} \tag{6.44}$$

The corresponding particle image displacement is given by

$$\boldsymbol{D} = \begin{pmatrix} M \cdot d_x \\ M \cdot d_y \end{pmatrix} \tag{6.45}$$

The intensity distribution of the interrogation region for the second exposure can be modeled as

$$I'(\boldsymbol{X}, \Gamma) = \sum_{j=1}^{N} P'(\boldsymbol{x}_j + \boldsymbol{d}) \cdot \tau(\boldsymbol{X} - \boldsymbol{X}_j - \boldsymbol{D}) \tag{6.46}$$

The particle image intensity distribution of the corresponding interrogation region for the first exposure is given by

$$\mathrm{I}(\boldsymbol{X},\Gamma)=\sum_{i=1}^{N}\mathrm{P}(\boldsymbol{x}_i)\cdot\tau(\boldsymbol{X}-\boldsymbol{X}_i) \tag{6.47}$$

The total image intensity field in the double-exposed region is therefore given by

$$\mathrm{I}''(\boldsymbol{X},\Gamma)=\mathrm{I}(\boldsymbol{X},\Gamma)+\mathrm{I}'(\boldsymbol{X},\Gamma)=\sum_{j=1}^{N}\left(\mathrm{P}(\boldsymbol{x}_j)\cdot\tau(\boldsymbol{X}-\boldsymbol{X}_j)+\mathrm{P}(\boldsymbol{x}_j+\boldsymbol{d})\cdot\tau(\boldsymbol{X}-\boldsymbol{X}_j-\boldsymbol{D})\right) \tag{6.48}$$

The autocorrelation function of the double exposed intensity field can be written as:

$$\begin{aligned}\mathrm{R}_{\mathrm{I}''}(\boldsymbol{s},\Psi,\boldsymbol{d})&=\left\langle \mathrm{I}''(\boldsymbol{X},\Gamma)\cdot\mathrm{I}''(\boldsymbol{X}+\boldsymbol{s},\Gamma)\right\rangle\\ &=\frac{1}{\mathrm{A_I}}\sum_{i,j}^{N}\mathrm{P}(\boldsymbol{x}_i)\cdot\mathrm{P}(\boldsymbol{x}_j)\int_{\mathrm{A_I}}\tau(\boldsymbol{X}-\boldsymbol{X}_i)\cdot\tau(\boldsymbol{X}-\boldsymbol{X}_j+\boldsymbol{s})\cdot\mathrm{d}\boldsymbol{X}\end{aligned} \tag{6.49}$$

where s is the separation vector in the correlation plane. By distinguishing the $i \neq j$ terms, which represent the correlation of different particle images and hence randomly distributed noise in the correlation plane (see Fig. 6.29), and the $i = j$ terms, which represent the correlation of the same particle images, it can be shown that the autocorrelation function, $R_{I''}$, can be decomposed into five terms:

$$\mathrm{R}_{\mathrm{I}''}(\boldsymbol{s},\Psi,\boldsymbol{d})=\mathrm{R_C}(\mathrm{s},\Psi,\boldsymbol{d})+\mathrm{R_F}(\mathrm{s},\Psi,\boldsymbol{d})+\mathrm{R_P}(\mathrm{s},\Psi)+\mathrm{R_{D^+}}(\mathrm{s},\Psi,\boldsymbol{d})+\mathrm{R_{D^-}}(\mathrm{s},\Psi,\boldsymbol{d}) \tag{6.50}$$

where R_C and R_F, both resulting from $i \neq j$ terms, form a random noise floor from which the signal peaks must be distinguished for a valid displacement estimate. $R_P(s, \Psi)$ is the self-correlation peak located in the center of the correlation plane (Fig. 6.29). It is caused by the correlation of each particle image with itself.

R_D+ and R_D- represent the correlation of images of particles in the first exposure with that of images of identical particles in the second exposure, hence they are related to the particle image displacement vector. They are symmetrically placed with respect to its central peak R_P. Since the particle image displacement vector could be related to the position of either correlation peaks, there is a 180 degree directional ambiguity. This arises from the inability to identify the particle images corresponding to the first and second exposures in a simple doubly exposed recording. The resulting directional ambiguity may be a serious limitation in the study of complex, reversing flow fields and it must be resolved by employing more sophisticated recording techniques and interrogation techniques.

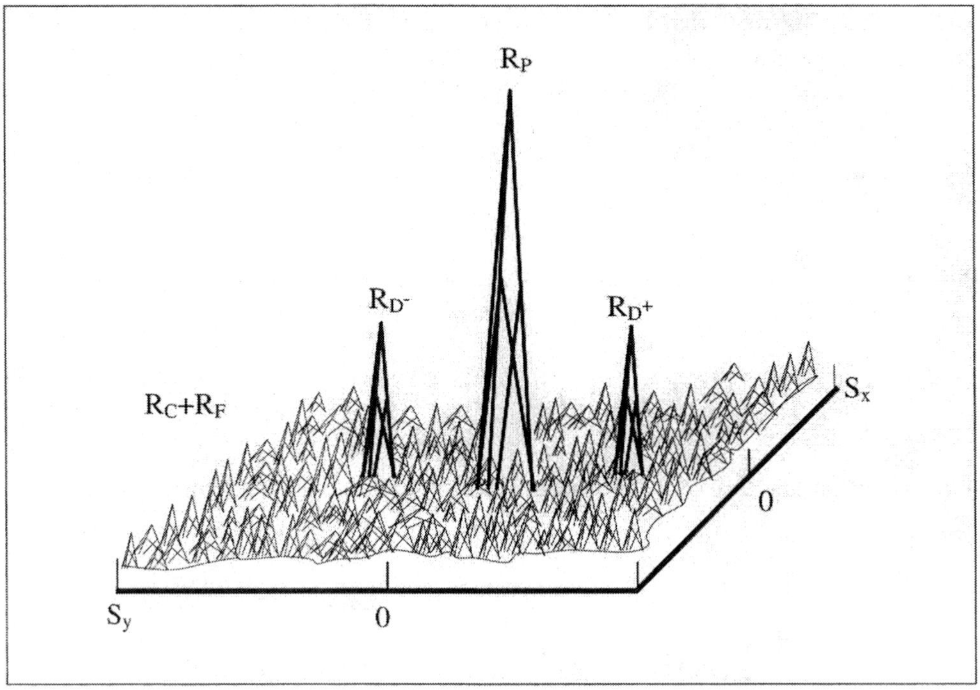

Fig. 6.29 Autocorrelation plane.

A particularly important case that should be avoided in the autocorrelation analysis of doubly exposed images is that when there are only two particle image pairs in the interrogation region. In this case, the signal peaks R_D+ and R_D- have the same height as the two largest spurious correlation peaks, R_C and R_F, so that resulting displacement vectors from these peaks appear equally likely, introducing additional directional ambiguities.

Because of the presence of non-correlation peaks in the correlation plane, the signal peaks should be of significant height so that they can be determined accurately. Since the signal peak is proportional to 2N, a sufficient number of particle images should be present in each interrogation region. In order to achieve good signal-to-noise ratios, Keane and Adrian [1990] recommended a minimum of 8 particles per interrogation region.

There is an upper limit on the number of particle images in the interrogation region, as well. As the number of particle images in a region increases, the signal peak height is always limited to a maximum value, whereas the likelihood of a number of spurious correlation peaks coinciding to give a spurious peak of significant height always increases.

6.3.4.4 Cross-Correlation Analysis of Double-Frame PIV Recordings

A PIV recording that is realized by the double-frame/single-exposure recording technique is evaluated by cross-correlation (see Fig. 6.30). In this case, the cross-correlation between two interrogation windows of the two recordings is calculated. It is desirable to offset the two windows according to an estimated mean displacement of the particles between the two illuminations [Vogt et al., 1996]. By offsetting the interrogation windows, the fraction of matched particle images to unmatched particle images is increased, thereby increasing the signal-to-noise ratio of the correlation peak. For each interrogation window, the cross-correlation function is calculated, and the position of the displacement peak is determined, which corresponds to the particle image displacement vector. The cross-correlation function is calculated by means of FFT algorithms.

As noted in the previous section, the intensity distribution in the interrogation region for the second exposure can be modeled as

$$I'(\boldsymbol{X},\Gamma)=\sum_{j=1}^{N} P'(\boldsymbol{x}_j+\boldsymbol{d})\cdot\tau(\boldsymbol{X}-\boldsymbol{X}_j-\boldsymbol{d}) \tag{6.51}$$

The particle image intensity distribution of the corresponding interrogation region for the first exposure is given by

$$I(\boldsymbol{X},\Gamma)=\sum_{i=1}^{N} P(\boldsymbol{x}_i)\cdot\tau(\boldsymbol{X}-\boldsymbol{X}_i) \tag{6.52}$$

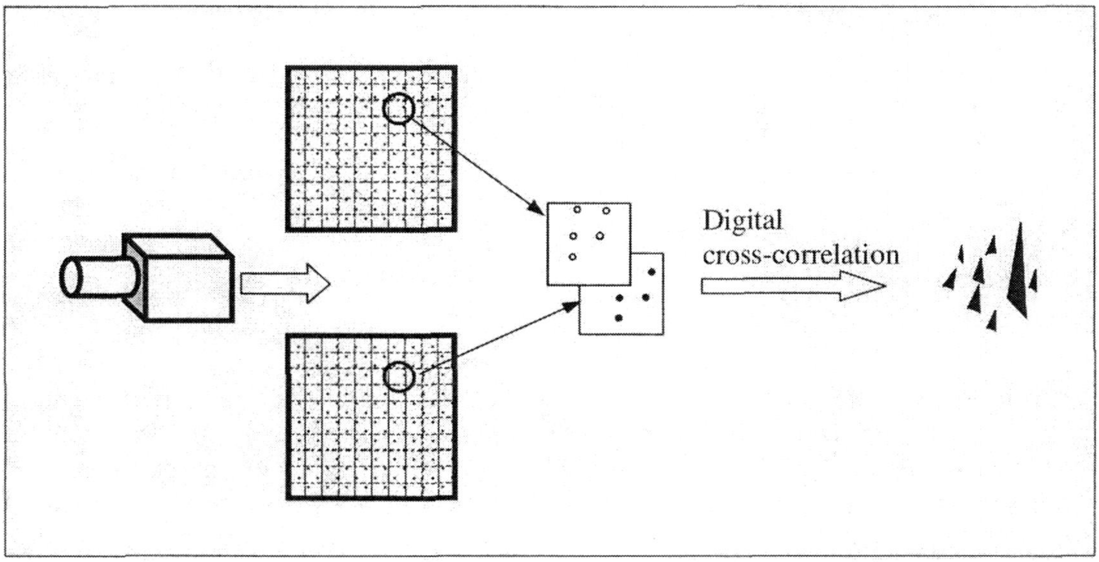

Fig. 6.30 Schematic diagram of the cross-correlation analysis of particle images.

The cross-correlation function of two interrogation areas can then be written as

$$R_{||}(\boldsymbol{s},\Psi,\boldsymbol{d}) = \langle I(\boldsymbol{X},\Gamma)\cdot I'(\boldsymbol{X}+\boldsymbol{s},\Gamma)\rangle \tag{6.53}$$

$$= \frac{1}{A_I}\sum_{i,j}^{N} P(\boldsymbol{x}_i)\cdot P(\boldsymbol{x}_j+\boldsymbol{d})\int_{A_I} \tau(\boldsymbol{X}-\boldsymbol{X}_i)\cdot\tau(\boldsymbol{X}-\boldsymbol{X}_j+\boldsymbol{s}-\boldsymbol{D})\cdot d\boldsymbol{X}$$

where $\boldsymbol{s}$ is the separation vector in the correlation plane. By distinguishing the $i \neq j$ terms, which respresent the correlation of different particle images and therefore randomly distributed noise in the correlation plane, and the $i = j$ terms, which contain the displacement information desired, we arive at the following expression:

$$R_{II}(\boldsymbol{s},\Psi,\boldsymbol{d}) = \frac{1}{A_I}\sum_{i\neq j} P(\boldsymbol{x}_i)\cdot P(\boldsymbol{x}_j+\boldsymbol{d})\int_{A_I}\tau(\boldsymbol{X}-\boldsymbol{X}_i)\cdot\tau(\boldsymbol{X}-\boldsymbol{X}_j+\mathrm{s}-\boldsymbol{D})\cdot d\boldsymbol{X}$$

$$+\frac{1}{A_I}\sum_{i=j} P(\boldsymbol{x}_i)\cdot P(\boldsymbol{x}_i+\boldsymbol{d})\int_{A_I}\tau(\boldsymbol{X}-\boldsymbol{X}_i)\cdot\tau(\boldsymbol{X}-\boldsymbol{X}_i+\mathrm{s}-\boldsymbol{D})\cdot d\boldsymbol{X}$$

or

$$R_{II}(\boldsymbol{s},\Psi,\boldsymbol{d}) = \sum_{i\neq j} P(\boldsymbol{x}_i)\cdot P(\boldsymbol{x}_j+\boldsymbol{d})\cdot R_\tau(\boldsymbol{X}_i-\boldsymbol{X}_j+\boldsymbol{s}-\boldsymbol{D}) \tag{6.54}$$

$$+R_\tau(\boldsymbol{s}-\boldsymbol{D})\sum_{i=j} P(\boldsymbol{x}_i)\cdot P(\boldsymbol{x}_i+\boldsymbol{d})$$

which can be decomposed into three parts:

$$R_{II}(\boldsymbol{s},\Psi,\boldsymbol{d}) = R_C(\boldsymbol{s},\Psi,\boldsymbol{d}) + R_F(\boldsymbol{s},\Psi,\boldsymbol{d}) + R_D(\boldsymbol{s},\Psi,\boldsymbol{d}) \tag{6.55}$$

where the first two terms R_C and R_F contribute to the background noise in the correlation plane as shown in Fig. 6.31, both resulting from the $i \neq j$ terms. $R_D(\mathbf{s}, \Psi, \mathbf{d})$ represents the component of the cross-correlation function that corresponds to the correlation of images of particles obtained from the first exposure with images of identical particles obtained from the second exposure ($i = j$ terms), i.e.,

$$R_D(s,\Psi,d) = R_\tau(s - D)\sum_{i=j} P(x_i) \cdot P(x_i + d) \tag{6.56}$$

Hence, this displacement correlation distribution reaches its peak at **s** = **D**, which means that the average particle image displacement vector **D** can be determined from the maximum of the displacement correlation distribution in the correlation plane. The sign of **D** uniquely defines the direction of flow in the interrogation.

The cross-correlation of two separate particle images offers additional advantages over conventional auto-correlation analysis. Because there is no self-correlation peak, displacements approaching zero may be measured, thus expanding the measurement dynamic range. With cross-correlation, the restriction of avoiding two particle image pairs is removed, so that lower mean seeding densities may be used. Furthermore, the use of higher seeding densities is permitted due to a reduced number of spurious correlation peaks in the cross-correlation plane. The reduced sensitivity of cross-correlation interrogation to seed density will be useful whenever large changes in seed density occur in the flow due to expansion, and more importantly, require less accurate seed density control.

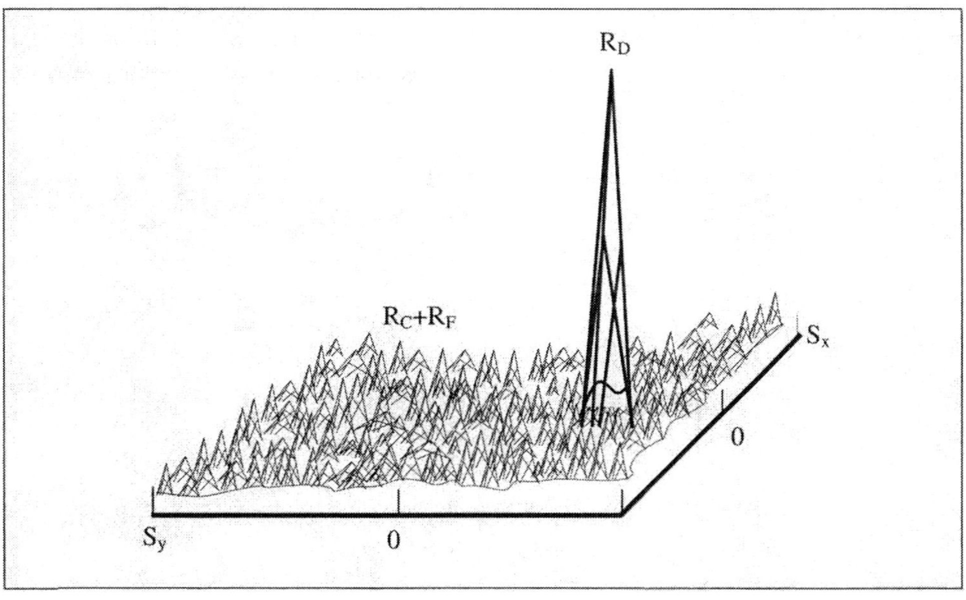

Fig. 6.31 Cross-correlation plane.

6.3.4.5 Digital Evaluation of Cross-Correlation and Autocorrelation

As noted above, the particle image displacement vector, **D**, can be determined from the signal peak positions in the correlation plane. Although correlation may be calculated directly using Eq. (6.49) and Eq. (6.53). In practice, the most efficient way of calculating correlation is carried out by means of FFT algorithms.

The correlation theorem states that the cross-correlation of two functions is equivalent to a complex conjugate multiplication of their Fourier transforms,

$$R_{II} \Leftrightarrow \hat{I} \cdot \hat{I}'^{*} \tag{6.57}$$

where $\hat{I}$ and $\hat{I}'$ are the Fourier transforms of the functions **I** and **I'**, respectively. In practice, the Fourier transform is efficiently performed using the FFT algorithm. The cross-correlation function can therefore be calculated by computing two two-dimensional FFTs on equal sized samples of the image, followed by a complex-conjugate multiplication of the resulting Fourier coefficients (see Fig. 6.32). The most common FFT implementation requires the input data (size of the interrogation window) to have a binary based dimension (e.g., 32 × 32 pixel, or 64 × 64 pixel samples).

Similar procedures are used in the calculation of the autocorrelation function except that both samples are the same (Fig. 6.32[a]). Unlike the cross-correlation function computed from different samples, the autocorrelation function will always have a self-correlation peak located at the origin and two symmetrically positioned displacement peaks with respect to the self-correlation peak.

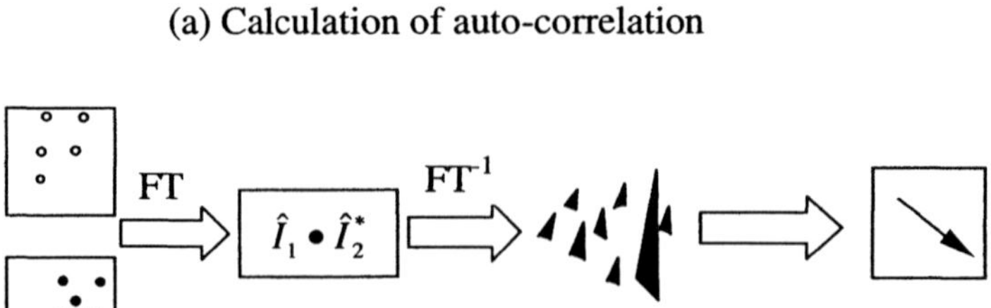

(a) Calculation of auto-correlation

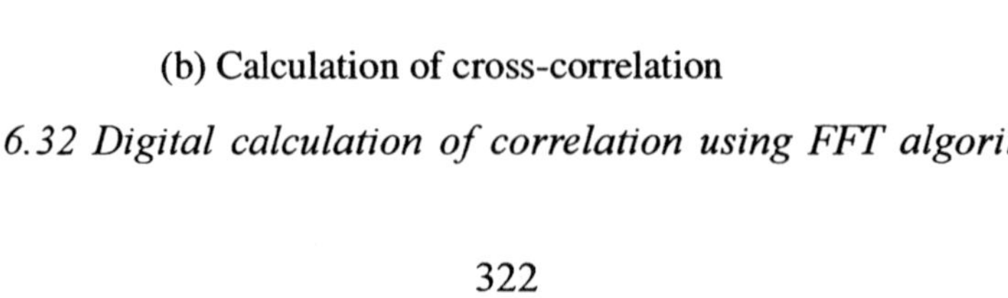

(b) Calculation of cross-correlation

Fig. 6.32 Digital calculation of correlation using FFT algorithms.

6.3.5 Removal of Directional Ambiguity

As noted previously, conventional photographic PIV recordings and their subsequent evaluation by autocorrelation suffer from 180 degree directional ambiguity. Although directional ambiguity may be addressed adequately by assuming bulk flow directions from prior knowledge of the flow structure, such a priori knowledge will not be available in many flows of practical importance.

A number of techniques have been proposed and developed in order to resolve the directional ambiguity. Amongst these are image shifting and pulse coding techniques.

Image shifting involves the introduction of a known and constant displacement in the position of the second exposure of particle images relative to the film. The principle of image shifting for removal of directional ambiguity is illustrated in Fig. 6.33. Two particles are imaged, one of which is moving to the right and the other to the left (reversed flow). By introducing an additional image shift, D_s, to those of these two particles, D_1, and D_2, the positions of the second exposure are always located in the positive direction, with respect to those of the first exposure, if D_s is selected to be greater than the reversed flow displacement. The true displacement values of these two particles are then obtained by subtracting D_s after the extraction of the positive displacement vectors for the PIV recording. In practice, the magnitudes of the two components of the image shift should be chosen to be larger than the most negative displacement vector components in the flow.

In addition to its use for removal of the directional ambiguity, image shift may be used to reduce the out-of-plane loss in conjunction with the reduced pulse separation time [Raffel et al., 1998]. The introduction of negative image shift can also lead to an increase in the resolution available for measuring turbulent fluctuations [Grant, 1997]. However, the magnitude of the measurement dynamic range is almost halved due to the reduced maximum particle image displacement.

Image shifting with conventional photographic recording systems has been achieved using both opto-mechanical and electro-optical methods. The most widely used image shifting technique is the rotating or oscillating mirror method first introduced by Adrian [1986]. It uses the motion of a mirror placed at a 45 degree to the camera optical axis to introduce the required image shift. Raffel et al. [1998] described the design and implementation of such a system. The practical difficulties of accurately phase-locking the mirror motion to that of an engine and synchronizing the mirror position with certain crank angle makes it difficult for the rotating mirror method to be applied to engine studies.

Electro-optical devices, involving polarized illumination, have been developed to obtain high-velocity shifts, where the image shift is produced using a birefringent crystal. Such devices rely on the use of a double pulsed laser, in which the polarization plane of the laser light is rotated by 90 degree between the two exposures. This can be achieved with high-pulse energy beams using commercial Pockels cells that are capable of rapid polarization switching. Difficulties may arise if significant depolarization occurs during the scattering process.

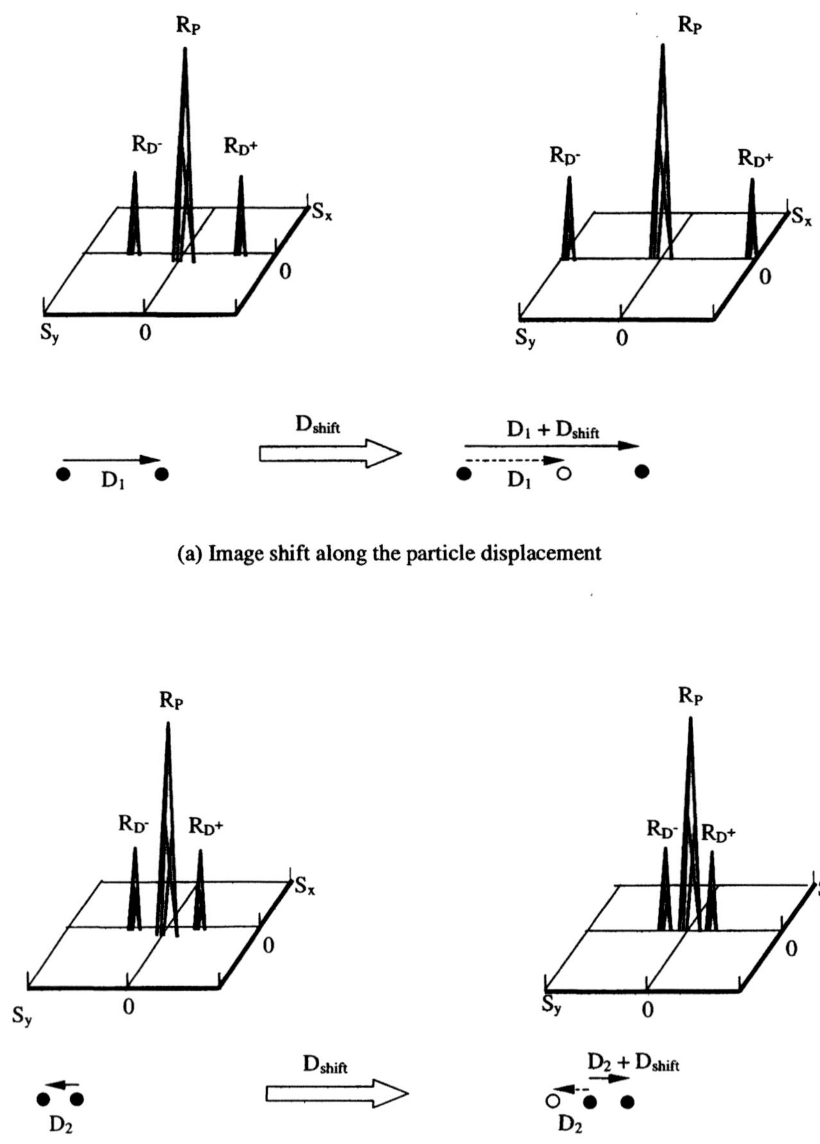

(a) Image shift along the particle displacement

(b) Image shift opposite to the particle displacement

Fig. 6.33 Removal of directional ambiguity by the image shifting technique.

Pulse coding or image labeling involves changing the appearance of the first and second particle images so that they can be distinguished uniquely at the interrogation stage. The most practical image labeling technique is achieved by color [Goss, et al., 1989]. In fact, the two-color method has been the only directional ambiguity removal technique applied to engine measurements [Nino et al., 1992; and 1993]. The first illumination was generated by the second harmonic of an Nd:YAG laser (532 nm) and the second one from the output of a pulsed dye laser (650 nm). The doubly exposed image was recorded onto a 35mm color film. Each interrogation region was illuminated alternately with red and green light and imaged onto a monochrome CCD camera separately. The resulting two images were then evaluated by the cross-correlation method. Considering the numerous benefits of cross-correlation over autocorrelation interrogation, the two-color method is particularly suited to resolving the directional ambiguity in conventional photographic PIV recordings in engine flows.

6.3.6 Optimization of a PIV System

As noted above, a number of experimental parameters are involved in successful PIV measurement. Having discussed the operation of a PIV system in detail, we now consider how to optimize PIV measurements.

A PIV system may be described in terms of the following properties:

- Spatial/temporal resolution
- Measurement dynamic range
- Accuracy
- Speed of data processing

The spatial resolution is important, since it determines the ability of the system to measure small spatial scales of flow structures. The spatial resolution is determined by the thickness of the laser sheet and the quality of the imaging optics. As noted previously, the laser sheet thickness is determined by the sheet forming optics and its thickness Δz is limited by the data drop-out rate due to the out-of-plane velocity component,

$$w \cdot \Delta t \leq \Delta z / 4$$

Therefore, a thicker laser sheet should be employed so long as there is sufficient scattered energy density on the recording medium.

Temporal resolution is limited by the speed at which one can advance the film or transfer the image electronically. Standard film camera is limited to a few frames per second. High-speed time resolved PIV recordings at 100 image pairs per second have been reported using a high repetition Cu vapor laser and a high-speed film drum camera [Stolz et al., 1992]. Typical cross-correlation cameras are capable of capturing up to 15 PIV image pairs per second, provided that the laser can be operated at such frequency.

The measurement dynamic range is determined by the pulse separation, imaging conditions, and dimension of interrogation regions. The minimum measurable displacement in the image plane is either limited by the effective image diameter given by Eq. (6.38)

$$d_i = \sqrt{M^2 d_p^2 + d_{diff}^2}$$

or the spatial resolution of the recording medium, (e.g., the pixel size of a CCD camera, d_{pixel})

Hence, the minimum velocity that can be resolved is given by

$$V_{min} = \Delta t \cdot d_i / M \quad \text{or} \quad V_{min} = \Delta t \cdot d_{pixel} / M \tag{6.58}$$

In order to optimize the correlation process, Keane and Adrian [1990] recommended that an upper limit for the particle displacement in double exposed images should be one quarter of the interrogation region length L. Therefore, the maximum velocity that can be measured given this restriction is

$$V_{max} = 0.25 \Delta t \cdot L/M \tag{6.59}$$

The dynamic range can then be expressed as a single ratio of these two quantities:

$$D_v = V_{max}/V_{min} = 0.25 L/d_i \quad \text{or} \quad D_v = V_{max}/V_{min} = 0.25 L/d_{pixel} \tag{6.60}$$

Typical values of the velocity dynamic range for autocorrelation PIV measurements are around 10. In engine flows, this limited measurement range prevents the simultaneous measurement of high- and low-velocity regions. Thus, erroneous velocity measurements can be expected in regions of the flow that fall outside the measurement range, and these will have to be removed in a systematic validation procedure. In practice, the limited velocity dynamic range may be further restricted at the lower end of the measurement dynamic range by enlarged particle images due to optical aberration of imaging optics and noise in the recording medium. Significant improvement in the measurement dynamic range up to 100 can be obtained when cross-correlation PIV measurement is used.

It is interesting to analyze the measurement dynamic range further when the particle image is diffraction limited, i.e., $d_i = d_{diff}$. A ssuming that the depth of field is equal to the light sheet thickness, then for a cubic interrogation volume we have

$$L/M = \delta_z = l = \Delta z \tag{6.61}$$

Substitution of Eq. (6.39) and Eq. (6.36) into Eq. (6.60) then gives the result

$$D_v \propto \frac{L}{d_i} = \frac{1}{1.22}\sqrt{\frac{l}{\lambda}} \tag{6.62}$$

Thus, the measurement dynamic range is a function of the interrogation volume size and the laser wavelength alone. Eq. (6.62) also shows that the dynamic range is independent of the imaging system magnification, assuming diffraction-limited imaging. This has several implications for the design of a PIV experiment. First, the magnification can be reduced to allow a larger area of flow to be recorded without compromising the dynamic range, as long as the recording medium is able to sufficiently resolve the particle image. Second, with sufficient scattering, a lower magnification allows the imaging system to be placed farther from the measurement section, limiting perspective errors and relaxing restrictions on experimental design.

The pulse separation Δt should be selected so that

$$\frac{(D_{max}/M)}{V_{max}} \leq \Delta t \leq \frac{(D_{min}/M)}{V_{min}} \tag{6.63}$$

As noted previously, the minimum measurable displacement D_{min} in the image plane is either limited by the spatial resolution of the recording medium (pixel size of a CCD camera) or the effective image diameter d_i. The upper limit for the particle displacement, D_{max}, should be one quarter of the interrogation region length, L.

Overall measurement accuracy in PIV is affected by a number of factors extending from the recording process all the way to the methods of evaluation, including recording medium noise, velocity gradients, out-of-plane particle motion, and particle motion outside the measurement dynamic range. The perspective error can be significantly reduced by placing the imaging system farther away from the measurement plane. The perspective error diminishes in proportion to the out-of-plane displacement, and hence can be reduced to a great extend if the measurement plane can be oriented with the predominant flow direction. The particle seeding density should be sufficiently high so that there should be 8 pairs of particles in each interrogation region, in order to increase the valid data rate of PIV evaluation.

Measurement accuracy in PIV evaluation can be assessed experimentally by using actual PIV recordings for which displacement data is known reliably. For example, PIV recordings of a quiescent flow were used in determining the measurement accuracy in the cross-correlation of single-exposed particle image pairs [Willert and Gharib, 1991].

For engine applications, ensemble-averaging over many engine cycles is required when there are large cyclic variations in the flow. The time it takes to record, process, and interrogate photographic PIV recordings significantly restricts the acquisition of ensemble-averaged flow data. The repetitive nature of photographic PIV systems makes such an endeavor unattractive for routine flow mapping and the long lead-times required make routine PIV studies unfeasible. This problem has now been solved by recent advances in digital recording and PIV evaluations. Real-time PIV systems are now available commercially with dedicated hardware correlation analyzers [Dantec Measurement Technology, 1998] or on-board processing using a PC

Table 6.2 Summary of PIV system variables and their optimization

Laser source	Light sheet thickness(Δz):	$\Delta z \geq 4 \cdot w \cdot \Delta t$; thicker sheet with sufficient spatial resolution.
	Pulse separation (Δt):	Autocorrelation: $D_{max} \leq 0.25$ L; cross-correlation: $D_{max} \leq 0.5$ L
Seeding particles	Particle size (d_p):	1 μm–10 μm; small particle to follow turbulent flow, large particle to increase scattered light.
	Seeding density:	≈ 6 particles in the interrogation region for improved valid data rate (signal-to-noise ratio).
Imaging optics	Magnification (M)	<1.0; smaller magnification for increased measurement area and reduced perspective error.[1]
	f-number (f#)	Selected so that depth-of-field $\delta_z \geq \Delta z$ and particle image diameter $D_i \leq d_{pixel}$.[1]
Cross-correlation CCD camera	Pixel size and No.:	The format of camera determines the spatial resolution of PIV recordings.
	Frame rate (30 fps):	The number of PIV image pairs per second is half the frame rate.
Interrogation regions	Size (L × L):	Larger L for increased dynamic range, smaller L for improved accuracy.
	Number:	Limited to a square matrix of $2^n \times 2^n$ by FFT algorithms used in digital correlation calculations.
Evaluation Scheme	Image shift between the two frames:	The two separately exposed images should be shifted by the estimated mean particle image displacement vector to reduce the in-plane loss of correlation.

[1] So long as there is sufficient light energy of scattered particle images on the recording medium.

computer with a fast image-acquisition card. The data rate is limited by either the cross-correlation CCD camera (15 PIV image pairs per second) or the repetition rate of the laser pulses.

In summary, successful PIV measurement demands the optimization of a number of parameters ranging from the illumination source and imaging optics, all the way to the evaluation of PIV recordings. The important parameters involved in the optimization of PIV measurement are summarized in Table 6.2. The basic procedures involved are as follows:

1. The design of a PIV experiment can begin with the selection of laser light sheet thickness, which should be sufficiently thick so that maximum out-of-plane displacement is less than a quarter of the light sheet thickness.
2. The required laser pulse separation should be chosen so that the maximum particle image displacement is approximately one-quarter and one-half of the interrogation region size for the autocorrelation and cross-correlation analysis, respectively.

3. The imaging optics should have a magnification factor much less than one, with the f-number setting selected so that the light sheet width falls within its depth-of-field. A longer distance between the imaging system and the object plane is preferred, so long as there is sufficient light energy of the scattered particle image on the recording medium.

4. The interrogation process bears important consequences for the measurement dynamic range and accuracy. Larger interrogation region increases dynamic range, but it compromises accuracy and spatial resolution. During interrogation, the interrogation windows corresponding to the two separately exposed particle images should be offset by the estimated mean particle image displacement vector for improved data yield.

In practice, the pulse separation will not be optimized without repeating each measurement using a large number of different pulse separations. The interrogation region size therefore has to be adjusted accordingly at the interrogation stage to accommodate the maximum particle displacements that are present. As a consequence, the offset required in the cross-correlation stage then must be optimized as well.

6.3.7 Post Processing of PIV Data

After evaluation of the PIV recordings, there are sometimes a certain number of incorrectly determined velocity vectors present due to noise or artifacts in the particle images. These spurious velocity vectors are referred to as outliers (see Fig. 6.34). In the simplest case, they can be detected by visual inspection of the magnitude and direction of velocity vectors and their deviation from their nearest neighbors. In practice, data validation algorithms can be implemented for the removal of these outliers, such as the global histogram operator and the dynamic mean value operator [Raffel et al., 1998].

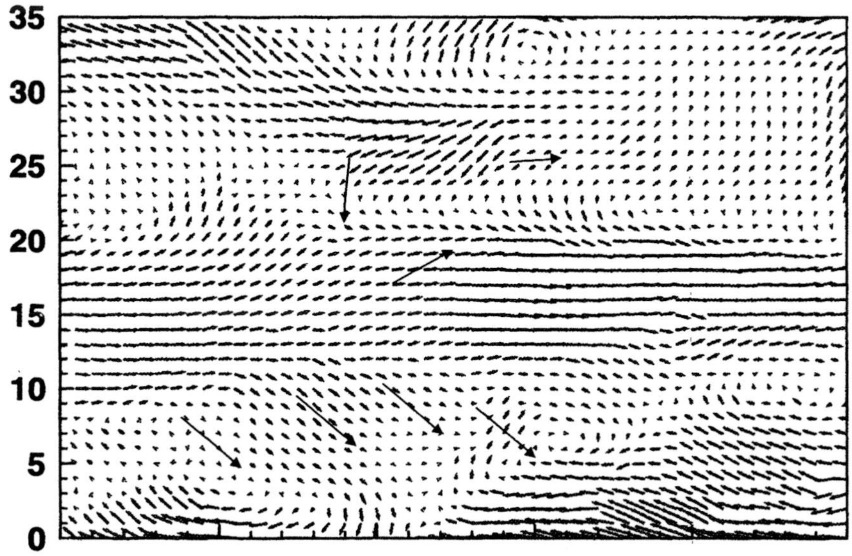

Fig. 6.34 Raw velocity vector map with outliers.

The removal of certain spurious velocity vectors leaves gaps in the vector field which must be filled. This normally involves interpolation using a weighted average of the surrounding data. If required, the continuity of the data can be further improved by smoothing using numerical low-pass filters.

Results of PIV measurements are normally presented as velocity vector maps. In order to understand the fluid mechanical features, ensemble-averaging of cyclic flows, and vector field operators (e.g., vorticity in order to detect tumble or swirl in the flow) are necessary. Animation of the PIV data is particularly useful for better understanding in the case of presenting time series of PIV recordings.

6.3.8 PIV Measurements in IC Engines

The first and yet the most comprehensive in-cylinder flow measurements using PIV were carried out in the late 1980s by Reuss et al. [1989]. The PIV experiments were performed in a specially constructed single-cylinder optical engine. The engine was equipped with a single poppet valve in order to maximize the size of a window in the cylinder head, through which the particle images were recorded. Images of TiO_2 particles from a single double-pulsed Nd:YAG laser were photographed with a 35 mm camera. The presence of strong axial swirl within the laser sheet allowed the velocity direction to be determined unambiguously using autocorrelation analysis. In addition to instantaneous velocity vector maps, large- and small-scale flow structures were analyzed by low-pass and high-pass filtering, respectively. The limited velocity range within the measurement region also allowed the experimental parameters to be optimized to minimize noise, so that spatial derivative quantities, such as vorticity and strain rate, could be evaluated. The same technique was also applied to a skip fired engine so that the velocity, vorticity, and strain measurements ahead of a propagating flame could be made [Reuss et al., 1990]. An electro-optical image shifting device was later incorporated into the PIV system in order to resolve the direction of flow velocity [Reuss et al., 1990].

The requirement to remove the directional ambiguity in photographic PIV recordings led to the development of a two-color PIV system by Nino et al. [1993]. In this work, particle images of two different colors were recorded onto a color film using Nd:YAG and pulsed dye lasers as the illumination sources. Cross-correlation of separately digitized red (scattering from the dye laser) and green (scattering from the Nd:YAG laser) images were employed to derive unambiguously the displacement vectors and their directions without a priori knowledge of the flow structure. A large cylinder head window allowed a complete flow field to be measured in a high-swirl engine under motored and fired conditions. Difficulty was encountered in the burned gas region due to the low seeding density.

While single PIV measurements per engine cycle are useful, continuous PIV recordings are required in the understanding of the development of flow structures and mixing processes as a function of crank angle. This was realized by means of a high-speed rotating drum camera and a copper vapor laser of high repetition rate [Stolz et al., 1992]. Due to the movement of the film between the two exposures an image shift of 0.25 mm was introduced into the particle images, which led to the removal of directional ambiguity of velocities up to 20 m/s. Successful motored and fired measurements in the vertical planes were made in a square piston optical engine at a speed of 200 consecutive PIV images per second.

The above PIV measurements were all carried out in specially designed optical engines which have atypical engine geometry of production engines. Reeves et al. [1994; 1996] reported the application of PIV measurements in a production geometry four-valve motored Hydra engine. The illumination was provided by a frequency doubled, dual oscillator, dual amplifier Nd:YAG laser, and particle images were recorded by a medium format Mamiya 645 camera (60 × 95 mm). The vertical laser sheets were introduced through the piston crown via a dielectric mirror, which was found superior to the transmission of laser sheets through side windows in minimizing stray reflections. PIV images along the center lines of an inlet and exhaust valve were obtained during the compression stroke. From the calculated velocity vector maps, the formation and destruction of a tumbling vortex were identified. An attempt was also made to correct for the small aberration in particle images, due to the curved surface of the transparent cylinder liner, by placing a specially designed corrective optical element between the camera lens and the engine cylinder. On the cylinder axis, corrected particle images were close to the diffraction-limited diameter afforded by the camera lens. It was also found that planes closer to the camera than the cylinder centerline gave unacceptable off-axis aberrations and should be avoided, with preference being given to planes farther away than the centerline.

In the last few years, the availability of digital cross-correlation cameras has dramatically simplified PIV measurements. As noted previously, cross-correlation analysis has a number of major advantages over autocorrelation: the removal of directional ambiguity, the significant increase in the measurement dynamic range, less sensitivity to out-of-plane movement, and less restricted seeding density requirement. Successful PIV measurements were carried out by Rouland et al. [1997] in a single cylinder four-valve optical engine with pent-roof combustion chamber. The PIV system used a standard CCD video camera to record the particle images onto two separate fields by operating in the field straddling mode. Since two images were recorded on odd and even fields in the video camera, respectively, the spatial resolution of the particle images was half the CCD array. The robustness of the cross-correlation PIV system allowed a large set of PIV images to be obtained in both the vertical and horizontal planes in the cylinder, from which large-scale and small-scale flow structures were determined by high-pass and low-pass spatial filtering. Application of the same technique to a square piston compression machine was also reported [Marc et al., 1997].

The application of a commercial digital PIV system to the intake manifold of a single-cylinder engine was reported by Gindele et al. [1997]. A full-frame progressive scan, interline transfer CCD camera was used to record two particle images onto two different frames of the camera. The cross-correlation was carried out by a dedicated processor, and velocity vectors were then transferred to a PC computer. Successful measurements were made at engine speed up to 5000 rpm.

6.4 Summary

In this chapter, the fundamentals of LDA and PIV and their applications to IC engines have been discussed. LDA has been applied to engine research for over two decades and provided much useful insight into the flow processes and their interactions with combustion within the cylinders of IC engines. The extension of the LDA setup to include the detection of the scattered signal phase permits the simultaneous measurement of fuel droplet size and velocity inside the cylinder; this will be described in Chapter 8. The advent of PIV systems has provided the ability to record instantaneous, cycle resolved, two-dimensional maps of velocity within extended areas in the cylinder. In both methods, flow-following particles have to be introduced into the engine's intake. The two techniques are complimentary and the introduction of fully digital PIV systems will see the application of more whole field velocity measurements in IC engines.

References

Adrian, R.J. (1984), "Scattering Particle Characteristics and Their Effect on Pulsed Laser Measurements of Fluid Flow: Speckle Velocimetry vs. Particle Image Velocimetry," *Applied Optics*, Vol. 23, No. 11, pp. 1690–1691.

Adrian, R.J. (1986), "Image Shifting Technique to Resolve Directional Ambiguity in Double-Pulsed Velocimetry," *Applied Optics*, Vol. 25, pp. 3855–3858.

Adrian, R.J. (1991), "Particle-Imaging Techniques for Experimental Fluid Mechanics," *Ann. Rev. Fluid Mech.*, Vol. 23, pp. 261–304.

Arcoumanis, C., Bicen, A.F., and Whitelaw, J.H. (1983), "Squish and Swirl-Squish Interaction in Motored Model Engines," *J. Fluids Engng*, Vol. 105, pp. 105–112.

Arcoumanis, C., Hu, Z., Vafidis, C. and Whitelaw, J.H. (1990), "Tumbling Motion: A Mechanism for Turbulence Enhancement in Spark-Ignition Engines," SAE Paper No. 900060.

Baby, X. and Floch, A. (1997), "Investigation of the In-Cylinder Tumble Motion in a Multi-Valve Engine: Effect of the Piston Shape," SAE Paper No. 971643.

Ball, W.F., Pettifer, H.F., and Waterhouse, C.N. (1983), "Laser Doppler Velocimeter Measurements of Turbulence in a Direct Injection Diesel Combustion Chamber," Paper C52/83, IMechE Publications.

Barker, D.B. and Fourney, M.E. (1977), "Measuring Fluid Velocities with Speckle Patterns," *Optics Lett.*, Vol. 1, No. 14, pp. 135–137.

Bicen, A.F., Vafidis, C., and Whitelaw, J.H. (1985), "Steady and Unsteady Airflow through the Intake Valve of a Reciprocating Engine," *J. Fluids Engng*, Vol. 107, p. 413.

Bopp, S., Durst, F., and Tropea, C. (1990), "In-Cylinder Velocity Measurements With a Mobile Fibre Optic LDA System," SAE Paper No. 900055.

Bopp, S., Vafidis, C., and Whitelaw, J.H. (1986), "The Effect of Engine Speed on the TDC Flowfield in a Motored Reciprocating Engine," SAE Paper No. 860023.

Burch, J.M., Tokarski, J.M. (1968), "Production of Multiple Beam Fringes from Photographic Scatters," *Optica Acta,* Vol. 15, Part 2, pp. 101–111.

Church, W. and Farrell, P.V. (1998), "Effects of Intake Port Geometry on Large Scale In-Cylinder Flows," SAE Paper No. 980484.

Clerk, D. (1921), "Cylinder Actions in Gas and Gasoline Engines," *SAE Journal*, Vol. 8, p. 523.

Coghe, A., Brunello, G., and Tassi, E. (1988), "Effects of Intake Ports on the In-Cylinder Air Motion Under Steady Flow Conditions," SAE Paper No. 880384.

Corcione, F.E. and Valentino, G. (1990), "Turbulence Length Scale Measurements by Two-Probe-Volume LDA Technique in a Diesel Engine," SAE Paper No. 902080.

Daneshyar, H. and Fuller, D.E. (1986), "Definition and Measurement of Turbulence Parameters in Reciprocating IC Engines," SAE Paper No. 861529.

Dantec Measurement Technology, (1998), *FlowMap PIV Systems*, Product Brochure.

Drain, L.E. (1980), *The Laser Doppler Technique*, John Wiley & Sons, New York.

Dudderar, T.D., Simpkins, P.G. (1977), "Laser Speckle Photography in a Fluid Medium," *Nature*, Vol. 270, pp. 45–47, 1977.

Durao, D.F., Lake, J., and Whitelaw, J.H. (1980), "Bias Effects in Laser Doppler Anemometry," *J. Phy. E: Scientific Instruments*, Vol. 14, pp. 442–445.

Durst, F., Melling, A., and Whitelaw, J.H. (1976), *Principles and Practice of Laser Doppler Anemometry*, Academic Press, New York.

Durst, F. and Krebs, H. (1984), " Adaptive Optics for IC Engine Measurements," *Experimental Measurements and Techniques in Turbulent Reactive and Non-Reactive Flows*, Vol. AMD 66, ASME.

Fansler, T.D. and French, D. (1988), "Cycle-Resolved Laser Velocimetry Measurements in a Reentrant-Bowl-in-Piston Engine," SAE Paper No. 880377.

Faure, M.A., Sadler, M., Oversby, K.K., Stokes, J., Begg, S.M., Pommier, L.S., and Heikal, M.R. (1998), " Application of LDA and PIV Technique to the Validation of a CFD Model of a Direct Injection Gasoline Engine," SAE Paper No. 982705.

Fraser, R.A. and Bracco, F.V.(1988), " Cycle-Resolved LDA Integral Length Scale Measurements in an IC Engine," SAE Paper No. 880381.

Gindele, J., Kech, J.M., Scholl, D.M., and Spicher, U. (1997), "Time Resolved Investigation of Unsteady Flow inside Inlet Manifolds and Characterization of Inlet Flow Behaviour," SAE Paper No. 972828.

Glover, A.R. (1986), "Towards Bias-Free Estimates of Turbulence in Engines," Proc. of 3rd Intl. Symp. on Applications of Laser Anemometry to Fluid Mechanics, Lisbon.

Glover, A.R., Hundleby, G.E., and Hadded, O. (1988), "The Development of Scanning LDA for the Measurement of Turbulence in Engines," SAE Paper No. 880379.

Goss, L.P., Post, M.E., Trump, D.D., and Sarka, B. (1996), "Two-Colour Particle Velocimetry," in ICALEO '89, 1989.

Grant, I. (1994), *Selected Papers on Particle Image Velocimetry*, SPIE Milestone Series MS99, SPIE Optical Engineering Press.

Grant, I. (1997), "Particle Image Velocimetry: A Review," *Proc. Inst. Mech. Engrs.*, Part C, Vol. 211, pp. 55–76.

Hadded, O. and Denbratt, I. (1991), "Turbulence Characteristics of Tumbling Air Motion in Four-Valve SI Engines and Their Correlation with Combustion Parameters," SAE Paper No. 910478.

Hall, M.J., Bracco, F.V., and Santavicca, D.A. (1986), "Cycle-Resolved Velocity and Turbulence Measurements in an IC Engine with Combustion," SAE Paper No. 860320.

Hilton, A.D., Roberts, J.B., and Hadded, O. (1991), " Autocorrelation Based Analysis of Ensemble Averaged LDA Engine Data for Bias-Free Turbulence Estimates: A Unified Approach," SAE Paper No. 910479.

Hoesel, W. and Rodi, W. (1977), "New Biasing Elimination Method for Laser Doppler Velocimeter Counter Processing," *Review of Scientific Instruments*, Vol. 48, pp. 910–919.

Jackson, N.S., Stokes, J., Sadler, M., Heikei, M.R., Faure, M., and Pommier, L. (1997), "Correlation of the Combustion Characteristics of Spark Ignition Engines with the In-Cylinder Flow Field Characterised Using PIV in a Water Analogue Engine," SAE Paper No. 971637.

Karimi, E.R. and Jane, P.A.H. (1993), "A Combustion System Investigation to Achieve Low Emissions on a Heavy Duty Diesel Engine," Proc. of AUTOTECH 93, pp. 143–149, Birmingham.

Keane, R.D. and Adrian, R.J. (1990), "Optimization of Particle Image Velocimeters. Part I: Double Pulsed Systems," *Meas. Sci. Tech.,* Vol. 2, pp. 963–974.

Khalighi, B. and Huebler, M.S. (1988), "A Transient Water Analogue of a Dual-Intake-Valve Engine for Intake Flow Visualisation and Full-Field Velocity Measurements," SAE Paper No. 880519.

Khalighi, B. (1990), "Intake-Generated Swirl and Tumble Motions in a 4-Valve Engine with Various Intake Configurations—Flow Visualisation and Particle Tracking Velocimetry," SAE Paper No. 900059.

Kuwahara, K., Watanabe, T., Takemura, J., Omori, S., Kume, J., and Ando, H. (1994), "Optimisation of In-Cylinder Flow and Mixing for a Centre-Spark Four-Valve Engine Employing the Concept of Barrel-Stratification," SAE Paper No. 940986.

Lading, L. (1987), "Spectrum Analysis of LDA Signals," in *The Use of Computers in Laser Velocimetry* (Ed. Pfeifer, H. and Jaeggy, B), Report R105/87, L'Institut Franco Allemand de Recherches de Saint Louis, Paper 20.

Lancaster, D.R. (1976), "Effects of Engine Variables on Turbulence in a Spark-Ignition Engine," SAE Paper No. 760159.

Lee, D.W. (1939), *A Study of Air Flow in an Engine Cylinder*, NACA Report 653.

Liou, T.-M. and Santavicca, D.A., (1985), "Cycle Resolved LDV Measurements in a Motored IC Engine," *Transactions of the ASME*, Vol. 107, pp. 232–240.

Liou, T.-M., Hall, D.A., Santavicca, D.A., and Bracco, F.V. (1984), "Laser Doppler Velocimetry Measurements in Valved and Ported Engines," SAE Paper No. 840375.

Lorenz, M. and Prescher, K. (1990), "Cycle Resolved LDV Measurements on a Fired SI Engine at High Data Rates Using a Conventional Modular LDV-System," SAE Paper No. 900054.

Marc, D., Boree, J., Bazile, R., and Charnay, G. (1997), "Tumbling Vortex Flow in a Model Square Piston Compression Machine: PIV and LDV Measurements," SAE Paper No. 972834.

Marko, K.A., Li, P., Rimai, L., and Ma, T. (1986), "Flow Field Imaging for Quantitative Cycle Resolved Velocity Measurements in a Model Engine," SAE Paper No. 860022.

McLaughlin, D.K. and Tiederman, W.G. (1973), "Biasing Correcting for Individual Realisation of Laser Anemometer Measurements in Turbulent Flows," *Phys. Fluids*, Vol. 16, pp. 2082–2088.

Meynart, R., (1980), "Equal Velocity Fringes in a Rayleigh-Bernard Flow by a Speckle Method," *Applied Optics*, pp. 1395–1386.

Morse, A.P., Whitelaw, J.H., and Yianneskis, M. (1980), "The Influence of Swirl on the Flow Characteristics of a Reciprocating Piston-Cylinder Assembly," *J. Fluids Engng.*, Vol. 194, pp. 291–299.

Neußer, H., Spiegel, L., and Ganser, J. (1995), "Particle Tracking Velocimetry—A Powerful Tool to Shape the In-Cylinder Flow of Modern Multi-Valve Engine Concepts," SAE Paper No. 950102.

Nino, E., Gajdeczko, B. F., and Felton, P.G. (1993), "Two-Colour Particle Image Velocimetry in an Engine with Combustion," SAE Paper No. 930872.

Obokata, T., Hanada, N., and Kurabayashi, T. (1987), "Velocity and Turbulence Measurements in a Combustion Chamber of SI Engine Under Motored and Firing Operations by LDA with Fibre-Optic Pickup," SAE Paper No. 870166.

Omori, S., Iwachido, K., Motomochi, M., and Hirako, O. (1991), "Effect of Intake Port Flow Pattern on the In-Cylinder Tumbling Air Flow in Multi-Valve SI Engines," SAE Paper No. 910477.

Pickering, J.D. and Halliwell, N.A.(1984),"Speckle Photography in Fluid Flows: Signal Recovery with Two-Step Processing," Applied Optics, Vol. 23, No. 8, pp. 1128–1129.

Pitcher, G., Wigley, G., and Jacobsen, G. (1988), "Benefit of Burst Spectrum Analysis for Laser Anemometry Studies in Internal Combustion Egnines," in *4th Int. Symp. On Applications of Laser Anemometry to Fluid Mechanics,* Lisbon.

Raffel, M., Willert, C.E., and Kompenhans, J. (1998), *Particle Image Velocimetry*, Springer, Heidelberg.

Rask, R.B. (1979), "Laser Doppler Anemometer Measurements in an Internal Combustion Engine," SAE Paper No. 790094.

Rask, R.B. (1981), "Comparison of Window, Smoothed-Ensemble, and Cycle-by-Cycle Data Reduction Techniques for Laser Doppler Anemometer Measurements of In-Cylinder Velocity," in *Fluid Mechanics of Combustion Systems*, Ed. by Morel, T., Lohmann, R.P., and Rackley, J.M., pp. 11–20, ASME.

Rask, R.B. (1984), "Laser Doppler Anemometer Measurements of Mean Velocity and Turbulence in Internal Combustion Engines," ICALEO 84, Laser Institute of America, Boston.

Reeves, M., Garner, C.P., Dent, J.C., and Halliwell, N.A. (1994), "Study of Barrel Swirl in a Four-Valve Optical IC Engine Using Particle Image Velocimetry," COMODIA 94.

Reeves, M., Garner, C.P., Dent, J.C., and Halliwell, N.A. (1996), "Particle Image Velocimetry Measurements of In-Cylinder Flow in a Multi-Valve Internal Combustion Engine," *Pro. Instn. Mech. Engrs*, Part D, Vol. 201, pp. 63–70.

Reuss, D.L., Adrain, R.J., Landreth, C.C., French, D.T., and Fansler, T.D. (1989), "Instantaneous Planar Measurements of Velocity and Large-Scale Vorticity and Strain Rate in an Engine Using Particle-Image Velocimetry," SAE Paper No. 890616.

Reuss, D.L., Bardsley, M.E.A., Felton, P.G., Adrain, R.J., and Landreth, C.C. (1990), "Velocity, Vorticity and Strain-Rate Ahead of a Flame Measured in an Engine Using Particle Image Velocimetry," SAE Paper No. 900053.

Reuss, D.L., Kuo, T., Khalighi, B., Haworth, D., and Rosalik, M. (1995), "Particle Image Velocimetry Measurements in a High-Swirl Engine Used for Evaluation of Computational Fluid Dynamics Calculations," SAE Paper No. 952381.

Rouland, E., Trinite, M., Dionnet, F., Floch, A., and Ahmed A. (1997), "Particle Image Velocimetry Measurements in High Tumble Engine for In-Cylinder Flow Structure Analysis," SAE Paper No. 972831.

Shack, D.H. and Reynolds, W.C. (1991), "Application of Particle Tracking Velocimetry to the Cyclic Variability of the Pre-Combustion Flow Field in a Motored Axisymmetric Engine," SAE Paper No. 910475.

Stolz, W., Kohler, J., Lawrenz, W., Meier, F., Bloss, W.H., Maly, R.R., Herweg, R., and Zahn, M. (1992), "Cycle Resolved Flow Field Measurements using a PIV movie technique in a SI engine," SAE Paper No. 922354.

Suen, K.O., Tindal, M.J., Yianneskis, M., and Paul, G.R. (1987), "The Developments of Swirl in the Cylinder of a Direct Injection Diesel Engine," Proc. 2nd Int. Conf. On Laser Anemometry-Advances and Applications.

Swords, M.D., Kalghatgi, G.T., and Watts, A.J. (1982), "An Experimental Study of Ignition and Flame Development in a Spark Ignition Engine," SAE Paper No. 821220.

Tindal, M.J., Cheung, R.S., and Yianneskis, M. (1988), "Velocity Characteristics of Steady Flows Through Engine Inlet Ports and Cylinders," SAE Paper No. 880383.

Tropea, C., Dimaczek, G., Kristensen, J., and Caspersen, C. (1988), "Evaluation of the Burst Spectrum Analyser LDA Signal Processor," in *4th Int. Symp. On Applications of Laser Anemometry to Fluid Mechanics,* Lisbon.

TSI Inc. (1998), *PowerView PIV System,* Product Brochure.

Vogt, A., Baumann, P., Gharib M., and Kompenhans, J. (1996), "Investigations of a Wing Tip Vortex in Air by Means of DPIV," Proc. 19th AIAA Advanced Measurements and Ground Testing Technology Conference.

Westerweel, J., Dabiri. D., and Gharb, M. (1997), "The Effect of a Discrete Window Offset on the Accuracy of Cross-Correlation Analysis of PIV Recordings," *Exp. Fluids*, Vol. 23, pp. 20–28.

Wigley, G. and Hawkins, M.G. (1978), "Three Dimensional Velocity Measurements by Laser Anemometry in a Diesel Engine Under Steady Flow Conditions," SAE Paper No. 780060.

Willert, C.E. and Gharib, M. (1991), "Digital Particle Image Velocimetry," *Exp. Fluids*, Vol. 10, pp. 181–193.

Witze, P.O. and Martin, J.K., (1986), "Cyclic Variation Bias in Spark Ignition Engine Turbulence Measurements," in *Laser Anemometry in Fluid Mechanics*, Ed. by Adrian R. J., et al.

Witze, P.O.(1980), "A Critical Comparison of Hot-Wire Anemometry and Laser Doppler Anemometry for IC Engine Applications," SAE Paper No. 800132.

Witze, P.O., Martin, J.K., and Borgnakke, C. (1984), "Conditionally-Sampled Velocity and Turbulence Measurements in a Spark Ignition Engine," *Combust. Sci. Technol.*, Vol. 36, pp. 301–317.

Zhao, H. (1989), *An Experimental Study of Methane Jet Diffusion Flames*, PhD Thesis, Leeds University.

Chapter 7

Measurement of In-Cylinder Mixture Distribution

Outline

7.1 Introduction

Mixture preparation and mixture distribution in SI engines are important for both fuel economy and exhaust emissions. With port-injected stoichiometric SI engines, accurate control of air/fuel ratio is required so that the three-way catalyst can operate effectively. In addition, the timing of fuel injection is known to influence combustion and pollutant formation through its effect on mixture preparation [e.g., Quader, 1982; Matsushita et al., 1985; Horie et al., 1992].

Furthermore, charge stratification can be employed to improve fuel economy and to reduce pollutant formation. For engines that operate with excess air (lean-burn) or EGR, a rich mixture around the spark plug is desirable to facilitate combustion [Kiyota et al., 1992; Lake et al., 1995; Kume et al., 1996].

To understand how to achieve good mixture distribution with stoichiometric SI engines, and the control of local charge stratification with stratified charge engines, it is important to be able to measure the fuel concentration and air/fuel ratio distribution in the combustion chamber in order to optimize the engine design.

Three optical techniques have been employed for in-cylinder fuel vapor concentration measurements. They are based on Rayleigh scattering, Raman scattering, and laser induced fluorescence. The underlying principles of these optical techniques have been introduced in Chapter 5. Their implementation and applications will be described in detailed in this chapter. The applications of non-optical techniques to in-cylinder fuel concentration measurements are discussed in Sections 6 (Fast FID) and 7 (High-Speed Sampling Valve). Techniques for the detection of in-cylinder liquid fuel distributions will be considered in Chapter 8.

7.2 Fuel Concentration Measurement by Laser Rayleigh Scattering (LRS)

7.2.1 Introduction

As discussed in Chapter 5, Rayleigh scattering is characterized by its simplicity and ease of implementation for gaseous species concentration measurements. It is the strongest of the molecular light scattering processes, and hence it has the potential for two-dimensional measurements. Despite the difficulties associated with the interference of Mie scattering from solid and liquid particles and spurious scattering from walls, laser Rayleigh scattering (LRS) has been successfully used for the measurement of gaseous fuel distribution in motored engines.

7.2.2 Principal of Fuel Concentration Measurement by LRS

The fundamental principle governing Rayleigh scattering has been discussed in Chapter 5. To derive the fuel concentration from the Rayleigh scattering signal, P_{fa}, of a binary fuel/air mixture, a calibration experiment is first performed for a known fuel mole fraction χ_{fo} in the binary mixture at a standard condition P_o and T_o. The resulting Rayleigh scattering signal is P_{fo}. The Rayleigh scattering from air only at the same condition, P_{ao}, is also required. From these three Rayleigh scattering values, the mole fraction of fuel vapor in a fuel/air mixture, χ_f, is given by (Eqs. [5.29] and [5.30]),

$$\chi_f = \chi_{fo} \frac{P_{fa}\dfrac{N_o}{N} - P_{ao}}{P_{fo} - P_{ao}} \tag{7.1a}$$

or

$$\chi_f = \chi_{fo} \frac{P_{fa}\dfrac{p_o}{p}\dfrac{T}{T_o} - P_{ao}}{P_{fo} - P_{ao}} \tag{7.1b}$$

It is clear from Eq. (7.1b) that the mole fraction of fuel, χ_f, can be obtained from the Rayleigh scattering signal, P_{fa}, if calibration signals P_{fo} and P_{ao}, are measured in advance. It is also necessary to know the temperature and pressure in order to derive the mole fraction of fuel from the Rayleigh signal. Generally, the pressure can be obtained with the use of a pressure transducer without any difficulty. However, measuring the temperature itself is a daunting task, which is one of the limiting factors in the application of Rayleigh scattering to IC engines. In some cases, however, with a judicious design of the diagnostic system and a suitable assumption of the flow under study, Rayleigh scattering can be used to obtain the concentration of a binary mixture system without the requirement of information on temperature (to be discussed later).

It can be seen from Eqs. (5.25), (5.26), and (7.1) that, if the values of σ_f and σ_a are close to each other, the Rayleigh signal will be independent of fuel mole fraction. Hence, fuel or simulated fuel should be selected to have a much larger Rayleigh scattering cross section in order to measure fuel concentrations. This can be normally achieved by use of heavy molecules, such as Freon or hydrocarbon fuels.

During the intake stroke, the temperature variation may be assumed to be small, and Eq. (7.1b) may be simplified to

$$\chi_f = \chi_{fo} \frac{P_{fa} \dfrac{p_o}{p} - P_{ao}}{P_{fo} - P_{ao}} \tag{7.2}$$

During the compression stroke, if the port-injected fuel in an SI engine is assumed to have evaporated completely by then, the fuel vapor mole fraction is given by [Zhao and Hiroyasu, 1993]

$$\chi_f = \chi_{fo} \frac{P_{fa} \dfrac{p_o}{p} \dfrac{T}{T_o} \dfrac{V}{V_c} - P_{ao}}{P_{fo} - P_{ao}} \tag{7.3}$$

where V_c is the cylinder volume at the start of the compression stroke, V the cylinder volume at the time of measurement. In order to determine the fuel vapor mole fraction in the combustion chamber during the compression stroke, the in-cylinder pressure, gas temperature, and scattered light intensity are required, as well as the calibration values at standard conditions.

7.2.3 Implementation of LRS

Fig. 7.1 shows schematically a typical setup for LRS experiments. Eq. (5.21) shows that the laser output should be vertically polarized in order to maximize the Rayleigh signal. The scattered light is collected at a right angle to enhance the spatial resolution. For a single-point measurement, the laser beam is focused into the test section so that the beam waist of the laser is viewed by a photomultiplier. For two-dimensional imaging, a laser sheet is introduced into the combustion chamber and imaged at a right angle by an intensified CCD camera. To suppress the spuriously scattered light from optical components, the laser beam is made to pass through pinholes or slits. After passing through the combustion chamber, the laser beam is dumped into a light trap. Synchronization between the laser, the detector, and the data acquisition system is normally achieved via a PC computer and a delay generator. Measurements can be made at a particular crank angle by triggering the system from a shaft encoder. Normalization of the scattered light signal to account for variations in laser power output can be effected by monitoring the laser pulse with a photodiode or power meter.

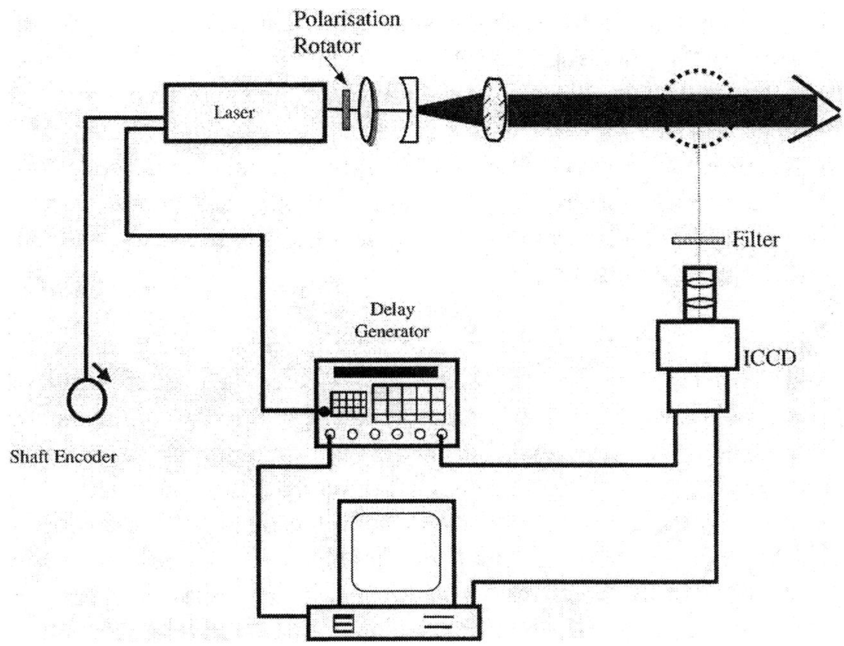

Fig. 7.1 Schematic of the optical setup for laser Rayleigh scattering.

As shown in Eqs. (7.2) and (7.3), if the Rayleigh scattered light intensities from the air and the standard premixed homogeneous air-fuel mixture are determined beforehand under a reference condition, the fuel concentration distribution can be derived from the detected scattered light from the fuel/air mixture, the cylinder volume, and the in-cylinder pressure and temperature. Normally, the calibration can be performed at room temperature and atmospheric pressure. During the intake stroke the gas temperature may be assumed to be constant, which would not cause any appreciable errors in estimating the mixture concentration. However, the gas temperature will vary during the compression stroke and may be estimated from the in-cylinder pressure measurement using the equation of state.

7.2.4 Minimization of Background Light in LRS

Since LRS is elastic, it suffers from interference of background light of the same wavelength or that of the continuous spectrum. This includes Mie scattering from particles, ambient light, scattered light from optics, entrance/exit optical windows, other solid surfaces, and flame radiation.

Since Mie scattering from particles is many orders of magnitude stronger than Rayleigh scattering from molecules, it poses serious interference with Rayleigh scattering measurements. For example, the Mie scattering cross section from a 1 μm particle is about 20 orders of magnitude higher than that of Rayleigh scattering from air. There are two procedures that can be used to reduce particle interference, as discussed in detail by Zhao and Hiroyasu [1993]. The first is the removal of particles in the measurement volume. This can be realized by supplying bottled air or compressed air with an appropriate filtering system for moderate flow rates. To suppress the lubricant-generated droplets, the engine should be operated with special piston rings. After the data have been collected, spikes of Mie scattering signals can be removed manually or automatically by software. However, sub-micron particles cannot be removed by the filtering system and the scattering from them will be more difficult to separate from the Rayleigh scattering from molecules. Post-data processing can become arbitrary, as the effect of the particle is multifaceted, random, and system-dependent.

Flame radiation can be reduced by the use of an interference filter centered at the laser wavelength and a shorter gating time for the detector. Moreover, because Rayleigh scattering from molecules is polarized, the flame luminosity can be halved by employing a polarization filter. However, background light yields the most severe limitation of the LRS technique. The magnitude of the background light determines the lower detection limit of the Rayleigh signal that can be detected by the detector. The lower the background level, the wider the dynamic range of the LRS system becomes. Furthermore, in IC engines the background light changes both spatially and temporally, which makes data reduction extremely complicated. The key to use the LRS technique is essentially the suppression of undesired light from all sources other than the molecules of interest.

To minimize the effect of background light, the optical system should be completely enclosed in a black box. To suppress spurious reflection of the incident laser beam during its propagation, slits, pinholes, and light traps are all extremely effective. The background light should be measured by imaging the scattered light from the gases with a comparatively smaller scattering cross section, such as helium, the scattering cross section of which is about 1.65% that of air. The signal derived in this way includes the noise in the detection system, stray light due to elastic scattering from the surroundings, and the Rayleigh scattering from the helium. As a result, background light intensity can be determined by subtracting the Rayleigh scattering from helium using the known cross sections of helium and air [Fourgette et al., 1986]. It should be noted that the background light must be measured with the same optical setup as that used in the actual experiment.

7.2.5 Application of LRS to IC Engines

The use of LRS to measure fuel concentration was first demonstrated by Arcoumanis et al. [1985] in a model engine. Freon-12 was injected through the center of an axisymmetrically located permanently open valve on the head of a reciprocating model engine motored at 200 rpm. This system was designed to study the fuel injection process in a diesel engine. Freon-12 was chosen due to its high-density ratio relative to air in order to simulate the density ratio between the vaporized diesel fuel and compressed air near the top dead center of the compression stroke in a diesel engine. The high scattering cross section of Freon-12 allowed time and space resolved fuel concentration measurements to be obtained. Because of the permanently open valve arrangement, this model engine was run with intake and exhaust stroke only. Because of the absence of a compression stroke, the complication of the time-varying density field was eliminated to facilitate the interpretation of the Rayleigh signals. To obtain more useful information from the LRS technique, Arcoumanis et al. [1989] and Green et al. [1989] extended the technique to a four-stroke diesel engine with a compression ratio of 3.5. In deriving the fuel concentration from the scattered light intensity, the gas temperature was assumed to be constant. The fuel mole fraction was measured by the LRS technique as a function of piston geometry, induction swirl, and nozzle-geometry.

In a series of papers, Kadota et al. [1990] and Zhao et al. [1990; 1991] described original applications of LRS to study the mixture formation process in a spark ignition engine. The initial work was carried out in a steady flow test rig with continuous injection of Freon-12 into the intake port during the intake stroke. The concentration measurements by LRS were then extended to a mixture in the presence of fuel droplets in the steady flow test rig with intermittent injection of liquid gasoline or Freon-113. They found that vapor concentration showed a large variation both temporally and spatially. The vapor concentration was also found to be dependent on the fuel injection pressure, duration, and fuels. Following the steady state measurements, Zhao et al. applied the LRS technique in a motored SI engine. The time histories of both mean and fluctuating fuel concentrations were obtained for both liquid fuels (gasoline, n-Pentane, n-hexane, and Freon-113) and gaseous fuel (Freon-12) injections. They found that the fuel concentration fluctuations consisted of a temporal variation in a specific cycle and a cyclic variation of the temporal mean concentration over many cycles.

Similar work was also reported by Arcoumanis et al. [1991], where fuel vapor concentration measurement was performed in a port-injected model SI engine. Their results showed that the fuel concentration in the combustion chamber was affected by the injection time and duration.

In combination with a spark plug with fiber optics, LRS was used by Lee and Foster [1995] to measure the fuel concentration in the vicinity of a spark plug gap in a single-cylinder spark ignition engine. The pressure, propane concentration, and initial flame development were measured simultaneously by a pressure transducer, LRS, and the fiber optics in the spark plug. An Argon-ion laser beam was focused near the spark plug with an effective beam waist diameter of approximately 75 μm. The scattered beam was collected at 90 degrees to the direction of beam propagation and passed through a 10 nm bandpass filter before it was detected by a photomultiplier. All inner surfaces of the combustion chamber were painted black so as to minimize the spurious scattered light. Any remaining background light from solid surfaces was subtracted from the scattered light. A threshold level was set at each measurement from an estimated maximum Rayleigh signal. All signals above this voltage level were rejected. Such discrimination against Mie scattering was done by visual inspection of signal amplitude and rise-time. Using this technique, they found that the variation in mixture concentration near the spark plug was not a major contributor to cycle-by-cycle variation in combustion for the homogeneous mixture cases. This probably is a result of the use of gaseous propane in their experiments.

Two-dimensional fuel vapor concentration was measured in a motored SI engine at 200 rpm by Zhao et al. [1993]. An Nd:YAG laser operating at the 2nd harmonic (532 nm) was employed to produce a laser sheet inside the combustion chamber. The scattered light from the measurement plane was detected through the optical window mounted in an extended piston crown by a gated intensified CCD camera. The engine was operated without lubrication to minimize the interference of oil mist and oil vapor with Rayleigh signals. Freon-12 and propane were used to study the mixing process of fuel and air. They found that the fuel vapor concentration distribution was inhomogeneous at the time of spark ignition.

Quantitative images of fuel vapor concentrations were obtained by Espey et al. [1994; 1997], in an evaporating and combusting diesel spray using planar LRS, in an optical single cylinder direct injection diesel engine under motored and firing conditions. Fig. 7.2 shows schematically the optical setup used. A frequency-doubled Nd:YAG laser sheet was introduced into the combustion chamber with its polarization perpendicular to the scattering plane in order to maximize the scattered signal collected in the vertical plane. The scattered images were acquired through a window mounted in place of one of the exhaust valves by a gated intensified CCD camera. A narrow bandpass filter (532 ± 5 nm) was placed in front of the camera to isolate the elastically scattered laser light.

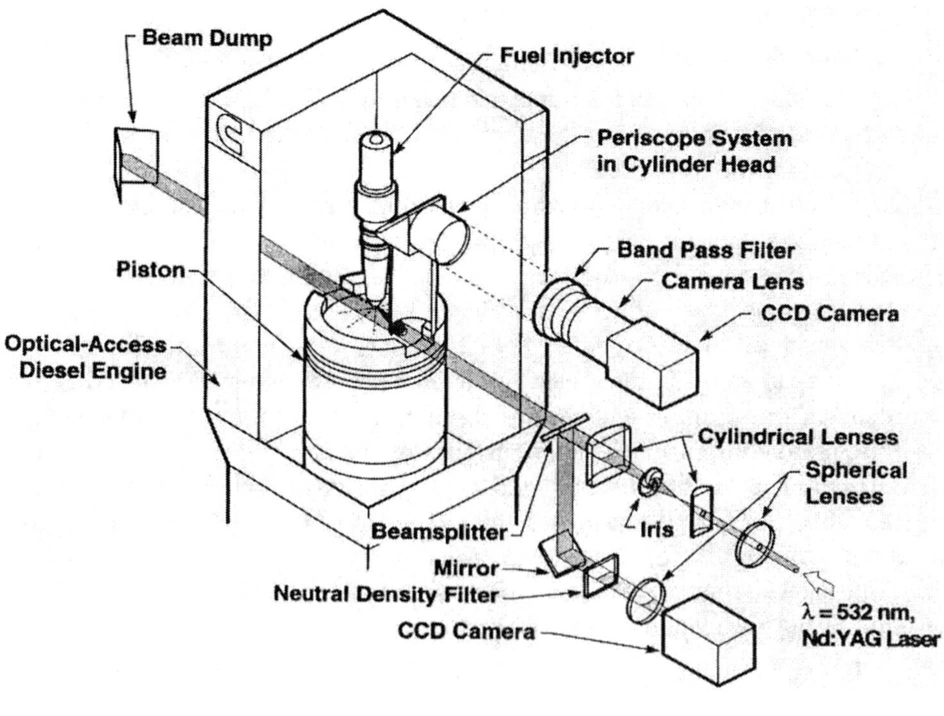

Fig. 7.2 Schematic of the optical setup used by Espey et al. [1997] for LRS measurements.

The fuel was supplied from an N_2-pressurized tank to avoid contamination by a normal fuel supply pump system. A mixture of the diesel reference fuels, heptamethylnonane and n-hexadecane, was used in the engine. The Rayleigh scattering technique was calibrated to obtain the two-dimensional quantitative fuel concentration measurements in the leading portion of the diesel spray, where the fuel was completely evaporated. Equivalence ratio was then deducted from the fuel vapor concentration using an adiabatic mixing assumption [Espey et al., 1997].

Since the Rayleigh scattering signal was directly related to the input laser power, simultaneous acquisition of the laser energy distribution was monitored through a second gated intensified CCD camera by splitting off a small fraction of the laser sheet. By normalization, the shot-to-short variation of laser energy output as well as the non-uniform distribution of the laser sheet, was removed. The interference from Mie scattering of particles was minimized by filtering the engine intake air and the fuel. Although background light was recorded and subtracted from the Rayleigh images, it could not be completely eliminated due to temporal variations in the background light. The uncertainties caused by background noise corresponded to as much as 0.3 equivalence ratio units. Other minor uncertainties included the accuracy of the Rayleigh scattering cross section (2.9%), the camera shot noise (0.7%), and the camera calibration curve (1.5%). Taking into account these uncertainties, Espey et al. estimated that the total uncertainty in their equivalence ratio images was 21% (0.4 units of equivalence ratio) for regions of equivalence ratio of 2 and 17% (0.8 units of equivalence ratio) for regions of equivalence ratio of 4.

Fig. 7.3 shows a sequence of LRS images of the leading portion of one of the diesel jets from an eight-hole injector. The left and right columns correspond to the reacting and non-reacting cases, respectively. These images show not only the effect of ignition and combustion on the diesel fuel jet, but also the limits of this diagnostic for obtaining quantitative fuel concentrations in the reacting jet. Before 4.5°CA after start of injection (ASI), the general appearance of fuel distributions of the reacting and non-reacting jets is very similar. After 4.5°CA ASI, the two sequences start to differ considerably due to the effects of combustion in the reacting jet, as combustion starts at about 4.5°CA ASI. Thus, the Rayleigh scattering signal can be related quantitatively to the fuel concentration and equivalence ratio until significant fuel breakdown takes place. Fuel breakdown due to high-temperature pyrolysis becomes significant after the start of combustion after 4.5°CA ASI, with the Rayleigh images of the reacting jet no longer being quantitative. For the non-reacting case, the Rayleigh scattering signal provides accurate 2-D images of a transient, vaporized fuel jet under realistic diesel conditions.

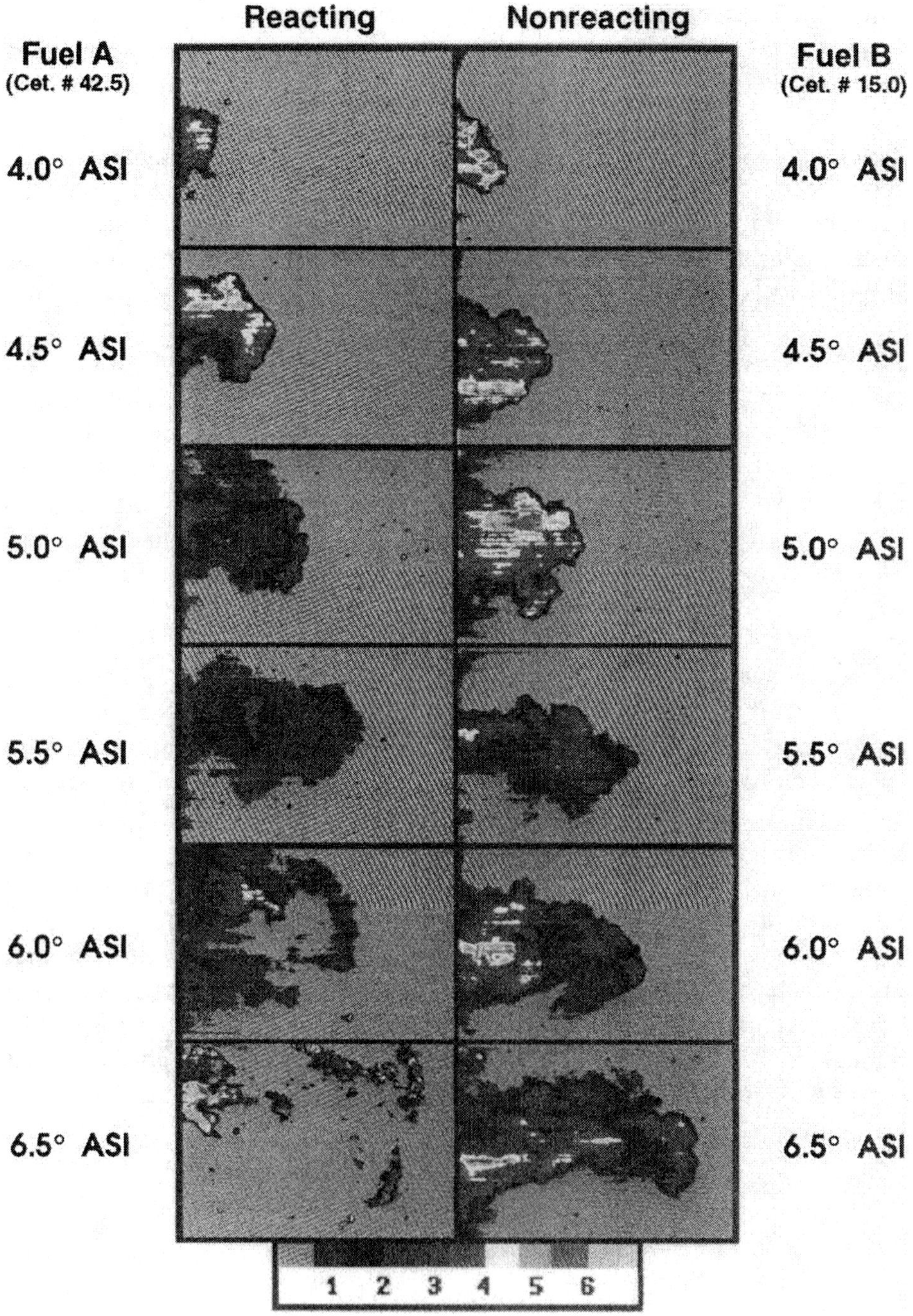

Fig. 7.3 Temporal sequence of equivalence ratio images of a diesel fuel jet [Espey et al., 1997].

7.3 Direct Air/Fuel Ratio Measurement by Spontaneous Raman Scattering (SRS)

7.3.1 Introduction

Raman scattering is the inelastic interaction between the incident light and the electronic-mediated vibrational-rotational modes of molecules, i.e., there is an energy exchange between the photons and scattering molecules. Hence, the scattered photons may gain or lose energy to the scattering molecules, so that the Raman signal has a different wavelength from the incident light. Spectral measurement of Raman scattered light serves to identify the molecular species. The intensity of Raman scattered light is dependent on the number of molecules that are thermally populated at particular rovibronic (rotational-vibrational) energy levels. Since Raman peaks are spectrally well separated from the incident light source, spurious scattered light can be discriminated against. Raman scattering has the advantage that the interaction time is very short (of the order of femtoseconds), so that quenching effects are not encountered. Quenching is a process in which the energy levels are altered in inelastic collisions with other molecules, leading to a large reduction in spectral emissions.

Raman scattering occurs with any laser excitation wavelength. It is capable of providing simultaneous multiple-species concentration measurements. Because spontaneous Raman scattering (SRS) does not involve electronic state transitions, it can measure the concentration of stable species, such as N_2, O_2, CO_2, etc., which are not possible with normal laser induced fluorescence techniques because these species require ultra-UV light to produce fluorescence. However, spontaneous Raman scattering is a very weak process and requires careful signal-to-noise consideration. Background noise from the scattering of unshifted laser wavelength can be a problem.

Spontaneous Raman scattering was first applied to internal combustion engines almost 20 years ago by Johnston [1979] and has recently been accorded renewed interest. With broadband detection, the Raman scattering technique permits simultaneous measurement of the concentrations of all major species of combustion: H_2O, CO_2, CO, N_2, O_2, and fuel, from which the air/fuel ratio and the residual gas content of the charge can be obtained simultaneously. Since air/fuel ratio is obtained from the ratio of two of the majority species, neither laser power fluctuations nor window fouling affect the measurement precision. Since SRS measures the air/fuel ratio directly, it can be applied to situations where residual gas or EGR gas exists in the cylinder, situations in which air/fuel ratio cannot be derived directly from fuel concentration measurement by LIF. Because the Raman signal is shifted from the incident laser wavelength, the interference from light scattered off walls can be largely eliminated and the accuracy of the measurement is typically limited by photon shot-noise. The drawback of the technique is the low signal strength, but this is becoming less of a problem with the commercially available high-power pulsed UV lasers and optical multi-channel detection systems with gated intensified CCD cameras.

7.3.2 Principle of SRS

In general, species concentration is measured from the Raman scattered light intensity in some spectrally integrated portion of the Raman spectrum. The spectral position of the Raman spectrum of a particular species is given by Eq. (5.40),

$$\nu = E_v / hc = \nu_0 \pm \left[\omega_e - 2\omega_e x_e (v+1) - \alpha_e J(J+1)\right] \tag{7.4}$$

where ν = the wave number, [cm^{-1}]
ν_0 = the excitation wave number, [cm^{-1}]
ω_e = the equilibrium vibrational frequency of the molecule, [cm^{-1}]
x_e = the anharmonicity constant of the vibrational mode of the molecule,
α_e = vibration-rotation interaction constant,
v = the vibrational quantum number of the initial energy state,
J = the rotational quantum number of the initial energy state.

For diagnostic purposes, Eq. (7.4) can be simplified to

$$\nu = \nu_0 \pm \nu_{vib} \tag{7.5}$$

The second term in Eq. (7.4) is replaced by the vibrational frequency, ν_{vib} in cm^{-1}, which is species specific. Table 5.1 lists vibrational frequencies [cm^{-1}] for molecules of interest in IC engines. The excitation wave number or frequency, ν_0 represents the incident light frequency in cm^{-1}. Equation (7.4) or Eq. (7.5) defines the spectral positions of the Q-branched vibrational Raman spectrum of a diatomic molecule. The plus and minus signs represent the anti-Stokes lines (frequency up-shifted) and Stokes lines (frequency down-shifted) respectively.

The Raman signal, P_r, from molecular species i, scattered into a solid angle Ω, is given by Eq. (5.42),

$$P_r = \eta_c I_i \Omega V_c N \left(\frac{\partial \sigma}{\partial \Omega}\right) \tag{7.6}$$

where η_c = the detection efficiency of the collection system
I_i = the incident laser intensity, (W/cm^2)
N = the number density of the Raman scattering gas molecule, (cm^{-3})
Ω = the solid angle of collection optics, (sr),
V_c = the collection volume, (cm^3)

The term, $(\partial\sigma/\partial\Omega)_i$, is the differential Raman scattering cross section of the molecule. For diatomic molecules, the scattering cross section can be expressed as the sum of contributions from molecules in different energy states, weighted by the vibrational quantum number v appropriate to each state. For a diatomic molecule, the anti-Stokes and Stokes fundamental ($\Delta v = \pm 1$) vibration Raman scattering cross sections are given by Eq. (5.43) and Eq. (5.44).

$$\left(\frac{\partial \sigma}{\partial \Omega}\right) \approx \left(\nu_0 \pm \nu_{vib}\right)^4 \frac{1}{1 - e^{(-hc\omega_e/kT)}} \tag{7.7}$$

where ν_{vib} is the characteristic vibrational frequency of the molecule, which determines the spectral position of vibrational Raman scattering. The exponential in the above equation indicates that, the higher the gas temperature is, the larger the Raman scattering cross section becomes as the higher energy vibrational states become more populated.

For polyatomic molecules, more complex expressions for the scattering cross section arise, but the conclusions drawn from this equation regarding the temperature dependence of the cross section are still valid.

Table 5.1 lists the Stokes fundamental vibrational Raman cross sections at room temperature for species of interest in IC engines. The cross sections were taken from the tabulation by Inaba and Kobayasi [1972] for an N_2 laser at 337 nm, except those of propane (C_3H_8), which were obtained from Schrotter and Klockner [1979]. The values at 248 nm and 308 nm (excimer lasers), 355 nm and 532 nm (Nd:YAG lasers) were calculated from the original values by scaling according to $\left(\nu_0 - \nu_{vib}\right)^4$. The scattering cross sections listed in Table 5.1 are calculated for the linearly polarized incident and scattered light perpendicular to the scattering plane as defined in Fig. 5.4. These cross sections are valid and unchanged for any scattering angle between the incident beam and scattering observation axes.

When a photon interacts with a molecule, energy exchange may occur between both the rotational and vibrational energy levels and the photon. This leads to rotational sidebands about the vibrational frequencies in the spectrum of scattered light, as shown in Fig. 5.3. The shape of the sidebands is affected by changes in temperature. Since Raman intensity is normally determined by integrating over only a portion of the sidebands, then this change in shape with changes temperature can influence the signal levels. In addition, the Raman scattering cross section is affected by temperature as shown by Eq. (7.7). These temperature effects on the Raman signal can be modeled by a bandwidth factor f(T), accounting for the fraction of the total Raman scattering actually detected. In general, Eq. (7.6) can be written as

$$P_r = \eta I_i N f(T) \tag{7.8}$$

where η is a factor dependent on the Raman cross section, wavelength, geometry, and optical collection efficiency, and is usually determined by experimental calibration. Equation (7.8) shows that the Raman signal is directly related to the concentration of Raman active species N when Raman scattering is detected at the Raman frequency of the species, defined by Eq. (7.5).

The bandwidth factor f(T) depends on the spectral location, shape, and bandwidth of the detection system, and the laser line spectral profile. For example, Eckbreth [1996] showed (Fig. 5.5) that for N_2 Raman scattering measurement, the temperature effect is small at low temperatures with a reasonable detection bandwidth (>2 nm). With a detection bandwidth of about 5 nm, the requirement of knowing temperature can be eliminated in order to obtain quantitative Raman scattering measurements.

In certain cases, the temperature effect may be ignored if the gas temperature is much lower than the characteristic vibrational temperature of the species to be measured, given by

$$T_v = \left(hc\omega_e / k\right)$$

The grouping hc/k has a value of 1.44 $K \cdot cm^{-1}$.

In general, species concentration is measured from the Raman scattered signal in some spectrally integrated portion of the Raman spectrum as defined in Eqs. (7.4) or (7.5). After determining η by calibration from Raman scattering in a known sample by independent measurements, the species concentration N can be calculated from the measured Raman scattering from Eq. (7.8).

7.3.3 Experimental Setup of SRS

Figure 7.4 shows two experimental setups for SRS experiments: one uses a single detector (Fig. 7.4[a]) and the other employs a multi-channel detector (Fig.7.4[b]). In SRS experiments, the selection of an appropriate laser has an important bearing on the signal-to-noise ratio, which is the limiting factor for Raman scattering. As shown in Table 5.1, the Raman cross section has a frequency scaling to the fourth power, i.e., $\left(\nu_0 - \nu_{vib}\right)^4$. Therefore, the laser wavelength greatly affects the signal strength of Raman scattering. However, the strong frequency scaling effect of Raman cross section is mitigated by the decreasing number of photons at higher frequencies (lower wavelength). Here, we compare the use of the 2nd harmonic output (532 nm) to that of the 3rd harmonic output (355 nm) of an Nd:YAG laser. The Raman cross section of N_2 at 355 nm (2.79×10^{-30} cm^2/sr) is about six times larger than that at 532 nm (0.46×10^{-30} cm^2/sr). Typical conversion efficiency from 2nd to 3rd harmonics is about 30%. As the laser wavelength decreases, the number of photons in each laser pulse is reduced by $\left(\nu_{532} - \nu_{vib}\right) / \left(\nu_{355} - \nu_{vib}\right)$ or 64%. Therefore, the total number of Raman photons at 355 nm is increased by 15% for 30% of the laser energy input. That is, if the same Nd:YAG laser is used, the Raman signal will be 15% stronger at its 3rd harmonic output than that obtainable at its 2nd harmonic output.

In recent years, excimer lasers have become more widely used in combustion diagnostics due to their higher power outputs in the UV region. A simple calculation shows that by going from 500 mJ at 532 nm of an Nd:YAG laser to 200 mJ at 248 nm of a KrF laser, the total number of Raman photons will be tripled.

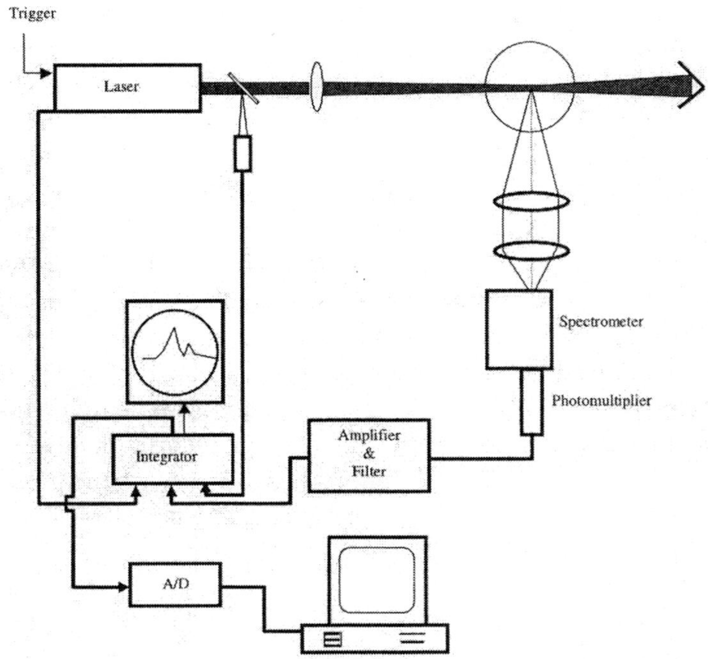

(a) Scanning spectrometer and photomultipler system

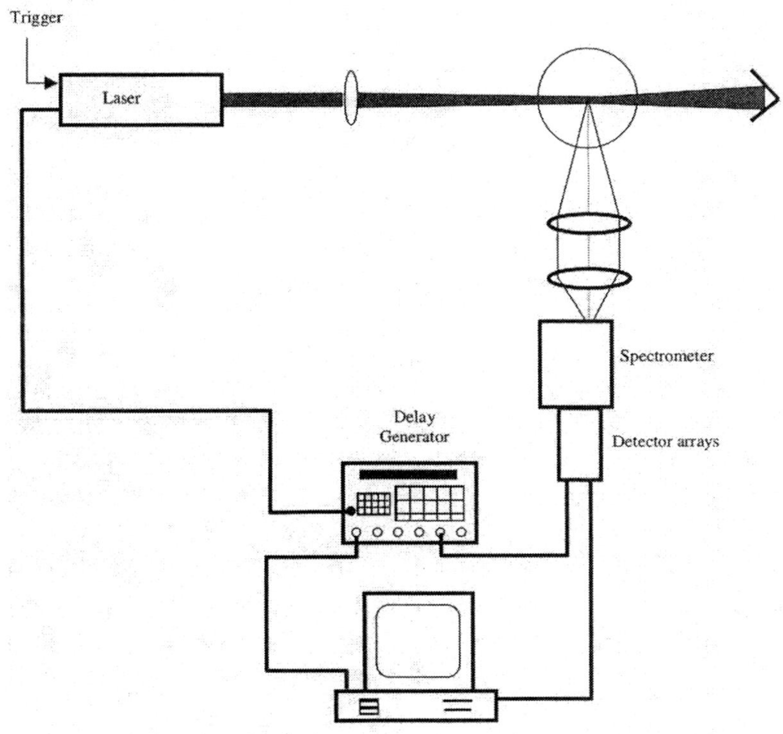

(b) Optical Multi-channel Analyser system

Fig. 7.4 Schematic of experimental setups for SRS experiments.

From the above discussion, lasers in the UV region seem to be particularly attractive for Raman scattering due to the large increase in the strength of the Raman cross section with increasing frequency. However, interference from laser induced fluorescence (LIF) can become more problematical in the UV region and care should be taken to avoid it. KrF lasers (248 nm) normally possess a relatively broad bandwidth on the order of 0.5 nm (80 cm^{-1}) which can excite a number of OH transitions between the ground and third (3, 0) vibrational band of the A-X system, as well as oxygen transitions in the Schumann-Runge system [Andresen et al., 1988; Wehrmeyer et al., 1992]. In this case, the fluorescence from OH and O_2 is the major source of interference with the Raman scattering signals in the absence of heavy hydrocarbons. To avoid the interference, the laser must be tuned to 248.62 nm and its spectral bandwidth narrowed by injection locking [Grünefeld et al., 1994a]. For XeCl at 308 nm, OH LIF from the lower vibrational bands (0, 0) and (1, 0) could pose a problem as well. In the presence of hydrocarbon fuels, extra care must be taken to avoid laser-induced fluorescence from certain fractions of fuels used. Since polarization during the Raman scattering process is preserved, interference from fluorescence emissions can be reduced by detecting the scattered light of the same polarization as the incident laser light.

Another aspect of Raman scattering must be considered when a laser source is selected. The line separation of the resulting Raman spectra is wider when using a longer excitation wavelength than when using a shorter wavelength laser. This can be seen in Table 5.1, where a comparison is made for several species at 248 nm and 308 nm (excimer lasers) and 355 nm and 532 nm (Nd:YAG lasers). This feature may become important in cases where several species are involved and their measurement and resolution are desired.

A pulsed laser is normally preferred in internal combustion engine applications. The short pulse width of a pulsed laser (<100 ns) enables instantaneous measurements to be made. More importantly, the use of a short pulse width combined with fast gated detection systems will suppress the background noise from the luminosity of combustion and will substantially increase the signal-to-noise ratio, which is of particular importance for SRS.

Since the Raman shift is comparatively small, a spectrometer is normally required to achieve the desired spectral resolution. Traditionally, the detection system consists of a monochromator and a photomultiplier as shown in Fig. 7.4(a). The Raman spectrum is scanned with the monochromator across the spectrum of interest. This method is lately being replaced with optical multi-channel detection. In an optical multi-channel detection system, the spectrometer's output is coupled to a photodiode array or a CCD camera, by which all frequency components of the spectrum can be recorded simultaneously, as shown in Fig.7.4(b). The use of an optical multi-channel detection system results in a savings in time and increased signal-to-noise ratio. If a two-dimensional array detector and a specially designed imaging spectrometer, as shown in Fig. 7.5, are used, one can make multi-point SRS measurements along a portion of the laser beam.

Standard spectrographs are optimized for the best spectral resolution at the expense of the vertical spatial resolution. In an imaging spectrograph, the optics are specially designed for

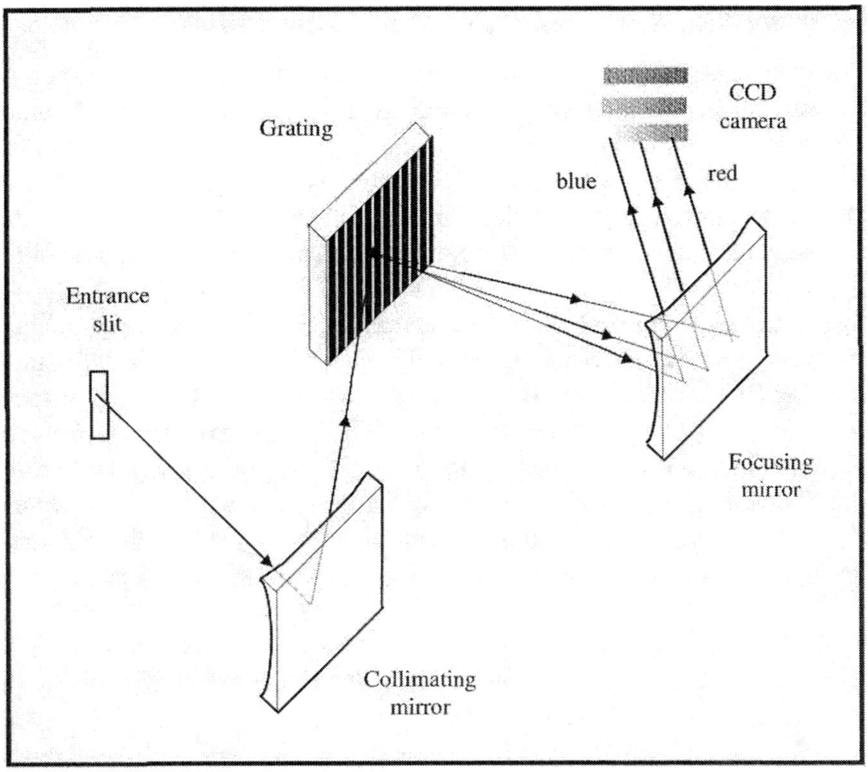

Fig. 7.5 Schematic of an imaging spectrograph.

point to point spectrographic imaging and enable many sources to be simultaneously diffracted and resolved as separate spectra on a CCD camera. As shown in Fig. 7.5, each point on the entrance slit produces a spectrum, and hence a series of spectra will be formed on the CCD detector. Thus, the horizontal axis in the CCD output corresponds to wavelength and contains information about the Raman spectrum from a single point, whereas the vertical direction of the image contains the spatial information along the entrance slit. If the CCD camera is intensified and gated, the Raman signal-to-noise ratio could further be increased. Scattered light is collected by a pair of achromatic lenses, the second of which should be selected to match the collection solid angle of the spectrograph. Sometimes multiple optical fibers replace the lenses to acquire spectra from different sources. Light at the incident laser wavelength can be minimized by placing a laser mirror before the entrance to the spectrograph. The laser mirror reflects light at the laser wavelength but transmits light at other wavelengths. There are basically two types of laser mirrors: the holographic notch filter and the dielectric notch filter. The holographic notch filter is preferred to the dielectric notch filter because it has much a higher optical density with a much narrower bandwidth and sharp edges, without the unwanted secondary leakage bands found in some dielectric notch filters.

7.3.4 Background Radiation and Signal-to-Noise Ratio

Because Raman signals are very weak, one must consider the background radiation and other sources of noises, which may interfere with the desired signal and render the Raman scattering useless. There are a number of sources that may contribute to the undesired background radiation. They include Rayleigh scattering, spurious scattering from optical windows and surfaces, Mie scattering from large particles, flame luminosity, soot incandescence, fluorescence from excited molecules, detector dark current, electrical noise, and detector shot noise.

The first three, Rayleigh, spurious, and Mie scattering, being of the same frequency as the incident laser beam, and thus spectrally separated from the desired signal, can be minimized by using proper filters (e.g., a dielectric laser mirror or holographic notch filter) or spectrographs used for the selection of the desired Raman signals. The interference from the broadband background radiation is particularly serious for SRS. For this reason, most of the applications of SRS measurements are restricted to intake and compression strokes or to premixed combustion. The fluorescence can be a serious problem, particularly if UV Raman scattering is employed. Proper selection of the laser system will eliminate this problem.

The detector dark current and electrical noise can be minimized by using coolers. The detector shot noise arises from the random process of photoelectron emission and can be the limiting factor for the detection of the Raman signal.

For a Poisson distribution of photoelectron arrival times, the number of signal photoelectrons equals the total count minus the background count, i.e., $N_s = N_{total} - N_b$. The Poisson distribution has a mean value of N_{total} with a standard deviation of $\sqrt{N_{total}}$. The signal-to-noise ratio may be written as

$$\frac{S}{N} = \frac{N_s}{\sqrt{N_s + N_b}} = \frac{\sqrt{N_s}}{\sqrt{1 + \frac{N_b}{N_s}}} \tag{7.9}$$

where N_s = the number of signal photoelectrons collected
N_b = the number of background photoelectrons

In the case of multiple pulse operation, the signal-to-noise ratio becomes:

$$\frac{S}{N} = \frac{n \cdot N_s}{\sqrt{n \cdot (N_s + N_b)}} = \frac{\sqrt{n \cdot N_s}}{\sqrt{1 + \frac{N_b}{N_s}}} \tag{7.10}$$

where n is the number of pulses used to obtain the averaged signal reading.

The experimental error is given by the reciprocal of S/N, that is

$$\text{Error}(\%) = \frac{\sqrt{1+\frac{N_b}{N_s}}}{\sqrt{n \cdot N_s}} \times 100\% \tag{7.11}$$

Equations (7.10) and (7.11) indicate that by increasing the number of averaged pulses, one can obtain a larger signal-to-noise ratio, or a lower measurement error.

In the following example, we assume that one is able to deliver 100 mJ of 532 nm laser radiation to the probe volume. The Raman signal, P, is obtained from Eq. (7.8) using the scattering cross sections given in Table 5.1. The number of photons collected is obtained by dividing the scattered light energy, P, b$\left[hc\left(\nu_0 - \nu_{vib}\right)\right]$. Then, the number of photoelectrons is estimated by multiplying the number of collected photons by system collection efficiency and detector quantum efficiency. If one desires a 0.5 mm spatial resolution, and employs an f#5 collection system, the solid angle of the collection optics, Ω, becomes 3.1×10^{-2} sr. The detector is assumed to have a 20% overall efficiency. The molecular number density 7.34×10^{19} cm^{-3} is used, corresponding to the value in a 10:1 compression ratio engine operating at part load (0.4 bar intake manifold pressure) and a 400 K charge temperature at the bottom dead center before compression. The amount of residual gas fraction used in the calculation is 10%. The calculated signal-to-noise ratios with different excitation wavelengths are presented in Table 7.1.

The values given in Table 7.1 are the shot-noise limited signal-to-noise ratios. In practice, the signal-to-noise ratios will be lower due to the background radiation, detector dark current, and electrical noise.

7.3.5 Data Analysis and Calibration

The Raman spectrum for a gas mixture consists of a number of spectral spikes of the scattering species. Fig. 7.6 shows a schematic representation of the Raman spectrum of a propane/air mixture. The spectral position of Raman scattering from each species is determined by the characteristic vibrational frequency of the scattering molecule, ν_{vib} relative to the laser wavelength, ν_0. The Raman signal strength is affected by the Raman cross section, the system calibration factor, concentration of the scattering species, the incident laser power, and a spectral function, as given by Eq. (7.8).

The Raman signal of each species is determined experimentally from the Raman spectrograph by integrating over its Raman spectral region. In Fig. 7.6, note that there are three Raman spectral regions associated with the fuel (C_3H_8), which result from its three vibrational modes: C–H, C–C, and CH_3. In this case, the integration over the large C–H region may be used to obtain the Raman signal of propane, as the other two are very small.

Table.7.1 Estimated signal levels for a mixture of gases at different excitation wavelengths

Species	Vibrational frequency (cm^{-1})	Volume fraction	Raman signal photoelectrons at different laser wavelengths				Signal-to-noise ratio at different laser wavelengths			
			248 nm	308 nm	355 nm	532 nm	248 nm	308 nm	355 nm	532 nm
N_2	2331	0.755	33104	16445	10169	2512	182	128	101	50
O_2	1556	0.182	10189	5107	3242	856	101	71	57	29
CO_2	1388	0.012	611	306	195	52	25	17	14	7
	1285		450	226	144	39	21	15	12	6
H_2O	3657	0.015	1558	735	445	96	39	27	21	10
C_3H_8	2890	0.036	10697	5172	3180	747	103	72	56	27
	1465		4286	2156	1373	364	65	46	37	19
	857		2525	1281	834	231	50	35	29	15

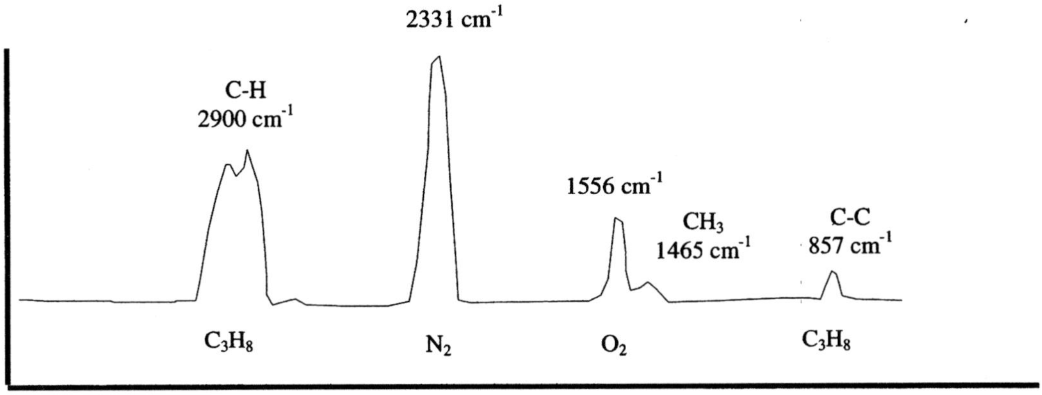

Fig. 7.6 Schematic representation of the Raman spectrum for a mixture of propane and air.

The mole fraction χ_i of each species can be determined from Raman scattering by

$$\chi_i = \frac{S_i / R_i}{\sum_i S_i / R_i} \tag{7.12}$$

where S_i is the Raman signal for species i obtained by integrating the rotational sidebands, which are included in each spectral peak and are not resolved in Fig. 7.6, as well as the central vibrational peak on the spectrum from the experiment. R_i is the relative system response factor obtained from a calibration procedure. From the mole fractions of oxygen/nitrogen and fuel, the air/fuel ratio can be determined. Information about the mole fractions of other major species can be used to measure the residual gas or EGR gas concentration.

Calibration may be performed in situ or in a calibration cell for various gases in the calibration mixture. The calibration signal at known conditions is first recorded for a calibration mixture of pure air, where the Raman signals from N_2 and O_2 are obtained by integrating over the rotational sidebands, as well as the central vibration peak. The N_2 system response, R_{N2}, is assigned a value of unity, from which the relative O_2 system response, R_{O2}, can be determined. Then a second calibration signal is recorded by adding fuel to the calibration mixture to obtain the relative system response R_{fuel}. Similarly, the relative system response factors are obtained for other species. When there is an overlap between the Raman spectral lines of different species, a more elaborate procedure is required [Miles and Dilligan, 1996]. For accurate calibration, the effect of temperature on the Raman scattering should be taken into account since temperature affects both the Raman spectral width and its cross section.

7.3.6 Applications of SRS to IC Engines

Using the SRS technique, Johnston [1979] obtained air/fuel ratio measurements in the Sandia's optical engine. The experiment was carried out in a motored, non-firing, direct injection stratified charge engine with gaseous propane fuel. The optical setup used is given in Table 7.2. Because of the low laser power output (5 W argon-ion laser) available, ensemble-averaged equivalence ratios over 100 cycles were measured from the ratio of Raman signals of C_3H_8 to that of N_2 at different crank angles at different radial positions. Large spatial variations in the value of air/fuel ratios were observed during and after fuel injection. The same experimental setup was later used to measure the air/fuel ratio near the spark plug electrode in the same engine by Johnston [1980].

Because of the problems associated with the small Raman scattering cross section, interference from other light scattering processes or flame luminosity can be serious for SRS. This is a problem particularly for the broad band background radiation. For these reasons, little progress was made in the development of SRS in the 1980s, except for the work reported by Sawerysyn et al. [1986]. In their work, Sawerysyn et al. used an Nd:YAG laser (532 nm) and a spatially resolved optical multi-channel detector to detect Raman scattering from different species simultaneously inside an SI engine. Spectral images were obtained for different crank angles, from which the spatial distribution of species N_2, H_2O, CO, CO_2, O_2, and fuel (C_3H_8) could be observed, along 15 mm of the laser beam with a spatial resolution of 1 mm. But the authors concluded that it was not possible to obtain quantitative single shot measurements at all because of the weak signals and the insensitive TV camera used in their experiments. Interference from oil fluorescence was observed after ignition.

Advances in lasers and optical multi-channel detectors in the last decade, however, have prompted several researchers to reinvestigate SRS as a useful technique for engine studies. In particular, the availability of intensified cooled CCD cameras, more powerful UV lasers, and the increasing interest in the fuel distribution in the cylinder of IC engines have lead to new developments in multi-point multi-species SRS measurements.

Heinze and Schmidt [1989] measured the air/fuel ratio distribution in a spray injected into a high-pressure and high-temperature chamber. Using a XeCl excimer laser (308 nm) and an optical multi-channel detector, they were able to determine air/fuel ratios from the recorded Raman spectrum of n-heptane and nitrogen. The presence of laser induced fluorescence from PAH due to thermal cracking of n-heptane introduced additional peaks into the spectrum. These PAH (polycyclic aromatic hydrocarbons) peaks in the recorded spectrum had to be removed, together with unsuppressed Mie scattering from droplets, before the air/fuel ratio could be determined from the spectral peaks of nitrogen and n-heptane. The authors claimed that the accuracy in the measured equivalence ratio was about ±8%. Quantitative, ensemble-averaged two-dimensional images of the air/fuel ratio in the spray were then obtained through a series of single-point SRS measurements, each of which was averaged over 50 measurements. The results showed that faster propagating fuel in the inner spray and at the spray tip played an important role in mixture formation.

Table 7.2 Summary of SRS measurements in IC engines reported in the literature

Laser source	Detection system	Spatial and temporal resolutions	Application and accuracy	References
Argon ion laser 488 nm 5.2 watts	F/1.5 collection optics A scanning spectrometer with a photomultiplier tube 1° CA gatewidth	Single point measurement Ensemble-averaged over 100 cycles	Air/fuel ratio measurements during compression stoke Gaseous propane fuel Motored engine (465 rpm) Error in equivalence ratios less than ±3–7%	Johnston, 1979; Johnston, 1980
Nd:YAG laser 532 nm 500 mJ (15 ns)	An optical multi-channel detector system with an intensified TV camera 40 ns gatewidth Schott OG 550 filter and laser mirror at 532 nm	1-D imaging over 15 mm Single cycle	Charge composition during compression stroke Gaseous propane fuel Firing engine (1660 rpm), Qualitative	Sawerysyn et al., 1986
Excimer laser 308 nm 100mJ (28 ns)	F2.4 collection optics, An optical multi-channel detector system with a 1-D photo-diode array, Schott WG 335 filter.	Single point measurement, ($0.1 \times 1 \times 1$ mm^3) Ensemble-averaged over 50 injections.	Air/fuel ratio in a fuel jet in a High T and P chamber, Liquid heptane fuel, Shot-noise limited accuracy: ±8%.	Heinze & Schmidt, 1989.
Non-seeded and seeded Nd:YAG lasers 532 nm 120 mJ (170 ns)	F3.2 collection optics An optical multi-channel detector system with a back-illuminated CCD Holographic notch filter	1-D spatially resolved measurements: (0.27 mm in dia. × 0.91 mm long) Ensemble-averaged as well as single cycle measurements	Mean concentration, fluctuation, and scalar length scales in charge composition Gaseous propane fuel Error in A/F ratios ≈ 1%.	Miles and Dilligan, 1996; Miles and Hinze, 1998; Hinze and Miles, 1999
Narrow band tunable excimer laser at 248.6 nm 150mJ (20 ns) linearly polarized	An optical multi-channel detector system with an gated ICCD camera A liquid butyl acetate long pass filter at 250–255 nm Polarized detection	Spatially averaged single cycle measurements: ($0.5 \times 10 \times 14$ mm^3) 1-D spatially resolved measurements: ($0.5 \times 1 \times 4$ mm^3) ensemble-averaged over 1000 cycles	Air/fuel ratios and residual gas contents during intake and compression strokes Isooctane fuel Overall error in A/F ratios less than 4%	Grünefeld et al., 1994a;1994b; 1995; Knapp et al., 1996; 1997

Measurements of major species in the Sandia transparent research engine were reported by Miles and Dilligan [1996] and Miles and Hinze [1998]. The experimental setup is given in Table 7.2. An imaging spectrograph detection system was employed in their study, which permitted one-dimensional line imaging of multi-species. The spectrograph entrance slit was orientated so that the slit was parallel to the laser beam. The light scattered from a portion of the laser beam was focused onto the entrance slit. In turn, the spectrograph formed multiple, spectrally dispersed images of the entrance slit onto a two-dimensional CCD array at the spectrograph exit. Thus, the vertical axis of the image on the CCD detector was the spatial extent of the measurement along the incident laser beam. It imaged several millimeters along the laser beam. In the same image, the horizontal axis corresponded to wavelength, and it contained information on the species concentration at a given spatial location.

The use of 532 nm as the excitation wavelength allowed the CO_2 spectral peak to be separated from the O_2 peak, and hence CO_2 content could be measured simultaneously with other species. As Table 5.1 shows, however, the employment of the longer excitation wavelength at 532 nm results in significant reduction in Raman signals. In order to deliver sufficient laser energy to the measurement volume, optical windows were arranged in a way that the laser beam passed them at Brewster's angle, at which the vertically polarized laser beam experienced little reflectivity. In addition, the laser beam expanded inside the window by the ratio of the reflective indices of window material and the surrounding gases, thereby allowing higher laser energy delivery. More importantly, the laser power was reduced by expanding the laser pulse to 170 ns by means of a pulse stretcher [Miles and Dilligan,1996].

When the same SRS apparatus was applied to a production-type single cylinder engine, the flush mounting of the windows precluded the use of Brewster windows. In order to reduce the risk of breaking windows, an injection seeded Nd:YAG laser was used [Hinze and Miles, 1999]. The seeded Nd:YAG laser offered a much better temporal pulse profile, resulting in fewer spikes in instantaneous power. The seeded laser also produced a better spatial profile of the beam, so there were fewer hot-spots in the beam that could damage the windows. The absence of Brewster windows called for a larger beam diameter at the windows, which was achieved at the expense of tighter focusing at the measurement location (0.27 mm in diameter). The smaller laser beam diameter improved the spatial resolution, but it resulted in occasional laser ignition at the focus due to the presence of particles in the charge. The interference of laser ignition on the recorded spectra had to be identified and removed before quantitative results could be obtained.

Mole fractions of H_2O, CO_2, O_2, and fuel (C_3H_8) were measured from the start of intake to the end of compression, from which the air/fuel ratio and the mole fraction of residual gases were calculated. Single-shot and cycle-resolved measurements were obtained from both the simple pancake type research engine [Miles and Hinze, 1998] and a four-valve DOHV engine, as well as from ensemble-averaged measurements.

SRS has been extensively used for measurements of in-cylinder species at Laser Laboratorium Göttingen. Their optical setup consisted of a tunable excimer laser, an imaging spectrometer, and a gated intensified CCD camera. The laser source was a specially designed tunable narrowband KrF excimer laser. In order to avoid LIPF interference of OH and O_2 around 248 nm, the laser output was tuned to 248.62 nm. A combination of linearly polarized laser beam and a half-wave plate placed in front of the spectrometer further suppressed the remaining O_2 fluorescence, because fluorescence was depolarized and vibrational Raman scattering was highly polarized. To attenuate Rayleigh and Mie scattering in their experiments, a long-pass liquid butyl acetate filter was used, which had a steep cut-off edge between 250 nm and 255 nm. This allowed complete recordings of Stokes Raman spectra of major species.

Three different optical setups were used in their experiments. The initial Raman scattering experiment was demonstrated by Grünefeld et al. [1994a; 1994b] in a four-cylinder SI engine with elongated pistons. The scattered light from the whole laser beam in the cylinder was detected through an optical window mounted in the piston, as shown in Fig. 7.7. The use of the large piston window permitted more scattered light to be collected, in order to increase the signal-to-noise ratio. The Raman signal was averaged over the whole probe volume (about 1.4 cm^3 in this case).

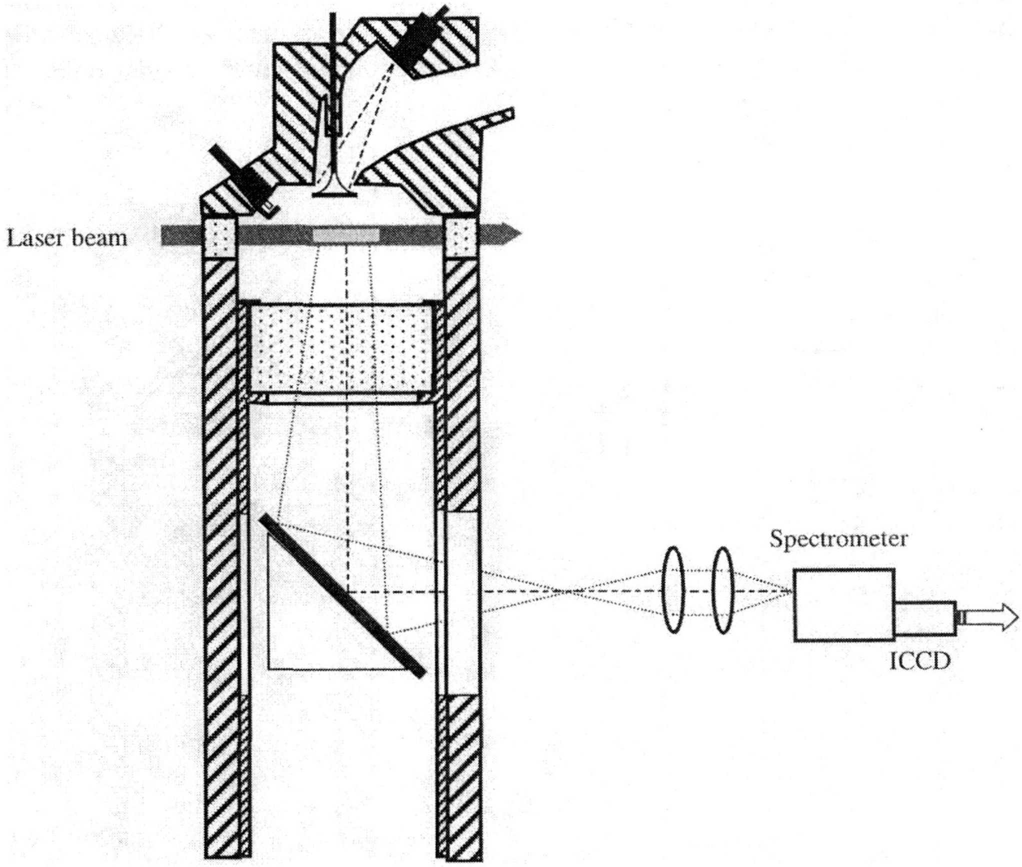

Fig. 7.7 SRS measurements in an engine with a transparent piston.

In order to apply the SRS technique to a production engine without extended piston, the back-scattered Raman light was collected through the laser entrance window [Grünefeld et al., 1995], as shown in Fig. 7.8. Since the Raman scattering cross section is independent of scattering angle (Eq. [7.7]), the Raman signal is not affected by the direction of detection. This setup required only two small windows placed in an aluminum plate between the cylinder head and cylinder block. One added advantage of the backscattering setup was that the signal was found to be less affected by oil or any other deposit on the window, because the intense laser beam burned out the deposit and kept a portion of the window clean. This was a particularly useful feature when they carried out experiments during cold start period when large amounts of fuel and combustion products condensed on the window. With the backscattering setup, however, Raman signals were averaged over the whole laser beam. In their experiments, a probe volume of $2 \times 20 \times 80$ mm^3, corresponding to about 5% of the charge volume at the time of measurement, was used. With this setup, Grünefeld et al. were able to measure the in-cylinder composition for the first 100 consecutive cycles at a fixed crank angle during cold start and warm-up periods.

Knapp et al. [1997] performed both spatially-averaged single-cycle and 1-D spatially resolved ensemble-averaged SRS measurements in a four-cylinder engine with a compression ratio of 10, instead of 6.5 as used in the previous studies. The scattered light was detected at a right angle through the third window mounted in a plate between the cylinder head and the engine

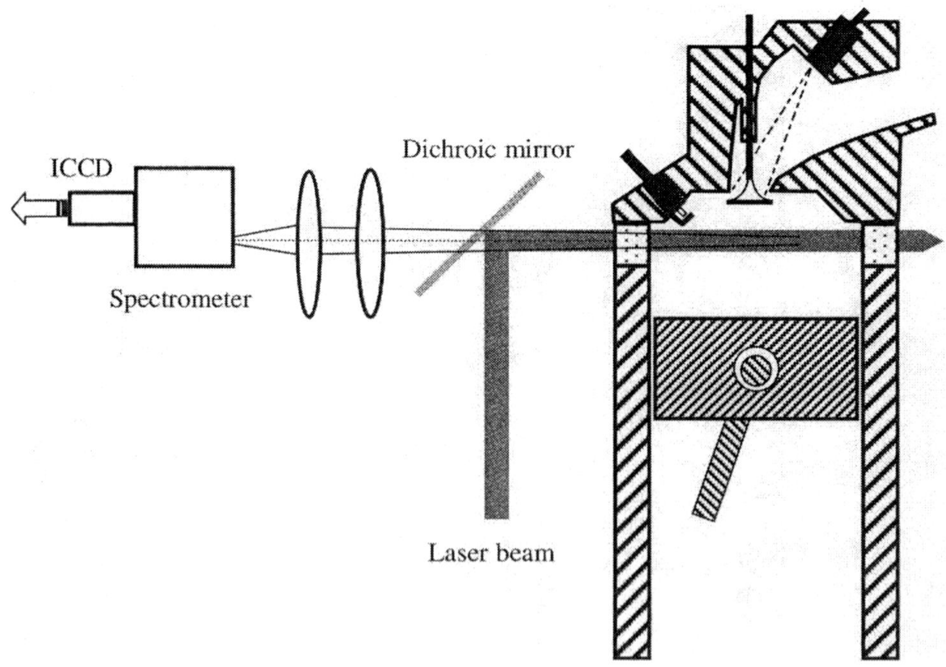

Fig. 7.8 SRS measurements in an engine with backward scattering.

block, as shown in Fig. 7.9. The spatially averaged ($0.5 \times 10 \times 14$ mm^3) measurements were used to detect the gas composition in the residual gas content under different engine operating conditions. For 1-D spatially resolved measurements, a plano-concave lens with 7 mm radius of curvature was used as the detection window in order to increase the field of view to compensate for the lower signal level encountered. The collection optics imaged a section of 40 mm along the laser beam (4 mm × 1 mm) onto the entrance slit of the spectrograph. The signal-to-noise level was further enhanced by averaging over 1000 cycles for the 1-D spatially resolved SRS measurements.

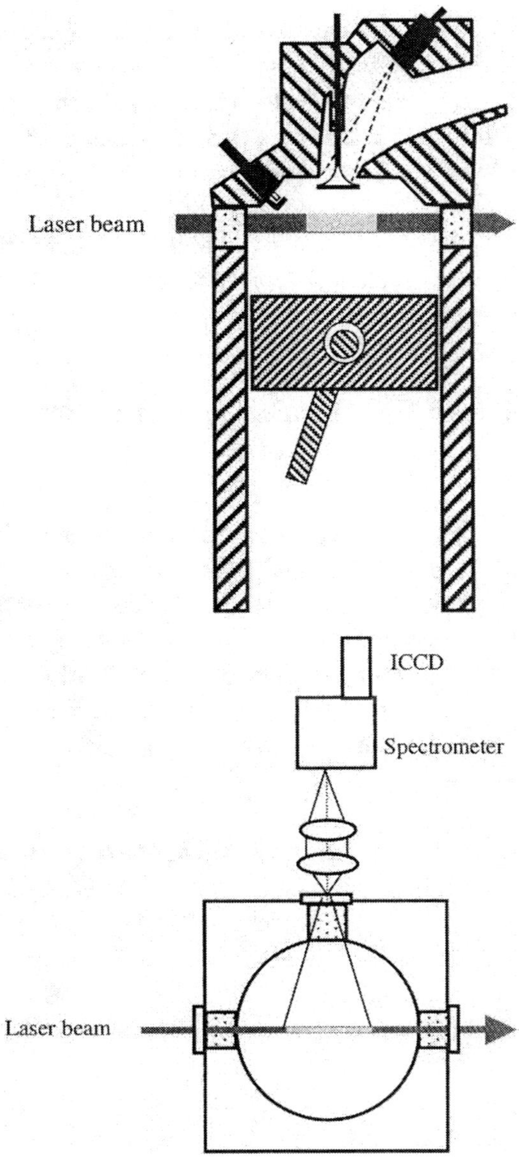

Fig. 7.9 Cycle resolved SRS measurements in a production engine.

In addition, Knapp et al. [1996] obtained air/fuel ratio measurements and NO LIF signals from separately recorded Raman spectrum and NO LIF signals. These measurements were carried out by firing the laser twice in the same engine cycle; one time before ignition and the other at 90°CA ATDC. An MgF_2 plate placed in front of the excimer laser produced a linearly polarized light perpendicular to the scattering plane. The scattered light passed through a spectrometer and then was separated into two orthogonally polarized components by a Glan prism, which were then detected by two gated ICCD cameras respectively. Since Raman scattering maintains its polarization, but fluorescence emissions are strongly depolarized, the horizontally polarized component of the scattered light is due to the fluorescence emission. The vertically polarized component of the scattered light contains both Raman scattering and fluorescence emission. Therefore, Raman scattering signals of fuel and nitrogen, and NO fluorescence could be determined from the two polarization components of the scattered light. The SRS results were confirmed to be better than 6%. Calibration measurements conducted by Knapp et al. also showed that the ratio of fuel Raman signal to that of nitrogen varied little with temperature up to 1100 K, so that the temperature effect could be ignored.

When the measured NO LIF signals were plotted against air/fuel ratios, they did not see any correlation between the two parameters in the same engine cycle. This was ascribed by authors to the small measuring volume involved.

7.4 Two-Dimensional Fuel Distribution Measurement by PLIF

7.4.1 Introduction

Laser induced fluorescence (LIF) is an electronic absorption and emission process that produces relatively strong signal intensity with high spatial resolution. The fluorescence signal is normally red-shifted from the excitation wavelength, hence it is easier to separate from stray scattered light. Fluorescence intensity is directly proportional to the molecular density, and it can be used to measure the concentration. The strong fluorescence signals allow two-dimensional imaging of species using a laser sheet. However, quenching effects can reduce fluorescence intensities, especially by oxygen at elevated pressures, in a way that depends on pressure and temperature, thereby making quantitative measurements in an engine difficult.

The early development of the laser induced fluorescence (LIF) method was driven by single-point measurements, but the applications of LIF to IC engines are dominated by multi-point planar imaging. Planar laser induced fluorescence (PLIF) may be viewed as a modern form of flow visualization. In common with methods such as Schlieren and shadowgraphy, PLIF is extremely useful for qualitative or semi-quantitative characterization of complex flow fields, providing spatially resolved information in a plane rather than integrated over a line-of-sight. The wide application of PLIF in IC engines is due to the strength of the fluorescence process, relative to its primary competitors, Rayleigh and Raman scattering, and to its species specificity. In the following sections, we will describe the basic physical processes involved in an LIF experiment. We will examine how LIF is adapted to IC engines, and how to select appropriate dopants, which influence the choice of a laser source and the interpretation of LIF signals. Then, experimental approaches to PLIF measurements will be considered. Finally, the utilization of PLIF for the instantaneous monitoring of in-cylinder fuel distribution will be described.

7.4.2 Principle

As discussed in Chapter 5, there are two regimes in which LIF can operate, depending on the relative laser intensity. If the laser intensity is strong enough, laser induced fluorescence will become independent of both the laser irradiance and the quenching. However, in engine applications, it is extremely difficult to reach the saturation regime due to the high-pressure environment inside the combustion chamber. Instead, LIF is in the so-called linear regime and its signal level is given by Eq. (5.61)

$$P_{flu} = \eta_c \Omega V_c f_1(T) \chi_{21} N I_{21} B_{21} \frac{A_{21}}{A_{21} + Q_{21}} \tag{7.13}$$

Eq. (7.13) indicates that the fluorescence signal is directly proportional to the mole fraction of the species of interest, χ_m, which forms the basis of LIF for species concentration measurements. Also note that in the linear regime, one normally needs to evaluate the quenching rate constant Q_{21} in order to make quantitative measurements of the species of interest. Because the quenching rate is sensitive to pressure, temperature, and composition, it is very difficult to make quantitative measurements of LIF in a higher-pressure and -temperature combustion environment.

To perform quantitative fluorescence measurements on a given species, several requirements must be satisfied. The first of these is that the molecule must have a known absorption and emission spectrum. Second, the molecule must have an absorption wavelength accessible to a laser source. Third, the rate of radiative decay of the excited state, which affects the fluorescence power directly, must be known. Fourth, excited state losses due to collisions, photo-ionization, and pre-dissociation, have to be taken into account. Some aspects of these requirements will be considered later. For a detailed discussion of quantitative LIF measurements for combustion diagnostics, readers are referred to the book by Eckbreth [1996] and the review paper by Kohse-Höinghaus [1994].

7.4.3 Adaptation of LIF for In-Cylinder Fuel Distribution Measurements

Three strategies can be used for fuel concentration measurements by LIF in IC engines: natural fluorescence from the fuel; fluorescence from a dopant molecule of matching properties to the fuel; and fluorescence from exciplex forming dopants. The first two strategies are used for fuels existing in either liquid form or gaseous form. A third approach is employed to study liquid fuel droplets and fuel vapor concentration simultaneously, which shall be dealt with in Chapter 8.

Some of the practical fuels may fluoresce when excited by UV light. The use of laser induced fluorescence from the fuel itself simplifies the experiments, but the interpretation of the results could be difficult. Because of the variation in fuel composition from batch to batch, the fluorescence signal may not be reproducible due to the lack of information on the photochemistry of the fluorescent component. Only some of the high-boiling point aromatics in the fuel are likely to fluoresce, which may not be representative of the fuel as a whole. Fluorescence has been observed from some hydrocarbon compounds, e.g., isooctane [Andresen et al., 1990] and n-heptane [Arnold et al., 1992]. This fluorescence may be due to impurities [Felton, 1996], and hence is subject to errors caused by variations in impurities from batch to batch. The natural fluorescence from fuel, however, can be used effectively in some applications [Fansler et al., 1995], particularly in liquid fuels [Felton et al., 1995].

Because the majority of fuels used in IC engines do not readily fluoresce with laser excitation in the visible and UV region, dopants have been used to mark the fuel concentration in the cylinder.

7.4.4 Selection of Fluorescence Dopants

Fluorescence tracers are used for indirect visualization of fuel concentrations in the cylinder. To obtain quantitative fuel distribution, several properties of the dopant must be considered:

- absorption of the available laser wavelength
- satisfactory fluorescence yield
- sufficiently low quenching properties
- non-toxic, stable, and soluble in fuel
- matching vaporization properties

In addition, the fluorescence should be sufficiently red-shifted to facilitate separation of elastically scattered laser light from the LIF signal.

The first requirement depends on the laser available. Two types of pulsed lasers are commonly employed, excimer lasers and Nd:YAG lasers with 3rd and 4th harmonic output, because of their high power outputs and short wavelengths. The second and third features directly affect the fluorescence signal level, delectability, and accuracy. The quenching rate of the fluorescence signal reduces the fluorescence signal level. Quenching can be particularly high by oxygen at high pressures in the cylinder, and can even prevent the use of air. Nitrogen sometimes has to be used to replace air, as in the use of laser induced exciplex fluorescence measurements. However, in one case, the quenching properties of a dopant by oxygen have been used to measure fuel/air ratios directly [Reboux et al., 1994; Reboux et al., 1996]; this will be discussed in Section 7.5. The last two properties will affect the accuracy of the measurement. In order to use the dopant as a tracer for the fuel, it must be soluble in the fuel. The matching boiling point helps to mimic the fuel vaporization process.

Based on the above requirements, a number of investigations have been conducted to find suitable dopants for SI engines. The required absorption wavelength in the visible and UV region restricts the dopants to two main categories: aromatic and carbonyl compounds. The fluorescence dopants used for LIF measurements are summarized in Table 7.3. The data of relative LIF intensity and the pressure and temperature effects in Table 7.3 were adapted from Fujikawa et al. [1997]. As shown in Table 7.3, aromatic compounds (e.g., Toluene) are strongly quenched by oxygen. Carbonyl compounds (compounds containing a –C=O bond) are less sensitive to oxygen quenching, and they can be excited by a variety of light sources, e.g., 3rd and 4th harmonics of Nd:YAG (355 nm and 266 nm) and excimer lasers (248 nm and 308 nm). Among the carbonyl compounds used are aldehydes (–C=O bond at one end), ketones (–C=O bond in the middle), and diones (two –C=O bonds). Diones, such as biacetyl (diacetyl, 2,3 butanedione), have longer excitation and fluorescence wavelengths. As shown in Table 7.3, acetone and 3-pentonone have higher fluorescence yield than aldehydes. In addition, ketones are preferred over aldehydes as dopants as they are less reactive under the high pressure and temperature environment, such as inside an engine. Ketones attack rubber or neoprene, and since fuel pumps, regulators, and fuel injectors, contain rubber or neoprene O-rings, the fuel supply system of an engine must be modified using ketone-resistant components such as stainless steel and Teflon.

7.4.5 Experimental Setup of PLIF

The typical setup of a PLIF system is shown in Fig. 7.10. A pulsed laser, normally with an output in the UV region, is used to excite the fluorescing molecule in the fuel at a specified crank angle. The laser light is formed into a sheet using combinations of cylindrical and spherical lenses. The fluorescence signal is imaged at a right angle using a gated ICCD camera. A band-pass filter may be placed in front of the camera to reject elastic scattering from the walls. A commercial lens can act as a filter as well as a lens, because it is anti-UV coated. A shaft encoder is used to generate the trigger pulse at a specified crank angle. A delay generator synchronizes the operation of the laser, the ICCD camera, and an image acquisition system via a PC computer. The intensifier should be gated for a short period of time (~100 ns) to discriminate against background light and combustion luminosity.

7.4.6 Calibration of PLIF Images

In order to obtain quantitative results, the raw LIF images must be post-processed and calibrated. The first process involves the removal of systematic errors present in the imaging system. This includes the removal of background and dark current signals from the recorded images, followed by corrections for non-uniformity in the responsivity of the detector and laser sheet intensity. Details on image processing and calibration have been discussed in Chapter 5. Other systematic errors, such as steering of the laser beam by density gradients, attenuation of the laser sheet across the imaged region, laser-produced perturbations, partial saturation, and radiative trapping (absorption of the fluorescence within the combustion gases) are often avoidable or negligible. But, these effects need to be considered in some applications.

Table 7.3 Summary of dopants used in PLIF measurements

Dopant and Fuel	T_{boil} (°C)	Absorption band (nm)	Emission band (nm)	Relative LIF intensity	Pressure effect** (%/MPa)	Temperature effect (%/1000 K)	Excitation source and Reference
$C_6H_5CH_3$, **toluene*** (5.0 vol. % in isooctane)	111	230–280	270–320	90 @ 248 nm 33 @ 266 nm	–80 @ 248 nm –90 @ 266 nm	–30 @ 248 nm 80 @ 266 nm	KrF excimer laser 248 nm, Reboux, et al., 1994; 1996.
$C_6H_5N(CH_3)_2$, **N, N dimethylaniline** (0.2 vol. % in isooctane)	193	200–320	310–420				KrF excimer laser 248 nm, Urushihara, 1996; Itoh, 1995.
$C_{16}H_{10}$, **fluoranthene** (1, 2-benzene) (<1.0 mass. % in isooctane)	solid 384	200–400	400–600				XeCl excimer laser 308 nm, Meyer et al., 1995; 1997.
C_2H_5CHO, **propionaldehyde** C_3H_7CHO, **butyraldehyde** C_5H_9CHO, **N-valeraldehyde** $C_6H_{13}CHO$, **heptaldehyde** C_6H_5CHO, **benzaldehyde** (9.1 vol. % in isooctane)	49 75 103 153 178	220–340 220–340	300–550 300–550	0.14 @ 248 nm 0.11 @ 248 nm			XeCl excimer laser 308 nm, Swindal et al., 1995.
CH_3COCH_3, **acetone**	56	200–330	300–600	1.0 @ 248 nm 1.0 @ 266 nm	60 @ 248 nm 5.0@ 266 nm	–25 @ 248 nm –7.0 @ 266 nm	XeCl excimer laser 308 nm, Wolff et al., 1994.
$C_2H_5COCH_3$, **2-butanone** (ethylmethylketone, EMK) (100%)	80	210–310	300–600				XeCl excimer laser 308 nm, Lawrenz et al., 1992.
$C_2H_5COC_2H_5$, **3-pentanone** (diethylketone) (2.0 –25.0 vol. % in isooctane)	102	220–340	330–600	1.3 @ 248 nm 1.3 @ 266 nm	70 @ 248 nm 3.0 @ 266 nm 0.0 @ 308 nm	–28 @ 248 nm –11 @ 266 nm 30 @ 308nm	Excimer lasers 248 nm, Johanssson et al., 1995. Nd:YAG laser 266 nm, Ghandhi and Bracco, 1995. Excimer lasers 308 nm, Berckmuller, 1996; 1997.
$CH_3COCOCH_3$, **biacetyl**, (2,3-butanedione) (2.0–6.0 vol. % in iso-octane)	88	200–480	440–510				Nd:YAG laser 355 nm, Baritaud and Heinze, 1992; Deschamps et al., 1994.
NO_2, **nitrogen dioxide** (0.2 vol. % in nitrogen)	gas	UV-visible	580 - 650				Nd:YAG laser 532 nm, Zhao et al., 1994.

* Toluene was used in FARLIF measurements.
** The pressure effects are valid up to 10 bar.

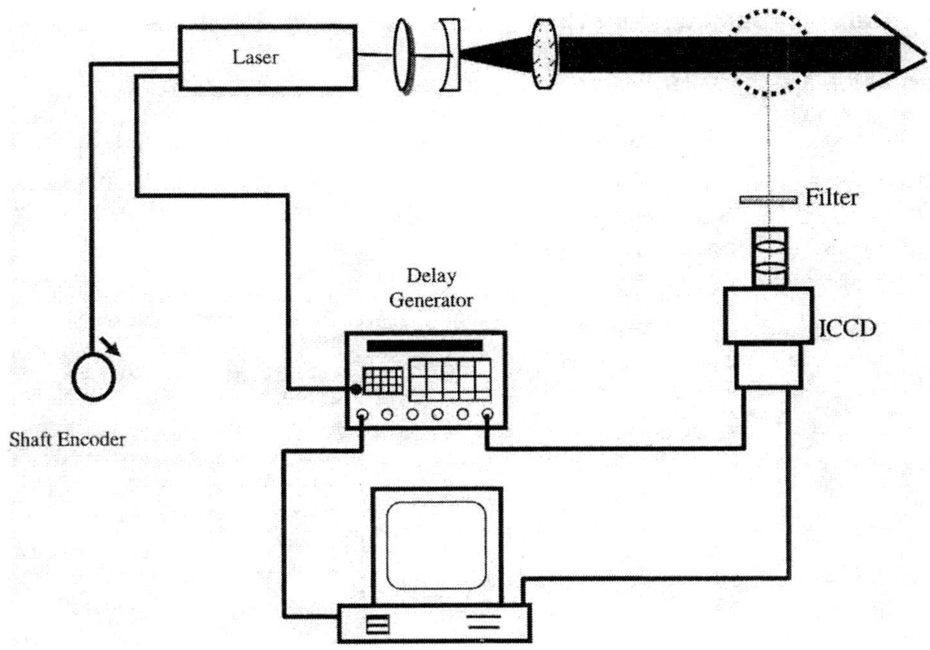

Fig. 7.10 Schematic of the optical setup for PLIF experiments.

The LIF images thus obtained represent the concentration of fluorescing molecules (dopant) at a particular condition. Since LIF intensity depends on temperature and pressure as well as dopant density, the LIF signal must be calibrated in order to obtain the fuel concentration and air/fuel ratios. Two calibration approaches can be adopted. The first involves in-situ calibration at each crank angle for different air/fuel ratios of homogeneous mixture, which means a calibration curve for each crank angle. The second approach is to relate laser induced fluorescence images to a calibration image taken at atmospheric pressure and ambient temperature. It can be shown [Berckmüller et al., 1994; 1996] that the effects of pressure and temperature on the LIF signal during the intake and compression strokes can be corrected for by a normalization factor, f_N

$$f_N = \frac{p_1 \cdot T_0 \cdot a + b \cdot T_1}{p_0 \cdot T_1 \cdot a + b \cdot T_0} \tag{7.14}$$

where p_0 and T_0 represent pressure and temperature at the calibration condition. p_1 and T_1 are the pressure and temperature of the gas in the cylinder when the LIF images are taken. The in-cylinder pressure can be obtained accurately from a pressure transducer. In-cylinder gas temperature is difficult to measure and is normally estimated. During the intake stroke, the gas temperature may be assumed constant. During the compression stroke, polytropic compression can be used to estimate the gas temperature. Constants a and b represent the effect of pressure and temperature on fluorescence yield and may be determined experimentally.

Table 7.4 Summary of PLIF experiments reported in the literature

Laser source and laser sheet (width × thickness)	Detection system and spatial resolution	Dopant and fuel	Applications and Reference
Nd:YAG 532 nm, 180 mJ 50 mm × 250 µm	ICCD camera 60 ns, >580 nm 0.23 × 0.23 × 0.25 mm	NO_2 (0.2 vol. %) N_2	N_2 injection to simulate fuel injection in a motored engine at low speeds. Zhao et al., 1994.
Nd:YAG 355 nm, 80 mJ 20 mm × 500 µm	ICCD camera 10 ms, 440–480 nm 0.25 × 0.2 × 0.5 mm	biacetyl (2–6 vol. %) isooctane	Effect of injection timing and engine speed on mixture homogeneity in a single-cylinder SI engine. Baritaud and Heinze., 1992, Deschamps et al., 1994
XeCl 308 nm, 150 mJ 27 mm × 1.0 mm	ICCD camera – – – – – –	fluoranthene (<1 mass %) isooctane	Effect of air/fuel ratio, engine speed, injection timing, and charge motion on mixture distribution in an SI engine Meyer et al., 1994; 1995; 1997.
XeCl 308 nm, – , 20 mm × 0.5 mm (10 mm × 0.5 mm)	ICCD camera – – – – – –	acetone (–) 3-pentanone (5 vol. %) isooctane	Effect of air/fuel ratio, injection timing, and EGR on mixture distribution in an SI engine. Andresen et al., 1990; Wolff et al., 1994; Knapp et al., 1997.
XeCl 308 nm, 220 mJ 76 mm × 0.3 mm	ICCD camera – – – – – –	EMK (100%)	Fluorescence characteristics of carbonyl compounds, imaging of air/fuel ratio in a square piston SI engine. Lawrenz et al., 1992.
KrF 248 nm, 400 mJ 69 mm × 2 mm	ICCD camera – , 280–400 nm. 0.19 × 0.19 × 2.0 mm	DMA (0.2 vol. %) isooctane	The effect of swirl and tumble flows on mixture distribution in a single-cylinder SI engine. Itoh et al., 1995; Urushihara et al., 1996; Hishinuma et al., 1996.
KrF 248 nm, 100 mJ 12 mm × 0.1 mm	ICCD camera 100 ns, visible range 0.19 × 0.19 × 0.1 mm	3-pentanone (0.2 vol. %) isooctane	The effect of injection timing and piston geometry on mixture homogeneity in a large SI engine. Johansson et al., 1995.
XeCl 308 nm, 150 mJ 50 mm × 0.25 mm	ICCD camera 200 ns, visible range 0.25 × 0.25 × 0.25 mm	3-pentanone (25.0 vol. %) isooctane	The effect of intake system and injection timing on mixture distribution in an SI engine. Berckmuller et al., 1996; 1997.
XeCl 308 nm, 60 mJ 20 mm × 0.2 mm	ICCD camera 2 µs, visible range. – – –	aldehydes (9.1 vol. %) isooctane	The effect of fuel boiling temperature on mixture homogeneity in an SI engine. Swindal et al., 1995.
Nd:YAG 266 nm, – , – – –	Intensified CID camera 150 ns, visible range. – – –	3-Pentanone (5.0 vol.%) isooctane	Fuel distribution in a special designed two-stroke direct injection gasoline engine. Ghandhi and Bracco, 1995.
Nd:YAG 266 nm, 10 mJ 50 mm × 0.5 mm	ICCD camera 200 ns, > 345 nm. 0.25 × 0.25 × 0.5 mm	gasoline fuel	Fuel distribution in a two-stroke direct injection gasoline engine. Fansler et al., 1995.
KrF 248 nm, 100 mJ 19 mm × 1 mm	ICCD camera 200 ns , > 280 nm. 0.19 × 0.19 × 1.0 mm	toluene (5.0 vol.%) isooctane	Fuel/air ratio LIF (FARLIF) measurements in a single-cylinder SI engine. Reboux et al., 1994; 1996.

7.4.7 In-Cylinder Fuel Distribution Measurements by PLIF

Because of its simplicity, species specificity, and strong signal level, PLIF has become the most useful technique for imaging fuel distributions in IC engines. Table 7.4 summarizes PLIF measurements in IC engines reported in the literature.

Following their LRS experiments, Zhao et al. [1994] measured cyclic variations in the in-cylinder mixture distribution by PLIF. Nitric dioxide (NO_2) was used as the fluorescence tracer. A mixture of N_2 and NO_2 was injected into the intake port via a solenoid valve to simulate the fuel injection and mixing process in a port-injected SI engine. The fluorescence from NO_2 was excited by a Nd:YAG laser at its 2nd harmonic (532 nm), and it was detected at 580 nm. From their measurements, Zhao et al. concluded that the in-cylinder mixture became homogeneous at the later stage of the compression stroke, but the cyclic variation of mixture homogeneity was still large. This type of experiment has provided useful information on the mixing process between the gaseous fuel and air in the cylinder. However, it is not representative of SI operation, since the vaporization characteristics of gasoline have an effect on mixture formation, as will be shown by the following studies.

Biacetyl has a boiling point of 88°C, close to that of isooctane (100°C), and it can be excited by 3rd harmonic output of an Nd:YAG (355 nm). When excited by a laser beam, biacetyl fluoresces from the first excited single state (lifetime: 50 ns, λ: 440–480 nm), and after an internal conversion it may also phosphoresce from a triplet state (lifetime: 1.2 ms, λ: 510–600 nm), or it may be collisionally de-excited (quenched). But the phosphorescence was found to be completely quenched and only fluorescence was observed [Baritaud and Heinze,1992; Deschamps et al. 1994]. Engine tests with pure nitrogen instead of air showed that less than 8% of the LIF signal was quenched by the oxygen in air. In their experiments, Baritaud and Heinze used a mixture of 2–6 % in volume biacetyl in isooctane in order to produce a strong fluorescence signal with negligible absorption. Because biacetyl decomposed slowly, a fresh mixture of isooctane and biacetyl was used for each experiment.

Fluoranthene was used as the fluorescence dopant in PLIF measurements by Meyer et al. [1994; 1995]. The experimental setup is summarized in Table 7.4. Fluoranthene was chosen as the dopant because of its superior fluorescence yield and very low sensitivity to quenching at high pressures. Although fluoranthene is solid below 110°C, it is readily soluble in fuel in low concentration (less than 1% by mass). Its ability to track the fuel is questionable since its boiling temperature is rather high (384°C at 1 bar). Meyer et al. [1997] studied the air/fuel distribution in an SI engine using PLIF of fluoranthene. The single-cylinder engine had two electromagnetically controlled intake valves, one for EGR gases and the other for the air/fuel mixture. The recirculated exhaust gases entered the combustion chamber with an outward swirl, and the air/fuel mixture entered into the center of the combustion chamber with a downward tumble. Fig. 7.11 was obtained when the inlet valve for the air/fuel mixture opened at 600°CA, when the other valve for EGR had just closed. The brightness in these images is proportional to the fuel concentration. The first sign of fuel appeared at 630°CA in the perpendicular images ($\perp$ IPD). At 645°CA, the swirl had lost its momentum, the tumble of fuel from the fully

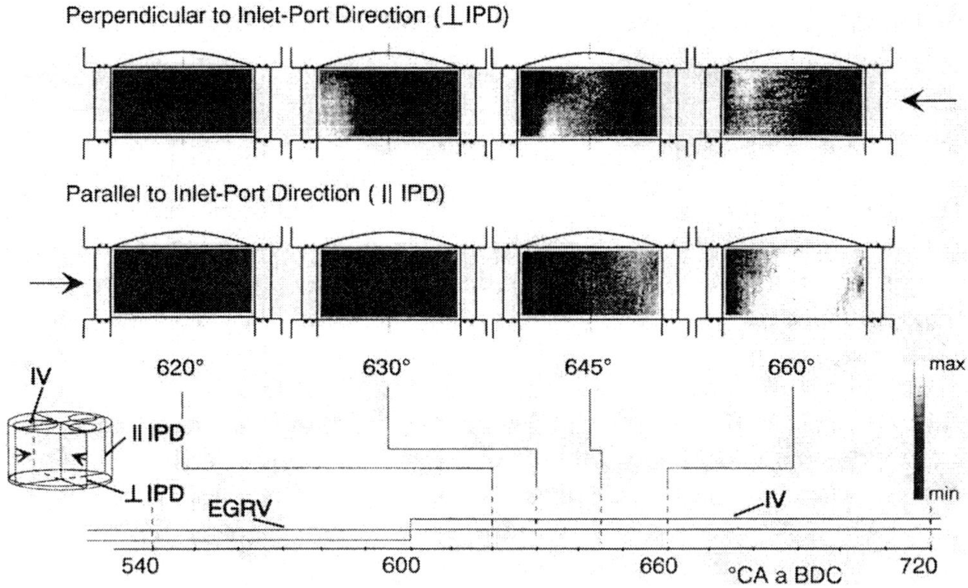

Fig. 7.11 In-cylinder fuel distribution with stratified EGR [Meyer et al., 1997].

opened valve began to dominate, as shown in the parallel images (|| IPD). At 645°CA, the downward tumble motion completely dominated the fuel distribution.

Initial application of PLIF measurements at Göttingen was reported by Andresen et al. [1990], where the fluorescence from isooctane was detected using a 248 nm laser. As discussed previously, isooctane normally does not fluoresce. The fluorescence observed by Andresen et al. was probably caused by impurities in the isooctane used [Felton, 1996]. In the subsequent experiments carried out at Göttingen, Wolff et al. [1994] employed acetone as the fluorescence dopant. More recently, PLIF measurements using 3-pentanone were also reported by the same research group [Knapp et al., 1997].

Lawrenz et al. [1992] conducted a series of experiments on carbonyl compounds to check their suitability as dopants using a XeCl excimer laser. The compounds investigated included: biacetyl, acetone, ethylmethylketone (EMK, 2-butanone), diethylketone, acetaldehyde, and valeraldhyde. They found that the ketones were insensitive to the partial pressure of air, whereas biacetyl fluorescence showed a strong dependence on the air pressure. They concluded that the biacetyl signal was too weak to be used in engines. However, Baritaud and Heinze [1992] reported no difficulty in the use of biacetyl as the fluorescing dopant. This discrepancy is likely caused by the absorption spectrum of biacetyl, which peaks around 355 nm. Despite its low boiling point (352 K), EMK was selected as the fluorescence dopant due to its strong fluorescence yield. Subsequent PLIF measurements in a square piston engine were carried out using pure EMK in order to boost fluorescence signals. Their results showed that substantial variations of fuel distribution were present in the cylinder at the time of ignition.

PLIF was extensively used to study in-cylinder fuel distributions in a single-cylinder optical engine with a transparent piston at Nissan Research Center [Urushihara et al., 1996; Hishinuma et al., 1996]. DMA (N,N-dimethylaniline) was used as the dopant in isooctane. When DMA concentrations of more than 2% by mass in isooctane were tested, both oxygen and DMA self-quenching were observed. Although 2% DMA by mass in isooctane was found to give the strongest fluorescence signal in the compression stroke, the strong absorption of laser light by the mixture in the vicinity of the optical window prevented the laser beam from passing through the combustion chamber. Eventually 0.2% DMA by volume in isooctane was used in their experiments. Because of the absorption by DMA, the two laser beams were introduced into the combustion chamber from opposite directions in order to reduce laser light attenuation. A KrF excimer laser at 248 nm was employed. The fluorescence peaked at 335 nm and was detected through a band-pass filter of 280 nm to 400 nm, which helped to block the scattered light of the laser beam and the fluorescence from quartz windows at 400 nm.

Although the use of isooctane in PLIF experiments helped to eliminate the absorption present in commercial gasoline fuels, isooctane represents only a small fraction of gasoline components. In order to carry out PLIF measurements under more realistic engine conditions, Itoh et al. [1995] developed a compound fuel that had a similar distillation curve to that of gasoline, but it was more transparent to the 248 nm laser beam than the normal gasoline. In-cylinder fuel distribution was obtained directed from fluorescence images of the compound fuel. The fluorescence signal arises from the aromatic components in the fuel. Significant difference was found in the mixing patterns when isooctane and the new compound fuel were used in a single-cylinder research engine.

PLIF of 3-pentanone in isooctane was used by Johansson et al. [1995] using a KrF excimer laser. Quantitative images of the fuel distribution within the engine were obtained through calibration with homogeneous known fuel/air mixtures in the cylinder. The authors claimed that the pulse-to-pulse spatial variations and the noise from the image intensifier yielded an error of about 10%. The background noise contributed another 3% error. The effects of quenching and absorption on PLIF were considered small. Therefore, the total error was estimated being around 15%. Using the PLIF technique and a pressure transducer, Johansson et al. established that there was a strong correlation between the mixture homogeneity and the combustion quality. They found that the inhomogeneity of the charge was reduced when a disc-shaped combustion chamber was replaced with a turbulence-generating geometry.

A similar experimental setup was used to study fuel distributions in a Honda VTEC single cylinder engine by Berckmüller et al. [1996]. A mixture of 75% isooctane and 25% 3-pentanone by volume was excited by an XeCl excimer laser. A calibration equation was employed to convert the LIF signals into quantitative fuel/air ratio measurements. Average fuel concentrations were recorded in four different planes below the spark plug during the intake and compression strokes. The recorded time history of the fuel distributions revealed the presence of an axial fuel distribution when one of the intake valves was disabled. The degree of stratification was found to be affected by the fuel injection timing. Subsequent studies by Berckmüller et al.

[1997] showed that the fuel concentration in a small region close to the spark plug had a dominating effect on the subsequent pressure development for overall lean mixtures.

Since commercial gasoline is a mix of different hydrocarbons, the distillation curve is quite broad. The boiling temperature of each component can vary from 50°C to 200°C, and it affects mixture preparation directly, because each component will evaporate differently. The difference between the hydrocarbons of different boiling temperatures was the subject of a study carried out by Swindal et al. [1995]. In their study, several aldehydes of boiling temperatures from 49°C to 178°C were employed to study the effect of fuel distillation properties on mixture preparation. Each dopant was mixed at a concentration of 9.1% by volume isooctane, and was excited by an XeCl excimer laser at 308 nm. A fairly long exposure time of 2 μs was used to record the fluorescence signal in the visible spectrum. Their results indicated that fractional distillation occurred during the intake, leading to non-uniform fuel distributions in the cylinder which were affected by the boiling point of each fuel component.

Continued pressure to reduce CO_2 emissions and the advent of relatively low-cost, electronically controlled direct fuel injection systems has created a resurgence of interest in direct injection gasoline engines. At low load, direct injection gasoline engines typically operate with a highly stratified fuel-air mixture. Improper fuel-air concentration (too rich or too lean) near the spark gap at the time of ignition causes misfires and partial burns. PLIF proves to be a very useful tool in visualizing the in-cylinder fuel distribution. The application of PLIF in direct injection two-stroke SI engines has been reported by Ghandhi and Bracco [1995] and Fansler et al. [1995]. The experimental setup used in these measurements can be found in Table 7.4. Ghandhi and Bracco studied the fuel distribution using 3-pentanone as the dopant in a specially designed two-stroke engine with six inlet and six exhaust ports radially positioned in the cylinder block. Fansler et al. visualized fuel distribution using PLIF from commercial gasoline. Strong fluorescence from the gasoline fuel was generated by a frequency quadrupled Nd:YAG laser (266 nm). From PLIF images, Fansler et al. found that fuel was mostly concentrated around the spark plug at the time of ignition and there were large cycle-to-cycle variations in the fuel concentration near the spark gap. They also noticed that the variation in fuel LIF intensity was consistent with the incidence of poor combustion as indicated by in-cylinder pressure analysis. As discussed previously, the use of the fluorescence signal from gasoline fuel is convenient, but the quenching characteristics of the fluorescing molecules in the fuel are unknown. Hence, this method is more suitable for qualitative visualization of fuel distribution.

7.5 Direct Fuel/Air Ratio Measurement by FARLIF

7.5.1 Principle

As discussed previously, laser induced fluorescence in the linear regime suffers from quenching losses. Quenching is particularly strong at elevated pressures, such as in the cylinder of an IC engine. This limits most in-cylinder LIF measurements to qualitative or semi-quantitative measurements.

However, quenching by oxygen molecules has been used by Reboux et al. [1994; 1996] to infer air/fuel ratios from laser induced fluorescence intensity signals. In the linear fluorescence regime, the fluorescence signal is given by Eq. (7.13)

$$P_{flu} = \eta_c \Omega V_c f_1(T) \chi_m N I_v B_{12} \frac{A_{21}}{A_{21} + Q_{21}}$$

which shows that the fluorescence is linearly proportional to laser intensity. At elevated pressures, the quenching rate Q_{21} of certain fluorescing molecules can be much higher than A_{21}, the above equation can be further simplified to

$$P_{flu} = \eta_c \Omega V_c f_1(T) \chi_m N I_v B_{12} \frac{A_{21}}{Q_{21}} \quad (7.15)$$

Since oxygen is a major species and its collisional cross section is much greater than that of nitrogen, the quenching rate Q_{21} of the in-cylinder charge is expected to be proportional to the oxygen concentration at constant temperature, i.e., $Q_{21} \propto [O_2]$. Hence, the above equation can be written as

$$P_{flu} = \eta_c \Omega V_c f_1(T) \chi_m N I_v B_{12} \frac{A_{21}}{[O_2]} \quad (7.16a)$$

or

$$P_{flu} = C_{cal} I_v \frac{[\text{Dopant}]}{[O_2]} \quad (7.16b)$$

where $\chi_m N$ has been replaced with the concentration of fluorescing dopant, [dopant], and a new calibration factor C_{cal} is introduced. C_{cal} is obtained for a particular system setup and the fluorescing dopant and its concentration. For a known concentration of dopant in the fuel, Eq. (7.16b) can be written as

$$P_{flu} = C_{cal} I_v \frac{[\text{Fuel}]}{[O_2]} \quad (7.17)$$

Therefore, the fluorescence signal is directly proportional to the fuel/air ratio at constant temperature, if oxygen is the major collision partner of the fluorescing dopant. Since the LIF signal is used to measure fuel/air ratios directly in this case, this measurement technique has been dubbed fuel/air ratio LIF or FARLIF.

When the normal LIF technique is used, the air/fuel ratio is calculated on the assumption that air is uniformly distributed in the mixture. However, when there is stratification in the cylinder charge, fuel/air ratios cannot be determined from fuel concentration alone using the normal LIF technique. In contrast, FARLIF measures directly the air/fuel ratio in such circumstances. In addition, the calibration of FARLIF needs to be carried out only at one sufficiently high pressure, whereas the fuel/air ratio measurement from normal LIF involves calibrations for different pressures. Because FARLIF needs to use dopants that are highly susceptible to quenching, its signal level will be lower.

7.5.2 Experimental Validation and Application of FARLIF

There are a number of prerequisites for FARLIF to be valid. Because Eq. (7.17) is derived from Eq. (7.13) for the linear regime, the fluorescence intensity should vary linearly with laser intensity. Since Eq. (7.17) is only valid, if the quenching rate Q_{21} becomes far greater than the Einstein coefficient, A_{21}, the quenching process should dominate the fluorescence signal. Finally, oxygen has to be the dominant quenching species.

In order to validate the above theoretical analysis, Reboux et al. [1994; 1996] carried out a series of experiments using toluene as the fluorescing dopant. A mixture of 5% toluene and 95% isooctane by volume was used as the test fuel. An excimer laser (KrF) was used to excite toluene. The fluorescence was red-shifted to 275 nm and detected through a long pass filter (Schott WG 280).

Experimental results showed that the fluorescence signal increased linearly with laser intensity. When pressure was greater than 3 bar, the fluorescence signal from toluene was found to be dominated by the quenching process. It was also found that the fluorescence signal from toluene was not quenched by isooctane nor self-quenched. In addition, the fluorescence signal was not affected by N_2, H_2O or CO_2. Therefore, the dominant collisional partner of excited toluene was confirmed to be oxygen.

Having validated Eq. (7.17) experimentally, Reboux et al. carried out direct fuel/air ratio measurements in an SI engine during the compression stroke. The calibration constant C_{cal} in Eq. (7.17) was obtained at 30°CA BTDC during the compression stroke from a homogeneous mixture with different equivalence ratios. The maximum difference between the equivalence ratios measured by a lambda sensor in the exhaust and FARLIF was 2%. Results showed that inlet manifold heating and injection closer to the inlet valve improved mixture homogeneity and reduced cycle-to-cycle variations in air/fuel ratios.

7.6 In-Cylinder Fuel Concentration Measured by Fast FID

The fast FID system has been described in detail in Chapter 3. The following sections discuss its applications to fuel concentration measurements.

7.6.1 Inlet Manifold Fuel Films

Ladommatos [1992b] and Ladommatos and Rose [1996] used the fast FID system to obtain estimates of the fuel mass residing on the inlet port of a port-injected gasoline engine (see Table 3.3, Set 7). The fast FID was installed in the No. 3 engine cylinder and sampled from the base of the spark plug. The following technique was used: With the engine operating at steady state, the No. 3 fuel injector was switched off. The FID signal plateau for successive engine cycles gradually decayed to zero over about 50 cycles. The decaying plateau level for each cycle was recorded and converted to fuel/air ratio using calibration data. The fuel entering the cylinder for each cycle after the injector was disabled was calculated from the fuel/air ratio and induced mass of air per cycle. These fuel quantities were then summed up. This total fuel mass was assumed to be equal to fuel mass residing on the No. 3 port wall when the injector was switched off. This assumption is based on the fact that after the injector was switched off, the cylinder was supplied with fuel coming only from the port wall. The results showed clearly that the mass of the port wall film was sensitive to the coolant temperature, degree of the injector fuel spray atomization, and injection timing.

A major source of imprecision of the above technique was the calibration of the FID system. This was carried out in situ, while the engine was operated at steady state over a wide range of air/fuel ratios. However, the calibration data required considerable extrapolation since the engine could not run at very lean air/fuel ratios. In addition, the calibration did not reflect the progressively diminishing level of combustion residuals after the No. 3 injector was switched off.

Shin et al. [1995] used a variant of the above method to study fuel transport from the inlet port to the cylinder of a port-injected spark-ignition engine. The fast FID tests were combined with flow visualization studies in the inlet port and estimates of indicated mean effective pressure. They used a single injection pulse to spray fuel into the port of the non-firing engine. Following this single injection event, the decay of the fuel equivalence ratio in the engine cylinder during the subsequent 100 engine cycles was monitored with a fast FID. Shin et al. found that it took about 10 engine cycles for most of the fuel to be transported into the engine cylinder. About 25 percent of the fuel entered the cylinder during the first one of these 25 cycles. The amount of fuel injected had to be increased eightfold before a combustible air-fuel mixture was formed in the cylinder during the first engine cycle.

7.6.2 Air Fuel Ratio Excursions during Fast Throttle Opening

Spark ignition engines when fitted with a three-way catalyst usually operate very close to the stoichiometric air/fuel ratio. When the engine undergoes a transient, for example a sudden increase in throttle opening, the air/fuel ratio must remain very close to the stoichiometric value throughout the transient in order to reduce pollutant exhaust emissions. However, during very fast throttle openings, the air fuel ratio is known to undergo a lean excursion which may last one or more engine cycles. Ladommatos [1992a], Rose et al. [1994], and Ladommatos and Rose [1996] used the fast FID to monitor, on a cycle-by-cycle basis, the air/fuel ratio during abrupt throttle openings in a production engine running on gasoline [see Table 3.3, Set 8]. The method involved measurement of the FID output plateau level for each successive engine cycle during the transient and a calibration curve (obtained in-situ) of air/fuel ratio against FID output plateau level. Their results show that following a fast throttle opening the air/fuel ratio becomes very lean (more than twice the stoichiometric value) for two or three engine cycles.

A major problem in carrying out the above transient tests was calibration of the FID system so that the signal obtained from the cylinder contents corresponded to the air/fuel ratio supplied to the cylinder. The above authors developed an in-situ calibration method which overcame this problem, but which was subject to significant imprecision. Briefly, this involved the engine operating at steady state (constant speed and throttle opening) while different *known* air/fuel ratios were supplied to the cylinder under investigation. The plateau level of the fast FID signal during the compression stroke was then plotted against the known air/fuel ratio to obtain a calibration curve (the brake load applied to the engine was adjusted to keep speed and throttle opening constant). This calibration curve applied only to a particular throttle opening (and, strictly speaking, load), because the amount of combustion residuals in the cylinder depend strongly on throttle opening. Therefore, calibration curves had to be obtained for each throttle opening of interest.

The transient results obtained by Ladommatos and Rose using fast FID are remarkably similar to those obtained by Li et al using an entirely different in-cylinder technique [Li et al, 1994a; 1994b]. Li et al. used a probe consisting of a tube of internal diameter 1.8 mm, which was installed on the engine cylinder (close to the spark plug) so that a continuous flow of sample could be extracted from the cylinder by means of a vacuum pump. Across the bore of the tube was placed a short length of thin (0.015 mm diameter) catalytic wire (platinum/iridium), which was maintained at a constant temperature (about 950 K). When air-fuel mixture flows over the catalytic wire, chemical energy released from hydrocarbon oxidation reduces the amount of electrical energy required to maintain the wire temperature constant. The reduction in the electrical energy requirement is related through calibration to the strength of the air-fuel mixture. The calibration procedure allowed for energy loss from the wire due to cooling of the hot wire by the flowing air-fuel stream (although the energy loss by heat is around one-tenth the value of the energy release due to fuel oxidation on the wire surface). Just downstream of the hot wire (about 3 mm) was placed a 0.3 mm diameter nozzle which was normally choked, and helped reduce the flow velocity fluctuations across the wire (although the flow sonic velocity changed with the fluctuating temperature of the cylinder contents). The transit time of this probe was very short (< 0.3 ms) owing to the fact that the hot catalytic wire was placed very close (2 mm) from the tube entrance. The close proximity to the tube entrance also reduced dispersion effects, which helped to lower the value of the time constant. The time constant also depended on the HC oxidation rate, which is normally very fast for catalytic surfaces at high temperature. Thus, the time constant is principally governed by the mass transport processes, involving the diffusion of fuel molecules to the wire surface (oxygen diffusivity is several times greater than that of large HC molecules with more than 5 carbon atoms).

7.6.3 In-Cylinder Mixture Inhomogeneity

Charlton et al. [1992] used a fast FID system to measure the continuous variation of mixture strength within the pre-chamber of a lean-burn natural gas engine during the intake and subsequent compression strokes. During bench tests on the system, it was found that an FID tube diameter of 0.3 mm was too large, causing the instrument output signal to be pressure-sensitive to abrupt increases in sample pressure. A reduction in FID tube diameter to 0.15 mm alleviated this pressure sensitivity. The early part of the fast FID signal, produced early in the compression stroke, was not considered meaningful, due to low sample pressures and poor instrument response (long time constant). The latter part of the signal was, however, used successfully to investigate, on an instantaneous basis, the dilution of the fuel in the engine pre-chamber as air was being forced into the pre-chamber during the latter part of the compression stroke. Theoretical results computed using a computational fluid dynamics (CFD) code agreed fairly well with the FID measurements obtained from the usable portion of the output signal. The results of Charlton et al. also show that the FID output signal responded correctly to changes in ignition timing. The abrupt drop in the FID signal when the mixture in the pre-chamber is consumed by the flame allows the time constant to be estimated at about 1 ms. Charlton et al. concluded that the fast FID system is useful for the measurement of mixture strength when this information is needed close to the point of ignition.

Sleightholme [1990] used a fast FID system to investigate mixture inhomogeneity within the single cylinder of a Ricardo E6 engine operating on propane at 1000 rpm and wide-open throttle. Sampling was far away from the spark plug. The FID sampling system did not include an expansion tube in "T" configuration, as shown in Figures 3.2 and 3.3. Instead, sample from the cylinder first entered a short length of 0.15 mm diameter tube. The sample then emerged into a larger-diameter tube where it expanded and its velocity was reduced. Finally, the sample exhausted from this larger diameter tube into an evacuated surge chamber (volume 7.5 cm^3). Just prior to exhausting into the surge chamber, a small portion of the sample was drawn into a 0.15 mm diameter tube and taken to the FID flame. This last 0.15 mm diameter tube performed the function of the FID tube, but was some three times as long as the FID tube used with the type of system illustrated in Figures 3.2 and 3.3. Sleightholme was able to obtain significant insight into the air-fuel mixing process in the cylinder. For example, it was possible to follow the mixing process on an instantaneous basis by observing the mixture strength at the sampling point during the induction stroke. Cycle-to-cycle variations in local mixture strength at the sampling point were also evident. Sleightholme was able to show quantitatively that there was a linear relationship between cycle-to-cycle variation in imep and cycle-to-cycle variability in mixture strength at the sampling point; however, it was not possible to reach a firm conclusion as to whether the cyclic variability in local mixture strength was due to variability in local inhomogeneity or due to variability in mixture strength supplied to the cylinder.

Meyer and Thring [1995] used a fast FID with two identical sampling systems to study the effects of mixture preparation on combustion. The two sampling probes, comprising 0.2 mm diameter transfer tubes each having a length of around 200 mm, were installed in the spark plug of a single-cylinder engine operated either on gasoline or natural gas at 1200 rpm and high load. The two sampling points were positioned at the base of the spark plug and were diametrically opposed. The results of Meyer and Thring with the engine running on gasoline show that the individual cycle HC concentrations measured at the two sampling points did not correlate particularly well, a phenomenon that was ascribed to poor vaporization and mixture stratification. In contrast, when the engine was run on natural gas the correlation between the HC concentrations at the two sampling points was a very good one. The poor correlation in the two measurements in the case of gasoline calls for a note of caution: the HC concentration measurements made at a single FID sampling point cannot always be assumed to be representative of HC concentration at other locations in the engine cylinder, even when the other locations are within a short distance (few mm) of the sampling point. Returning to the results of Meyer and Thring, other interesting findings include a fairly good inverse correlation between the averaged HC concentration at the spark plug and the ignition delay (crank angle degrees for 0–10% of mixture burned).

7.6.4 HC Concentration in the Combustion Products

Summers [1996] carried out experiments in a single-cylinder engine (Ricardo Hydra) fitted with a mass-produced road-vehicle cylinder head. The aim of the experiments was to measure the in-cylinder concentration of unburned HC in the combustion products. The engine ran on pre-vaporized gasoline at 1040 rpm. A problem encountered by Summers was pressure fluctuation in the CP chamber during engine operation, causing great flow rate fluctuations through the FID tube. Fortuitously, it was possible to make the in-cylinder measurements during a period in the engine cycle when the pressure drop across the FID tube was stable for a sufficiently long period (see also Section 3.4.2.1).

Peckham and Collings [1994] studied the evolution of unburned HCs emerging (outgassing) from in-cylinder crevices during the expansion stroke. Two in-cylinder sampling arrangements were used. For the first, the sampling point was attached to the moving piston so that unburned HCs emerging from the piston top land crevice (volume 500 mm^3) could be sampled. A length flexible sample transfer tubing conveyed the sample to the fast FID head; this was referred to as the "flying" sampling arrangement (summarized in Table 3.3, Set 9). The second sampling arrangement comprised an artificially created crevice having a volume of 1 mm^3 and located in the engine cylinder head far away from the spark plug; this was referred to as the "static" sampling arrangement (Table 3.3, Set 10). Peckham and Collings were able to resolve individual events on from the instantaneous fast FID signal recorded during an engine cycle. For example, the "flying" sampling arrangement revealed that high HC concentrations from the piston land crevice build up near the cylinder wall during the expansion stroke. The "static" sampling arrangement enabled observation of unburned HCs emerging from the artificial crevice during the expansion stroke. Although it was possible to link these observations events in the engine cycle (e.g., exhaust valve opening) care was needed because the FID signal time axis was distorted due to variable sample transit time. In particular, great care was required with the "flying" sampling arrangement, because the large transfer tube length (450 mm) resulted in large transit time variations (7 to 18 ms). In another series of tests, Peckham and Collings

[1992] used the same fast FID system and similar techniques to monitor the unburned HC from three sampling points located in the cylinder head, adjacent to the exhaust valve seat, and in the exhaust port, respectively.

7.7 In-Cylinder Fuel Measurement by a High-Speed Sampling Valve

A Fast FID probe provides cycle-resolved measurements of hydrocarbons in the cylinder. If measurements of multiple species and/or time (crank angle) resolved measurements are required, a high-speed intermittent sampling valve can be used. A detailed discussion on high-speed sampling valves can be found in Chapter 4. In the following sections, their applications to measure fuel concentration in the cylinder are given.

7.7.1 Local Mixture Strength at the Spark Plug

In spark ignition engines, the air/fuel ratio within the electrode gap of the spark plug is an important factor, which can affect combustion performance. In order to investigate the cycle-by-cycle fluctuations in mixture strength at the gap, several investigators combined the sampling valve with the spark plug.

For example, Matsui et al. [1979] used a needle-type sampling valve which was ingeniously modified so that it performed the additional function of a spark plug. The tip of the sampling valve is shown in Fig. 7.12. A carrier gas was supplied close to the needle valve seat in order to dilute the sample. The carrier gas performed a number of important functions. First, dilution of the in-cylinder mixture with the carrier gas reduced the HC concentration, bringing it closer to the operating range of exhaust HC analyzers. Second, the carrier gas allowed cyclic resolution of the mixture samples to be obtained, by sampling one engine cycle every few seconds. In this way, the carrier gas had the dual function of transporting the samples to the FID analyzer as well as separating the individual samples. This allowed a degree of cycle resolution of the mixture strength in the engine cylinder. Matsui at al. conducted systematic experiments that showed that the entire path of the diluted sample had to be maintained at a temperature greater than 150°C in order to avoid condensation of the heavier gasoline fractions. An initial problem encountered by Matsui et al. [1979] was a sudden variation in the carrier gas flow rate every time the sampling valve opened. This problem was caused by the abrupt inrush of sample into the valve, upon valve opening, which generated a pressure wave that caused variation in the flow rate of the carrier gas. This problem was overcome by installing a surge tank and a flow-control valve in the carrier gas supply tubing upstream of the sampling valve, and by judicial choice of the tubing length.

Another potential problem investigated by Matsui et al. was the effect on the combustion process of removing the quantity of sample between the spark plug electrodes. This was done by comparing the 0–10% burn duration, obtained from heat release analysis, with and without sampling. Although an effect on this duration was detectable, it was considered small when sampling occurred before the spark discharge.

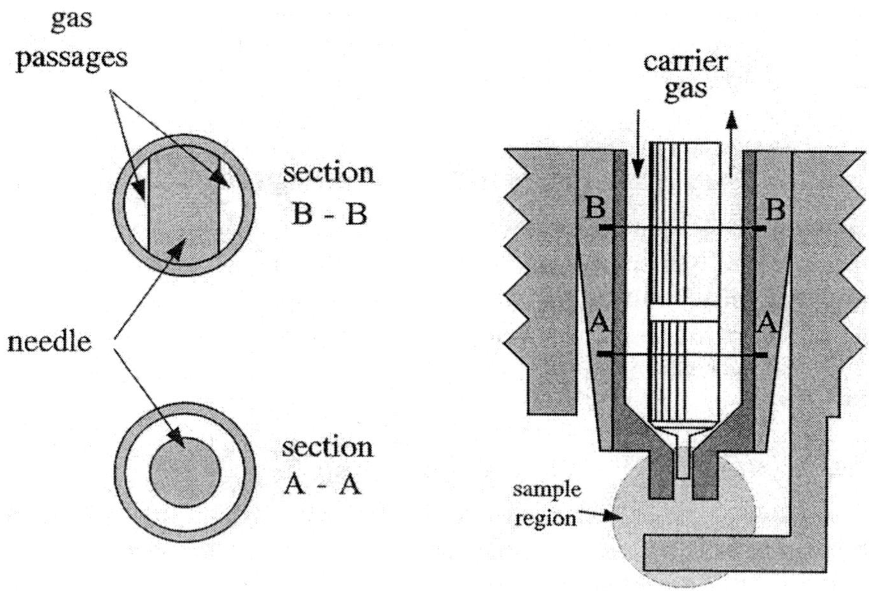

Fig. 7.12 Collection of hydrocarbons from the engine cylinder using carrier gas.

Ayusawa et al. [1978] measured the un-ignited air/fuel ratio at the spark plug gap by means of a needle-type valve, similar to that used by Matsui et al. [1979]. One two-stroke and two four-stroke engines were used, running on gasoline. The valve was used, both, in multi-cycle and single-cycle modes. In the multi-cycle mode, the valve was operated continuously once every 10 to 15 consecutive engine cycles and the samples from the different cycles were all mixed together. Thus, the measured air/fuel ratio was the average value for many cycles. In the single-cycle mode, the valve was operated only once, in order to obtain a sample from just a single cycle. Helium was fed close to the valve seat, where it entrained and diluted the sample from the single valve opening. The carrier gas containing the diluted sample then emerged from the valve into a loop of sample line and was subsequently transferred to a heated stainless steel syringe. Analysis of the samples was carried out by gas chromatography. Ayusawa at al. also found that the entire sample transfer and storage system had to be maintained at 150°C in order to avoid condensation of the heavier gasoline fractions.

A significant problem encountered by Ayusawa et al. [1978] was the sealing of the sample-storage syringes to prevent ingress of air, which would have greatly distorted the O_2 concentration in the sample. O_2 was measured for the calculation of the air/fuel ratio. Use of stainless steel as the syringe material, Teflon seals, and pressurization of the syringe seals with helium solved this problem. Another consideration during the experiments by Ayusawa et al. was the sampling timing. If the sampling timing was too close to the ignition timing, then the mixture did not ignite. Thus, sampling of the uni-ignited mixture was carried out 5 to 10°CA before spark discharge.

Quader [1982] also used a sampling valve that performed the additional function of the spark plug, in order to investigate the effects of mixture preparation and the use of EGR in a multi-cylinder gasoline engine with port injection.

7.7.2 Determination of Air/Fuel Ratio at the Spark Plug in a Direct Injection Gasoline Engine

7.7.2.1 Gas Sampling Procedure

Sadler et al. [1997] used a poppet-type sampling valve, developed at Brunel Engines Laboratory, to measure the air/fuel ratio at the spark plug of a direct injection gasoline engine. In this type of engine, the mixture strength is highly stratified, and of great interest is the air/fuel ratio at the spark plug at the time of spark discharge. The spark plug was replaced by the tip of the sampling valve and the engine was motored at 1500 rpm, having previously heated the engine to operating temperature using electric heaters. When sampling late in the compression stroke, an uni-ignited mixture sample of about 3 cm^3 per cycle was obtained (at ambient conditions), equivalent to 2.25 L/min. It was not possible to analyze this uni-ignited mixture sample by means of normal exhaust emissions gas analyzers because, first, a minimum sample flow rate of around 30 L/min is usually required by such equipment and, second, because the concentration of HC in the uni-ignited mixture was far greater than that found in exhaust gases. The uni-ignited gas sample was, therefore, first diluted with nitrogen before being analyzed by normal exhaust gas analyzers. The sample handling system is illustrated in Fig. 7.13.

Referring to Fig. 7.13, the sampling procedures used by Sadler et al. [1997] were as follows. The sampling valve was first operated continuously with the sample exhausted to waste (exhaust), in order to ensure that the sample lines were filled with sample. Control valve A in Fig. 7.13 was then opened for a brief period in order to divert some sample to the sample bag, which was previously half-filled with nitrogen. Once the required amount of sample entered the bag, the sample flow was discontinued. Finally, more nitrogen was added so as to fill the sample bag and to mix the sample evenly. Thus, the sample was diluted by nitrogen with a dilution ratio that could be varied between 30 and 30,000, depending on how much sample was allowed to flow into the bag. (As will be shown in the next section, it is not necessary to know the dilution ratio in order to calculate the gravimetric air/fuel ratio of the sample extracted from the engine cylinder.) The gas analyzers were then purged with nitrogen, and the sample bag was emptied through the analyzers so that the concentrations of HC and O_2 in the sample bag were measured. The gravimetric air/fuel ratio in the engine cylinder, at the spark plug, was then calculated as described in the following section.

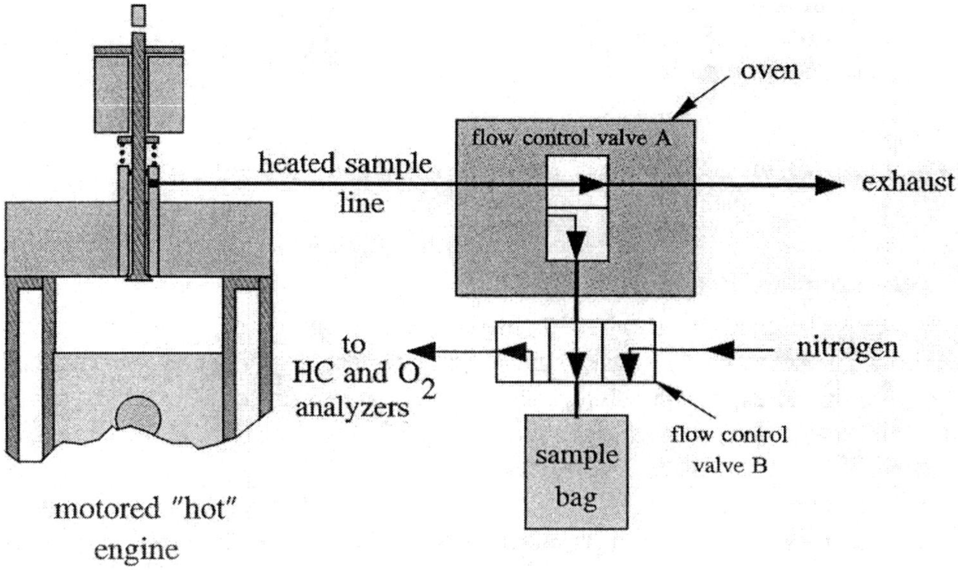

Fig. 7.13 Sampling of fuel-air mixture from a motored direct injection gasoline engine.

7.7.2.2 Calculation of Air/Fuel Ratio

Let,

$[HC]_{c3}$ = measured molar (i.e., by volume) HC concentration in the sample bag in terms of ppmC_3 (This implies that the HC analyzer has been calibrated with propane, C_3H_8.)

$[HC]_{c1}$ = molar concentration of HC in the sample bag in terms of ppmC_1

$[HC]$ = molar concentration of HC in the sample bag (%)

$[O_2]$ = measured molar oxygen concentration in the sample bag (%)

y = H/C ratio of the fuel (atoms of hydrogen/atoms of carbon per molecule of fuel; for gasoline, H/C ratio is normally within the range 1.8 to 2, and for diesel fuel within the range 1.7 to 1.9)

(A/F) = air/fuel ratio by mass in the diluted mixture
= air/fuel ratio by mass in the engine cylinder

As mentioned above, the task is to calculate the gravimetric air/fuel ratio, A/F, at the point of sampling in the cylinder using measurements of $[HC]_{c3}$ and $[O_2]$ made with the system shown in Fig. 7.13.

First, convert the measured concentration of the fuel in the sample bag from ppm propane (i.e., ppmC_3) to concentration in ppmC_1,

$$[HC]_{c1} = 3[HC]_{c3} \quad (\text{ppmC}_1)$$

Now, convert from ppm to %,

$$[HC] = 3\ [HC]_{c3} \times 10^4 \quad (\%)$$

Without combustion, $[CO_2]$, [CO], $[H_2O]$, [NO], and $[NO_2]$ are set to zero, then,

$$\left(\frac{A}{F}\right) = 4.773\left(\frac{M_a}{M_f}\right)\frac{[O_2]}{[HC]} \tag{7.18}$$

Substituting for $[HC] = 3[HC]_{c3}\ 10^{-4}$ (%), $M_a = 28.96$, and $M_f = 12.01 + 1.008\ y$

$$\left(\frac{A}{F}\right) = 4.773 \times \frac{28.96}{12.01 + 1.008y} \times 10^4 \times \frac{[O_2]}{3[HC]_{C3}} \tag{7.19}$$

Fortunately, the dilution ratio of the air-fuel mixture in the inert gas, such as nitrogen, need not be known. The reason for this is that the volume of nitrogen in the sample bag cancels out when the ratio of the measured concentrations, $[O_2]/[HC]_{C3}$, is used in the above expression.

7.7.3 Mixture Stratification

Over the years, gasoline engine designers have employed various engine designs that deliberately cause the mixture strength in the engine cylinder to stratify. In this way, the mixture strength is greater close to the spark plug and weaker elsewhere. This mixture distribution can, for example, enhance the establishment of the combustion flame, particularly when a weak overall air-fuel mixture is used, or when diluents such as recirculated exhaust gas are used.

Quader [1982] employed a combination of axial swirl and timed port-injection to obtain mixture stratification within the engine cylinder, with a richer mixture being present near the cylinder head. A spark ignition engine operating on isooctane was used. In order to establish the extent of mixture stratification, Quader used an electromagnetically operated poppet valve to obtain mixture samples close to the cylinder head, prior to spark discharge. The sample was conveyed to an FID analyzer via heated tubing.

Johnston and Green [1982] carried out extensive studies of the air/fuel ratio in the engine cylinder of an optically accessible experimental engine running on propane at 500 rpm. The air-fuel mixture was highly stratified within the engine cylinder due to a combination of axial swirl and timed in-cylinder fuel injection (i.e., direct gasoline injection). Two techniques were used for the measurement of the air-fuel mixture strength. The first employed an inward-opening needle-type sampling valve with a minimum sampling period of 1 ms. The second used a laser Raman scattering spectroscopic system. The sample gas extracted from the engine cylinder by mean of the sampling valve was transferred to an FID system via a heated sample line. Three sample valve orientations and a number of valve tip penetrations within the cylinder were used. The equivalence ratios measured at different crank angles by means of the sampling valve and Raman spectroscopy were in close agreement. However, insertion of the sampling valve caused significant disturbances to the aerodynamic flow field within the combustion chamber.

Ozasa et al. [1991] also used a gas sampling technique to investigate the effect of combustion chamber shape on in-cylinder mixture strength in the case of a lean-burn spark ignition engine. Kano et al. [1998] used a sampling valve to analyze mixture formation in a direct injection gasoline engine. The outward-opening sampling valve was electromagnetically actuated and had a maximum lift of 0.2 mm. The sampling valve took 0.4 ms to reach this maximum lift from the closed position. The inlet to the sampling valve was positioned at the end of a tube having inner diameter of 1.7 mm. The other end of the tube was positioned within the combustion chamber. When the sampling valve opened, mixture from the cylinder flowed along the tube and entered the sampling valve. In general, extra care is needed in using this type of sampling arrangement, because the storage capacity of the tube could lead to sample contamination. When the sampling valve opens, the initial sample flowing into the valve will be the contents of the tube. These contents will be a combination of: (a) mixture that entered the tube during the compression stroke before the sampling valve opened and (b) residual mixture that was left in the tube at the end of the previous sampling valve opening. As small a tube internal volume as possible is therefore desirable. However, minimizing the tube volume is constrained by the need to keep the sample flow through the tube unrestricted while the sampling valve is open.

7.7.4 Mixture Distribution between Cylinders

Several investigators used sampling valves in order to obtain the mixture strength in the cylinders of multi-cylinder engines. In this way, maldistribution of the mixture between the engine cylinders could be revealed. A sample of the in-cylinder air-fuel mixture is normally extracted prior to spark discharge, and analyzed so that the air/fuel ratio in a particular cylinder can be determined.

This technique was used by Bartholomew et al. [1938] in order to determine the mixture strength prior to ignition from each of the eight cylinders of an automotive spark ignition engine running on gasoline. Bartholomew et al. used a needle-type sampling valve which was modified so that it performed the additional function of a spark plug. The valve and spark plug were combined because the only access to the combustion chamber was through the spark plug of the engine, which was installed in a vehicle. Bartholomew et al. collected the sample by allowing it to fill an evacuated tube. Others using similar techniques to study the mixture distribution between cylinders include Cooper et al. [1959], who used an electromagnetically actuated sampling valve, also incorporating the spark plug; and, Donahue et al. [1950], who used mechanically actuated poppet-type valves in a four-stroke six-cylinder gasoline engine operating at 1000 rpm.

7.7.5 Scavenging Efficiency

In two-stroke engines, the complete engine cycle is completed in a single engine revolution. A critical aspect of two-stroke engine operation is the scavenging process. This involves the removal of the combustion products from the engine cylinder and their replacement with fresh combustible mixture. Careful design of the engine ports and piston geometry is necessary in order to ensure that as great amount of fresh mixture as possible is trapped in the engine cylinder.

Isonomy and Yanagihara [1962] measured the scavenging efficiency of small (0.05 to 0.125 liter displacement) two-stroke motorcycle ported engines running up to 9000 rpm. Isonomy and Yanagihara used both poppet and needle-type sampling valves, and employed a technique of skipping several engine cycles between sampling. Thus, cylinder contents were sampled once every 100 or 200 cycles, first, to ensure adequate sampling valve response and, second, to allow engine running to recover after extracting a large proportion of the in-cylinder gas during the sampled cycle. In this type of engine, large maldistributions of the fresh air-fuel mixture and combustion residuals in the engine cylinder can exist. Therefore, Isonomy and Yanagihara considered it necessary to extract as much of the in-cylinder contents as possible, after the exhaust port had closed, so as to ensure that the calculated scavenging efficiency was a representative value for the whole cylinder. For example, in the case of a 0.05 liter engine, between 25 and 50% of the trapped mass of the cylinder contents was extracted once every 100 or 200 engine cycles.

7.8 Summary

The three optical techniques employed to measure fuel vapor concentrations, LRS, SRS, and LIF, share similar optical setups. High-power pulsed lasers are invariably used in order to obtain adequate signals. For most imaging applications, scattered light is detected with intensified solid state array photodetectors, such as ICCD cameras, as opposed to photographic films, because of their linear response curve, high sensitivity, and fast processing speed.

In principle, all three optical techniques described in this chapter can be used to measure fuel vapor concentration in the combustion chamber of an engine. The fuel concentration measurements are often carried out in order to obtain air/fuel ratio distributions in the cylinder. Hence, it is preferred that experimental results be expressed in terms of air/fuel ratios. LRS and LIF signals are directly related to the fuel concentration. Homogeneous air distribution is assumed when the fuel concentration is converted into air/fuel ratios. On the other hand, SRS and FARLIF measure directly the air/fuel ratio, which is particularly useful when stratified charge is present in the cylinder.

Laser induced fluorescence (LIF) is the most widely used technique for in-cylinder fuel distribution measurements. One of its most attractive features is the ease of acquiring instantaneous, two-dimensional images of fuel concentration in the cylinder of an IC engine because of its strong signal level. The red-shifted frequency of fluorescence emission enables the spurious scattered light from particles or walls to be eliminated. The most severe drawback of LIF is its sensitivity to collision quenching. This not only reduces the available signal intensity, but also complicates quantitative interpretation of experimental results. However, collision quenching during LIF measurement can be utilized to measure fuel/air ratio directly, such as in FARLIF.

Raman scattering is a very weak process. In most cases, SRS is limited by the signal-to-noise ratio. High laser intensity and a very sensitive detector are required to obtain SRS measurements. Most major species can be detected by SRS. The high-pressure environment in an IC engine actually facilitates the implementation of SRS, because it increases species concentration and hence the signal-to-noise ratio. The advantage of SRS is its capability of simultaneous detection of multiple major species, from which the fuel/air ratio is directly measured and residual gas concentration can be calculated. When an optical multi-channel detector, such as an ICCD array coupled to an imaging spectrometer, is used, multi-point and multi-species measurements can be obtained in IC engines.

Rayleigh scattering is a much stronger scattering process than Raman scattering. The optical setup is rather simple. Because LRS has the same wavelength as the incident laser light, it suffers from interference from Mie scattering and spurious scattered light from surrounding walls. This has limited most of LRS to motored engine operations with the extreme precaution of removing particles from the environment. LRS is more suitable to density measurement than concentration measurement, since it is not species specific. For most concentration measurements, LIF can be used to replace LRS.

The major advantage of a fast FID and a high-speed sampling valve over optical techniques is their easy adaptation to multi-cylinder production engines. The fast FID technique provides cycle resolved fuel concentration, and the high-speed sampling generates time-resolved multi-species measurements. However, both techniques are characterized by limited temporal and spatial resolutions.

References

Andresen, P., Bath, A., Gröger, W., Lülf, H.W., Meijer, G., and ter Meulen, J.J., (1988), "Fluorescence Imaging Inside an Internal Combustion Engine Using Tunable Excimer Lasers," *Appl. Opt.* Vol. 27, pp. 365–378.

Andresen, P., Meijer, G., Schlüter, H., Voges, H., Koch, A., Hentschel, W., Oppermann, W., and Rothe, E. (1990), "Fluorescence Imaging Inside an Internal Combustion Engine Using Tunable Excimer Lasers," *Applied Optics*, Vol. 29, pp. 2392–2402.

Arcoumanis, C., Green, H.G., and Whitelaw, J. (1985), "Velocity and Concentration Measurements in a Model Diesel Engine," *Expts. Fluids* 3, 270–276.

Arcoumanis, C., Whitelaw, J.H., and Wong, K.Y. (1989), "Gaseous Simulation of Diesel Engine Sprays," SAE Paper No. 890793.

Arnold, A., Dinkelacker, F., Heitsmann, T., Monkhouse, P., Schäfer, M., Sick, V., Wolfrum, J., Hentschel, W., and Schindler, K. (1992), "DI Diesel Engine Combustion Visualized by Combined Laser Techniques," *Proc. 24th Symposium (International) on Combustion*, pp. 1605–1612.

Arcoumanis, C. and Enotiadis, A.C. (1991), "In-Cylinder Fuel Distribution in a Port-Injected Model Engine Using Rayleigh Scattering," *Expts. Fluids* 11, 375–387.

Ayusawa, T., Nemoto, T., Koo, Y., and Jo, S.H.(1978), " Relationship Between Local Air-Fuel Ratio and Combustion Character in Spark Ignition Engines," SAE Paper No. 780147.

Baritaud, T.A. and Heinze, T.A. (1992), " Gasoline Distribution Measurements with PLIF in a SI Engine," SAE Paper No. 922355.

Bartholomew, E., Chalk, H., and Brewster, B. (1938), "Carburetion, Manifolding and Fuel Antiknock value," *SAE Transactions*, Vol. 33, pp. 141–156.

Berckmüller, M., Tait, N., and Greenhalgh, D. (1996), "The Time History of the Mixture Formation Process in a Lean Burn Stratified Charge Engine," SAE Paper No. 961929.

Berckmüller, M., Tait, N. and Greenhalgh, D. (1997), "The Influence of Local Fuel Concentration on Cyclic Variability of a Lean Burn Stratified Charge Engine," SAE Paper No. 970826.

Berckmüller, M., Tait, N. Lockett, R., and Greenhalgh, D. (1994), "In-Cylinder Crank Angle Resolved Imaging of Fuel Concentration in a Firing Spark Ignition Engine Using Planar Laser Induced Fluorescence," *Proc. 25th Symposium (International) on Combustion*, The Combustion Institute, pp. 151–156.

Charlton, S.J., Jager, D.J., Tawfig, M.E. (1992), "In-Cylinder Measurement of Mixture Strength in a Turbocharged Natural Gas Engine," *Proc. Instn. Mech. Engrs.*, Vol. 206, Part A, pp. 167–175.

Cheng, W.K., Hamrin, D., Heywood, J.B., Hochgreb, S., Min, K.D., and Norris, M. (1993), "An Overview of Hydrocarbon Emissions Mechanisms in Spark Ignition Engines," SAE Paper No. 932078.

Cooper, D.E., Courtney, R.L., and Hall, C.A.(1959), "Radioactive Tracers Cast New Light on Fuel Distribution," *SAE Trans.*, Vol. 67, pp. 619–639.

Daniel, W.A. (1957), "Flame Quenching at the Walls of an Internal Combustion Engine," *Proc. 6th Symposium (International) on Combustion*, The Combustion Institute.

Daniel, W.A. and Wentworth, J.T.(1962), "Exhaust Gas Hydrocarbons—Genesis and Exodus," SAE Paper No. 48613.

Deschamps, B., Snyder, R., and Baritaud, T. (1994), "Effect of Flow and Gasoline Stratification on Combustion in a 4-Valve SI Engine," SAE Paper No. 941993.

Donahue, R.W. and Kent, R.H. (1950), "A Study of Mixture Distribution," *SAE Quart. Trans.*, Vol. 4, No. 4, pp. 546–558.

Eckbreth, A.C. (1996), *Laser Diagnostics for Combustion Temperature and Species*, 2nd Edition, Gordon and Breach Publishers.

Espey, C., Dec, J., Litzinger, T.A., and Santavicca, A. (1994), "Quantitative 2D Fuel Vapour Concentration Imaging in a Firing DI Diesel Engine Using Planar Laser Induced Rayleigh Scattering," SAE Paper No. 940682.

Espey, C., Dec., J., Litzinger, T., and Santavicca, D. (1997), "Planar Laser Rayleigh Scattering for Quantitative Vapour Fuel Imaging in a Diesel Jet," *Combustion and Flames*, Vol. 109, Number 1/2, pp. 65–86.

Fansler, T.D., French, D.T. and Drake, M.C. (1995), "Fuel Distribution in a Firing Direct Injection Spark Ignition Engine Using Laser Induced Fluorescence Imaging," SAE Paper No. 950110.

Felton, P.G. (1996), "Laser Diagnostics for Direct-Injection Gasoline Engines," Paper presented at IMechE International Seminar on Lean Burn Combustion Engines, London.

Felton, P.G., Kyritsis, D.C., and Fulcher, S.K. (1995), "LIF Visualisation of Liquid Fuel in the Intake Manifold During Cold Start," SAE Paper No. 952464.

Fourgette, D.C., Zurn, R.M., and Long, M.B. (1986), "Two Dimensional Rayleigh Thermometry in a Turbulent Nonpremixed Methane Hydrogen Flame," *Combust. Sci. Technol.* Vol. 44, 307–317.

Fox, J.W., Min, K.D., Cheng, W.K., and Heywood, J.B. (1992), "Mixture Preparation in a SI Engine with Port Fuel Injection During Starting and Warm Up," SAE Paper No. 922170.

Fujikawa, T., Hattori, Y., and Akihama, K. (1997), "Quantitative 2-D Fuel Distribution Measurements in an SI Engine Using Laser-Induced Fluorescence with Suitable Combination of Fluorescence Trace and Excitation Wavelength," SAE Paper No. 972944.

Ghandhi, J.B. and Bracco, F.V. (1995), "Fuel Distribution Effects on the Combustion of a Direct Injection Stratified Charge Engine," SAE Paper No. 950460.

Green, H.G., Whitelaw, J.H., and Wong, K.Y. (1988), "Air Fuel Mixture Characteristics of Reciprocating Engines," *Combust. Sci. Technol*. 59, 225–246.

Grünefeld, G., Beushausen, V., Andresen, P., and Hentschel, W. (1994b), "A Major Origin of Cyclic Energy Conversion Variations in SI Engines: Cycle by Cycle Variations of the Equivalence Ratio and Residual Gas of the Initial Charge," SAE Paper No. 941880.

Grünefeld, G., Beushausen, V., Andresen, P., and Hentschel, W. (1994a), "Spatially Resolved Raman Scattering for Multispecies and Temperature Analysis in Technically Applied Combustion Systems: Spray Flame and Four Cylinder In Line Engine," *Appl. Phys. B.*, Vol. 58, pp. 333–342.

Grünefeld, G., Knapp, M., Beushausen, G., Andresen, P., Hentschel, W., and Manz, W. (1995), "In Cylinder Measurements and Analysis on Fundamental Cold Start and Warm Up Phenomena of SI Engines," SAE Paper No. 952394.

Hardalupas, Y., Taylor, A.M.K.P., Whitelaw, J.H., Ishii, K., Miyano, H., and Urata, Y. (1995), "Influence of Injection Timing on In Cylinder Fuel Distribution in a Honda VTEC-E Engine," SAE Paper No. 950507.

Heinze, T. and Schmidt, T. (1989), "Fuel Air Ratios in a Spray, Determined Between Injection and Autoignition by Pulsed Spontaneous Raman Spectroscopy," SAE Paper No. 892102.

Heinze, P. and Miles, P. (1999), "Quantitative Measurements of Residual and Fresh Charge Mixing in a Modern SI Engine Using Spontaneous Raman Scattering," SAE Paper No. 1999-01-1106.

Hishinuma, H., Urushihara, T., Kaduho, A. and Itoh, T. (1996), "Development of a Technique for Quantifying A/F Ratio Distribution Using LIF Image Processing," *JSAE Review*, Vol. 17, pp. 355–359.

Horie, K., Nishizawa, K., Ogawa, T., Akazaki, S., and Miura, K. (1992), "The Development of a High Fuel Economy and High Performance Four Valve Lean Burn Engine," SAE Paper No. 920455.

Inaba, H. and Kobayasi, T. (1972), "Laser Raman Radar—Laser Raman Scattering Methods for Remote Detection and Analysis of Atmospheric Pollution," *Opto-Electronics*, Vol. 4, 101–123.

Isonomy, T. and Yanagihara, S. (1962), "Gas Sampling Valve for Measuring Scavenging Efficiency in High-Speed Two-Stroke Engines," *SAE Transactions T-47*, pp. 420–433.

Itoh, T., Kakuho, A., Hishinuma, H., Urushiahara, T., Takagi, Y., Horie, K., Asano, M., Ogata, E., and Yamasita, T. (1995), "Development of a New Compound Fuel and Fluorescent Tracer Combination for Use with Laser Induced Fluorescence," SAE Paper No. 952465.

Johansson, B., Neij, H., Alden, M., and Juhlin, G. (1995), "Investigations of the Influence of Mixture Preparation on Cyclic Variations in a SI Engine Using Laser Induced Fluorescence," SAE Paper No. 950108.

Johnston, S.C. (1979), "Precombustion Fuel/Air Distribution in a Stratified Charge Engine Using Laser Raman Spectroscopy," SAE Paper No. 790433.

Johnston, S.C. (1980), "Raman Spectroscopy and Flow Visualisation of Stratified Charge Engine Combustion," SAE Paper No. 800136.

Johnston, S.C. and Green, R.M. (1982), "Engine Hydrocarbon Concentration Measurements: A Comparison Between a Sampling Probe and Raman Spectroscopy," SAE Paper No. 820499.

Kadota, T., Zhao, F.-Q., and Miyoshi, K. (1990), "Rayleigh Scattering Measurements of Transient Fuel Vapour Concentration in a Motored Spark Ignition Engine," SAE Paper No. 900481.

Kano, M., Saito, K., Basaki, M., Matsushita, and S., Gohno, T. (1998), "Analysis of Mixture Formation of Direct Injection Gasoline Engine," SAE Paper No. 980157.

Kiyota, Y., Akishino, K., and Ando, H. (1992), "Concept of Lean Combustion by Barrel Stratification," SAE Paper No. 920678.

Knapp, M., Beushausen, V., Hentschel, W., Manz, P., Grunefeld, G., and Andresen, P. (1997), "In Cylinder Mixture Formation Analysis with Spontaneous Raman Scattering Applied to a Mass Production SI Engine," SAE Paper No. 970827.

Knapp, M., Luczak, A., Beushausen, V., Hentschel, W., Manz, P., and Andresen, P. (1996), "Polarisation Separated Spatially Resolved Single Laser Shot Multispecies Analysis in the Combustion Chamber of a Realistic SI Engine with a Tunable KrF Excimer Laser," *Proc. 26th Symposium (International) on Combustion*, The Combustion Institute, pp. 2589–2596.

Kohse-Höinghaus, K. (1994), "Laser Techniques for the Quantitative Detection of Reactive Intermediates in Combustion Systems," *Prog. Energy Combust. Sci.*, Vol. 20, pp. 203–279.

Kume, T., Iwamoto, Y., Iida, K., Murakami, M., Akishino, K., and Ando, H. (1996), "Combustion Control Technologies for Direct Injection SI Engine," SAE Paper No. 960600.

Ladommatos, N. (1992a), "Cyclically Resolved Measurements of Hydrocarbons in the Cylinders of Internal-Combustion Engines by Means of a Fast Flame-Ionisation Detector," *J. Inst. Energy*, Vol. LXV, pp. 94–101.

Ladommatos, N. (1992b), "Investigation of Gasoline Films Accumulating in the Inlet Ports of Spark-Ignition Engines," *J. Inst. Energy*, Vol. LXV, pp. 102–104.

Ladommatos, N. and Rose, D. (1995), "Monitoring of the Mixture Strength in the Cylinders of a Port-Injected Gasoline Engine During Transient Operation," *Proc. Instn. Mech. Engrs.*, Vol. 209, Part D, pp. 127–134.

Ladommatos, N. and Rose, D. (1996), "On the Causes of In-Cylinder Air Fuel Ratio Excursions During Load and Fuelling Transients in Port-Injected Spark-Ignition Engines," SAE Paper No. 960466.

Lake, H., Stokes, J., and Pendlebury, K.J. (1995), "Development Experience of a Multi-Cylinder CCVS Engine," SAE Paper No. 950165.

Lawrenz, W., Köhler, J., Meier, F., Stolz, W., Wirth, R., Bloss W., Maly, R., Wagner, E., and Zahr M. (1992), "Quantitative 2D LIF Measurements of Air/Fuel Ratios During the Intake Stroke in a Transparent SI Engine," SAE Paper No. 922320.

Lederman, S. (1977), "The Use of Laser Raman Diagnostics in Flow Fields and Combustion," *Prog. Energy Combust. Sec.*, Vol. 3, pp. 1–34.

Lee, K. and Foster, D. (1995), "Cycle-by-Cycle Variations in Combustion and Mixture Concentration in the Vicinity of Spark Plug Gap," SAE Paper No. 950814.

Li, C.F., Dual, Q.C., Maroteaux, F., and Murat, M. (1994a), "Time-Resolved Measurement of Fuel Transient Behaviour and Cycle to Cycle Variation of Local Fuel/Air Equivalence Ratio in Gasoline Engines," SAE Paper No. 940989.

Li, C.F., Murat, M., Jullien, J., and Zhu, M.J. (1994b), "A Fast-Response Catalytic Hot-Wire Sonic Nozzle Probe to Measure Fuel Concentration in a Fuel/Air Mixture," *Measurement Science and Technology*, 5, pp. 151–156.

Matsui, K., Tanaka, T., and Ohigashi, S.(1979), "Measurement of Local Mixture Strength at Spark Gap of SI Engines," SAE Paper No. 790483.

Matsushita, S., Inoue, T., Nakanishi, K., Kato, K., and Kobayashi, N. (1985), "Development of the Toyota Lean Burn Combustion System," SAE Paper No. 850044.

Meyer, J., Haug, M., Schreiber, M., and Unverzagt, S. (1994), "Quantitative Fuel Distribution in a Spark Ignition Engine," *Proc. COMMODIA 94*, pp. 95–105.

Meyer, J., Graul, W., Kiefer, K., Thiemann, J., and Landry, M. (1997), "Study and Visualisation of the Fuel Distribution in a Stratified Spark Ignition Engine with EGR Using Laser Induced Fluorescence," SAE Paper No. 970868.

Meyer, J., Haug, M., Schreiber, M., and Unverzagt, S. (1995), "Controlling Combustion in a Spark Ignition Engine by Quantitative Fuel Distribution," SAE Paper No. 950107.

Meyer, R.C. and Thring, R.H. (1995), "Mixture Preparation Measurements," SAE Paper No. 950069.

Miles, P. and Dilligan, M. (1996), "Quantitative In-Cylinder Fluid Composition Measurements Using Broadband Spontaneous Raman Scattering," SAE Paper No. 960828.

Miles, P. and Hinze, P. (1998), "Characterisation of the Mixing of Fresh Charge with Combustion Residuals Using Laser Raman Scattering with Broadband Detection," SAE Paper No. 981428.

Ozasa, T., Suzuki, S., Kondo, H., Takahashi, H., Kataoka, M., and Oshima, Y. (1991), "Effect of Combustion Chamber Shape on a Lean-Burn SI Engine—Measurement of In-Cylinder Mixture Strength Using a Gas Sampling Technique," *JSAE Review*, Vol. 12, No. 2, pp. 4–11.

Peckham, M. and Collings, N. (1992), "Study of Engine Wall Layer Hydrocarbons with a Fast-Response FID," SAE Paper No. 922237.

Peckham, M. and Collings, N. (1994), "Static and 'Flying' FID sampling of In-Cylinder Crevice Hydrocarbons," IMechE Seminar "Measurement and Observation Analysis of Combustion in Engines," London.

Quader, A.A. (1982), "The Axially Stratified Charge Engine," SAE Paper No. 820131.

Reboux J., Puechberty, D., and Dionnet, F. (1994), "A New Approach of Planar Laser Induced Fluorescence Applied to Fuel/Air Ratio Measurement in the Compression Stroke of an Optical SI Engine," SAE Paper No. 941988.

Reboux J., Puechberty, D., and Dionnet, F. (1996), "Study of Mixture Inhomogeneities and Combustion Development in a SI Engine Using a New Approach of Laser Induced Fluorescence (FARLIF)," SAE Paper No. 961205.

Rose, D., Ladommatos, N., and Stone, R. (1994), " In-Cylinder Mixture Excursions in a Port-Injected Engine Using Fast Throttle Opening," SAE Paper No. 940382.

Sadler, M., Stokes, J., Edwards, S.P., Zhao, H., and Ladommatos, N. (1997), "Optimisation of the Combustion System for a Direct Injection Gasoline Engine Using a High-Speed In-Cylinder Sampling Valve," *Proc. Intern. Semin. on Euro IV Challenge—Future Technologies and Systems*, IMechE.

Sawersyn, J.P., Sochet L., Desenne, D., Crunelle-Cras, M., Grase, F., and Bridoux, M. (1986), "A Study of Spatial Distribution of Molecular Species in an Engine by Pulsed Multichannel Raman Spectroscopy," *Proc. 21st Symposium (International) on Combustion*, pp. 491–496, The Combustion Institute.

Schrotter, H.W. and Klockner, H.W. (1979), in *Raman Spectroscopy of Gases and Liquids*, Ed. by A. Weber, pp. 123–166, Springer-Verlag, Berlin.

Seitzman, J.M. and Hanson, R.K. (1993), *Instrumentation for Flows with Combustion*, Ed. by A.M. K.P. Taylor, pp. 405–466, Academic Press.

Shin, Y., Min, K., and Cheng, W.K. (1995), "Visualisation of Mixture Preparation in a Port-Fuel Injection Engine During Warm-Up," SAE Paper No. 952481.

Sleightholme, G.R. (1990), "In-Cylinder Measurements of Charge Inhomogeneity in a Spark-Ignition Engine," SAE Paper No. 900484.

Summers, T. (1996), *Fast-Response FID Measurement of SI Engine Residual Gas Hydrocarbon Concentration,* PhD Thesis, Cambridge University.

Swindal, J.C., Dragonetti, D.P., Hahn, R., Furman, P.A., and Acker, W.P. (1995), "In-Cylinder Charge Homogeneity during Cold Start Studied with Fluorescent Tracers Simulating Different Fuel Distillation Temperatures," SAE Paper No. 950106.

Urushihara, T., Nakata, T., Kaduhou, A., and Takagi, Y. (1996), "Effects of Swirl and Tumble Motion on Fuel Vapour Behaviour and Mixture Stratification in Lean Burn Engines," *JSAE Review,* Vol. 17, pp. 239–244.

Wehrmeyer, J.A., Cheng, T-S., and Pitz, R.W. (1992), "Raman Scattering Measurements in Flames Using a Tunable KrF Laser," *Appl. Opt.* Vol. 31, pp. 1495–1504.

Wolff, D., Beushausen, V., Schlüter, H., and Andresen, P. (1994), "Quantitative 2D-Mixture Fraction Imaging Inside an Internal Combustion Engine Using Acetone Fluorescence," *Proc. COMMODIA 94*, pp. 445–450.

Zhao, F-Q, Kadota, T., and Takemoto, T. (1990), "Rayleigh Scattering Measurements of Fuel Vapour Concentration Fluctuation in a Motored Spark Ignition Engine," *Proc. Int. Symp. COMODIA* 90, 377–382.

Zhao, F.-Q., Kadota, T., and Takemoto, T. (1991), "Temporal and Cyclic Fluctuation of Fuel Vapour Concentration in a Spark Ignition Engine," SAE Paper No. 912346.

Zhao, F. and Hiroyasu, H. (1993), "The Applications of Laser Rayleigh Scattering to Combustion Diagnostics," *Prog. Energy Combust. Sci*. Vol. 19, pp. 447–485.

Zhao, F.-Q., Taketomi, M., Nishida, K., and Hirorasu, H. (1993), "Quantitative Imaging of the Fuel Concentration in a SI Engine with Laser Rayleigh Scattering," SAE Paper No. 932641.

Zhao, F.Q., Taketomi, M., Nishida, K., and Hiroyasu, H. (1994), "PLIF Measurements of the Cyclic Variation of Mixture Concentration in a SI Engine," SAE Paper No. 940988.

Chapter 8

In-Cylinder Liquid Fuel Measurement

Outline

8.1 Introduction

In a diesel or gasoline direct injection (GDI) engine, the fuel is directly injected into the combustion chamber, where flow is highly turbulent. Fuel injection takes place in a very short period of time. Both liquid droplets and fuel vapor exist in the combustion chamber. The characterization of liquid fuel atomization and vaporization is extremely important since these factors directly affect the engine's performance and emissions. For a port injected spark ignition engine, the amount of liquid fuel droplets is negligible at the time of spark discharge. However, it has been found that the atomization and vaporization processes affect the vapor fuel distribution inside the combustion chamber at the time of combustion. During cold start, the temperature of the cylinder walls is too low to effectively vaporize the liquid fuel. As a result, in port injection SI engines, liquid fuel can enter the cylinder during intake, and remain in the cylinder both as droplets dispersed in the gas mixture and as liquid films on walls. The large droplets and fuel film on the cylinder walls will then contribute to the large uHC emissions during the cold-start.

The simplest and most utilized optical technique for fuel spray characterization is the direct imaging of sprays. Photographic and, more recently, digital imaging using a solid state camera have been used to characterize automotive fuel sprays. Optical imaging of sprays has been reviewed recently by Chigier [1991], in which direct imaging, high-speed cinematography, and holography, were discussed in detail. Direct imaging of fuel sprays with back-illumination has been used extensively to measure spray geometry, penetration depth, wall impingement, and subsequent combustion in diesel combustion engines [e.g., Kamimoto et al., 1987; Yoshikawa et al., 1989; Rao et al., 1992; Kosaka et al., 1992; Winklhofer et al., 1993]. Most of these in-cylinder spray measurements were carried out together with combustion imaging, and they will be described in detail in Chapter 9.

In addition to the general spray structure, it is important to know the droplet size distributions in the spray, because the droplet size distribution has a direct effect on the fuel atomization and vaporization processes. Optical techniques for droplet sizing can be divided into direct imaging techniques and light-scattering techniques.

The simplest optical droplet sizing technique is the direct imaging of fuel droplets by photographic films or solid state cameras. These images can provide information on the general structure of a fuel spray as well as droplet size distribution. The droplet size distribution is determined by counting and sizing the diameter of the droplets from the recorded images. The direct imaging technique normally involves a strobe light of short duration in order to freeze the flow in the fuel spray. Fig. 8.1 illustrates a suitable optical geometry for spark photography of sprays. Alternatively, the spark source can be replaced by a pulsed laser or a modulated CW laser, which has a much shorter pulse duration and much higher pulse energy. A photographic film is preferred when high-resolution images are required, whereas a solid state camera provides easy and automated data processing. This technique can only be applied to relatively dilute regions of the droplet field. Also, because only one plane of the droplet field at a time can be photographed, care must be taken not to include the image of the droplets out of focus into the size analysis. There is also a trade-off between the minimum droplet size that can be resolved and the detectable area by the direct imaging technique. Detection of smaller droplets requires larger magnification, which leads to smaller detectable area in the droplet field. In the extreme cases, the detectable droplet field may be restricted to a few millimetres.

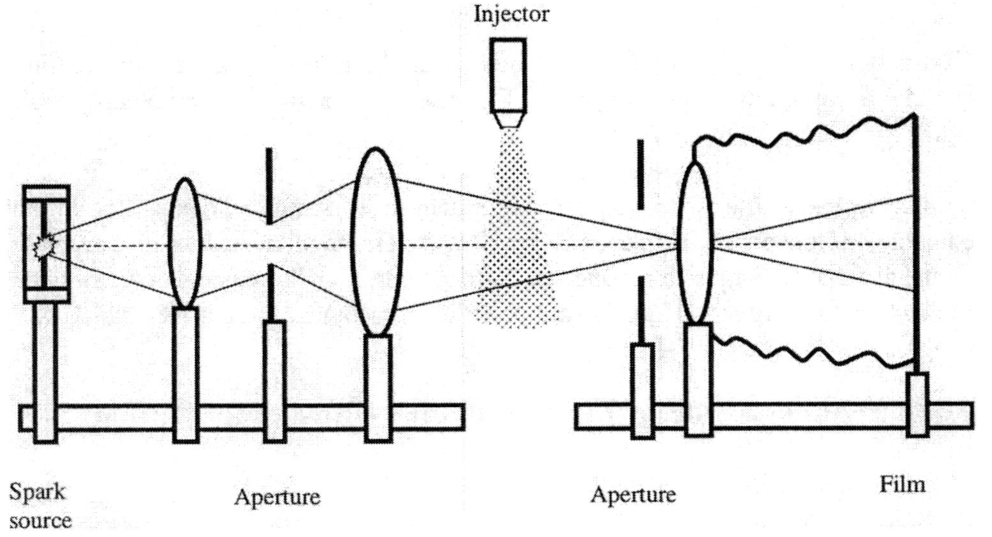

Fig. 8.1 Optical setup for the back illuminated photography of sprays.

Photographic techniques can be extended into three-dimensional measurements by holographic recordings. Holograms provide three-dimensional pictures of an object, which can be examined in much the same way as a photograph. A complete review of particle and flow field holography was given by Trolinger [1991].

There are two approaches to light scattering measurements: integral methods and particle counting. The integral methods, typically performed by the Fraunhofer diffraction, measure a large numbers of droplets simultaneously. The resulting intensity comes from integration over space, and a spatial average is obtained. This implies moderate to high droplet concentrations and a significant sample volume. On the other hand, the particle counting techniques, such as a laser phase Doppler analyzer (PDA), droplets are viewed one at a time. This implies low to moderate droplet concentrations and a small sample volume. The higher the concentration, the smaller the volume needs to be. The size distribution by PDA measurements is built up over time and a temporal average is obtained. The temporal distribution also provides a measure of mass flux of droplets of different sizes. The spatial and temporal size distributions can be related to each other, if the droplet size-velocity relationship is known.

In a fuel spray, where both liquid droplets and fuel vapour may be present at the same time, it is desirable to be able to detect them simultaneously. This can be achieved by laser induced exciplex fluorescence (LIEF) method. Though elastic (Mie) scattering or fluorescence signal can be obtained simultaneously from both the liquid and vaporized fuel, the signal from the liquid phase is many orders of magnitude stronger than that from the fuel vapor, which limits the elastic scattering or LIF techniques to the detection of liquid only. The LIEF technique offers significant improvement in the dynamic range because the fluorescence from the liquid and the vapor phase can be separated in wavelength and detected separately.

Laser Extinction/Absorption (LEA) is the other technique that is capable of simultaneous detection of fuel vapour and liquid droplets. LEA may also generate information of droplet size in the spray.

This chapter begins with a review of the two optical drop-sizing techniques, the Fraunhofer diffraction method and phase Doppler analyzer (PDA). This is followed by a discussion of two-dimensional imaging techniques for liquid fuel distributions. Finally, techniques for simultaneous measurements of fuel vapor and liquid droplets are discussed, including LIEF and LEA.

8.2 Liquid Droplet Sizing by the Fraunhofer Diffraction Method

8.2.1 Fraunhofer Diffraction

The Fraunhofer diffraction method derives the droplet size information from the angular distribution of the elastically scattered radiation. When a relatively large opaque particle ($d \gg \lambda$) is illuminated by a beam of parallel monochromatic light, the Fraunhofer diffraction pattern can be observed on a screen at a great distance from the particle. This pattern is characterized by a series of alternating dark and bright rings that relate to the location of zero and peak intensities, as shown in Fig. 8.2. In practice, the Fraunhofer diffraction pattern is observed at the focal plane using a lens.

The mathematical description of the Fraunhofer diffraction pattern for a particle of diameter d can be shown to be

$$I = I_0 \left(\frac{2J_1(X)}{X} \right)^2 \tag{8.1}$$

where I_0 is the intensity at the center of the diffraction pattern, $J_1(X)$ is the first order spherical Bessel function, and X is a dimensionless size parameter given by

$$X = \frac{\pi d R}{\lambda f} \tag{8.2}$$

where f is the focal length of the collection lens, and R is the radial distance in the detection plane (focal plane) as measured from the optical axis.

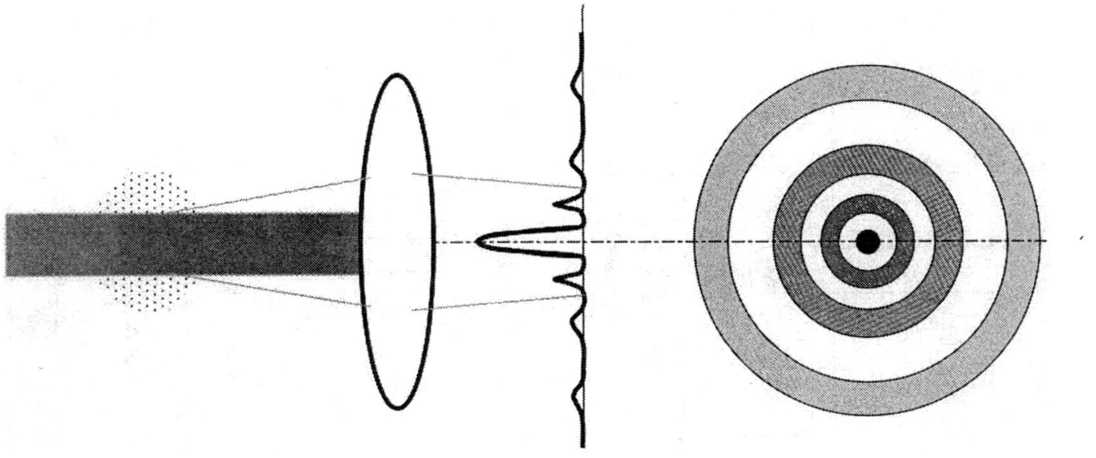

Fig. 8.2 Radial intensity distribution of the Fraunhofer diffraction pattern.

As shown by Eq. (8.1), the diffraction pattern is not sensitive to refractive index or shape, and it is unique for a particle or a collection of particles of different sizes. It is therefore possible to deduce the size distribution of a collection of particles from their diffraction patterns.

8.2.2 Implementation of the Fraunhofer Diffraction Method

The basic apparatus of the Fraunhofer diffraction method is illustrated in Fig. 8.3. A parallel beam of monochromatic light, readily from a laser (typically a low-power He-Ne laser), traverses a cloud of droplets or particles. The transmitted and scattered light is collected by a lens, and its intensity distribution is observed in the focal plane. The intensity distribution can be measured by traversing a photodetector across the focal plane. This is time-consuming and it is preferable to use an array of detectors.

An apparatus developed by Swithenbank et al. [1977] has been used extensively and adopted commercially by Malvern Instruments. It collects the light diffracted by a particle using a set of concentric photodetector rings. The light energy within any ring of radii R_1 and R_2 can be shown to be, [Swithenbank et al., 1977]

$$P_{R_1,R_2} = P\left[J_0^2(X_1) + J_1^2(X_1) - J_0^2(X_2) - J_1^2(X_2)\right] \tag{8.3}$$

where J_0 is the zero-order spherical Bessel function, and P is the energy falling on the particle. For N particles of the same size, the total energy falling on the particles is given by

$$P = NI_0\pi\left(\frac{d}{2}\right)^2 \tag{8.4}$$

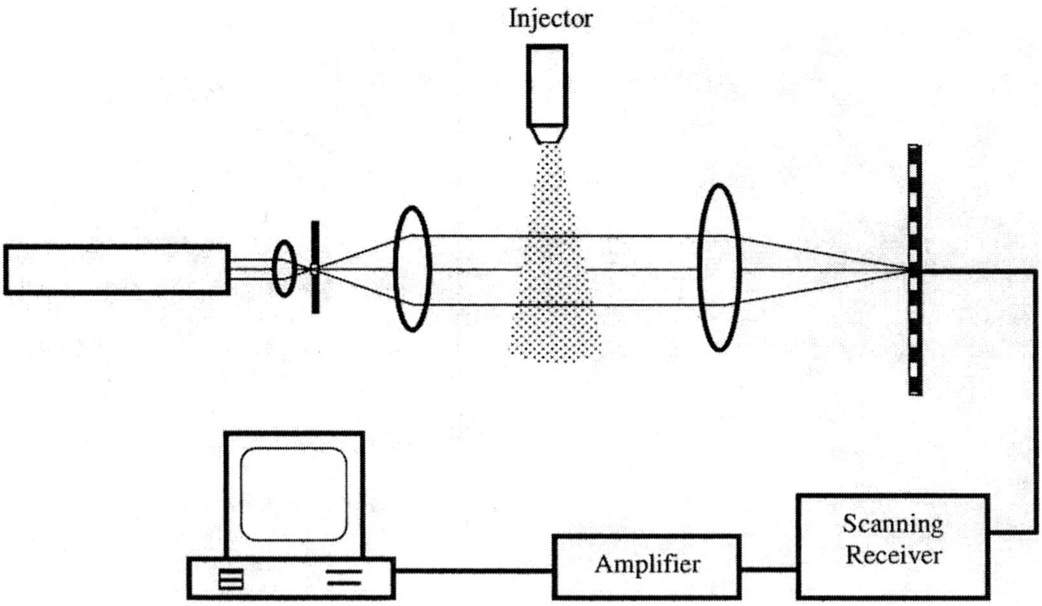

Fig. 8.3 Experimental setup for the Fraunhofer diffraction technique.

If a collection of particles of different sizes is present, then the light energy falling on any ring in the focal plane is the sum of those from individual particles:

$$P_{R_1,R_2} = C\sum_{j=1}^{M} \frac{W_j}{a_j}\left[J_0^2(X_{j,1}) + J_1^2(X_{j,1}) - J_0^2(X_{j,2}) - J_1^2(X_{j,2})\right] \quad (8.5)$$

where C is a constant, W_i is the weight fraction of the particles of diameter d_i, and M is the total number of particle size ranges. Theoretically, the size distribution may be obtained from the measured data by inverting Eq. (8.5). However, this is difficult to do in practice. An alternative approach to the inversion problem is to solve a discretized version of Eq. (8.5).

$$P(I) = W(J)T(I,J) \quad (8.6)$$

where P(I) and W(J) are the light energy and weight distributions, respectively, and T(I, J) contains the coefficients that define the light energy distribution for each size range.

Unfortunately, Eq. (8.6) cannot be inverted directly. For this reason, least-squares-minimization-of-error solutions are adopted, where a form for W(J) is assumed and iteratively adjusted until the sum of squared errors in P(I) is minimum. This process can be expressed as

$$\sum_{I=1}^{N_d}\left[P_m(I)-W(J)T(I,J)\right]^2=\min \tag{8.7}$$

where $P_m(I)$ is the measured light energy distribution, and N_d is the total number of detector rings. Several distributions, which are typically two-parameter functions, have been used, the most common being Rosin-Rammler and log-normal. The two parameters in these functions are varied until a "best fit" to the measured data is found. This approach is rather limited. At best, only an approximation to the true size distribution is achieved unless it fits the assumed form exactly. Further, if the size distribution is multi-modal, the assumption of a single-peak will be a very poor representation.

To overcome this difficulty, Malvern Instruments has developed a "modal-independent" software to perform the direct inversion of the diffraction measurements. The principle behind this has not been made clear, but there is evidence that it performs satisfactorily [Young and Bachalo, 1988].

The ratio of light intensity measured at the center diode, before and after the droplet field is introduced into the field, gives the fraction of the light transmitted by the particles. This ratio is related to the drop concentration by the Beer-Lambert law:

$$\ln\left(\frac{I}{I_0}\right)=-\tau L \tag{8.8}$$

where L is the optical path length, and τ is given by

$$\tau=2\sum_{j=1}^{M}N_j\pi\left(d_j/2\right)^2 \tag{8.9}$$

where N_j is the number of droplets in the size range j per unit volume, and d_j the mean drop diameter of the jth size class. Thus, when the size distribution is known, the light attenuation can be used to determine the droplet concentration. The volume fraction of droplets is given by

$$C_V=-\frac{\ln(I/I_0)}{3L\sum_{j=1}^{M}\left(V_j/d_j\right)} \tag{8.10}$$

where V_j is the drop volume distribution that results from the size distribution measurement. For a low-density particle field, Eq. (8.10) cannot be used because of small attenuation. The volume fraction of droplets can be, instead, evaluated from the total scattered light energy measured by the detector:

$$C_V = C'' \sum_{i=1}^{N_d} P(i) \left(L \sum_{i=1}^{N_d} \sum_{j=1}^{M} V(i) T(i,j) \right)^{-1} \tag{8.11}$$

where P(i) is the total scattered light energy collected from all detector rings other than the central diode, C″ is a system constant and is determined by calibration.

Using different focal lengths for the receiving lens, droplets in the size range 0.2–2000 mm can be measured. The instrument can be calibrated using particles of known size or calibration reticles that have a large numbers of discs of well-determined size photographically etched onto glass.

The method is relative in that it relies on the shape of the scattering pattern rather than the actual intensities. Thus, it is independent of variations in incident intensity. At low level of drop concentration, scattering from dust on lenses or windows is a particular problem. In order to minimize scattering by dust particles, the collection lens and any windows should be kept clean. Practically, the minimum extinction appears to be around 5–10%, as recommended by Malvern Instruments.

If the concentration is too high, multiple scattering occurs and alters the diffraction pattern. Special procedures developed over the years can extend the applicable range of the diffraction method to extinction in excess of 90%.

If measurements are carried out in a high-temperature environment, deflections due to density changes will cause the central spot to move. Over a finite time, this will have the effect of increasing the spot size and thus reduce the maximum measurable drop size. One way around this difficulty is to gate the readings so that they are only taken when the spot is central. The readings need to be taken in a time that is short compared to the inverse of the fluctuation frequency.

The Fraunhofer diffraction-based instruments are well established as a means for spray characterization. Since the diffracted light is integrated over the measurement volume along the laser beam, it has limited spatial resolution, although an Abel inversion scheme can be used with the assumption of an axisymmetric spray to procure spatially resolved measurements. The method is particularly prone to minor fluctuations in the refractive index, which can cause significant beam deflections. It is very difficult to combine the method with other techniques to produce simultaneous size and velocity measurements of the drops.

8.3 Droplets Sizing and Velocity Measurements by a Phase Doppler Analyzer

8.3.1 Introduction

The laser phase Doppler analyser (PDA) requires that an interference fringe pattern be created within a small volume in the spray. As a droplet moves through this volume, a scattered light signal is generated which is modulated in both time and space. The temporal modulation is related to the velocity of the droplet, while the spatial frequency is related to the droplet size. PDA requires that no more than one droplet be in the probe volume at any moment, and hence it is not well suited to high-density sprays. The PDA technique cannot provide size distributions in a single transient injection. After a number of years of development and refinement, the method has received wide acceptance as a reliable means of characterizing sprays. It can provide size and velocity data, number density, volume flux, and time resolved information.

8.3.2 The Phase Doppler Principle

The PDA technique is a particle size measurement technique based on the detection of the phase of the light scattered from an individual particle. The scattering of light by homogeneous dielectric spheres of arbitrary size is described most rigorously by the Lorenz-Mie theory. However, the geometrical optics approach will be employed here, which uses simple ray optics and provides a ready insight into the scattering phenomena used in the phase Doppler method.

Fig. 8.4(a) shows light rays incident upon the lower half of a sphere. At the first interface, part of the light is reflected from the surface of the sphere and these rays are referred to as $p = 0$ (reflected rays). Light rays transmitted and refracted by the sphere are designated as $p = 1$ rays (1st order refracted rays). Rays reflected from the internal surface and refracted in the backward direction are $p = 2$ rays (2nd order refracted rays).

In order to compute the phase from ray optics, it is common to reference the actual ray to a hypothetical ray scattered without a phase lag at the center of the sphere, as shown in Fig. 8.4(b). It is apparent that the reflected wave from the first surface experiences a shortcut of length Δs relative to the hypothetical wave, and hence a phase difference between the two beams will result:

$$\Delta\sigma_0 = \frac{\Delta s}{\lambda} \tag{8.12}$$

Similarly, the first refracted ray undergoes a change of phase relative to the hypothetical wave.

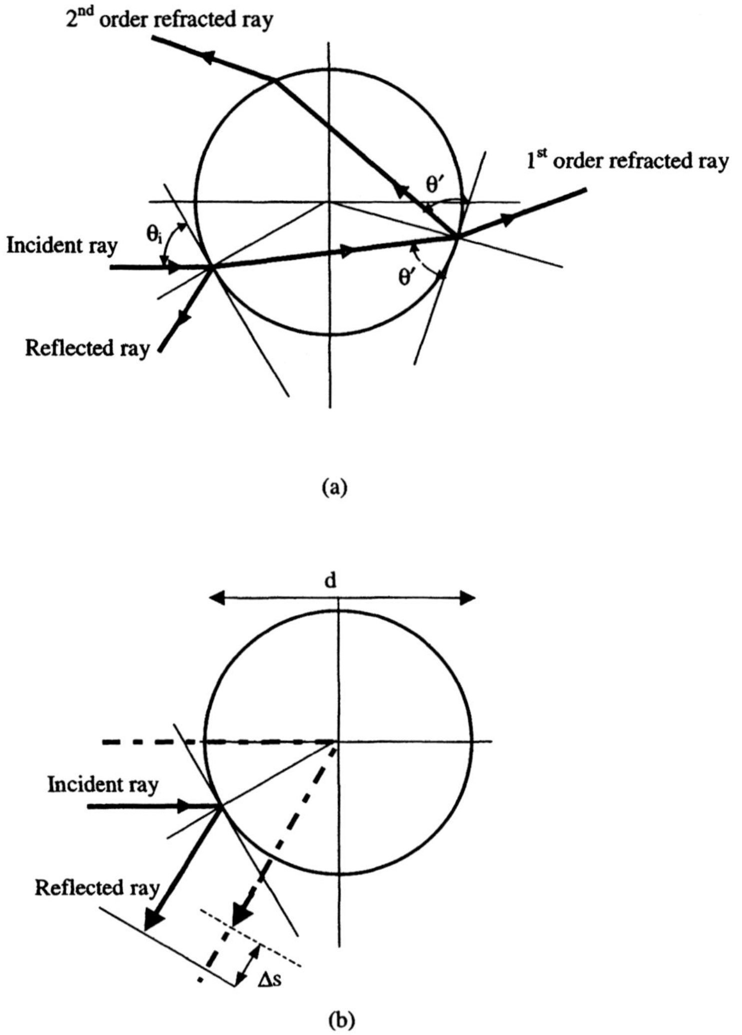

Fig. 8.4 Geometrical ray trace of incident ray through a sphere.

The general expression for the optical path length of a light ray passing through the sphere relative to the hypothetical wave was derived by Bachalo [1980],

$$\Delta s = 2a\left(\sin\theta_i - p\cdot m\cdot \sin\theta'\right) \tag{8.13}$$

The resulting phase difference is given by

$$\Delta\sigma = \frac{2\pi d}{\lambda}\left(\sin\theta_i - p\cdot m\cdot \sin\theta'\right) \tag{8.14}$$

where $m = n - ik = m_2/m_1$ is the relative refractive index of the sphere to the surrounding medium, $a = \pi d/\lambda$ is the size parameter, and θ_i and θ' are the angles between the surface tangent and the incident and refracted rays, respectively. The value of the integer p depends on the nature of ray (p = 0 for reflection; p = 1 for 1st refraction; p = 2 for 2nd refraction).

Equation (8.14) illustrates that the phase difference is proportional to the diameter of the spherical particle, d. The question is now how to measure this phase change. Fig. 8.5(a) shows the solution based on a standard dual-beam LDA setup. The rays from each beam are incident upon the particle at different angles, and therefore reach the common points on the receiver by different optical paths. The relative phase shift due to the differing optical paths is described by

$$\varphi = \frac{2\pi d}{\lambda}\left[\left(\sin\theta_{1,i} - \sin\theta_{2,i}\right) - pm\left(\sin\theta_1' - \sin\theta_2'\right)\right] \tag{8.15}$$

where the subscripts represent beams 1 and 2. Since all the angles are fixed by the transmitting and receiving optics, the phase shift only changes as a result of the particle diameter, d.

The phase difference produces an interference fringe pattern on the detector with an intensity distribution as shown in Fig. 8.5(b), and it can be described as

$$I \propto I_1 + I_2 + 2\sqrt{I_1 \cdot I_2}\,\cos\left[2\pi\nu_D t + \varphi\right] \tag{8.16}$$

Equations (8.15) and (8.16) illustrate that the spatial frequency (φ) of the fringe pattern is linearly related to the size of the sphere, and the temporal frequency of the fringe pattern is the Doppler difference frequency ν_D in the LDA technique (Chapter 6). Therefore, the particle size can be theoretically determined from the phase difference or spatial frequency if it can be measured. Since there is no reference for the phase difference, the position of the particle in the beams has to be known if the absolute value of the phase of the scattered light is to be determined. The solution is to measure the phase difference between the signals from two different detectors by means of an electronic phase difference processor. The velocity of the sphere is measured simultaneously from the temporal frequency ν_D of the scattered signal, as discussed in Chapter 6.

In order to determine the parameters affecting the phase shift, it is necessary to introduce the coordinates shown in Fig. 8.6. The coordinates are arranged with the z-axis in the direction of the bisector of the two incident beams. The x-axis is in the plane of the beams, and the y-axis completes the coordinate system. The beams are incident in the x-z plane with an angle $\pm\alpha/2$ to the z-axis. The polarization is described by the angle ξ from the y-axis ($\xi = 0°$ perpendicular; $\xi = 90°$ parallel). The scattering direction is described by the off-axis angle ϕ and the elevation

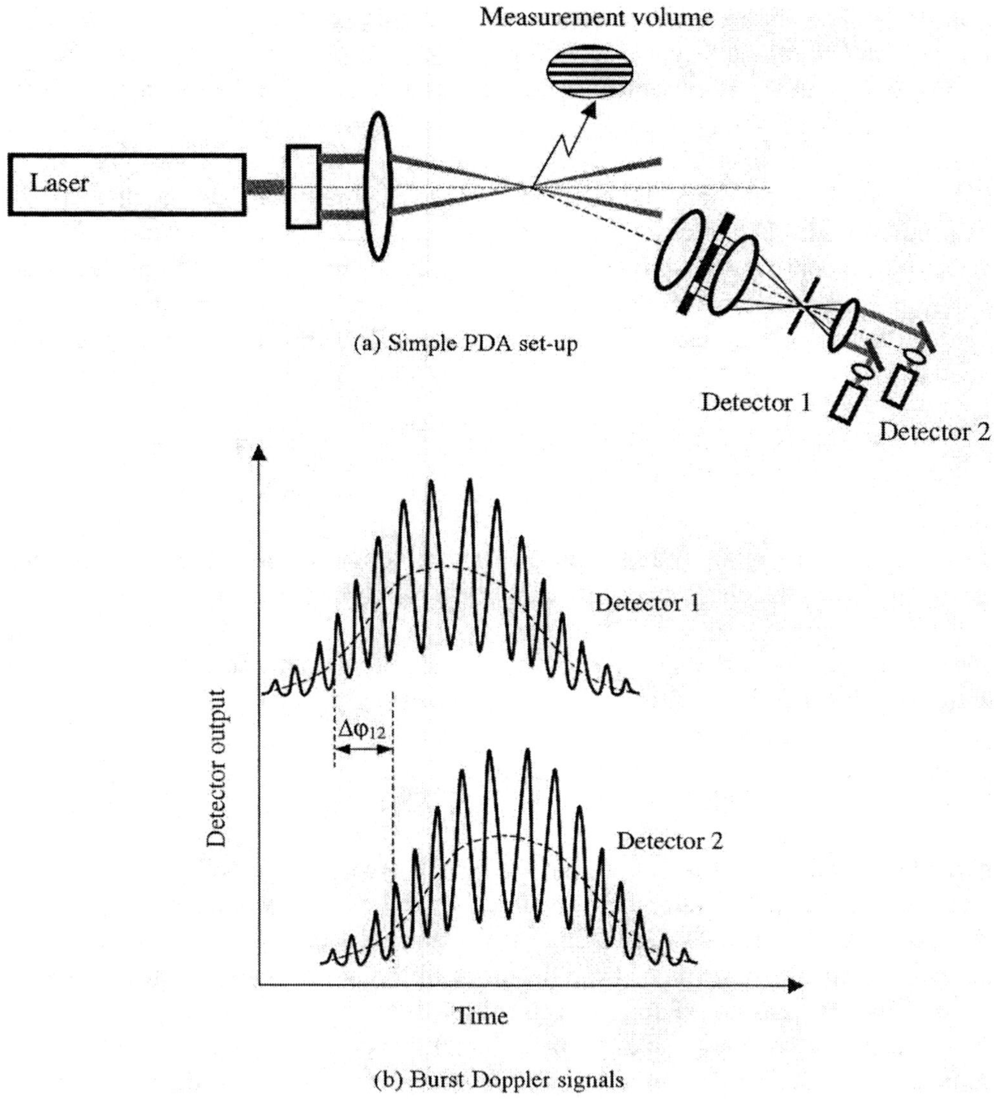

Fig. 8.5 Optical geometry of a phase Doppler analyzer and its burst signal output.

angle ψ. The phase difference given by Eq. (8.16), in terms of the geometry of the two laser beams, can then be determined from geometric optics. In the case of reflection (p = 0), it is given by

$$\Delta\varphi^0 = \frac{\sqrt{2}\pi m_1 d}{\lambda}\left[\sqrt{1+\sin\phi\sin\psi\sin\frac{\alpha}{2}-\cos\phi\cos\frac{\alpha}{2}}-\sqrt{1-\sin\phi\sin\psi\sin\frac{\alpha}{2}-\cos\phi\cos\frac{\alpha}{2}}\right] \tag{8.17}$$

The phase difference for the 1st order refracted beam is given by

$$\Delta\varphi^1 = \frac{2\pi d}{\lambda}\left\{\sqrt{1+m^2-\sqrt{2}m\sqrt{1+\sin\phi\sin\psi\sin\frac{\alpha}{2}+\cos\phi\cos\frac{\alpha}{2}}}\right.$$
$$\left.-\sqrt{1+m^2-\sqrt{2}m\sqrt{1-\sin\phi\sin\psi\sin\frac{\alpha}{2}+\cos\phi\cos\frac{\alpha}{2}}}\right\} \tag{8.18}$$

The phase difference for higher order refracted rays cannot be expressed in a closed form and must be found by an iterative procedure. In general, both reflected and refracted light will reach the detector, and the phase-diameter relation thus depends on the relative amount of each and can be a highly oscillating function. One of the important considerations in performing phase Doppler based measurements is to select an observation direction and a detector aperture so that the phase-diameter relationship is a linear relation as given either by Eq. (8.17) or Eq. (8.18).

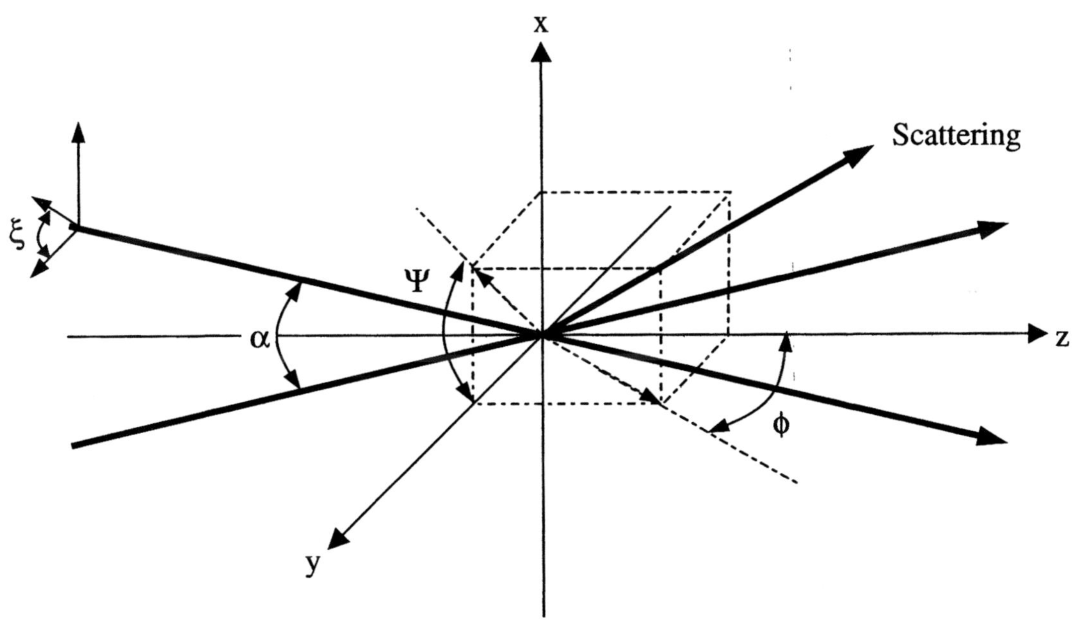

Fig. 8.6 Definition of scattering geometry.

8.3.3 Implementation of the Phase Doppler Method

The simplest PDA configuration is shown in Fig.8.5(a), which is very similar to that of a dual-beam LDA, except that two detectors are required to obtain the information that will allow the determination of the particle size. In this system, the laser beam is split into two beams of equal intensity. The beams are then focused and made to intersect using a transmitting lens. As the particle passes through the beam intersection region, it produces a scattered interference fringe pattern that appears to move past the detectors. Doppler burst signals are produced by each detector, as shown in Fig. 8.5(b). The frequency of each burst signal is used to determine the particle velocity. The phase shift between the two Doppler burst signals, $\Delta\varphi_{12}$, can then be determined by measuring the time delay, τ_{12}, between the zero-crossings of the high-pass filtered signals from detectors 1 and 2, and dividing by the measured Doppler period, τ_D:

$$\Delta\varphi_{12} = \frac{\tau_{12}}{\tau_D} \cdot 2\pi \qquad (8.19)$$

where the measured phase differences are averaged over all the cycles in the Doppler burst signal.

Measurements of the phase shift are then related to the particle size using the relationships given by Eq. (8.17) or Eq. (8.18), depending on whether the reflection or the 1st refraction component of the scattered light is detected. With parallel polarization, the light intensity scattered by refraction from a liquid droplet in air will be approximately 80 times that scattered by reflection when the droplet is subject to a uniform illumination. By careful selection of the light collection angles for the detectors, a linear relationship derived from Eq. (8.18) can be obtained, as shown in Fig. 8.7, assuming the 1st refraction is the only scattering component received by the detectors. In Fig. 8.7, the slope of the line can be altered by changing the optical parameters, including the laser beam intersection angle (α), collection angle (ϕ and ψ), laser wavelength (λ). From Eq. (8.18), it is also noted that the linear relationship is different for droplets of different index of refraction.

The preferred light collection position for the receiver is at an off-axis of from 30–40 degrees (ϕ) and in a plane (y-z plane, $\Psi = 0$) that passes through and is orthogonal to the intersection of the two laser beams. Two identical apertures on a single lens are placed symmetrically on either side of the lens center in the y-z plane.

In practice, three detectors are used in the measurement of phase differences to determine the droplet size, as shown in Fig. 8.8. These three detectors produce two pairs of redundant measurements which are then compared to estimate the quality of the measurements. More importantly, the introduction of the third detector extends the size range and sensitivity of the PDA system. In the design of the PDA system shown in Fig. 8.8, these detectors are spaced in an approximate 3 to 1 spacing. The resulting phase versus diameter curves are plotted in Fig. 8.9. The phase difference between the signals from the closely spaced detectors, D1 and

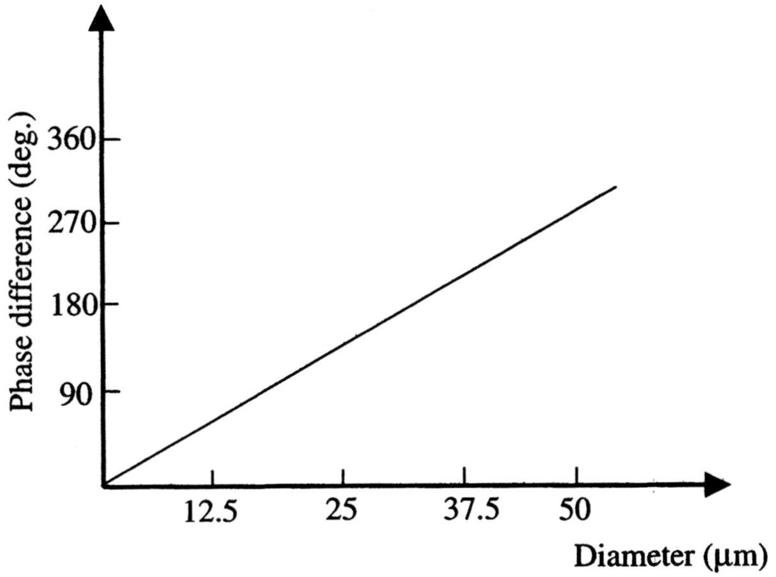

Fig. 8.7 Phase difference and droplet size relationship.

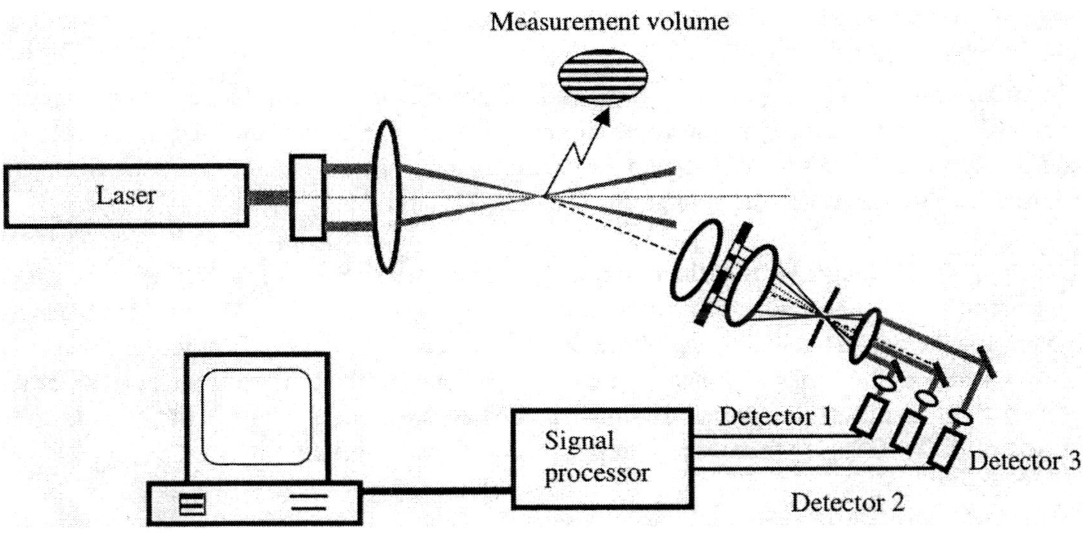

Fig. 8.8 Schematic of a PDA system with three detectors.

D2, follows the dashed line. The phase between the signals for the detectors with the larger spacing, D1 and D3, follows the solid lines with steeper gradient. The outputs of D1 and D2 produce the phase measurement, φ_{12}, and those of D1 and D3 yield the phase measurement, φ_{13}. With this arrangement, the phase may be measured for detector separations that extend over several fringes (one fringe corresponds to a measured phase shift of 2π). Using the

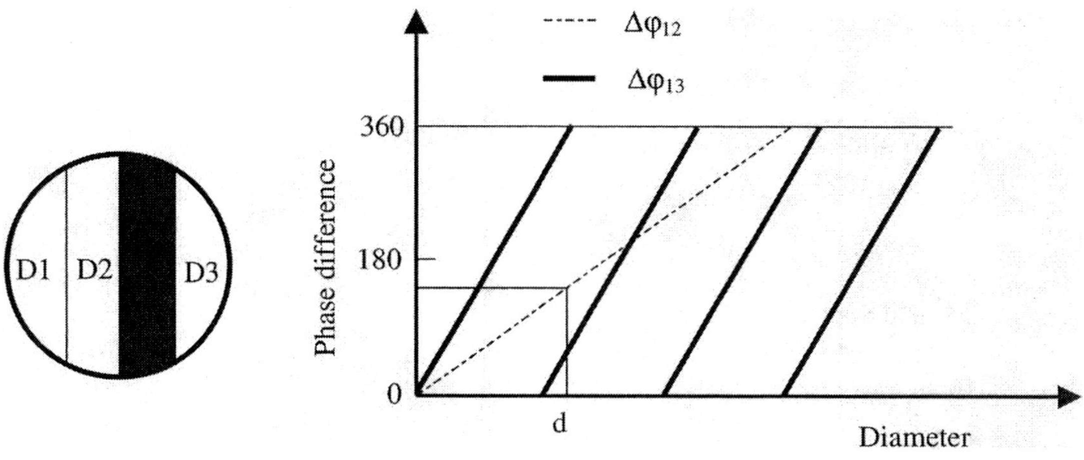

Fig. 8.9 Phase difference and droplet diameter relationship with three detectors.

phase measurement of φ_{12} allows the determination of the cycle in which the measurement of φ_{13} took place. For example, for a droplet of size d, the value of φ_{12} indicates that φ_{13} is made on the 2nd cycle of phase, which means that there is an entire interference fringe between the D1 and D3 detectors. And the measurement on detector D3 extends on to the second fringe between the detectors.

The two measurements, φ_{12} and φ_{13}, are weighted according to their relative sensitivities and averaged to obtain the mean phase value for that particle. Obviously, the measurement of D1 and D3 has the greatest sensitivity, with the change in phase with a change in droplet size being approximately three times the change for detectors D1 and D2.

The Doppler burst signals from the two photodetectors are shown in Fig. 8.5(b), from which measurements must be made to find at least three quantities, namely the Doppler frequency of a burst for the droplet velocity, the phase difference between the two Doppler bursts for the droplet size, and the number of measurements per second (or the time between bursts) for the droplet flux. At least four types of signal processor have been used for the simultaneous measurement of Doppler frequency and phase, and they are based on

(i) measuring zero crossing by a counter;
(ii) auto- and cross-correlation functions by a correlator or computer;
(iii) auto- and cross-covariance functions by a covariance processor;
(iv) discrete Fourier transform (DFT) methods by a digital signal processor or computer.

Because of their low signal-to-noise ratio threshold and high data acquisition rate, covariance and DFT methods have been adopted commercially in the processing of PDA signals. A detailed discussion on these four methods is given by Heitor et al. [1993].

8.3.4 Data Reduction and Presentation

The PDA technique is a single particle counting technique and as such provides the velocity and size measurement for each droplet. The results are normally presented as droplet size distributions in bar charts together with the corresponding velocity distribution, as shown in Fig. 8.10. Different mean diameters can be derived from the individual droplet sizes to correlate with the physical process being investigated. Both Lefebvre [1989] and Sowa [1992] gave detailed discussions on representative mean droplet diameters and their interpretation. The commonly used mean diameters are the Sauter mean diameter (SMD) and the volume mean diameter (VMD), as given by

$$\text{SMD}: D_{32} = \frac{\sum N_i d_i^3}{\sum N_i d_i^2} \tag{8.20}$$

$$\text{VMD}: D_{30} = \frac{\sum N_i d_i^3}{\sum N_i} \tag{8.21}$$

where N_i is the number of droplets in size class i, and the summation is taken over all the particle size classes.

The PDA technique has the ability to measure the drop number density and volume flux, which are important parameters in the spray atomization process. Volume flux is determined from the measured volume mean diameter, D_{30}, sampling cross sectional area, A, and the number of droplets counted, N, in the sampling time interval, τ, as given by

$$F = \frac{\pi}{6} D_{30}^3 \frac{N}{A \cdot \tau} \tag{8.22}$$

where

$$N = \sum_{i=1}^{n} N_i$$

As mentioned previously, size distributions are highly skewed in the larger size classes with a relatively small population, it is important that a large number of particles be measured to obtain good statistical representation and hence accurate determination of D_{30}. Furthermore, since the volume flux is proportional to the D_{30} cubed, the accurate determination of this quantity is extremely important in realizing good volume flux measurements.

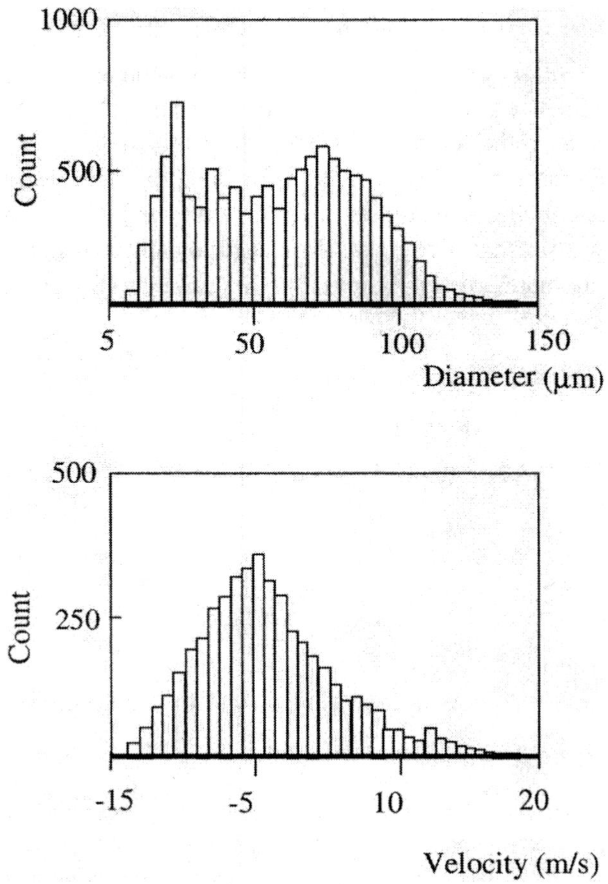

Fig. 8.10 Typical PDA results showing the droplet size and velocity distribution.

It should be noted that the volume flux measured by the PDA technique is a temporal distribution. In contrast, the estimation of the volume flux from a photograph would yield a spatial distribution.

The number density N_d is calculated from the measured number of particles N passing the sampling cross section. In practice, the particle number density is determined from the transit time (or residence time), t, of each particle, as it passes the sample volume relative to the total sampling time, τ, of each measurement, and the sample volume Ω_s:

$$N_d = \frac{1}{\tau}\sum_i \frac{\sum_j t_{i,j}(d_i)}{\Omega_s(d_i)} \tag{8.23}$$

where $t_{i,j}(d_i)$ is the residence time of the jth droplet within the droplet size class d_i. The sample volume, $\Omega_s(d_i)$, is written as a function of particle size because of the Gaussian distribution of the laser beam and the droplet trajectory through the cross section of the two beams [Bachalo et al., 1991]. For a Gaussian distribution of the illumination beam, smaller particles require a greater illumination to produce a detectable signal. Thus, the effective diameter of the sampling volume is smaller for small particles than the large particles. Therefore, reliable measurement of particle number density does not depend on accurate sizing of the particles but is very dependent on reliable determination of the measurement volume and counting of the particles. For one-dimensional flow, the effective diameter of the sampling volume can be determined from the transit time and velocity of the particle:

$$D_s(d_i) = u(d_i) \cdot t(d_i) \qquad (8.24)$$

Alternatively, it can be measured by counting cycles in the Doppler burst and using fringe spacing. For each size class, an accumulation of fringe counts is collected. Since particles will pass on all trajectories through the laser beam, the trajectories leading to the maximum number of fringe counts in the signal indicate that these particles passed normally through the diameter of the beam. The product of this fringe count and the fringe spacing then provides the effective diameter of the sampling volume. This procedure is followed for each particle size class.

For three-dimensional flows, the effective diameter of the sampling volume may be better determined by the signal amplitude method. A maximum signal amplitude is detected when a droplet passes through the center of the Gaussian beam, whereas the minimum detectable amplitude occurs when a droplet travels in a trajectory away from the peak intensity of the Gaussian beam. The ratio of the maximum (I_{max}) and minimum (I_{min}) signal amplitude measured amongst each droplet class (d_i) can then be used to find the effective diameter (d_{samp}) of the sampling volume for that droplet size class, which is given by

$$\ln\left(\frac{I_{max}(d_i)}{I_{min}}\right) = \frac{d_{samp}^2}{3r_{laser}} \qquad (8.25)$$

where r_{laser} is the laser beam waist diameter at $1/e^2$. Normally the minimum signal amplitude is defined by the threshold signal-to-noise ratio (SNR).

8.3.5 Optimization of a PDA System

The potential source of errors in a PDA system is the detection of signals produced by the wrong light scattering component. From Eqs. (8.17) and (8.18), it is noted that different relationships of phase and diameter are applicable to reflection and refraction. Because most PDA systems are designed to make use of the 1st refraction component of the light scattered, the presence of the reflection component must be identified and removed, otherwise serious errors will occur.

The presence of reflection for a PDA system optimized for the refraction detection is due to the problem of trajectory-dependent light scattering. As shown in Fig. 8.11, if a droplet that is relatively large compared to the focused beam waist diameter passes on certain trajectories, the peak intensity of the beam will fall on a point on the droplet and be reflected to the detector. Although 1st refraction is a much more efficient process than reflection, on trajectories shown here, the incident intensities at the points on the droplet scattering light by refraction and reflection may result in the dominance of the refraction being lost. On some trajectories, the scattered light may be dominated by the reflection from large droplets. In these cases, the three-detector arrangement is used to remove the reflection component by determining the direction of the fringe pattern movement. From geometric optics, it is known that reflection produces a pattern that will move in the opposite direction of light scattered by refraction. Additional measures are required to reject the occurrences of unwanted reflection components. For example, adaptive threshold validation has been developed to minimize this error by means of setting a minimum detectable intensity or signal amplitude for each drop size class.

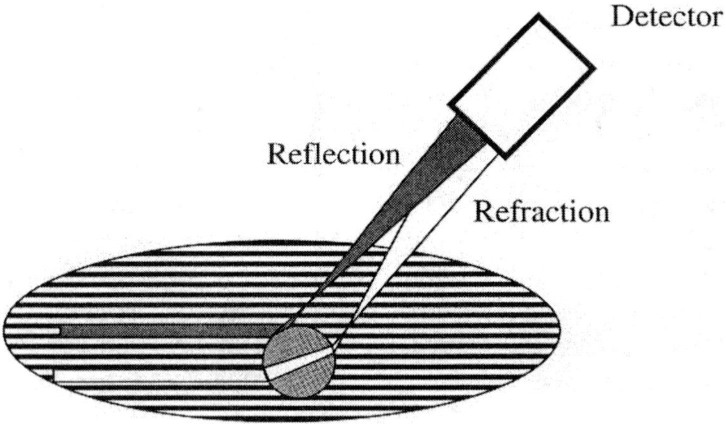

Fig. 8.11 Illustration of reflection interference with PDA measurements.

Another problem that occurs in PDA measurements is that large droplets are more likely to be detected than smaller ones. When using laser beams with Gaussian beam intensity distributions, small particles will need to cross the beams closer to the central peak intensity to produce a detectable signal. While larger particles can pass further out on the edges of the Gaussian beam and still be detected, because particles scatter light in proportional to the diameter squared. This bias toward large droplets must be accounted for in the data reduction.

For an accurate measurement of droplet size, PDA requires knowledge of the refractive index of the droplet and the surrounding gases. Practical liquid fuels are a mixture of many hydrocarbons, so the density and refractive index, m = n − ik, changes according to fuel composition. The imaginary refractive index, k, which is determined by the light absorption in the liquid, is normally negligible for most hydrocarbons in the visible range. The real refractive indices for some common fuels and fuel components are given in Table 8.1. There are two points that need to be made about Table 8.1: The first is that diesel or gasoline fuels are multicomponent, with each component having different physical properties. The second is that commercial fuels may contain dyes, and therefore exhibit very high absorption and fluorescence, which will modify drastically the angular distribution of the refracted light intensities.

The variation of refractive index with temperature is indirectly obtained from the Eykman relationship, which is derived empirically and specifically for hydrocarbons, as given below

$$\left(\frac{n^2-1}{n+0.4}\right)=\rho\cdot \text{const.} \tag{8.26}$$

where fuel density, ρ, as a function of temperature is readily available.

Table 8.1 Refractive indices for fuels and liquids

Liquid	Refractive index
Water	1.33
Heptane	1.39
Isooctane	1.39
Benzene	1.50
Toluene	1.50
Diesel	1.45
Ethanol	1.36

The theory underlying the PDA size measurement technique described in the preceding sections is only applicable to spherical scatters. It is the surface tension that keeps droplets spherical. However, as the size increases or its velocity relative to the continuous phase, the droplet will deform and ultimately break up. The Weber number, the ratio of the momentum force on the droplet to the surface tension forces, must be less than one for surface tension to dominate and the droplet to remain spherical. In general, non-spherical droplets are rejected in the data reduction stage by means of their departure from sphericity, described by

$$\varphi_{12}+\varphi_{23}+\varphi_{31}=0 \tag{8.27}$$

Therefore, if PDA is to be used in the region of a spray where atomization is poor and liquid jet and ligaments are present, serious measurement errors may occur.

For an optimized PDA system, the following quantities are claimed by manufacturers:

- Droplet size from 0.5 to 10,000 μm (spherical)
- Dynamic size range of 50:1 at one optical setup
- Size accuracy of 1%
- Droplet velocity up to 300 m/s
- Sample volume determined for each size class
- Particle time-of-arrival 0.6 microsecond resolution
- Maximum droplet number density 10^6/cc
- Maximum date rate 50,000 samples/s

8.3.6 Application of PDA to IC Engines

Over the last few years, the PDA technique has become the most widely used method for characterizing fuel sprays. The sizes and velocities of the fuel droplets downstream from the gasoline port injectors have been measured by a number of researchers using this technique. At Wayne State University, the PDA technique has been extensively used to characterize the spray atomization processes from various gasoline port injectors [Lai et al., 1994; Kelly-Zion et al., 1995; Zhao et al., 1995]. Nemecek et al. [1995] studied the spray drop size and velocity in the intake port of a SI engine using the PDA technique. They found that the atomization process within the intake system was similar to that in a quiescent environment except for the large scale structures appearing at the end of fuel injection. For open-valve injection, the intake flow accelerated the fuel droplets of all sizes. However, small droplets could decelerate in the intake port for closed-valve injection. The droplet entrainment effect of the intake air flow was also demonstrated by Vannoble et al. [1992] using the PDA technique. They found that the intake air flow caused smaller drops to deviate toward the upper region of the manifold, while larger ones continued with little change in their trajectory. Similar results were reported by Brenn et al. [1995] for port-fuel gasoline injection in a quiescent flow chamber and cross-flow environment.

The advances in PDA data acquisition and processing techniques have made it possible to investigate pulse-to-pulse variations of individual injection events, by providing data rates of several thousands per second. Amer and Lai [1995] observed clear evidence of pulse-to-pulse variations in the spray droplet velocity exiting the injector. Injection-to-injection velocity variation of a gasoline injector for two-stroke marine engines was also measured by Ikeda et al. [1997] using a very high data rate PDA system.

Liquid fuel behavior in the cylinder of firing SI engines has been investigated by a number of researchers using the PDA technique [Posylkin et al., 1994; Hardalupas et al., 1995; Meyer and Heywood, 1997]. Posylkin et al. found that the closed-valve injection allowed droplets to impinge on the valve and its seat, as well as on the valve stem and the wall of the port. This resulted in the evaporation of a fraction of the fuel, which was directly related to the residence time of the fuel. Thus, the amount of liquid fuel entering the cylinder increased with engine speed because of the reduced residence time. Measurements of droplets in a single cylinder

Honda VTEC engine by Hardalupas et al. revealed that open-valve injection resulted in a cloud of liquid droplets, the location of which in the cylinder was affected by the injection timing. Liquid fuel behavior in the cylinder of a square piston optical engine was studied by Meyer and Heywood using the PDA technique. Measurements of liquid fuel droplets during start-up showed that the amount of liquid fuel entering the cylinder reached a peak value due to strip atomization of liquid fuel with closed-valve injection. With open-valve injection, smaller amount of liquid fuel entered the cylinder at start-up due to enhanced fuel atomization through injection into the moving charge. However, under warmed-up conditions, the situation reversed so that more liquid fuel was transported into the cylinder with the open-valve injection.

Diesel spray represents a severe challenge to PDA measurements. It is optically thick, so light transmission through or into the spray is very low. The extremely high drop density requires the employment of a very small measurement volume, so that there is only one droplet present in the measurement volume. The spray close to the injector can be very poorly atomized with liquid jet and ligaments dominating the flow processes. The presence of liquid jet or ligaments can refract the incident laser light to prevent direct transmission and the formation of a measurement volume. The transient, high-pressure nature of the spray also brings experimental difficulties with very low duty cycles for data arrival and very high velocities. Despite these difficulties, measurements are possible and have been performed by a number of groups. Most of these measurements have been carried out on non-evaporating diesel sprays at atmospheric conditions [e.g., Arcoumanis et al., 1993; Tinaut et al., 1993; Desantes et al., 1998; Levy et al., 1998].

More recently, PDA measurements have been successfully carried out in a high-pressure and high-temperature combustion bomb by Schünemann et al. [1998]. In order to limit the occurrence of more than one droplet in the measurement volume under the high droplet density conditions, the measurement volume diameter was reduced to 70 μm by means of a beam expander in the transmitting optics. The sphericity of each droplet was checked by a three-detector receiving optics system using Eq. (8.19). In order to minimize the effect of varying refractive index, due to temperature change, Schünemann et al. changed the receiving optics (scattering angle) to 69 degrees, at which the phase-diameter relationship showed good linearity and was insensitive to changes of the fuel's refractive index for the 1st order refraction. The trade-off was the much reduced signal level than that at 30 degrees. After optimizing the receiving optics, Schünemann et al. measured the droplet size and velocity distribution in a diesel spray from a common rail injection system in the combustion chamber at 45 bar and 380°C.

8.4 Two-Dimensional Techniques for Liquid Fuel Distribution Measurements

8.4.1 Introduction

As discussed previously, early studies on the structure of fuel sprays have generally been carried out using direct photography, Schlieren and shadowgraphy. A major limitation of these methods is that they generate information integrated along a line of sight through the flow. An alternative approach is to use a laser sheet to illuminate a plane in the spray or a cloud of droplets. These laser sheet based methods are capable of giving spatially resolved information with excellent temporal resolution.

8.4.2 The Mie Scattering Method

The simplest method for two-dimensional imaging of sprays and liquid fuel droplets is to use Mie scattering of a laser sheet by the liquid droplets themselves. Mie scattering refers to the light scattering process when a particle is of diameter much larger than the wavelength or the size parameter $a = \pi d/\lambda >> 1$. For visible wavelengths, Mie scattering occurs from particles larger than 0.5 μm. The scattering intensity from a group of particles of diameter d is given by [Van de Hulst, 1957]

$$I_{sca}(n,\theta,d) = I_0 \cdot C_n \cdot f(n,\theta) \cdot d^2 \qquad (8.28)$$

where n is the real part of the refractive index of the particle, θ is the scattering angle, and C_n is the number density of particles. F(n, θ) is a complicated function of the refractive index, particle shape, and scattering angle.

The experimental setup for Mie scattering is shown in Fig. 8.12. A laser beam is focused into a sheet by a combination of spherical and cylindrical lenses. The scattered light is detected at a right angle by a camera. The resulting images are then processed to subtract the background noise and corrected for non-uniformity in the laser sheet. When combustion emission is present, it can be rejected using a narrow-band interference filter centered at the laser wavelength, and synchronized and gated recording. Because of the strong signal from Mie scattering, a wide choice of cameras is available from standard 35 mm SLR cameras to CCD cameras. Solid state cameras are preferred because of their instant feedback and easy post-processing. Nd:YAG and excimer lasers are normally used. If a high-repetition pulsed laser is used, it is possible to use high-speed film cameras or high-speed solid state cameras to make high-speed movies [Fansler et al., 1995].

The Mie scattering technique described above is limited to qualitative visualization of the liquid fuel distribution, because the scattered light intensity depends on both the drop concentration and size distribution. In order to overcome this limitation, Hodges et al. [1991] employed the ensemble-scattering polarization ratio method, which was originally proposed by Cavaliere et al. [1992] in their studies of diesel fuel distribution in a single-cylinder optical engine. The drop size and concentration are determined from the relative intensity of two orthogonal polarizations of the scattered radiation. The scattered light is simultaneously imaged through two orthogonally polarized filters, and then the two images are ratioed. The assumption in the technique is that depolarization of the incident light should be negligible. The authors found the depolarization effects within the spray core were large, thereby preventing measurement of the drop size in that part of the spray in the cylinder [Hodges et al., 1991]. Reliable quantitative measurement of the mean drop size and concentration was therefore limited to the peripheral regions of the diesel spray.

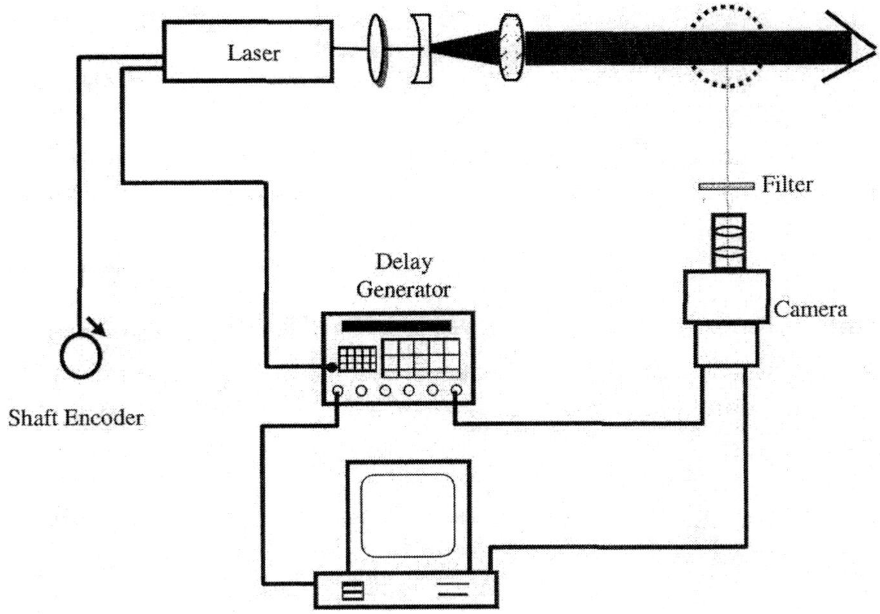

Fig. 8.12 Experimental setup for two dimensional Mie scattering imaging of fuel sprays.

8.4.3 Planar Laser Induced Fluorescence (PLIF) Imaging

Another technique for two-dimensional imaging of sprays and liquid fuel droplets is to use fluorescence of a laser sheet by fuels or fluorescent dopant. The principles and implementation of PLIF imaging have been discussed in Chapter 7. The experimental setup for this technique is shown in Fig. 7.10. Detection of fluorescence from practical fuels such as Diesel or gasoline is a very attractive approach from the standpoint of simplicity in implementation, but it introduces difficulties in interpretation of the signals. Since the fluorescing components are likely to be high-boiling point aromatics, the fluorescence signal will not be representative of the fuel as a whole. In addition, the photochemistry will not be well characterized and will probably not be reproducible from batch to batch. Fluorescence from a dopant molecule of matching boiling point to the practical fuel is therefore a preferred way of recording PLIF images of liquid fuel.

8.4.4 Mie Scattering vs. LIF for Spray and Liquid Fuel Distribution Imaging

Mie scattering imaging of sprays and fuel droplets requires no dopant and is insensitive to the wavelength used, thus a wide choice of light sources and cameras is possible. LIF imaging of fluorescing dopant requires ultraviolet light sources. Mie scattering gives a much stronger signal than fluorescence and can be imaged using photographic film or unintensified cameras. However, Mie scattering is biased toward larger droplets because of its second order size dependence. LIF can be used to produce high-contrast images of sprays by rejecting the spurious scattered light from walls and dusts at the laser wavelength.

Since Mie scattering of liquid droplets is many orders of magnitude stronger than Rayleigh scattering from vaporized fuel molecules, the Mie scattering method only gives information on the liquid phase. When both liquid droplets and fuel vapor are present in the laser sheet, the fluorescence signal from liquid fuel will be several orders of magnitude greater than that of the fuel vapor. In order to overcome this difficulty, laser induced exciplex fluorescence (LIEF) was developed for simultaneous imaging of both liquid and vaporized fuel distribution in the cylinder.

8.4.5 Planar Imaging of Sprays and Liquid Fuel Distribution in the Cylinder

Applications of Mie scattering and PLIF techniques include spray characterization, studies of port-injected fuel in the intake manifold and cylinder, in-cylinder spray and liquid fuel measurements.

Spray characterization of various gasoline injectors was investigated using PLIF [Zhao et al., 1995]. The interaction between the flow and fuel spray in the intake port of a port-injection SI engine was investigated by Wensing et al. [1997] using Mie scattering. The geometric structure of the liquid fuel spray entering the cylinder was visualized by Meyer et al., [1998] using fluorescence signals of acetone added to the fuel. The effects of fuel volatility on the evaporation process and unburned hydrocarbon emissions were investigated by means of adding fluorescing dopants to the gasoline fuel [Dawson and Hochgreb, 1998]. The fluorescing dopants, including acetone, 3-pentanone, and aromatics, were selected to represent the low-, middle-, and high-boiling fractions of fuel, respectively.

To better understand the highly turbulent diesel spray structure and evaporating process, several studies have been conducted using Mie scattering of liquid fuel [Felton et al., 1987; Hodges et al., 1991; Baritaud et al., 1994; Espey and Dec, 1995]. Fig. 8.13 illustrates a temporal sequence of liquid fuel images obtained from the Sandia optical diesel engine [Cannan et al., 1998]. The experimental setup is shown in Fig. 8.14. These images were selected from twelve different engine cycles. For all images in Fig. 8.13, the laser sheet traveled up the fuel jet axis (Fig. 8.14b) from right to left, primarily illuminating the fuel jet in the 3 o'clock position. Because some of the liquid jets traveled beyond the width of the laser sheet, some jets appeared shorter than others for 2 °CA ASI (after the start of injection). In fact, all jets attained similar penetration for a given crank angle throughout the injection event [Espey and Dec, 1995]. By following the images sequence in Fig. 8.13, it can be seen that fuel injection did not appear in the first image, corresponding to the timing just before the start of injection needle motion. Injection actually commenced between the first and the second images. From 0.5 °CA to 3.5 °CA ASI, the liquid fuel jet penetrated rapidly, reaching out to roughly 25 mm in length by 3.5 °CA ASI. Beyond 3.5 °CA ASI, the liquid extent stabilized at a relatively constant length despite the fact that the fuel vapor continued to penetrate across the combustion chamber, as discussed in Chapter 6 [Espey et al., 1997].

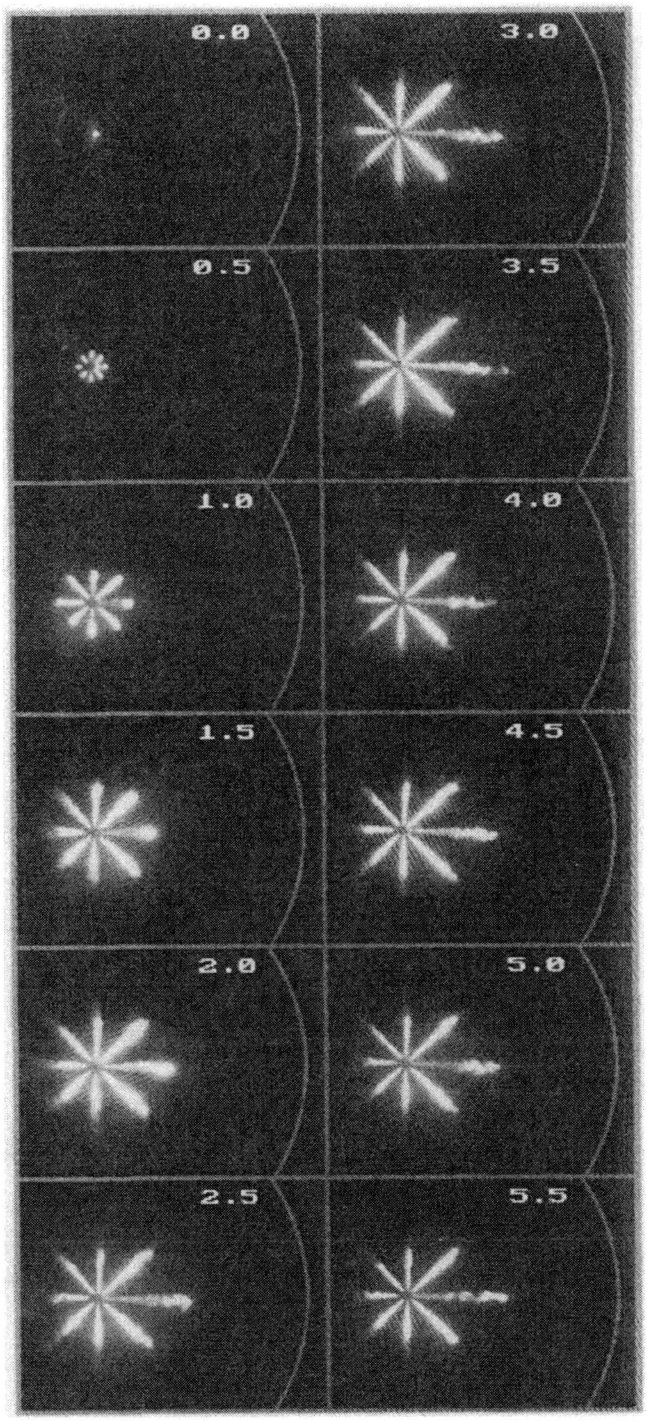

Fig. 8.13 Temporal sequence of Mie-scattering images of liquid fuel jets, the crank angle degrees after the start of injection (ASI) and after BDC are given in the upper right-hand corner and beside of each image, respectively. [Canaan et al., 1998].

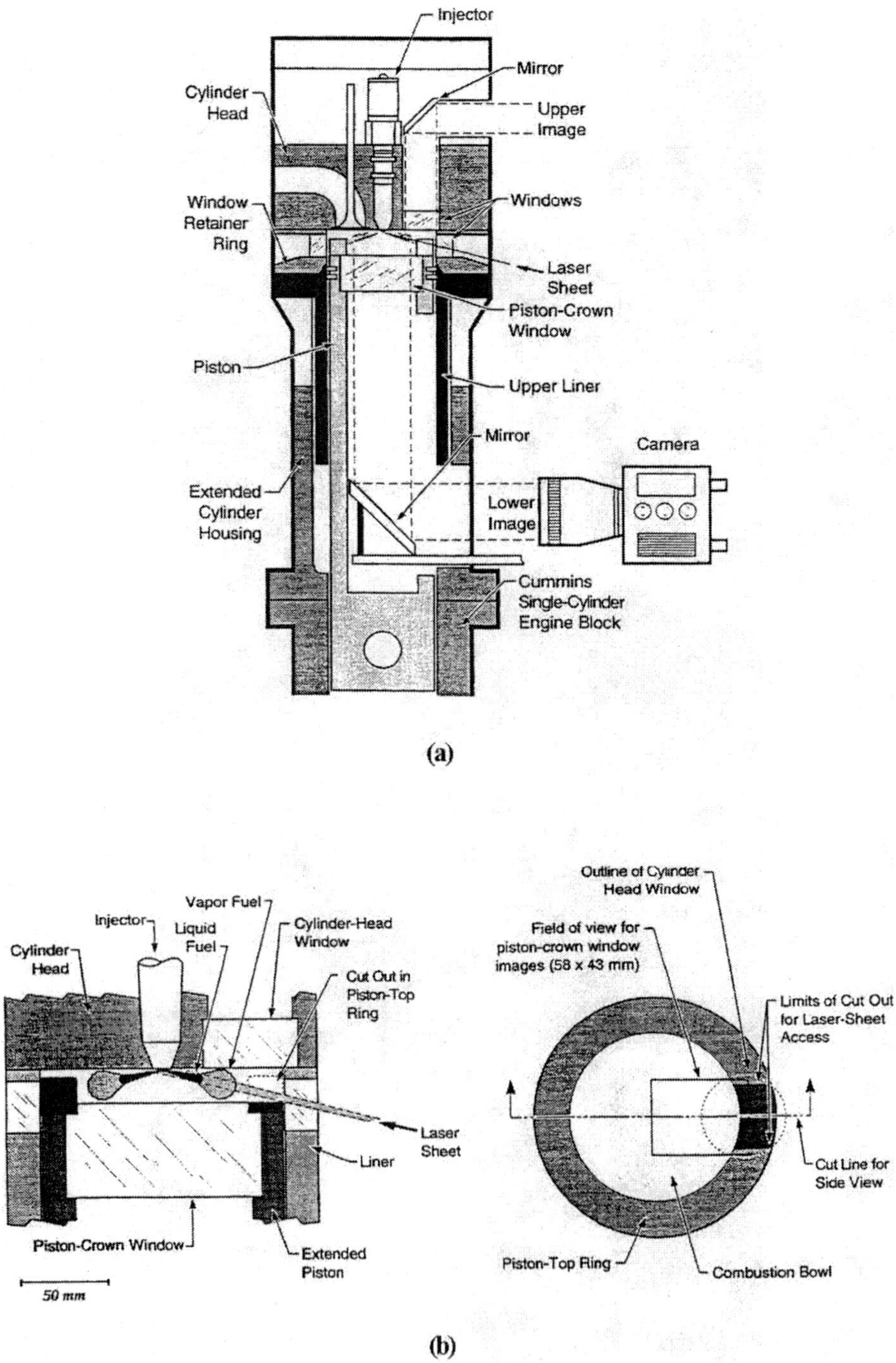

Fig. 8.14 (a) Schematic of the optical diesel engine. (b) Optical configuration of the combustion chamber with single cutout in the piston-top ring for laser access.

All of the above experiments were carried out using the 2nd output of an Nd:YAG laser or an XeCl excimer laser. Because of their slow repetition rates, only one image could be obtained from each engine cycle. In order to overcome this difficulty, a copper vapor laser, together with a high-speed film camera or a high-speed video camera, was used to record continuous Mie scattering images of fuel drops over many consecutive engine cycles, and direct images of ignition and combustion, in a direct injection two-stroke SI engine [Fansler et al., 1995].

In addition to planar fluorescence imaging, LIF has been used to observe liquid fuel in the intake port [Felton et al., 1995] and in the cylinder [Witze and Green, 1997] by means of an expanded laser beam. In order to observe the liquid fuel behavior in the intake port of a port injection SI engine, Felton et al. used an expanding Nd:YAG laser operating in the third harmonic mode (355 nm) as the light source to illuminate the inside of the manifold. This was done via a small quartz periscope inserted through the wall of the manifold. Imaging of the natural fluorescence of a gasoline fuel was observed through an endoscope, which was also mounted in the manifold. The endoscope allowed fluorescence imaging of liquid fuels at different regions of the manifold to be recorded. Using this technique, they were able to compare the liquid fuel behavior of a pintle-type injector and an air-assisted injector, and they found that the coarse spray from the pintle-type injector contributed significantly to exhaust μHC emissions during the cold-start.

In order to observe the liquid films in the cylinder during a simulated cold-start, Witze and Green [1997] introduced an expanded Nd:YAG laser beam (355 nm) through the transparent piston crown in a single-cylinder SI engine with an extended cylinder block. The resulting natural fluorescence of the gasoline fuel was recorded by an intensified CCD camera, through the transparent piston crown and a dichroic mirror. The dichroic mirror was placed between the cylinder block and the rest of transmitting and receiving optics, and it reflected the laser beam into the cylinder and transmitted the fluorescence onto the camera. A limitation of this imaging technique is that it is a line-of-sight method, which means that it is not possible to explicitly distinguish between liquids located on the cylinder head and those on the window in the piston crown. However, it is often possible to infer which surface the liquid films are on from their shape. Using this technique, Witze and Green were able to study the effect of injection timings and coolant temperatures on the in-cylinder liquid fuel distribution during a simulated cold-start.

8.5 Simultaneous Visualization of Fuel Vapor and Liquid Fuel by LIEF

8.5.1 Principle

When natural fuel fluorescence or a single fluorescence dopant is used to visualize liquid fuel and fuel vapor at the same time, it is not possible to separate spectrally the fluorescence from the liquid phase or vapor. This is because the absorption and fluorescence spectra of organic molecules dissolved in nonpolar solvents, such as typical fuels, are virtually identical to the spectra of the same molecules in the vapor phase. Since the liquid fluorescence signal is many orders of magnitude greater than the vapor fluorescence signal, it is difficult to detect fuel vapor in the presence of liquid fuel due to the limited dynamic range of photodetectors. To overcome this difficulty, laser induced exciplex fluorescence (LIEF) was developed by Melton and Verdieck [1984;1985]. In this method, a fluorescent molecule reacts in an excited state, M*, with another molecule, G, to form a second fluorescent species (M-G)*, i.e.,

$$M^* + G \Leftrightarrow (M\text{-}G)^* \tag{8.29}$$

The newly formed fluorescence species (M-G)*, called excited state complex or exciplex, is bound in the excited state rather than in the ground state. The emission from the exciplex is red-shifted with respect to that of M*. The reaction between M* and G to form (M-G)* is reversible and in some cases it is possible to adjust the concentration of species G so that the exciplex (M-G)* is the dominant emitter in the liquid phase and the monomer M* is the dominant emitter in the vapor phase. This is because of the fact that, for the liquid phase, the equilibrium of the reaction can be driven to the right side (increasing exciplex) with a large concentration of G, but the exciplex is not stable in the vapor phase due to the low density and high temperature.

The emission peaks of the excited monomer, M*, and the exciplex, (M-G)*, should be sufficiently separated for emitting species M* and (M-G)* to act as markers for the vapor and liquid phases, respectively. If the spray is optically thin and if the species M co-evaporates with the fuel, then the emission from these markers, separated by appropriate filters, will be directly related to the amount of fuel in the vapor and liquid phases. In the absence of quenching effects, the fluorescence intensity is directly proportional to the fuel concentration, allowing qualitative interpretation of results or potentially quantitative measurements with proper calibration.

Quantification of the LIEF signal is limited by several factors. The equilibrium of exciplex formation is temperature dependent, which means that higher temperatures reduce the number of exciplex events. As a result, exciplex fluorescence from a hot liquid phase is weaker at higher temperature. In addition, the emission from the excited monomer increases, leading to a spectral overlap of fluorescence from the liquid phase and vapor. Another problem arises because of the temperature dependence of the fluorescence yield. Furthermore, the exciplex fluorescence is severely quenched by oxygen. Quenching of the vapor phase fluorescence by oxygen can be a serious problem, particularly at high pressures. Quenching of the liquid fluorescence by oxygen is much less of a problem, because the dissolved oxygen can be removed from the fuel prior to the experiment, and any ambient oxygen will not have enough time to diffuse into the liquid droplet during the short droplet lifetime. The quenching of oxygen can be minimized by performing experiments in a nitrogen atmosphere. Quantitative measurements may also be hampered by the necessity to match the boiling temperatures of the particular fuel used and the exciplex generating dopants.

The N,N,N′,N′ tetramethyl-p-phenylene diamine (TMPD)/naphthalene (Np) system was the first exciplex system developed for diesel fuels [Melton and Verdieck,1984]. The principal features of LIEF, as discussed above, can be illustrated by this exciplex system. A schematic summary of the photophysics of the TMPD/naphthalene (Np) exciplex system is shown in Fig. 8.15. Fig. 8.16 shows the fluorescence spectra of the two phases in a mixture of 90% n-decane, 9% naphthalene, and 1% TMPD by mass. The vapor phase fluorescence is dominated by the excited monomer (TMPD*) emission, while the exciplex ([Np + TMPD]*) emission dominates the condensed phase emission. Although the liquid and vapor phase emissions peak at 480 nm and 390 nm, respectively, an overlap in spectrum between the two fluorescence signals exists over a wavelength range from 380 nm to 510 nm. This may be attributed to the small emission of the monomer in the liquid phase and the slight formation of the exciplex in the vapor phase. Normally, the absolute quantity of the fluorescence intensity from the vapor phase is quite small compared with the emission from the liquid phase, because the fluorescence intensity depends on the mass of the excited fluorescent matter. If a long-pass filter is employed to cut off at 520 nm, the cross-talk from the vapor phase can be eliminated. By contrast, the fluorescence signal near 390 nm is always contaminated with liquid phase fluorescence emission. In a region where both liquid droplets and fuel vapor exist in an evaporating spray, the cross-talk from the liquid phase is unavoidable. It will be difficult to obtain quantitative vapor concentration measurements in this region.

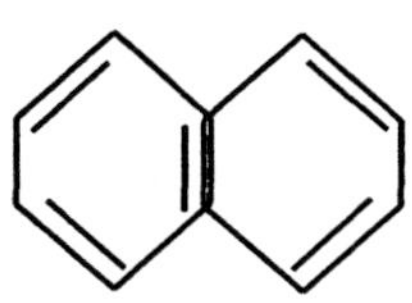

Naphthalene

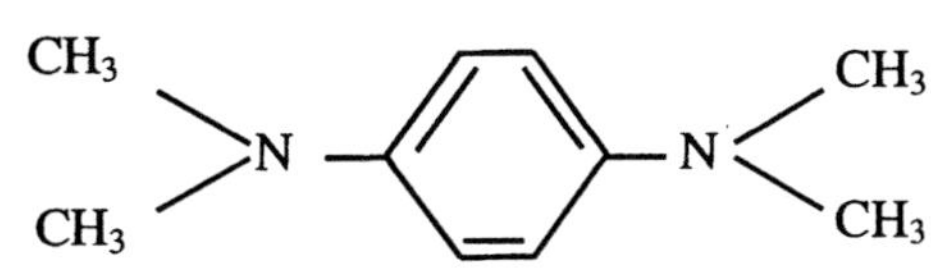

TMPD

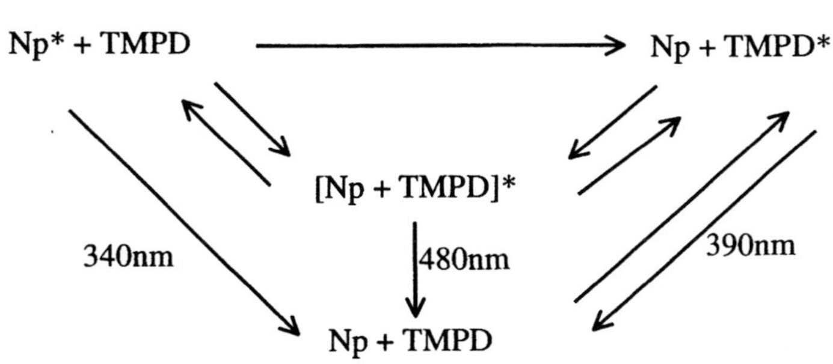

Fig. 8.15 Illustration of the naphthalene and TMPD exciplex system.

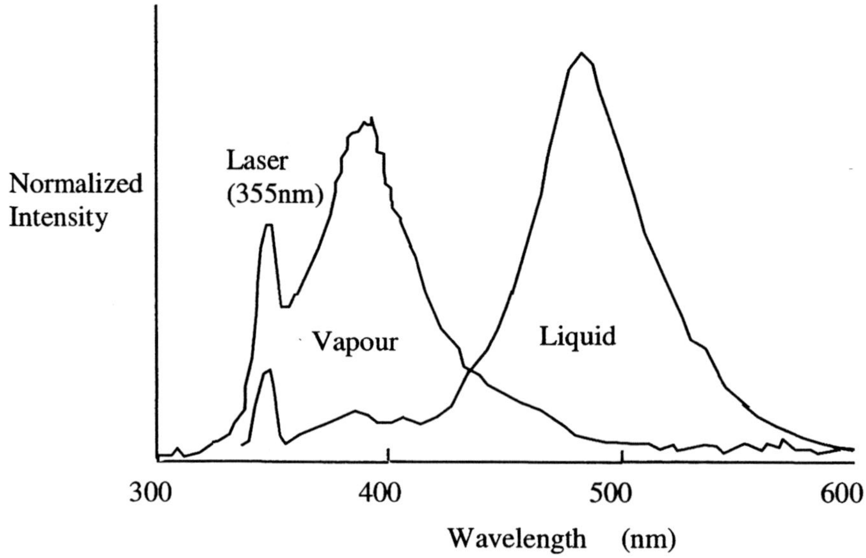

Fig. 8.16 Naphthalene and TMPD fluorescence spectra.

As noted previously, the reaction between the excited monomer TMPD* and the naphthalene molecule Np to form the exciplex [Np + TMPD]* is reversible. By adjusting the concentration of naphthalene, the exciplex [Np + TMPD]* can be the dominant emitter in the liquid phase, and the dissociated species TMPD* will be the dominant emitter in the vapor phase.

8.5.2 Experimental Setup for LIEF

Fig. 8.17 shows the experimental setup for planar LIEF. In order to carry out PLIEF measurements, fluorescence signals from both liquid fuel and fuel vapor must be detected simultaneously by two separate cameras, or by one camera with an image doubler in front of the camera. Spectral filters are placed in front of each camera in order to select the spectral regions of liquid and vapor fluorescence, respectively. The selection of the spectral filters depends on the monomer and exciplex fluorescence spectra. For the fluorescence spectra shown in Fig. 8.14, for example, a broad bandpass filter (410 ± 9.5nm) and a narrow bandpass filter (532 ± 1 nm) have been used to separate the vapor and liquid fluorescence signals [Senda et al., 1997]. However, because of the overlap in the liquid and vapor emission spectra (Fig. 8.16), some cross-talk will still be present in the exciplex images of liquid fuel and fuel vapor. When a single camera is used, a half-width neutral density filter is also required to attenuate the stronger fluorescence signal from the liquid fuel. The experimental setups and dopants employed for LIEF measurements are summarized in Table 8.2.

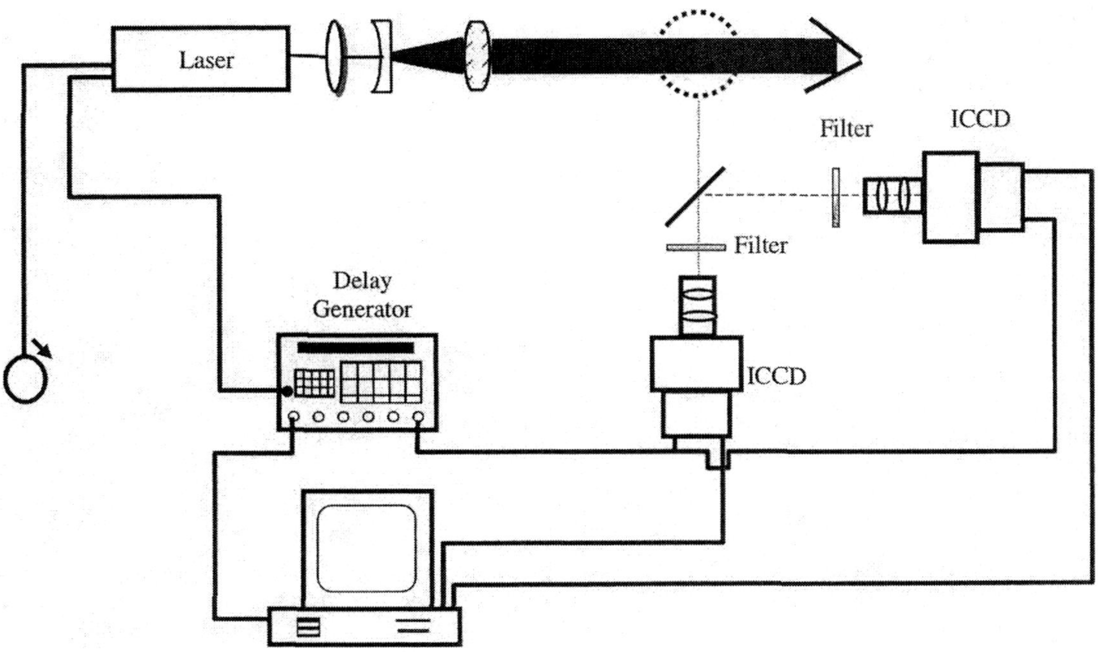

Fig. 8.17 Experimental setup for planar laser induced exciplex fluorescence (PLIEF).

8.5.3 Quantitative Analysis of Fuel Vapor Concentration Using PLIEF

As discussed in the previous section, quantification of the LIEF signal is limited by several factors. The fluorescence signals from both exciplex (liquid phase) and excited monomer (vapor phase) are temperature dependent. The fluorescence can be quenched severely by oxygen, which is particularly serious for the vapor phase, which can be minimized by performing experiments in a nitrogen atmosphere. Thirdly, the exciplex fluorescence is affected by the vapor concentration, i.e., it is self-quenched. When the LIEF is applied to a dense spray, the attenuation of the laser beam also needs to be taken into account.

In the absence of procedures that relate the measured fluorescence intensities to vapor and liquid concentrations, quantitative interpretation of the images is not possible. Melton [1988] developed procedures based on the calculation of light absorption, quantum yields for emission, refraction within droplets, etc. The procedures and the photophysical parameters required are summarized in a "user's manual." However, the photophysical function/parameter methods described in the user's manual are difficult to perform and may not be very accurate. Rotunno et al. [1990] developed a direct calibration method for both the liquid and vapor phases. A known amount of liquid or vapor, located at the plane to be imaged, is irradiated, and the fluorescence intensity is measured. If all excitation and detection parameters are kept constant, the ratio of the measured intensities to the known liquid or vapor concentration can be used as a calibration factor for subsequent spray experiments.

Table 8.2 Summary of exciplex dopants and PLIEF experimental setups reported in the literature

Laser source	Exciplex dopants (monomer/exciplex forming dopant)	Detection system spectral filters	Application and Reference
Nd:YAG 266 nm, 80 mJ	TMPD/naphthalene (1.0%, 2.5% mass in cetane)	Camera with colored film	Diesel spray in a confined volume, Melton and Verdieck [1984; 1985].
Nd:YAG 355 nm	TMPD/naphthalene (1.0%, 9.0% mass in decane)	Intensified solid state camera Vapor: 400 ± 12 nm; Liquid: >450 nm	Diesel spray in a motored two-stroke engine with N_2 Bardsley et al. [1988]; Campbell et al. [1995].
Nd:YAG 355 nm, 60 mJ	TMPD/naphthalene (0.5%, 5.0% mass in n-tridecane)	Intensified CCD camera Vapor: 350–390 nm; Liquid: >460 nm	Diesel spray in an RCM filled with N_2 Yeh et al. [1993].
Nd:YAG 355 nm, 100 mJ	TMPD/naphthalene (1.0%, 10% mass in heptane)	Intensified CCD camera Vapor: 400 ± 12 nm; Liquid: >435 nm	Diesel spray in a motored four-stroke engine with N_2 Hodges et al. [1991].
Nd:YAG 355 nm, 60 mJ	TMPD/naphthalene (1.0%, 10% mass in n-tridecane)	Intensified CCD camera Vapor: 410 ± 10 nm; Liquid: 532 ± 1 nm	Diesel spray impingement in a constant volume filled with N_2 Senda et al. [1992; 1994; 1997].
Nd:YAG 355 nm, 200 mJ	TMPD/naphthalene (1.0%, 9.0% mass in n-octane)	Intensified CCD camera Vapour: 400 ± 8 nm; Liquid: 480 ± 10 nm	Mixture injection SI engine filled with N_2 Fujimoto and Tabata [1993]; Tabata et al. [1995].
Nd:YAG, 355 nm, 40 mJ	TMPD/naphthalene (1.0%, 10% mass in decane)	Intensified vidicon camera Vapor: 400 ± nm; Liquid: 500 ± nm	Fuel spray from an air-assist injector inside a motored two-stroke engine with N_2 Diwakar, et al. [1992].
XeCl 308 nm	TMPD/naphthalene (in isooctane)	Intensified CCD camera Vapor: 2xWG355, UG11; Liquid: WG360, GG455, GG475	Liquid/vapor fuel distribution in an SI engine running with isooctane Knapp et al. [1996].
Nd:YAG 266 nm	fluorobenzene/ DEMA (0.5%, 9.0% vol. in hexane)	Intensified CID camera Vapour: 295 ± 11nm; Liquid: 400 ± 12 nm	Fuel spray from an air-assist injector inside a motored two-stroke engine with N_2 Ghandhi et al. [1994].
XeCl 308 nm	DMA/naphthalene (5.0%, 5.0% mass in gasoline)	Intensified CCD camera Vapour: 340 ± 10nm; Liquid: >380 nm	Quantitative fuel vapour concentration in a SI engine Schimizu et al. [1992].
XeCl 308 nm	DMA/1,4,6-TMN (5%, 7% mass in isooctane)	Vapour: λ_{peak} = 355 nm; Liquid: λ_{peak} = 405 nm	New exciplex seeding compound Kim et al. [1996].
KrF 248 nm, – –	TEA/benzene (2.0%, 2.9% in isooctane)	Intensified CCD camera Vapour: 307 ± 12nm; Liquid: > 325 nm	Fuel distribution in the intake port of an SI engine Münch et al. [1996].

Since an evaporating diesel spray produces a highly dense vapor cloud, the absorption of the laser beam by the fuel vapor needs to be considered, as well as the quenching effects due to temperature and concentration. Senda et al. [1997] developed a calibration and correction procedure for quantitative measurements of fuel vapor concentration in an evaporating diesel spray using LIEF. The procedure is as follows.

The fuel spray is divided into n cells as shown in Fig. 8.18. From Eq. (5.61), the fluorescence intensity from the fluorescing molecules (monomer) in a cell, j, of the spray can be written as

$$I_{flu} = \eta_c f_1(T)\chi_m NKI_a \tag{8.30}$$

where K is introduced to represent the last term in Eq. 5.61, $(B_{12} \cdot A_{21}/(A_{21}+Q_{21}))$, and I_a is the laser absorbed by the monomer in the region. I_a is related to the incident laser intensity to the cell, I_j, by Lambert-Beer's law

$$I_a = I_j[1 - \exp(-\varepsilon C_j l)] \tag{8.31}$$

where ε = the molar absorption cross section for the incident wavelength, [m^2/mol]
C_j = the molar concentration of the dopant in the jth cell, [mol/m^3]
l = the dimension of the cell, [m].

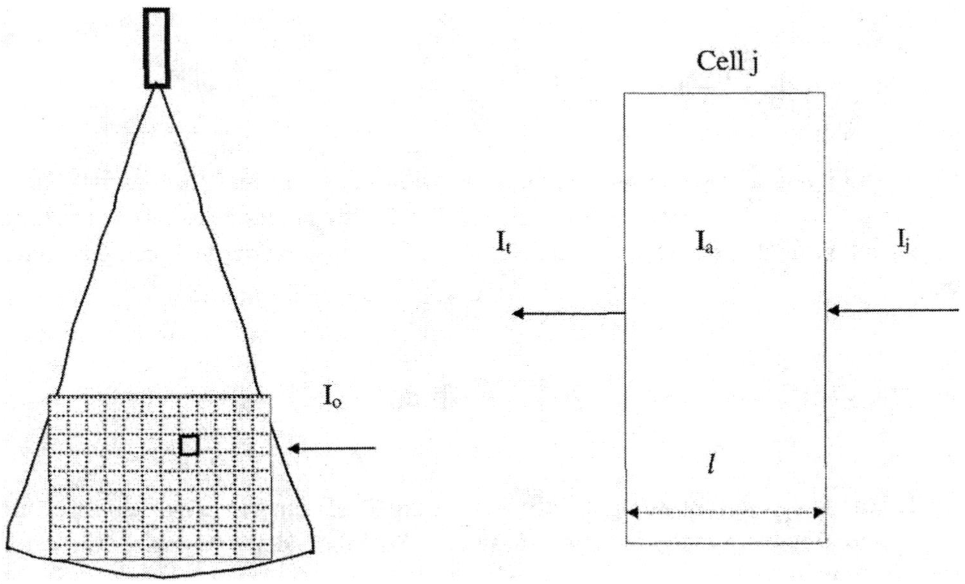

Fig. 8.18 Schematic of the theoretical model for the quantitative vapor concentration analysis with correction for light absorption.

Since the laser output is attenuated as it passes through the upstream regions, the incident laser intensity to the cell, j, is given by

$$I_j = I_0 \exp\left(-\varepsilon l \sum_{i=1}^{i=j-1} C_i\right) \quad (8.32)$$

Substituting Eqs. (8.31) and (8.32) into Eq. (8.30), the fluorescence signal is given by

$$I_{flu} = \eta_c f_1(T)\chi_m NKI_0 \exp\left(-\varepsilon l \sum_{i=1}^{i=j-1} C_i\right)\left[1-\exp\left(-\varepsilon C_j l\right)\right] \quad (8.33)$$

Note that Eq. (8.33) is derived assuming that the laser light is attenuated only by vapor absorption. If liquid droplets are present, the light scattering by droplets must be considered (to be discussed in a later section). Therefore, Eq. (8.33) is valid only in the completely vaporized regions.

In order to correct for the fluorescence loss due to self-quenching (concentration quenching) and temperature quenching, Senda et al. [1997] measured the fluorescence intensity as a function of TMPD concentration and temperature. Their results show that the concentration quenching can be modeled by a linear relationship, i.e.,

$$I_{flu} = \eta f_1(T) K_c I_0 \exp\left(-\varepsilon l \sum_{i=1}^{i=j-1} C_i\right)\left[1-\exp\left(-\varepsilon l C_j\right)\right] \quad (8.34)$$

where K_c is included to represent the concentration effect on the fluorescence intensity. In addition, the temperature effect $f_1(T)$ was found to follow an exponential form. Taking these two quenching effects into account, the fluorescence signal from vaporized monomer can be written as

$$I_{flu} = \eta K_c I_0 \exp\left(-\varepsilon l \sum_{i=1}^{i=j-1} C_i\right)\left[1-\exp\left(-\varepsilon l C_j\right)\right]\left[K_{t1}\exp\left(-K_{t2}T_m\right)\right] \quad (8.35)$$

where K_{t1} and K_{t2} are the temperature quenching coefficients, and T_m is the fuel vapor mixture temperature. Hence, the last term in Eq. (8.35) models the temperature effect on the fluorescence intensity.

In the case of an evaporating fuel spray, the fuel mixture temperature T_m is lower than that of ambient gas. The mean temperature of the mixture in the cell j can be estimated from the bulk mean temperature of liquid and ambient gas by

$$T_m = \frac{-C_v\,[C_{fl}(T_{sat} - T_{l0}) + h_f] + \rho_a C_a T_a + C_v C_{fv} T_{sat}}{C_v C_{fv} + \rho_a C_a} \tag{8.36}$$

where C_v, = mass concentration of the fuel vapor, [kg/m^3]
C_{fl}, = specific heat of the liquid fuel, [kJ/kg·K]
C_{fv} = specific heat of the fuel vapor, [kJ/kg·K]
C_a = specific heat of the ambient gas, [kJ/kg·K]
T_{l0} = initial temperature of the liquid fuel, [K]
T_{sa} = saturation temperature of the fuel, [K]
ρ_a = density of ambient gas, [kg/m^3]
h_f = latent heat of fuel vaporization, [kJ/kg]
T_a = temperature of the ambient gas entrained into the spray, [K].

T_a can be obtained by averaging the mixture temperature at the adjacent outside zone and the ambient gas temperature.

Thus, using Eqs. (8.34), (8.35), and (8.36), both the mass concentration of the fuel vapor, C_v, which is calculated from C_j knowing the dopant concentration, and averaged mixture temperature T_m for each cell are obtained quantitatively with an assumption that the saturated temperature and the evaporating rate for the fuel and the dopant are the same. The calculation procedure is summarized in Fig. 8.19.

Since there is an overlap in wavelengths between the fluorescence signals from the vapor and liquid phases of fuel, quantitative measurement of fuel vapor concentration is limited to the region where there are few fuel droplets.

8.5.4 Application of PLIEF for Diesel Spray Visualization

Since the boiling temperatures of TMPD (260°C) and naphthalene (218°C) correspond to the middle boiling points of diesel fuels, these substances are particularly suited to the study of diesel sprays. Initial PLIEF work using TMPD/naphthalene as the exciplex compound was realized using an Nd:YAG laser at its 4th harmonic output (266 nm) as the excitation source [Melton and Verdieck, 1984; Melton and Verdieck, 1985]. Later on, the 3rd harmonic at 355 nm was adopted by Bardsley et al. [1988], due to much reduced absorption by both TMPD and naphthalene at 355 nm. The fuel used was a mixture of 90% decane, 9% naphthalene, and 1% TMPD by mass. The vapor phase fluorescence was detected at 400 nm, while a long-pass filter (>450 nm) was employed to isolate the fluorescence signal from the liquid phase. Other details can be found in Table 8.2. Quenching of the vapor phase fluorescence by oxygen was found to be very strong at high pressures in the engine. Consequently, the

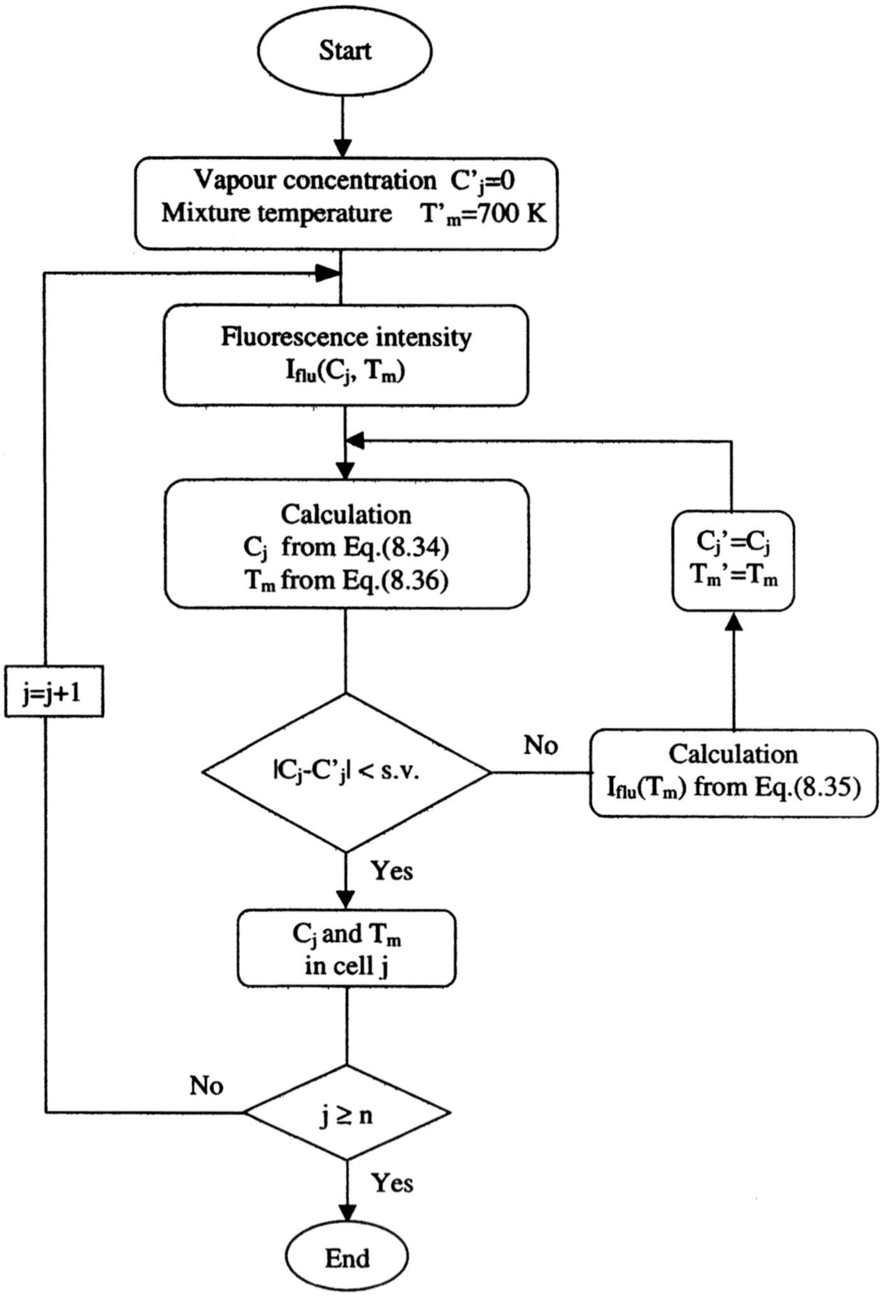

Fig. 8.19 Flow chart for the quantitative calculation of the fuel vapor concentration and spray temperature.

experiments were carried out in a motored two-stroke engine with N_2 rather than air. Their results showed that the vaporized fuel traveled further in a broader cone than the liquid fuel, and the fuel vapor was deflected and convected by the swirling flow in the cylinder.

A similar setup was used to study the interaction between the pilot injection and main injection in a motored two-stroke engine with N_2 [Campbell et al., 1995]. The optical system used is given in Table 8.2. Their results suggested that pilot injection affected the in-cylinder flow field, particularly in the near wall region, in such a way that less liquid fuel accumulation occurred with pilot injection.

A rapid compression-expansion machine was employed by Yeh et al. [1993] in order to obtain two-dimensional images of the fuel vapor concentration. They investigated the suitability of TMPD/naphthalene dopants for LIEF in a high-temperature and high-pressure environment. No thermal decomposition was found during the 10 ms operation period of the rapid compression-expansion machine. The fluorescence intensity was found to be approximately proportional to TMPD vapor concentration at temperatures between 500 K and 800 K and a pressure of 20 bar. Using the LIEF technique, they obtained two-dimensional images of an evaporating spray in the rapid compression-expansion machine filled with nitrogen.

Instantaneous images of liquid fuel and fuel vapor of a diesel jet inside a motored engine were obtained by Hodges et al. [1991]. Exciplex images of liquid phase emission revealed the presence of oscillations in the direction parallel to the jet axis. However, no such structures were observed in the gas phase emission.

In a series of PLIEF experiments, Senda et al. [1992;1994; 1997] investigated the impingement process of an evaporative diesel spray upon a wall surface in a high-temperature and high-pressure constant-volume bomb. Quantitative analysis of fuel vapor concentration in the diesel spray was obtained, following the calibration and correction procedure described in the previous section. They found that TMPD was stable as a vapor tracer in the high-temperature environment during the 80 ms experiment. Because of the spectral overlap between the exciplex and excited monomer and interference of light scattering due to liquid droplets, quantitative vapor concentration was obtained only in the completed vaporized region. No attempt was made to quantify the fuel droplet concentration, which would require the droplet size distribution and its number density in order to calculate the mass concentration of the liquid fuel droplets. From the PLIEF images, they were able to measure the change in vapor concentrations in radial distance from the spray axis and the temporal change in fuel vapor mass and volume. They observed that as the fuel spray was impinging on a flat surface, the fuel vapor was reflected backward at peripheral regions, as well as along the wall in the radial direction, due to the wall jet vortex formed at the edge of the wall spray flow.

It should be noted that all of the above studies were conducted in a nitrogen atmosphere in order to avoid vapor phase fluorescence quenching by oxygen.

8.5.5 Applications of PLIEF to SI Engines

Using the PLIEF technique, Fujimoto and Tabata [1993] studied the mixture distribution inside the combustion chamber of an SI engine with a swirl control intake system. The details of their experimental setup are given in Table 8.2. A pulsed Nd:YAG laser (355 nm) was used to excite the fluorescing exciplex (TMPD/naphthalene in N-octane). The experiments were carried out under motoring conditions with nitrogen gas to prevent the quenching of the fuel vapor fluorescence by oxygen. They found that under the high swirling condition, fuel vapor was spread around the peripheral region of the combustion chamber in a doughnut-like shape in the later stage of the compression stroke. Under the low swirling condition, a high concentration of fuel vapor was found on the exhaust valve side due to the fuel impingement on that side.

Also using the PLEF technique, Tabata et al. [1995] investigated the mixture distribution inside the combustion chamber of a mixture injected SI engine. A fuel/air mixture was injected directly into the combustion chamber at the beginning of the compression stroke from a small mixture chamber in the cylinder head. LIEF images show that a high concentration of fuel vapor was preserved during the compression stroke and reduced fuel impingement on cylinder walls, which led to a stratified charge and hence improved combustion stability under lean operating conditions. The engine was reported to be capable of operating at air/fuel ratios up to 40:1.

Using a similar experimental setup, Diwakar et al. [1992] studied the liquid and vapor fuel distributions from an air-assist injector in a two-stroke direction injection gasoline engine.

The above studies involved the use of PLIEF in a motored SI engine with N_2. In order to study the effect of fuel injection timing on mixture distributions and combustion, Knapp et al. [1996] applied PLIEF to a firing SI engine. The quenching of vapor fluorescence by oxygen was partially compensated for by the use of a high-power XeCl excimer laser with a shorter excitation wavelength. Their results showed that, for open valve injection in a port injected SI engine, the liquid fuel droplets were responsible for the inhomogeneous in-cylinder fuel distribution, which was consistent with the PDA measurements reported by Hardalupas et al. [1995].

The boiling point of TMPD, which dominates the vapor phase fluorescence, is 260°C. Thus, its ability to trace the evaporation of gasoline fuel is questionable. To overcome the problem, several research groups are searching for LIEF dopants with matching properties to gasoline fuels.

Initial work on gasoline-based exciplex compounds focused on mixtures of fluorobenzene, triethylamine (TEA), and isooctane, following the initial work by Melton [1993]. But fluorobenzene was found to evaporate much faster than the isooctane despite the fact that the boiling point of fluorobenzene (85°C) is close to that of isooctane (98°C) [Law, 1989].

Ghandhi et al. [1994] found that in a mixture of fluorobenzene/diethylmethylamine (DEMA) and hexane, co-evaporation was more closely achieved between fluorobenzene and hexane. Since the boiling temperature of hexane (68°C) is low, the results obtained using this mixture will be representative of only the lighter components in gasoline fuels. The LIEF of fluorobenzene/DEMA was excited by an Nd:YAG laser at its 4th harmonic output (266 nm). The emissions of the monomer (fluorobenzene) at 290 nm and exciplex at 380 nm were sufficiently shifted to allow discrimination with interference filters. They found that the vapor phase fluorescence from fluorobenzene was severely affected by temperature quenching, which limited their results to the qualitative interpretation of the images. Further study by Lee and Gajdeczko [1996] also showed that the cross-talk from Mie scattering onto the vapor phase signal could also be a problem in dense regions of the spray. This was later resolved by adding a WG 305 Schott glass filter and increasing the concentration of fluorobenzene to 2%. Because the fluorobenzene fluorescence was quenched by oxygen, the studies were limited to motored engines with nitrogen.

A mixture of 5% N,N-dimethylaniline (DMA), 5% naphthalene, and 90% gasoline by mass was employed by Shimizu et al.[1992] to observe both liquid and vapor phase concentrations of fuel. DMA was found to account for 85% of the fluorescence at λ_{max} = 355 nm in the vapor phase, when excited at 308 nm by an XeCl excimer laser. The fluorescence of the liquid phase from the exciplex formed by DMA and naphthalene was red-shifted to λ_{max} = 410 nm. Because DMA was used to trace the vapor phase of the fuel, the true vapor concentration of fuel may be underestimated in the early stages of evaporation, because its high boiling point (193°C) corresponds to the upper end of the gasoline distillation curve. The vapor phase fluorescence signal was calibrated under motoring conditions for different equivalence ratios at different crank angles. The authors claimed an accuracy of 3% in equivalence ratios from vapor phase fluorescence signals. However, the pulse-to-pulse variations in the laser output were not considered, which would have been an additional source of errors. Only qualitative imaging of liquid fuel in the cylinder was attempted. It is worth mentioning that all of the measurements were carried out under firing conditions. Two laser sheets were introduced into the combustion chamber from opposite directions to minimize the effect of light attenuation, which was caused by the liquid droplets and affected the quantitative vapor concentration measurement during the early stage of the intake stroke. They found that the evaporation of liquid droplets was enhanced by the introduction of swirling flow during the intake stroke, resulting in a more homogeneously distributed mixture at the time of spark ignition.

Kim et al. [1996] reported a new exciplex mixture suitable for tracking the vaporization of gasoline fuel. The exciplex mixture consists of 5% N,N-dimethylaniline (DMA) and 7% 1,4,6-trimethylnaphthalene (TMN) in 88% isooctane by mass. The excited DMA dominates the vapor phase fluorescence. Because DMA has a boiling temperature of 193°C it will simulate the heavy components in gasoline. When excited at 308 nm, the vapor phase fluorescence can be detected at a peak of 350 nm, while the liquid phase fluorescence peaks at 405 nm. The exciplex formation was found to be stable at elevated temperatures (145°C), as was the spectral separation between monomer and exciplex emissions.

Münch et al. [1996] developed another new exciplex mixture, triethylamine (TEA)/benzene. The boiling points of benzene and TEA are 81°C and 89°C, respectively, which match closely to that of isooctane (99°C). When a mixture of 2.9% benzene and 2 % TEA in isooctane was excited by a KrF excimer laser at 248 nm, a red shifted exciplex fluorescence was observed in the liquid phase with a maximum around 350 nm, while the vapor phase fluorescence peaked at 290 nm. Fuel evaporation inside the inlet manifold was investigated using the PLIEF technique. Incomplete fuel film evaporation was observed at the inlet valve opening and large local and cyclic variations in the fuel vapor concentrations were found to be present throughout the cycle in the intake manifold.

8.6 Simultaneous Measurement of Fuel Vapor and Fuel Droplets by the Laser Extinction/Absorption (LEA) Technique

8.6.1 Introduction

As discussed in the previous section, LIEF can be used to visualize the fuel atomization and vaporization process. However, it is difficult to make a reliable quantitative analysis from LIEF measurements. For example, estimation of the absolute liquid phase concentration by LIEF requires detailed knowledge of temperature-dependent spectral emission from exciplex dopant in the spray, which is difficult to implement. In addition, because the gas phase fluorescence is effectively quenched by oxygen, most of the experiments have to be carried out in a nitrogen environment.

8.6.2 The Principle of the LEA Technique

Chraplyvy [1981] was the first to propose the use of two-wavelength laser extinction (scattering)/ absorption to measure fuel vapor concentration and liquid fuel droplets simultaneously in an evaporating spray. The principle of the two-wavelength laser extinction and absorption (LEA) technique is as follows.

In the presence of liquid droplets, the light beam will be attenuated by both the absorption of vapor and liquid fuel and scattering of liquid droplets, when traversing through an evaporating spray. Two separate measurements must be made in order to obtain the vapor concentration: a transmission measurement at a non-absorbing wavelength (λ_T) to determine the droplet optical thickness, and a transmission measurement at an absorbing wavelength (λ_A) to determine the joint vapor and droplet optical thickness. The difference between these two measurements can then be used to determine the vapor concentration.

At an absorbing wavelength λ_A, the laser intensity $I(\lambda_A)$ transmitted through a mixture of vapor and droplets is related to the incident laser intensity $I_o(\lambda_A)$ by

$$I(\lambda_A) = I_o(\lambda_A)\exp\left[-\alpha(\lambda_A)C_{vap}L\right]\exp\left[-\int_0^L K_{ext}(\lambda_A)dx\right] \tag{8.37}$$

where $\alpha(\lambda_A)$ = the vapor absorption coefficient at λ_A, [m^2/mol]
C_{vap} = the vapour concentration, [mol/m^{-3}]
L = the optical path length, [m]
K_{ext} = the extinction coefficient for a cloud of droplets, [m^{-1}]

The first exponential term in Eq. (8.37) represents the light attenuation by the vapor, and the second exponential term is the light extinction due to droplets, which includes both the scattering loss and absorption.

From light scattering theory [Bohren and Huffman, 1983], the extinction coefficient for a cloud of particles is related to the number, size, and properties of individual droplets by

$$K_{ext}(\lambda_A) = -\frac{\pi}{4} C_n \int_0^{\infty} Q_{ext}(\lambda_A) N(D) D^2 dD \tag{8.38}$$

where C_n = the number density of droplets, [m^{-3}]
$Q_{ext}(\lambda_A)$ = the extinction efficiency of a droplet at wavelength λ_A
D = the droplet diameter, [μm]
N(D = the droplet size distribution function defined such that $\int N(D)\, dD = 1$.

Determination of the vapor concentration C_{vap} requires two extinction measurements: one at the absorbing wavelength (λ_A) as depicted by Eq. (8.37), and the other one at a non-absorbing wavelength (λ_T). The extinction of light at the non-absorbing wavelength is caused purely by the scattering by the droplets

$$I(\lambda_T) = I_0(\lambda_T) \exp\left[-\int_0^L K_{sca}(\lambda_T) dx \right] \tag{8.39}$$

The scattering coefficient, K_{sca}, for a cloud of particles is related to the number, size, and properties of individual droplets by

$$K_{sca}(\lambda_T) = -\frac{\pi}{4} C_n \int_0^{\infty} Q_{sca}(\lambda_T) N(D) D^2 dD \tag{8.40}$$

where $Q_{sca}(\lambda_T)$ is the scattering efficiency of a droplet.

At the same wavelength, the extinction efficiency is the sum of the scattering efficiency and absorption efficiency, i.e.

$$Q_{ext}(\lambda)= Q_{sca}(\lambda)+ Q_{abs}(\lambda) \tag{8.41}$$

The values of extinction and scattering efficiencies Q_{ext} and Q_{sca} are complex functions of the particle size and the refractive index of the particle. Exact expressions for these values do not exist, except for some special cases. Perhaps the most important exactly soluble problem in the theory of absorption and scattering by small particles is that for a sphere of arbitrary radius and refractive index as described by the Lorentz-Mie theory [Mie,1908].

Equations (8.37) and (8.39) may also be written in terms of the optical thickness τ:

$$I(\lambda_A) = I_0(\lambda_A)\exp\left[-\tau(\lambda_A)_{vap}\right]\exp\left[-\tau(\lambda_A)_{drop}\right] \tag{8.42}$$

$$I(\lambda_T) = I_0(\lambda_T)\exp\left[-\tau(\lambda_T)_{drop}\right] \tag{8.43}$$

If Eq. (8.43) is substituted in Eq. (8.42), the following expression for the line-of-sight averaged vapor concentration results:

$$\begin{aligned} C_{vap} &= \frac{1}{\alpha(\lambda_A)L}\left[-\tau(\lambda_T)_{drop}\left(\frac{\tau(\lambda_A)_{drop}}{\tau(\lambda_T)_{drop}}\right) - \ln\left(\frac{I}{I_0}\right)_{\lambda_A}\right] \\ &= \frac{1}{\alpha(\lambda_A)L}\left[\ln\left(\frac{I}{I_0}\right)_{\lambda_T} R - \ln\left(\frac{I}{I_0}\right)_{\lambda_A}\right] \end{aligned} \tag{8.44}$$

where R is simply the ratio of the drop optical thickness at the two wavelengths.

Note that the size distribution of droplets is required in order to obtain the extinction coefficient $K_{ext}(\lambda_A)$ and scattering coefficient $K_{sca}(\lambda_T)$. This may be obtained from visible laser scattering experiments using the Fraunhofer diffraction technique, as discussed in Section 8.2.

The mass concentration of fuel droplets per unit volume C_d [kg/m^3] is related to the first term in Eq. (8.44), and is given by

$$C_d = -\frac{2}{3L}\rho_f \frac{D_{32}}{Q_m} \ln\left(\frac{I}{I_0}\right)_{\lambda_T} \tag{8.45}$$

where ρ_f is the liquid fuel density, and Q_m is the mean extinction efficiency for a polydispersed particle cloud

$$Q_m = \frac{\int_0^\infty N(D)Q(D)D^2 dD}{\int_0^\infty N(D)D^2 dD} \tag{8.46}$$

and D_{32} is known either as the volume/surface mean diameter or Sauter mean diameter

$$D_{32} = \frac{\int_0^\infty N(D)D^3 dD}{\int_0^\infty N(D)D^2 dD} \tag{8.47}$$

Thus, simultaneous measurements of fuel vapor concentration and liquid droplets can be achieved by the two-wavelength LEA method, if the droplet size distribution is obtained from the Fraunhofer diffraction technique. Under certain situations, however, the ratio of optical thickness R may be approximated as one, as will be discussed later.

8.6.3 Implementation of the LEA Technique

Fig. 8.20 depicts schematically the experimental setup of an LEA system. Two lasers are used, one of which is tuned to the absorption wavelength of the fuel used, while the other is used to detect the presence of liquid droplets. A fortuitous overlap of absorption bands for most hydrocarbons in a homologous series, as shown in Fig. 8.21, allows He-Ne laser radiation at 3.39 μm to be used to probe the fuel vapor. Alternatively, a dopant may be used which has a matching boiling point to the fuel, and has an absorption band accessible to a laser source. Since most hydrocarbon vapor does not absorb visible radiation, a visible He-Ne laser (0.632 μm), or any other laser with visible output, is suited for scattering measurements to determine drop-size distributions. Thus, using this dual wavelength system (visible and infrared or UV), the liquid and vapor phases of the fuel can be discriminated and quantified.

As discussed in the previous section, even if only the vapor concentration is to be measured in the presence of liquid droplets, the knowledge of droplet size distribution is needed, which

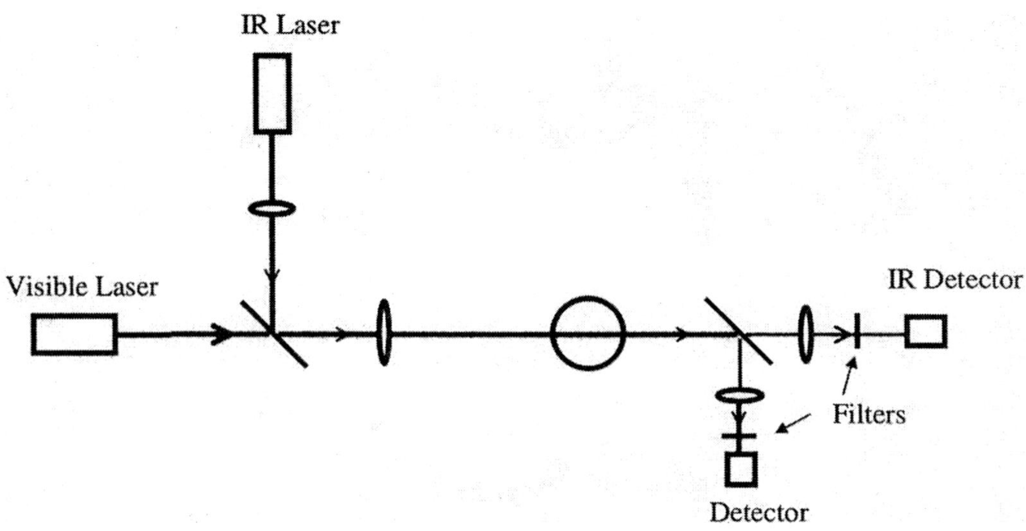

Fig. 8.20 Schematic of the optical setup for the dual wavelength LEA technique.

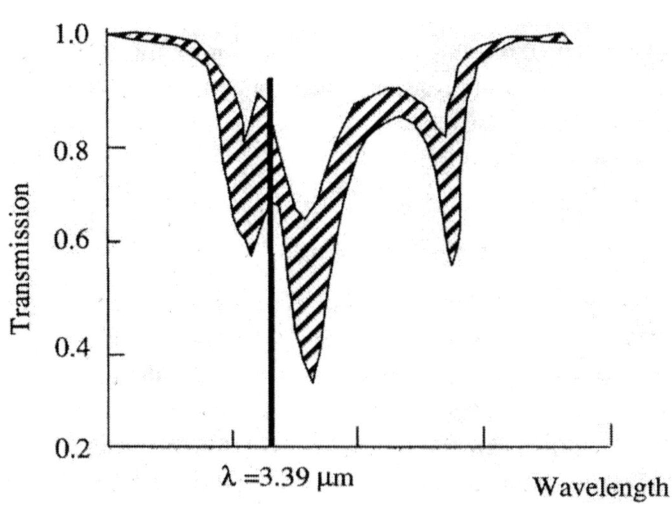

Fig. 8.21 Hydrocarbon absorption spectrum.

complicates the measurement. A significant simplification of Eq. (8.44) can be achieved, given certain constrains on the drop size and optical setup. Drallmeier [1994] estimated that the ratio of the optical thickness, R, approaches one if drop-size distributions have an area mean diameter D_{20} greater than 20 μm, where D_{20} is defined as

$$D_{20} = \left[\frac{\int_0^\infty N(D) D^2 dD}{\int_0^\infty N(D) dD} \right]^{1/2} \tag{8.48}$$

In addition, according to light scattering theory [Van de Hulst, 1957; Kerker, 1969; and Bohren and Huffman, 1983], the extinction coefficient Q_m approaches a constant value of 2 as the droplet size becomes much larger than the wavelength. Therefore, the mass concentration of liquid droplets can be simplified to

$$C_d = -\frac{1}{3L} \rho_f D_{32} \ln\left(\frac{I}{I_0}\right)_{\lambda_T} \tag{8.49}$$

The Sauter mean diameter, Q_{32}, can be determined experimentally by the Fraunhofer diffraction technique. Alternatively, it may be estimated from [Kamimoto et al., 1989],

$$D_{32} = \frac{0.63 Q_{ext} M_f}{\rho_f \sum \left\{ \log\left(\frac{I_0}{I}\right)_{\lambda_T} \Delta S \right\}} \tag{8.50}$$

where ΔS is the unit projection area. M_f is the total mass of injected fuel within the spray. $\Sigma\left[\log(I_0/I)_{\lambda_T} \Delta S\right]$ is the summation of the transmissivity by fuel droplets in the entire spray.

8.6.4 Application of LEA to Evaporating Fuel Sprays and IC Engines

Chraplyvy [1981] developed the first dual wavelength LEA system, in order to measure the vapor concentration inside an evaporating n-heptane fuel spray. The vapor absorption and liquid scattering were measured at 3.39 μm and 0.6328 μm, respectively. The droplet size distributions were deduced from the forward scattered light distributions using a Malvern ST-1800 analyser. A similar setup was used by Drallmeier and Peters [1991] to measure the vapor flux, droplet concentration, and size in an isooctane fuel spray. Radial distributions of vapor concentration in the spray were obtained from tomographic line-of-sight infrared extinction measurements. They found that the vapor concentration was relatively insensitive to the measured drop size distribution, but it was most sensitive to the measurement of the infrared extinction ratio. An upper limit for the overall error in the vapor concentration measurement was estimated as 15% over the spray radius.

Following the initial development work by Winklhofer and Plimon [1991], a dual wavelength LEA system was used to obtain quantitative measurements of fuel vapor distributions in a gasoline research engine [Winklhofer et al., 1992]. In order to simply the experimental setup and data interpretation, quantitative fuel vapor concentration measurements were carried out in the absence of liquid droplets. The visible He-Ne laser was only used to check the presence of liquid fuel in the cylinder. The attenuation of the infrared He-Ne laser was monitored with an HgCdTe detector. Since laser beam attenuation by absorption was continuously monitored, the information obtained was characterized by high temporal resolution, which provided cycle-to-cycle variation measurements, as well as mapping of the temporal development of in-cylinder fuel distribution. The sensitivity of their system was suited to cover near stoichiometric fuel concentration levels ($\phi = 0.5$–2.0), and signal-to-noise ratios allowed the resolution of mixture variations on the order of 5–10%.

Skippon et al. [1996] developed a dual wavelength infrared LEA system. The infrared radiation from a tungsten-halogen lamp was used to illuminate a cylindrical volume of 12 mm diameter in the combustion chamber, through a silica fiber and collimating optics. After the combustion chamber, the attenuated beam was transmitted through another fiber before it was split into two infrared bands at 2.07 μm and 2.31 μm by means of a beam splitter and two interference filters. The infrared radiation was then detected by two room-temperature lead sulfide (PbS) detectors. The infrared radiation at 2.31 μm was absorbed by the band associated with two C–H bond vibration modes, while there was no absorption at 2.07 μm. Measurements using this band are thus sensitive to species containing $-CH_3$ and $-CH_2$ groups. According to Eq. (8.44), the fuel vapor concentration was obtained from the ratio of the intensities at the two wavelengths, assuming that R in Eq. (8.44) could be approximated as one. In order to improve the signal-to-noise ratio, phase sensitive detection was included in the dual wavelength LEA system, by modulating the source of the infrared radiation and employing two synchronously demodulated lock-in amplifiers for the two detectors. The technique was used to measure the effect of fuel distillation characteristics on the response of the in-cylinder air/fuel ratio to a simple fuel shut-off transient. They found that the delay between the fuel shut-off and in-cylinder air/fuel ratio response increased with increasing boiling point of the component.

The most comprehensive application of the LEA technique was carried out by Suzuki et al. [1993]. Fig. 8.22 shows the experimental setup employed for the simultaneous measurement of the fuel vapor concentration and liquid droplets in an evaporating diesel spray. In this study, α-methylnaphthalene was used as the test fuel. The *para* absorption band of the α-methylnaphthalene was probed by a pulsed dye laser at 280 nm. The light extinction component was measured for the visible output (560 nm) of the dye laser before the frequency doubling crystal. The two laser beams were expanded to parallel beams of 100 mm diameter before they traversed the diesel spray in a high temperature and pressure combustion chamber. The transmitted light beams were then separated into two paths and detected by two CCD cameras (Fig. 8.22). Absorption and extinction images were then processed to find out the equivalence ratio distributions of fuel vapor and droplets in the diesel spray by a computerized tomographic transformation method. The mass fuel vapor concentration, liquid fuel mass concentration, and Sauter mean diameter of fuel droplets, were calculated from the extinction and absorption using Eqs. (8.44), (8.49), and (8.50). Calibration was conducted before the experiments to find the relationship between fuel vapor concentration and absorption coefficient for different temperatures and pressures. The accuracy of this measuring technique was claimed to be better than 11% in their concentration measurements.

Fig. 8.23 shows the results of the concentration distributions (equivalence ratio) of liquid fuel and fuel vapor in the impinging fuel spray at 2.4 ms after the start of fuel injection at a injection pressure of 196 bar. The impingement took place 25 mm away from the injector. At this particular time, the fuel droplets were not present at the peripheral regions of the impinging spray. However, as shown in Fig. 8.23, the dispersion pattern of the fuel vapor in the impinging spray showed a raised curve shape at the peripheral region around the impinging point. Similar images were also obtained for the diesel spray at different times after the start of fuel injection for both

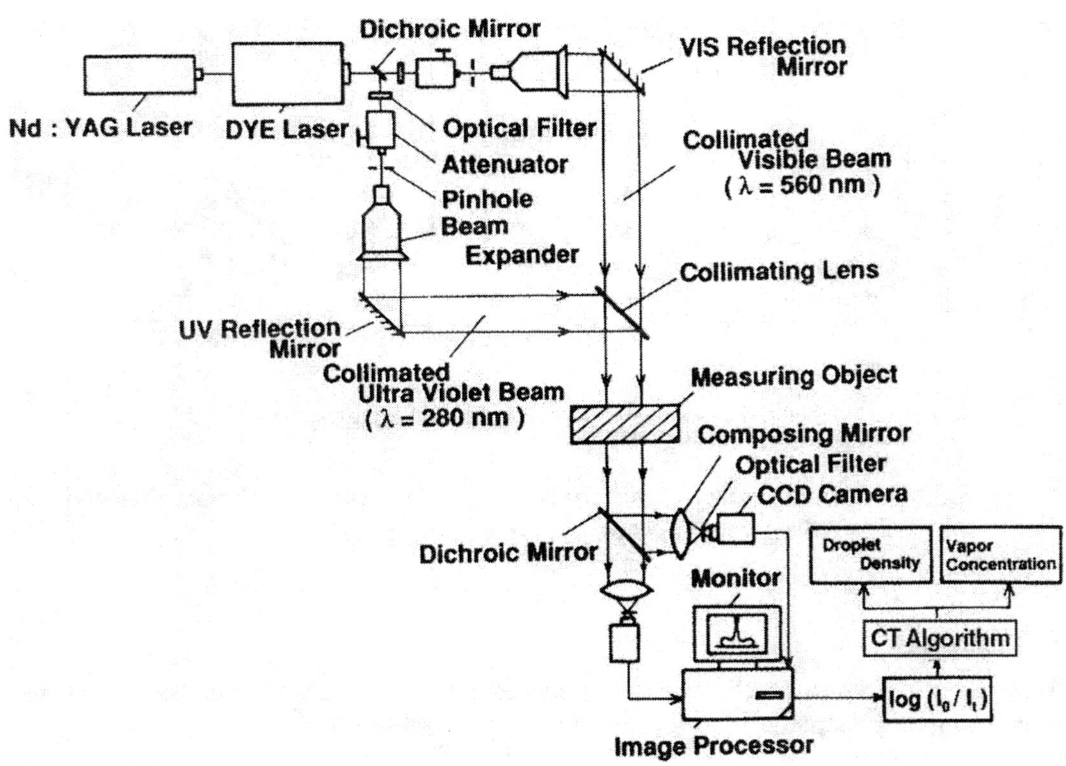

Fig. 8.22 Experimental setup for the LEA imaging system used by Suzuki et al.[1993].

vertically and non-vertically impinged fuel sprays. The resulting images were then used to study in detail the time history of fuel vaporization processes. Studies were also carried out on the interaction of fuel spray and combustion chamber geometry using the same experimental setup by Suzuki et al. [1994].

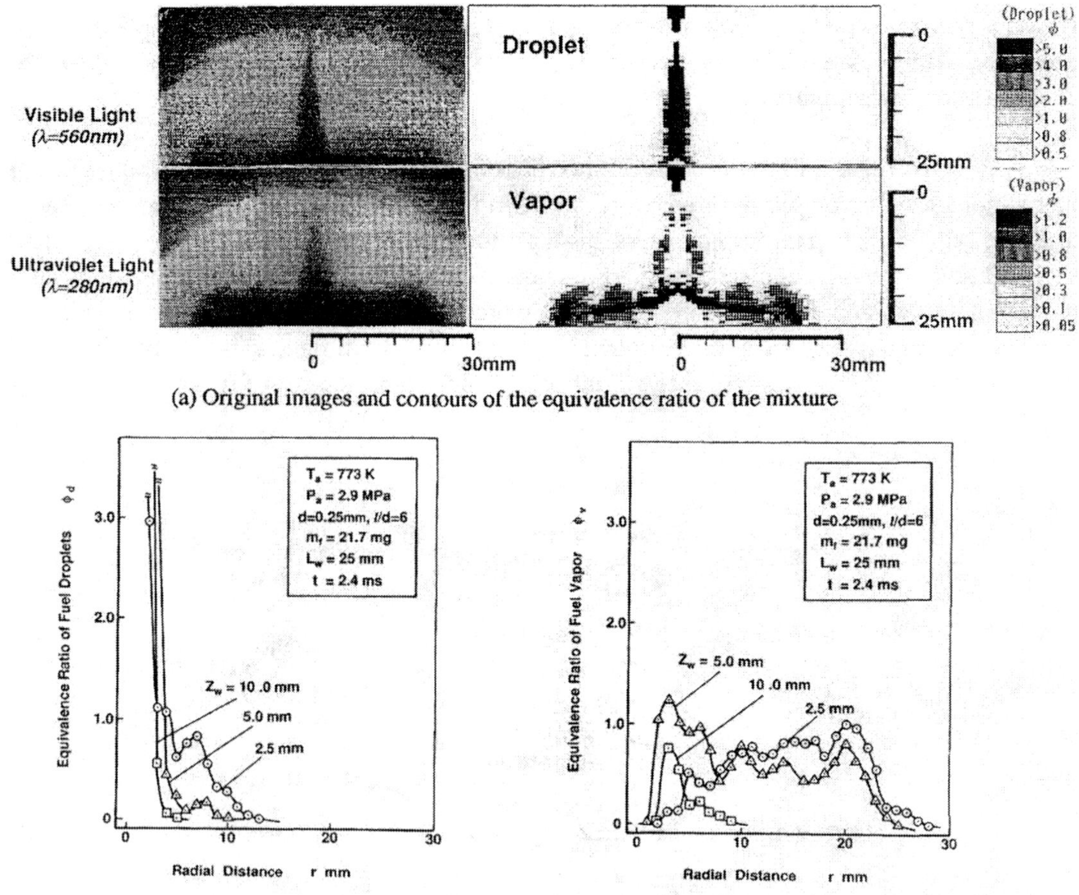

(a) Original images and contours of the equivalence ratio of the mixture

(b) Radial distributions of the equivalence ratio of the mixture

Fig. 8.23 Two-phase mixture concentration distributions of an impinging spray at $t = 2.4$ ms after the start of injection, [Suzuki et al., 1993].

8.7 Summary

Three groups of optical techniques for sprays and liquid fuel distribution measurements are discussed in this chapter and they are summarized in Table 8.3.

Table 8.3 Summary of optical techniques for sprays and liquid fuel distribution measurements

Technique	Applications	Comments
Back illuminated photography	Spray structure in and outside engines	Qualitative line-of-sight visualization Sensitive to liquid fuel only
Schlieren and shadowgraph	Spray atomization and vaporization in and outside engines	Qualitative line-of-sight visualization Sensitive to both liquid and vapor phases
Fraunhofer diffraction	Liquid drop sizing in sprays	Spatially averaged size distribution Medium to high drop concentration
Phase Doppler analyzer	Liquid drop sizing and mass flux measurements in sprays and engines	Spatially resolved size distribution Temporal averaged size distribution
Mie scattering	2-D imaging of sprays and liquid fuel distribution	Simple experimental setup Sensitive to large droplets
Planar LIF	2-D imaging of sprays and liquid fuel distribution	Excellent contrast Sensitive to liquid drops
Planar LIEF	2-D imaging of liquid and vaporized fuel distribution	Separate liquid and vapor fluorescence signals Severe quenching of vapor fluorescence
Laser extinction/absorption	Simultaneous detection of fuel vapor and drops concentration	Quantitative two-phase measurements Line-of-sight technique

Spray structure and liquid drops can be directly observed with the aid of a back illumination source. The sensitivity of the direct photography to the vaporized fuel will be improved if a Schlieren or shadowgraph setup is used. Laser sheet techniques are particularly suited for spatially resolved measurements of spray and liquid fuel distributions. Images of liquid fuel distribution can be obtained with excellent contrast if LIF is used.

Droplet sizing can be realized by either the Fraunhofer diffraction method or the PDA technique. The former provides a line-of-sight and spatially averaged size distribution, while the latter yields a single point and temporally averaged size distribution. The PDA technique has the advantage of good spatial resolution, and is capable of simultaneous measurements of drop size and mass flux.

Simultaneous measurements of fuel vapor and liquid fuel can be performed by LIEF or LEA techniques. LIEF shares many of the features associated with LIF. LIEF has been used to obtain two-dimensional images of fuel vapor and liquid droplet distributions in IC engines. The fluorescence from liquid fuel and fuel vapor are separated spectrally by means of an exciplex forming compound. The effect of quenching is worsened in LIEF because of the limited choice of exciplex forming compounds. The quenching of fluorescence from fuel vapor has limited most LIEF applications to a nitrogen environment in a motored engine.

Dual wavelength LEA is another optical technique that is capable of simultaneous detection of fuel vapor and liquid droplets. The LEA technique can be used to simultaneously measure fuel vapor concentration, liquid droplet concentration, and droplet sizing. It is not affected by the presence of oxygen, and hence can be applied to conditions that are more realistic. LEA provides line-of-sight measurement along the laser beam. Tomographic techniques are required in order to obtain spatially resolved measurements.

References

Amer, A. and Lai, M.-C. (1995), "Time-Resolved Measurements of a Transient Gasoline Spray," SAE Paper No. 950509.

Arcoumanis, C., Chang, J.C., Morris, T. (1993), "Spray Characteristics of Single and Two-Spring Diesel Fuel Injectors," SAE Paper No. 930922.

Bachalo, W.D. (1980), "Method for Measuring the Size and Velocity of Spheres by Dual-Beam Light-Scatter Interferometry," Applied Optics, Vol. 19, Part 3, p. 363.

Bachalo, W.D., Rosa, A., and Sankar, S.V., (1991), "Diagnostics for Fuel Spray Characterization," *Combustion Measurement,* Ed. by N. Chigier, pp. 229–277.

Bardsley, M., Felton, P., and Bracoo, F., (1988), "2-D Visualisation of a Hollow-Cone Spray in a Cup-in-Head, Ported, I.C. Engine," SAE Paper No. 880521.

Baritaud, T.A., Heinze, T.A., and Le Cos, J.F. (1994), "Spray and Self-Ignition Visualisation in a DI Diesel Engine," SAE Paper No. 940681.

Bohren , C.F. and Huffman, D.R. (1983), *Absorption and Scattering of Light by Small Particles*, John Wiley & Sons.

Brenn, G., Domnick, J., Dorfner, V., Durst, F., and Tremel, P. (1995), "Unsteady Gasoline Injection Experiments: Comparison of Measurements in Quiescent Air and in a Model Intake Port," SAE Paper No. 950512.

Campbell, P., Sinko, K., and Chehroudi, B. (1995), "Liquid and Vapour Phase Distributions in a Piloted Diesel Fuel Spray," SAE Paper No. 950445.

Canaan, R.E., Dec, J.E., Robert, M.G., and Daly, D.T. (1998), "The Influence of Fuel Volatility on the Liquid-Phase Fuel Penetration in a Heavy-Duty DI Diesel Engine," SAE Paper No. 980510.

Cavaliere, A., Ragucci, R., D'Alessio, A., and Noviello, C. (1992), *Analysis of Diesel Sprays through Two-Dimensional Laser Light Scattering, Proc. of 22nd Symposium (International) on Combustion,* The Combustion Institute.

Chigier, N., (1991), "Optical Imaging of Sprays," *Prog. Energy Combust. Sci.*, Vol. 17. pp. 211–262,

Chraplyvy, A.R., (1981), "Nonintrusive Measurements of Vapour Concentrations Inside Sprays," *Applied Optics*, Vol. 20, No. 15, pp. 2620–2624.

Cousyn, B., Posylkin, M., Vannobel, F., and Whitelaw, J.H. (1995), "Droplet Characteristics in Two Cylinders of a Firing Spark Ignition Engine," SAE Paper No. 952466.

Dawson, M. and Hochgreb, S. (1998), "Liquid Fuel Visualisation Using Laser Induced Fluorescence During Cold Start," SAE Paper No. 982466.

Desantes, J., Arregle, J., Pastor, J., and Delage, A. (1998), " Influence of the Fuel Characteristics on the Injection Process in a DI Diesel Engine," SAE Paper No. 980802.

Diwakar, R., Fansler, T., French, D., Ghandhi, J., Dasch, C., and Hefflfinger, D. (1992), "Liquid and Vapour Fuel Distributions from an Air-Assist Injector—An Experimental and Computational Study," SAE Paper No. 920422.
Drallmeier, J.A. and Peters, J.E. (1991), "Experimental Investigation of Fuel Spray Vapour Phase Characterisation," *Atomisation and Sprays*, Vol. 1, pp. 63–88.
Drallmeier, J.A. (1994), "Hydrocarbon-Vapour Measurements in Fuel Sprays—A Simplification of the Infrared Extinction Technique," *Applied Optics*, Vol. 33, No. 30, pp. 7175–7179.
Espey, C. and Dec, J. (1995), "The Effect of TDC Temperature and Density on the Liquid-Phase Fuel Penetration in a DI Diesel Engine," SAE Paper No. 952456.
Espey, C., Dec., J., Litzinger, T., and Santavicca, D. (1997), "Planar Laser Rayleigh Scattering for Quantitative Vapour Fuel Imaging in a Diesel Jet," *Combustion and Flames*, Vol. 109, Number 1/2, pp. 65–86.
Fansler, T.D., French, D.T., and Drake, M.C. (1995), "Fuel Distributions in a Firing Direct-Injection Spark-Ignition Engine using Laser-Induced Fluorescence Imaging," SAE Paper No. 950110.
Felton, P.G., Kyritsis, D.C., and Fulcher, S.K. (1995), "LIF Visualisation of Liquid Fuel in the Intake Manifold during Cold Start," SAE Paper No. 952464.
Felton, P.G., Mantzaras, J., Bardsley, M.E.A., and Bracco, F.V. (1987), "2-D Visualisation of Liquid Fuel Injection in an Internal Combustion Engine," SAE Paper No. 872074.
Fry, M., Nightingale, C., and Richardson, S. (1995), "High-Speed Photography and Image Analysis Techniques Applied to Study Droplet Motion Within the Porting and Cylinder of a 4-Valve SI Engine," SAE Paper No. 952525.
Fujimoto, M. and Tabata, M. (1993), "Effect of Swirl Rate on Mixture Formation in a Spark Ignition Engine Based on Laser 2-D Visualisation Technique," SAE Paper No. 931905.
Ghandhi, J.B., Felton, P.G., Gajdeczko, B.F., and Bracco, F.V. (1994), "Investigation of the Fuel Distribution in a Two Stroke Engine with an Air-Assisted Injector," SAE Paper No. 940394.
Hardalupas, Y., Taylor, A., and Whitelaw, J.H. (1995), "Influence of Injection Timing on In-Cylinder Fuel Distribution in a Honda VTEC-E Engine," SAE Paper No. 95050.
Heitor, M.V., Starner, S.H., Taylor, A.M.K.P., and Whitelaw, J.H. (1993), "Velocity, Size and Turbulent Flux Measurements by Laser Doppler Velocimetry," in Instrumentation for Flows with Combustion, ed. by Taylor, Academic Press.
Hodges, J., Baritaud, T., and Heinze, T. (1991), "Planar Liquid and Gas Fuel and Droplet Size Visualisation in a DI Diesel Engine," SAE Paper No. 910726.
Ikeda, Y., Hosokawa, S., Sekihara, F., and Nakajima, T. (1997), "Cycle-Resolved PDA Measurement of Size-Classified Spray Structure of Air-Assist Injector," SAE Paper No. 970631.
Kamimoto, T., Yokota, H., and Kobayashi, H. (1987), "Effect of High Pressure Injection on Soot Formation Processes in a Rapid Compression Machine to Simulate Diesel Flames," SAE Paper No. 871610.
Kamimoto, T., Yokota, H., and Kobayashi, H. (1989), "A New Technique for the Measurement of Sauter Mean Diameter of Droplets in Unsteady Dense Sprays," SAE Paper No. 890316.
Kelly-Zion, P.L., DeYoung, C.A., Peters, J.E., and White, R.A. (1995), "In-Cylinder Fuel Drop Size and Wall Impingement Measurements," SAE Paper No. 952480.

Kerker, M. (1969), *The Scattering of Light and Other Electromagnetic Radiation*, Academic Press.Kerker, M. (1969), *The Scattering of Light and Other Electromagnetic Radiation*, Academic Press.

Kim, J., Golding, B., Schock, J., Keller, P., and Nocera, D., "Exciplex Fluorescence Visualisation Systems for Pre-Combustion Diagnosis of an Automotive Gasoline Engine," 1996, SAE Paper No. 960826.

Knapp, M., Luczak, A., Beushausen V., Hentschel, W., and Andresen, P. (1996), "Vapour/Liquid Visualisation with Laser Induced Exciplex Fluorescence in an SI Engine for Different Injection Timings," SAE Paper No. 961122.

Kosaka, H., Won, Y., and Kamimoto, T. (1992), "A Study of the Structure of Diesel Sprays," SAE Paper No. 920107.

Lai, M.-C., Zhao, F.-Q., Amer, A.A., and Chue, T.-H. (1994), "An Experimental and Analytical Investigation of the Spray Structure from Automotive Port Injectors," SAE Paper No. 941873.

Law, C.K. (1989), "Some Recent Advances in Droplet Combustion," Proc. of the Third International Colloquium on Drops and Bubbles, AIP Conference Proceedings, 197, pp. 321–337.

Lee, C.F. and Gajdeczko, B.F. (1996), "Quantitative Exciplex Measurements of the Vapour Concentration in Air-Assisted Sprays," Extended Abstract in Proceedings of 9th Annual Conference on Liquid Atomisation and Spray Systems, ILASS-Americas 96, pp. 297–01.

Lefebre, A.H. (1989), *Atomisation and Spray*, Taylor and Francis.

Levy, N., Amara, S., Champoussin, J.-C. (1998), "Simulation of a Diesel Jet Assumed Fully Atomised at the Nozzle Exit," SAE Paper No. 981067.

Melton, L.A. (1993), "Exciplex-Based Vapour/Liquid Visualisation Systems Appropriate for Automotive Gasolines (A Users Manual)," Final report to Texas Higher Education Coordinating Board, Energy Research Applications Program, Contract 27.

Melton, L.A. (1988), "Quantitative Use of Exciplex Based Vapour/Liquid Visualisation Systems (A Users Manual)," Final Report of Army Research Office Contract DAAL 03-86-k-0082, Jan.

Melton, L. and Verdieck, J.(1985), "Vapour/Liquid Visualisation for Fuel Sprays," *Combust. Sci. and Tech.*, Vol. 42, pp. 217–222.

Melton, L. and Verdieck, J. (1984), "Vapour/Liquid Visualisation in Fuel Sprays," *Proc. of 20th Symposium (Int.) on Combustion*, pp. 1283–1290..

Meyer, R. and Heywood, J. (1997), "Liquid Fuel Transport Mechanisms into the Cylinder of a Firing Port Injected SI Engine During Start Up," SAE Paper No. 970865.

Meyer, R., Yilmaz, E., and Heywood, J.B. (1998), "Liquid Fuel Flow in the Vicinity of the Intake Valve of a Port-Injected SI Engine," SAE Paper No. 982471.

Mie, G. (1908), "Beitrage zur Optik Truber Medien Speziell Kollodaler Metallosungen," Ann. Phys., Vol. 25, pp. 377–445.

Münch, K., Kramer, H., and Leipertz, A. (1996), "Investigation of Fuel Evaporating Inside the Intake of a SI Engine Using Laser Induced Exciplex Fluorescence with a New Seed," SAE Paper No. 961930.

Nemecek, L.M., Wagner, R.M., and Drallmeier, J.A. (1995), "Fuel Droplet Entrainment Studies for Minimization of Cold-Start Wall-Wetting," SAE Paper No. 950508.

Posylkin, M., Taylor, A., Vannobel, F., and Whitlaw, J.H. (1994), "Fuel Droplets Inside a Firing Spark Ignition Engine," SAE Paper No. 941989.

Rao, K.K., Winterbone, D.E., and Clough, E. (1992), "Laser Illuminated Photographic Studies of the Spray and Combustion Phenomena in a Small High Speed DI Diesel Engine," SAE Paper No. 922203.

Rotunno, A.A., Winter, M., Dobbs, G.M., and Melton, L.A. (1990), "Liquid Visualisation of Fuel Sprays," *Combustion Science and Technology*, Vol. 71, pp. 247–261.

Shimizu, R., Matumoto, S., Furuno, S., Murayama, M., and Kojima, S. (1992), "Measurement of Air Fuel Mixture distribution in a Gasoline Engine Using LIEF Technique," SAE Paper No. 922356.

Schunemann, E., Fedrow, S., and Leipertz, A. (1998), "Droplet Size and Velocity Measurements for the Characterisation of a DI-Diesel Spray Impinging on a Flat Wall," SAE Paper No. 982545.

Senda, J., Kanda, T., Kobayashi, M., and Fujimoto, H. (1997), "Quantitative Analysis of Fuel Vapour Concentration in Diesel Spray by Exciplex Fluorescence Method," SAE Paper No. 970796.

Senda, J., Kobayashi, M., Tanabe, Y., and Fujimoto, H. (1994), "Visualisation and Quantitative Analysis of the Fuel Vapour Concentration in Diesel Spray," *JSAE Review,* Vol. 15, pp. 149–156.

Senda, J., Fukami, Y., Tanabe, Y., and Fujimoto, H. (1992), "Visualisation of Evaporative Diesel Spray Impinging Upon Wall Surface by Exciplex Fluorescence Method," SAE Paper No. 920578.

Skippon, S., Nattrass, S., Kitching, J., Hardiman, L., and Miller, H., (1996) "Effects of Fuel Composition on In-Cylinder Air/Fuel Ratio During Fuelling Transients in an SI Engine, Measured using Differential Infrared Absorption," SAE Paper No. 961204.

Sowa, (1992), "Interpreting Mean Drop Diameters," *Atomization and Sprays*, Vol. 2, Part 1.

Suzuki, M., Nishida, K., and Hiroyasu, H. (1994), "Trajectory and Dispersion of Vapour and Drop Clouds in Evaporating Diesel Spray Injected into Model Combustion Chamber," *COMMODIA 94*, pp. 199–204.

Suzuki, M., Nishida, K., and Hiroyasu, H. (1993), "Simultaneous Concentration Measurement of Vapour and Liquid in an Evaporating Diesel Spray," SAE Paper No. 930863.

Swinthenbank, J., Beer, J.M., Taylor, D.S., Abbott, D., and McGreath, G.C., (1977), "A Laser Diagnostic Technique for the Measurement of Droplet and Particle Size Distribution," *Prog. Astronaut. Aeronaut.*, Vol. 53, pp. 421–427.

Swinthenbank, J., Cao, J., and Hamidi, A. (1991), "Spray Diagnostics by Laser Diffraction," *Combustion Measurement*, Ed. by N. Chigier, pp. 179–227.

Tabata, M, Kataoka, M., Fujimoto, M., and Noh. Y. (1995), "In-Cylinder Fuel Distribution, Flow Field, and Combustion Characteristics of a Mixture Injected SI Engine," SAE Paper No. 950104.

Tinaut, F., Castro, F., Melgar, A., Sanchez, M.-L., and Gimenez, B. (1993), "Visualisation and Measurement of Speed and Size in Diesel Sprays," SAE Paper No. 933370.

Trolinger, J.D., (1991), "Particle and Flow Field Holograph," in *Combustion Measurements*, Ed. by Chigier, N., pp. 51–89.
Van de Hulst, H.C. (1957), *Light Scattering by Small Particles*, Chapman and Hall, London.
Vannobel, F., Dementhon, J.B., and Robart, D. (1992), "Phase Doppler Anemometry Measurements on a Gasoline Spray inside the Inlet Port and Downstream of the Induction Valve: Steady Flow Conditions," *Proc. 6th International Symposium on Laser Techniques and Applications in Fluid Mechanics*, Ed. by Heitor, M.V., Maeda, M., and Whitelaw, J.H., Springer-Verlag.
Wensing, M., Munch K., and Leipertz, A. (1997), "Characteristics and Application of Gasoline Injectors to SI Engines by Means of Measured Liquid Fuel Distributions," SAE Paper No. 972947.
Winklhofer, E. and Plimon A. (1991), "Monitoring of Hydrocarbon Fuel-Air Mixtures by Means of a Light Extinction Technique in Optically Accessed Research Engine," *Optical Engineering*, Vol. 30, No. 9, pp. 1263–1268.
Winklhofer, E., Fraidl, G.K., and Plimon, A. (1992), "Monitoring of Gasoline Fuel Distribution in a Research Engine," *Proc. Instn. Mech. Engrs.* Vol. 206, pp. 107–115.
Winklhofer, E., Fuchs, H., and Philipp, H. (1993), "Diesel Spray Combustion—An Imaging Analysis," SAE Paper No. 930862.
Witze, P.O. and Green, R.M. (1997), "LIF and Flame-Emission Imaging of Liquid Fuel Films and Pool Fires in an SI Engine During a Simulated Cold Start," SAE Paper No. 970866.
Yeh, C., Kamimoto, T., Kobori, S., and Kosaka, H. (1993), "2-D Imaging of Fuel Vapour Concentration in a Diesel Spray via Exciplex-Based Fluorescence Technique," SAE Paper No. 932652.
Young, B.W. and Bachalo, W.D. (1988), "The Direct Comparison of Three In-Flight Droplet Sizing Techniques for Pesticide Spray Research," *Optical Particle Sizing: Theory and Practice*, Ed. by Gouesbet, G. and Grehan, G., pp. 483–497, Plenum Press, New York.
Yoshikawa, S., Fursawa, R., Arai, M., and Hiroyasu, H. (1989), "Optimizing Spray Behaviour to Improve Engine Performance and to Reduce Exhaust Emissions in a Small DI Diesel Engine," SAE Paper No. 890463.
Zhao, F.-Q., Yoo, J.-H., and Lai, M.-C. (1995), "The Spray Characteristics of Dual-Stream Port Fuel Injectors for Applications to 4-Valve Gasoline Engines," SAE Paper No. 952487.

Chapter 9

Engine Combustion and Flame Visualization

Outline

9.1 Introduction

Since internal combustion engines rely on the conversion of chemical energy into mechanical power through combustion of fuel and air mixture in the cylinder, it is paramount for engine designers to be able to optimize the combustion process for maximum power output, maximum efficiency and yet minimum pollutant emissions. For this reason, since the very early years of IC engine efforts had been devoted to techniques that were able to "see" the flame inside the engine.

Among the techniques first developed for engine combustion visualization are high-speed direct photography and Schlieren/shadowgraph techniques. Their implementation and application form the first part of this chapter. This is followed by an introduction to the laser sheet techniques for spatially resolved two-dimensional images of flame structure.

The substantial engine modification required by optical imaging techniques has lead to the development of multiple probe techniques for engine combustion diagnostics. The implementation of these techniques and the interpretation of their output are discussed in this chapter.

9.2 Visualization of Engine Combustion

9.2.1 Introduction

Historically, visualization of engine combustion was first carried out by direct recording of luminous combustion, followed by the introduction of the Schlieren technique for autoignition and knock studies. Visualization involves the formation of visible images directly or indirectly by the action of light on sensitive materials. Therefore, light and recording of the images formed are essential to any forms of visualization. Light can be generated from the object itself, if it is luminous. Otherwise, a light source has to be deployed to illuminate the object to be imaged. Traditionally, photographic films were the only means of recording images by chemical processes induced by the action of light. Photoelectric devices are now replacing photographic films in some applications. Due to the rapid nature of engine combustion, each image must be taken with a very short exposure time in order to freeze any fluid motion or combustion. This can be realized by intermittent recordings at high shutter speeds, so that one image is recorded from one cycle and a sequence of images as a function of crank angles is acquired over many engine cycles. The other involves recording of hundreds of images or more per second continuously so that a sequence of images is obtained within the same engine cycle. We refer to the former method of recording as high-speed single-shot imaging and the latter as high-speed continuous recording, though both methods have been described as high-speed photography in the literature.

9.2.2 Light Source for High-Speed Visualization

A Schlieren/shadowgraph system, as to be discussed later, requires a light source. In order to observe non-luminous events in a combustion chamber, such as fuel sprays inside an IC engine, an illumination light source is needed. Fig. 9.1 illustrates one typical experimental setup used to visualize diesel combustion in a single-cylinder research engine. The light from a light source reflects off the cylinder walls and illuminates the fuel spray inside the combustion chamber. The illuminated fuel spray is then imaged onto a camera located on the other side of the engine.

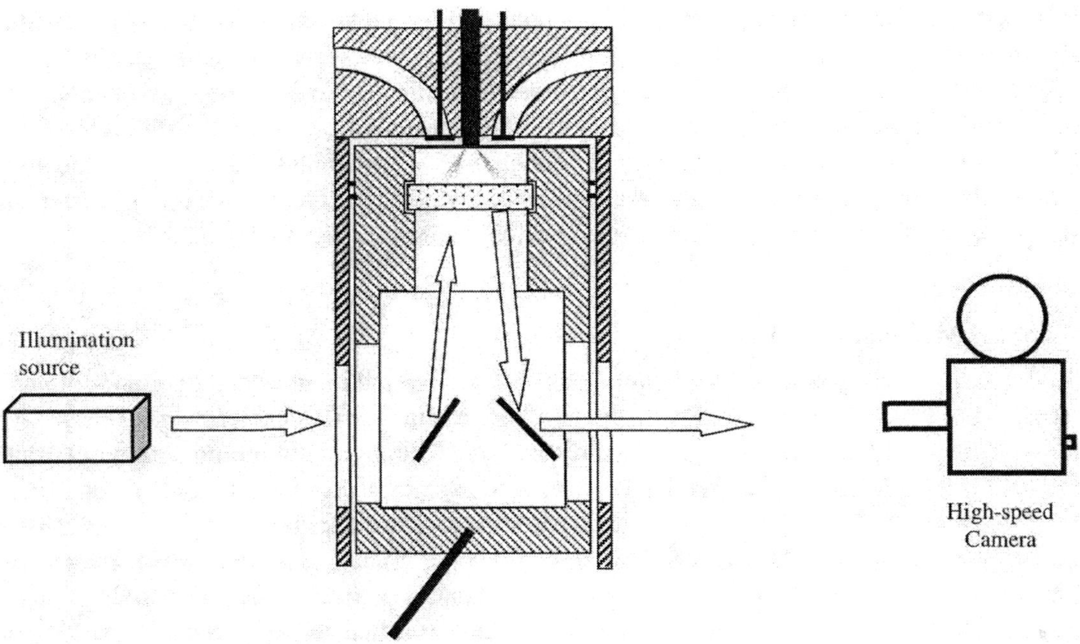

Fig. 9.1 Experimental setup for spray and combustion visualization in a DI diesel engine.

High-speed visualization requires large quantities of light over a very shot time duration. The principal characteristics that must be considered for the selection of a light source are its brightness, spectral distribution, and repetition rate.

Light sources can be of either continuous or pulsed nature. Pulsed light sources are preferred in the single shot mode where a light pulse of very shot duration defines the time-resolution of the system and permit a less fast camera to be used. Because energy is delivered within a short period of time, a pulsed light source provides more light than an equivalent continuous light source for the high-shutter speed operation.

There are two types of continuous lamps in the visible spectrum: Quartz tungsten halogen (QTH) lamps, and DC arc lamps. QTH lamps have significant advantages over arc lamps of high total visible output, large source area, and a smooth continuum without the sharp peaks of arc lamps. The majority of pulsed lamps are xenon arc lamps with microsecond pulses. Nanosecond flash-lamps are also available with a repetition rate up to 100 kHz.

Lasers offer much more brightness than the conventional light sources mentioned above. Pulsed lasers, such as pulsed Nd:YAG lasers, are better suited for single-shot operations where the time resolution may be determined by the pulse width of the laser output (<20 ns). However, they are of limited use for high-speed continuous recordings due to their low repetition rate (typically, 10 or 20 Hz). When CW lasers, such as a He-Ne laser and an argon-iron laser, are used as the illuminating light source for continuous high-speed visualization, the exposure time is determined by the speed of the camera (1/no. of frames per sec.). For example, at 10,000 frames/sec, the exposure time is 100 μs. In some applications, this may not be short enough to freeze the fluid motion. A copper vapor laser or a high-repetition rate Nd:YAG laser, on the other hand, generates laser pulses less than 30 ns and a few millijoules at a repetition rate up to tens of kHz. When synchronized with a high-speed continuous recording system, extremely sharp images of fuel spray and combustion can be recorded at very high speeds.

9.2.3 High-Speed Cameras

Traditionally, the majority of visualization has been carried out by means of photographic film recordings. More recently, solid state cameras, particularly CCD cameras, have been used for both single and continuous high-speed recordings. Although photographic material offers the benefit of high resolution and high data acquisition rates, the processing time is long and it has a nonlinear response curve. The instant feedback of the captured image in a solid state camera permits quick adjustment of the optical setup. When only qualitative images are required, such as flame visualization, a solid state camera output can be connected to a high-quality video recorder, and hence unlimited images can be stored and analyzed off-line. A solid state camera is more sensitive than film, and it has a linear response to the incident light. If a solid state camera is equipped with a gated intensifier, ultra-fast (>5 ns) imaging can be achieved at very low light conditions. However, since CCD cameras are normally limited to a frame rate of 50/60Hz or lower, they are more suited to single shot applications.

For high-speed continuous recordings, there are three types of high-speed cameras: high-speed cine film cameras, electronic imaging cameras, and high-speed video CCD cameras. High-speed cine film cameras are among the most commonly used. The rotating prism camera is one of them. In this type of camera, a continuously moving band of film is fed from a roll over a drum with pins and then onto a take-up roll, as shown in Fig. 9.2(a). Portions of film are held stationary during exposure in the same way as in still cameras. A rotating prism is inserted between the lens and the film. The combination of the rotating prism and the moving film allows framing rates more than 25,000 pictures per second. The images from the lens are synchronized with the film motion and optical compensation is achieved by rotating optical components. They normally use 16 mm films so that the images are quite small, approximately 16 mm on the diagonal. Resolution is quite good and typically on the order of 40-50 line pairs per mm, while the high-quality 35 mm camera has a definition of approximately 90 line pairs per mm. With a half-height head in the camera, speeds can be doubled but with an image size half of the original one. Nearly two-thirds of a 400 foot film on the roll may be used while the film is accelerating from zero to the required rotation speed. Filming at higher speeds allows filming at a constant speed for only several fractions of a second. Some of these cameras have a second entry lens so that other parameters displayed on an oscilloscope can be recorded simultaneously with flame images for accurate time reference. Some of the other cameras are equipped with two LEDs: one is driven by the built-in clock and the other by an external clock that may be driven by a shaft encoder output. The timing marks printed on the edge of

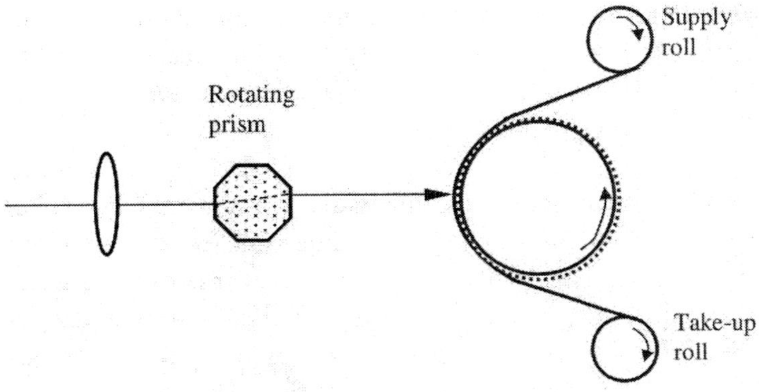

(a) a high-speed rotating prism camera

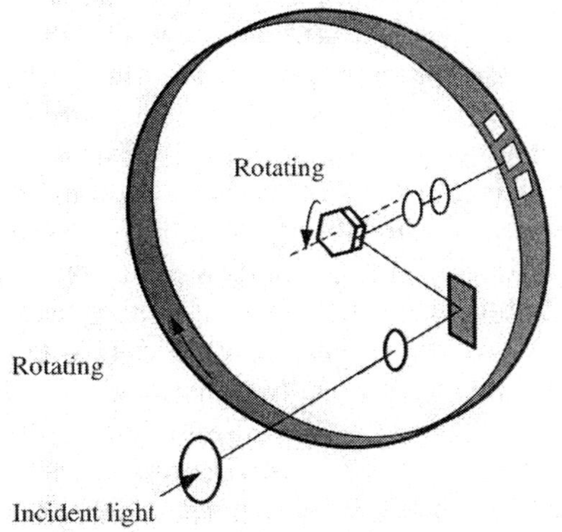

(b) a high-speed rotating drum camera

Fig. 9.2 High-speed film cameras.

the film by these LEDs provide excellent reference to other parameters measured at the same time (e.g., cylinder pressure).

For even higher film speed, rotating drum cameras can be used. A rotating drum camera has a film track that lies the inside cylinder of the rotating drum (Fig. 9.2[b]). A special film cassette is used to load a specified length of film into the inside perimeter of the drum. The drum is powered by an electric motor, controlled by an electronic control unit or variable power source. A rotating mirror is mechanically coupled to the drum. During exposure, successive faces of the mirror reflect light through relay lenses at regular intervals. The relay lenses focus the object into images on the moving film, creating the regularly spaced discrete frames.

Rotating drum cameras can produce records at up to 200,000 frames per second. In faster models, the drum rotates in a housing that has to be vacuumed to reduce air friction. This allows the drum to reach higher speeds with less power, and reduces temperature rise and the slight image resolution loss due to air turbulence.

The major element of electronic imaging cameras is an image converter tube, which converts photons to an array of electrons analogous to the image. This electron image is electrostatically focused, deflected, and shuttered to produce a record. A phosphor screen in the image converter tube reproduces the optical image by converting electrons back to photons, which can then be recorded on film or a CCD camera. The non-mechanical manipulation of the image allows for extremely fast recording speeds and provides great flexibility in the timing and duration of exposures.

In the Hadland 700 series Imacon cameras, 8 images are produced on the phosphor screen and then recorded by a Polaroid camera. The slowest speed is 10,000 frames per second and the fastest is 20 million frames per second. The speed is changed by means of a plug-in unit. For any framing speed, each frame is the size of a 35 mm negative, that is approximately 35 mm across the diagonal. The definition of Imacon cameras depends on the speed, but is approximately 10 to 15 line pairs per mm. The Hadland Imacon 468 camera takes pictures at rates from 1000 frames per second up to 100 million frames per second, but only 6 images can be recorded at one time. In this camera, each picture can be individually set so that the exposure of each channel and the inter-frame time of each channel can be entirely different and independent of any other one. The pictures are stored on a PC in a TIFF format file so that they can be easily transferred to other software packages for further analysis.

The high-speed video cameras are CCD cameras operating from 200 to 4,500 frames per second. The images produced can either be stored on videotape or directly in solid state memory. The NAC 400 high-speed video runs at 200 frames per second with a full height picture on the monitor. The pictures are recorded on standard VHS tape. The Kodak Ektapro EK uses Kodak's own special tape, which runs at high speed. Speeds of at up to 1000 frames per second can be obtained with a full aspect ratio picture of 240(H) × 200(V) pixels. The maximum recording time is 30 seconds. The Kodak HS camera stores pictures directly in solid state memory and records at up to 4,500 frames per second with a full-size picture and up to 40,500 frames per second with reduced-size pictures. The number of pictures that can be stored varies with the size of the memory. Typically, 3000 pictures can be stored so that, at 4,500 frames per second, one can store pictures for two thirds of a second.

Many of these high-speed cameras are equipped with synchronized strobe light of short duration (a few microseconds), which is good for freezing the action and improving the sharpness of the pictures. For combustion studies, a high-repetition pulsed laser is preferred because of its higher energy output and extremely short pulse duration.

For direct visualization of weak luminous events, an image intensifier can be used with these high-speed cameras. The image intensifier improves a camera's sensitivity by thousands of times, and it is capable of extremely short exposure times (nanoseconds) by gating on-and-off electronically. There is a compromise between the gain and the speed at which the intensifier can be operated. For high-speed continuous recordings, a very rapid decay phosphor is employed with a reduced gain in hundreds. The use of an intensifier will also compromise the spatial resolution of the imaging system.

9.2.4 Schlieren/Shadowgraph Techniques

Direct visualization of combustion and flame propagation is effective when the flame is luminous. In order to observe non-luminous events in the cylinder, such as spray and flow field, Schlieren/shadowgraph techniques can be used because these techniques are sensitive to density changes. The terms of shadowgraph and Schlieren have been sometimes used interchangeably, but there are differences between the two. The optical setup of a Schlieren system is similar to that for shadowgraph, with the exception that a knife edge or aperture is used to block part of the light through the test section. Fig. 9.3 shows the setup of a conventional Schlieren system. Because of the obstruction, the Schlieren technique is sensitive to the first derivative of density, whereas the shadowgraph technique responds to the second derivative of density. In general, the Schlieren technique gives greater contrast than the shadowgraph technique, and hence it has been preferred method for flow visualization. For engine applications, the density gradients are high enough to give good contrast in a shadowgraph image. The shadowgraph technique

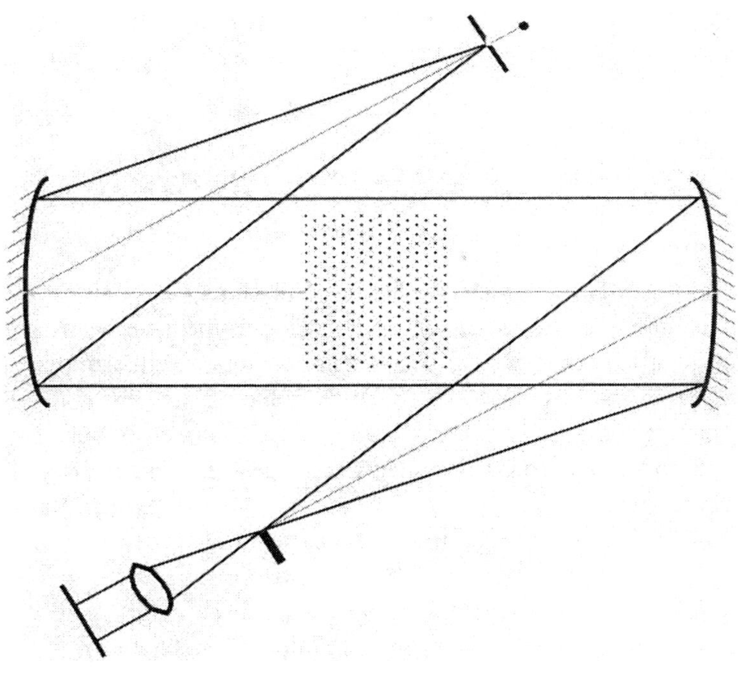

Fig. 9.3 Typical setup of a Schlieren system with concave mirrors.

is far less sensitive to the quality of the optical components than the Schlieren technique, leading to considerable cost savings. More importantly, when a double-pass setup (see Fig.9.5) is used, the rocking of the mirrored piston surface deflects the return image relative to the knife-edge. The deflection of the resulting images makes the Schlieren system difficult to use.

Fig. 9.4 shows one typical laser shadowgraph system. Traditionally, a high-intensity lamp has been used as the light source. The use of a laser as the light source may be advantageous in some applications. The monochromatic nature of the light avoids the need for expensive achromatic lenses. An additional advantage of monochromatic light is found when studying high luminosity combustion: Using a narrow band-pass filter, one can eliminate the emission from the combustion; using color photography, one can often distinguish combustion emissions from the shadowgraph. Finally, the use of a laser as the light source yields maximum image sharpness. One disadvantage of the laser light is that it enhances background noise due to interference of the light scattered from foreign matters in the test sections. When white light source is used, the interference effect is absent.

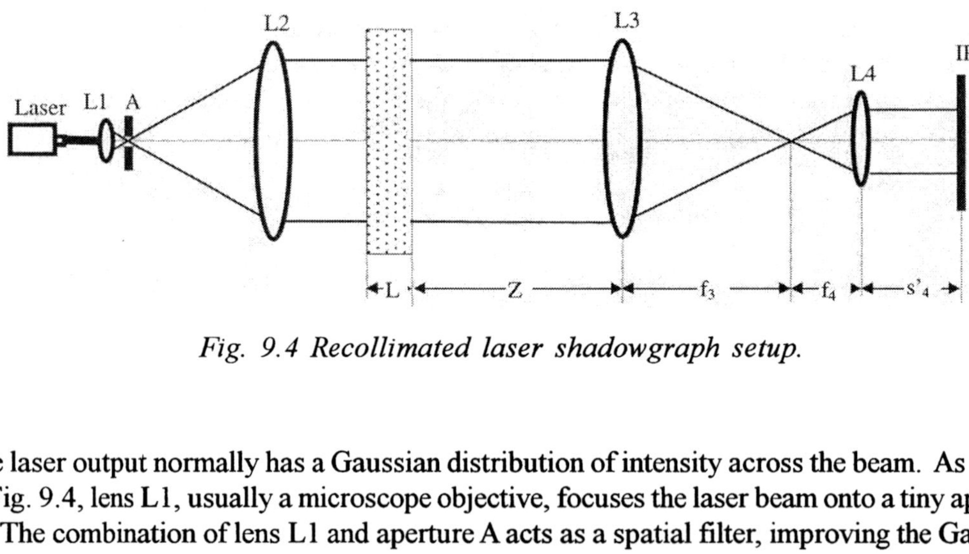

Fig. 9.4 Recollimated laser shadowgraph setup.

The laser output normally has a Gaussian distribution of intensity across the beam. As shown in Fig. 9.4, lens L1, usually a microscope objective, focuses the laser beam onto a tiny aperture A. The combination of lens L1 and aperture A acts as a spatial filter, improving the Gaussian intensity distribution of the laser beam, as well as reducing the physical size of the "point" source. The expanding beam exiting the aperture is recollimated by lens L2, which must be larger in diameter than the maximum dimension of the desired field of view. The parallel light is passed through the test section, and then focused by a third lens, L3. Lens L4 could be the camera lens or one used to form a real image on a translucent screen.

Such setup as shown in Fig. 9.4 requires two optical windows mounted at each end of the test section. When only one optical access is available, a double-pass shadowgraph system can be used as shown in Fig. 9.5(a). Component changes are the elimination of one large lens, since a single lens serves the function of both L2 and L3, and the addition of a beam splitter, BS. The use of a beam splitter permits the incident and return light to be coincident. This is a desirable feature, but it comes at the expense of light intensity at the image plane. Each pass through the beam splitter results in approximately 50% loss in light, such that the efficiency of the system is less than one-quarter that of a single-pass system. Alternatively, the incident beam is tilted slightly such that the reflected beam can be separated from the incident beam, as shown in Fig. 9.5(b). This system permits a less powerful light source to be used, but at the expanse of sensitivity.

For both setups shown in Figures 9.4 and 9.5, the change in light intensity ΔI for a tiny area, dxdy, at the image plane IP can be shown to be

$$\frac{\Delta I}{I} = -\frac{\lambda}{n_0}\int_0^L \left(\frac{\partial^2 n}{\partial x^2} + \frac{\partial^2 n}{\partial y^2}\right) dz \qquad (9.1)$$

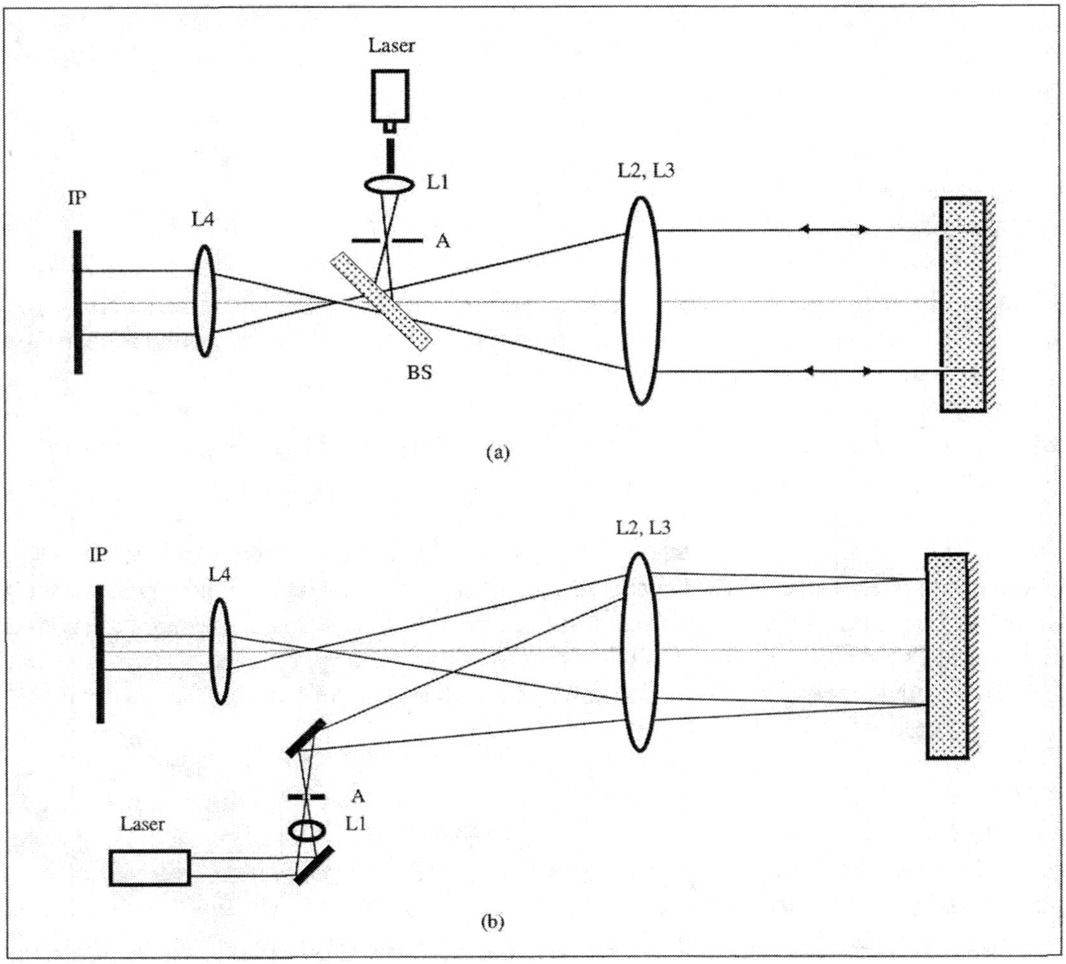

Fig. 9.5 Double-pass recollimated laser shadowgraph setup.

where n is the refractive index, n_0 the refractive index at a reference temperature and pressure. The sensitivity of this shadowgraph system, λ, is given by

$$\lambda = z + (1 - M)\frac{f_3}{M} + \frac{s_4'}{M^2}$$

and the magnification factor M can be shown to be

$$M = \frac{f_4}{f_3}$$

Note that the image is inverted. The sensitivity of both systems can be varied by moving only the image plane. This proves to be very convenient for engine experiments where the bulk of setup is normally fixed. In the double-pass setup, the light passes through the test section twice, and sensitivity to density gradients within the test section is doubled. The upper integration limit in Eq. (9.1) becomes 2L.

9.2.5 Engine Combustion Visualization

9.2.5.1 SI Engines

Due to lack of large optical windows strong enough to survive engine combustion in the 1920s, Glyde [1930] mounted six button-like windows in a line traversing the cylinder head of an L-head SI engine. To facilitate interpretation, these windows were viewed through a slit in a revolving disk above the cylinder head, which acted as a stroboscope. They found that a bright yellowish flash appeared in all windows simultaneously, but not until a normal flame had traveled almost all the way across the chamber.

Withrow and Boyd [1931] advanced Glyde's design by replacing these windows with a slit window, as shown in Fig. 1.22. They used a revolving-drum camera having a rotational axis parallel to the window so that a record of flame-travel distance versus time could be produced. Marvin and Best [1931] extended Glyde's design by distributing 31 such windows over the area of the cylinder head. This provided sufficient spatial resolution to plot flame position as a function of crank angle.

The first successful implementation of a quartz window to cover the entire cylinder head was realized five years later by Withrow and Rassweiler [1936] in an L-head engine. Using a high-speed cine film camera, they obtained up to 30 pictures during a single combustion event at 500 frames/s or 2.4°CA at 900 rpm. These pictures revealed for the first time the initiation of combustion as a flame kernel at the spark plug, followed by growth of the kernel into a wrinkled flame front. They found that knock started from autoignition at one or more isolated points located in the end gas ahead of the flame front. Combined with in-cylinder pressure measurements, Withrow and Rassweiler [1938] found from these combustion movies that increments in combustion pressure rise were proportional to the corresponding increments of mass fraction burned. Based on these experimental observations, they developed a method of calculating mass fraction burned profile from the in-cylinder pressure measurements, as described in detail in Chapter 2.

Combustion of leaner or diluted mixtures in an SI engine is characterized by flames of weak luminosity. While investigating the effects of charge dilution on combustion by high-speed photography, Nakanishi et al. [1975] found that a high-speed film camera was only suited to rich mixture combustion with bright yellow luminance. In order to increase combustion radiation, they added copper oleate or copper acetate to the fuel. However, they found that the flame was dramatically altered by the presence of these seedings. Eventually, they resorted to a specially designed high-speed camera (though little information was given of the camera) operated at 800 frames/s to record the weak blue flames with diluted mixtures.

In order to visualize the low-intensity blue-colored flames in a lean-burn engine, Nakamura et al. [1989] employed an SIT (silicon intensified target) camera. Images produced on the fluorescent screen of an Imacon™ camera were recorded directly by the SIT camera. Using the system, Nakamura et al. were able to record eight images of the initial flame kernel in a single engine cycle.

Because of the weak luminance of the initial flame kernel, the direct imaging of its development requires extremely sensitive film or an intensified camera. Bates [1989] employed two intensified CCD (ICCD) video cameras to record instantaneous flame images from two directions in a single cylinder engine with a sapphire liner and transparent piston. One pair of images was recorded at a preset crank angle in each engine cycle. In order to prevent the ICCD cameras from being overexposed to the stronger radiation during the main combustion period, the intensifier gains had to be adjusted as the recordings took place later in the flame propagation period.

Because they are very sensitive to density changes, Schlieren/shadowgraph techniques are particularly suited for studies of flame initiation and development. Ask et al. [1994] studied the flame kernel development in a single-cylinder SI engine using a Schlieren system similar to that shown in Fig. 9.4. Two optical windows were installed in a specially designed cylinder head. A halogen lamp was used as the light source. A circular aperture in front of the lamp and a circular cutoff aperture produced an axisymmetric Schlieren image. The Schlieren images of initial flame development were recorded by either a high-speed film camera for continuos recording lasting several combustion cycles or a CCD video camera for intermittent recordings.

Using the double-pass shadowgraph system shown in Fig. 9.5(a), stunning images of flame structures (Fig. 9.6) were obtained from the Sandia optical engine (Fig. 1.23) [Smith, 1982]. An argon-ion laser was used as the light source in the shadowgraph system. An acousto-optic modulator (AOM) was employed to strobe the laser light synchronously with the engine. The expanded laser beam passed through the engine combustion chamber and reflected off a mirror bonded to the piston top. A 35mm camera for single-shot photography or a cine film camera for high-speed photography was used to record shadowgraphs of flow and flame. In the initial experiment, shadowgraphs were directly imaged onto the camera. Because of the rocking of the piston mirror, shadowgraph images moved around. In the worst case, this motion was so large as to deflect the image partially off the film. This motion was a maximum at the focal plane of the source image. A mylar screen was therefore placed behind lens L4, as shown in Fig. 9.5, to form a real image. The image on the mylar screen was in turn focused by the camera lens onto the film. By recording shadowgraph images indirectly through the mylar screen, the image movement on the film was minimized. Using a laser shadowgraph system, Smith studied the effects of flow field, spark location, and air/fuel ratios on flame propagation and flame structure. Fig. 9.6 shows a set of enlarged shadowgraphs of a 9 mm diameter section of the propagating flame in the cylinder, viewed normal to the flame surface. This shows that increasing engine speed convoluted the thin flame front, and the length scales became smaller as turbulence levels in the unburned charge went up. The thin reaction sheet flame front was multiply connected. This is consistent with the fact that flame speed increases with engine speed.

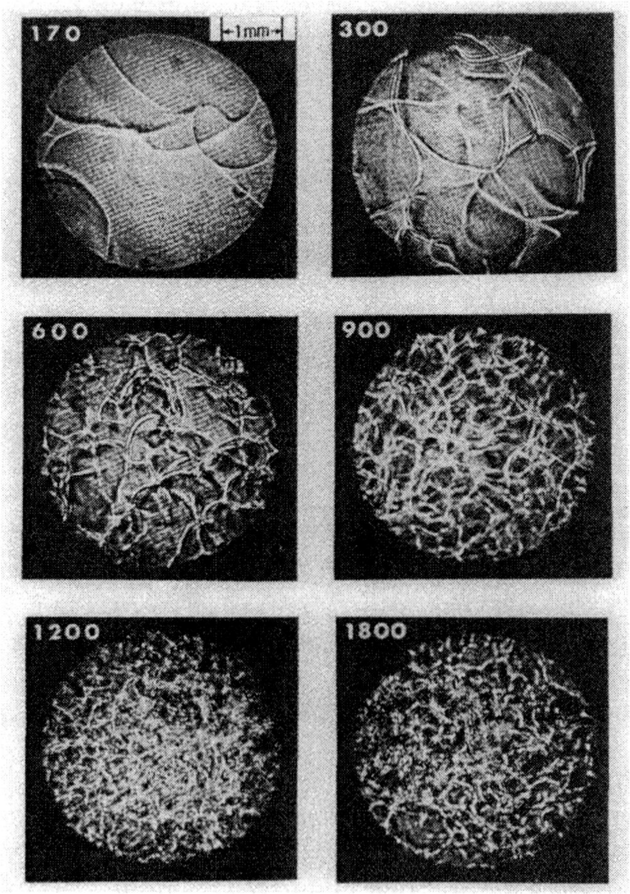

Fig. 9.6 Shadowgraphs of flames propagating toward the viewer at different engine speeds [Smith, 1982].

In addition to observing the initial flame development in an SI engine, Schlieren/shadowgraph techniques have been used extensively in autoignition and knock studies. Because of their sensitivity to density variations, they are effective in detecting autoignition sites in the end gas region and the pressure waves generated during knocking combustion.

Miller [1947] reported photographic studies of knock in an overhead-valve engine using a high-speed Schlieren system. Most of the photographs were taken with a special cine film camera at 40,000 frames/s. Limited studies were performed with a different cine camera capable of exposing 200,000 frames/s. Miller concluded that autoignition in the end gas sometimes did and sometimes did not cause a detonation wave. He also concluded that autoignition of the end gas was not essential to the creation of a detonation wave, because cases were observed where the wave originated in the gas region was already enveloped by the normal flame. This research stirred the controversy between supporters of the autoignition theory and detonation-wave theory of knocking combustion. The autoignition theory states that autoignition in the end gas induces rapid heat release and a pressure wave that is responsible for the characteristic sound of knocking combustion. The detonation wave theory suggests that a detonative wave is established by an accelerating flame front near the end of its travel, which results in the characteristic gas vibrations accompanying knock.

High-speed Schlieren photographic studies on knocking combustion were also reported by Male [1949] and Ball [1955]. Recent studies of similar kind include Smith et al. [1984] in the Sandia engine (Fig. 1.23) and Konig and Sheppard [1990] in a two-stroke SI engine (Fig. 1.20).

Rapid compression machines have been widely used in autoignition studies. Fig. 9.7 shows a series of shadowgraphs of knocking combustion obtained from a rapid compression and expansion machine (RCEM) [Hayashi et al., 1984]. These pictures were taken using a double-pass laser shadowgraph system, similar to that shown in Fig. 9.5, at 100,000 frames/s through a quartz window at the top of the RCEM. A mixture of air and n-butane was compressed and ignited by a spark plug. The spark plug, in each case, was located at the bottom of the image.

The images in Fig. 9.7 were taken under heavy knock conditions. The first three images are characterized by an arc-shaped flame front originating from the spark plug located at the bottom of the image. The flame front separates a crescent-moon-like unburned region appearing at the top of these images from the dark burned gas region. In these images, the small black spots and the dark region at the top were caused by scratches or flaws in the optical window. The first autoignition site can be seen at 1210 μs as an extended dark region, accompanied by a pressure rise measured at this location by a pressure transducer. From t = 1220 to 1250 μs, the autoignition region expands rapidly in the unburned gas region (the bright region in these pictures), and a pressure wave is generated. The pressure wave is more visible in the burned gas region and reaches maximum sharpness at the end of this period. The wave observed in the frames t = 1240 to 1250 μs was confirmed to be a shock wave, as its speed was measured at about 1700 m/s, approximately 2.8 times faster than the sonic velocity in the unburned gas

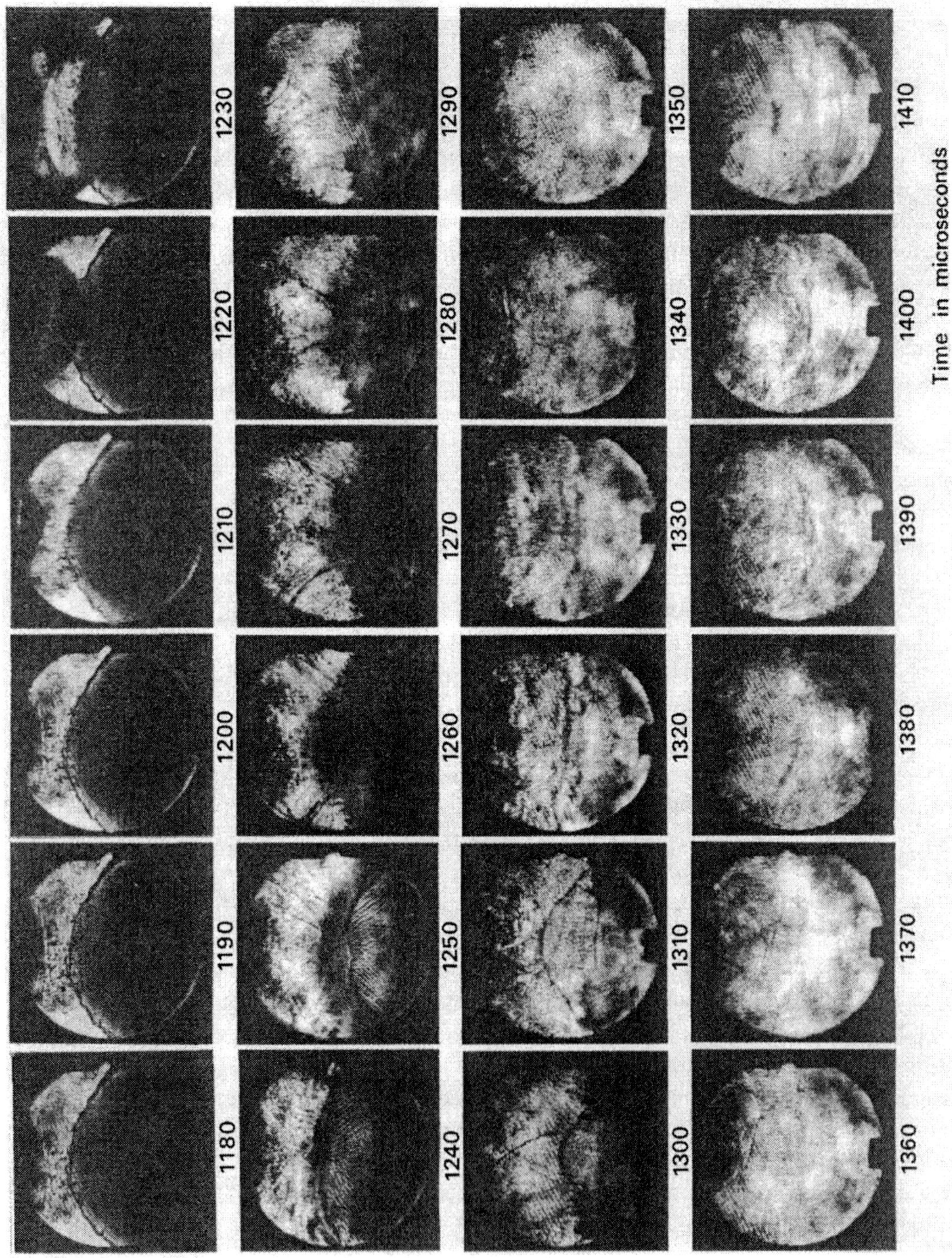

Fig. 9.7 Shadowgraphs of knocking combustion taken at 100,000 frames/s [Hayashi et al., 1984].

region. Subsequent pictures reveal that the shock wave is reflected off the opposite wall, and is degraded into a pressure wave that undergoes a round-trip process, which is responsible for the characteristic frequency of knocking combustion.

9.2.5.2 Diesel Engines

Combustion visualization in a diesel engine is often carried out in conjunction with sprays. Fig. 9.1 and Fig. 9.8 illustrate two ways of visualizing spray and combustion through a transparent piston. Observation through the top and side of the combustion chamber is depicted in Fig. 9.9 and Fig. 9.10, respectively. The use of an illumination source is necessary for the visualization of spray.

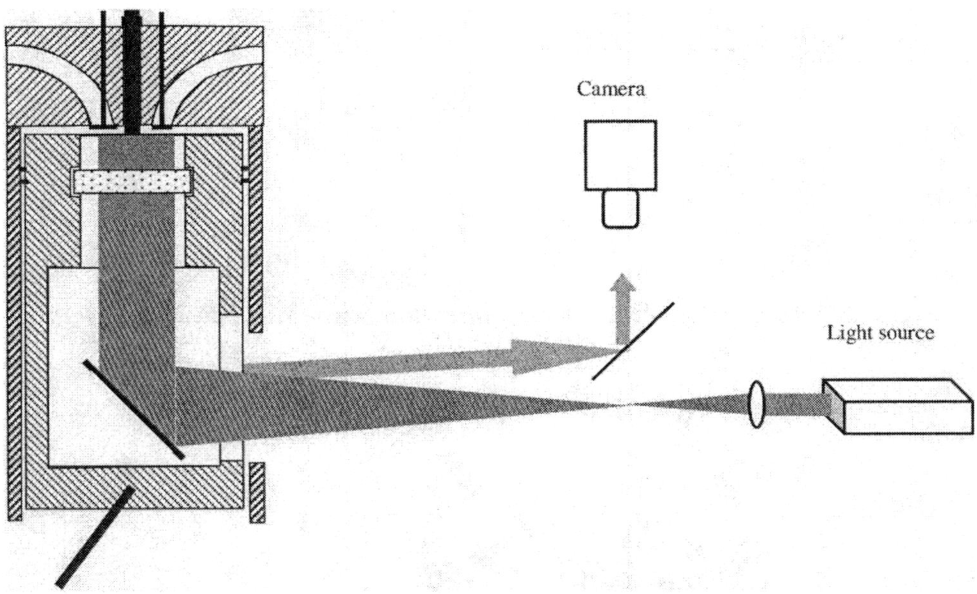

Fig. 9.8 Spray and combustion visualization through the piston with one port.

The first high-speed photographs of diesel combustion in a DI diesel engine were carried out by Rothrock and Waldon [1935]. Although the direct injection configuration they used was atypical of modern production engines, their photographs did reveal the three stages of diesel combustion: an ignition delay period, a period of uncontrolled combustion (premixed combustion), and a period of controlled combustion (diffusion combustion). Several photographic studies of diesel combustion were reported in IDI diesel engines during the 1960s [Nagao and Kakimoto, 1962; Alcock and Scott, 1962; Scott, 1969]. Excellent color photographs of diesel combustion in various forms of combustion chambers were presented by Scott [1969] and reproduced by Heywood [1988]. The combustion chamber investigated included the quiescent chamber, deep piston bowl chamber with swirl and multiple sprays, M.A.N single spray diesel engine,

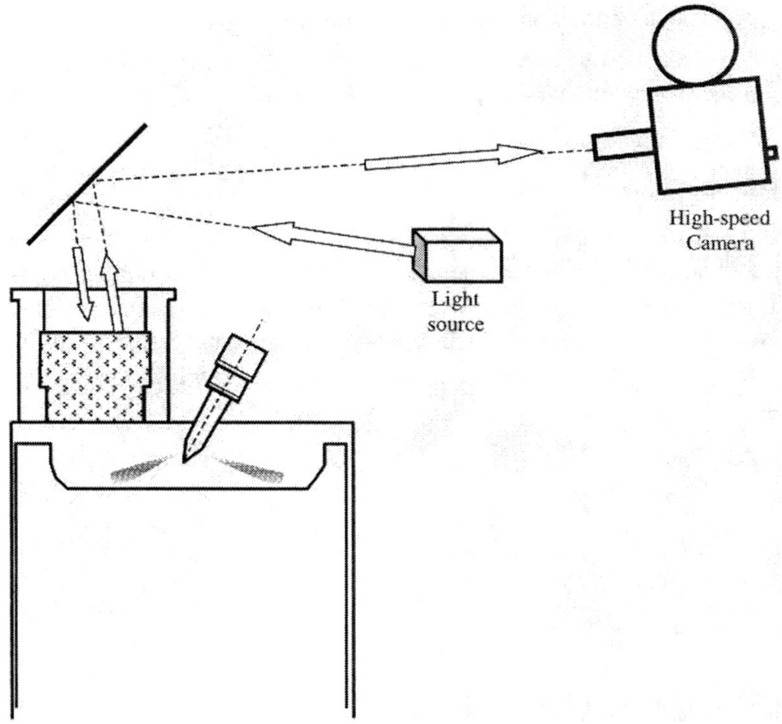

Fig. 9.9 Spray and combustion visualization through the cylinder head.

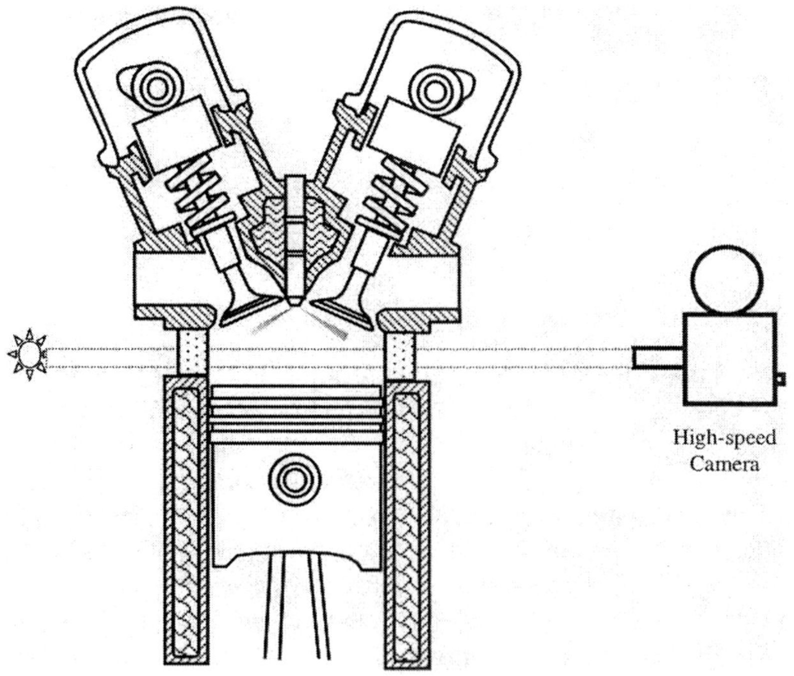

Fig. 9.10 Spray and combustion visualization with back illumination from the side.

and the prechamber and main chamber in a Ricardo Comet IDI diesel engine. During the premixed combustion phase, the flame was of too low luminosity to be recorded. 2% copper oleate was added to the fuel so that the faint blue flames turned bright green due to the emission of copper hydroxide formed under high temperature. As combustion proceeded, flames were dominated by brilliant white-yellow flame from the burning of soot particles, followed by dull yellow and red color as the flame temperature dropped [Heywood, 1988].

A number of recent combustion visualization studies in DI diesel engines have been performed in conjunction with flame temperature measurements using the two-color method, which will be discussed in Chapter 10 [e.g., Zhang et al., 1993; Arcoumanis et al., 1994; Winterbone et al., 1994; Chaves et al., 1996; Shiozaki et al., 1998]. The combined use of combustion visualization and the two-color method provides researchers with a means of studying the interaction of fuel injection and subsequent combustion.

In order to study diesel combustion in an IDI engine, Nishida et al. [1986] developed an optical system that was capable of simultaneous recording of combustion in both combustion chambers, as shown in Fig. 9.11. A xenon lamp was used to illuminate both chambers. A pair of images from each chamber was recorded by the same frame of the high-speed cine film camera. A similar setup was also used by Hotta et al. [1998] in their study of an IDI diesel engine.

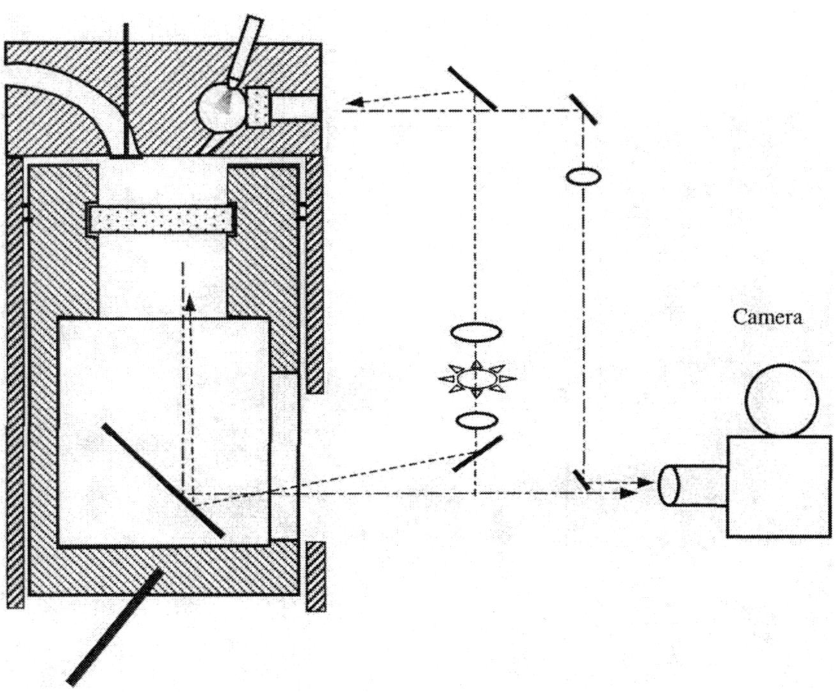

Fig. 9.11 Simultaneous visualization of combustion in the prechamber and main combustion chamber of an IDI diesel engine.

Because of their high sensitivity to density changes, high-speed Schlieren or shadowgraph techniques are particularly suited to studies of fuel spray vaporization and fuel/air mixing in the prechamber of an IDI diesel engine where two windows can be easily installed. From high-speed Schlieren photographs taken from the prechamber of an IDI engine, Lyn and Valdmanis [1962] concluded that the establishment of large-scale swirl in the prechamber also produced small-scale turbulence, which accelerated the mixing of fuel vapor and air during the ignition delay period. A similar setup was also employed by Allan and Khin [1970] to investigate the injection, evaporation, and mixing of fuel with air, as well as combustion in the swirl chamber in a modified CFR engine. Browne et al. [1986] measured the liquid core length of diesel fuel, toluene, ethanol, and sunflower oil in the prechamber of an IDI engine by means of high-speed single-shot shadowgraphs.

In order to take advantages of solid state cameras for their instant feedback and large dynamic ranges, Chaves et al. [1996] developed a multi-CCD camera system for the visualization of spray development in the Volkswagen's multi-cylinder DI optical engine. The experimental setup is shown in Fig. 9.12. Uniform illumination was provided by a 30 watt halogen bulb and a frosted glass. The reflected light was split into eight beams by an eight-faced reflecting pyramid and imaged onto eight CCD chips installed around it. A programmable sequencer controlled the delay time between the CCD chips, and hence the framing speed. The shutter speed of each CCD chip could be individually altered with a minimum value of 1 μs. Using this system, they were able to obtain a sequence of 8 images at 50,000 frames/s from a single engine cycle.

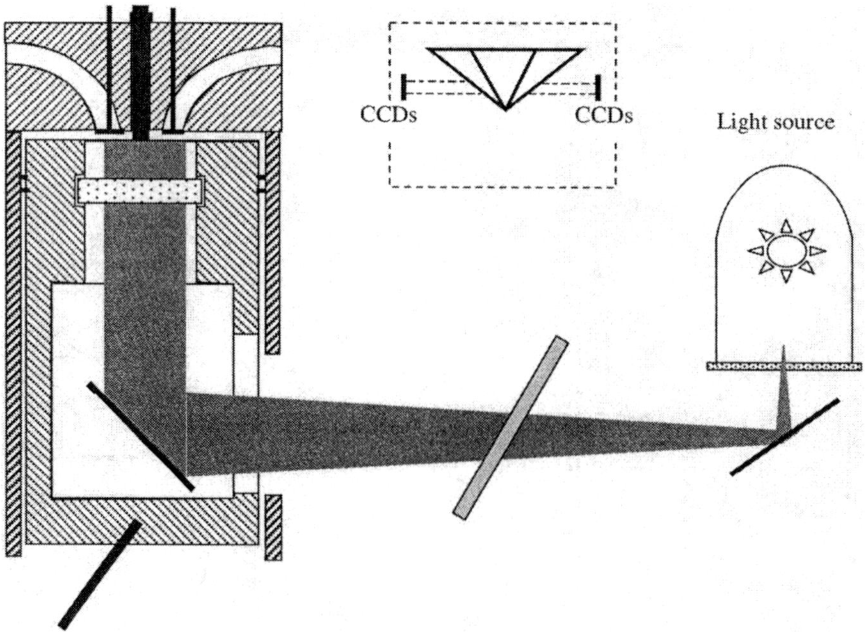

Fig. 9.12 Spray and combustion visualization using a specially designed high-speed camera.

All of the above mentioned test results were obtained through large optical accesses, which required substantial modification to the engine and sometimes led to configurations atypical of real engines. In order to overcome this difficulty, Cartellieri and Werlberger [1984] developed an endoscope-based imaging system. The novelty of the technique is the requirement of very small optical access to the engine. The key component of the system is an endoscope which can be fitted into an engine through a hole of pressure transducer size. The endoscope is a rigid rod-lens scope with an insertion tube having an outer sheath diameter of 5 mm. Since the optical components can only withstand temperatures of 110°C, a cooling system is provided within the endoscope to prevent it from being damaged by overheating. The image from the endoscope is then transmitted to a high-speed camera through an optical linkage as shown in Fig. 9.13. The linkage consists of four rigid rod-lens relay tubes linked together at the ends by means of swivel joints. This system was later used in the study of fuel injection and combustion events in a single-cylinder research engine [Werlberger and Cartellieri, 1987] and a four-cylinder production DI diesel engine [Karim, 1989].

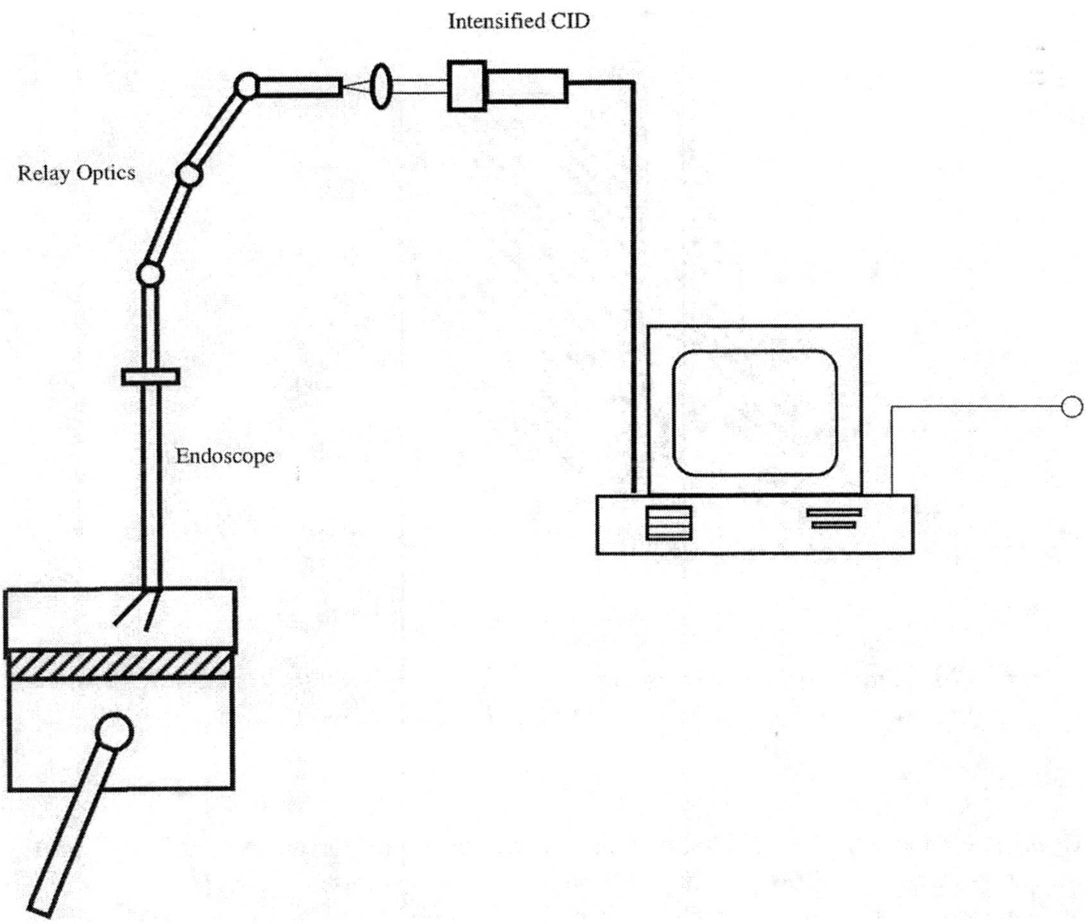

Fig. 9.13 Schematic of an endoscope-based imaging system.

Fig. 9.14 shows the installation of an endoscope system in a large single-cylinder DI diesel engine [Kong et al., 1995]. Optical access for the imaging system was achieved through one of two exhaust valves in the engine. The valve seat insert, the valve guide, and the valve were replaced by an adapter, a stainless probe, and window assembly of an AVL Type 510 system, as shown in Fig. 9.13. The adapter and probe were fixed to the engine head. The adapter was press-fitted into the valve to seal in-cylinder gases. The probe had an outer thread to secure it to the adapter and an inner thread to secure the window assembly. In order to provide a larger field of view of the combustion chamber, sections of the adapter and the probe were machined away. The optical system provided a line of sight of 30° from its axis with an 80° field of view, which allowed for the visualization of one whole spray plume of a six hole injector. The images were then transmitted via the relay optics to an optical coupler, which focused the light from the optical linkage to the input cathode of an image intensifier. The output of the image intensifier was then imaged onto a charge-injected device (CID) camera before it was digitized and stored in a PC computer.

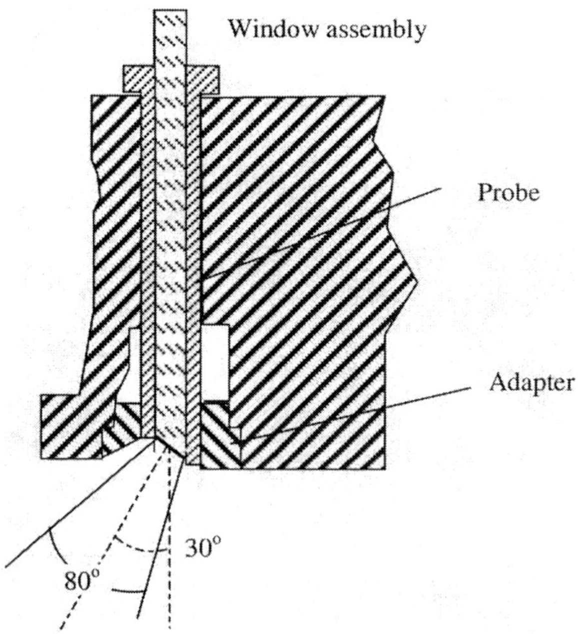

Fig. 9.14 Installation of the optical window of an endoscope system through one exhaust valve.

Using this endoscope-based system, Kong et al. obtained an extensive set of combustion images for different injection pressures, timings, and injection patterns. Ricart and Reitz [1996] applied the same system in their studies of pilot injection. The experimental results were then used to validate CFD predictions.

9.3 Two-Dimensional Flame Imaging by Mie Scattering

9.3.1 Introduction

Both direct imaging and shadowgraph techniques provide visualization of flow field and flames integrated over the line-of-sight along the optical path. Two-dimensional visualization techniques are preferred in some applications where spatial resolution is required. Two-dimensional visualization is realized by imaging just a thin slice of the turbulent flame illuminated by a sheet of laser light. The experimental setup, as shown in Fig. 9.15, is similar to that for PLIF, as discussed in Chapter 7. In practice, both PLIF and Mie scattering have been employed in two-dimensional flame visualization. The former involves imaging of OH radicals in the flame front and burned gas region. This will be discussed in Chapter 13. In this section, we will concentrate on two-dimensional imaging techniques based on Mie scattering.

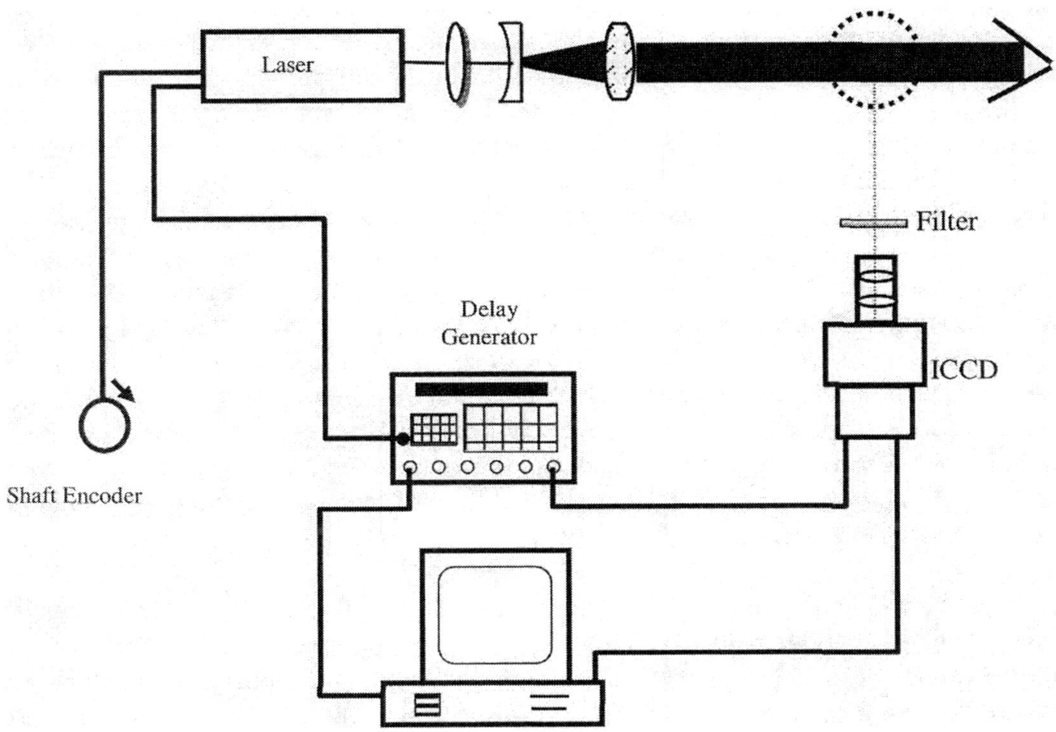

Fig. 9.15 Experimental setup for two-dimensional laser sheet imaging.

9.3.2 Principles and Implementation

A thin sheet of laser light is introduced into the cylinder through optical windows and scattered by particles in the mixture. These particles are introduced into the cylinder with incoming air during the intake stroke. A two-dimensional detector or camera captures the Mie scattering from these particles at 90° through a third window. For seeding levels sufficiently low to eliminate multiple scattering, the scattered light intensity is proportional to the local number density of particles in the plane illuminated by the laser sheet. In the burned gas region, the seed number density is substantially reduced by the burn-out of seeding (if oil droplets are used), or by the expansion of seeding (if solid particles are used). Therefore, a reduction in the scattered light intensity allows the boundary between the burned and unburned gases to be imaged. The flame front appears as an interface between the bright and dark areas, corresponding to unburned and burned gases respectively.

The 2nd harmonic of a high-energy pulsed Nd:YAG laser is normally employed as the light source because of its high power and relatively low cost. Its green output coincides with the high spectral response part of most imaging devices. A high-repetition pulsed laser, together with a high-speed film drum camera, is ideally suited for continuous high-speed visualizations.

The seeding of the flow field is critical in two-dimensional Mie scattering experiments. Since determination of the flame front relies on the variation in the scattered light intensities between the unburned and burned gases, seeding particles in the burned gas region should lead to drastically lower intensity of the scattered light by smaller scattering cross sections and/or reduced seeding number densities. Previous studies using TiO_2 showed that this material exhibited a large and rapid reduction in its scattering cross sections when traversing the flame front. Dry TiO_2 particles are available commercially in sizes of microns, and can be seeded directly into the intake flow. Alternatively, they can be generated online by passing gaseous $TiCl_4$ into the intake charge, which reacts with water vapor in the air, forming TiO_2 particles. In addition to the toxicity problem, the presence of HCl can cause extensive wear and corrosion in the cylinder. Micro-balloons or some other solid particles may be used as seeding particles, but they tend to scratch optical windows after a period of use. Oil droplets and smoke particles can be easily generated from commercial devices and added continuously to the intake air. The burning of oil droplets and smoke particles provides a sharper boundary between the unburned and burned gas regions. More information regarding particle/droplet seeding systems can be found in Chapters 6 and 8.

Because Mie scattering is many orders of magnitude stronger than LIF or Rayleigh scattering, less sensitive detectors are sufficient to record the images of scattered light. A narrow bandpass filter is normally placed in front of the camera so that only the scattered laser light is imaged onto the detector to minimize the interference of flame radiation. The use of a gated intensifier is also desirable to reduce the interference from flame emission. In order to improve the contrast or the signal-to-noise ratio of images, background subtraction should be adopted in the production of final images.

9.3.3 Visualization of Flame Propagation in SI Engines

Baritaud and Green [1986] and Zur Loye et al. [1987] were among the first to apply the two-dimensional Mie scattering technique to study the flame structure in SI engines. Both studies were carried out using the 2nd harmonic of a pulsed Nd:YAG laser and an intensified solid state camera. Submicron seeding particles were generated online by introducing gaseous $TiCl_4$ into the intake charge via a venturi immediately upstream of the intake valve. This was done to provide high enough levels of seeding at relatively higher engine speed. Serious window fouling was experienced during these experiments. Deposition on the side and top windows limited the number of flame images to two or three before the windows needed cleaning [Baritaud and Green, 1986]. Nevertheless, they obtained, for the first time, two-dimensional images of flame structures in an SI engine, which revealed flame front structures of scales as small as 0.2 mm. Evidence of increasing flame front wrinkles was observed, together with peninsulas and pockets of unburned mixture in the burned gas region.

Ziegler and his co-workers [1988] extended the technique to high-speed continuous recordings of flame front structures in a transparent square piston SI engine. The output of a copper vapor laser was formed into a thin laser sheet. Both the green (510.6 nm) and yellow (578.2 nm) lines of the copper vapor laser were employed in order to increase the energy output of each laser pulse. The scattered light from smoke particles was recorded by a high-speed film camera through a broad bandpass filter. The laser and camera were synchronized and operated at 6 kHz. The photographs on the film were then digitized using a CCD camera. As the light intensity changed rapidly from the unburned gases to the burned gases by a factor of four, the flame front was readily determined by a threshold in the image processing software. The flame contours thus determined were analyzed using fractal analysis and complex Fourier transform procedures. Radii of curvature, wrinkling factors, and turbulent burning velocities were determined at different engine speeds and crank angles.

Mantzaras et al. [1988] extended a single laser sheet technique to a 3-D visualization technique, in which scattered light from four spatially separated laser sheets were simultaneously imaged. The four laser beams were the 2nd and 3rd harmonic beams of an Nd:YAG laser (532 nm and 355 nm), along with the two strongest beams of the first Stokes (683 nm) and first anti-Stokes (436 nm) from a hydrogen Raman cell pumped by the 2nd harmonic. The four laser sheets were parallel and separated by 0.9 mm to 1.5 mm. The light from the four sheets was scattered by TiO_2 particles and detected by an intensified diode array. The collection optics included a quadrupling multi-vision prism, with four different color filters in front of each of its facets, and a video lens, allowing imaging areas from each of the four planes to occupy a quarter (44×44 pixels) of the 100×100 diode array. Using this technique, Mantzaras et al. found that the islands of unburned gases behind the flame front visible in 2-D images were in fact sections through "fingers" of unburned mixture interconnected to the main unburned gas regions on different planes within the combustion chamber.

A similar setup was also employed by Hicks et al. [1994] to characterize the flame structure and turbulence scales in a two-stroke SI engine. The four laser beams were generated from the 2nd and 3rd harmonics of an Nd:YAG laser, and the first Stokes lines of these two harmonics through a hydrogen Raman cell. Using this technique, Hicks et al. found that islands of unburned gases behind the flame front were associated with encirclement by large-scale structures rather than partial-burn or total quenching due to flame stretch.

9.4 Flame Propagation Measurements by Multiple Ionization Probes

9.4.1 Introduction

Optical imaging of combustion in an engine, whether it is realized by direct visualization or laser sheet techniques, requires extensive modification in the engine design. Most imaging techniques are difficult to apply to actual production engines. In order to minimize engine modifications, multiple probe techniques have been developed to measure flame propagation in IC engines.

9.4.2 Multi-Point Ionization Probe Head Gasket

Ionization probes have been used for many years to detect the reaction zone in a flame, as well as in FIDs (flame ionization detectors) as described in Chapter 3. In its simplest form, an ionization probe consists of a single electrode and a ground probe body. When hydrocarbons are burned, radicals with positive charges are generated. When a DC voltage is applied to the insulated electrode, a current flows between the electrode and probe body. Because of the greater mobility of electrons, a positive voltage is commonly applied to the electrode or probe. The strength of the current, and thus the detected signal, depends on the magnitude of the bias voltage, the probe surface area, and the ion density in burned gases. A simple electronic circuit can turn this current flow to an easily detectable voltage signal. The sudden rise in the signal output indicates the arrival of the flame front.

Curry [1963] performed an outstanding demonstration of the use of ionization probes in engine combustion measurements. Forty-nine ionization probes were installed in the cylinder head and piston of an SI engine. Signals from the piston-mounted probes were taken out through spring-loaded pins mounted in brush assemblies. These probes were adapted from steel-jacketed thermocouples. The thermocouple wires acted as the electrodes, while the steel jacket acted as the ground. These probes were flush-mounted in the cylinder wall so that they did not perturb the flow. The arrival time of the flame at each probe was monitored on a cycle-resolved basis, permitting the shape of the burned volume to be estimated as a function of time. Curry's experiment proved the value of ionization probes in the study of flame propagation, but it involved extensive machining of the cylinder head in order to accommodate these probes.

Witze and Bopp [1991] simplified the installation requirement of ionization probes by incorporating multiple ionization probes in a head gasket. Bare steel wires were sandwiched between two sheets of paper gasket material with the ends positioned flush with the bore cutout, such that, when installed, the current flow was from the electrodes to either the engine head or cylinder block. Eight probes were equally spaced around the cylinder. A positive biased voltage of 300 V was applied to the electrodes. A signal-conditioning unit carried out charge-to-voltage conversion and signal amplification. The flame arrival time was determined when a pulse generator was triggered at a predetermined level by the incoming signal. This design proved to be adequate for a single-cylinder research engine, because the engine had a low compression ratio and was fired only every fourth engine cycle. The durability of this gasket design was not seriously tested.

In order to use multiple ionization probes in the head gasket of a production engine, Nicholson and Witze [1993] further developed the technique by using printed-circuit-board technology. A double-sided printed circuit board (PCB) was used as the base material. Tracks were etched onto one surface of the PCB to form precisely located ionization probes as shown in Fig. 9.16. The underside of the PCB was left intact and used as the ground electrode. Each probe could be several millimeters wide by less than 0.1 mm thick. Shield tracking was placed between the ion probe trackings in order to eliminate the cross-talk that may occur. A layer of insulating material was sandwiched between the PCB and the cylinder head to create a high impedance gap. Tests with a glass-fiber-reinforced polyimide resin board material [Russ et al., 1997] suggested that the gasket could be used for extensive periods at part load (the limit being signal degradation due to deposits). However, durability at full load, even at low speeds, could be measured in seconds or minutes. Jackson et al. [1996] were able to extend the lifetime of their instrumented head gasket to several hours of full-load operation at 2000 rpm without failure, by using a glass-fiber-reinforced PTFE sheet, which could withstand high temperature, and by reducing the thickness of the underside of the PCB board for enhanced heat transfer from the board to the engine block.

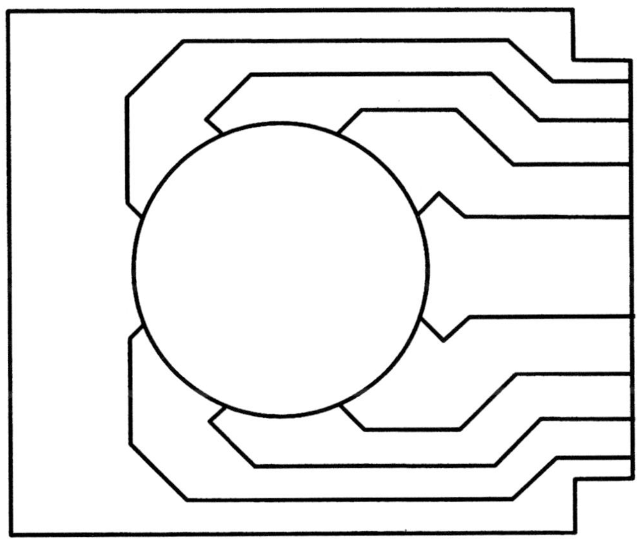

Fig. 9.16 Ionization probes instrumented head gasket on a printed circuit board.

9.4.3 Data Acquisition and Analysis

Fig. 9.17 shows the electronic circuit used by Nicholson and Witze [1993] for their ionization probes. A single 250 V DC power supply was used for these probes. Voltage dividers reduced the actual bias voltage supplied to the probes to 150 V, which was designed to buffer the individual ionization probe signals from cross-talk through the common supply voltage. The circuitry converted the ionization current into voltages and provided an amplified output. High-pass filters were incorporated into the circuitry in order to remove the DC component. The flame arrival times at each probe were then registered when the ionization signals were above a preset threshold. Nicholson and Witze found that 10% of the difference between the maximum value of the ionization signal and the background value was a good choice of threshold.

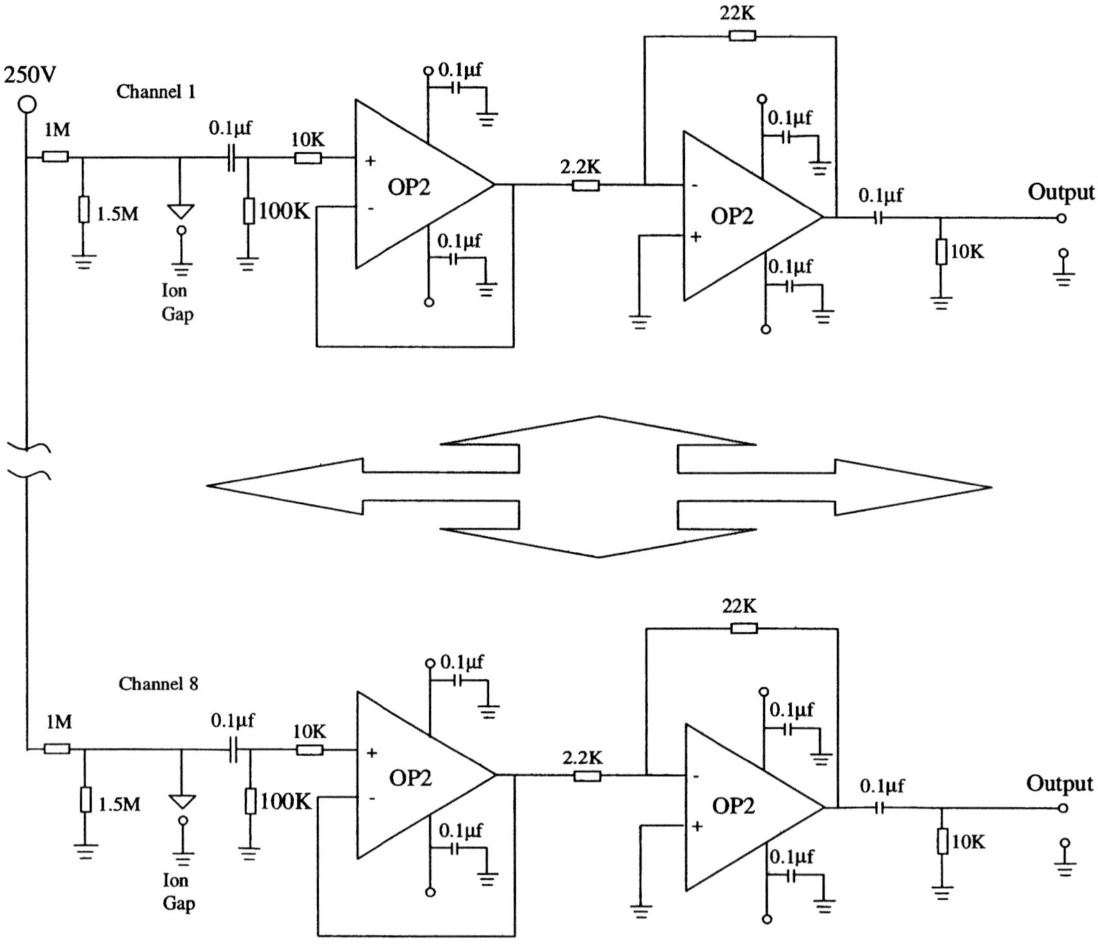

Fig. 9.17 An electronic circuit for multiple ionization probes.

The flame arrival times can be presented in a polar plot, as shown in Fig. 9.18, where the flame arrival time, the interval from ignition until flame arrival at the sensors, is given as a polar plot superimposed on a schematic diagram of the engine cylinder. The inner circle in solid line represents the cylinder bore, and the ionization probe measurements are shown as arrows originating at the spark plug location and intersecting the bore at the location of the individual sensors. The line at the tip of the arrows is a smooth curve fit to the flame arrival times. This approach permits the direction and distance of the flame path to each ionization probe to be displayed. The three circles of dashed lines represent the locations of a spherical flame front at 10, 20 and 30 degrees crank angle after ignition.

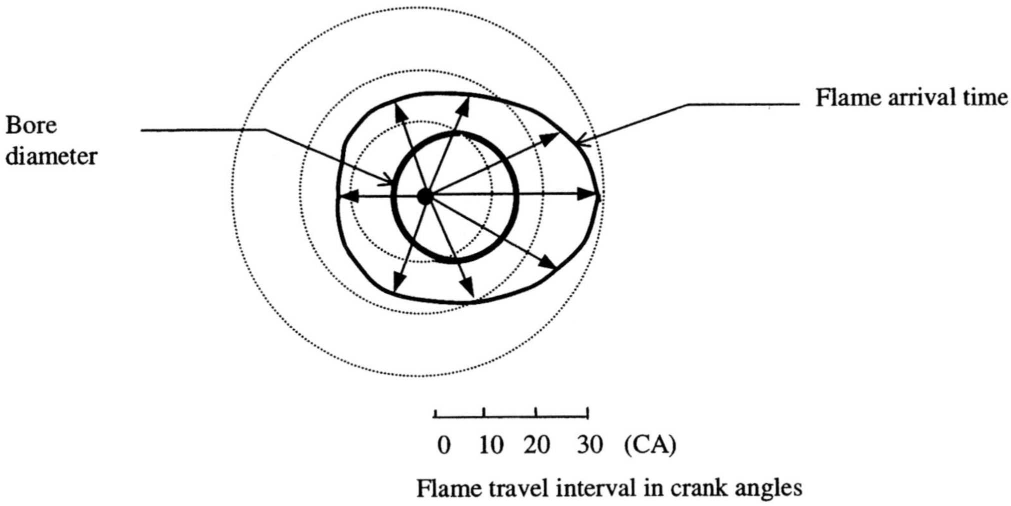

Fig. 9.18 Flame arrival time represented in a polar diagram.

However, for an engine designer or development engineer to locate the end gas region in the combustion chamber or investigate the effect of charge motion on flame propagation, flame contours will be easier to interpret than polar plots of flame arrival times. The flame arrival times obtained by an ionization probe at the end of the flame travel process, however, are not easily related to flame contours, because the flame propagation speed varies in time and direction. To draw the flame contours at any given time, one has to know how the flame evolves from the point of ignition both in space and time. Two approaches have been used to derive the flame contours from ionization probe measurements.

Russ et al. [1997] produced flame contours by using the flame arrival information from the ionization probes with burned gas volume information from the pressure trace. First, the relationship between mass fraction burned and flame radius for the test engine was obtained using a two-zone engine simulation program, assuming a spherical flame growth. Then the horizontal component of the calculated flame radius was calculated, which represents the flame radius projected on the piston surface. At the time of flame arrival for each probe, a ratio of the actual flame radius to the horizontal component of calculated flame radius could be found. This ratio was then used to scale the predicted flame radius of the flame in the direction of each probe. Therefore, an estimated relationship between the calculated mass fraction burned from the pressure trace and true flame radius in the direction of each probe was established, from which the flame contours could be drawn. The limitation of this approach is that the scale factor, between the predicted and measured flame radius, is invariant throughout the flame propagation process.

Hadjiconstantinou and Heywood [1995] developed a model that converted flame arrival times at eight electrode locations into approximate flame contours at any given time. The model was based on the assumption that the flame evolution curve for any engine could be expressed in the form of $r/r_0 = f(t/t_0)$, i.e., the distance traveled by the flame in any direction normalized by a known distance r_0, was given by a function of the time to travel there, t, normalized by the time, to, taken to travel to r_0. The function, f, was obtained from analysis of flame propagation data in the literature, which were scaled to account for the different geometry of combustion chambers. It was also assumed that the flame evolution obeyed this relation in every direction. Given eight arrival times, t_i, and the distances of the ionization probes from the point of ignition, r, they solved for the eight t_0's (the time it took for the flame to reach the midpoint distance in the eight different directions) by setting r_0 to the midpoint distance. Then, for any given time t, the distances traveled in the eight probe directions were calculated, and flame contours at different times were determined by joining these flame locations smoothly using a spline fit.

9.4.4 Multi-Ionization Probe Measurements of Flame Propagation

In the first application of an instrumented head gasket to flame propagation measurements, Witze [1989] obtained cycle-resolved flame arrival times from eight ionization probes located around the perimeter of a single-cylinder engine. These probes were embedded in the head gasket between the cylinder head and the cylinder block. The results were presented in polar diagram as shown in Fig. 9.18. To demonstrate the technique, measurements were carried out for uniformly spaced and clustered arrangements of ionization probes, and for differing conditions of fluid motion, spark location, spark plug configuration, and equivalence ratios. The same instrumented gasket was also used to infer fluid motion by using one of the ionization probes as the ignition site, with the ensuing flame front serving as a flow marker. Using this technique, Witze was able to measure the swirling gas flow and flame propagation velocities in the engine.

Following initial development work by Nicholson and Witze [1993], a number of research groups have reported the use of the instrumented PCB head gasket in production engines. Park et al. [1997] investigated the effect of combustion chamber design on the flame propagation and knocking combustion in an SI engine by means of an ionization gasket, pressure transducer, and flame visualization. The end-gas region determined by ionization probes compared well with flame imaging measurements. Jackson et al. [1996] studied the flame propagation in a Ricardo CCVS stratified EGR combustion system by means of a 16 channel ionization probe head gasket. With glass-fiber-reinforced PTFE sheet, the instrumented head gasket lasted over 11 hours of full load running at 1800-2000 rpm with detonation without failure.

In additional to its direct measurements of flame arrival times, multiple ionization probes have been employed to infer flame contours, as discussed in the previous section. Russ et al. [1997] investigated flame development in a production engine with two different intake configurations.

Flame contours at 2%, 10%, and 50% mass-fraction-burned crank angles generated from the ionization probe outputs were used to study cycle-to-cycle variations in the flame shape and speed in the two engines.

Ionization probe derived flame contours were also employed by Hadjiconstantinou et al. [1996] to investigate sources of unburned hydrocarbon emissions from a lean burn SI engine. Combined with a fast FID, they found that partial burn cycles with very lean mixtures were caused by either a globally slow burning flame that had not completed combustion, or a quenched region that developed toward the end of the combustion process.

9.5 Combustion Studies by Multiple Optical Fibers

9.5.1 Introduction

Fiber optic technology offers an alternative to the large optical window that is required for direct imaging of combustion process. The principles of fiber optics can be found in many textbooks. An optical fiber can be considered to be a highly efficient light guide that transmits light over long distance. The fiber itself consists of a core and a cladding, over which protective layers called buffer and jacket are used, as shown in Fig. 9.19. The refractive index of the core is higher than that of the cladding. Light entering the fiber at angles within its acceptance cone is internally reflected off of the core/cladding interface, and thus is guided by the fiber. Step-index fiber optics are normally used as the light guide because of their high transmission. The acceptance cone angle of a fiber depends on its construction and material. For a silica/silica optical fiber, the acceptance cone angle is about 25°.

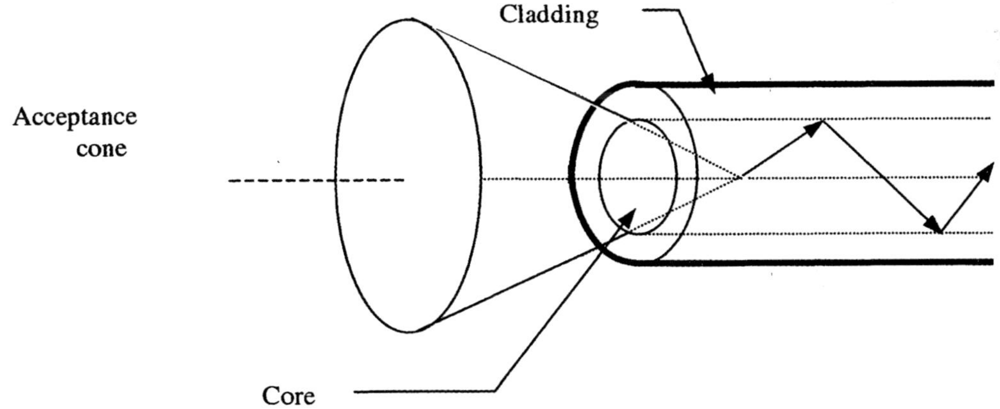

Fig. 9.19 Schematic of an optical fiber.

Spicher and co-workers [1984; 1986; 1987; 1988; 1991] solved the optical access problem by installing as many as 240 optical fibers around the combustion chamber, including the cylinder wall, and cylinder head. In addition, non-contact transmission was employed to direct the light output of fibers mounted in the piston to the detection system. The flame luminosity detected by each fiber was monitored with a high-speed transient recorder, and this information was subsequently used to reconstruct the three-dimensional evolution of the burned gas volume. Each optical probe was screw-mounted in the walls, as shown in Fig. 9.20. To protect the optical fibers from the influence of in-cylinder pressure and temperature, a sapphire rod was glued inside the optical probe. The recess in the wall or the gap between the sapphire rod and the optical fiber restricted the field of view to 2°–12°.

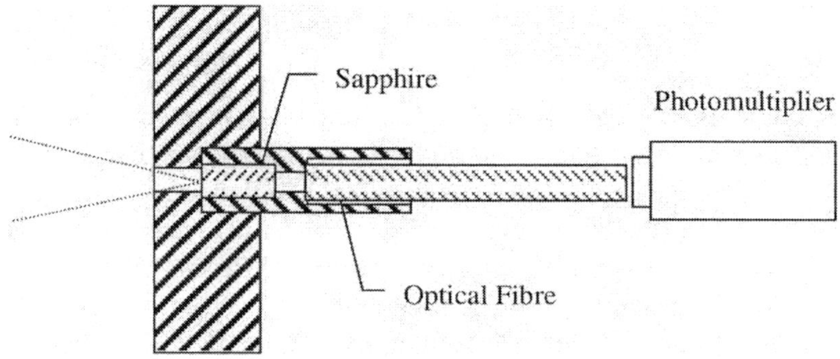

Fig. 9.20 Mounting of an optical probe [Spicher and Velji, 1984].

9.5.2 Multiple Optical Fibers Spark Plug

9.5.2.1 Installation of Multiple Optical Fibers in a Spark Plug

The approach adopted by Spicher et al. requires substantial modifications to the engine. In order to simplify the experimental work involved, Witze et al. [1988] installed eight optical fibers directly into a standard spark plug as shown in Fig. 9.21. This setup is particularly suited for measurements of flame kernel development. When the optical fiber tip was flush mounted, the field of view of each optical probe was determined by the acceptance cone angle of the fiber, which was about 22° for the silica/silica fiber used in their setup.

Fig. 9.22 shows an improved design by Meyer et al. [1993]. The central electrode of the probe, part of a commercially available spark plug, was inserted in a specially machined housing. This housing held eight symmetrically spaced optical fibers on an 11 mm diameter ring. The fibers were protected from the hot gas by press-fitted sapphire windows, and the other end of each fiber was connected to a photomultiplier tube. Once the photomultiplier output exceeded a preset threshold, the flame arrival time was registered by sending an electrical pulse to a digital timer board. Salvat et al. [1994] found that post-processing the recorded signals produced more consistent flame arrival times, particularly under operating conditions near the combustion limit.

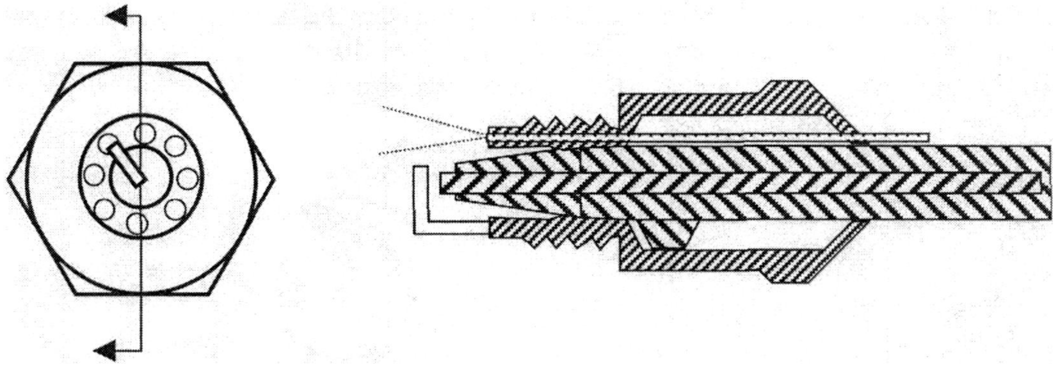

Fig. 9.21 Schematic of an optical fiber instrumented spark plug.

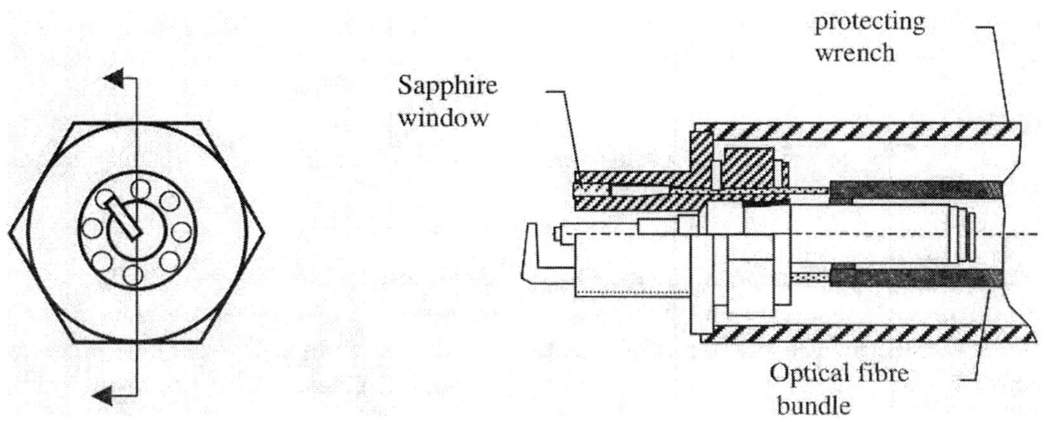

Fig. 9.22 Improved design of an optical fiber instrumented spark plug.

9.5.2.2 Data Processing and Analysis of Optical Fiber Outputs

Two flame kernel models have been employed to infer the flame kernel growth rate and convection velocity based on the information obtained from the optical probes. One is an ellipse fit model, developed by Kerstein and Witze [1990], and the other one is a cubic spline interpolation approach, developed by Lord et al. [1993].

In the ellipse model, an ellipse fit applies a nonlinear least square algorithm to fit the group of eight flame front velocity vectors extending radially from the ignition point with an ellipse, as shown in Fig. 9.23(a). The minor axis of the ellipse is interpreted as the flame kernel growth rate in a radial direction away from the ellipse centroid. The major axis is interpreted as the sum of growth rate and convection velocity, from which a kernel convection velocity may be drawn. In determining the ellipse dimensions, and in turn the kernel convection velocity, the algorithm repeatedly weights the earliest arrival times most heavily. Kerstein and Witze [1990] considered that those early arrival times should provide the best representation of the convection velocity.

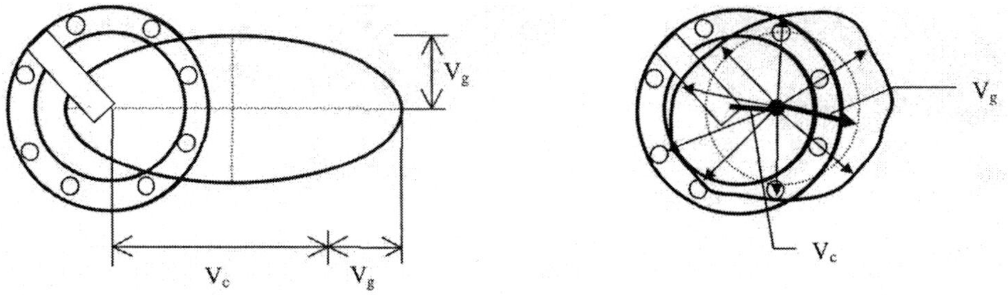

(a) Ellipse model

Growth rate (V_g) is defined by the semi-minor axis of the ellipse. Convection velocity (V_c) is derived from the distance from the spark gap to the farthest point on the ellipse and V_g.

(b) Cubic spline model

Growth rate (V_g) is defined by the average distance from flame centroid to flame front. Convection velocity (V_c) is determined by the displacement of the flame controid from the spark gap.

Fig. 9.23 Flame kernel development models used in the data analysis of fiber outputs.

The cubic spline model, as shown in Fig. 9.23(b), first defines eight vectors extending radially from the point of ignition. The magnitude of each vector is determined by multiplying the reciprocal of the corresponding arrival time by the fixed radial distance from the spark gap to the fiber probe. This group of vectors defines the flame front expansion rate away from the point of ignition. The cubic spline interpolation algorithm fits the group of eight flame front velocity vectors extending radially from the ignition point with a cubic spline, as shown in Fig. 9.23(b). Advantages of the spline fit include its ability to pass through all eight data points and to account for geometrical irregularities in flame kernel perimeter, which are evident from combustion photographs.

In the cubic spline model, kernel growth rate and convection velocity rely on the calculation of the flame area centroid, and are determined as follows:

(a) First, a characteristic time, Δt, is defined as the average flame front arrival time over all eight fibers for the given cycle,

$$\Delta t = \frac{1}{8}\sum_{i=1}^{8}\left(t_i - t_{ign}\right) \tag{9.2}$$

Here, t_i is the flame arrival time at the ith fiber, and t_{ign} is the time of spark ignition.

(b) Secondly, the location of the centroid, (X_c, Y_c), of the area enclosed by the interpolated flame perimeter is calculated by the numerical integration,

$$X_c = \frac{\int_A r_c \cos(\theta) dA}{\int_A dA} \tag{9.3}$$

$$Y_c = \frac{\int_a r_c \sin(\theta) dA}{\int_A dA}$$

Where r_c is the radial distance from origin to centroid of incremental wedge, and A the total area enclosed in flame perimeter. The displacement of the flame centroid from the spark gap is thus known, and the kernel convection velocity vector (V_c) and convection direction (θ_c) for a given cycle can be calculated by

$$V_{cx} = \frac{X_c}{\Delta t} \tag{9.5}$$

$$V_{cy} = \frac{Y_c}{\Delta t} \tag{9.6}$$

$$|V_c| = \sqrt{V_{cx}^2 + V_{cy}^2} \tag{9.7}$$

$$\theta_c = \arctan\left(\frac{V_{cy}}{V_{cx}}\right) \tag{9.8}$$

(c) Finally, the flame kernel growth rate, S_g, is calculated from the distance from the centroid to the interpolated flame perimeter in eight equally spaced directions,

$$S_g = \frac{1}{8}\sum_{k=1}^{8} Se(k) \tag{9.9}$$

A similar cubic spline model was also used by Bianco et al. [1991] to study flame kernel development in an SI engine. The definition of the kernel growth rate in their model was based on the area expansion rate with respect to perimeter, while the model described by Eqs. (9.3)–(9.9) features the average rate of radial flame front expansion in eight specified directions. The convection velocity was calculated similarly to the previous model.

9.5.2.3 Flame Shape Measurements Using Ionization Probes and Fiber Optics

As noted previously, multiple ionization probes measure the flame arrival times at the perimeter of the combustion chamber, while the multiple optical fibers in a spark plug are particularly suited for the study of flame kernel development. Based on the flame arrival times at the fiber optic instrumented spark plug and those at the head gasket ionization probes, the development of the flame radius in each direction from the center of the spark plug toward each head gasket ionization probe was estimated using a cubic spline interpolation [Salvat et al., 1994]. The flame contour for a given time was then drawn from the spline interpolation. This curve fitting methodology produced reasonable flame shape contours, from which Salvat et al. were able to study the cycle-to-cycle variations in the combustion rate, as well as the effect of different flow characteristics on flame development and end-gas region geometry and location.

A similar study was also reported by Meyer et al. [1993], using head gasket ionization probes and spark plug fiber optics.

9.5.3 AVL Tomographic Combustion Analysis System

The multi-channel optical fiber system developed by Spicher et al. provides spatially resolved information about flame radiation intensity detected by a large number of optical fibers, each viewing a small sector of the combustion chamber. The optical probes, however, must be distributed around the combustion chamber and installed in the cylinder head, liner, and piston. A fiber optic probe instrumented spark plug avoids engine modifications, but the spatial information is limited to the few points accessed by the probes. To overcome these shortcomings, Philipp et al. [1995] developed a tomographic combustion analysis (TCA) system, which detects the flame propagation via fiber optic probes embedded in a cylinder head gasket.

The TCA system obtains flame images in a way different from the multiple optical fiber probes techniques described previously. The optical probes in Spicher's or Witze's systems are mounted either radially along the cylinder walls or parallel to each other. They detect the arrival time of the flame in the field of view of each probe, from which a flame contour is calculated. The TCA system still relies on the detection of flame radiation along the field of view of each probe, but the probes are arranged in such a way that the field of view of different probes crosses each other as shown in Fig. 9.24. The flame image is then calculated from the outputs of all the probes by means of a tomographic reconstruction technique, which has been widely used in medical diagnostics. The number of crossing points defines the spatial resolution of the TCA system.

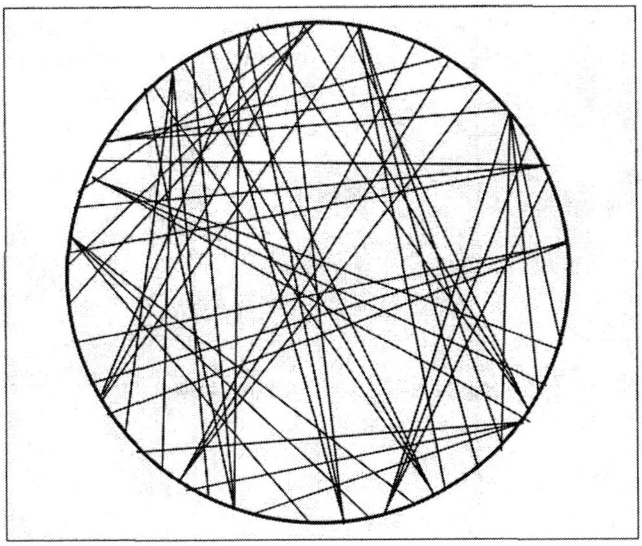

Fig. 9.24 Illustration of optical paths defined by the fiber optics in the AVL TCA system.

The TCA system setup is shown in Fig. 9.25. It comprises an instrumented head gasket with multiple optical fiber probes. A standard system configuration has 100 optical fibers, which are divided into groups of up to five optical fibers. Each group shares a convex lens which is used to reduce the field of view of each fiber from 30° to 4°, as shown in Fig. 9.26. When such a system is applied to an 80 mm bore combustion chamber, the spatial resolution is about 8 mm wide and 4 mm thick. The individual sensor channel comprises a front optic element, fiber optic transmitter, photodiode, and amplifier. UV transmitting optical fibers and UV sensitive photodiodes, with a high dynamic range, are used to cover the whole radiation spectrum during the combustion process. The signal outputs from the amplifiers are recorded on a multi-channel transient recorder. The temporal resolution can be selected according to the application. In standard applications, flame radiation can be recorded for 64 consecutive cycles for a period from 30° CA BTDC to 50° CA ATDC at a resolution of 0.3° CA. To study knocking combustion, the typical crank angle resolution required is about 0.05° CA, and the range of crank angles is limited to 25 CAs (10° CA BTDC to 15° CA ATDC) if 64 consecutive cycles are to be recorded. The flame images are obtained offline from the probe outputs using a tomographic image reconstruction algorithm on a Unix workstation. The reconstructed flame images can then be either sent to a printer for printing or downloaded to a video recorder.

The TCA system was demonstrated both in a single-cylinder research engine and in a multi-cylinder production engine [Philipp et al., 1997]. The optical fibers and the micro-optical lens elements were implanted in a 2.5 mm thick steel sheet, which was sandwiched between a pair of laminated gasket sheet layers that were part of a standard head gasket in an SI engine. The space required by the additional thickness of the optical gasket was gained by machining the face planes of the cylinder block and/or the cylinder head, depending on the engine and measurement requirement. Flame propagation under both normal and knocking combustion conditions were measured, from which the end-gas location and autoignition centers could be identified.

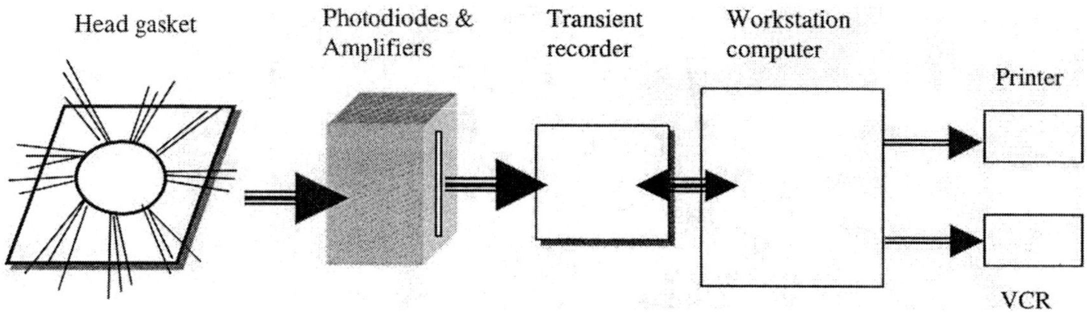

Fig. 9.25 AVL tomographic combustion analysis (TCA) system.

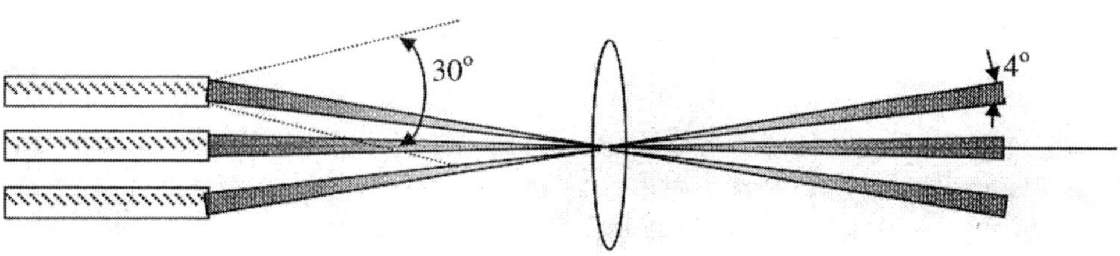

Fig. 9.26 Micro-optic component for limiting the field-of-view of fiber optics.

9.6 Summary

Visualization of combustion and flame in IC engines is often required for better understanding of the nature of turbulent flame propagation in premixed mixtures, the mechanisms of end-gas autoignition/knocking combustion, and spray diffusion flame combustion. Visualization using specially designed research engines has been widely used in the study of engine combustion processes because of its simplicity and easy interpretation. The availability of the endoscope system makes it possible to accomplish the task with much less engine modification. Two-dimensional Mie scattering provides high spatial and temporal resolution images of flame contours that are particularly useful for turbulent combustion modeling. Multiple probe approaches to combustion visualization have been developed because of the increased need for the techniques that can be readily applied to production engines with minimum engine modifications. Combined with reasonably sound flame propagation models, these multiple probe techniques can be used effectively in the measurement of both normal and abnormal combustion processes in SI engines. A summary of these engine combustion visualization techniques is given in Table 9.1.

Table 9.1 Summary of engine combustion visualization techniques

Technique	Application	Comment
High-speed photography	Fuel spray imaging Flame propagation measurements Diesel combustion visualization	Simple setup *Large optical windows* *Limited to luminous combustion*
Endoscope system	Fuel spray imaging Diesel combustion visualization	Minimum engine modification *Restricted and slightly distorted view*
Shadowgraph and Schlieren	Flow field imaging Spray evaporation and mixing Flame kernel and flame propagation Autoignition sites and knocking combustion	Multi-purpose applications Most sensitive imaging technique *Two large line of sight windows required*
2-D Mie scattering	Flame front contour in an SI engine Turbulent combustion modeling	Good spatial and temporal resolution *Large windows, seedings required*
Head gasket ionization probes	Flame arrival time (flame contours) End-gas location	Suitable for production engines *Limited to flame arrival measurements*
Multiple fiber optics Probes in walls	Flame arrival time Flame contours Knocking combustion	Suitable for production engines *Complicated installation*
Spark plug mounted fiber optics	Flame kernel development in a SI engine	Suitable for production engines *Limited to flame kernel measurement*
AVL TCA system	Flame propagation Knocking combustion	Suitable for production engines *Limited resolution (8 × 4 mm)*

References

Alcock, J. and Scott, W. (1962), "Some More Light on Diesel Combustion," Proc. Instn. Mech. Eng., No. 5, pp. 179–191.

Allan, A.B. and Khin, A. (1970),"Investigation of Combustion Phenomena in a Swirl Chamber Compression Ignition Engine Using Schlieren Techniques," SAE Paper No. 700500.

Arcoumanis, C., Hu. Z., and Nagwaney, A. (1994), "Spray and Combustion Visualization in Direct Injection Diesel Engines," IMechE Seminar on Measurement and Observation Analysis of Combustion in Engines, London.

Ask, T.O., Almas, T., Valland, H., and Paulsen, H. (1994), "Study of the Initial Flame Growth in a Spark Ignition Engine," COMMODIA 94, Japan, 69–83.

Bates, S. (1989). "Flame Imaging Studies of Cycle-by-Cycle Combustion Variation in a SI Four-Stroke Engine," SAE Paper No. 892086.

Ball, G.A., (1955), "Photographic Studies of Cool Flames and Knock in an Engine," Proc. 5th Symposium (International) on Combustion, The Combustion Institute.

Baritaud, T.A. and Green, R. (1986), "A 2-D Flame Visualisation Technique Applied to the IC Engine," SAE Paper No. 860025.

Bianco, Y., Cheng, W., and Heywood, J. (1991), "The Effects of Initial Flame Kernel Conditions on Flame Development in SI Engine," SAE Paper No. 912402.

Browne, K.R., Partridge, I.M., and Greeves, G. (1986), "Fuel Property Effects on Fuel/Air Mixing in an Experimental Diesel Engine," SAE Paper No. 860223.

Cartellieri, W.P. and Werlberger, P. (1984), "Investigation of Combustion in High Speed Direct Injection Diesel Engines Using a Novel High Speed Combustion Photography Technique," Proc. 20th FISTA Congress, Vienna, Paper 845008.

Chaves, H., Hentschel, W., Obermeier, F., and Stasicki, B. (1996), "In Cylinder High Speed and Stroboscopic Video Observation of Spray Development in a DI Diesel Engine," SAE Paper No. 961206.

Curry, S. (1963), "The Relationship between Flame Propagation and Pressure Development during Knocking Combustion," SAE Paper No. 630095

Glyde, H.S. (1930), "Experiments to Determine Velocities of Flame Propagation in a Side Valve Petrol Engine," J. Inst. Petroleum Technologists, Vol. 16, 756-776.

Hadjiconstantinou, N. and Heywood, J. (1995), "A Model for Converting SI Engine Flame Arrival Signals into Flame Contours," SAE Paper No. 950109.

Hadjiconstantinou, N., Min, K., and Heywood, J. (1996), "Relation Between Flame Propagation Characteristics and Hydrocarbon Emissions Under Lean Operating Conditions in Spark Ignition-Engines," 26th Symposium (International) on Combustion, The Combustion Institute, Naples, Italy.

Hayashi, T., Taki, M., Kojima, S., and Kondo, T. (1984), "Photographic Observation of Knock with a Rapid Compression and Expansion Machine," SAE Paper No. 841336.

Heywood, J.B., *Internal Combustion Engine Fundamentals*, McGraw-Hill, 1988.

Hicks, R.A., Lawes, M., Sheppard, C.G. W., and Whitaker, B.J. (1994), "Multiple Laser Sheet Imaging Investigation of Turbulent Flame Structure in a Spark Ignition Engine," SAE Paper No. 941992.

Hotta, Y., Nakakita, K., Inayoshi, M., Ogawa, T., Sato, T., and Yamada, M. (1998), "Combustion Improvement for Reducing Exhaust Emissions in IDI Diesel Engines," SAE Paper No. 980503.

Jackson, N.S., Stokes, J., Lake, T.H., Sapsford, S.M., Heikal, M., and Denbratt, I. (1996), "Understanding the CCVS Stratified EGR Combustion System," SAE Paper No. 960837.

Karim, E.R. (1989), "High-Speed Photography of Fuel Spray and Combustion Events in a Production Diesel Engine and Combustion Bomb," Proc. Instn. Mech. Engrs, Vol. 203, pp. 269–281.

Kerstein, A.R. and Witze, P.O. (1990), "Flame-Kernel Model for Analysis of Fiber-Optic Instrumented Spark Plug Data," SAE Paper No. 900022.

Kong, S.-C., Ricart, L., and Reitz, R. (1995), "In-Cylinder Flame Imaging Compared with Numerical Computations," SAE Paper No. 950455.

Konig, G. and Sheppard, C.G.W. (1990), "End Gas Autoignition and Knock in a Spark Ignition Engine," SAE Paper No. 902135.

Lord, D.D., Kim, Y., Anderson, R.W., and Brehob, D.(1993), "The Effect of Charge Motion on Early Flame Kernel Development," SAE Paper No. 930463.

Lyn, W.T. and Valdmanis, E. (1962), "The Application of High-Speed Schlieren Photography to Diesel Combustion Research," J. of Photo. Sci., Vol. 10, pp. 74–82.

Male, T. (1949). "Photographs at 500,000 Frames per second of Combustion and Detonation in a Reciprocating Engine," Proc. 3rd Symposium on Combustion, 721–726.

Mantzaras, J., Felton, P.G., and Bracco, F.V. (1988), "3-D Visualisation of Premixed Charge Engine Flames: Islands of Reactants and Products; Fractal Dimensions; and Homogeneity," SAE Paper No. 881635.

Marvin, C.F. and Best, R.D. (1931), "Flame Movement and pressure Development in an Engine Cylinder," NACA Report 399.

Meyer, R., Kubesh, J.T., and Shahed, S.M. (1993), "Simultaneous Application of Optical Spark Plug Probe and Head Gasket Ionisation Probe to a Production Engine," SAE Paper No. 930464.

Miller, C.D. (1947), "Roles of Detonation Waves and Autoignition in Spark Ignition Engine-Knock as Shown by Photographs Taken at 40,000 and 200,000 Frames per sec," SAE Trans. Vol. 1, 98–143.

Nagao, F. and Kakimoto, H. (1962), "Swirl and Combustion in Divided Combustion Chamber Type Diesel Engines," SAE Trans., Vol. 70, pp. 680–696.

Nakamura, A., Ishii, K., and Sasaki, T., (1989), "Application of Image Converter Camera to Measure Flame Propagation in SI Engine," SAE Paper No. 890322.

Nakanishi, K., Hirano, T., Inoue, T., and Ohigashi, S. (1975), "The Effects of Charge Dilution on Combustion and Its Improvement-Flame Photograph Study," SAE Paper No. 750054.

Nicholson, D. and Witze, P. (1993), "Flame Location Measurements in a Production Engine Using Ionisation Probes Embodied in a Printed-Circuit-Board Head Gasket," SAE Paper No. 930390.

Nishida, K., Hiroyasu, H., Matsuoka, T., and Yamauch, H. (1986), "Characterisation of Combustion Processes in the Prechamber and Main Chamber of an Indirect Injection Diesel Engine by High-Speed Photography," SAE Paper No. 861187.

Park, D., Kim, H., and Kim, W. (1997), "A Study of Flame Propagation for Different Combustion Chamber Configurations in an SI Engine," SAE Paper No. 970876.

Philipp, H., Fraidl, G.K., and Winklhofer, E. (1997). "Flame Visualisation in Standard SI Engines-Results of a Tomographic Combustion Analysis," SAE Paper No. 970870.

Philipp, H., Plimon, A., Fernitz, G., Hirsch, A., Fraidl, G. K., and Winklhofer, E. (1995), "A Tomographic Camera System for Combustion Diagnostics in SI Engines," SAE Paper No. 950681.

Ricart, L. and Reitz, R. (1996), "Visualisation and Modelling of Pilot Injection and Combustion in Diesel Engines," SAE Paper No. 960833.

Rothrock, A.M. and Waldon, C.D. (1935), "Some Effects of Injection Advance Angle, Engine-Jacket Temperature, and Speed on Combustion in a Compression-Ignition Engine," NACA Technical Report No. 525.

Russ, S., Peet, G., and Stockhausen, W. (1997), "Measurements of the Effect of In-Cylinder Motion on Flame Development and Cycle-to-Cycle Variations Using an Ionisation Probe Head Gasket," SAE Paper No. 970507.

Salvat, O.P., Cheng, A.S., Cheng, W.K., and Heywood, J.B.(1994), "Flame Shape Determination Using an Optical-Fibre Spark Plug and a Head-Gasket Ionisation Probe," SAE Paper No. 941987.

Scott, W.M. (1969), "Looking in on Diesel Combustion," SAE Trans., Vol. 78, pp. 33–48.

Shiozaki, T., Nakajima, H., Yokota, H., and Miyashita, A., (1998), "The Visualisation and Its Analysis of Combustion Flame in a DI Diesel Engine," SAE Paper No. 980141.

Smith, J. R. (1982), "The Influence of Turbulence on Flame Structure in an Engine," Flows in Internal Combustion Engines, Ed. by T. Uzban, ASME, New York, 67–72.

Smith, J.R., Green, R.M., Westbrook, C.K., and Pitz, W.J. (1984), "An Experimental and Modeling Study of Engine Knock," Proc. of 20th Symposium (International) on Combustion, pp. 91–100.

Spicher, U. and Velji, A. (1984), "Measurements of Spatial Flame Propagation and Flow Velocities in a Spark Ignition Engine," 20th Symposium (International) on Combustion, The Combustion Institute.

Spicher, U. and Kollmeier, H. (1986), "Detection of Flame Propagation During Knocking Using Simultaneously High Speed Schlieren Cinematography and Multi-Optical Fibre technique," SAE No. 861532.

Spicher, U., et al.(1987), "An Experimental Study of Combustion and Fluid Flow in Diesel Engines," SAE Paper No. 872060.

Spicher, U. and Schmitz, G., and Kollmeier, H. (1988), "Application of a New Optical Fibre Technique for Flame Propagation Diagnostics in IC Engines," SAE Paper No. 881637.

Spicher, U., Kroger, H., and Ganser, J. (1991), "Detection of Knocking Combustion Using Simultaneously High Speed Schlieren Cinematography and Multi Optical Fibre Technique," SAE Paper No. 912312.

Werlberger, P. and Cartellieri, W.P., "Fuel Injection and Combustion Phenomena in a High Speed DI Diesel Engine Observed by Means of Endoscopic High Speed Photography," SAE Paper No. 870097, 1987.

Winterbone, D.E., Yates, D.A., Clough, E., Rao, K.K., Gomes, P., and Sun, J.-H. (1994), "Combustion in High-Speed Direct Injection Diesel Engines-A Comprehensive Study," Proc. Instn. Mech. Engrs, Part C, Vol. 208, pp. 223–239.

Withrow, L. and Boyd, T.A. (1931), "Photographic Flame Studies in the Gasoline Engine," Ind. Eng. Chem., Vol. 23, pp. 539–547.

Withrow, L. and Rassweiler, G.M. (1936), "Slow Motion Shows Knocking and Non-Knocking Expositions," SAE Trans., Vol. 39, pp. 297–303.

Withrow, L. and Rassweiler, G.M. (1938), "Motion Pictures of Engine Flames Correlated with Pressure Cards," SAE Trans., Vol. 43, pp. 185–204.

Witze, P. (1982), "The Effect of Spark Location on Combustion in a Variable Swirl Engine," SAE Paper No. 820044.

Witze, P. (1989), "Cycle-Resolved Multipoint Ionisation Probe Measurements in a Spark Ignition Engine," SAE Paper No. 892099.

Witze, P. and Bopp, S. (1991), "Investigation of In-Cylinder Fluid Motion Using a Head Gasket Instrumented with Ionisation Probes," SAE Paper No. 910719.

Witze, P. and Green, R. (1993), "Determining the Location of End-Gas Autoignition Using Ionisation Probes Installed in the Head Gasket," SAE Paper No. 932645.

Witze, P. and Vilchis, F.R. (1981), "Stroboscopic Laser Shadowgraph Study of the Effect of Swirl on Homogeneous Combustion in a Spark-Ignition Engine," SAE Paper No. 810226.

Witze, P., Hall, M., and Bennett, M. (1990), "Cycle-Resolved Measurements of Flame Kernel Growth and Motion Correlated with Combustion Duration," SAE Paper No. 900023.

Witze, P., Hall, M., and Wallace, J. (1988), "Fibre-Optic Instrumented Spark Plug for Measuring Early Flame Development in Spark Ignition Engines," SAE Paper No. 881638.

Zhang, L., Minami, T., Takatsuki, T., and Yokota, K., (1993), "Analysis of the Combustion of a DI Diesel Engine by Photograph Processing," SAE Paper No. 930594.

Ziegler, G.F.W., Zettlitz, A., Meinhardt, P., Herweg, R., Maly, R., and Pfister, W. (1988), "Cycle-Resolved Two-Dimensional Flame Visualisation in a Spark-Ignition Engine," SAE Paper No. 881634.

Zur Loye, A.O. and Bracco, F.V. (1987), "Two-Dimensional Visualisation of Premixed-Charge Flame Structure in an IC Engine," SAE Paper No. 870454.

Chapter 10

Diesel Combustion Temperature and Soot Loading Measurement by the Two-Color Method

Outline

10.1 Introduction

One of the characteristics of diesel combustion is its high luminosity caused by the thermal radiation of soot particles at high temperature. To take advantage of its luminous nature, the two-color method has been developed and used to measure diesel combustion temperature since 1940s [Uyehara et al. 1946 and 1947; Lyn, 1957]. In this chapter, the theory and implementation aspects of the two-color method will be discussed in detail. Then, particular features of the full-field two-color method are considered. This chapter concludes with some examples of the applications of the two-color method to in-cylinder temperature and soot concentration measurements.

10.2 Theory of the Two-Color Method

10.2.1 Flame Temperature Measurement

10.2.1.1 Definition of Flame Temperature

During diesel combustion, both solid soot particles and gaseous combustion products are present. However, diesel combustion is dominated by the intense radiation from soot particles. Since the two-color method utilizes the thermal radiation from soot particles, it measures directly the temperature of soot particles. The temperature of combustion gases is not directly measured. However, it can be shown that the temperature difference between the two temperatures is negligible (<1 K) when the ambient gas and soot particles have attained thermal equilibrium [Schack, 1925], a condition that can be attained in about 10^{-5} to 10^{-6} seconds within the cylinder [Baker and Ryder, 1961]. In the absence of surface reaction on the soot particles, it can also be proven that the soot particles can faithfully follow the surrounding gases temperature as it changes with time.

From the above discussions, it may be assumed that the soot particle and combustion gas temperatures are approximately the same. In the rest of this chapter the flame temperature refers to the soot particle temperature.

10.2.1.2 Principle of Two-Color Temperature Measurement

In the two-color method, the thermal radiation at two different wavelengths is detected, and the flame temperature is then determined from their ratio by eliminating an unknown factor.

The intensity of radiation from a blackbody varies with wavelength, and depends on the temperature of the blackbody. This is described mathematically by Planck's equation,

$$E_{b,\lambda}(T) = \frac{C_1}{\lambda^5 \left[e^{(C_2/\lambda T)} - 1 \right]} \tag{10.1}$$

where $E_{b,\lambda}$ = monochromatic emissive power of a blackbody at temperature T, ($W\ m^{-3}$)
λ = wavelength, (μm)
T = temperature, (K)
C_1 = the first Planck's constant
= $3.7418 \times 10^{-16}\ W\ m^2$
C_2 = the second Planck's constant
= $1.4388 \times 10^{-2}\ m\ K$

The monochromatic emissivity of a non-blackbody is defined as

$$\varepsilon_\lambda = \frac{E_\lambda(T)}{E_{b,\lambda}(T)} \tag{10.2}$$

where $E_\lambda(T)$ and $E_{b,\lambda}(T)$ are the monochromatic emissive power of a non-blackbody and a blackbody respectively, at the same temperature T and wavelength λ. In other words, ε_λ is the fraction of the blackbody radiation emitted by a surface at wavelength λ.

In the two-color method, an apparent temperature T_a is introduced, defined as the temperature of a blackbody that will emit the same radiation intensity as a non-blackbody at temperature T. T_a is also known as the brightness temperature. From the definition of T_a, it follows that

$$E_{b,\lambda}(T_a) = E_\lambda(T)$$

Combining the definition for monochromatic emissivity with that of the apparent temperature gives:

$$\varepsilon_\lambda = \frac{E_{b,\lambda}(T_a)}{E_{b,\lambda}(T)} \tag{10.3}$$

Combining Eq. (10.1) and Eq. (10.3), the monochromatic emissivity is given theoretically by

$$\varepsilon_\lambda = \frac{e^{C_2/\lambda T} - 1}{e^{C_2/\lambda T_a} - 1} \tag{10.4}$$

In practice, ε_λ is estimated for soot particles using the empirical correlation developed by Hottel and Broughton [1932],

$$\varepsilon_\lambda = 1 - e^{\left(-KL/\lambda^\alpha\right)} \tag{10.5}$$

where K = an absorption coefficient,
= proportional to the number density of soot particles
L = the geometric thickness of the flame along the optical axis of the detection system

The value of the parameter α depends on the physical and optical properties of the soot in the flame. The selection of the α value will be discussed later. Matsui et al. [1979] carried out validation studies of the above correlation by simultaneous measurements of the soot emissivity at three wavelengths in a diesel engine. They concluded that at least in the visible range this is a correct functional relation between emissivity and wavelength.

Combining Eq. (10.4) and Eq. (10.5) gives,

$$KL = -\lambda^{\alpha} \ln\left[1 - \left(\frac{e^{(C_2/\lambda T)} - 1}{e^{(C_2/\lambda T_a)} - 1}\right)\right] \qquad (10.6)$$

The unknown product KL can be eliminated by rewriting the above equation for two specific wavelengths, λ_1 and λ_2

$$\left[1 - \left(\frac{e^{(C_2/\lambda_1 T)} - 1}{e^{(C_2/\lambda_1 T_{a1})} - 1}\right)\right]^{\lambda_1^{\alpha_1}} = \left[1 - \left(\frac{e^{(C_2/\lambda_2 T)} - 1}{e^{(C_2/\lambda_2 T_{a2})} - 1}\right)\right]^{\lambda_2^{\alpha_2}} \qquad (10.7)$$

The actual flame temperature T is independent of wavelength, while the apparent temperature T_a and, possibly, the parameter α vary with wavelength.

It can be seen that this equation can be solved for the flame temperature, T, provided the apparent temperature T_{a1} and T_{a2} for the flame are known at the two wavelengths λ_1 and λ_2 These two apparent flame temperatures can be measured at these two wavelengths using a calibrated two-color optical pyrometer system. In fact, this is the purpose of the two-color system: to provide instantaneous measurements of the apparent temperatures T_{a1} and T_{a2} for the flame at two chosen wavelengths, λ_1 and λ_2. The choice of the two wavelengths, at which T_{a1} and T_{a2} are monitored, will be discussed in Section 10.3.2.

10.2.2 KL Factor and Soot Concentration

Once the value of the flame temperature, T, has been determined using Eq. (10.7), substitution of T back into Eq. (10.6) gives an estimate for the value of the product KL, which is proportional to the soot concentration. In addition, if some assumptions are made, an estimate of the volumetric density of soot, C_v (the volume of particles in a unit of space), and soot gravimetric density, C_m (mass of soot per unit volume), can be obtained.

It will be shown in Chapter 11 that the volumetric density of soot particles, C_v, is given by

$$C_v = \frac{1}{6\pi L \, \mathrm{Im}\left\{\dfrac{m^2-1}{m^2+2}\right\}} \frac{KL}{\lambda^\alpha} \tag{10.8}$$

where m is the complex refractive index of the soot particle. Thus, using the estimate of KL factor it is possible to obtain an estimate of the volumetric density of soot particles. Of course, the values of L, α, and m are needed.

The gravimetric density of soot particles, C_m, can also been obtained if a value of soot density ρ is known or can be estimated. The estimation of the soot number density, i.e., the number of soot particles per unit volume, requires knowledge of the size distribution of soot particles, which is normally difficult to obtain.

10.3 Implementation

A typical arrangement for the implementation of the two-color method in a diesel engine is shown in Fig. 10.1. In this system, a window (usually made of quartz or sapphire for measurements involving infrared radiation) allows optical access to the engine combustion chamber.

The radiation emitted by a cloud of incandescent soot particles passes through the optical window and the aperture, and then is guided by a fiber-optical bundle to a bifurcation where it is split into two beams. Each beam then passes through a bandpass filter which only allows radiation within a narrow wavelength range (usually 5 nm to 20 nm) to go through. The instantaneous signals generated by the photodiodes are amplified and the resulting voltages at each wavelength are then recorded by a data acquisition system.

It should be pointed out that the aperture inserted between the optical window and the receiving fiber defines the field of view of the system to a narrow cone that lies within the combustion chamber. The use of an aperture increases the spatial resolution, but the instrument sensitivity is reduced as a result of less radiation reaching the receiving optics. Another important function of the aperture is to reduce the possibility of leaving the field of view not filled. When only part of the field of view is filled with combustion gas, the brightness temperature of the flame will be underestimated.

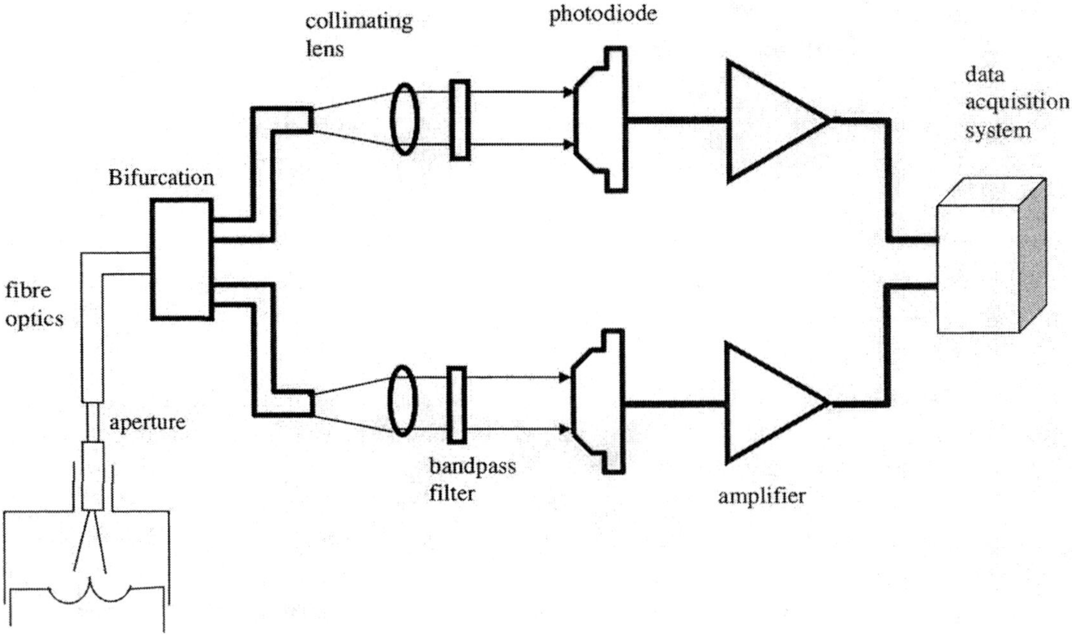

Fig. 10.1 Schematic layout of a two-color system.

10.3.1 Selection of α

From Equations (10.6), (10.7), and (10.8), it can be seen that estimation of the flame temperature, the KL factor, and the volumetric density of soot particles, depends on the values chosen for the parameter α. Thus, it is necessary to consider the choice of a suitable value for α.

The value of α is a function of a number of factors. It depends on the light wavelength, the soot particle size, and the refractive index of soot. Fuel type may also have an influence on the value of α [Gaydon and Wolfhard, 1970]. Before selecting the value of α, it should be noted that when the two wavelengths are selected to be in the visible region, then the choice of the value of α is less critical to the estimation of the flame temperature. This is because the flame temperature given by Eq. (10.7) is not sensitive to the value of α. This point is clearly shown in Fig. 10.2 which is calculated for $\alpha = \alpha_1 = \alpha_2$. It is also noticed from Fig. 10.2 that for wavelengths in the infrared region the influence of the value of α is significant.

A similar effect of α on the estimation of KL is observed. As shown in Fig. 10.3, the estimation of KL is less sensitive to the value of α if the two wavelengths are chosen in the visible spectrum, whereas the sensitivity is rather large when these two wavelengths are in the infrared region.

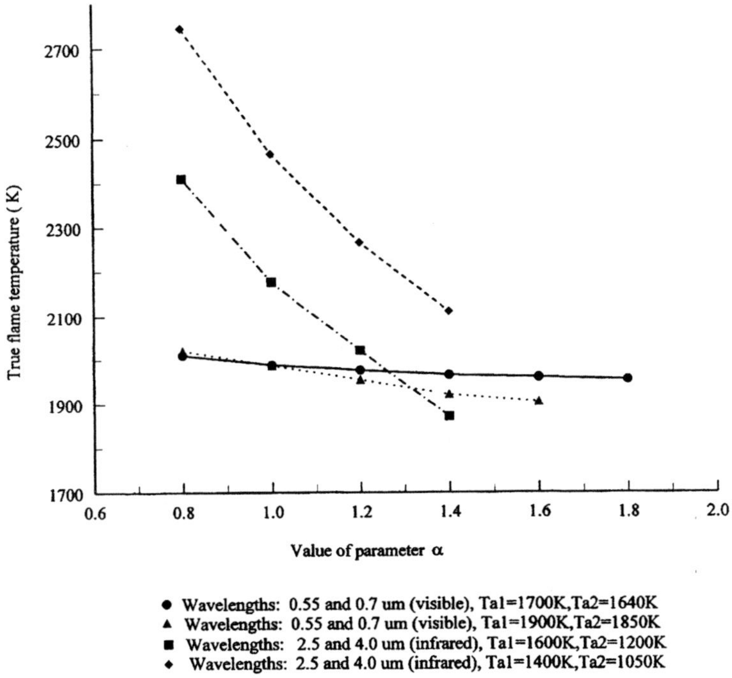

Fig. 10.2 The effect of the value of parameter α on the estimated temperature.

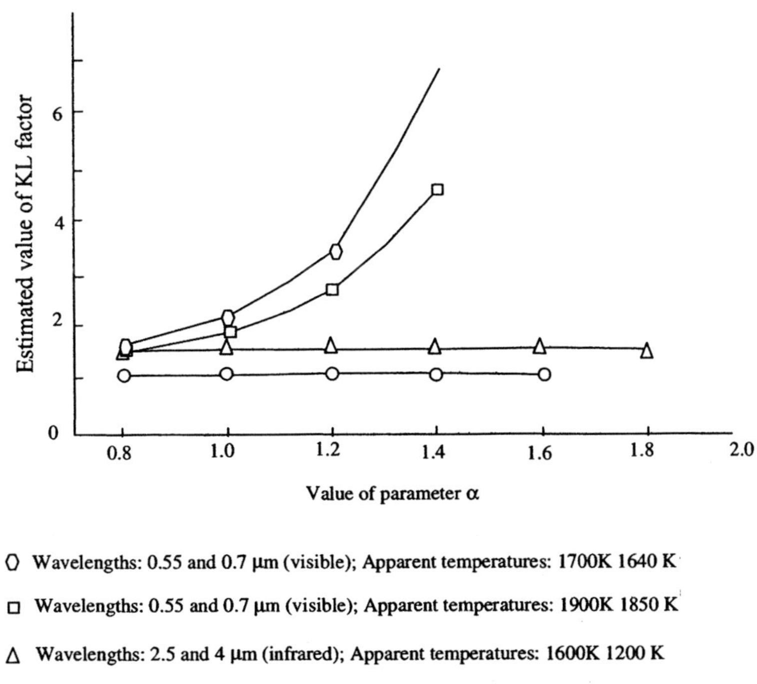

Fig. 10.3 The effect of the value of parameter α on the estimated KL factor.

Table 10.1 summarizes the values of α used in the literature. From these values, it can be recommended that a fixed value of 1.39 would be appropriate for most fuels when using visible wavelengths. For the infrared region, the choice of the value of α is less clear-cut, but a fixed value close to unity appears to be appropriate for diesel fuel flames for the wavelength range of 0.8 to about 7 μm.

Table 10.1 Compilation of values of the parameter α in the Hottel and Broughton correlation for soot emissivity

Value of parameter α				
Visible wavelengths	Infrared wavelengths		Fuel or flame type	Reference
1.39	0.95	($\lambda > 0.8\mu m$)	steady luminous flame	Hottel and Broughton [1932]
1.38	0.91–0.97	($\lambda = 2–4\mu m$)	diesel engine soot	Matsui et al. [1980]
1.39	-		diesel engine soot	Yan and Borman [1988]
-	0.94–0.96	($\lambda > 0.8\mu m$)		Liebert and Hibbard [1970]
-	0.89, 1.0	($\lambda =1–7\mu m$)	amyl acetate	Siddall and McGrath [1963]
-	0.77	"	kerosene	
-	0.94, 0.95	"	benzine	
-	0.93	"	candle	
-	0.96, 1.14, 1.25	"	furnace samples	
-	1.06	"	petrotherm	
-	1.00	"	propane	
-	$\alpha = 0.91+0.28 \ln \lambda$	"	various fuels	
1.43	-		acetone	Rossler and Behrens [1950]
1.39	-		amyl acetate	
1.29	-		coal gas/air	
1.23	-		benzene	
1.14	-		nitrocellulose	
0.66 - 0.75	-		acetylene/air	

10.3.2 Selection of Wavelengths

It will be recalled that an estimate of the flame temperature can be made using Eq. (10.7). However, solution of Eq. (10.7) requires measurements of the apparent temperatures of the flame at two wavelengths. The choice of wavelengths for the two-color method involves a number of considerations.

In general, visible and infrared wavelengths have been used in the two-color method. Visible wavelengths are preferred for a number of reasons. First, if the wavelengths are chosen in the visible part of the spectrum, the sensitivity of the measuring system to changes in flame temperature will be greater because the rate of change of spectral radiance with respect to temperature, dI_λ/dT, is larger in the visible than in the infrared region, as shown in Fig. 10.4. Also, there will be a greater difference in signal output from the system at the two visible wavelengths, due to the fact that the spectral radiance with respect to wavelength, $dI_\lambda/d\lambda$, is higher in the visible region for temperatures around 1000 to 2000 K, as shown in Fig. 10.4. This makes the effect of signal-to-noise ratio on the calculation of flame temperature less important. The larger separation of two wavelengths used will also improve the sensitivity of the two-color system, provided that the α parameters for the two wavelengths are well known.

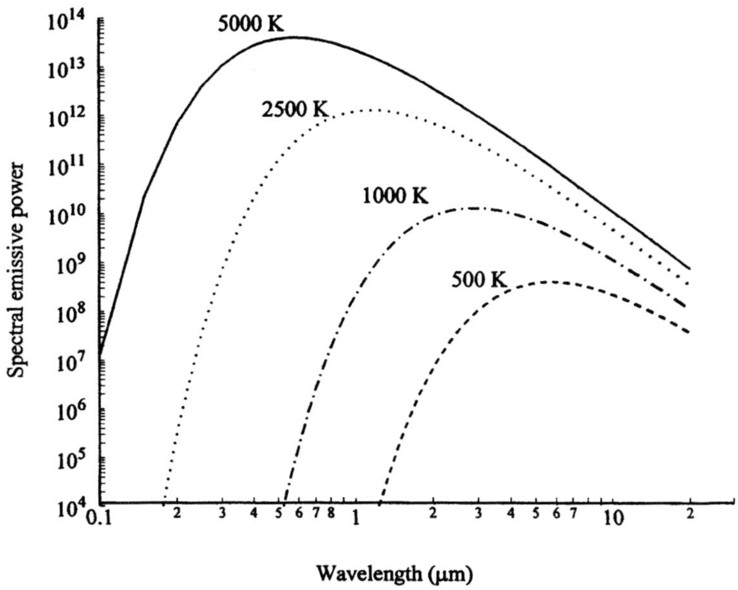

Fig. 10.4 The spectral radiance of a blackbody as a function of temperature.

As shown in Fig. 10.2 and Fig. 10.3, estimation of temperature and KL factor is not very sensitive to the choice of the value of the parameter α, when the two wavelengths are chosen to be in the visible region.

During combustion, strong band emissions from gaseous species can be present, as well as thermal radiation from soot particles. In the reaction zones of flames, many radicals, such as OH, CH, C_2, HCO, NH, NH_2 may be formed and give appreciable emission in the visible and near ultraviolet regions. In the infrared region, wavelengths must be chosen carefully to avoid radiation or absorption from CO_2, CO, water vapor, and fuel vapor. Fig. 10.5 shows the spectra from various gaseous species in the visible and infrared region.

Photodiodes with high responsivity in the visible and near infrared regions are readily available commercially. Also, there are commercially available optical fibers for use as light guides which have a high transmittance in the visible and near infrared regions.

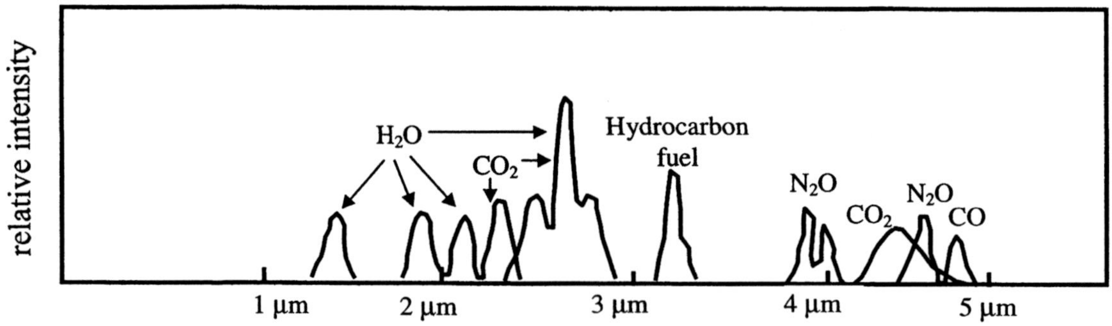

Fig. 10.5 Emission spectra in the infrared region for various gas molecules.

Table 10.2 shows a compilation of the wavelengths used by some investigators in the application of the two-color method.

10.3.3 Calibration

It will be recalled that the two-color method measures the apparent temperatures of the flame T_{a1} and T_{a2} at wavelengths λ_1 and λ_2. These apparent temperatures can then be used in Eq. (10.7) to obtain the actual flame temperature. Since the two-color system will produce initially

Table 10.2 Wavelengths used by various investigators in the application of the two color method

Wavelength (nm)					
Visible			Infrared		Reference
λ_1	λ_2	λ_3	λ_1	λ_2	
549	750	-	2300	3980	Matsui et al. [1980]
529	624	738	-	-	Matsui et al. [1979]
550	700	850	-	-	Yan and Borman [1988]
-	-	-	750	1000	Peterson and Wu [1986]
460	640	-	-	-	Kawamura et al. [1989]
550	650	-	-	-	Shakal and Martin [1994]
450	640	-	-	-	Arcoumanis et al. [1995]
581	631	-	-	-	Winterbone et al. [1994]
540	643	-	-	-	Miyamoto, et al. [1996]
600	-	-	-	1000	Bertoli et al. [1996]
470	650	-	-	-	Shiozaki et al. [1996]

only voltages $V_{\lambda 1}$ and $V_{\lambda 2}$ corresponding to the radiation from the flame at the two wavelengths, the system must be calibrated so that the two voltages correspond to the apparent flame temperatures. This can be done using a source of blackbody radiation.

Using a source of blackbody radiation, a calibration curve can be generated in the form of a plot of voltage output against blackbody temperature for each wavelength. Once the calibration curves have been produced for the two wavelengths, the instantaneous voltages can be converted to the apparent temperatures T_{a1} and T_{a2}. The flame temperature, T, and the KL factor can then be obtained from Eq. (10.7) and (10.6), respectively.

Often, a blackbody may not be readily available In these circumstances, it may be more convenient to use a calibrated tungsten lamp. The lamp calibration is often carried out at a wavelength of 650 nm only, but it is sufficiently accurate at other wavelengths close to 650 nm. For wavelengths very different from 650 nm, a new calibration curve will be required for the tungsten lamp at that wavelength. Another difficulty when using a tungsten lamp for calibration is that the size of the tungsten ribbon filament may be too small to illuminate the whole of the field of view of the two-color system. This problem can be overcome by using a lens as shown in Fig. 10.6. The focal length and the position of the optical lens determines the size of the image of the tungsten ribbon D′, which should at least be the same as the diameter of the receiving optical fiber or detector. The diameter of the lens should be chosen so that the field of view in the calibration is the same as that as defined by the aperture in the probe.

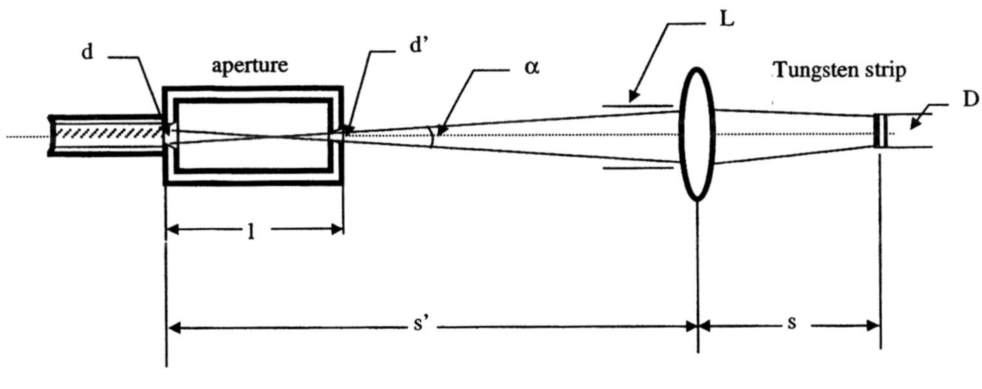

Angle α must be smaller than the acceptance angle of the optical fibre
Angle α, dimensions ***d, l, D***, and focal length f are specified
Calculate d' to give required angle α from $d' = 2\,l\,\tan(\alpha/2) - d$
Calculate distance *s'* from $s' = f\,[1+(d/D)]$ where d/D is the magnification ratio
Calculate distance *s* from $s = f\,[1+(D/d)]$
Calculate lens aperture ***L*** from $L = [\,2\,s'\,\tan(\alpha/2)] - d$
Note: if aperture is designed with both holes of equal diameter, then $d = d'$, in which case d is specified and l is calculated from $l = d/\tan(\alpha/2)$, to give the required value of angle α

Fig. 10.6 Optical setup for the calibration of a two-color system using a tungsten ribbon lamp.

When an optical lens is used, some of the radiation from the tungsten lamp will be lost through reflection and absorption by the lens. If the transmittance of the lens is τ, the voltage output V from a photodiode will be lowered by $(1 - \tau)$ compared to what it would be without the lens for a given value of T. Thus in constructing the calibration curve for the two-color system, the quantity V/τ should be plotted against the apparent temperature, since when the system is operational in an engine without the lens the voltage output would be somewhat higher.

10.3.4 Data Acquisition and Analysis

The output from a two-color system is normally converted into voltages by the photodiodes before subsequent calculations are carried out, in order to obtain the flame temperature and soot loading factor. A high-speed data acquisition system is used to collect the amplified signal outputs from the photodiodes on a crank angle basis using the output from a shaft encoder as the clock for data acquisition. Large cycle-by-cycle variation in radiation, however, is present in diesel combustion. The radiation signals can be very irregular because of the motion of soot clouds. The irregularity in radiation signals can also be caused by the random location of ignition and combustion. It has been found that the large cyclic variation of the radiation signals gives a large variation in the results of the KL factors. The flame temperature, though, was found to be less affected by the cycle-by-cycle variation. Thus, some form of ensemble-average is required to obtain the temperature and KL measurement from the two-color method.

There are two ways to obtain an ensemble-averaged temperature and KL factor. In the first approach, which Matsui et al. [1979] used, the voltage outputs corresponding to the radiation intensity at the two wavelengths at each crank angle for 1000 cycles are ensemble-averaged. These averaged voltage outputs for the two wavelengths are then used in the calculation of the flame temperature and KL factor. An alternative approach is to use an ensemble-averaged temperature and KL factor. In this method, the signal for each cycle is used to calculate the temperature variation at each degree crank angle during a particular engine cycle. Then, the individual cycle temperatures are ensemble-averaged over many engine cycles. Because the flame temperature and KL factor are not linearly related to the radiation intensity, it is expected that these two approaches may not generate the same results. Yan and Borman [1988] compared these two methods and found that the difference between them was small for KL factor and temperature during most of the combustion period. In practice, the first approach is preferred, as it is computationally faster and reduces the influence of noise on the calculation of temperature.

After conversion of signal outputs into apparent temperatures at the two wavelengths, the true flame temperature can be calculated using Eq. (10.6). Eq. (10.6) is highly nonlinear and can be solved by the method of Newton iteration. However, this method will not always give a convergent solution for this equation, especially when the two apparent temperatures are very close, i.e., the monochromatic emissivities are near unity. As a consequence, a calculation scheme has been developed by Yan and Borman [1988]. This involves the estimation of the apparent temperature at the first wavelength, T_{a1}, for an assumed T and measured T_{a2}, and an iteration process until the calculated T_{a1} is within 0.1°C of the measured T_{a1}, each iteration being calculated for a new estimated T.

The numerical scheme is illustrated in Fig. 10.7. It is based on the observation that the variation of T_{a1} with T at a given T_{a2} is approximately linear for a short temperature range, as shown by rearranging Eq. (10.7),

$$\frac{1}{T_{a1}} = \frac{\lambda_1}{C_2} \ln\left[1+\left(\frac{e^{(C_2/\lambda_1 T)}-1}{A}\right)\right] \qquad (10.9)$$

where

$$A = 1 - \ln\left[1-\left(\frac{e^{(C_2/\lambda_2 T)}-1}{e^{(C_2/\lambda_2 T_{a2})}-1}\right)\right]^{\alpha_2/\alpha_1}$$

It is also true that the flame temperature is greater than its apparent temperatures, T_{a1} and T_{a2}. The detailed procedures to determine the flame temperature are summarized as follows:

1) Assume that the true flame temperature lies between points 1 and 2 in the figure. Point 2 should therefore be selected to be greater than the true flame temperature. Because the flame temperature in a diesel engine is around 2500 K, $T_2 = 3000$ K will be a good value. The temperature at point 1, T_1, is assigned to the measured T_{a2} at the longer wavelength, which should always be lower than the true flame temperature.
2) Evaluate the first apparent temperature at point 2, $T_{a1,2}$, with T_2 and T_{a2} using Eq. (10.9). T_{a2} should be greater than $T_{a1,m}$, which is the measured apparent temperature at the shorter wavelength.
3) Use $T_{a1,m}$ to estimate the new flame temperature by linear interpolation

$$T_2^{new} = T_1 + \frac{(T_2 - T_1)(T_{a1,m} - T_{a21})}{(T_{a1,2} - T_{a21})} \qquad (10.10)$$

4) Use the newly estimated flame temperature T_2^{new} together with T_{a2}, to evaluate a new $T_{a1,2}$.
5) Compare $T_{a1,2}^{new}$ with $T_{a1,m}$. If the difference is less than 0.1°C, T_2^{new} is the solution to the true flame temperature for given $T_{a1,m}$ and T_{a2}; otherwise, replace $T_{a1,2}$ with $T_{a1,2}^{new}$, and repeat steps 3, 4, and 5 outlined above until a convergent solution is found. A few iterations are normally sufficient.

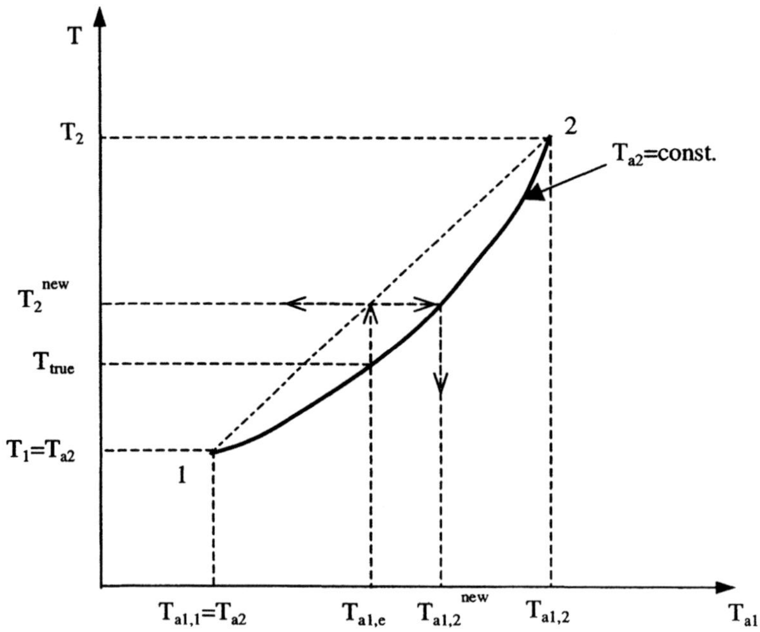

Fig. 10.7 Schematic of the numerical scheme for solving Eq. (10.7).

10.4 Accuracy

10.4.1 Effect of Soot Deposition on Window

During engine tests, soot accumulated on the optical window will reduce its transmissivity and cause an error in the estimation of flame temperature and soot concentration. Matsui et al. [1980] found that in a DI diesel engine, a reduction of 14% in transmissivity at $\lambda = 0.55$ μm resulted an error of around 1% in the flame temperature and less than 5% in the estimate of the soot concentration. As shown by Eq. (10.7), the true flame temperature is mainly determined by the ratio of flame radiation at the two wavelengths, while the KL factor is directly proportional to the absolute amount of flame radiation. The loss in transmissivity would therefore have a much smaller effect on the measured flame temperature than KL factors.

In order to minimize the effect of the soot deposition on two-color measurement results, Yan and Borman [1988] developed a self-cleaning sapphire window arrangement, which was able to keep the window surface free of soot even after several hours of operation.

10.4.2 Wall Reflections

Note that the radiation detected by a two-color system includes not only the direct radiation from the flame (soot + gaseous species), but also the radiation reflected off the walls, and the thermal radiation by the wall surfaces. Because the surface temperatures of walls are much lower than the soot particles, their effect can be ignored in the visible region. In the infrared region, significant radiation from the walls is present, and the effect of this needs to be taken into account. Using visible radiation, Matsui et al. [1980] found that the effect of reflected radiation from an opposite wall on the estimation of temperature was very small (on the order of 2 or 3 K), while the error in the determination of KL factor was at most 10% at very low soot concentrations, falling to 2% at high soot concentrations. When infrared wavelengths are used, the error in the estimated temperature ranged from 50 to 150 K. In addition, the error in the estimated value of the KL factor could be as high as 50%.

10.4.3 Effects of Non-Uniform Temperature and Soot Distributions

The two-color method will only give a true flame temperature if a spatially uniform temperature exists along the line of sight. The non-uniformity of temperature and soot concentration will affect the physical meaning of the flame temperature as well as the KL measured by this technique. Detailed studies were carried out on the effects of non-uniform temperature and soot distribution on two-color results by Matsui et al. [1979; 1980] and by Yan and Borman [1988]. Their findings can be summarized as follows:

1) The effect of soot concentration non-uniformity on the measured temperature is less severe than the effect of uneven temperature on the KL factor. If a severe non-uniform distribution of temperature exists, the KL factor will be underestimated.
2) The temperature measured by the two-color method is higher than the arithmetic mean temperature, because the rate of change of monochromatic intensity with temperature becomes greater with increasing temperature.
3) The flame temperature closer to the window will have a larger effect on the measured temperature than will those of flames farther away, i.e., the measured flame temperature will be biased toward that near the window.
4) The measured flame temperature will be closer to the maximum flame temperature if the maximum values of temperature and soot concentration are coincident with the line of sight.

10.4.4 Overall Accuracy

One general conclusion that emerges from the above discussions is that the two-color method is particularly well-suited for accurate temperature measurements. On the other hand, soot loading measured by the two-color method should be used with caution, because of its proneness to interference by window fouling and non-uniform soot distributions along the optical path of measurements.

10.5 Full-Field Imaging of Temperature and Soot Distributions

10.5.1 Introduction

In the previous sections, the salient features associated with the two-color method have been discussed. The two-color system described so far measures the flame temperature in a small control volume as defined by the field of view of the probe. Thus, it is essentially a "line-of-sight" measurement. In diesel engine combustion, however, it is desirable to obtain the two-dimensional temperature and soot distribution. This can be achieved by the use of an imaging device as the detector in a two-color system. The imaging device can be either a conventional film camera or a solid-state camera. As will be shown in the next two sections, the use of a solid-state camera considerably reduces the overhead associated with long and tedious film development and data analysis processes. However, a high-speed film camera is still the preferred way to obtain consecutive high-speed two-color images, because of its superior spatial resolution and speed.

10.5.2 Two-Color Image Recordings by Films

10.5.2.1 Principles

Fig. 10.8 shows the experimental setup for high-speed measurement of two-dimensional temperature and KL factor distributions in a diesel engine [Ahn et al., 1981]. The flame radiation is directly recorded by a high-speed camera with a negative color film. After recording the flame images, the density at each point on the film is measured by a densitometer by illuminating the film with a white light source. The density at each point is measured for two different colors by placing an interchangeable color filter between the light source and the film. The ratio of the densities for the two colors at each point on the film is then used to obtain the flame temperature and KL factor distributions in the cylinder. Negative color films are preferred for full-field recordings, because of their wider dynamic range in density than positive films.

To realize the two-color method by means of a film, it is necessary to find out the relationship between the density of the film and the apparent temperature.

The relationship between the density of the film, D_λ, at a given wavelength, λ, and the magnitude of exposure, E_x, is given by

$$D_\lambda = \gamma_\lambda \log(E_x) + \text{const.} \tag{10.11}$$

where γ_λ is a constant of the film used, which depends, among other things, on the process and the chemical used for the film development. E_x is proportional to the product of the radiation intensity and the exposure time.

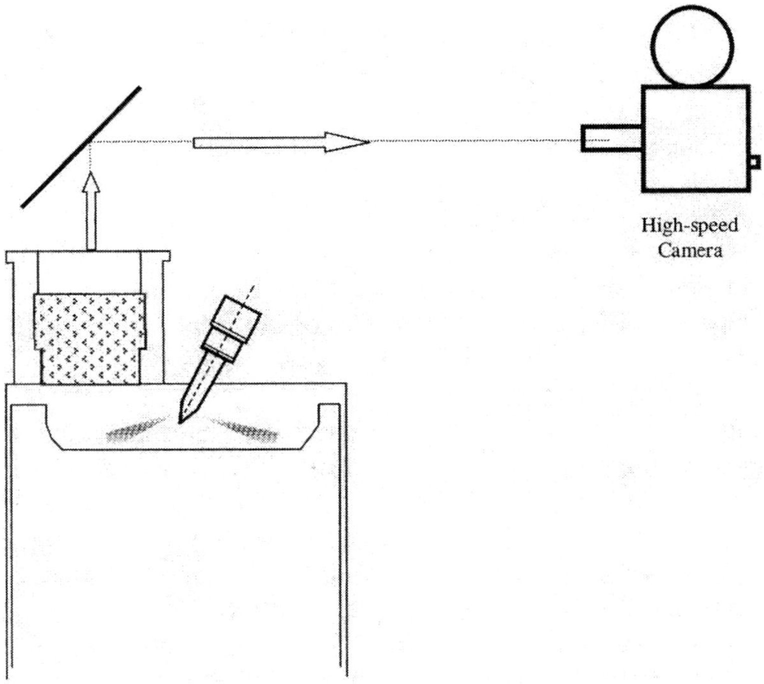

Fig. 10.8 Experimental setup for the high-speed recording of two-color images [Ahn et al., 1981].

The density of the film can then be expressed as follows:

$$D_\lambda = \gamma_\lambda \log(\int \varepsilon_\lambda E_b S_\lambda \tau_\lambda P_\lambda d\lambda) + \text{const.} \quad (10.12)$$

where ε_λ = emissivity of the soot particles
E_b = the blackbody radiation power
S_λ = the spectral sensitivity of the film
τ_λ = the spectral transmissivity of the filter
P_λ = the spectral sensitivity of the photomultiplier of the densitometer

For narrow bandwidth interference filters, ε_λ, S_λ, τ_λ, and P_λ are constants and the replacement of E_b by Eq. (10.1) leads to the following formula:

$$D_\lambda = -\frac{\gamma_\lambda}{2.3026}\frac{C_2}{\lambda T_a} + \text{const} \quad (10.13)$$

The calibration constants in Eq. (10.13) can be obtained by exposing the film to a source of blackbody radiation. The exposed film is then illuminated by white light and recorded by a photodetector at a particular wavelength. A device developed for this purpose is a densitometer, which is essentially a photomultiplier tube. An easy way to do the job, however, is to use a

CCD camera. If a color CCD camera is available, all that is required is to shine white light through the film and leave the camera to sort out the colors. In the absence of a color CCD camera, a monochrome camera can be used with filters attached to it.

10.5.2.2 Film Analysis Using Offline Calibration

Theoretically, two calibration curves at wavelengths λ_1 and λ_2 can be obtained from a source of blackbody radiation. The flame temperature can then be determined using Eq. (10.7), once the apparent temperatures, T_{a1} and T_{a2}, have been obtained by Eq. (10.13) from the density readings of flame images at two different wavelengths λ_1 and λ_2.

In order to obtain the apparent flame temperature at each position on the film, the relationship between the apparent flame temperature and the film density at each wavelength, i.e., the calibration curves as shown in Fig. 10.9, needs to be known. However, it is not possible to alter the apparent temperature of a blackbody source during high-speed recording, so other means must be explored. According to Eq. (10.13), if the film calibration constant, γ_λ, and the offset value, *const*, are known, the apparent temperature is uniquely determined by the film density, assuming that the wavelength at which the film density is measured is well defined. γ_λ and *const* can be determined by recording the radiation of a blackbody source at a known temperature through a set of neutral density filters, then plotting the resulting film density against the logarithmic value of exposure, as shown in Fig. 10.10.

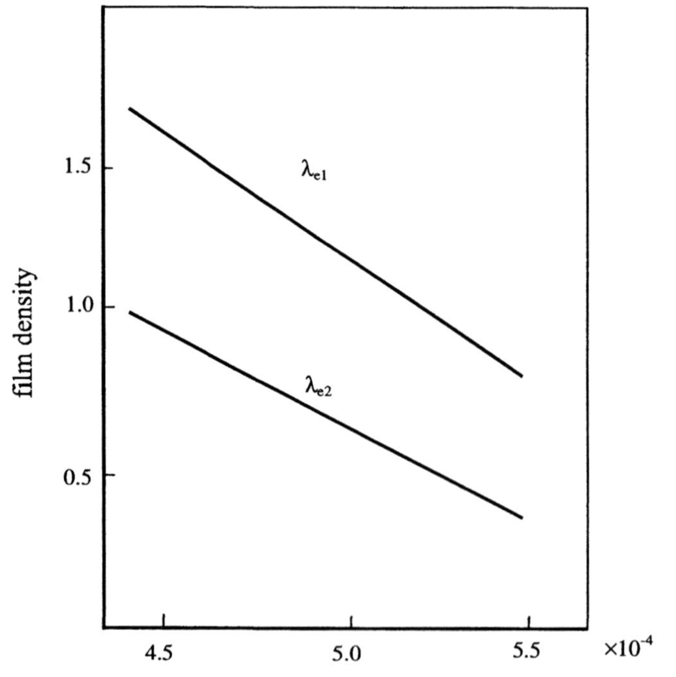

Fig. 10.9 Film density plotted against the reciprocal of the apparent temperature.

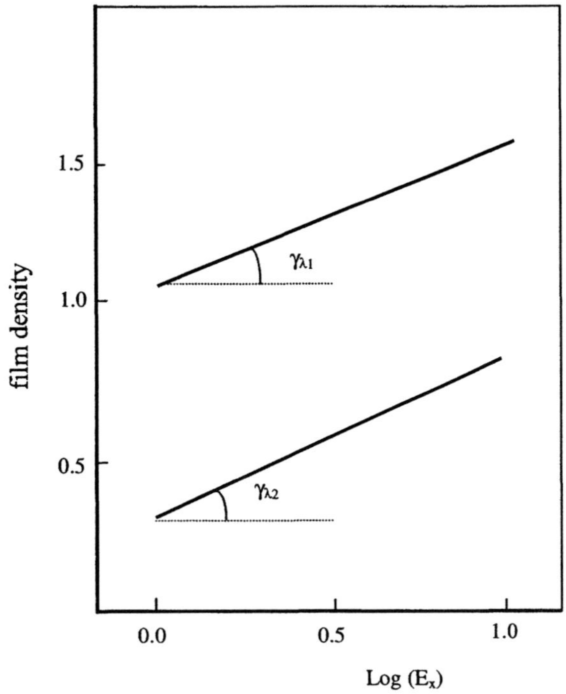

Fig. 10.10 Film density plotted against the logarithmic value of exposure.

γ_λ and *const* can be obtained either offline or online. In both cases, the calibration images of blackbody radiation and combustion images need to be recorded on the same film, in order to minimize the effect of different film and film processing on film densities. In the offline approach [Ahn et al., 1981], recording of the calibration source is done after the experiments. During the experiment, the high-speed camera starts to roll the film at 6000 fps when the engine is motored. As the film reaches the end of its first 50 feet, a magnetic shutter is forced to open, and the fuel injection starts simultaneously. Thus, the flame images are recorded on the second half of a 100 foot piece of film. The film is then loaded in the reverse direction, and the images of a calibration source are recorded on the remaining 50 feet of film. The calibration images are taken at 6000 fps through the detached optical window and a disc rotating at 3800 rpm in which 10 different neutral density filters are installed.

For the online approach, the blackbody radiation and diesel combustion are recorded simultaneously [Kobayashi et al., 1992]. The experimental setup is shown in Fig. 10.11. Both images of diesel combustion and those of the calibration source are recorded simultaneously by a 16 mm high-speed film camera at 6000 fps, through a set of rotating neutral density filters. The rotating disk with 8 different neutral density filters rotates at 3000 rpm. The exposed film is digitized by a color CCD camera through a red color filter and blue color filter, respectively. The color-filtered images of the blackbody radiation are then used to calibrate the flame temperature measurement obtained from the color-filtered images of diesel combustion on the same film.

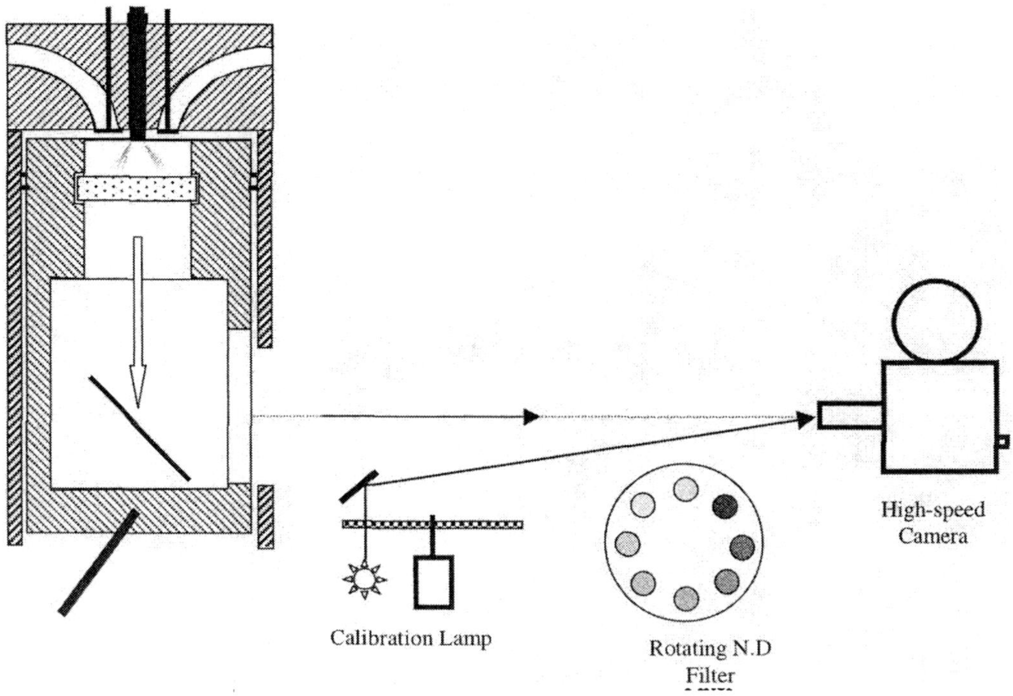

Fig. 10.11 Experimental setup for high-speed two-color experiments with online recording of the calibration source.

10.5.2.3 Film Analysis with Online Calibration

To overcome the difficulty of ensuring consistent color balance of the film, both during exposure and during the subsequent processing, both a "live" measurement of the temperature at one point in the cylinder and high-speed combustion photographs over the whole field of view were taken [Winterbone et al., 1994].

As shown in Fig. 10.12, a single point two-color system and a high-speed camera for two-dimensional temperature measurements are used simultaneously. This requires having optical access to the cylinder through two independent optical paths. In one optical path, a single point two-color measurement is performed using a calibration camera to focus the light onto two photomultipliers via a bifurcated fiber. The other optical path is used to take high-speed combustion photographs. The single point measurement provides a continuous reading of apparent temperature with crank angles. The film provides full-field color images of the combustion process over the whole chamber at discrete times, defined by the framing rate of the high-speed camera.

Each image on the film is digitized twice by a monochrome CCD camera through two interference filters at 581 nm and 631 nm. And each time the image is converted into a computer image with 512 × 512 pixels and 256 grey levels.

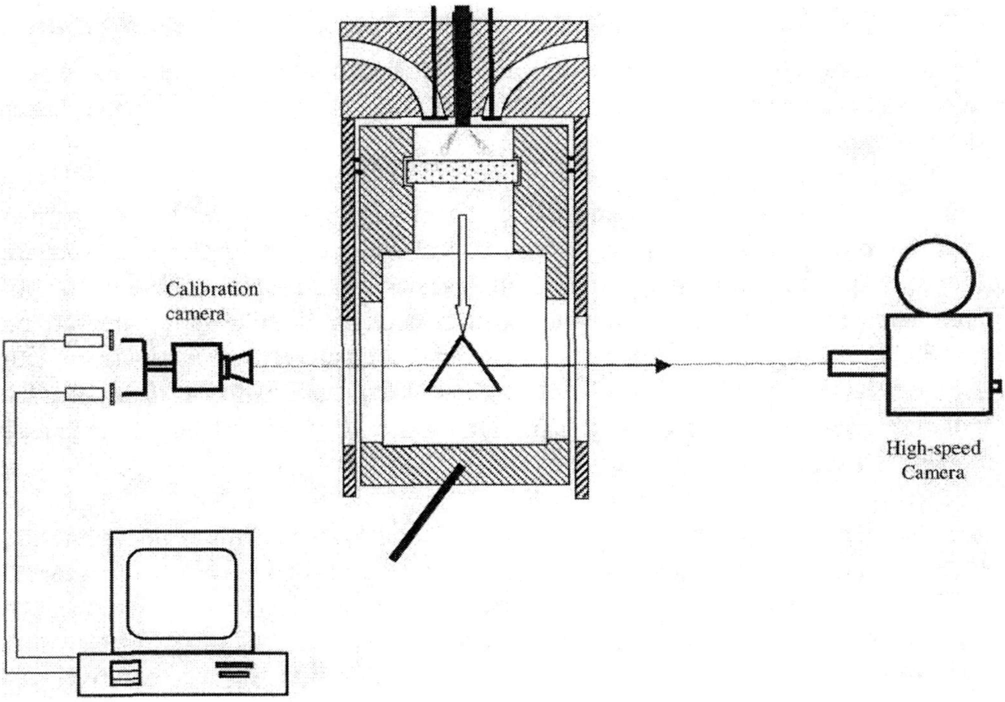

Fig. 10.12 Experimental setup for high-speed recording of two-color images with online calibration [Winterbone et al. 1994].

First, the grey scale of those pixels, corresponding to the live calibration point, is measured and averaged. This process is repeated for each image on the film for each wavelength. The apparent temperatures obtained at various crank angles from the calibration camera, combined with the grey scales at the respective crank angles of the pixels of the computer images, produce an apparent temperature vs. grey scale correlation for the digitized images. A similar correlation is obtained for the other wavelength as well.

The next step is to extend this single piece of information to the whole combustion chamber. This is achieved by applying the apparent temperature vs. grey scale correlation obtained above across the digitized image. The apparent temperature of each pixel is found for each wavelength. Once the two apparent temperatures for each pixel are known, the flame temperature and soot concentration (KL factor) for each pixel are calculated from Eq. (10.7) and Eq. (10.7).

Calibration curves (grey level vs. apparent temperature) have to be determined for each film because the film constant varies between film batches and with development. In the results of Yates and Gomes [1993], the calibration curves were rather scattered. The flame measurements showed their two-color temperatures were about 80 K higher than those measured by the thermocouples.

10.5.3 Full-Field Temperature Measurement Using Solid-State Cameras

While the use of a high-speed film camera provides high-resolution time-resolved full-field temperature and soot measurement, it requires lengthy film development and processing and tedious data analysis.

A solid-state camera can be viewed as consisting of a two-dimensional array of photosensors (pixels), which converts the incident photons into electrons. The output from a solid-state camera can be easily digitized into a number of grey scales, which can be calibrated to obtain the apparent flame temperature. In this case, the discussion on the single point measurement applies readily to the full-field measurement with solid-state cameras. In comparison with a high-speed film camera, the use of a solid-state camera dispenses with the requirement for online calibration and tedious film development. The most widely used solid-state cameras are charged-coupled devices (CCD).

Two approaches have been adopted for full-field two-color systems with monochrome CCD cameras, as shown in Fig. 10.13. In the first approach, a single CCD camera is used (Fig. 10.13[a]). Two images at the selected wavelengths are recorded on the camera side-by-side by placing an image doubler in front of the camera and two band-pass filters mounted between the image doubler and the camera [Shakal and Martin, 1994]. Alternatively, two images at the selected wavelength can be recorded separately by two cameras, each of which has a color filter placed in front of it [Arcoumanis et al., 1995].

The drawback of a CCD camera is that it has a lower resolution than film, and normally can take only one image per cycle, although a high-speed CCD camera, such as Kodak "EktaPro HS Solid-State Motion Analyser," has been used to obtain multiple images in the same cycle. Its resolution is limited to 128 × 128 pixels [Arcoumanis et al., 1995]. Alternatively, multiple two-color images can be obtained using a framing streak camera [Shiozaki et al, 1996].

The full-field two-color systems discussed so far are designed to take photographs through a large optical window, which involves substantial modification to engines. To overcome this difficulty, Shakal and Martin [1994] developed an endoscope type imaging system that requires minimum engine modification. The system setup is shown in Fig. 10.14. A coherent fiber-optical bundle is used as the means for light transmission. A coherent fiber-optical bundle contains many single optical fibers, the relative positions of which are arranged to be the same at both ends so that image transmission is possible. The use of fiber-optical bundles also allows a much wider field of view than viewing through a window directly. The field of view is limited to 45 degrees by a lens mounted at the tip of the bundle on the engine side. The spatial position of the field of view is further determined by a small mirror attached to the end of the fiber-optical bundle, as shown in Fig. 10.14(a). The whole bundle is placed 1 mm away from the optical window mounted in a probe adapter, which prevents the fiber-optical bundle from being exposed to the harsh combustion environment. The dimension of the optical access port is 14 mm in diameter. The fiber-optical bundle has a diameter around 8 mm.

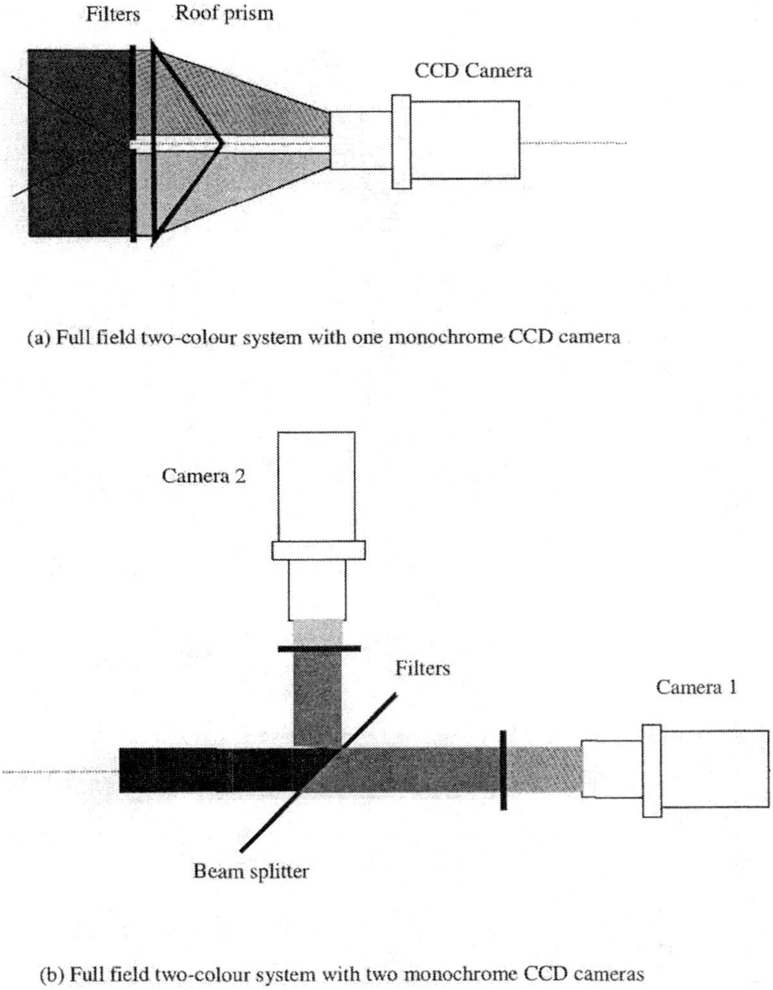

Fig. 10.13 Optical layout for full-field two-color measurements using solid-state cameras.

As shown in Fig. 10.14(b), the output from the other end of the fiber bundle is magnified by a 10X-microscope objective lens. The magnified image is than split into two side-by-side images via an image doubler. Two 10 nm bandwidth filters at 550 nm and 650 nm are used for the two-color measurements. Use of a gateable intensified solid-state camera is necessary because of the low radiation received by the camera, due to transmission losses in the optical system. The two-images recorded on the same frame can then be used to obtain the full-field temperature and KL factor measurements using the two-color method.

Alternatively, the combustion images can be transmitted to the intensified camera by an endoscope system, which comprises an optical window, a long glass rod probe, and a four-piece cardanically linked transmission arm [Bakenhus and Reitz, 1999].

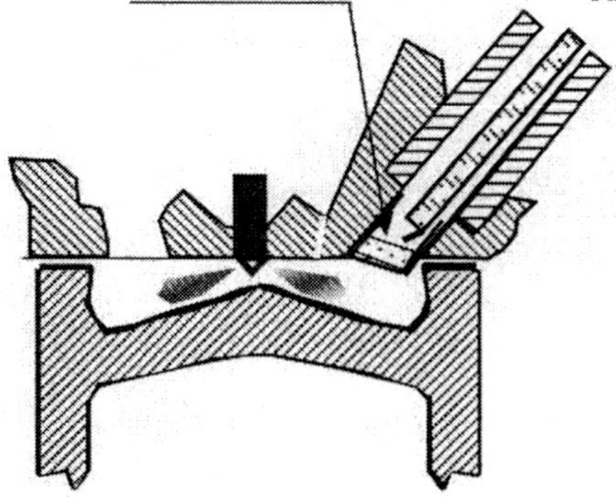

(a) Probe adapter with fibre optical bundle installed

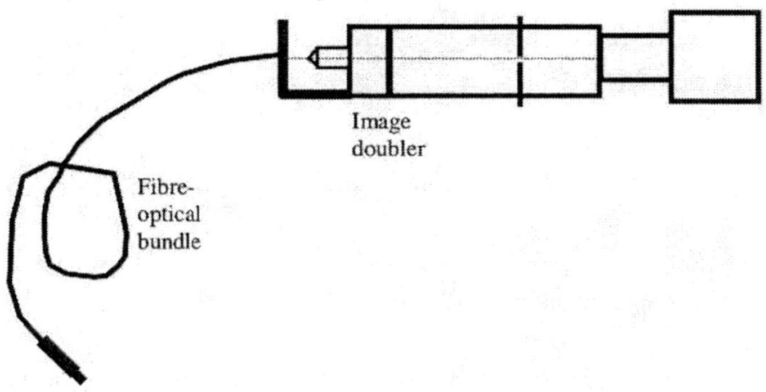

(b) Optical set-up of the imaging system

Fig. 10.14 Experimental setup for the full-field temperature measurements using fiber optics [Shakal and Martin, 1994].

10.6 Application of the Two-Color Method to IC Engines

As stated at the beginning of this chapter, original work on the development and application of the two-color method was carried out by Uyehara et al. in 1940s. The technique has since been used to measure diesel combustion flame temperature in diesel engines [e.g., Lyn, 1957]. The technique was revived by Matsui et al. [1979; 1980; 1982], who carried out the most extensive evaluation of the two-color technique and extended the technique to provide semi-quantitative soot concentration measurements. Since then, the two-color method has become the most used technique for diesel flame temperature measurements because of its simplicity, low cost, and relative ease of application. Because of the large number of applications of the two-color method to diesel combustion studies, only selected examples are presented here.

In additional to the comprehensive development work on the two-color method and its application to diesel engines by Matsui et al. [1979;1980;1982], Yan and Borman [1988] discussed several aspects of the two-color method and its application to a single-cylinder heavy-duty DI diesel engine. The two-color system consists of a trifurcated fiber optical bundle and a specially designed self-cleaning sapphire rod window, which stays clean under steady state full load conditions. Fig. 10.15 shows some typical in-cylinder temperature and KL factor results for different engine loads. As shown in Fig. 10.16, when the load increases, the soot concentration during the earlier part of combustion decreases. It is also noticed that the second bump on the curves of KL factor is more pronounced for the higher load. This was attributed by the authors to later burning and/or probably poor quality of end of injection. The flame temperature increases slightly with load. While the others [Matsui et al., 1980, Matsui et al., 1982, Peterson and Wu, 1986] reported that the peak flame temperature decreased with load. However, for the injection equipment used in Yan and Borman's experiments, the injection timing was advanced with the load, while the end of injection stayed approximately constant. The results shown in Fig. 10.14 are therefore consistent with the other measurements, if the timing is taken into account.

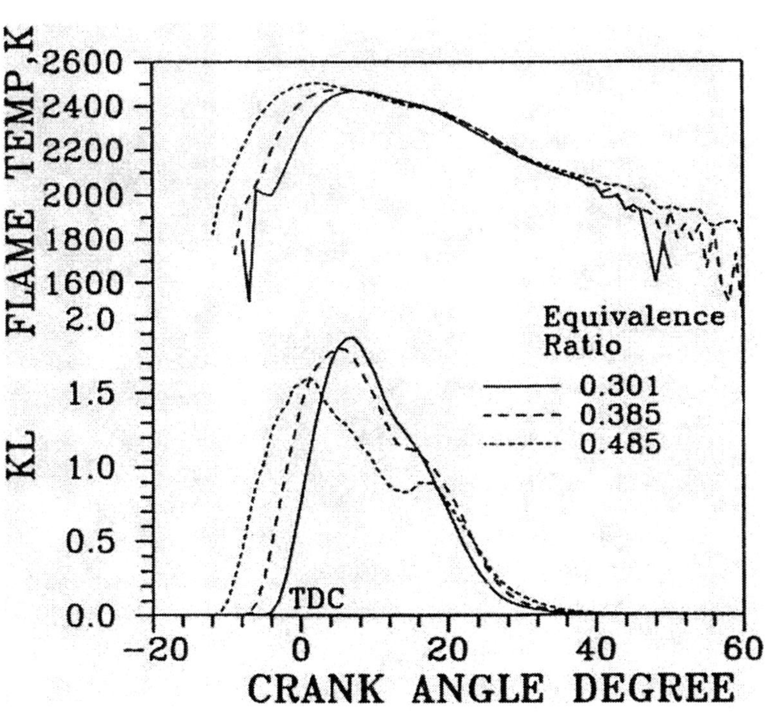

Fig. 10.15 Flame temperature and KL factors at different loads [Yan and Borman, 1988].

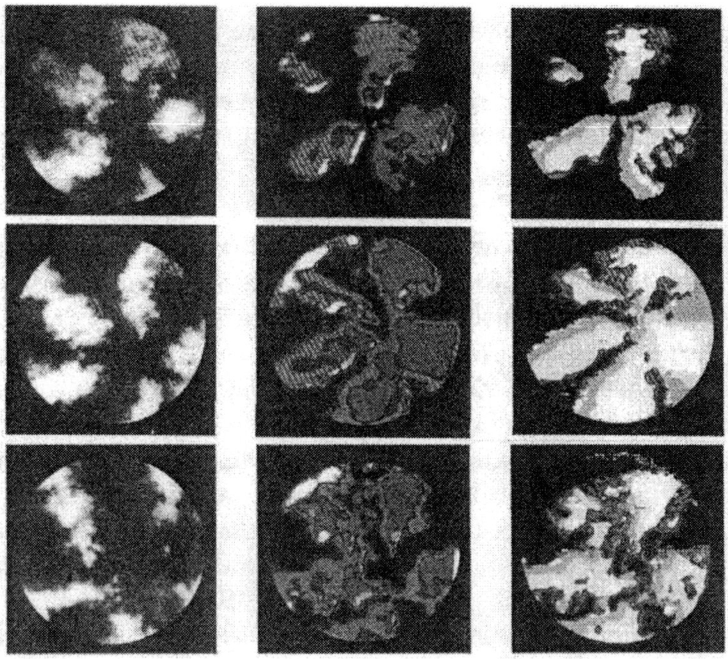

top row 5.04 °CA ATDC; middle row 7.20 °CA ATDC; bottom row 9.36 °CA ATDC
(a) flame images, temperature and KL factor distribution with 0% EGR

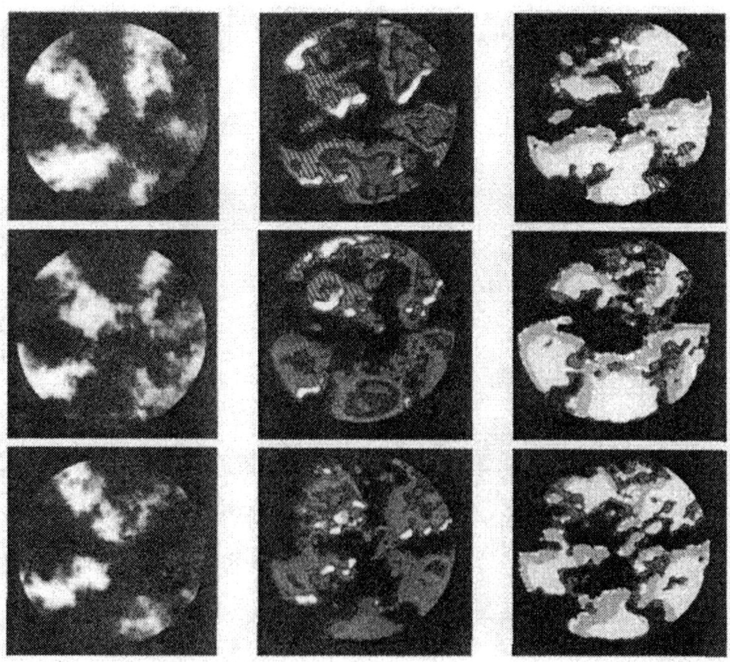

top row 7.2 °CA ATDC; middle row 9.36 °CA ATDC; bottom row 11.52 °CA ATDC
(b) flame images, temperature and KL factor distribution with 30% EGR

Fig. 10.16 Flame images and two-color results from a DI diesel engine [Arcoumanis et al., 1995].

Bertoli et al. [1996] investigated the effect of fuel properties on in-cylinder soot concentration in a DI diesel engine using the two-color method. They found that for pure paraffin fuels the cetane number had a strong effect on the measured KL factor, whereas the aromatic fuel content had a strong influence on the soot loading only for lower (<58) cetane number fuels.

Two-color temperature measurements were also used to investigate diesel combustion in the pre-chamber of indirect injection diesel engines [Nagase et al., 1985; Peterson and Wu, 1986]. Shakal and Martin [1994] studied the effects of auxiliary fuel injection on combustion in a two-stroke DI diesel engine using a full-field two-color system as shown in Fig. 10.14. Temperature measurements indicated that the presence of pilot injection resulted in a high-temperature region (>2300K) around the injection nozzle at the start of main injection. Neither temperatures nor soot concentrations were detectable during the two-stage ignition process of the port fuel injection, indicating low-temperature premixed combustion.

Miyamoto, et al. [1996] investigated the effect of oxygenated additives on combustion temperature and soot formation in a DI diesel engine. The combustion images were recorded using a high-speed film camera. Exhaust analysis showed that significant smoke and particulate suppression could be achieved without increases in NO_X with every oxygenated additive. These results were confirmed by the full-field two-color measurements, which showed that KL factors decreased during combustion while the flame temperature distribution was not affected with the addition of the oxygenated additive.

Full-field two-color temperature and KL factor distributions in a DI diesel engine were also measured by Kobayashi et al [1992]. The engine speed was fixed at 1000 rpm and a high-pressure injection system was employed. Combustion photographs were taken through an transparent piston crown by a rotating prism-type 16 mm camera at 6000 fps. Higher temperatures were observed in the central portions of the combustion chamber. Higher KL factors were observed at the ends of the spray plumes. Other regions of higher KL factors were found to persist through the combustion process and possibly contributed to exhaust emissions. In general, they found that the flame temperature increased with injection pressure, but the soot was reduced in the case of high-pressure injection.

The effect of EGR on flame temperature was studied by Arcoumanis et al. [1995] and Shiozaki et al. [1996]. Using a single intensified CCD camera, Arcoumanis et al. obtained combustion temperature and KL factor distributions in one of four optically accessible cylinders of a DI diesel engine. Two sets of images in Fig. 10.16 were obtained for 0% and 30% EGR rates, respectively, when the engine was running at 1000 rpm without load. The images at each crank angle were recorded from different cycles. From these two sets of images, it is apparent that the presence of EGR reduced flame temperature. When EGR was added to the cylinder, the detectable area of sooting flame was reduced while the KL factor values with each area remained approximately the same. This indicates a reduction of soot with increasing EGR that is not in line with the expected values. This is likely to be caused by the limited dynamic range of the CCD camera used, as well as the inaccuracy in KL factor measurements by the two-color method.

There are also other applications of the full-field two-color method. For example, Winterbone et al. [1994] examined the effect of swirl on combustion. Shiozaki et al. [1996] identified a dependence of OH radical emissions on flame temperature.

Kawamura et al. [1989] extended the application of the two-color method to an SI engine. Since the premixed combustion in an SI engine is generally much less luminous that the diffusion flame in a diesel engine, fine alumina particles (0.3 mm) were mixed with the intake air. The burning of these particles allowed some qualitative burned gas temperatures to be obtained. In order to extend this technique to quantitative temperature readings, the emissivity or the factor α has to be known as a function of temperature and wavelength.

10.7 Summary

The two-color method is the most widely used technique for diesel combustion temperature measurements. It is capable of instantaneous and continuous flame temperature measurements in a diesel engine. It is simple to set up and can be achieved with minimum optical access to the engine. It is now routinely used in diesel engine combustion studies.

In addition, the two-color method is capable of detecting soot loading in a diesel engine. However, the measured KL factor should be treated with caution, because its values are easily affected by window fouling, surface reflections, and non-uniform soot distributions.

The two-color method is a line-of-sight measurement method. It gives a true flame temperature assuming a spatially uniform temperature along the line of sight. The non-uniformity of temperature and soot concentration will influence the physical meaning of the flame temperature, as well as the KL factor measured by this technique.

References

Ahn, S., Matsui, Y., Kamimoto, T., and Matsuoka, S. (1981), "Measurement of Flame Temperature Distribution in a DI Diesel Engine by Means of Image Analysis of Nega-colour Photographs," SAE Paper No. 810183.

Arcoumanis, C., Bae, C., Nagwaney, A., and Whitelaw, J. (1995), "Effect of EGR on Combustion Development in a 1.9L DI Diesel Optical Engine," SAE Paper No. 950850.

Arcoumanis, C., Cutter, P., Folkes, D., and Tabaczynski, R. (1996), "Spray, Combustion, and Heat Transfer Studies in a Ricardo Hydra Direct Injection Diesel Engine," IMechE Seminar Publication—Lean Burn Combustion Engines.

Bakenhus, M. and Reitz, R. (1999), "Two-Colour Combustion Visualisation of Single and Split Injections in a Single-Cylinder Heavy-Duty DI Diesel Engine Using an Endoscope-Based Imaging System," SAE Paper No. 1999-01-1112.

Baker, H.D. and Ryder, E.A., *Temperature Measurement in Engineering*, Vol. 2, John Wiley and Sons (1961).

Bertoli, C., Beatrice, C., Stasio, S., and Giacomo, N. (1996), "In-Cylinder Soot and NOx Concentration Measurements in DI Diesel Engine Fed by Fuels of Varying Quality," SAE Paper No. 960832.

Gaydon, A. and Wolfhard, H. (1970), *Flames*, Chapman and Hall, London.

Hottel, H. and Broughton, F. (1932), "Determination of True Temperature and Total Radiation from Luminous Gas Flames," *Ind. and Eng. Chem.*, 4-2, pp. 166–174.

Kawamura, K., Saito, A., Yaegashi, T., and Iwashita, Y. (1989), "Measurement of Flame Temperature Distribution in Engines by Using a Two-Colour High Speed Shutter TV Camera System," SAE Paper No. 890320.

Kobayashi, S., Sakai, T., Nakahira, T., Komori, M., and Tsujimura, K. (1992), "Measurement of Flame Temperature Distribution in DI Diesel Engine with High Pressure Fuel Injection," SAE Paper No. 920692.

Liebert, C.H. and Hibbard, R.R. (1970), NASA TN D-5647.

Lyn, W.T. (1957), "Diesel Combustion Study by Infrared Emission Spectroscopy," *J. Inst. Petrol.*, 43–398 p. 25.

Matsui, Y., Kamimoto, K., and Matsuoka, S. (1979) "A Study on the Time and Space Resolved Measurement of Flame Temperature and Soot Concentration in SI Diesel Engine by the Two-Colour Method," SAE Paper No. 790491.

Matsui, Y., Kamimoto, K., and Matsuoka, S. (1980), "A Study on the Application of the Two-Colour Method to the Measurement of Flame Temperature and Soot Concentration in Diesel Engines," SAE Paper No. 800970.

Matsui, Y., Kamimoto, K., and Matsuoka, S. (1982), "Formation and Oxidation Processes of Soot Particulates in a DI Diesel Engine—An Experimental Study via the Two-Colour Method," SAE Paper No. 820464.

Miyamoto, N., Ogawa, H., Arima, T., and Miyakawa, K. (1996), " Improvement of Diesel Combustion and Emissions with Addition of Various Oxygenated Agents to Diesel Fuels," SAE Paper No. 962115.

Nagase, K., Funatsu, K., Muramatsu, Y., and Kawakami, M.(1985), "An Investigation of Combustion in Internal Combustion Engines by Means of Optical Fibres," SAE Paper No. 851560.

Peterson, R. and Wu, K. (1986), "The Effect of Operating Conditions on Flame Temperature in a Diesel Engine," SAE Paper No. 861565.

Rossler, F. and Behrens, H. (1950), "Bestimung des Absorptionskoeffizieten von Russteilchen Verschiedenser Flammen," *Optik*, V. 6, p. 145.

Schack, A.Z. (1925), "Temperaturdifferenz zwischen Flame und Kohlenstoffteilchen," *Z. Techn. Physik.*, 6–10, pp. 530–540.

Shakal, J. and Martin, J. (1994), "Imaging and Spatially Resolved Two-Colour Temperature Measurements through a Coherent Fiberoptic: Observation of Auxiliary Fuel Injection Effects on Combustion in a Two-Stroke DI Diesel," SAE Paper No. 940903.

Shiozaki, T. et al. (1996), "The Analysis of Combustion Flame Under EGR Conditions in a DI Diesel Engine," SAE Paper No. 960323.

Siddall, R.G. and McGrath, I.A. (1963), "The Emission of Luminous Flames," *Proc. of 9th Symposium on Combustion*, pp. 102–110.

Uyehara, O.A., Myers, P.S, Watson, K.M., and Wilson, L.A. (1946), "Flame-Temperature Measurements in Internal Combustion Engines," *Trans. ASME*, Vol. 68, p. 17.

Uyehara, O.A., Myers, P.S, Watson, K.M., and Wilson, L.A. (1947), *"Diesel Combustion Temperatures—The Influence of Operating Variables,"* *Trans. ASME,* Vol. 69, pp. 465–477.

Winterbone, D., Yates, D., Clough, E., Rao, K., Gomes, P., and Sun, J. (1994), "Combustion in High-Speed Direct Injection Diesel Engines—A Comprehensive Study," *Proc. Instn. Mech. Engrs.,* Vol. 208, 223.

Yan, J. and Borman, G. (1988), "Analysis and In-Cylinder Measurement of Particulate Radiant Emissions and Temperature in a Direct Injection Diesel Engine," SAE Paper No. 881315.

Yates, D.A. and Gomes, P.C. (1993), Proc. of Experimental and Predictive Methods in Engine Research and Development, IMechE seminar publication.

Chapter 11

In-Cylinder Soot Measurements

Outline

11.1 Introduction

Diesel engines offer significant fuel economy advantages over other power plants used for road transportation. However, ever more stringent emission legislation is being introduced for diesel engines. To meet the upcoming legislation, various techniques have been developed to study soot formation and removal processes in a diesel engine.

The light extinction method is a simple and accurate way to measure the in-cylinder soot concentration. For soot particle sizing, both light scattering and extinction are used. The principle and implementation aspects of each technique are described first, followed by examples of results from the applications of these techniques. The descriptions of light scattering theory will be kept to minimum. For a complete description of light scattering from small particles, readers are directed to books by Van de Hulst [1957], Kerker [1969], and Bohren and Huffman [1988].

Laser induced incandescence is a relatively new but power tool for two-dimensional soot distribution measurements. Its principle and operational considerations will be discussed in Section 11.4. This is followed by a discussion of the optical techniques suitable for two-dimensional soot distribution measurements. Finally, applications of in-cylinder sampling techniques to in-cylinder soot analysis will be presented.

11.2 Soot Concentration Measurement by the Light Extinction Method

11.2.1 Principle of the Light Extinction Method

When a cloud of particles is placed in a beam of light radiation, as shown in Fig. 11.1, the intensity of the light reaching the detector will be reduced. In this case, we say that the presence of the particles has resulted in extinction (attenuation) of the incident light. The attenuation of light is caused both by scattering by the particles and by absorption in the particles, i.e.,

Extinction = Scattering + Absorption.

The light intensity after passing through the particle cloud is given by Lambert-Beer's law

$$I = I_0 \exp\left(-\int_0^L K_{ext} dx \right) \quad (11.1)$$

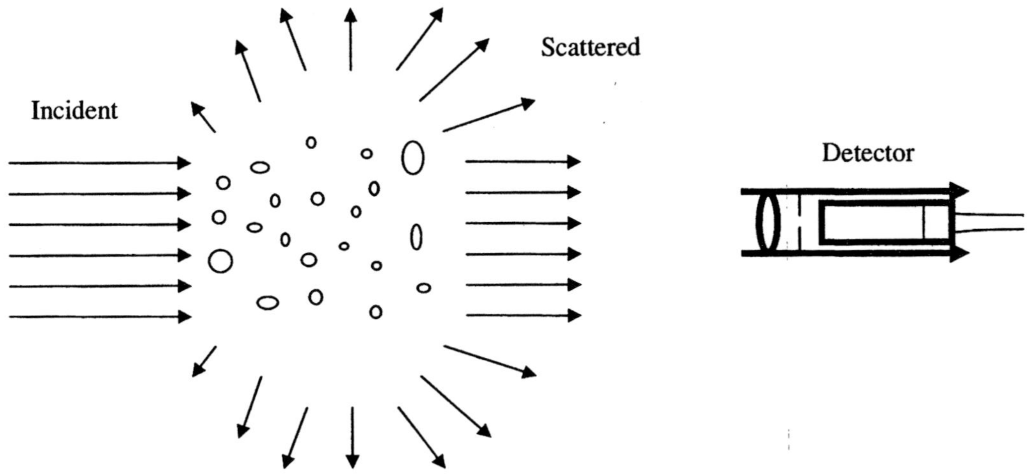

Fig. 11.1 Light extinction and scattering by a collection of particles.

where I_0 = the incident light intensity, $[W\ m^{-2}]$
K_{ext} = the extinction coefficient for a cloud of particles, $[m^{-1}]$
L = the path length of the light beam, [m]

From light scattering theory, the extinction coefficient for a cloud of particles is related to the number, size, and properties of individual particles by

$$K_{ext} = \frac{\pi}{4} C_n \int_0^\infty Q_{ext} N(D) D^2 dD \tag{11.2}$$

where C_n = the number density, $[m^{-3}]$
Q_{ext} = the extinction efficiency of a particle
D = the particle diameter, [μm]
$N(D)$ = the particle size distribution function defined such that $\int N(D)dD = 1$.

The value of extinction efficiency of each particle, Q_{ext}, is a complex function of its size and refractive index. The exact expressions for Q_{ext} do not exist except for some special cases. Perhaps the most important exactly soluble problem in the theory of absorption and scattering by small particles is that for a sphere of arbitrary radius and refractive index as described by the Mie theory [1908]. Various approximate techniques are available, which can lead to simpler expressions for calculation, provide physical insight and allow solutions for situations not open to rigorous analysis. One of the most common approximate methods is Rayleigh scattering. This applies to small particles with $x \ll 1$ and $x|m - 1| \ll 1$, where m is the complex refractive index, $x = \pi D/\lambda$ is the particle size parameter.

Soot particles, especially those encountered in diesel engine combustion, have a size distribution over a range from a few nanometers to 70 nm. The shape of the individual particle is roughly spherical, although non-spherical agglomerates may form in the exhaust, which may take the form of a long chain or a cluster. When visible light with wavelength ranging between 0.4 μm and 0.7 μm is used, sub-micron soot particles lie in the regime $x|m| << 1$. For example, at wavelength of 0.5 μm and a typical refractive index of $m = 1.6 - i0.6$, we find that $x \leq 0.063$ or $x|m| \leq 0.107$ for $D \leq 10$ nm. For larger soot particles with $D \leq 70$ nm, $x \leq 0.44$ or $x|m| \leq 0.75$. According to Kerker et al. [1978] the Rayleigh approximation is valid to within 1% for $x|m| < 0.2$ and to within 10% for $x|m| < 0.5$. Thus, we may apply the Rayleigh approximation to individual soot particles, but not to any but the smallest agglomerates. The agglomerates are not spherical so that, strictly speaking, the Mie theory cannot be used. However, most researchers have tended to use the spherical assumption because of its simplicity. Their results should be considered in that light. In addition, because the composition of soot varies, the refractive index is uncertain. This leads to errors in the calculation of scattering properties and the analysis of experimental results.

Within the Rayleigh scattering regime, the extinction efficiency is given by

$$Q_{ext} = 4x\,\mathrm{Im}\left(\frac{m^2-1}{m^2+2}\right) + \frac{8}{3}x^4\,\mathrm{Re}\left(\frac{m^2-1}{m^2+2}\right) \tag{11.3}$$

The first term in Eq. (11.3) is the absorption efficiency, Q_{abs}, and the second is the scattering efficiency, Q_{scat}. Thus, for Rayleigh scatterers, the scattering efficiency is proportional to x^4, whereas the absorption efficiency is proportional to x. Since $x << 1$ in the Rayleigh regime, absorption dominates where it exists at all, i.e., $Q_{abs} >> Q_{scat}$. Since soot is heavily absorbing, the extinction efficiency of a soot particle can be estimated accurately by

$$Q_{ext} = Q_{abs} = 4x\,\mathrm{Im}\left(\frac{m^2-1}{m^2+2}\right) \tag{11.4}$$

So, Lambert-Beer's law becomes

$$I = I_0 \exp\left(-\pi^2 \frac{L}{\lambda}\,\mathrm{Im}\left(\frac{m^2-1}{m^2+2}\right) C_n \int_0^{\infty} N(D) D^2 dD\right) \tag{11.5}$$

The volumetric density of particles, C_v (the volume of particles per unit volume), is related to the number density of particles, C_n (the number of particles in a unit of space) by

$$C_v = \frac{\pi}{6} C_n \int_0^\infty N(D) D^3 dD \tag{11.6}$$

which can be obtained from the extinction measurement by

$$C_v = -\frac{2}{3L} \frac{D_{32}}{Q_m} \ln\left(\frac{I}{I_0}\right) \tag{11.7}$$

where Q_m is the mean extinction efficiency for a polydispersed particle cloud

$$Q_m = \frac{\int_0^\infty N(D) Q(D) D^2 dD}{\int_0^\infty N(D) D^2 dD} \tag{11.8}$$

D_{32} is known as either the volume/surface mean diameter or the Sauter mean diameter

$$D_{32} = \frac{\int_0^\infty N(D)\, D^3 dD}{\int_0^\infty N(D)\, D^2 dD} \tag{11.9}$$

For particles of size parameter $x < 0.5$, the difference is less than 5% between the mean extinction efficiency, Q_m, defined by Eq. (11.8) and the extinction efficiency, $Q(D_{32})$, of a particle of Sauter mean diameter D_{32}, regardless of the size distribution function [Kontani and Gotoh, 1983 and 1986]. The error will be increased to 10% if the size parameter approaches 1.

Substituting Q_m with $Q(D_{32})$, the volumetric density of the particles (also known as the volume fraction), Eq. (11.7) becomes

$$C_v = -\frac{\lambda}{6\pi L \, \mathrm{Im}\left\{\frac{m^2-1}{m^2+2}\right\}} \ln\left(\frac{I}{I_0}\right) \tag{11.10}$$

It is obvious that the particle volume fraction can be calculated from the ratio of I/I_0 without any knowledge of particle size distributions. The only approximation will be the value of refractive index to be used. This could be the largest source of error as the refractive index varies with wavelength. Chang and Charalampopoulos [1990] has measured experimentally the variation of refractive indices with wavelength and compared them with the more widely quoted earlier measurements by Dalzell and Sarofim [1969] and the theoretical results of Lee and Tien [1981]. Followings are the refractive indices from Chang and Charalampopoulos, together with the value from Lee and Tien:

Wavelength $\lambda(\mu m)$	**Refractive Index** $m = n - ik$
0.337	1.48 – i 0.77
0.532	1.75 – i 0.61
0.633	1.80 – i 0.58
0.810	1.85 – i 0.57
1.064	1.91 – i 0.59

$m = n - ik = 1.90 - i0.55$; (Lee and Tien)

A simple calculation using Eq. (11.10) shows that the soot volume fraction will be 33% lower if the refractive index is changed from $m = 1.9 - i0.55$ to $m = 1.4 - i0.55$. Thus, the value of refractive index has a significant effect on the measured value of soot concentration.

If the particle density is known or assumed, the soot mass concentration C_m can then be obtained by $C_m = C_v\rho$.

11.2.2 Implementation of the Light Extinction Method

The setup of the light extinction measurement is rather straightforward. All that is required is a laser and a photodetector, as shown in Fig. 11.2. A very narrow-band interference filter (FWHM [full width at half maximum] = 1.0 nm) is normally placed in front of the photodetector to suppress the influence of flame radiation on the measured light intensities. Using such a filter, the flame radiation reaching the detector can be reduced to an amount on the order of 0.1% of the laser beam. Light deflection due to hot combustion gases can be minimized to a large extent by using a large detection area detector and a focusing lens placed between the detector and measurement volume. The interference of scattering with extinction is quite a common source of error when the light extinction measurement is carried out for large particles. It is less a problem for measurements involving soot particles, because the first lobe of scattering maximum is quite far from the optical axis. In addition, in the case of soot, absorption far outweighs scattering. For soot particles with a diameter $D < 0.1$ μm, or $x < 0.5$ in the visible, using Eq. (11.14) given in the Section 11.3.1, the estimated ratio of the absorbed power to that scattered arriving the detector is greater than 10,000 if a 1 cm aperture is used at a distance of 1 m. Thus, in the case of soot, aperture size may be neglected as a problem.

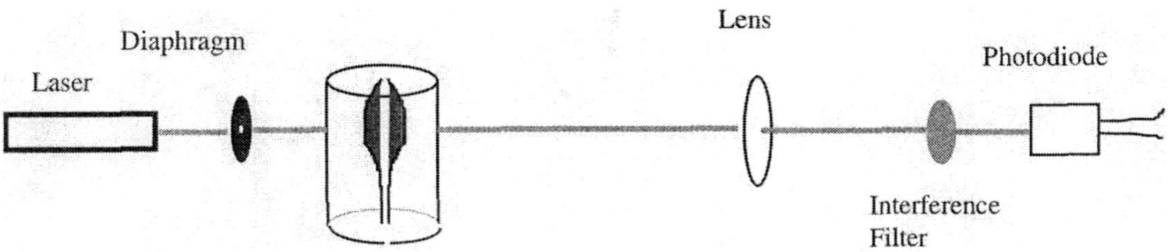

Fig. 11.2 Schematic of the setup for the light extinction measurement.

Since light can be attenuated due to scattering and absorption by molecules and particles, the total transmitted light through the combustion chamber suffers from extinction by air, dust, and any liquid droplets, as well as soot particles. This is illustrated in Fig. 11.3. When the engine is motored, the light intensity is slightly reduced due to Rayleigh scattering from compressed air in the cylinder. If the engine is motored with ambient air, the second peak of the dotted line may appear during the expansion stroke because of the presence of water droplets from condensation of water vapor in the air [Hentschel and Richter,1995]. Another source of interference with light extinction measurements is the liquid fuel droplets, as shown in Fig. 11.3. It is possible to obtain the light extinction curve (dashed line) due to liquid fuel droplets in a nitrogen motored engine with fuel injection. If the fuel droplet curve is subtracted from the measured total extinction curve, however, the soot concentration will be underestimated at the beginning of the measurements because the actual fuel vaporization process will be faster in a firing engine. Hentschel and Richter [1995] found that the vaporization was normally completed a few degrees CA after the end of injection under the high pressure and temperature environment in a modern diesel engine. It can therefore be assumed that extinction soon after the end of the injection is caused by soot particles only.

11.2.3 Application of the Light Extinction Method to Diesel Engines

11.2.3.1 Single Point Light Extinction Measurements

The light extinction method was used by Wiartalla et al. [1995] to study soot formation in a high pressure and temperature combustion chamber. The measurements were carried out in a matrix of locations in the fuel spray. The mass of soot in different layers below the nozzle, as

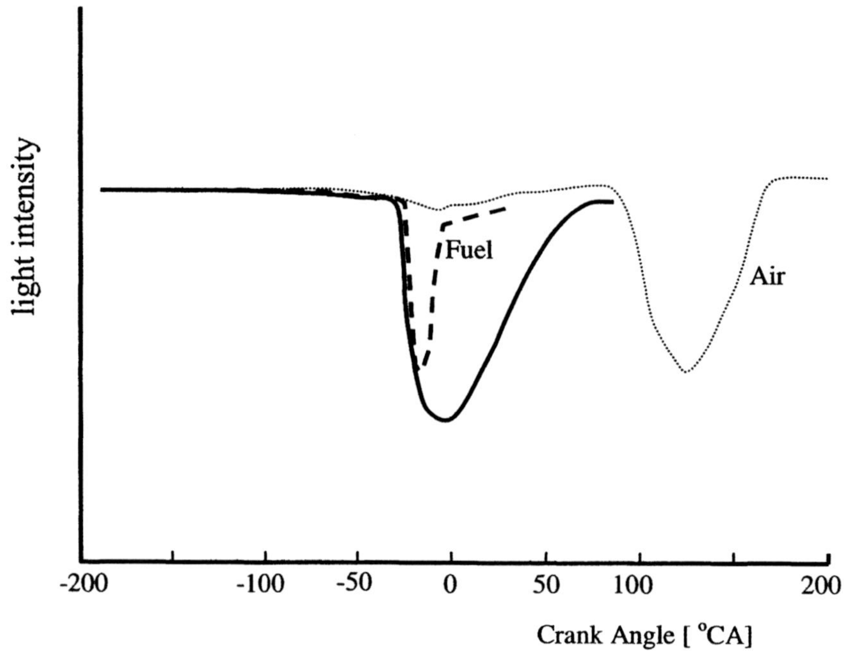

Fig. 11.3 Sources of interference in the light extinction measurement in a diesel engine

well as the total mass of soot in the spray, was then calculated as a function of time from the measured data. The effect on soot formation of injection system parameters, such as injection pressure, injection rate, nozzle hole diameter, and spray injection angle, was investigated. They found that higher overall injection rates led to much less soot formation in the spray, but high injection rates in the early phase of injection lead to higher soot formation in the spray. A large overlapping of injection and combustion caused strong secondary soot formation, indicated by a second peak of high soot mass concentration close to the injection nozzle. From these observations, the authors suggested that an optimum injection history should be characterized by an increasing injection rate during the injection process and a sharp cutoff of injection. In their calculation, a refractive index $m = 1.90 - i0.45$ and a soot density of 1.8 g/cm^3 were used.

Hentschel and Richter [1995] applied the light extinction technique to a high-speed DI diesel engine. Laser beams from an He-Ne laser and an argon ion laser were introduced into the combustion chamber via a bifurcated optical fiber. The transmitted laser beams were then separated back to the two wavelengths by another bifurcated fiber. A combination of a narrow bandpass filter, short pass filter, and neutral density filter was employed in each path to suppress the flame radiation. Outputs from the two photodiodes were recorded by a transient recorder. They found that increasing EGR extended the ignition delay period and delayed soot formation and oxidation. The peak soot concentration was found a few crank angle degrees after autoignition, which increased with increasing EGR rates. The soot oxidation rate was reduced with increasing amount of EGR in the cylinder due to reduced combustion temperature and oxygen concentration.

In addition to carrying out theoretical analysis on the light extinction method, Kontani and Goto [1983; 1986] attempted to measure the size of soot particles using the spectral extinction or two-wavelength dispersion quotient method, which assumes that the ratio of extinction efficiency at two wavelengths ($Q_{ext}(\lambda_1) / Q_{ext}(\lambda_2)$) is a simple function of particle size. Unfortunately in the Rayleigh regime, $Q_{ext}(\lambda_1)/Q_{ext}(\lambda_2)$ is independent of particle size. Therefore, no information on particle size could be retrieved from their spectral extinction measurements.

11.2.3.2 Visualization of In-Cylinder Soot Distribution by Light Extinction

In the previous section, the soot volume fraction is measured along a line-of-points on the narrow laser beam through the soot cloud inside the combustion chamber. A matrix of measurement points is required to build up an image of soot distribution, which is tedious and time-consuming. In addition, because measurements have to be carried out at different times, only ensemble-averaged soot distributions can be measured. Therefore, it is desirable to measure the soot volume fraction at multiple points simultaneously.

It seems logical to think that infinite numbers of extinction measurements can be accomplished by means of an imaging device and an expanded light source, as shown in Fig. 11.4. An expanded parallel beam passes through the soot cloud before they reach an imaging device. The light intensity will be reduced in the region of the soot cloud due to extinction by soot particles. This setup is very similar to that used for Schlieren or shadowgraphy. However, the system shown in Fig. 11.4 measures the light attenuation due to extinction only when the soot particles are suspended in a quiescent and isothermal environment. However, when there are variations of the refractive index in surrounding gases, part of the beam will be deflected. Consequently, the Schlieren effect will be recorded, as well as extinction caused by soot particles.

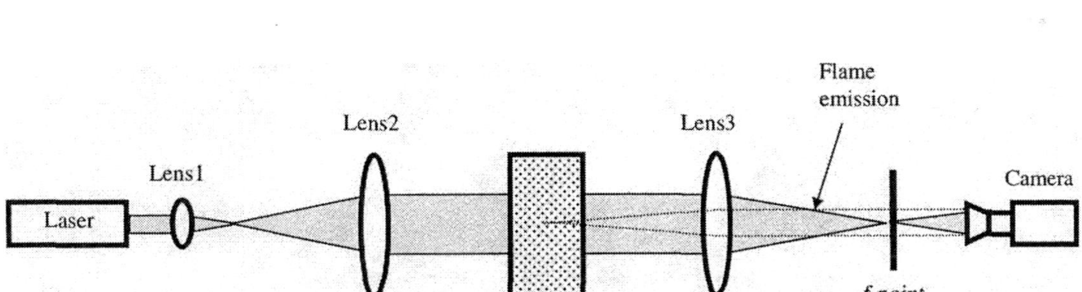

Fig. 11.4 Idealized setup for infinite numbers of light extinction measurements.

The effect of changes in refractive index or the Schlieren effect on soot visualization was studied by Nakakita et al. [1990]. They compared three optical setups for soot visualization. The results are shown in Fig. 11.5. The images were taken from the swirl chamber of an IDI diesel engine, which had two quartz windows at each end, by a high-speed cine camera at 4800 frames per second. The images in the top row (Fig. 11.5[a]) were obtained with a Schlieren system, which was similar to the optical setup shown in Fig. 11.4 and comprises a 6 mm diameter circular stop at the focal point. These images show fuel sprays, fuel/air mixture, flame-front, and luminous flame regions. But it was impossible to tell where the soot cloud is.

Fig. 11.5(b) shows the photographs recorded without the circular stop at the focal point, i.e., they are shadowgraphs of the mixture in the swirl chamber. With the removal of the circular stop, the sensitivity of those shadowgraphs is lower than that of the Schlieren photographs. The density gradients in the mixture are still visible. Hence, dark regions in these images at 2 and 4 °CA ATDC are due to the density gradients in the mixture as well as attenuation by the soot cloud. It is still difficult to separate the soot cloud from changes in density gradients.

Fig. 11.5(c) shows the images taken with a uniformly diffused light through a diffuser (a frosted glass plate), using the experimental setup shown in Fig. 11.6. Flame emission was block by a narrow bandpass filter. In setting up the system in Fig. 11.6, the camera was

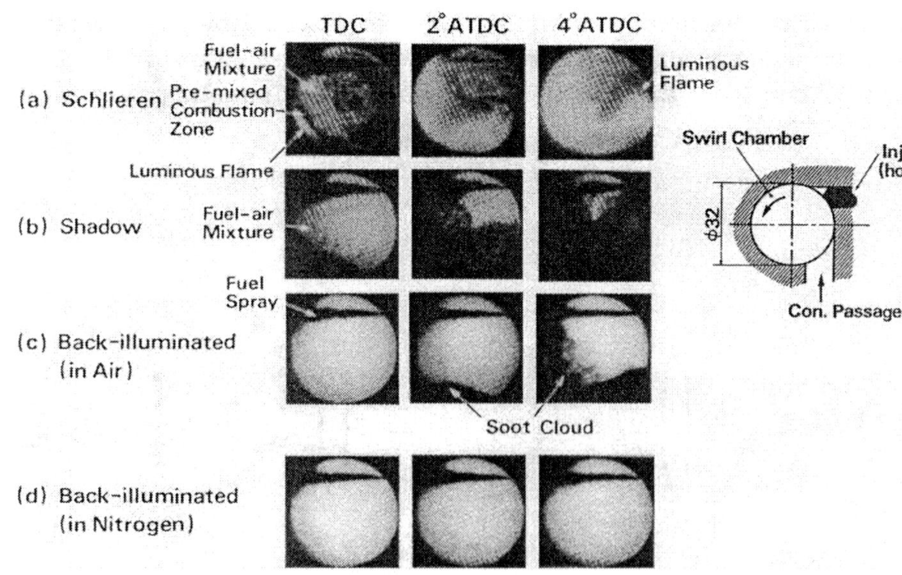

Fig. 11.5 Schlieren, shadowgraph, and light-extinction photographs of diesel combustion [Nakakita et al., 1990].

focused to the measurement section so that the shadowgraphic effect was minimized. The photographs with this back-illuminated method do not show the variations cased by the density gradients or temperature gradients seen in Fig. 11.5(a) and Fig. 11.5(b). The Schlieren effect caused by the change in the refractive index of the mixture was suppressed by the use of diffused light. The dark areas are therefore caused by extinction alone. Further comparison with photographs taken in the swirl chamber with a nitrogen motored engine reveals that the top shadow was due to the extinction by liquid fuel droplets and the other shadow areas were caused by the presence of a soot cloud.

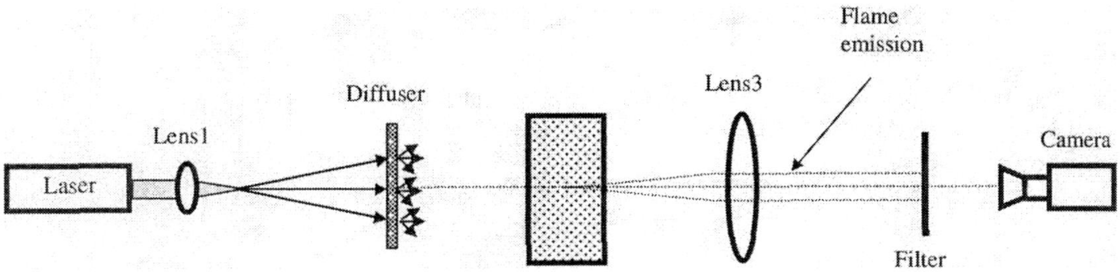

Fig. 11.6 Optical setup for soot visualization with back-illumination.

Having established the appropriate optical setup for soot visualization, Nakakita and his co-workers examined the soot formation process in the swirl chamber of a two-stroke diesel engine with a single-hole injector. They found that the main soot area was much smaller than the luminous flame area, particularly in the initial soot formation process. The main soot cloud first appeared in the tip region of fuel/air mixture downstream of the ignition position immediately after ignition, and was then carried away by the gas flow. The soot cloud formed in the high temperature region at the outer region of the fuel-rich zone.

Earlier work on soot visualization was reported by Greeves and Meehan [1975]. Their experimental setup is shown in Fig. 11.7. A lens collimated the UV light from a mercury arc lamp. The beam was slightly converging. Three images were recorded simultaneously on a black and white film by a high-speed cine camera. The three images included: (1) a UV image of the step wedge for calibration, (2) a UV absorption image of the soot cloud, and (3) an image of the luminous flame. The optical density at each point of the UV absorption image was measured by a micro-densitometer. The attenuation of light was then obtained by comparison with that on the UV calibration image. The third image provided useful information on the location of flame. The optical setup was applied to the swirl chamber of a Ricardo E16 engine operating at 366 rpm. Limited results were reported on the spatial distribution of soot in the swirl chamber with a single-hole injector. Because their experimental setup for soot concentration measurements was essentially the same as that for shadowgraph, it is likely that their results could have been affected by the density and temperature gradients present in the combustion mixture.

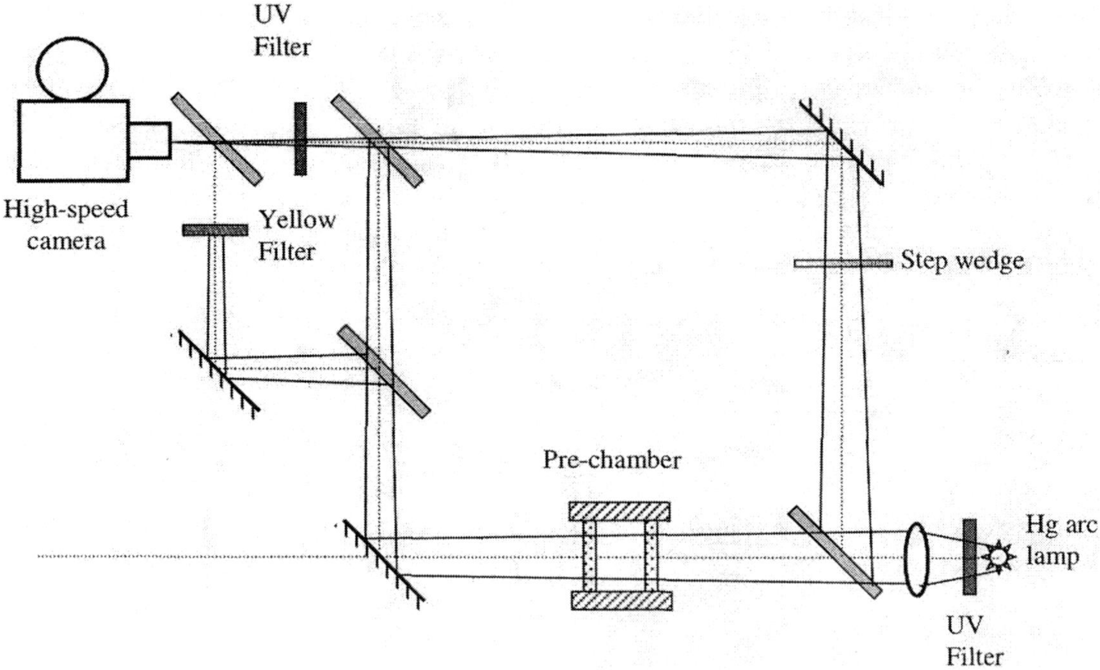

Fig. 11.7 Optical setup for simultaneous combustion imaging and soot extinction measurements [Greeves and Meehan, 1975].

Chang et al. [1985] studied soot formation and combustion in a rapid compression machine. Both quantitative point measurement of soot concentration by the light extinction method and soot visualization were used. The soot visualization was realized by a Schlieren system. The Schlieren stop used was a circular aperture. They found that the sensitivity of the system to density gradient in the flame was minimized when the diameter of the aperture was increased to 10 mm. Their experimental results showed that the soot particles first appeared near the tip of the flame and reached its peak value immediately after injection, which was similar to what Nakakita et al. observed. They found that the soot was concentrated in the high temperature but fuel-lean region, which was attributed to the fact that evaporated fuel required time to be pyrolyzed as it traveled through the burning fuel-rich region toward the flame tip.

Both Kontani and Gotoh [1986], and Miyamoto et al. [1990] studied soot formation using an optical setup similar to that shown in Fig. 11.4. A small aperture was placed at the focal point, which rendered the setup essentially a Schlieren system. As discussed at the beginning of this section, such a setup will lead to images including both the Schlieren effect and attenuation due to the soot cloud. Their soot concentration results, therefore, could have been affected.

A two-dimensional light extinction method was also reported by Bäcker and Wiartalla [1994]. A back-illuminated system similar to that shown in Fig. 11.6 was employed. To compensate for the loss in the light intensity by the presence of a diffuser screen, an argon ion laser (2 W) was used as the light source. The transmitted light was filtered by a bandpass filter to suppress the influence of flame radiation on the measured light intensity, and it was recorded with a high-speed cine film camera. The film was then digitized with a video camera and the transmission was calculated for each pixel by normalizing the intensity after the start of injection to the intensity just before the start of injection.

11.3 Soot Particle Sizing by Scattering and Extinction

11.3.1 Principle

Since light scattering from soot particles falls in the Rayleigh regime, the scattering polar diagram (scattering intensity as a function of scattering angle) is not sensitive to size. Therefore, optical particle sizing methods based on the scattering intensity variation cannot be used. Neither can information be obtained from wavelength variation, such as spectral extinction or the two-wavelength dispersion method where one can infer particle size from the ratio of extinction coefficients at two different wavelengths. In such circumstances, Van de Hurst [1957] proposed a method based on the ratio of scattering to extinction from small particles. The principle of the scattering and extinction method is as follows.

In the presence of a particle, the scattering intensity is given by

$$I_{sca} = \frac{3I_0}{8\pi r^2} K_{sca} (\sin^2 \phi + \cos^2 \theta \cos^2 \phi) \tag{11.11}$$

where ϕ = the angle between the plane of measurement and the direction of polarization of the incident wave,

θ = the scattering angle,

r = the distance between the measurement point and the detector.

For Rayleigh scatters ($x << 1$) the scattering coefficient K_{sca} is given

$$K_{sca} = \frac{2}{3}\frac{\pi^5}{\lambda^4}\left|\frac{m^2 - 1}{m^2 + 2}\right|^2 \int_0^{\infty} D^6 C_n N(D)\, dD \tag{11.12}$$

To find the power received at the detector, P_{sca}, I_{sca} must be integrated over the aperture of the detector, i.e.,

$$P_{sca} = V \int_A I_{sca} dA$$

where V = scattering volume;
A = detector area.

From Eq. (11.1), the power detected by an extinction detector is

$$P_{ext} = AI_0 \exp\left(-\int_0^L K_{ext} dx\right)$$

As discussed in the Section 11.2.1 (Eq. [11.3]), for Rayleigh scatters, the scattering efficiency is proportional to x^4, whereas the absorption efficiency is proportional to x. Since soot is heavily absorbing and $x << 1$, the extinction is dominated by absorption

$$K_{ext} = K_{abs} = \frac{\pi^2}{\lambda} \mathrm{Im}\left(\frac{m^2-1}{m^2+2}\right)^2 \int_0^\infty D^3 C_n N(D) dD \tag{11.13}$$

Both K_{sca} and K_{ext} contain the unknown soot concentration, C_n, which can be eliminated by taking the ratio K_{sca}/K_{ext},

$$K_{sca}/K_{ext} = \frac{2}{3}\left(\frac{\pi}{\lambda}\right)^3 \frac{\left|\frac{m^2-1}{m^2+2}\right|^2 \int_0^\infty D^6 N(D) dD}{\mathrm{Im}\left|\frac{m^2-1}{m^2+2}\right| \int_0^\infty D^3 N(D) dD} \tag{11.14}$$

Introducing a measure of the particle size, D_{63}

$$D_{63} = (\bar{D}^6 / \bar{D}^3)^{1/3}$$

with

$$\bar{D}^6 = \int_0^\infty D^6 N(D)\, dD$$

$$\bar{D}^3 = \int_0^\infty D^3 N(D) dD$$

Then, Eq. (11.14) becomes

$$K_{sca}/K_{ext} = \frac{2}{3}\left(\frac{\pi}{\lambda}\right)^3 \frac{\left|\frac{m^2-1}{m^2+2}\right|^2}{\text{Im}\left|\frac{m^2-1}{m^2+2}\right|} \frac{\bar{D}^6}{\bar{D}^3} = \frac{2}{3}\left(\frac{\pi}{\lambda}\right)^3 \frac{\left|\frac{m^2-1}{m^2+2}\right|^2}{\text{Im}\left|\frac{m^2-1}{m^2+2}\right|} D_{63}^2 \tag{11.15}$$

Therefore, a measure of size, D_{63}, can be obtained from the measurement of the scattered light and light extinction. This method will give the true size for a monodispersed cloud of particles. However, it will yield a measure of the size that is biased by the size distribution when the particle cloud is polydispersed. There is not enough information to recover the size distribution function. A measure of the soot concentration can be obtained from either the extinction or scattered power, using D_{63} as the particle size. In any case, the measured concentration $C_{n,sca}$ from the scattered power or $C_{n,ext}$ from extinction is different from the true concentration C_n because of

$$K_{sca} \propto C_n \bar{D}^3 = C_{n,sca}(D_{63})^6$$

$$K_{ext} \propto C_n \bar{D}^3 = C_{n,ext}(D_{63})^6$$

The other factor that affects the accuracy of this method is the refractive index, m, because the ratio of scattering to extinction involves

$$\left|(m^2-1)/(m^2+1)\right|^2 / \text{Im}[(m^2-1)/(m^2+2)]$$

For the range of refractive indices suitable for soot as listed previously, the ratio can vary by a factor of 2 or more and is an important unknown.

The scattering and extinction method has been mostly used to study soot formation in flames. A typical apparatus for light scattering and extinction measurements is shown in Fig. 11.8. The light source is commonly a laser because of its high brightness, excellent collimation, and high degree of linear polarizability. The transmitted beam passes to an extinction detector. Due to temperature gradients and movement of the gas, the transmitted beam is continually in motion. A simple lens placed in front of the detector is used to minimize the movement of the transmitted beam. In this way, deflections arising at the flame are brought back to the optic axis of the system, and the laser beam always arrive at the same point on the detector. The use of a lens also concentrates the flame radiation onto the detector.

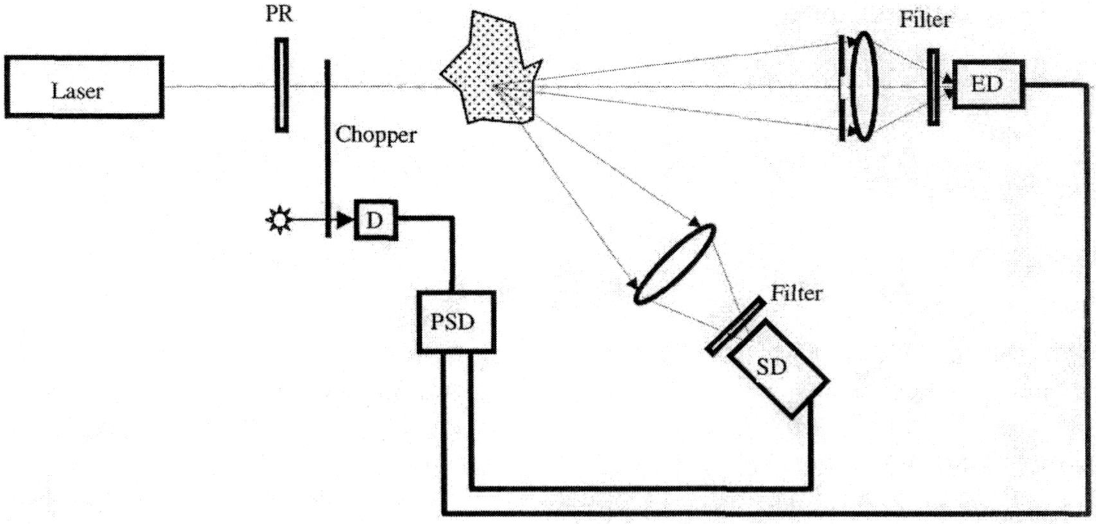

PR: polarization rotator; SD: scattering detector; ED: extinction detector;
D: photodetector; PSD: phase sensitive detection.

Fig. 11.8 General optical setup for light scattering and extinction measurements.

The scattered light is collected by a lens and focused onto a detector at a small angle to the forward direction. The scattering volume is determined by the intersection of the laser beam and the image of the aperture placed in front of the detector. In order to increase the signal-to-noise ratio of the scattering signal, phase-sensitive detection can be used by means of an optical chopper and lock-in amplifiers.

11.3.2 Adaptation and Application of the Light Scattering and Extinction Technique to a Diesel Engine

To apply the scattering and extinction method with the optical setup shown in Fig. 11.8 in a diesel engine normally requires three optical windows. Tree and Foster [1994] designed an ingenious optical probe that was capable of transmitting laser light and receiving the scattered light simultaneously, dispensing with the need for three windows. The use of such a probe also avoided the beam steering problem associated with transmitting and receiving optics. The schematic of the probe used in shown in Fig. 11.9.

The output from an argon ion laser (488 nm + 514 nm) was introduced into the cylinder through a transmitting window in the probe via an optical fiber (Fig. 11.9). Flame radiation and scattered light were collected through a sapphire window and transmitted into a trifurcated optical fiber bundle which split light into three equal portions. Light in each fiber then passed through filters of 488 ± 1 nm, 514 ± 1 nm, and 850 ± 25 nm, respectively. The scattered laser light at

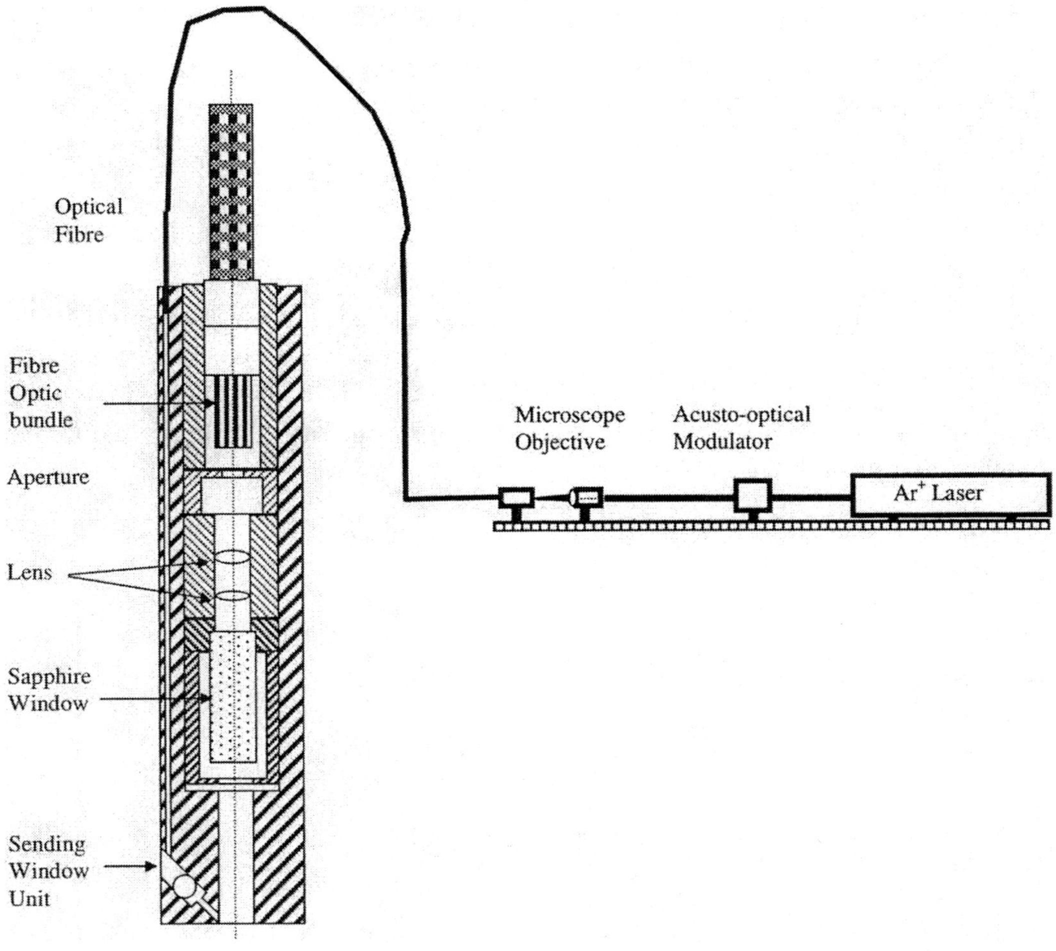

Fig. 11.9 The optical probe and experimental setup for combined light scattering/extinction and scattering/radiation measurements in a diesel engine [Tree and Foster, 1994].

488 nm and 514 nm was detected by photomultiplier tubes, while the near IR radiation at 850 nm was measured with a photovoltaic cell. The measurement volume for light scattering was defined by the intersection of the laser beam and the viewing cone angle, which was approximately 5 mm long and 1.5 mm wide. It should be noted that flame radiation was detected from a volume along the line-of-sight as defined by the cone angle, which was larger than the scattering volume.

Two different methods were used by Tree and Foster in order to obtain the diameter, number density, and soot volume fraction in the combustion chamber of a DI diesel engine. The first was the light scattering/extinction method and the second was the radiation/scattering method.

11.3.2.1 The Scattering and Extinction Technique

When scattered light is detected from the measurement volume within a cloud of soot particles, as shown in Fig. 11.9, the incident light intensity will be reduced by extinction on its path in, and the scattered light will be similarly reduced on its path out. If this extinction effect is taken into account, then the scattered power detected is given by

$$P_{sca} = V \exp\left(-\int_{L_1} K_{ext} dx\right) \exp\left(-\int_{L_2} K_{ext} dx\right) \int_A I_{sca} dA \tag{11.16}$$

where L_1 and L_2 are the path lengths in and out, and it is understood that I_{sca} is the scattered intensity for the incident intensity I_0.

Hence, the scattered power detected at each laser wavelength is

$$P_{sca,\lambda_1} = V \exp\left(-\int_L K_{ext,\lambda_1} dx\right) \int_A I_{sca,\lambda_1} dA \tag{11.17}$$

$$P_{sca,\lambda_2} = V \exp\left(-\int_L K_{ext,\lambda_2} dx\right) \int_A I_{sca,\lambda_2} dA \tag{11.18}$$

where L is the optical path length as defined by the intersection volume of the laser beam and the viewing cone angle. For monodispersed soot particles, using the scattering and extinction coefficients given by Eqs. (11.12) and (11.13), Equations (11.17) and (11.18) become

$$P_{sca,\lambda_1} = C_{cal} f(m) \exp\left(-C_n D^3 L/\lambda_1\right) C_n D^6 / \lambda_1^4 \tag{11.19}$$

$$P_{sca,\lambda_2} = C_{cal} f(m) \exp\left(-C_n D^3 L/\lambda_2\right) C_n D^6 / \lambda_2^4 \tag{11.20}$$

where C_{cal} = calibration constant, including the transmission characteristics of optics and the sensitivity of the detector,

$f(m)$ = a function of the refractive index of soot, given by

$$f(m) = \exp\left(-\mathrm{Im}\left(\frac{m^2-1}{m^2+2}\right)\right)\left|\frac{m^2-1}{m^2+2}\right|^2 \tag{11.21}$$

From Eqs. (11.19) and (11.20), the size of the soot particles can be obtained by measurement of the scattered light at two laser wavelengths. If the calibration constant, C_{cal}, is known, the number density of soot, C_n, can also be calculated. The calibration can be done by placing the optical probe in front of water before and after each engine measurement.

The light scattering and extinction method will only works when there is a sufficient amount of soot in the measurement volume. If the soot cloud were completely opaque, there would be no scattering. On the other hand, there would not be enough extinction if the soot cloud were transparent. Fortunately, during the combustion period in a diesel engine, the amount of soot allows both scattering and extinction to produce enough signals.

11.3.2.2 The Radiation and Scattering Technique

In this method, two radiation signals are detected together with the scattering signal from one of the laser beams. The scattering signal includes both scattering and extinction from the laser beam. The two radiation signals are given by Planck's law

$$P_{rad,\lambda_1} = \varepsilon_\lambda \frac{C_1}{\lambda_1^5\left[\exp(C_2/\lambda_1 T)-1\right]} \tag{11.22}$$

$$P_{rad,\lambda_2} = \varepsilon_\lambda \frac{C_1}{\lambda_2^5\left[\exp(C_2/\lambda_2 T)-1\right]} \tag{11.23}$$

From $\varepsilon_\lambda = 1-\exp\left(-K_{ext}L/\lambda^a\right)$ and Eq. (11.13), the above equations can be generalized as

$$P_{rad,\,\lambda 1} = f(C_n\, D^3/\,\lambda_1,\, \lambda_1 T) \tag{11.24}$$

$$P_{rad,\,\lambda 2} = f(C_n\, D^3/\,\lambda_2,\, \lambda_2 T) \tag{11.25}$$

From two radiation signals the flame temperature and quantity KL_{eff} ($\sim C_n D^3 L_{eff}$) can be determined using the two-color method described in Chapter 10. The effective optical length, L_{eff}, is defined by the receiving optics and is less than the distance from the cylinder head to the piston. If this effective length is known, the soot diameter and number density can be found from Eqs. (11.20), (11.24), and (11.25).

Tree and Foster estimated the value of L_{eff} from the radiation and scattering signals. The effective optical length, L_{eff}, was approximated by matching the radiation/scattering soot volume fraction at a crank angle of peak soot concentration and was found to be 0.6 times the length from the piston to the cylinder head. This was only done at one engine condition.

11.3.2.3 Implementation

The optical probe was positioned in place of one of the exhaust valves in a single-cylinder direct injection diesel engine. The laser beam was pulsed on and off every other cycle. This allowed every other cycle to produce radiation signals only when the cycles in between contained both radiation and scattering signals. An ensemble-average of 200 cycles was taken, 100 cycles with the beam on and the other 100 cycles with the beam off. An ensemble-averaged scattering signal was obtained by subtracting the ensemble-averaged value with the beam off from that with the beam on. As a result, three sets containing the ensemble-averaged radiation signals at the three wavelengths of 488, 514, and 850 nm and two sets of scattering signals of the soot particles at wavelengths of 488 and 514 nm were generated. The soot particle diameter, number density, and soot volume fraction were determined from these five sets of data by the scattering/extinction and radiation/scattering methods.

11.3.2.4 Results

Fig. 11.10 shows the results for soot volume fraction, soot diameter, and number density at 1300 rpm and 0.5 equivalence ratio. The flame temperature is also shown as calculated from the radiation signals. Analysis of the raw data by Tree and Foster suggested that the soot diameter calculation was only valid in the region between 4 and 16 °CA ATDC. The large soot volume fraction occurring between −11 and 4 °CA ATDC in the scattering and extinction method was attributed to interference from fuel droplets and low signal-to-noise ratios.

As shown in Fig. 11.10, between 4 and 16 °CA ATDC the soot diameter increased slightly to a peak then dropped rapidly. As the soot diameter peaked, the soot volume fraction dropped slowly and continuously. Detailed examination of the figure reveals that the diameter was decreasing in size very rapidly between 11 and 16 °CA ATDC even though there were still large measurable soot volume fractions, indicating a change in soot diameter and not simply a lack of soot particles.

Also shown in Fig. 11.10 is the soot number density. Owing to the high level of uncertainty involved in the result, the soot number density results were not considered in the paper by Tree and Foster.

In addition to the detailed study of the sooting process at this speed and equivalence ratio, measurements were also made at engine speeds of 1000 rpm and 1700 rpm and equivalence ratios 0.2 to 0.5. The measured peak soot diameter was found to be in the range of 30–50 nm. Soot volume fraction was found to be in the range of $2.5–6.5 \times 10^{-5}$. The peak soot diameter, soot volume fraction, and soot temperature at the measurement location were found to increase as equivalence ratio increased, although the effect of speed was inconclusive.

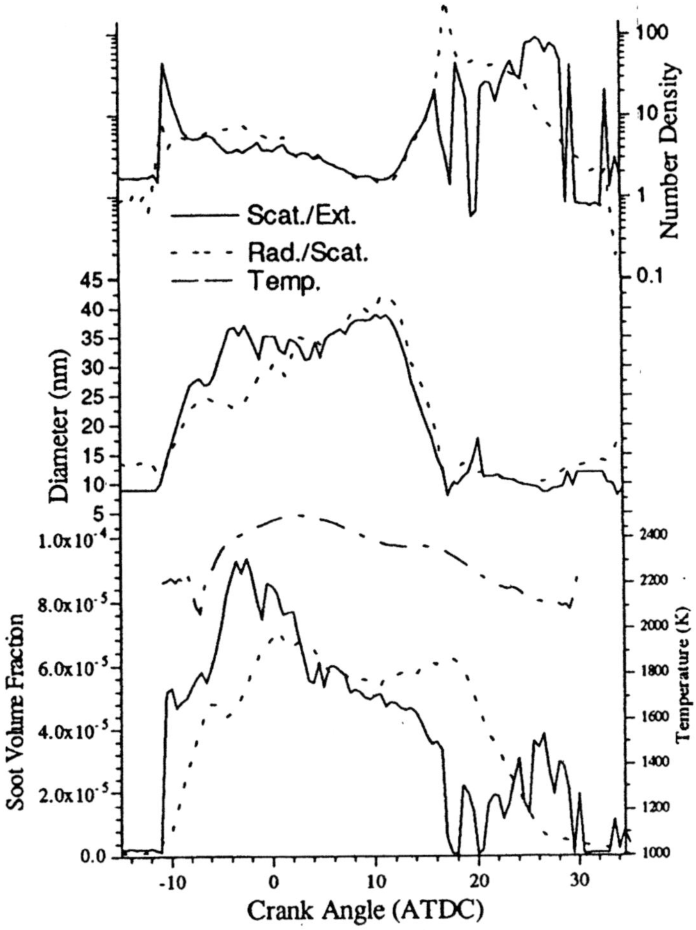

Fig. 11.10 Soot volume fraction, diameter, number density, and temperature at 1300 rpm, f = 0.5 equivalence ratio [Tree and Foster, 1994].

There were two sources of errors that rendered the scattering and extinction method more quantitative than the radiation and scattering method. One was the approximation of the effective optical length in the radiation measurement. The other was that radiation from soot outside the scattering measurement volume was detected.

The uncertainty in the absolute soot diameter measured was estimated to be approximately 30% plus the uncertainty caused by the unknown refractive index, using the scattering and extinction method. The radiation and scattering method involved an assumed path length in calculating the soot volume fraction and soot diameter, as well as the uncertainties in the scattering and extinction method.

11.4 Laser Induced Incandescence (LII)

11.4.1 Introduction

The light extinction method described in Section 11.2 measures the soot concentration integrated along the line-of-sight, and hence lack spatial resolution. Uneven soot distribution along the line-of-sight cannot be resolved. This difficulty can be overcome by means of laser sheet illumination. Two different types of laser-sheet or two-dimensional soot imaging techniques have been developed and used to obtain planar soot distributions in both optically accessible diesel engines and in diesel engine-like environments. The two techniques are laser induced scattering (LIS) and laser induced incandescence (LII). The principle of LIS from soot particles has been discussed in Section 11.3. In the following sections, the principle and implementation of LII will be discussed in detail. Visualization of two-dimensional soot distribution by LIS and LII in diesel combustion systems will be considered in Section 11.5.

11.4.2 Principle of Laser Induced Incandescence

Laser induced incandescence was first noticed by Eckbreth [1979] as an interference signal in coherent anti-Stokes Raman scattering (CARS) experiments in particle-laden flames.

Laser induced incandescence (LII) is generated when a cloud of soot is irradiated with an intense laser light. The soot particles are heated to a temperature well above the surrounding gas temperature due to the absorption of laser energy. The heated soot particles subsequently emit blackbody radiation corresponding to the elevated soot particle temperature. If laser fluences are high enough, the vaporization of small carbon fragments, typically C2 and C3, from the surfaces will take place once the vaporization temperature of carbon of 4000 K is reached. Both laser induced incandescence and laser induced fluorescence from the vaporized material have been suggested previously by Dasch [1984a, 1984b, and 1984c], Melton [1984], and Bengtsson and Alden [1995] to study soot volume concentrations in flames. LII is more widely used because it is less susceptible to spurious signals, e.g., those from other channels of C2 excitation or subtle changes in the nature of the vaporized material, which may be associated with soot particle aging. It is also observed that the LII signal, in general, dominates over that from fluorescent emission of the vaporized carbon.

11.4.2.1 Energy and Mass Balance Equations

The absolute temperature of the heated soot particles is determined by the rate of laser energy absorbed, the rate of conductive heat transfer to the surrounding medium, soot vaporization, and the energy loss due to blackbody radiation. Both Dasch [1984a] and Melton [1984] gave derivations for laser induced vaporization with a number of simplifications. A more thorough theoretical analysis of the laser induced incandescence technique was given by Hofeldt [1993]. The analysis is based on a transient energy balance on a particle, which is small compared to the wavelengths of absorbed and emitted radiation,

$$m_s \frac{d(C_s T_s)}{dt} - \frac{H_v}{M_v}\frac{dm_s}{dt} = I_o C_{abs} - hA_s(T_s - T_\infty) - \int_0^\infty A_s \varepsilon_{s,\lambda} E_{b,\lambda}(T_s)d\lambda + \int_0^\infty C_{abs}\varepsilon_{w,\lambda} E_{b,\lambda}(T_w)d\lambda \tag{11.26}$$

where m_s = mass of the soot particle, [kg]
C_s = specific heat of the soot particle, [J/(kg·K)]
T_s = temperature of the soot particle, [K]
H_v = enthalpy of vaporization of the soot particle, [J/mole]
M_v = molecular weight of vaporized material
I_0 = laser intensity, [W/m^2]
C_{abs} = absorption cross section, [m^2]
h = heat transfer coefficient, [W/(m^2·K)]
A_s = surface area of the soot particle, [m^2]
= πD^2
T_∞ = temperature of the surrounding medium, [K]
$\varepsilon_{s,\lambda}$ = spectral emissivity of the soot particle
$E_{b,\lambda}$ = blackbody spectral irradiance, [W/(m^2·μm)]
$\varepsilon_{w,\lambda}$ = spectral emissivity of the wall

Starting from the left-hand side, the terms in Eq. (11.26) are, respectively, the energy storage, the heat of vaporization, the laser power absorbed, the rate of heat transfer to the medium, the rate of energy loss by radiation, and the rate of radiative absorption from surroundings.

If radiation from the walls is neglected, using appropriate kinetic theory, Eq. (11.26) can be rewritten as

$$\frac{dT_s}{dt} = \frac{I_0 A_{abs}}{C_s} - \frac{6h(T_s - T_0)}{\rho_s C_s D_s} - \frac{6N_v H_v}{\rho_s C_s D_s N_{Av}} - \frac{3}{2\rho_s C_s D_s}\int_0^\infty Q_{abs} E_{b,\lambda}(T_s)d\lambda \tag{11.27}$$

where ρ_s = density of the soot particle, [kg/m^3]
A_{ab} = mass specific absorption cross-section, [m^2/kg]
= C_{abs}/m_s
N_v = the molecular diffusion flux, [m^{-2}/s]
N_{Av} = Avogadro's number, [#/kmol]
Q_{abs} = absorption efficiency of the soot particle, which is equal to the spectral emissivity of the soot particle, $\varepsilon_{s,\lambda}$
= $C_{abs}/(\pi D^2/4)$

The continuity equation for the particle mass is used to derive the above equation, which, assuming the particle surface is essentially stationary and the vapor is carried away from the particle by diffusion, is given by

$$\frac{dm_s}{dt} = -\frac{N_v \pi D_s^2 M_v}{N_{Av}} \tag{11.28}$$

As shown in Eq. (11.28), the change in soot particle temperature is affected by the particle diameter, which needs to be known when vaporization becomes important. For a spherical particle,

$$\frac{dm_s}{dt} = \frac{\pi}{2} \rho_s D_s^2 \frac{dD_s}{dt} \tag{11.29}$$

Combining Eqs. (11.28) and (11.29) yields

$$\frac{dD_s}{dt} = -\frac{2 N_v M_v}{N_{Av} \rho_s} \tag{11.30}$$

Thus, Eq. (11.27) and Eq. (11.30) are coupled nonlinear equations that must be solved simultaneously if vaporization is important, in order to obtain the time-dependent particle size and temperature distributions.

The molecular diffusion flux, N_v, is related to the flux, N_{vK}, for free molecule conditions ($K_n >> 1$) and the flux under continuum conditions, N_{vC}, by

$$\frac{1}{N_v} = \frac{1}{N_{vK}} + \frac{1}{N_{vC}} \tag{11.31}$$

The molecular and continuum fluxes can be calculated from kinetic theory [Chapman and Cowling, 1970] and mass transport relations [Bird et al., 1960],

$$N_{vK} = \beta n \left(\frac{RT_s}{2\pi M_v} \right)^{1/2} \left(x_{vs} - x_{v\infty} \right) \tag{11.32}$$

$$N_{vC} = 2n \frac{D_{AB}}{D_s} \frac{\left(x_{vs} - x_{v\infty} \right)}{\left(1 - x_{vs} \right)} \tag{11.33}$$

where β = evaporation coefficient
= 0.9 (Experimental values spanned 0.6 to 1.0 [McCoy and Cha, 1973].)
R = gas constant
n = molecular number density, $[m^{-3}]$
x_{vs} = mole fraction of carbon vapor at the surface of the soot particle
$x_{v\infty}$ = mole fraction of carbon vapor in gas far from the soot particle
D_{AB} = mass diffusivity of carbon vapor in gas, $[m^2/s]$

Since the mass flow rate term is only important when vaporization is significant, the mole fraction in the vapor phase can be obtained from the Clausius-Clapeyron equation

$$x_{vs} = \exp\left(H_v \frac{(T_s - T^*)}{RT_s T^*} \right) \tag{11.34}$$

which is invalid if the particle is superheated beyond the vaporization temperature, T*, and for temperatures far different from T*.

In order to solve Eq. (11.27), the absorption efficiency, Q_{abs}, and the heat transfer coefficient, h, are also required. Although Q_{abs} can be completely evaluated with computer programs based on Mie theory [e.g., Bohren and Huffmann, 1983], there are two regimes where the value of Q_{abs} can be easily evaluated. For small particles in the Rayleigh regime, Q_{abs} is a linear function of particle diameter D (Eq. [11.4]). For very large particles ($D >> \lambda$), Q_{abs} is independent of particle size. A complicated relationship exists between Q_{abs} and the particle diameter when the particle size is comparable to the radiation wavelength.

Since the soot particle size is on the order of or smaller than the mean-free-path in the surrounding gas, the heat transfer coefficient, h, is given by [McCoy and Cha, 1973]

$$h = \frac{2k_\infty}{D_s(1 + GK_n)} = \frac{2k_\infty}{D_s + G\lambda_{mfp}} \tag{11.35}$$

where K_n is the Knudsen number, λ_{mfp} is the mean-free-path in the gas, and G is a factor that depends on geometry,

$$G = \frac{8f}{\alpha(\gamma + 1)} \tag{11.36}$$

where f is the Eucken factor [Chapman and Cowling, 1970] and α the thermal accommodation factor ($\alpha = 0.9$).

From Eq. (11.27), it can be seen that when a short intense laser pulse is applied, the soot particle temperature rises rapidly during the duration of the laser pulse as the soot particle absorbs the laser energy. The heat loss through the conductive (the second term) and radiative heat transfer (the last term) to the surrounding medium is much smaller than the laser energy absorption rate (the first term). As the soot particle temperature approaches the soot vaporization point (approximately 4000 K), the temperature rise is severely hampered by the energy consumed in the vaporization of soot particles. After the laser pulse is switched off, the temperature of the soot particle gradually decreases due to conductive and radiative heat losses.

11.4.2.2 LII Signal and Its Relation to Soot Volume Fractions

As a blackbody, the soot particle radiates according to Planck's equation. The detector signal, S, collected over a given solid angle, Ω, from a distribution of soot particles, can be written as

$$S = V_{mv} \frac{\Omega}{4\pi} \int_0^{\tau} C_n(t) W(t) \int_0^{\infty} N(D_s) \int_0^{\infty} C_{abs} E_{b,\lambda}(T_s) d\lambda \, dD_s dt \tag{11.37}$$

where V_{mv} = measurement volume, [m^3]
$C_n(t)$ = soot particle number density, [m^{-3}]
$W(t)$ = signal windowing function
$N(D_s)$ = normalized size distribution, i.e., $\int N(D_s) dD_s = 1$

The emissivity of the soot particles is assumed equal to the absorption efficiency, Q_{abs}. Equations (11.27) and (11.30) may be numerically integrated to obtain the soot temperature (T_s) and its diameter (D_s). For the sake of simplicity, we have neglected laser intensity variations and assumed that any fluorescence from excited-state vapor molecules, which may be generated if a particle is vaporized, is not significant in the wavelength band being examined.

The time integral is included to allow for the variation of a laser pulse that is a function of time, integrating detectors, and/or post processing of time-resolved signal. A unit weight delta function $\delta(t = \tau)$ can be used to replace W(t) to generate the signal at any time $t = \tau$. For integrated measurements, the exposure time may be adjusted depending on illumination pulse length, particle response time, particle velocity, detector response time, and background level. Independent size calibration may be necessary to determine the optimum gating window if temporal integration is to be employed without sensitivity to the particle size. On the other hand, it might be possible to infer information about soot particle size distribution if the temporal profile of the signal is recorded.

Computations of LII in response to an idealized laser pulse based on the above blackbody radiation laws and the soot particle energy balance were carried out by Melton [1984] and Tait and Greenhalgh [1992].

In the limit of high laser power and maximum soot particle temperature near its vaporization point, according to Melton, Eq. (11.27) becomes

$$I_0 \frac{Q_{abs}}{4} = \frac{H_v N_v}{N_{Av}} \tag{11.38}$$

For small particles (D < λ), the absorption efficiency is given by Eq. (11.4). The soot temperature can be approximated by

$$\frac{1}{T} \approx \frac{1}{3915}\left[1 - \frac{1}{23.9}\log\left(I_0 \frac{\pi D}{\lambda} \mathrm{Im}\left|\frac{m^2-1}{m^2+2}\right|\right)\right] \tag{11.39}$$

Inserting this result into Eq. (11.37) we have approximately

$$S = C_{cal}\int_0^{\tau} C_n(t) W(t) \int_0^{\infty} N(D_s) \int_0^{\infty} D_s^x \, d\lambda \; dD_s dt \tag{11.40}$$

where C_{cal} = calibration factor
x = exponential factor of dependence of LII signal on soot particle size
= $(3 + 0.154\lambda^{-1})$

For λ between 0.4 and 0.7μm, for example, the LII signal is proportional to the mean soot diameter raised to the power of 3.38 to 3.22, or approximately to the soot volume fraction, i.e.,

$$S_{LII} \propto C_v \approx C_n \cdot D^3 \tag{11.41}$$

Eq. (11.41) forms the basis of soot volume fraction measurements by the LII technique.

11.4.3 Implementation of LII

Fig. 11.11 shows the optical setup for LII measurements. The output from a high-power pulsed laser is expanded into a light sheet by a combination of spherical and cylindrical lenses. The LII signal is detected perpendicularly to the laser sheet by a gated intensified CCD camera. The low LII signal-to-noise ratio requires the use of a gateable camera with gating down to nanoseconds. A UV-enhanced camera lens, typically a Nikon 105 mm UV lens, is used. As will be discussed in the next section, the best compromise between the absolute intensity of the LII signal and the ratio of the LII signal to the background luminosity is obtained at a detection wavelength of about 400 nm. A set of low-pass filters or a UV narrow bandpass filter is

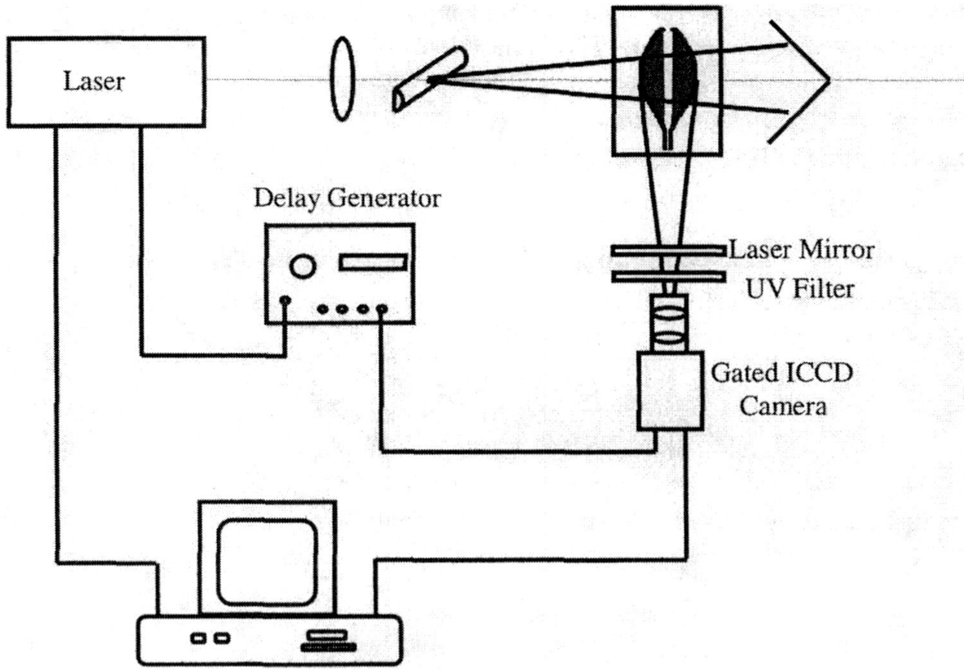

Fig. 11.11 Schematic of the 2-D laser induced incandescence setup.

required to restrict the detection range of the LII signal. One or two laser mirrors are used to block off any elastic scattering from soot particles or any surfaces. Under optimum conditions, a signal-to-noise ratio of no more than 1% may be achieved.

The intensity of an LII signal depends on a number of factors. The dominant factor is the temperature of the laser-heated soot particles, which is determined by the rate of laser energy absorption and heat loss through conduction, radiation, and soot vaporization. The detection wavelength will also affect the LII signal intensity and its signal-to-noise ratio. These factors must be considered in the application of LII for quantitative measurement of soot volume fraction.

11.4.3.1 Selection of Excitation and Detection Wavelengths

The excitation and detection wavelengths must be selected to optimize the LII signal-to-noise ratio. When visible light is used, most soot particles in diesel engines scatter light in the Rayleigh region, where the particle absorption efficiency varies approximately inversely with wavelength at fixed refractive index and fixed size, as shown in Eq. (11.4). Hence, a shorter excitation wavelength will be more effective in heating up the soot particles. However the broadband fluorescence from polycyclic aromatic hydrocarbons (PAHs) may interfere strongly with the LII signal if a UV excitation is to be used. When the 4th harmonic of an Nd:YAG laser (266 nm) was used [Cignoli et al., 1994], for example, strong broad-band fluorescence from PAHs was observed around 400 nm. This imposed severe interference with the LII signal. The use of the second harmonic of an Nd:YAG laser at 532 nm is a good compromise between the excitation efficiency and signal-to-noise ratio. With excitation at 532 nm, any fluorescence from PAHs or other molecules will be at longer visible wavelengths and near the IR region. Because the LII signal is normally observed at around 400 nm, interference from fluorescence can be minimized by using a low-pass filter.

In the LII technique, the temperature difference between the laser-heated and non-heated soot particles is used to reject the background luminosity from the soot not in the probe volume. In a diesel engine or sooting flame, the temperature of non-heated soot particles is about 2200 K or less, whereas the temperature of laser-heated soot particles can reach 4000K. Assuming that blackbody radiation takes place on soot surfaces, we can compare the blackbody radiation at 4000 K to that at 2200 K, as shown in Fig. 11.12. Fig. 11.12 is calculated assuming that the thickness of the laser sheet is two orders of magnitude less than that of a soot cloud in the field of view of the detector. It is seen that the ratio of radiation from laser-heated soot particles to that of non-heated soot particles increases as the detection wavelength becomes shorter.

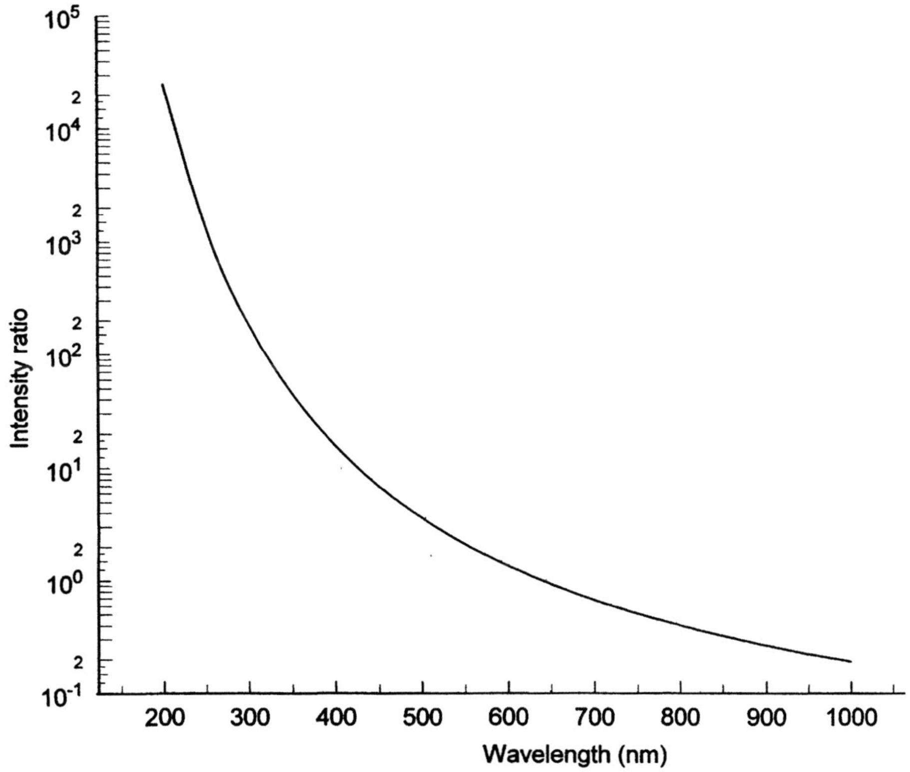

Fig. 11.12 Ratio of blackbody radiation from heated soot at 4000 K to that from soot at 2200 K.

Therefore, the detection of LII signal at shorter wavelengths is preferred in order to reject unwanted luminosity. On the other hand, the overall intensity of the collected light will also decrease as the cutoff wavelength is reduced. The LII signal is normally detected around 400 nm to achieve a good signal-to-noise ratio. A low-pass filter or a narrow bandpass filter is normally used. Since the LII signal-to-noise ratio is very low (<1%), a small amount of elastically scattered laser light from soot, dust, and wall passing through the filter may cause serious interference with the LII signal. A laser mirror, which reflects light at the laser wavelength but allows other wavelengths to pass, will be needed to block the elastically scattered laser light. A holographic notch filter can be used as a very effective laser mirror because of its extremely high blockage at the laser wavelength.

11.4.3.2 Effect of Laser Output

The laser output affects the intensity of a LII signal in two ways: laser fluence and laser beam shape. Fig. 11.13 illustrates the effect of laser fluence on the LII signal for a soot particle of radius 100 nm [Tait and Greenhalgh, 1993]. Below a certain laser fluence value, no

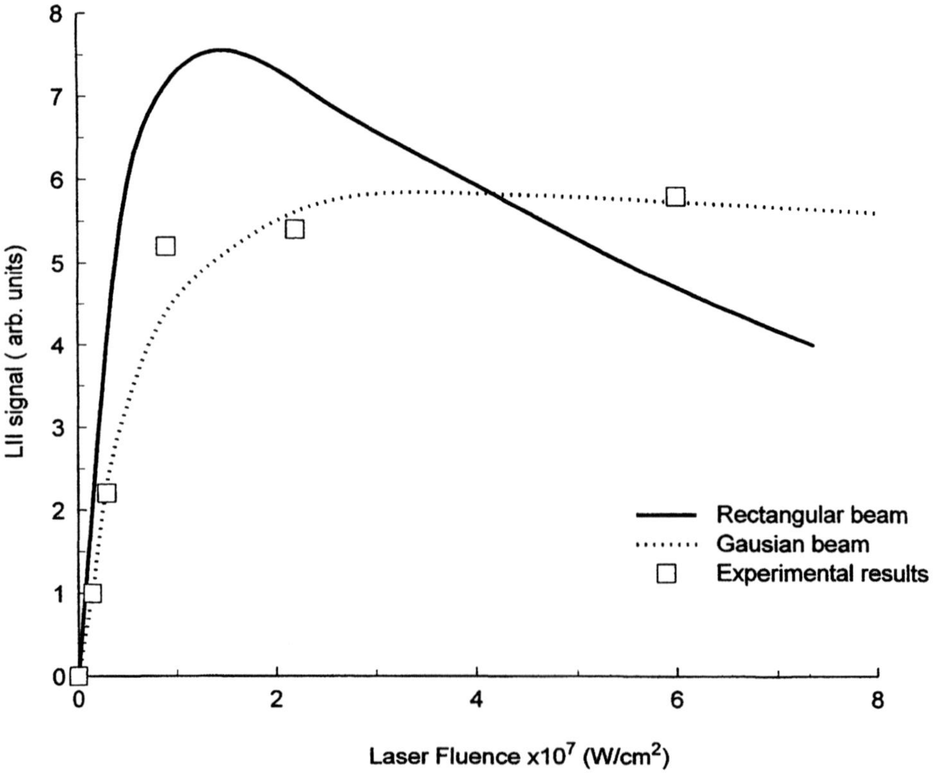

Fig. 11.13 Effect of laser fluence on the LII signal level, adapted from Tait and Greenhalgh [1993].

observable LII signal is detected. Just above the laser fluence threshold, the LII signal exhibits a rapid increase with increased laser fluence. This is because the soot particle temperature increases as a function of the laser fluence, which causes a corresponding increase in the LII signal.

For a rectangular beam, the signal decreases at very large laser fluences due to the soot particles losing significant mass. When a Gaussian profile laser beam is used, calculation predicts the LII signal to be nearly independent of laser power above the threshold fluence (about $10 MW/cm^2$). This can be explained by the fact that for a Gaussian profile beam, the laser intensity is high in the middle and gradually falls toward the edge of the beam. Increasing laser energy results in an increase in the effective laser beam diameter or the LII probe volume, in which the laser fluence exceeds the threshold of the LII signal. A similar result was also reported by Quay et al. [1994].

11.4.3.3 Selection of the Temporal Resolution

Since the LII signal overlaps with flame radiation, synchronization between the laser output and detection of the LII signal is required. Unlike elastic scattering (Rayleigh or Mie scattering)

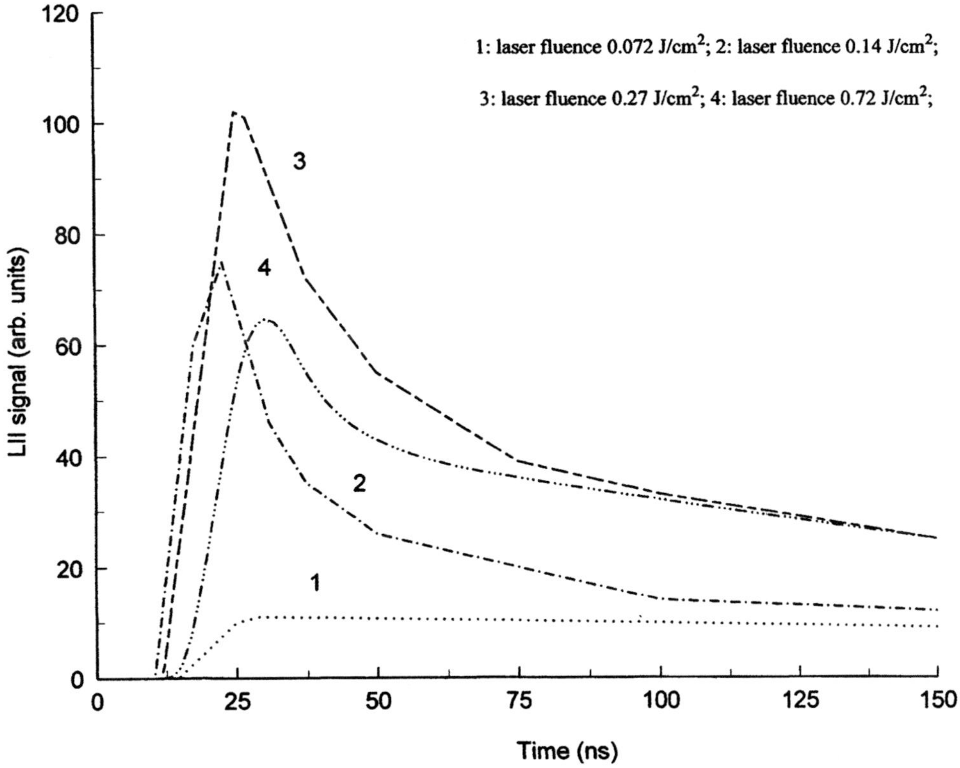

Fig. 11.14 Effect of laser fluence on the temporal profile of the LII signal, adapted from Ni et al. [1995].

which lasts as long as the laser pulse, laser induced incandescence can last a few hundred nanoseconds. Fig. 11.14, shows the temporal profiles of LII signals for different laser fluences [Ni et al., 1995]. As can be seen from Fig. 11.14, when the laser fluence was not saturated, the rise and the decay time of the LII signal was long. There was a dramatic decrease in the LII decay time as the laser fluence was increased because of the reduction in soot particle diameter by vaporization. The peak LII signal increased with laser fluence initially and then became lower for very high laser fluence. Thus, when the LII signal is to be detected, the detector gate should be delayed by few nanoseconds with respect to the laser pulse because of the time required for LII signal to reach its peak value.

The effect of particle size on LII signal decay time was also studied by Ni et al. [1995]. Fig. 11.15 shows the temporal profile of the LII signal at heights of 10 and 30 mm above an ethene-air laminar diffusion flame. Spatially resolved particle-size measurements on this flame by Santoro et al. [1983] indicated that the mean particle diameter (D_{63}) of soot particles at these two heights were 64 nm and 160 nm, respectively. As shown in Fig. 11.15, the rising edges of LII temporal profiles for soot particles of different sizes were closely matched, consistent with the theoretical prediction that in the Rayleigh regime, all particles, regardless of their sizes, increase in temperature at the same rate. However, the LII signal from the smaller soot underwent a faster decay time than the larger soot. This is because the heat loss was dominated by convection and radiation losses in this region and the smaller particles had a larger surface-to-volume ratio and hence cool faster.

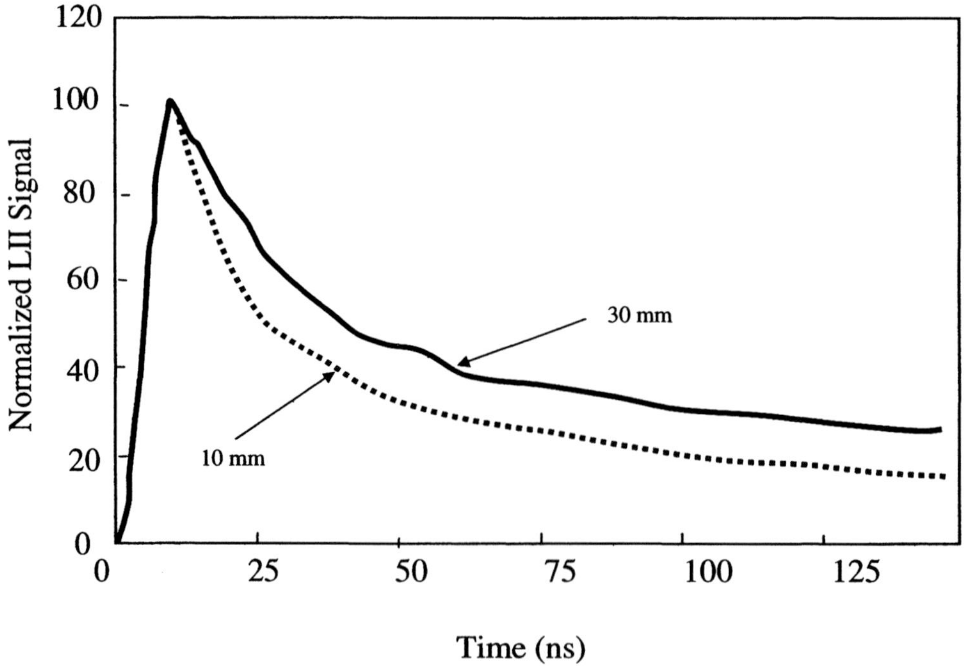

Fig. 11.15 Effect of particle size on the temporal profile of the LII signal, adapted from Ni et al. [1995].

Because the LII signal is more closely related to the soot volume fraction at its peak value when soot particles approach vaporization temperature, and because its rise time is independent of the particle size, the LII signal should be integrated up to its peak. According to Fig. 11.14, a gate width of 30 ns should be adopted.

Additional considerations arise from the laser beam profile and reflected light off of walls. Since a laser beam is typically characterized by a Gaussian intensity distribution, the LII signal resulting from the edge of the laser beam is characterized by a longer rise time. Delayed detection will include more contributions from low laser fluence. Because the low laser fluence does not heat the soot particles to near vaporization temperature, the dependence of LII signal on particle size to $D^{3+0.1154/\lambda}$ is no longer valid and soot volume fraction cannot be accurately determined from LII signals. The same consideration applies to the laser beam reflected from surfaces, which results in erroneous soot volume fraction measurements because of its reduced laser fluence and longer LII rise time.

Therefore, even though the LII signal could have a lifetime as long as a few hundred nanoseconds, a short gate width should be used with minimum delay from the laser pulse. For their experimental setup, Ni et al. [1995] estimated that an error could be as high as 10% due to the size effect when a gate width of 50 ns was used. The temporal resolutions, together with other experimental information, are given in Table 11.1 for LII measurements reported in the literature.

11.4.3.4 Calibration of LII for Soot Volume Fraction Measurements

Since the LII technique involves the absolute measurement of incandescent light, calibration of the LII signal is required in order to obtain the soot volume fraction. This can be carried out in a sooting laminar flame. The LII signal at a single spatial point is equated with the known value of the soot volume fraction, which can be obtained by the tomographic inversion method after the line-of-sight values of soot volume fraction in several angular directions have been measured using the laser extinction technique [Santoro et al., 1983 and 1987].

A simple calibration procedure was adopted in the authors' laboratory for a laminar diesel flame. This calibration is performed based on the principle of Beer-Lambert's law of extinction of light through a cloud of soot particles (Eq. [11.1]). According to Eq. (11.4), the extinction coefficient for monodispersed soot particles is given by

$$K_{ext} = \frac{6\pi}{\lambda}\,\mathrm{Im}\left(\frac{m^2-1}{m^2+2}\right) C_V \qquad (11.42)$$

Table 11.1 Summary of experimental setups used for LII measurements

Excitation wavelength	Detection wavelength	Detector	Reference
Nd:YAG 532 nm	<450 nm, two 450 nm low-pass filters One BG39 Schott-glass short-pass filter A 532 nm laser mirror	Gated intensified CID camera (240 × 388) 70 ns gate width Gated intensified CCD (380 × 480), 50 ns gate width	Dec, et al., 1991; Dec, 1992; Dec and Espey, 1992; Dec and Espey, 1995.
Nd:YAG 532 nm	A 400±70 nm bandpass filter A 532 nm laser mirror	Gated intensified lens coupled CCD camera (768 × 480), 50 ns gate width	Alatas, et al., 1993.
Nd:YAG 532 nm	A 450 ± 25 nm bandpass filter	Gated intensified CCD camera (384 × 578), 18 ns gate width	Ni, et al., 1995.
Nd:YAG 532 nm	A 400 ± 66 nm bandpass filter	Gated intensified lens coupled CCD camera, 40 ns gate width	Kosaka, et al., 1995.
Nd:YAG 532 nm	A 410 ± 70 nm bandpass filter A 532 nm laser mirror	Gated intensified CCD camera (384 × 578), 20 ns gate width	Wiltafsky, et al., 1996.
Excimer 308 nm		Gated intensified CCD camera (600 × 400)	Tait and Greenhalgh 1992; 1993.
Nd:YAG 266 nm	A 400 ± 35 nm bandpass filter Time delayed detection	Gated intensified CCD camera (576 × 578)	Cignoli, et al., 1994.

Since the LII signal is proportional to the soot volume fraction, scaled by a constant of proportionality K_{cal}, the integrated soot volume fraction along a line-of-sight in the flame is given by

$$\int_0^L C_V dx = K_{cal}\int_0^L S_{LII}(x)\,dx = K_{cal}\sum_{pixel=1}^{n} S_{LII}(i)\Delta x \tag{11.43}$$

where L is the optical path length, and Δx the size of a pixel. The soot distribution profile, $S_{LII}(i)$ is determined from the LII images.

Using Eqs. (11.1) and (11.43) and writing out the imaginary part in Eq. (11.42) explicitly, it can be shown that

$$K_{cal}\Delta x \sum_{pixel=1}^{n} S_{LII}(i)\frac{6\pi}{\lambda}\left[\frac{6nk}{\left(n^2+k^2+2\right)^2+4n^2k^2}\right]=\ln\left(\frac{I_0}{I}\right) \tag{11.44}$$

Thus, the calibration factor can be determined from the light extinction measured along a line in the flame corresponding to the row of pixels (pixel 1 to pixel n) in the CCD detector. It can then be used to convert the LII signal in each pixel into the soot volume fraction.

Another effect that requires consideration before the true soot volume fraction can be determined is the trapping of the LII signal, due to its extinction by the soot field between the laser sheet and the detection system. A procedure for accurately calibrating and correcting the LII signals for signal trapping was reported by Choi and Jensen [1998] for an axisymmetric flame. However, quantitative LII measurements in IC engines have not yet been realized.

11.4.3.5 Soot Particle Sizing Using LII and LIS

We have shown that the LII signal is proportional to the soot volume fraction $C_v \approx C_n \cdot D_s^3$. On the other hand, the LIS signal from soot particles is proportional to $C_n \cdot D_s^6$ according to the Rayleigh approximation for soot particles. Therefore, from the measured LII and LIS intensities one can calculate the relative value of particle size

$$D_{63} = \left(D_s^6 / D_s^3\right)^{1/3} \approx \left(I_{scat} / I_{LII}\right)^{1/3} \tag{11.45}$$

and the number density of soot particles

$$C_n \approx I_{LII}^2 / I_{scat} \tag{11.46}$$

11.6 Two-Dimensional Soot Imaging in Diesel Engines

11.5.1 Introduction

Both LIS and LII have been used to measure the soot distribution in diesel engines. Laser induced scattering, either Mie scattering or Rayleigh scattering, depending on the particle size, is frequently used to study the soot formation process in flames. The LIS signal is much stronger than the LII signal.

LIS can be used to map the soot distribution in laminar and turbulent flames using gaseous fuels. The flame luminosity is suppressed by a fast shutter speed and employment of a narrow bandpass filter. The application of elastic scattering to a diesel engine, however, is more problematic since the elastically scattered light from liquid droplets (Mie scattering) cause significant interference in identifying soot particles during the early stage of combustion. Parasitic scattering from windows and walls imposes another serious interference with elastic scattering. Lee et al. [1990] attempted to distinguish between soot particles and liquid droplets by comparing the elastic scattering intensities of vertically and horizontally polarized laser light. This approach was reasonably successful, but it complicated the experiment significantly because it was necessary to acquire two images of different polarization simultaneously. In addition, post-processing of the images to separate the signals from soot particles and liquid droplets was a lengthy process.

There are other disadvantages of using elastic scattering for soot measurements. Elastic scattering of soot particles in the cylinder is in the Rayleigh regime. The Rayleigh scattering signal is proportional to the particle diameter to the sixth power. Therefore, the scattered light favors zones with larger particle size. A strong Rayleigh scattering signal may be a result of very high soot number density, large particle size, or a combination of both. Consequently, interpretation of the results obtained by LIS is difficult.

The LII technique overcomes most of the problems associated with the elastic scattering technique. The LII signal is broadband. It is therefore possible to receive the light signal at a wavelength away from the laser wavelength with a narrow bandpass filter. Although the LII signal-to-noise ratio is low, an intensified CCD camera can be used to obtain sufficient detection sensitivity. The flame luminosity can be suppressed by synchronized gating. Because the LII signal originates from heated soot particles, interference from liquid droplets is avoided. Most importantly, the LII signal is proportional to the soot volume fraction. The dependence of LII on soot particle size will be very small when care is taken in gating during the experiment. Furthermore, interference from the elastically scattered light from combustion chamber walls is avoided because the LII signal is collected at a different wavelength from the laser wavelength.

11.5.2 Application of Two-Dimensional Soot Imaging to Diesel Engines

Lee et al. [1990] applied planar laser induced scattering to visualize soot distribution in a specially designed DI diesel engine. The fuel injection nozzle was mounted in an extended piston. The combustion chamber was formed by a quartz ring and a window in the cylinder head for maximum optical access. The fuel used was a 50/50 mixture of isooctane and tetradecane. A skip-fire mechanism was employed to reduce window fouling by soot deposition. In order to separate the liquid fuel droplets from soot particles in the resulting images, two images for vertically and horizontally polarized scattered light were recorded by two cameras, simultaneously and separately. Soot particles were identified as regions with low polarization ratios between the two images. This is because the larger liquid droplets behave like Mie scatters and the smaller soot particles like Rayleigh scatters. When a polarized laser beam is used, only vertically polarized light will be scattered by soot particles at a right angle. Consequently, soot particles will only appear in the images of the vertically polarized light.

Using this technique, Lee and his co-workers studied the spatial and temporal variations of soot during the combustion period. They found that for the high-swirl case, soot was uniformly distributed in the combustion bowl. Without swirl, soot was detected mainly near the wall and persists into the expansion stroke. The main difficulties encountered were soot deposition on the window and multiple scattering from the fuel droplets during the fuel injection period. The same experimental setup was later extended by Alatas et al. [1993] to include LII measurements.

Two-dimensional imaging techniques were used extensively at Tokyo Institute of Technology to study diesel spray flames in a rapid compression machine. Initially, the two-dimensional distribution of a soot cloud in the rapid compression machine was measured by the LIS technique [Won et al., 1991]. A cross section of the flame axis was illuminated by a thin laser sheet from an Nd:YAG laser (532 nm). The temporal variation of scattered light images was obtained for different injection pressures. Fig. 11.16 shows the main features of the diesel flame in the rapid compression machine, measured by the LIS technique.

As shown in Fig. 11.16, the scattered light appeared first in the periphery of the flame tip and it appeared U-shaped (point 1). After the scattering appeared in the periphery of the flame tip, scattering was observed near the flame center, but the intensity was weak at first. The amount of scattered light upstream from the center was fairly uniform at the early stage of combustion (point 2), but later it was broken into a number of small clusters on the order of several millimetres (point 3). The intensity of these clusters increased as they traveled downstream. This was followed by the break-up of the peripheral U-shaped region (point 4) as the flame moved further downstream. It was also found that increasing the injection pressure lowered soot concentration in the flame. The LIS technique was later extended to obtain the

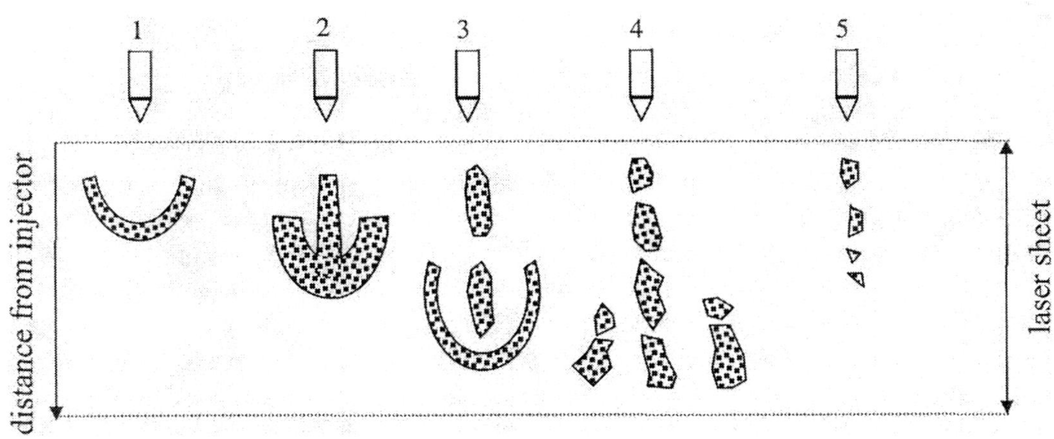

Fig. 11.16 Schematic representation of the temporal variation of the soot distribution in a diesel flame in a rapid compression machine, adapted from Won et al. [1991].

velocity vectors of soot clouds in the diesel flame, by employing laser sheets from a double-pulsed Nd:YAG laser [Won et al., 1992].

During their experiments, Won et al. found that the laser sheet was strongly attenuated as it passed through the soot cloud. They also pointed out that a large soot number density or large soot particles, or a combination of both, could have contributed toward the high scattering intensity. This was attributed to the fact that the LIS signal was proportional to the soot number density, C_n, and the size of the soot particles to the sixth power, D^6.

In order to measure the growth rate of the soot particles, Kosaka et al. [1995] carried out simultaneous recording of LIS and LII images of soot particles, from which the soot concentration, their size distribution, and their relative number density were determined in the rapid compression machine (Fig. 11.17). From Fig. 11.17 it can be seen that soot particles of larger diameter were distributed in the flame head region. It is also noticed that there was a reciprocal correlation between the soot particle number density and the particle sizes, i.e., as the soot particles grew their number density was reduced.

Comprehensive results have been reported by Dec and his co-workers [Dec et al., 1991, 1992a, 1992b, and 1995] using two-dimensional laser sheet diagnostics. The schematic layout of the optical setup for the simultaneous LII and LIS imaging is shown in Fig. 11.18.

Experiments were conducted in a single-cylinder, direct injection, four-stroke diesel engine. Two windows located between the cylinder head and cylinder block allowed the laser sheet to pass through horizontally or along the axis of the fuel jet as shown in Fig. 11.18. A window in the cylinder head replaced one of the two exhaust valves to obtain a view of the squish region and the outer portion of the combustion bowl. A unit electronic fuel injector was mounted in the cylinder head and oriented so that one of the eight sprays was aimed directly under the detection optical window.

A frequency doubled (532 nm) Nd:YAG laser was used for the LII and LIS measurements. A combination of cylindrical and spherical lenses produced a laser sheet approximately 300 μm thick and 25 mm wide in the engine. A half-wave plate was used to rotate the polarization of the laser, making it horizontal (perpendicular to the scattering plane) to maximize the LIS signal collected at a right angle. An edge filter was used to split the optical signal onto two cameras.

LII images were recorded by an intensified CCD camera. Using a combination of 450 nm short-wavelength-pass and colored glass filters and a camera-intensifier gate width of 50 ns, the contribution of background luminosity to the LII images was found to be negligible. A 532 nm laser-line mirror (reflection at 532 nm) was necessary to reject the elastic scattering of the laser beam from particles and walls. Since the laser power density used was above the threshold of LII (10 MW/cm^2), the relative intensities in an LII image provided information about the relative soot volume fractions.

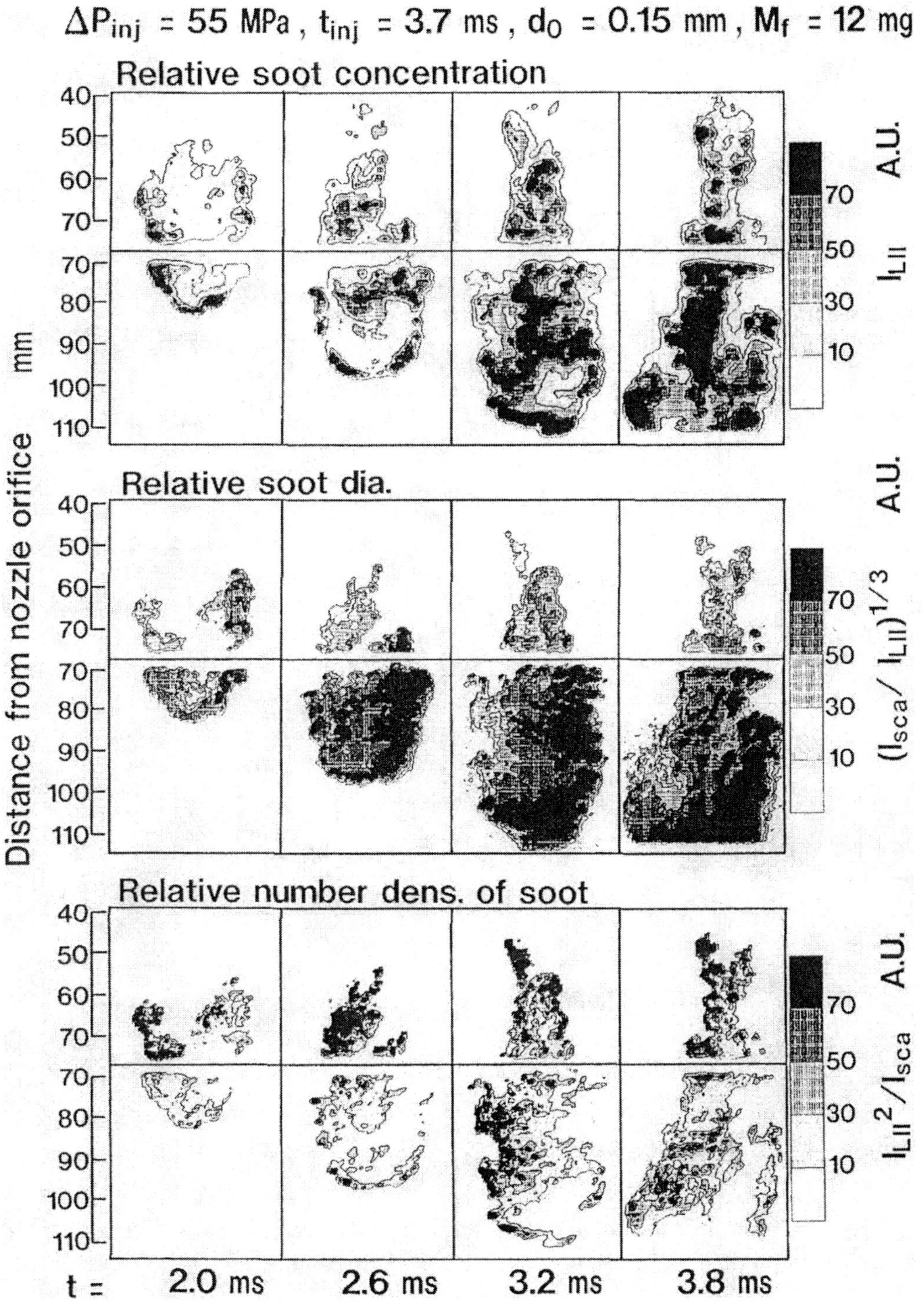

Fig. 11.17 Distributions of the relative soot concentration, relative soot diameters, and relative number density of soot particles in a diesel flame in a rapid compression machine [Kosaka et al., 1995].

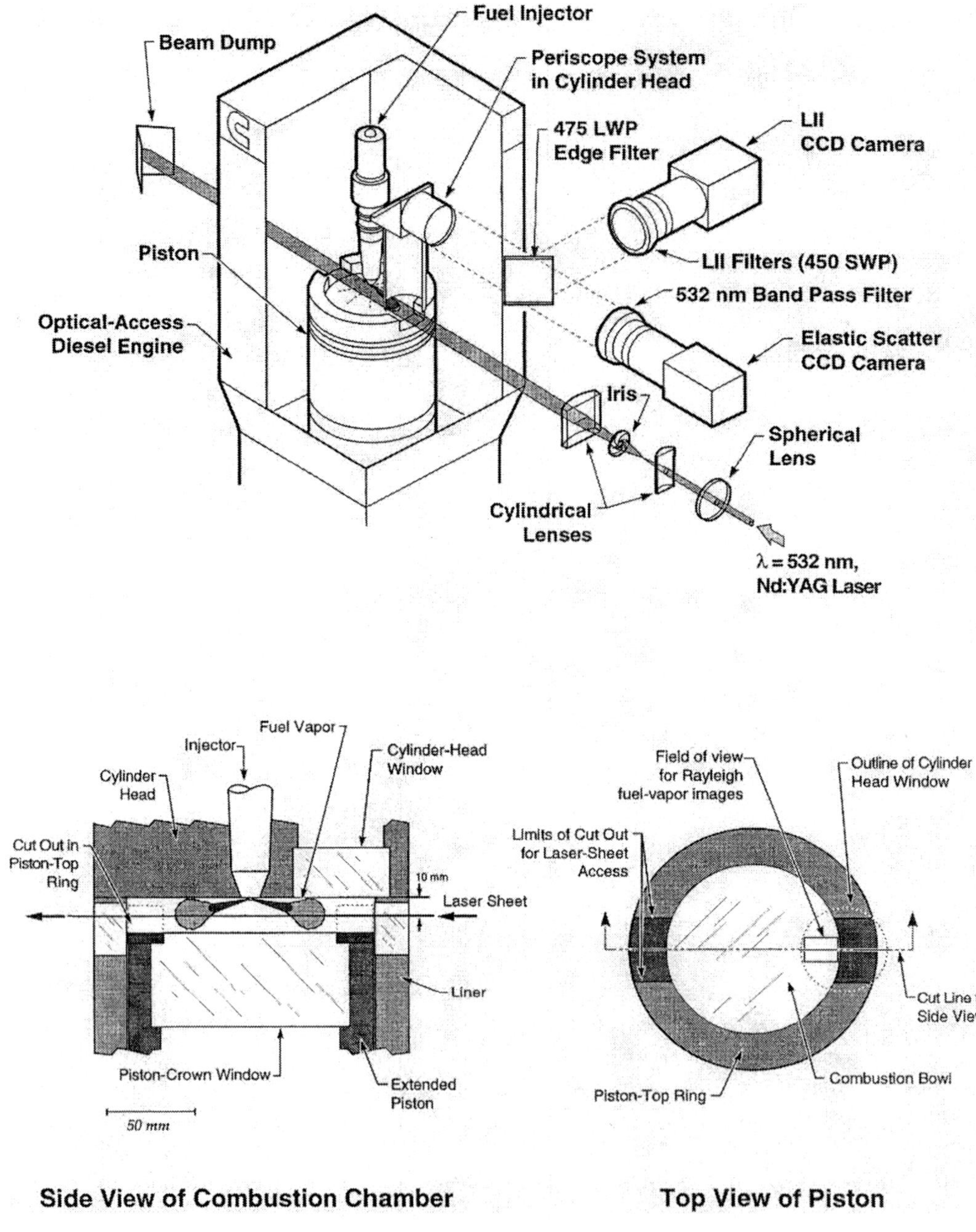

Fig. 11.18 Schematic of the optical setup for simultaneous LII and LIS imaging, Dec and Espey [1995].

LIS images were recorded simultaneously by the other intensified CCD camera. A 100 ns gatewidth and a 532 ± 10 nm bandpass filter were used to suppress the interference from flame radiation and LII, although spurious scattered light from walls could not be removed, and it added to the noise in the LIS images.

To overcome the severe beam attenuation immediately after the start of combustion with normal diesel fuels, low-sooting fuels were used. To minimize the rate of window fouling and to avoid overheating, the engine was operating at very low load and fired once every 10th or 20th engine cycle, at which the images were taken. The images or images pairs were acquired in sets of 12, from 12 separate cycles.

Fig. 11.19 shows a temporal sequence of simultaneous LII and LIS images for the low sooting fuel. The laser sheet was horizontal in the plane 9 mm below the cylinder head and the left edge of the image was 27 mm (along the fuel jet axis) downstream of the injector. The first soot was detected at 6 °CA after the start of injection (ASI) with LII only, appearing near the leading edge of the fuel jet. At 6.5 °CA ASI, the LII image shows more soot being formed across the fuel jet with the highest soot concentration in the leading portion of the fuel jet. From 7.0 °CA ASI through the end of the sequence at 8.5 °CA ASI, the LII images in Fig. 11.19 show a progressive development of the soot distribution. Soot occurred throughout the majority of the cross section of the leading portion of the jet, and its concentration continued to grow as indicated by the intensifier gas on the top right-hand corner. Although soot concentrations were generally higher toward the leading part of the jet, the soot concentration was approximately the same in the center and at the edge of the jet at any axial location. In contrast, there was no sign of soot in LIS images until 7.0 °CA ASI. The presence of soot in LIS images was concentrated at the peripheral region of the jet. Since the LIS signal was likely biased to regions of larger soot particles, the LIS images show that large particles are generally confined to the peripheral regions during these early stages of diesel combustion. There were more large soot particles at the front of the jet, but few in the central region.

11.6 In-Cylinder Particulate Sampling and Analysis

11.6.1 Introduction

There are several examples in the literature of the extraction of particulates from the cylinder of diesel engines using sampling valves. In general, extraction of particulates requires care so that particulates are not deposited onto the walls of the valve passages and onto the walls of the sample transfer system. In addition, a method of separating the particulates in the sample from the gas flow is required. By far, the most common method of separating the particulates is to deposit them on fiberglass filters.

Because of the small sample sizes allowed within the short sampling periods by a fast sampling valve, the samples are generally ensemble-averaged over many cycles. In contrast with local

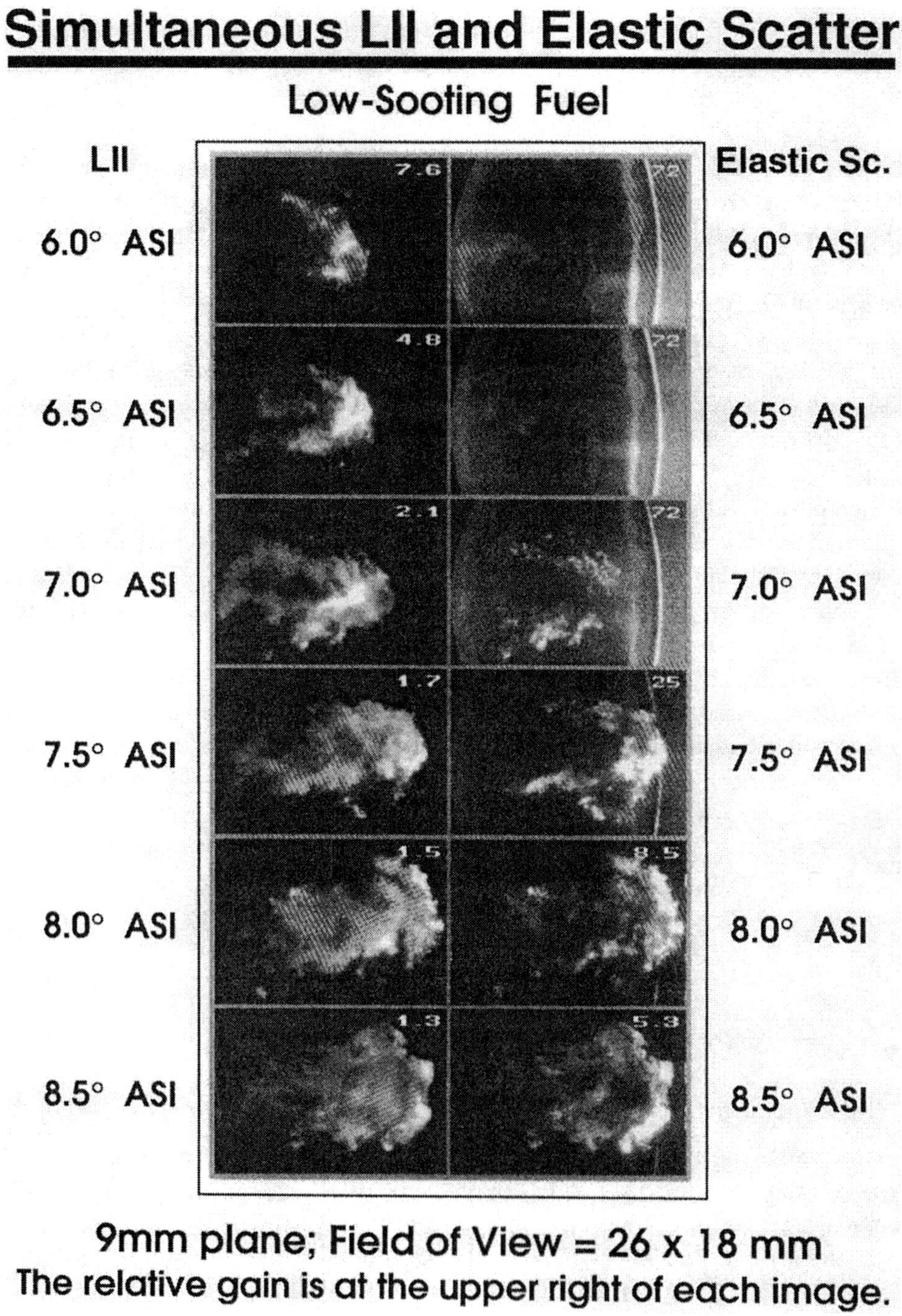

Fig. 11.19 Simultaneous LII and LIS images of diesel flames in a DI diesel engine, Dec and Espey [1995].

sampling by a fast sampling valve, the total cylinder dumping method introduced in Chapter 4 gives the spatially averaged soot concentration of the cylinder at a given crank angle.

In the following sections, applications of fast sampling valves for soot concentration measurements are described, followed by examples using total cylinder dumpling techniques.

11.6.2 Use of Carrier Gas for Particulate Sampling

Norris-Jones et al. [1984] measured in-cylinder particulates at various locations both within and outside the spray of a diesel engine by means of a needle-type sampling valve. The tip of the needle valve used is shown in Fig. 11.20. It can be seen from the figure that a carrier gas was used, partly in order to increase the flow velocity through the valve to avoid precipitation of the particulates, and partly to aid quenching of the sample close to the valve seat. The carrier gas was supplied close to the valve seat via the hollow valve stem, as shown in the figure. Kato et al. [1997] used air as the carrier gas in order to dilute the sample in a similar manner to Norris-Jones et al. The air carrier was delivered close to the valve seat, again, in order to help quench the sample entering the valve and in order to avoid precipitation of the particulates. The ratio of the sample volume to the air volume was determined by measuring the concentration of CO_2 in the undiluted sample, in the diluted sample, and in the carrier air, using gas chromatography. Bäcker et al. [1997] and Lepperhoff and Houben [1990] used nitrogen as the carrier gas, also delivered close to the valve tip. Bäcker et al. [1997] recommend a dilution ratio of about 10:1. Shimazaki et al. [1996] used different carrier gases during sampling studies in a two-stroke DI diesel engine. The choice of carrier gas used depended on the type of chromatographic technique employed to analyze the gas sample. Helium was used as the carrier gas when TCD gas chromatography was employed to analyze the sample for H_2, O_2, N_2, CO, and CO_2. Nitrogen was used as the carrier gas when FID gas chromatography was employed to analyze the sample for CH_4, C_2H_2, and C_2H_4.

11.6.3 Deposition of Particulates on Filters

Norris-Jones et al. [1984] used a fiberglass filter to collect the particulates. The filter was held in a filter holder, as shown in Fig. 11.20. The net mass of the particulates collected on the filter was obtained by weighing the conditioned filter before and after deposition, which is a standard technique commonly applied to exhaust particulates. Several other investigators used filtering techniques similar to those used by Norris-Jones et al. For example, Duggal et al. [1978] also collected the soot from a diesel engine cylinder onto glass fiber filters mounted on a holder made of stainless steel. Again, the mass of the particulates was obtained by weighing. The sample volume was measured with a gas flow meter positioned after the soot filter holder, and this enabled the soot density (in g/m^3 of sample gas) to be obtained. Other investigators using similar filtering and weighing techniques for particulate collection from diesel engines include Bäcker et al. [1997], Lepperhoff and Houben [1990], and Fujiwara et al. [1984].

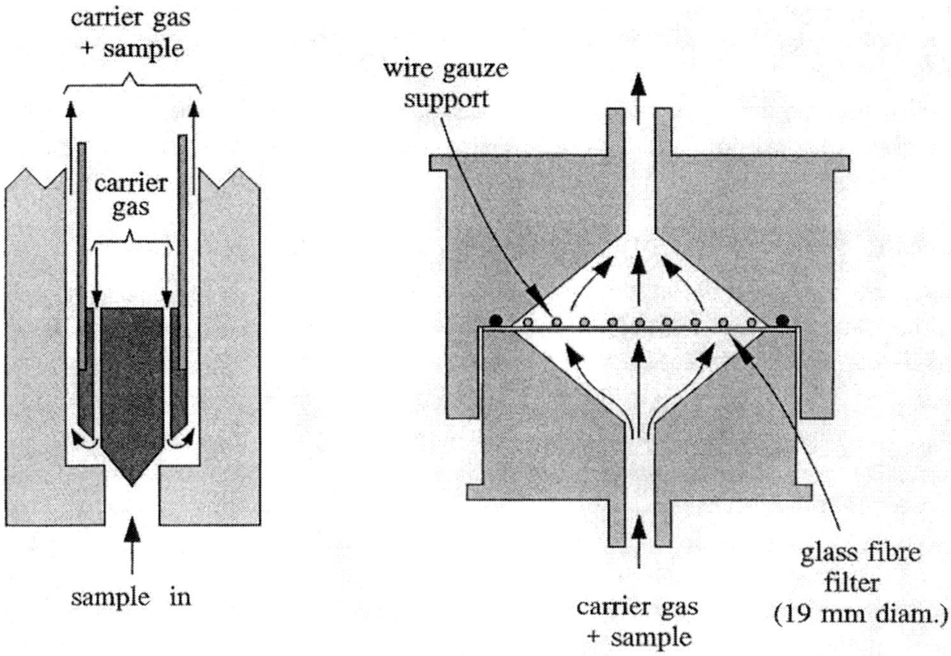

(a) the sampling valve

(b) the sampling filter holder

Fig. 11.20 Collection of particles from the engine cylinder using carrier gas.

Kato et al. [1997] carried out extensive particulate sampling from the combustion chamber of a DI diesel engine. The particulates were collected onto glass fiber filters which were coated with PTFE (Pallflex TX40 HI 20-WW), and the deposited mass was obtained by means of the standard weighing method. The particulate mass was then split into soluble and insoluble organic fractions using standard techniques commonly used for exhaust particulates. Sampling was carried out once every 11 engine cycles or so as to minimize the effect of sample extraction on engine operation.

Shimazaki et al. [1996] collected soot onto filters, but did not obtain the particulate mass by weighing. Instead, they measured the optical reflectance (Bosch number) of the particulates deposited onto the filters. A technique similar to that of Shimazaki et al. was also used by Aoyagi et al [1980], who carried out extensive sampling studies of soot in the combustion chamber of DI diesel engine. Particulates were collected onto filter papers held in a filter holder similar to the one shown in Fig. 11.20. The Bosch smoke number for each filter was measured by means of a Bosch smoke meter (optical reflectance), and converted to smoke density (g/m^3). This conversion made use of the measured amount of sample gas and, one assumes, an appropriate correlation between the Bosch smoke number and the weight of particulate on the filter. It is interesting to note that both Shimazaki et al. [1996] and Aoyagi et al. [1980] used evacuated vessels positioned after the filter holder, into which the sample gas emerged. These vessels were evacuated prior to each sample being collected. The vessel used by Shimazaki et al. had a volume of 330 cm^3 while that used by Aoyagi et al had a volume of 155 cm^3.

When sampling particulates from the combustion chamber of a DI engine, Bäcker et al. [1997] arranged for the valve tip to be heated by high-temperature oil so as to maintain the tip at 350°C in order to avoid soot deposition on the walls and condensation of unburned hydrocarbons [see also Lepperhoff and Houben, 1990].

11.6.4 Analysis of Soot Morphology

Some investigators collected soot from the engine cylinder onto specially prepared filters, and then analyzed the soot morphology. For example, Fujiwara et al. [1984] collected soot in this manner from a pre-chamber diesel engine using an electromechanically actuated needle-type sampling valve. The soot deposited onto the filters was then examined by means of a transmission electron microscope (magnification 100,000) to determine the particle size distribution. The diameter of some 2,000 to 3,000 particles was measured at each sampling condition. Particle diameters ranged from 5 to 55 nm, the most frequent size being around 20 nm. Fujiwara et al. [1984] also obtained the concentrations of CO, CO_2, and O_2 in the gas sample by means of gas chromatography. They then proceeded to compare the soot density (mass of soot collected on the filter per unit volume of sample gas) with the oxygen concentration in the sample gas. Fujiwara et al found that oxidation of soot took place when the local O_2 concentration was greater than about 5%. Also, oxidation took place more rapidly early during the combustion, when the gas temperature was high.

Ning et al. [1991] also used transmission electron microscopy to examine soot extracted from the combustion chamber of a DI diesel engine. The soot was collected on specially prepared glass fiber filters (HG-6.7 with trapping efficiency of 99.9999% for particulates having 300 nm diameter). Onto each of these filters was mounted a microscopy copper specimen grid supporting a 20 Å carbon film (2.5 mm diameter).

Bäcker et al. [1997] obtained gas samples from a DI diesel engine at a sampling location that was adjacent to one of the fuel sprays and upstream of the swirl direction. Bäcker et al. used an electrical aerosol analyzer to measure the diameter of soot particles in cylinder gas samples extracted during the diffusion burning phase. The diameter of the particles ranged from 10 to 1000 nm. However, 95% of the particles had a diameter less than 200 nm, while around 50% had a diameter less than 40 nm. According to Bäcker et al., soot particles with diameters less than 200 nm could follow the sample gas flow faithfully with negligible slip. Bäcker et al. found that the soot density at the sampling point reached a peak level when almost half of the injected fuel was burned.

11.6.5 Particulate Density when Carrier Gas is Used

As explained in an earlier section, when sampling particulates from the engine cylinder, a carrier gas is sometimes used to dilute the gas sample from the engine cylinder. This carrier gas is mixed, close to the valve seat, with the sample from the cylinder, and the resulting mixture is then passed through a filter system to extract the particulates [e.g., Kato et al., 1997; Norris-Jones et al., 1984].

The particulate density is defined as the mass of particulates in the sample gas per unit volume of sample gas at atmospheric conditions. It should be noted that the volume of the sample gas used to calculate the particulate density should exclude the carrier gas volume.

When a carrier gas is used, the particulate density can be calculated using one of the following two methods.

The first method requires measurements of:

- Particulate mass collected on the filter (M_{pm} g)
- Volume of carrier gas supplied to the sampling valve at atmospheric temperature and pressure conditions (V_{gas} m^3)
- Volume of diluted sample (mixture of carrier gas plus sample from the cylinder) at atmospheric temperature and pressure conditions (V_{ds} m^3)

Obviously, volumes should be measured at the same pressure and temperature.

The concentration of particulate matter in the sample gas (C_{pm} g/m^3) can then be calculated simply from:

$$C_{pm} = M_{pm} / \left(V_{ds} - V_{gas} \right)$$

The second method requires measurements of:

- Particulate mass collected on the filter (M_{pm} g)
- Volume of diluted sample (mixture of carrier gas plus sample from the cylinder) at atmospheric temperature and pressure conditions (m^3)
- Molar (volumetric) concentration of CO_2 in the undiluted sample from the cylinder ($[CO_2]_{us}$ %)
- Molar (volumetric) concentration of CO_2 in the diluted sample (carrier gas plus sample from cylinder) ($[CO_2]_{ds}$ %)
- Molar (volumetric) concentration of CO_2 in the carrier gas prior to mixing with sample ($[CO_2]_{gas}$ %)

Volumes should, of course, be measured at the same pressure and temperature. Also, let:

V_{us} volume of undiluted gas sample from the cylinder but at atmospheric temperature and pressure (m^3)
R_d dilution ratio
$= V_{ds}/V_{us}$

By definition,

$$C_{pm} = M_{pm} / V_{us}$$

Also,

$$V_{us} = V_{ds} / R_d$$

where R_d is given by,

$$R_d = \left\{[CO_2]_{us} - [CO_2]_{gas}\right\} / \left\{[CO_2]_{ds} - [CO_2]_{gas}\right\}$$

Therefore,

$$C_{pm} = M_{pm} R_d / V_{ds}$$

11.6.6 Measurements of Soot Concentration by the Total Cylinder Dumping Technique

The total cylinder dumping technique was originally developed by Voiculescu and Borman [1978] to study NO histories in a DI diesel engine. Kittelson et al. [1988] adapted the technique for measurements of the in-cylinder soot concentration as a function of crank angle. The experimental detail has been given in Chapter 4. Results obtained at engine speeds of 1000 and 1500 rpm showed that soot concentration peaked between 15 and 30 CA degree after top dead center (CAD ATDC) and approached exhaust levels by 40–60 CAD ATDC. At low load ($\phi = 0.4$), the exhaust soot concentration was only 15% of the peak in-cylinder value. At high load ($\phi = 0.4$), the exhaust value increased to 60–80% of the peak in-cylinder concentration. This suggested that oxygen availability later in the cycle is a critical factor in determining exhaust soot concentrations in a diesel engine.

Time resolved primary and agglomerate particle size distribution measurements were also obtained from the same experimental setup [Luo et al., 1989]. They found that the mean diameter of the primary particles decreased from 30 nm at the start of the combustion to 23 nm in the exhaust, while the agglomerates increased from 120 nm to 294 nm during the same process. These results indicated that agglomeration and oxidation occurred simultaneously during the combustion process.

11.7 Summary

In this chapter, a range of techniques is considered for in-cylinder soot measurements. The light extinction method is characterized by its simplicity and accuracy. It provides instantaneous soot concentration measurements. The light extinction technique can be extended to the full-field imaging of soot distribution in a diesel engine. However, care should be taken of the optical setup so that interference from the Schlieren or shadowgraphic effects is minimized. The light extinction method can be extended to obtain particle size information by measuring the scattered light as well as transmitted light, but its application to a diesel engine has proved difficult.

Laser induced incandescence (LII) is a relatively new technique. The LII signal is directly proportional to the soot volume fraction. Compared with laser induced scattering (LIS), it is not biased by the particle size. Since LII is a broadband emission, a bandpass filter can be used to eliminate interference from the spurious scattered light from surfaces, liquid droplets, and dusts, which causes noise in the LIS signal. LII is particularly suited for two-dimensional imaging of soot volume fraction distributions in a diesel engine. The combined use of LII and LIS measurements can provide information on soot particle size, as well as soot volume fraction in a diesel engine. Beam attenuation by soot particles can hamper the use of LII in heavy sooting flames.

In-cylinder sampling by a fast sampling valve provides spatially resolved ensemble-averaged soot concentration measurements, while the total cylinder dumping technique permits spatially averaged soot concentration of the cylinder at a given crank angle. Both techniques allow the extracted samples to be further analyzed with regard to the size and structure of soot particles present in the combustion chamber, which is not possible by other techniques.

References

Alatas, B., Pinson, A., Litzinger, T., and Santavicca, D. (1993), "A Study of NO and Soot Evolution in a DI Diesel Engine Via Planar Imaging," SAE Paper No. 930973.

Aoyagi, Y., Kamimoto, T., Matsui, Y., and Matsuoka, S. (1980), "A Gas Sampling Study on the Formation Processes of Soot and NO in a DI Diesel Engine," SAE Paper No. 800254.

Bäcker, H. and Wiartalla, A. (1994), "Application of Light Extinction Methods for the Investigation of Soot Formation During Combustion in a Model Chamber," *COMODIA 94*, pp. 565–570, Japan.

Bäcker, H., Hild, O., and Lepperhoff, G. (1997), "Investigation on Particle Size Distribution in the Cylinder of a Passenger Car DI Diesel Engine Using a Fast Gas Sampling Valve," SAE Paper No. 970875.

Bengtsson, P. and Alden, M. (1995), "Soot Visualization strategies Using Laser Techniques: Laser Induced Fluorescence in C2 from Laser Vaporized Soot, and Laser Induced Soot Incandescence," *Applied Phys*. B60, pp. 51–59.

Bird, R.B., Stewart, W.E., and Lightfoot, E.N. (1960), *Transport Phenomena*, John Wiley and Sons, New York, pp. 510, 528, and 648.

Bohren, C.E. and Huffman, D.R. (1988), *Absorption and Scattering of Light by Small Particles*, Wiley-Interscience, New York.

Chang, H. and Charalampopoulos, T. (1990), "Determination of the Wavelength Dependence of Refractive Indices of Flame Soot," *Proc. R. Soc. Lond. A*, 430, pp. 577–591.

Chang, Y.J., Kobayashi, H,. Matsuzawa, K., and Kaminoto, T. (1985), "A Photographic Study of Soot Formation and Combustion in a Diesel Flame with a Rapid Compression Machine," *Proc. of COMODIA*, pp. 149–157.

Chapman, S. and Cowling, T.G. (1970), *Mathematical Theory of Non-Uniform Gases*, 3rd ed., Cambridge University Press, p. 249.

Choi, M.Y. and Jensen, K. (1998), "Calibration and Correction of Laser-Induced Incandescence for Soot Volume Fraction Measurements," *Combustion and Flames*, Vol. 112, pp. 485–491.

Cignoli, F., Benecchi, S., and Zizak, G. (1994), "Time-Delayed Detection of Laser Induced Incandescence for the Two-Dimensional Visualization of Soot in Flames," *Applied Optics*, Vol. 33, No. 24, pp. 5778–5782.

Dalzell, W.H. and Sarofim, A.F. (1969), "Optical Constants of Soot and Their Application to Heat Flux Calculations," *Trans. ASME J. Heat Transfer*, Vol. 91, pp. 100–104.

Dasch, C. (1984a), "Continuous-Wave Probe Laser Investigation of Laser Vaporization of Small Soot Particles in a Flame," *Appl. Opt.*, 23, pp. 2209–2215.

Dasch, C. (1984b), "New Soot Diagnostics in Flames Based on Laser Vaporisation of Soot," *Proc. 20th Symposium (International) on Combustion*, The Combustion Institute, pp. 1231–1237.

Dasch, C. (1984c), "Spatially Resolved Soot Absorption Measurements in Flames using Laser Vaporization of Particles," *Opt. Lett.*, Vol. 9, pp. 214–216.

Dec, J., Zur Loye, A., and Siebers, D. (1991), "Soot Distribution in a DI Diesel Engine Using 2-D Laser Induced Incandescence Imaging," SAE Paper No. 910224.

Dec, J. (1992), "Soot Distribution in a DI Diesel Engine Using 2D Imaging of Laser Induced Incandescence, Elastic Scattering and Flame Luminosity," SAE Paper No. 920115.

Dec, J. and Espey, C. (1992), "Soot and Fuel Distributions in a SI Diesel Engine via 2D Imaging," SAE Paper No. 922307.

Dec, J. and Espey, C., (1995), "Ignition and Early Soot Formation in a DI Diesel Engine Using Multiple 2D Imaging Diagnostics," SAE Paper No. 950456.

Duggal, V.K., Priede, T., and Khan, I.M. (1978), "A Study of Pollutant Formation within the Combustion Space of a Diesel Engine," SAE Paper No. 780227.

Eckbreth, A.J. (1979), "Effects of Laser-Modulated Particulate Incandescence on Raman Scattering Diagnostics," *Appl. Phys.*, Vol. 48, pp. 4473–4479.

Fujiwara, Y., Fukazawa, S., Tosaka, S., and Murayama, T. (1984), "Formation of Soot Particulates in the Combustion Chamber of a Pre-Combustion Chamber Type Diesel Engine," SAE Paper No. 840417.

Greeves, G. and Meehan, J.O. (1975), "Measurement of Instantaneous Soot Concentration in a Diesel Combustion Chamber," *Proc. of IMechE Conference on Combustion in Engines*, pp. 73–82.

Hentschel, W. and Richter, J. (1995), "Time-Resolved Analysis of Soot Formation and Oxidation in a Direct Injection Diesel Engine for Different EGR-Rates by an Extinction Method," SAE Paper No. 952517.

Hofeldt, D. (1993), "Real-Time Soot Concentration Measurement Technique for Engine Exhaust Streams," SAE Paper No. 930079.

Kato, S., Takayama, Y., Sato, G.T., Tanabe, H., and Sato, T.G. (1997), "Investigation of Particulate Formation of DI Diesel Engine with Direct Sampling from Combustion Chamber," SAE Paper No. 972969.

Kerker, M. (1969), *The Scattering of Light*, Academic Press, New York.

Kerker, M., Scheiner, P., and Cooke, D.D.J. (1978), "The Range of Validity of the Rayleigh and Thomson Limits for Lorentz-Mie Scattering," *Opt. Soc. Am.*, Vol. 68, pp. 135–237.

Kittelson, D., Pipho, M., Ambs, J., and Luo, L. (1988), "In-Cylinder Measurements of Soot Production in a Direct-Injection Diesel Engine," SAE Paper No. 880344.

Kontani, K. and Gotoh, S. (1983), "Measurement of Soot in a Diesel Combustion Chamber by Light Extinction Method and In-Cylinder Observation by High-Speed Shadowgraphy," SAE Paper No. 831291.

Kontani, K. and Gotoh, S. (1986), "Effects of Particle Size Distribution on Soot Particle Measurement by Transmissive Light Extinction Method," SAE Paper No. 861234.

Kosaka, H., Nishigaki, T., Kamimoto, T., and Harada, S. (1995), "A Study on Soot Formation and Oxidation in an Unsteady Spray Flame via Laser Induced Incandescence and Scattering Techniques," SAE Paper No. 952451.

Ladommatos, N., Balian, R., Horrocks, R., and Cooper, L. (1996), "The Effect of Exhaust Gas Recirculation on Soot Formation in a High-Speed Direct-Injection Diesel Engine," SAE Paper No. 960841.

Lee, S. and Tien, C. (1981), "Optical Constants of Soot in Hydrocarbon Flames," *Proc. 18th Symposium (International) on Combustion*, The Combustion Institute, pp. 1159–1166.

Lee, W., Solbrig, C., Litzinger, T., and Santoro, R. (1990), "Planar Laser Light Scattering for the In-Cylinder Study of Soot in a Diesel Engine," SAE Paper No. 902125.

Lepperhoff, G., and Houben, M.(1990), "Soot Formation and Oxidation in IDI and DI Diesel Engine," SIA Congress, SIA No 90097, Lyon, 13–14 June 1990.

Luo, L., Pipho, M, Ambs, J., and Kittelson, D. (1989), "Particle Growth and Oxidation in a Direct-Injection Diesel Engine," SAE Paper No. 890580.

McCoy, B.J. and Cha, C.Y. (1973), "Transport Phenomena in the Rarefied Gas Transition Regime," *Chem. Engr. Sci*. Vol. 29, pp. 381–388

Melton, L. (1984), "Soot Diagnostics Based on Laser Heating," *Applied Optics*, Vol. 23, No. 13, 2201.

Mie, G. (1908), "Beitrage zur Optik truber Medien speziell kolloidaler Metallosungen," *Ann. Phys.*, Vol. 25, pp. 377–445.

Miyamoto, N., Ogawa, H., Goto, N., and Sasak, H. (1990), "Analysis of Diesel Soot Formation Under Varied Ignition Lag with a Laser Light Extinction method," SAE Paper No. 900640.

Nakakita, K., Nagaoka, M., Fujikawa, T., Ohsawa, K., and Yamaguchi, S. (1990), "Photographic and Three Dimensional Numerical Studies of Diesel Soot Formation Process," SAE Paper No. 902081.

Ni, T., Pinson, J., and Santoro, R. (1995), "Two-Dimensional Imaging of Soot Volume Fraction by the Use of Laser Induced Incandescence," Applied Optics, Vol. 34, No. 30, 7083.

Ning, M., Yuan-Xian, Z., Zhen-Huan, S., and Hu Guo-Dong (1991), "Soot Formation, Oxidation and Its Mechanism in Different Combustion Systems and Smoke Emission Patterns in DI Diesel Engines," SAE Paper No. 910230.

Norris-Jones, S.R., Hollis, T., and Waterhouse, C.N.F. (1994), "A Study of the Formation of Particulates in the Cylinder of a Direct Injection Diesel Engine," SAE Paper No. 840419.

Quay, B., Lee, T., Ni, T., and Santoro, R.J. (1994), "Spatially-Resolved Measurements of Soot Volume Fraction Using Laser Induced Incandescence," *Combustion and Flame* , Vol. 97, pp. 384–392.

Santoro, R.J., Semerjian, H.G., and Dobbins, R.A. (1983), "Soot Particle Measurements in Diffusion Flames," *Combustion and Flame*, 51, pp. 203–218.

Santoro, R.J., Yeh, T.T., Horvath, J.J., and Semerjian, H.G. (1987), "The Transport and Growth of Soot Particles in Laminar Diffusion Flames," *Combust. Sci. Technol.*, Vol. 53, pp. 89–115.

Shimazaki, N., Hatanaka, H., Yokota, K., and Nakahira, T.(1996), "A Study of Diesel Combustion Process Under the Condition of EGR and High-Pressure Fuel Injection with Gas Sampling Method," SAE Paper No. 960030.

Tait, N. and Greenhalgh, D. (1993), "PLIF Imaging of Fuel Fraction in Practical Devices and LII Imaging of Soot," *Ber. Bunsenges, Phys. Chem.*, Vol. 97, No. 12, pp. 1619–1625.

Tait., N. and Greenhalgh, D. (1992), "2D Soot Field Measurements by Laser Induced Incandescence," *Proceedings of the Optical Methods and Data Processing in Heat and Fluid Flow,* pp. 185–193, IMechE, London.

Tree, D. and Foster, D. (1994), "Optical Measurements of Soot Particle Size, Number Density, and Temperature in a Direct Injection Diesel Engine as a Function of Speed and Load," SAE Paper No. 940270.

Van de Hulst, H.C. (1957), *Light Scattering by Small Particles*, Chapman and Hall, London.

Voiculescu, I.A. and Borman, G.L. (1978), "An Experimental Study of Diesel Engine Cylinder-Averaged NOx Histories," SAE Paper No. 780228.

Wiartalla, A., Bäcker, H., and Durnholz, M. (1995), "Influence of Injection System Parameters on Spray Development Combustion and Soot Formation by Optical Measurement Techniques in a Model Combustion Chamber," SAE Paper No. 950233.

Wiltafsky G., Stolz W., Köhler, J., and Espey, C. (1996), "The Quantification of Laser Induced Incandescence for Planar Time Resolved Measurements of the Soot Volume Fraction in a Combusting Diesel Jet," SAE Paper No. 961200.

Won, Y.H., Kamimoto, T., and Kosaka, H. (1992), "A Study on Soot Formation in Unsteady Spray Flames via 2D Soot Imaging," SAE Paper No. 920114.

Won, Y.H., Kamimoto, T., Kobayashi, H., and Kosaka, H. (1991), "2-D Soot Visualisation in Unsteady Spray Flame by Means of Laser Sheet Scattering Technique," SAE Paper No. 9101223.

Zhao, H., Lowry, G., and Ladommatos, N. (1996), "Time-Resolved Measurements and Analysis of In-Cylinder Gases and Particulates in Compression-Ignition Engines," SAE Paper No. 961168.

Chapter 12

In-Cylinder Gas Temperature Measurement

Outline

12.1 Introduction

Pressure, temperature, and density are properties that define the thermodynamic state of the cylinder gas. Cylinder pressure is normally uniform and can be readily measured with a piezoelectric pressure transducer as described in Chapter 2. Charge density can be easily calculated from the cylinder volume and the amount of gases trapped in the cylinder. The gas temperature during compression can be considered uniform except for a thermal boundary layer very close to the cylinder walls. The average gas temperature can be estimated at a given crank angle from the equation of state,

$$T = \frac{pV}{nR}$$

However, that is as much as we can know about the gas temperature without measurement. As soon as combustion starts, a sharp temperature gradient will form between burned and unburned gases. Even within the burned gas, there is a temperature gradient from the first burned element to the last. There is also a temperature gradient from the bulk gas across the thermal boundary layer to the chamber walls.

Combustion gas temperature plays a significant role in NO_x formation and its subsequent emissions. Understanding autoignition kinetics requires accurate measurements of end-gas temperature history. In addition, cylinder gas temperature is needed to assess convective heat transfer and radiative heat loss to the walls and the thermal stress of wall materials.

Since thermocouples typically lack the necessary time response or durability, in-cylinder gas temperature measurements have been dominated by optical techniques. Before the advent of laser-based diagnostics, all temperature measurements were carried out using line-of-sight techniques. Thus, the classification of temperature measurement techniques according to their spatial resolution seems to be a natural choice from both a historical and a technical point of view. The measurement of combustion temperature in a diesel engine by means of the two-color method has been discussed separately in Chapter 10 because it is often used to measure the in-cylinder soot loading at the same time.

In this chapter, the line-of-sight techniques for temperature measurements are described first. Spatially and temporally resolved laser diagnostics will then be introduced.

The line-of-sight temperature measurement techniques to be introduced in the following sections are classified into three types according to their operational principles. The first group of techniques may be termed radiation thermometers, as they are based on thermal radiation laws. The other group consists of spectroscopic techniques that measure gas temperatures from the intensities of spectral lines.

The use of spectroscopic techniques is limited to optically thin gases because of the self-absorption. On the other hand, radiation thermometers are suited to optically thick gases where the emissivity or absorptivity of the gas can be measured accurately.

The velocity-of-sound method is the only non-optical technique used for the measurement of in-cylinder gas temperatures. It offers the advantages of compact design and limited modification required to the engine.

Temperature measurements by spontaneous Raman scattering (SRS), laser Rayleigh scattering (LRS), coherent anti-Stokes Raman scattering (CARS), and laser induced fluorescence (LIF) have been developed in the recent years. The common feature of these laser-based techniques is their capability for non-intrusive, in situ measurements with high spatial and temporal resolution. Therefore, these techniques offer the unique opportunity to accurately measure in-cylinder gas temperature distributions and their effect on subsequent combustion and pollutant formation processes. More recently, two-dimensional temperature measurements have become possible using Rayleigh scattering or LIF, with the latter offering a greater potential to be applied to more realistic engine conditions.

The spectroscopic principles of these laser-based methods have been introduced in Chapter 5. In this chapter, the emphasis will be on their use for in-cylinder gas temperature measurements.

12.2 Radiation Thermometry

12.2.1 Principles of Radiation Thermometry

We are all familiar with the dull red glow of coals in a fire and the bright white glow of incandescent lamps. We know that the whiter and brighter an object appears, the hotter it is. This is in fact the very basic principle on which most of the optical techniques are based. They are normally referred to as radiation thermometers or pyrometers.

Because our eyes are only sensitive to a very narrow range of the radiation spectrum (0.4–0.7 μm), shown as the visible spectrum in Fig. 12.1, most of us associate thermal radiation only with objects that are hot (>500°C) and dangerous. At the other extreme, when they reach temperatures above 1500°C, objects become so bright that our eyes have difficulty observing them comfortably.

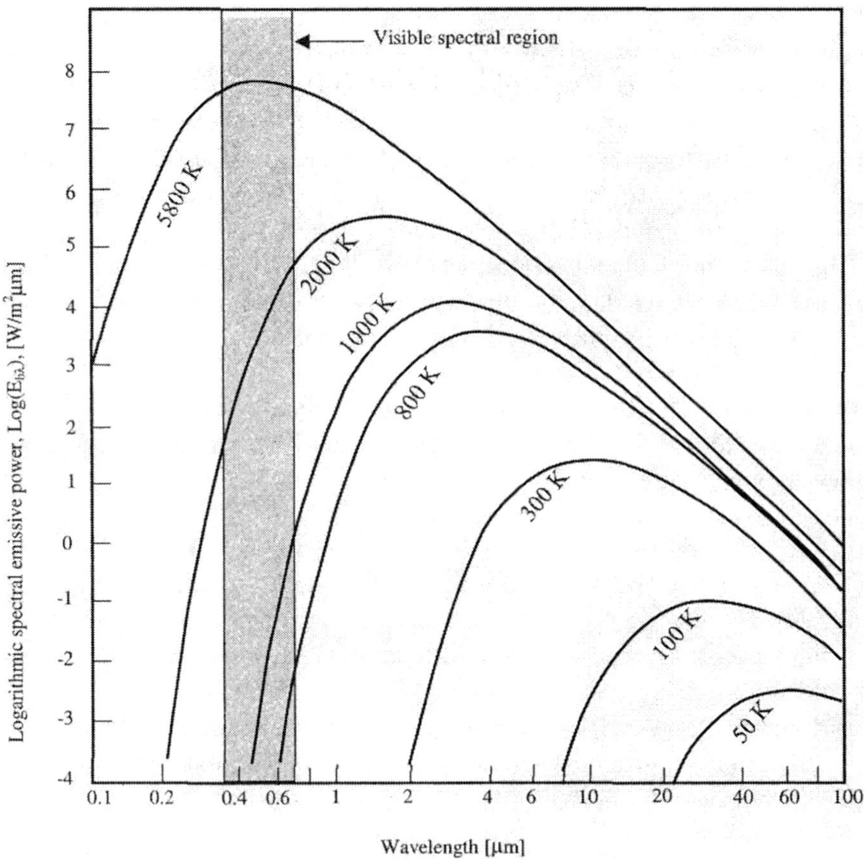

Fig. 12.1 Monochromatic blackbody emissive power distribution as a function of wavelength.

In order to determine temperature from the emission of radiation as shown in Fig. 12.1, we need to introduce the concepts of spectral radiance and the blackbody radiation.

Spectral radiance, measured as the energy emitted by a surface per unit area, per unit wavelength, per unit solid angle, is the technical term for the brightness of an object. It is advantageous to use radiance, as it is independent of the distance to the object and the size of the object. In addition, in an ideal optical system of lenses, the radiance of an object is constant. Instruments that measure radiance allow us to infer temperatures at a distance and, if necessary, to use close-up or telephoto lenses.

A blackbody is simply a perfect emitter and absorber of radiation. In 1900, Planck derived the mathematical description of the distribution of blackbody radiation curves shown in Fig. 12.1,

$$E_{b,\lambda}(T) = \frac{C_1}{\lambda^5\left[\exp(C_2/\lambda T) - 1\right]} \tag{12.1}$$

where $E_{b,\lambda}$ = monochromatic emissive power of a blackbody at temperature T, [W m^{-3}]
λ = wavelength, [μm]
T = temperature, [K]
C_1 = the first Planck's constant
= 3.7418×10^{-16} W m^2
C_2 = the second Planck's constant
= 1.4388×10^{-2} m K

For several practical reasons, most radiation thermometry is carried out at wavelengths in the visible or near infrared region. In this portion of the spectrum, $(C_2/\lambda T) > 4.6$, and Planck's law is approximated to 1% or better by Wien's law:

$$E_{b,\lambda}(T) = \frac{C_1}{\lambda^5 \exp(C_2/\lambda T)} \tag{12.2}$$

However, practically all objects emit less radiation than a blackbody at the same temperature. The fraction of full radiation that is actually emitted is called the monochromatic emissivity, ε_λ, of the object. Thus, the radiation of a non-blackbody is given by

$$E_\lambda(T) = \varepsilon_\lambda \frac{C_1}{\lambda^5\left[\exp(C_2/\lambda T) - 1\right]} \tag{12.3a}$$

or

$$E_\lambda(T) = \varepsilon_\lambda \frac{C_1}{\lambda^5 \exp(C_2/\lambda T)} \tag{12.3b}$$

The monochromatic emissivity, ε_λ, is normally dependent on temperature, wavelength, and concentration, and it is the most important and yet the most limiting factor for accurate determination of gas temperature using radiation thermometry. If the emissivity, ε_λ, is independent of wavelength, the emitting object is considered to be a grey body.

Another important concept in radiation thermometry is the apparent temperature or brightness temperature, which has been introduced in Chapter 10:

$$E_{b,\lambda}(T_a) = E_\lambda(T)$$

Thus, the apparent temperature is the temperature of a blackbody that will emit the same amount of radiation as a non-blackbody at temperature T. Combining Eqs. (12.2) and (12.3b), it can be shown that

$$\frac{1}{T} = \frac{1}{T_a} + \frac{\lambda}{C_2} \ln \varepsilon_\lambda \tag{12.4}$$

In practice, the amount of radiation detected by a thermometer is given by

$$P = \int_{\lambda_1}^{\lambda_2} \eta_A \tau_\lambda S_\lambda E_\lambda(T) d\lambda \tag{12.5}$$

where η_A = the geometrical factor of the optical system
τ_λ = the spectral transmission of the optical system
S_λ = the spectral response of the detector

Hence, the gas temperature can be determined from the detector output if the emissivity, the geometrical characteristics, the spectral transmission, and the spectral response of the system are known. This is the theoretical basis of radiation thermometers.

For gases that are not opaque, the absorptivity becomes more relevant. According to Kirchhoff's law, the monochromatic emissivity, ε_λ, is equal to the monochromatic absorptivity for the gas, α_λ. For gases, the absorptivity is normally expressed as

$$\alpha = 1 - I/I_o = 1 - e^{-K_{abs}L} \tag{12.6}$$

where K_{abs} = the absorption coefficient (cm^{-1})
L = the optical path length (cm)
I_o = the light intensity without absorption
I = the light intensity after absorption

The absorptivity, and hence the emissivity of a gas, depends on the gas composition, concentration, and temperature, as well as the gas thickness (the KL factor).

The application of radiation thermometry to in-cylinder gas temperature measurements relies upon the detection of thermal radiation of gases present in the cylinder. Elementary gases such as O_2, N_2, and dry air have a symmetrical molecular structure, and they neither absorb nor emit radiation unless they are heated to such extremely high temperatures that they become ionized plasmas and electronic energy transformations take place. On the other hand, gases such as CO_2, CO, H_2O, and hydrocarbons, are asymmetric in one or more of their modes of vibration so that they absorb and emit radiation in limited infrared ranges (called bands). The most important gases for thermal radiation detection are CO_2 and water vapor (H_2O).

As discussed in Chapter 5, changes in energy levels due to changes in vibrational frequency manifest themselves as a strong peak at the wavelength corresponding to the vibrational transformation, accompanied by multiple rotational energy changes slightly above or below the peak. This process results in absorption or emission bands. The shape and width of these bands depend on the temperature and pressure of the gas, while the magnitude of the monochromatic absorptivity is primarily a function of the thickness of the gas layer. The absorption spectrum of water vapor shown in Fig. 12.2 illustrates the complexity of the process. The most important absorption/emission bands for water vapor lie between 1.7 and 2.0 μm, 2.2 and 3.0 μm, 4.8 and 8.5 μm. The absorption spectrum of CO_2 is shown in Fig. 12.3, which identifies the most important absorption/emission bands as being between 2.6 and 2.8 μm, 4.2 and 4.5 μm. One or two of these bands are used for gas temperature measurements by radiation thermometry.

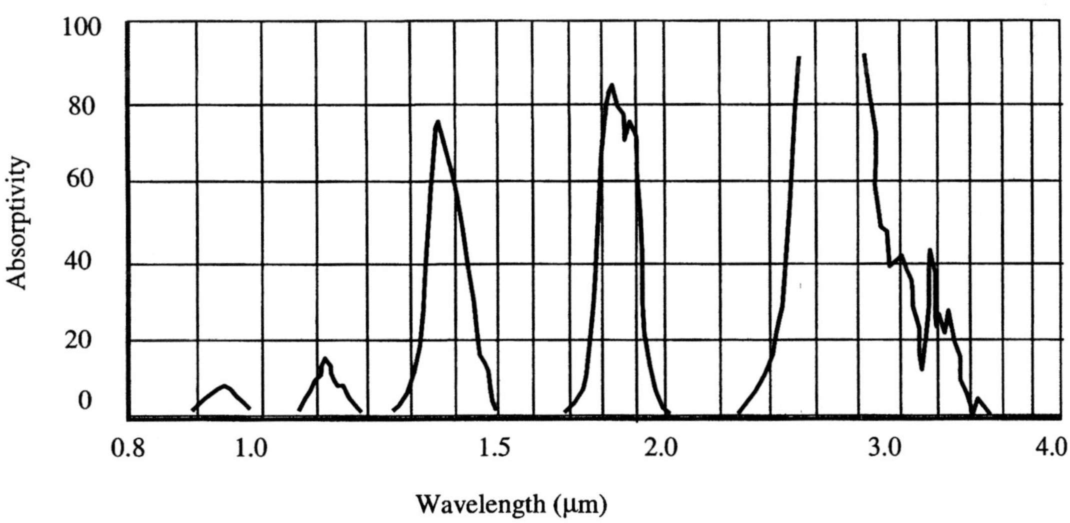

Fig. 12.2 Absorption bands of water vapor.

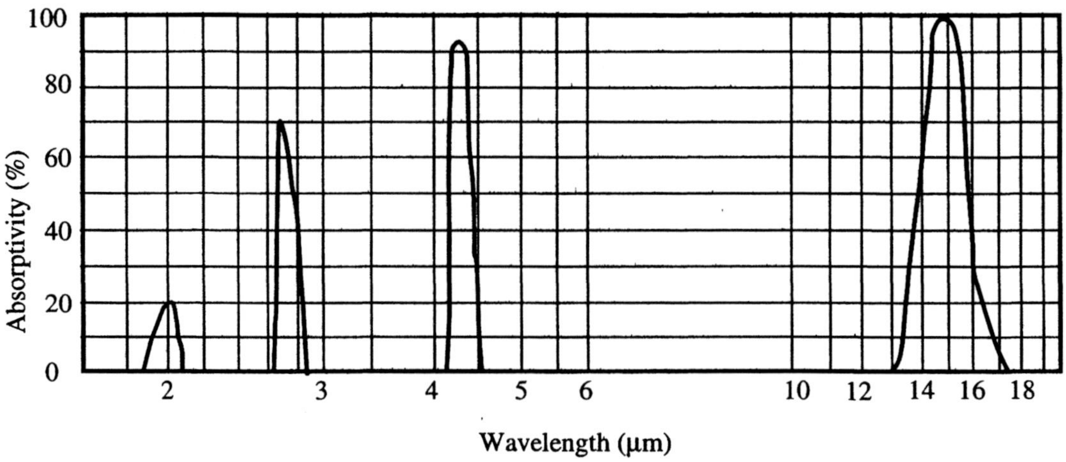

Fig. 12.3 Absorption bands of carbon dioxide.

When both CO_2 and H_2O exist in a mixture and the radiation is detected from an overlapping spectral band (e.g., around 4.5 μm), the emissivity of the gases will not simply be the sum of the two but will have a lower value due to absorption/emission in overlapping wavelength bands. The precise method of calculating the absorptivity or emissivity of a gas mixture is quite complex. A comprehensive treatment of this subject is beyond the scope of this text, and the reader should consult Hottel and Egbert [1942], Tien [1968], and Siegel and Howell [1993], for details on the subject. On the other hand, for engine applications the value of emissivity is normally determined experimentally because of the number of unknown parameters involved.

It is worth noting that both radiation thermometry and spectroscopic techniques to be introduced later are applicable to gases in thermal equilibrium. In transient flows, the relaxation time for local thermal equilibrium depends on the collision frequency and sometimes exceeds a characteristic flow time, giving rise to a non-equilibrium condition. However, while the relaxation time is short in comparison with the flow residence time, the assumption of local thermal equilibrium is good. This is usually the case at or above atmospheric pressure, as in an IC engine. Non-equilibrium conditions, on the other hand, can exist in combustion gases. Slow or incomplete combustion can prevent combustion gases from reaching equilibrium. Intermediate products, such as OH radicals, may become chemically excited and large departures from equilibrium in reaction zones will occur, which could lead to erroneous high temperature readings.

Accurate temperature measurements by radiation thermometry will be obtained only when the radiation is of thermal origin. Excess radiation, such as that which might result from chemiluminescence, should be excluded in the selection of detection wavelengths in order to minimize its interference with accurate temperature measurement.

12.2.2 Basic Components and Classification of Radiation Thermometry

Although radiation thermometry can be realized in different ways, as will be discussed later, it is important to recognize that every system consists of three basic components: a reference source, optics, and a detector, as shown in Fig. 12.4(a).

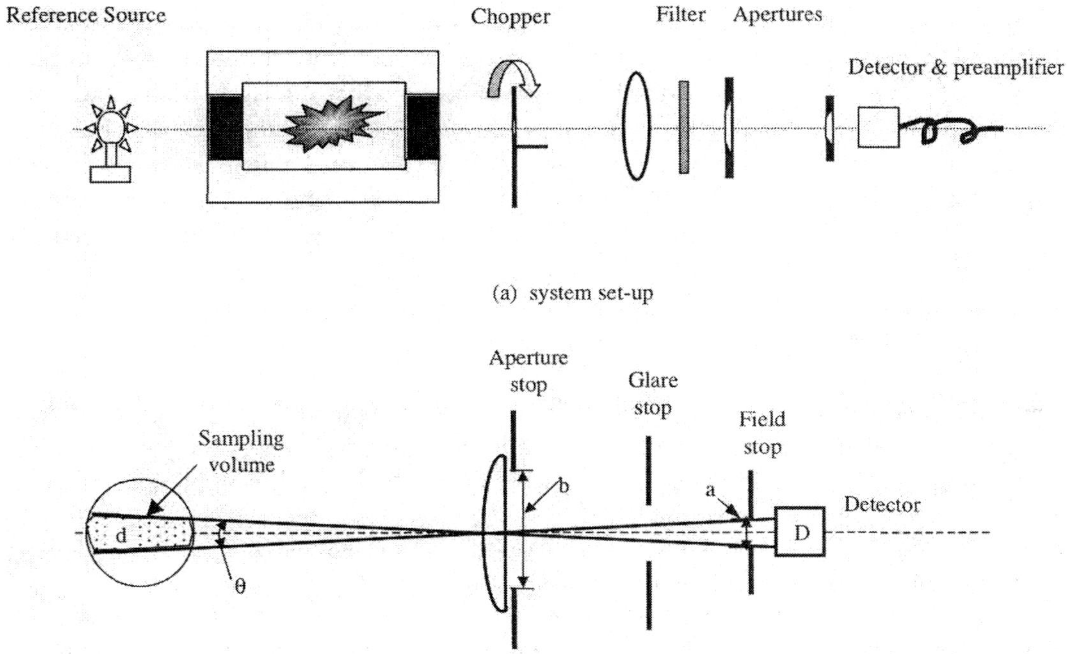

(b) detection optics

Fig. 12.4 Schematic of a radiation thermometry system.

The detector converts the amount of radiation into a measurable electrical signal. For gas temperature measurements in IC engines, the detector is normally a photon type device. Such devices convert the incident photons to either electrical current flow or voltage which can then be processed by amplifiers and other electronics into a display or recording. Photomultipliers are much more sensitive than any other detectors in the near UV and visible regions, and are characterized by their very fast response times. Silicon photodiodes are widely used in the visible and near infrared region due to their low cost and ease of operation. In the infrared (IR) region of 2 μm to 5 μm, PbS and PbSe detectors are available. For better performance in this region and beyond, the HgCdZnTe family of detectors can be employed. These IR detectors are normally cooled to improve their performance.

The optics serves two functions. First, it defines the spectral region of the detection by means of spectral filters or spectrometers. Second, it defines the angular field of view of the pyrometer

(Fig. 12.4[b]), θ. The field of view, θ, determines the spatial resolution of the system, and is defined by the use of a field stop and an aperture stop of diameter a and b, respectively, as shown in Fig. 12.4b. The use of optical lenses ensures that only a certain part of the gases or flame is imaged onto the detector. The focal length and the position of the optical lens determine the size of the image, D, of the emitting object, d.

The optical windows and lenses should be selected to be transparent in the spectral region of operation. For many applications, infrared radiation is detected. A few optical materials are transparent in the infrared region. Calcium fluoride has a high transmission (>90%) over the spectral range from 0.2–8.0 μm. It is slightly soluble in water and susceptible to thermal shock, and hence is not suitable for high temperature and pressure applications. However, calcium fluoride lenses can be used as an alternative to the more expensive sapphire ones. Germanium and zinc selenide are transparent up to 16.0 μm. Their transmission without optical coating is much lower. Sapphire offers excellent mechanical and optical properties as a window material. It has a transmission band that extends from the UV throughout the visible and near IR regions to 5.5 μm. The drawback is its relatively high cost, and its relatively large reflection loss without anti-reflection coating.

Finally, a reference source is normally required for performing the required temperature measurements. A reference source of blackbody radiation should be used, although a calibrated tungsten filament lamp may be used as a low-cost alternative. The temperature of the tungsten filament can be adjusted easily by adjusting the electrical current flow through the filament. A calibration curve at a particular wavelength can be obtained for the apparent temperatures of the filament as a function of the supplying current.

The two main classifications of radiation thermometers are based on the wavelength selection principle and on the method of operation. Classifications by wavelength selection principle are as follows:

(a) Total radiation thermometers. They are sensitive to the major part of the thermal radiation spectrum. In practice, the spectral response range of the detector or window will always limits the detection of radiation to a certain band of wavelengths.
(b) Broadband radiation thermometers. This type is sensitive to radiation over a broad band of wavelengths. The spectral range is usually restricted by the short wavelength cutoff of Planck's law and the long wavelength cutoff of the detector or window.
(c) Narrow band or monochromatic radiation thermometer. The response of this type of thermometer is limited to a single narrow band of wavelengths by the inclusion of an optical filter.
(d) Two-color or ratio thermometers/pyrometers. These instruments attempt to compensate for errors that arise from the emissivity of a real object by measuring the ratio of the spectral radiance in two narrow bands at different wavelengths. It is, however, necessary to assume that the source behaves as a grey body or to assume a value for the ratio of the spectral emissivity at the two wavelengths. The two-color method can be extended to multi-wavelength detection.

Alternatively, radiation thermometry can be classified based on its method of operation as direct-reading and comparison thermometers. As the name implies, direct-reading thermometers relate the output signal from the radiation detector directly to the temperature of the radiating source. This type is less commonly used than the comparison type thermometer, in which the unknown source temperature is obtained by comparison with a stable reference source, normally a small tungsten filament lamp as part of the thermometer. The comparison mode can be achieved by either employing a null detection system that gives a zero output when the test and reference sources match each other, or the test and reference sources are compared by chopping the radiation from them so that each of them alternately illuminates the detector. With phase-sensitive detection, AC amplification and filtering techniques may be used to reduce the effects of thermal noise, interference, and drift.

12.2.3 Monochromatic Radiation Thermometry

12.2.3.1 The Chopped Method (The Emission-Absorption Method)

Consider a reference source, gas volume, optics, and detector arranged as shown in Fig. 12.4. In the absence of gas, radiation received by the detector from the reference source through a narrow bandpass filter centered at λ is given by:

$$P_1 = \eta_c \tau_{w1} \tau_{w2} \varepsilon_b E_{b,\lambda}(T_b) \tag{12.7}$$

where η_c includes the geometric factor and spectral response of the detection system and τ_{w1} and τ_{w2} are the window transmission of windows 1 and 2, respectively. ε_b is the emissivity of the reference source.

With the gas present at a given temperature between the reference source and the detector, the radiation received by the detector includes radiation from the reference and the gas volume, and is given by

$$P_2 = \eta_c \tau_{w2} \left\{ \tau_{w1} \varepsilon_b E_{b,\lambda}(T_b)\left[1 - \alpha_g(\lambda, p, T_g)\right] + E_{g,\lambda}(T_g)\varepsilon_g(\lambda, p, T_g) \right\} \tag{12.8}$$

If a chopper is placed between the reference source and the gas, the radiation from the gas alone will be detected

$$P_3 = \eta_c \tau_{w2} \varepsilon_g(\lambda, p, T_g) E_{g,\lambda}(T_g) \tag{12.9}$$

Combining Eqs. (12.7), (12.8), and (12.9) together with Kirchhoff's law ($\alpha_g = \varepsilon_g$) leads to

$$\ln\left[\frac{P_1 + P_3 - P_2}{\varepsilon_b \tau_{w1} P_3}\right] = \frac{C_2}{\lambda}\left(\frac{1}{T_g} - \frac{1}{T_b}\right) \tag{12.10}$$

The chopped method can be applied to an IC engine by detecting the reference source radiation, P_1, during the intake stroke when the emission of gas is negligible. If this radiation is absorbed by more than a negligible amount by the gas during intake, the absorptivity $\alpha_g(\lambda, p_1, T_{g1})$ must be considered so that Eq. (12.7) becomes

$$P_1 = \eta_c \tau_{w1} \tau_{w2} \varepsilon_b \left[1 - \alpha_g\left(\lambda, p_1, T_{g1}\right)\right] E_{b,\lambda}\left(T_b\right) \tag{12.11}$$

Eqs. (12.8), (12.9), and (12.11) can then be combined as follows:

$$\ln\left[\frac{P_1 + P_3 - P_2}{\varepsilon_b \tau_{w1} P_3} \frac{1}{\left(1 - \dfrac{\alpha_g\left(\lambda, p_1, T_{g1}\right)}{\alpha_g\left(\lambda, p_2, T_{g2}\right)}\right)}\right] = \frac{C_2}{\lambda}\left(\frac{1}{T_{g2}} - \frac{1}{T_b}\right) \tag{12.12}$$

where T_{g2} is the gas temperature measured during compression or combustion.

Eq. (12.12) forms the basis of the emission-absorption method as applied to an IC engine. It shows that this method is capable of recording radiation data for a gas-temperature history as a function of crank angle on an individual cycle basis. Because it is difficult to obtain accurate values of the absorptivity of the gas under measurement conditions, the second term on the left-hand side of Eq. (12.12) should be ideally zero or negligible. This can be achieved by performing the calibration at a point during the intake stroke when the absorption is minimal.

12.2.3.2 The Null Method

Considering the simplified setup of Fig. 12.4 where the windows are absent, the radiation from a blackbody source received by the detector without and with the presence of gas is given, respectively, as follows:

$$P_1 = \eta_c E_{b,\lambda}\left(T_b\right) \tag{12.13}$$

$$P_2 = \eta_c \left\{ E_{b,\lambda}(T_b)\left[1-\alpha_g(\lambda,p,T_g)\right] + E_{g,\lambda}(T_g)\varepsilon_g(\lambda,p,T_g) \right\} \tag{12.14}$$

When the radiation from the reference source P_1 and that from the gas and source P_2 become equal, then Eqs. (12.13) and (12.14) can be equated and written in the form

$$E_{b,\lambda}(T_b)\alpha_g(\lambda,p,T_g) = E_{g,\lambda}(T_g)\varepsilon_g(\lambda,p,T_g) \tag{12.15}$$

According to Kirchhoff's law, the emissivity of the gas is the same as its absorptivity. Hence, Eq. (12.15) becomes

$$E_{b,\lambda}(T_b) = E_{g,\lambda}(T_g)$$

or

$$T_b = T_g$$

This equality establishes the conditions for determining the gas temperature at the null point: The radiation from reference source P_1 is the same as the radiation received at the detector from the gas media and reference source P_2.

In deriving the null point, Eqs. (12.14) and (12.15) do not include the effect of the windows. The radiation from both windows can be minimized by cooling. In addition, transmission characteristics of the window between the reference source and the gas does affect the null point and should be considered. In addition, when the null method is applied to an IC engine, the reference source radiation is detected during the intake stroke. The effect of gas absorption during the intake process should be considered. Effects of the window's transmission and gas absorptivity can be analyzed by referring to Eq. (12.12). At the null point, Eq. (12.12) becomes

$$\ln\left[\frac{1}{\varepsilon_b \tau_{w1} P_3}\frac{1}{\left(1-\dfrac{\alpha_g(\lambda,p_1,T_{g1})}{\alpha_g(\lambda,p_2,T_{g2})}\right)}\right] = \frac{C_2}{\lambda}\left(\frac{1}{T_{g2}}-\frac{1}{T_b}\right) \tag{12.16}$$

which gives the corrected gas temperature at the null point.

The null method is a point-by-point technique; that is, one gas temperature is measured at a particular time or crank angle, and a certain amount of time is required to build up a temperature-time history. Unlike the chopped method, by which the gas temperature can be measured as a function of time for the entire engine cycle with a given source temperature, the null method generates the gas temperature time history from different cycles, as it requires the adjustment of source temperature for different crank angles.

It should be pointed out that the radiation laws on which radiation thermometry is based apply to radiation of thermal origin only. Chemiluminescence, which is light emission from an electronically excited state due to chemical reaction, should be avoided when the gas temperature is measured by a radiation thermometer by proper selection of the detection wavelength range.

12.2.3.3 Application of Radiation Thermometry to IC Engines

In an SI engine, the combustion process is characterized by the propagation of a premixed flame burning with near or leaner than stoichiometric mixtures. Hence, there is very weak radiation in the visible spectrum. To combat this difficulty, solid particles may be introduced into the cylinder to produce intense radiation in the visible spectrum for direct flame photography and measurements of combustion temperature in SI engines. The first burned gas temperature measurement was achieved by the commonly known sodium line-reversal method by Withrow and Rassweiler [1932] more than sixty years ago.

The sodium line-reversal technique is a practical implementation of the null method. The spectral wavelengths employed in this case are the D lines of sodium. A brief description of the technique is given as follows.

When light from a bright background source is passed through sodium vapor, the two yellow lines of sodium vapor at 0.5890 μm and 0.5896 μm appear dark, in absorption, against the continuous spectrum from the background. It can be shown using Kirchhoff's law that when light from a bright background source giving a continuous spectrum is passed through a flame containing sodium vapor, the sodium lines will appear either in absorption, as dark lines against the continuum, or as bright lines, standing out brighter than the continuum according to whether the brightness temperature of the background source is higher or lower than the flame temperature. When the brightness temperature of the background and the flame temperature are exactly the same, the lines become invisible, having the same brightness as the background. This is shown by Eq. (12.15).

Thus, if one can vary the brightness of the background until the sodium lines just disappear, then the flame temperature is determined to be equal to the brightness temperature of the background. The background can be either a blackbody radiation source or a calibrated light source. Note that the brightness temperature of an object is the temperature of a blackbody for which the spectral radiance has the same value as for the object considered, as defined by Eq. (12.4). The brightness temperature of the calibration source is obtained from prior calibration.

Rassweiler and Withrow [1935] measured the average temperature in the burned gas region using the sodium-line reversal technique. Their experimental setup was similar to that shown in Fig. 12.4. A calibrated tungsten lamp was used as the light source. The radiation from the lamp through the cylinder gas was detected at the sodium lines by a spectroscope. The temperature of the calibrated lamp filament was adjusted until the sodium lines in the spectroscope reversed from emission to absorption, thus establishing the line-of-sight average temperature of the burned gas. The use of an optical chopper enabled the measurement of temperature at a particular crank angle. Using this technique, they found that the burned gas near the spark plug was a couple of hundred degrees hotter than the gas at the far end of the combustion chamber, due to the compression of the earlier burned gases.

The sodium line reversal method produces spatial and time averaged temperature measurements of burned gases. Because of the tedious procedure involved, it is more suited to stationary situations.

To overcome the limitation of the line-reversal technique, the absorption-emission method was adopted by a research group at the University of Wisconsin [Millar, et al., 1953] to give a continuous recording of the burned gas temperature. In a similar setup to Fig. 12.4, light from a molten zirconium arc lamp passed through the combustion chamber onto the entrance slit of a spectrograph. The radiation at one of the sodium lines was then measured by a photomultiplier at the output of the spectrograph. The attenuated intensity of the transmitted light was used to measure the absorbance (emissivity) of the cylinder gas. When the chopper obscured the light from the zirconium lamp, the photomultiplier measured radiation from the cylinder gas alone. Combination of the two readings enabled the temperature of the flame to be determined from Eq. (12.12).

In order to use the sodium line to measure the burned gas temperature, sodium has to be introduced into the combustion chamber. Millar et al. [1953] described a procedure of introducing a small amount of sodium into fuel. Sodium methoxide was first dissolved in methyl alcohol to form a solution. This solution was mixed with equal parts of acetone. The mixture was then mixed 10% by volume with the test fuel. After thorough mixing, the fuel was allowed to stand and the undissolved sodium mixture decanted. In the end, only 1% sodium was left in the fuel.

The introduction of sodium or any other solid particles is not desirable in the study of engine combustion due to complication of the seeding process and interference of the particles with the combustion process. Spectral radiation of sodium D lines occurs due to the presence of combustion, and hence cannot be used to measure the unburned gas temperature. On the other hand, CO_2 and water vapor are naturally present in both unburned gas and burned gas regions and emit/absorb strongly in the infrared region. Thus, radiation thermometry based on the detection of infrared radiation from CO_2 and/or water vapor superseded that based on sodium lines.

A comprehensive evaluation of an infrared monochromatic thermometer was carried out by Burrows et al. [1961] in order to measure the end-gas temperature in an SI engine. The experimental setup shown in Fig. 12.4 was used. Radiation from a blackbody source passed through sapphire windows and the combustion chamber before it was focused onto a gold-doped germanium detector. Diaphragms were employed to eliminate internal reflections off of cylinder walls. An interference filter was used to restrict the radiation to be detected to a wavelength band of 2.5 to 2.9 μm, which corresponds to the emission-absorption bands of water vapor and carbon dioxide. Both the null method and the chopped technique were tested. Calibration tests carried out in a steady flow device indicated that the null method produced more consistent temperature readings than the chopped method, compared with those obtained from a thermocouple probe. No obvious reasons were found for this discrepancy. Therefore, they adopted the null method for their temperature measurements. However, this had the disadvantage of requiring a number of engine cycles to obtain a temperature time history.

The temperature measurements were carried out in both a specially designed pocketed combustion chamber, for comparison tests with the velocity-of-sound technique, and a modified L-head combustion chamber, to compare the results with those obtained by the iodine-absorption thermometer [Burrows et al., 1961]. It was found that the infrared thermometer gave a temperature reading weighted toward the peak temperature in the gas, whereas the velocity-of-sound system measured a gas temperature closer to the mass-averaged temperature. The infrared thermometer was also found to perform better than the iodine-absorption thermometer.

Temperature data obtained by the infrared thermometer under motored and fired conditions showed an increase in the gas temperature with engine speed. The greatest temperature differences occurred near TDC where the heat transfer effect was largest. The infrared thermometer also showed that the use of lower octane fuel resulted in higher end-gas temperatures.

Infrared thermometry was also used to measure the burned gas temperature in the main combustion chamber of a divided-chamber spark ignition engine [Ryu and Asanuma, 1984; Sakurauchi, et al., 1987; Takagi et al., 1990]. The experimental setup is similar to that shown in Fig. 12.4. The infrared radiation from CO_2 gas was selected by a bandpass filter centered around 4.3 μm and focused onto a cooled InSb detector by calcium fluoride lenses. The initial measurements were carried out using the chopped method, where both the absorption and emission of CO_2 gas were measured [Ryu and Asanuma, 1984; Sakurauchi et al., 1987]. Later, the absorptivity of CO_2 gas as a function of crank angles was evaluated from

$$\alpha_{CO_2} = 1 - \frac{P_2 - P_3}{P_1}$$

where P_1, P_2, and P_3 are the radiation signal from the blackbody radiation source, the total radiation from the blackbody and the gas volume, and that from the gas volume alone, respectively.

They found that, among the three filters tested, the absorptivity of CO_2 gas was almost constant (0.88 ± 0.02) at a wavelength of 4.3 μm at an air/fuel ratio of 15:1 when the CO_2 concentration exceeded 3% [Takagi, 1990]. Based on this observation, burned gas temperature was measured by the direct emission method. After calibration against the blackbody, the gas temperature was inferred from Eq. (11.9) assuming a constant emissivity of 0.88. Comparison of gas temperature measured by absorption-emission and emission thermometers showed good agreement.

Temperature measurement by emission thermometer offers the advantage that only one window is required. However, the measurement of emissivity itself often proves to be very difficult. In the absorption-emission method (the chopped method), the emissivity factor may be ignored if the reference source radiation can be detected without the gas absorption as shown in Eq. (12.12).

12.2.4 Two-Color Thermometer/Pyrometer

The radiation from a volume of emitting medium detected at the wavelength λ is, according to Eq. (12.9), given by

$$P_\lambda = \eta_c \tau_w \varepsilon E_\lambda(T) \tag{12.17}$$

If the emissivity ε is known, and the detector output can be compared with that of a blackbody at the same wavelength, then the gas temperature T can be determined at once. Unfortunately, the emissivity is not generally known. It is possible, however, to cancel the need for that parameter or reduce its influence if measurements are made at two wavelengths rather than at one. The ratio of detector outputs at two different wavelenghts is given by

$$\frac{P_{\lambda 1}}{P_{\lambda 2}} = \frac{\eta_{c1}\varepsilon_{\lambda 1}}{\eta_{c2}\varepsilon_{\lambda 2}}\left(\frac{\lambda_2}{\lambda_1}\right)^5 \frac{\left(e^{(C_2/\lambda_2 T)}\right)-1}{\left(e^{(C_1/\lambda_1 T)}\right)-1} \tag{12.18}$$

where the spectral response factor η_c can be readily determined from calibration with a blackbody radiation source. If the radiation source behaves like a grey body, the emissivity ratio will be one and the radiation body temperature can then be obtained from Eq. (12.18). Otherwise, the emissivity ratio must be determined or some law for the dependence of emissivity on other parameters must be assumed. The two-wavelength radiation method, or two-color method, has been discussed in detail with regard to its application to the diesel combustion temperature measurements in Chapter 10. The emissivity of soot particles is assumed to vary as a simple power, α, of the wavelength as given by the Hottel and Broughton correlation [1932].

$$\varepsilon_\lambda = 1 - e^{\left(-KL/\lambda^{\alpha}\right)}$$

The application of the two-color method for measurement of gas temperature will be discussed in this section.

At General Motors, Agnew [1960] developed a two-color system for measuring the end-gas temperature in an SI engine. The cylinder head included an end-gas pocket with two opposite windows. The optical path length between the two windows was approximately 73 mm. Radiation from the end-gas region was focused onto the entrance slit of a monochromator and detected by a lead sulfide detector. Due to the absence of luminous particles in an SI engine, water vapor was taken as the radiating substance, and the wavelengths chosen were those of two water vapor bands at $\lambda_1 = 1.89$ and $\lambda_2 = 2.55$ microns. Through the use of a monochromator, radiation intensities at the two wavelengths were detected from different cycles though at the same crank angle. Thus, the temperature readings represent cycle-averaged values. An optical chopper was placed in front of the monochromator to provide a zero reference when the radiation was blocked and to prevent the extremely intense hot flame emission from reaching the detector by blocking radiation soon after TDC on the power stroke. The self-absorption by water vapor present between the window and detector was minimized by flushing the spectrometer with nitrogen continuously.

The crucial aspect of this technique lies in the accuracy of the emissivity ratio. In this study, the emissivity ratios were determined under motored conditions with air from measured radiation intensity ratios and temperatures calculated from measured gas pressure, air flow, and exhaust gas measurements, using the ideal gas law. The emissivity ratios were determined as a function of the optical path. The results showed a rather large effect of water vapor concentration on the measured emissivity ratios.

The major difficulty associated with this two-color method is accurate determination of the emissivity ratios. It is a recognized fact that the radiation temperature corresponds more closely to the maximum gas temperature, whereas the temperature calculated from pressure and the ideal gas law should correspond more closely to the mass-averaged temperature. Because the difference between the two can be as much as 10%, it seems that the emissivity ratio used may be better determined by means of the emission-absorption measurements at the two wavelengths.

Another difficulty with this method of temperature measurement is related to the fact that the emissivity ratio is critically dependent on the optical path of the gas and water vapor concentration. Because water vapor is produced in the end-gas region during the autoignition process, an iterative correction technique was devised by Agnew [1960] to account for the variation of emissivity ratio due to extra water vapor. The iteration procedure, as shown in Fig. 12.5, can be summarized as follows:

1. Determine the calibration curves for
   (a) the emissivity ratio as a function of optical path length;
   (b) the radiation intensity as a function of the optical path length at a known temperature that is close to the ones to be measured.
2. Choose a reference point from curve (b) at which the temperature, radiation intensity, and optical path length are all known.
3. Calculate the temperature using the emissivity ratio determined from curve (a) for an optical path length at the reference point.
4. Calculate the change in intensity from the reference point due to the difference in temperatures determined in step 3 and at the reference point using Planck's law.
5. Calculate the actual intensity change from the reference and subtract the change due to temperature alone (determined in step 4) to work out the difference in intensity due to optical path length.
6. From the difference in intensity due to optical path length and curve (b), determine the change in optical path length from the reference.
7. From the optical path length so determined, find a new value of emissivity ratio from curve (a) so that the true temperature is calculated from Eq. (12.18).

Despite these difficulties, through careful calibration and extensive evaluation, Angew was able to measure the end-gas temperature history during the compression stroke and early part of the combustion stroke in the engine. A strong influence of cylinder wall temperature on end-gas temperature was identified. The end-gas temperature was found to increase with increasing inlet temperature, coolant temperature, spark advance, and decreasing fuel octane number.

Compared with the monochromatic absorption-emission technique (the chopped method), the two-wavelength method requires only one window instead of two. The effect of window fouling on the accuracy of temperature measurements is minimized using the two-color method as the intensity ratio of two lines is measured.

12.3 The Velocity-of-Sound Method

12.3.1 Principle of the Velocity-of-Sound Technique

The velocity of sound in gases is given by the equation

$$c = \sqrt{\gamma R T} \tag{12.19}$$

where c = the speed of sound, [m/s]
γ = the ratio of the specific heats for the gas
T = the absolute gas temperature, [K]
R = the specific gas constant, [kJ/kg·K]

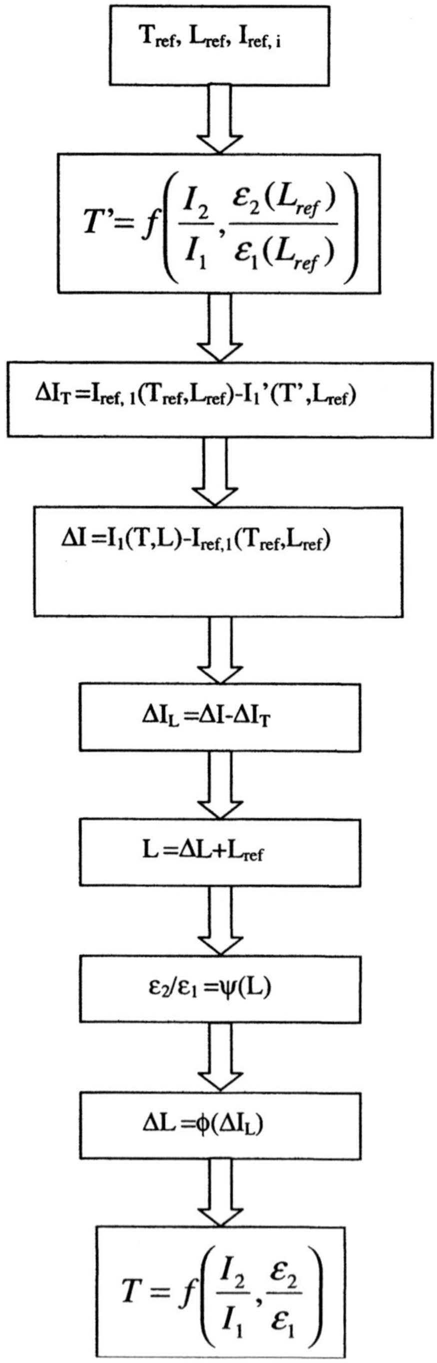

Fig. 12.5 Flow chart for the correction of the optical path length for the two-color gas temperature measurement.

The gas constant R is given by

$$R = \frac{\tilde{R}}{M}$$

where $\tilde{R}$ is the universal gas constant (8.31 KJ/kmol·K) and M is the molecular weight of the gas mixture.

The basic procedure for the velocity of sound measurement is rather straightforward. Two transducers are placed across the intended sound path of a known length. A sonic signal is sent across the gas volume, and the received signal is recorded. The time of flight is determined from the received signal, and an averaged velocity of sound is then calculated from,

$$c = \frac{d}{t}$$

where d = the sound path length as defined by the distance between the two transducers
t = the time of flight of sound between the two transducers.

To convert the measured velocity of sound into temperature, a calibration curve based on Eq. (12.19) is required prior to the start of experiments, which can be calculated from the values of γ and R for the gas mixture. The value of γ and molecular weight M of a mixture of known composition can be determined from the thermodynamic properties of each component [e.g., Heywood, p. 130, 1988; Ely and Huber, 1990]. For burned gas mixtures, chemical equilibrium calculations are required to determine the gas composition before the value of γ and M can be ascertained. A NASA equilibrium program is readily available for this purpose [e.g., Gordon and McBride, 1973]. A computer program for calculating properties of equilibrium combustion products, designed specifically for use in internal combustion engine applications, has been developed by Olikara and Borman [1975].

For unburned mixtures at atmospheric or lower pressures, the pressure has no effect on the velocity of sound. The speed of sound, however, is modified at elevated pressures. Because of the tedious nature of the calculation of sound velocity at elevated pressures, Livengood et al. [1958] plotted the ratio of sound velocities at a certain pressure and 0.01 bar as a function of pressure and used it to correct for the pressure effect on the velocity of sound.

12.3.2 Transmission of Sound and Selection of Acoustic Frequency

When sound travels through a medium, much of its energy is lost at the boundaries and within the medium. The reflection of sound at a boundary is given by [Kleppe, 1989],

$$R = \frac{4r_1r_2}{(r_1 + r_2)^2}$$

where r_i is the acoustic impedance of the medium I, which is defined as

$$r_i = \rho_i c_i$$

where ρ_i = the density of the ith medium
c_i = the speed of sound in the ith medium

When two transducers are employed in the velocity-of-sound temperature measurement, sound energy will experience a reflection loss twice as the sound signal is transmitted from transducer to cylinder gas and back to transducer.

Major loss occurs within the gas medium due to viscous losses, heat conduction losses, losses associated with molecular exchange of energy, and losses from turbulence. Therefore, the total absorption coefficient can be expressed as [Kleppe, 1989]

$$\alpha = \alpha_c + \alpha_m + \alpha_t$$

where the sum of the first two terms for air can be calculated as a function of temperature and pressure from a program [Kleppe, 1989]. The absorption coefficient due to turbulence α_t has to be determined experimentally for the particular flow system under investigation.

In the cylinder, the mixture composition varies due to stratification of fresh charge, residual gas, and fuel vapor throughout the cycle and from cycle to cycle. The gas temperature and pressure also changes rapidly during one engine cycle. Therefore, sound transmission varies accordingly throughout the engine cycle.

The sound transmission losses are strongly affected by the frequency used. Fig. 12.6 shows the frequency dependencies of absorption over a distance of 8 cm for different temperatures and at pressure of 1 bar. The peak of the absorption occurs around 1750 K. The position of the peak will move slightly to lower temperatures with increasing pressure. Thereby, attenuation will impose severe limitations on the operating frequency, especially if the burned gas temperature is to be measured.

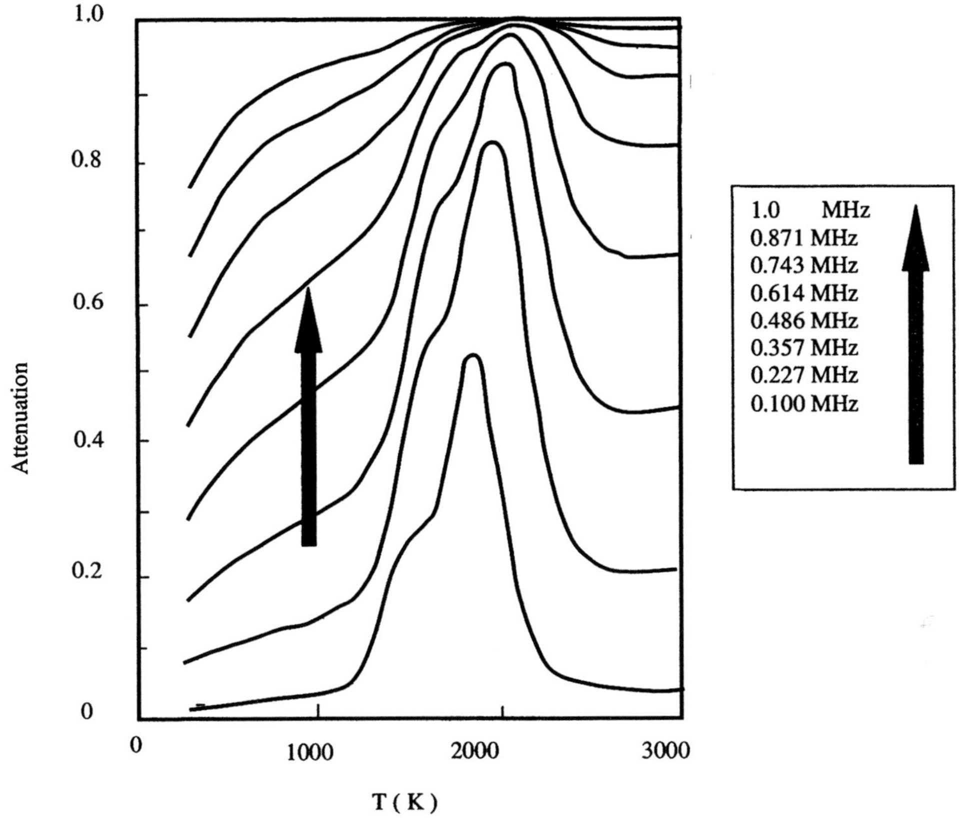

Fig. 12.6 Attenuation losses as a function of frequency over an 8 cm path.

12.3.3 Accuracy of the Velocity-of-Sound Method

The temperature inferred from Eq. (12.19) is averaged along the sound path between the two transducers. In the cylinder, there is a non-uniform gas temperature distribution. In the unburned region, there is a cool gas boundary layer near the cylinder wall and a core gas. In the burned gas region, there is a temperature gradient in the core gas region due to the compression of earlier burned gas volume as well as the thermal boundary near the cylinder wall. For such a non-uniform temperature distribution, the mass-averaged temperature is defined as,

$$T_{mass,V} = \frac{\int_V \rho T dV}{\int_V \rho dV}$$

Bauer et al. [1997] investigated the effect of boundary layer and temperature gradient on the calcualtion of the velocity-of-sound temperature. They found that the velocity-of-sound temperature deviated from the mass-averaged temperature by less than 0.5% in the unburned gas region. For the burned gas case, T_{sound} exceeded $T_{mass,\ v}$ by about 3% for a boundary layer thickness of 2 mm and a temperature difference of 180 K in the burned gas core region. Thus, the gas temperature measured from the velocity-of-sound can be considered very close to the mass averaged temperature.

To infer the gas temperature from the velocity of sound, values of γ and R must be found, and for this, the mixture composition must be known. Mixture composition in the cylinder is approximately known through measurement of the exhaust air/fuel ratio. However, uncertainty is introduced through the residual gas concentration, which is not readily measured. Bauer et al. found that a 20% residual gas concentration resulted in a 1% increase in dT/T for a stoichiometric propane and air mixture at 500 K. The effect of residual gas is more pronounced at temperatures above 1500 K.

The ratio of specific heats is affected by the temperature, as well as the mixture composition. The effect of γ on the temperature measurement can be assessed by differentiation of Eq. (12.19) with respect to γ,

$$dT = T \times \frac{-d\gamma}{2\gamma}$$

In the unburned gas region, the value of γ is about 1.35. If γ is changed from 1.3 to 1.35, a measurement error less than 2% would occur. In the burned gas region, the value of γ is much lower and difficult to predict accurately. Chemical equilibrium models are required to determine the values of γ and R. An iterative procedure may be used to infer the gas temperature from the velocity-of-sound measurement.

12.3.4 Implementation of the Velocity-of-Sound Method

The experimental setup consists of several components as shown in Fig. 12.7. A shaft encoder is used to trigger the transceiver at a specified crank angle. When triggered, the pulse generator sends out a high-voltage burst to the transmitting transducer on the right. The output of the receiving transducer is amplified before it is compared with the driving signal to the transmitting transducer.

Each transducer consists mainly of an oscillating crystal and a metal rod/plate to protect the crystal from the hot gases in the cylinder. Fig. 12.8 and Fig. 12.9 show the velocity-of-sound apparatus as used by Livengood et al. [1958] and Bauer et al. [1997]. Two facing holes are normally required in the cylinder head/block for the pair of transducers. Fig. 11.8 shows an alternative arrangement used by Livengood et al., where the brass coupling rod in the transmitter

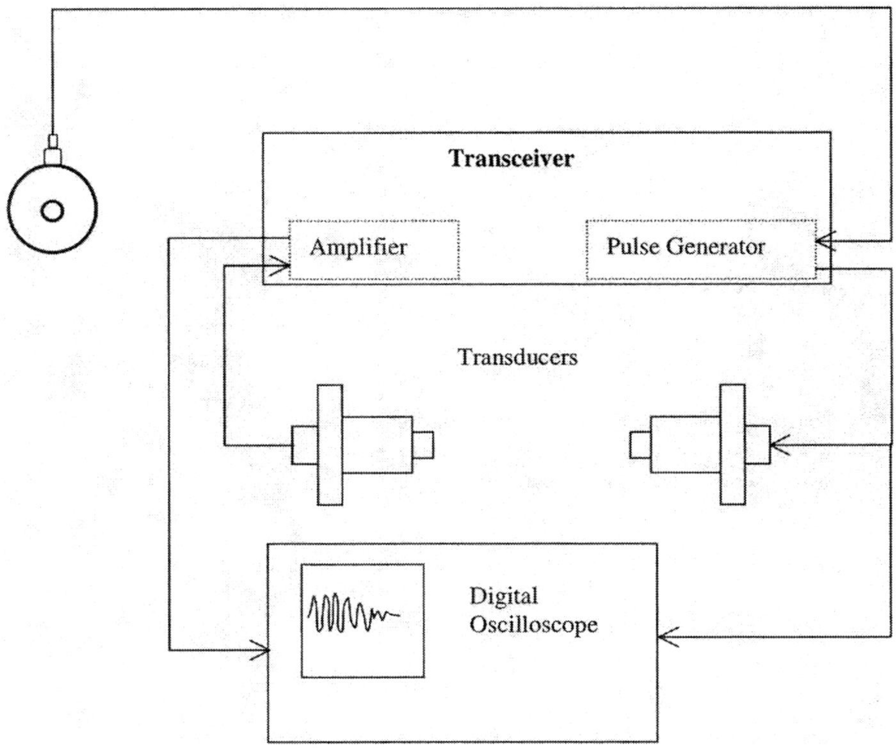

Fig. 12.7 Block diagram of a velocity-of-sound system.

protruded into the combustion chamber so that the temperature of the gas volume between the two orthogonally arranged transducers was measured.

The front faces of the transducers should be designed to be as large as possible to maximize the acoustic output and reception. However, the transducer housing is limited by the available access to the combustion chamber. The dimension of the coupling rod/plate should be chosen to match the nearest harmonic mode of the oscillating crystal and be thin to reduce transmission loss in the rod. In order to protect the crystal from overheating, the coupling rod is normally cooled by a built-in cooling circuit as shown in Fig. 12.8 or by an internal cooling jacket in the cylinder block (Fig. 12.9).

During engine operation, sound is generated by the vibration of many engine components. To receive a discernible sound signal through the gas volume, it is necessary to minimize sound coupling through solid sound paths at the interface between the transducers and the engine block. At the same time, sealing is required for the high cylinder pressure at the front surfaces of the transducers. O-rings have been used for this purpose by Livengood (Fig. 12.8) and Bauer et al. (Fig. 12.9). As cylinder pressure increases, the O-rings are compressed in the

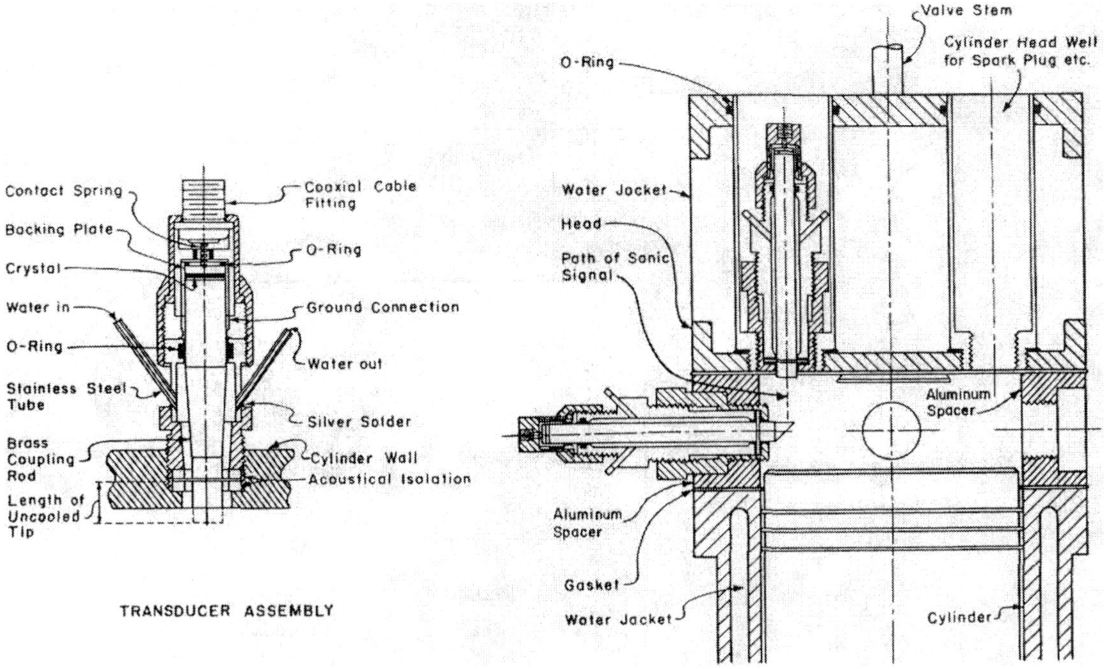

Fig. 12.8 Velocity-of-sound transducers and their application to a combustion chamber [Livengood et al., 1954].

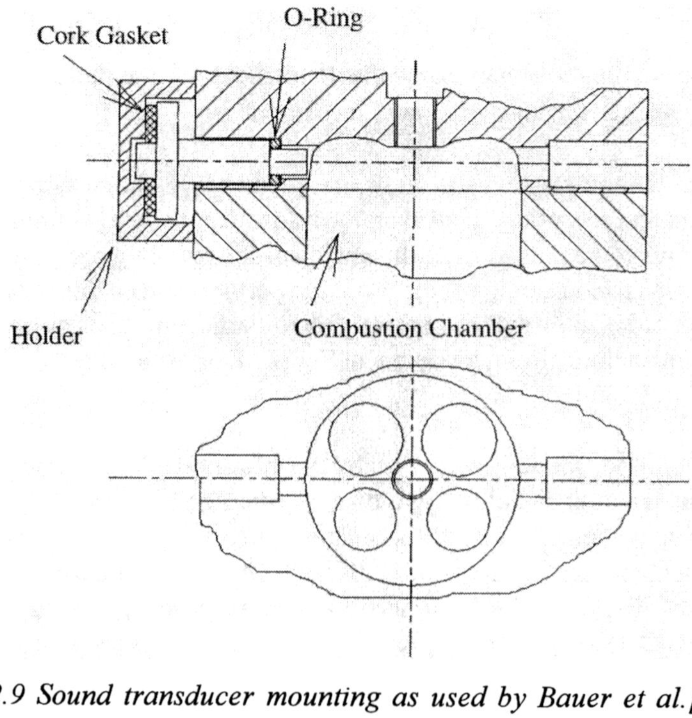

Fig. 12.9 Sound transducer mounting as used by Bauer et al.[1997].

axial direction and expanded in the radial direction to seal the cylinder gas. With such an installation, the seal contact force is proportional to the pressure load to be sealed. Because cross-talk increases with the in-cylinder pressure, signal strength also increases due to higher gas pressure. At low cylinder pressures, signal-to-noise ratio improves significantly. The insulation of the contact surfaces at the back of the transducers by cork disk behaves in a manner analogous to that of the O-ring seal in the front.

The generation and detection of the sound wave must be optimized. Fig. 12.10 shows a pair of transmitted and received signals. A gated high-voltage electric packet is used to generate a sending signal. High-voltage electric pulses cause the crystal in the transmitter to vibrate. This mechanical motion is conducted to the rod, which then starts a sound-wave pulse across the gas medium. On reaching the detector, the wave stimulates mechanical vibrations in the rod of the receiver, which are then converted to electric oscillations by the crystal in the receiver. The input drives the received signal up in amplitude for the duration of the input burst, and the received signal amplitude decreases thereafter. The peak of the received signal corresponds to the end of the input burst. The time difference, then, corresponds to the flight time of the signal plus a small constant delay in the electronic circuit.

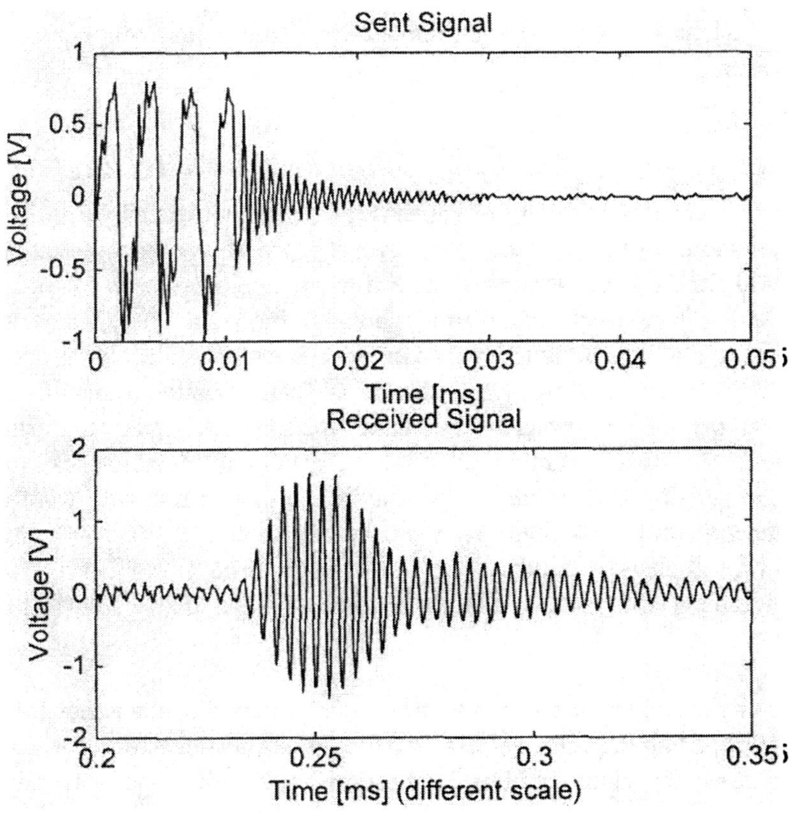

Fig. 12.10 Example of transmitted and received sonic signal traces [Bauer et al., 1997].

The selection of sound frequency has a large effect on the signal strength and accuracy of the measurement. As shown in Fig. 12.6, the signal frequency should be as low as possible to reduce attenuation. On the other hand, high frequencies will lead to smaller errors, because the measurement accuracy is on the order of half a wavelength. A 2.5 MHz frequency ultrasound was used by Livengood et al. for the measurement of the end-gas temperature with a rather short sound path of 15 mm. As a compromise, Bauer et al. adopted a frequency of 320 kHz due to the fairly long sound path of 80 mm and intended high-temperature applications.

Higher amplitude and longer burst width of the driving voltage are desirable for increased signal strength. However, a limit will be reached when the cross-talk begins to rise with the burst width.

Before transducers are applied to an engine, they should be calibrated prior to each experiment to determine the time delay from signal arrival at the transmitter to the signal peak in the received signal. Subsequent time measurements are made at the signal peak to determine the time of flight through the gas medium.

The velocity-of-sound technique provides time-resolved gas temperature measurements at a particular crank angle with a time resolution depending on the sound frequency and sound path, typically less than 1°CA. A complete temperature-time history will require measurements from many cycles.

12.3.5 Application of the Velocity-of-Sound Method to IC Engines

The initial development of the velocity-of-sound method was motivated by a desire to understand knocking combustion. In an effort spanning several years, the velocity-of-sound method was developed at MIT for the measurement of the end-gas temperature in an SI engine [Livengood et al., 1954; 1958]. The experimental setup is shown in Fig. 12.8. Here, an oscillating barium titanate ceramic crystal was mounted on the outer end of a water-cooled brass rod. Electric pulses from a 2.5 MHz oscillating circuit caused the crystal on the transmitter to vibrate. A fairly short sound path (15 mm) was used to reduce the sound losses due to the high-frequency signal employed. The low temperature encountered also facilitated the use of high-frequency sound for better measurement accuracy. Comparative engine test data were obtained on a specially designed combustion chamber with a pocketed end-gas region using the infrared thermometer and velocity-of-sound system by Burrows et al. [1961]. Excellent agreement was found between the two systems, though the former tended to give slightly higher readings than the latter.

An important contribution to the understanding of knocking combustion was made by Gluckstein and Walcutt [1961] using the velocity-of-sound method. They measured end-gas temperature just prior to incipient knocking combustion for many fuels and under a variety of operating conditions. The measurements were conducted in a CFR engine with a modified combustion chamber. The transducers were mounted on the sides of a tapering ellipsoidal pocket of the combustion chamber. The paramount importance of the end-gas compression time and final gas temperature, that is, the temperature of the end-gas at the time of autoignition, was well established by this work. Gluckstein and Walcutt's work showed that the end-gas temperature at the detonation limit was nearly constant at 1020 ± 22°C for a wide variety of fuels over a considerable range of inlet temperatures with compression ratio as the dependent variable.

The velocity-of-sound method was recently developed further by Bauer et al. [1997] for temperature measurements of both unburned and burned gases. The two transducers were mounted opposite to each other across the triangular portion of a pent-roof cylinder head (Fig. 12.9). Uncertainties due to the time-of-flight detection, gas composition, and the impact of gas temperature distribution, were analyzed. For unburned cases, the sound temperature was found to be very close to the mass-averaged gas temperature. The effect of mixture composition variation was considered negligible. The largest error stemmed from the inaccuracy in the time-of-flight detection and was estimated to be 2%. Uncertainties were larger for burned gas cases, due to shorter time-of-flight and larger temperature gradient. An aggregate uncertainty of approximately 5% was claimed. However, the in-cylinder measurements were restricted to compression and early expansion strokes. Low-pressure and high-temperature conditions during the remainder of the cycle were too detrimental to receive a signal.

The same experimental setup was also evaluated for exhaust gas temperature measurement by Bauer et al. In the exhaust, the wind effect, due to the presence of bulk gas movement, blows the sound signal in the downstream direction. For a given exhaust gas velocity u, the deflection angle is $b = \tan^{-1}(u/c)$. Because of the wind effect and possible interference by exhaust gas dynamics, measurements of exhaust gas temperature were excluded during the blow-down period before TDC of the exhaust stroke.

12.4 Spectroscopic Temperature Measurement Techniques

12.4.1 Rotational Temperature

12.4.1.1 Theory

The intensities of spectral lines can be used for the measurement of gas temperature. These line emissions arise from transitions between two energy states, as discussed in Chapter 5, according to

$$\nu_{nm} = (E_n - E_m)/h \tag{12.20}$$

where ν_{nm} is the frequency of the light emission resulting from a transition between energy states E_n and E_m, and h is the Planck's constant.

The emission intensity of an individual line depends on the transition probability for the spontaneous emission for that particular line, also known as the Einstein coefficient, and on the population in the excited energy state. The intensity of a particular line is thus given by an expression of the form

$$I_{nm} = \eta_c A_{nm} N_n h\nu_{nm} \tag{12.21}$$

where η_c is a calibration factor depending on the experimental setup. A_{nm} is the Einstein coefficient for spontaneous emissions from the energy level n to level m. N_n is the population of the species in the upper level n.

For molecules, rotational lines are normally used in spectroscopic temperature measurement. Spectra resulting from pure rotational transitions alone are usually too far in the infrared for useful application. Spectra resulting from simultaneous rotational and vibrational transitions are also in the infrared regions where measurements are difficult. However, for simultaneous electronic, vibrational, and rotational transitions, the rotational fine structure may appear in the visible or ultraviolet wavelength ranges where high resolution and rapid precision measurements are most convenient. One such example is the emission from the OH radical. OH radicals are abundantly available during the combustion of hydrocarbon fuels. They are easily excited to the first electronic state at high temperatures due to collision, and this is followed by spontaneous radiation as they move from the excited state to the ground state. The emission lies in the near UV region (306 nm), and can be easily detected by a spectroscopic detection system.

For diatomic molecules, such as OH radicals, the energy due to rotation is given by Eq. (5.1)

$$E_n = J(J+1)Bhc \tag{12.22}$$

The population in the excited energy state in thermal equilibrium is given by the Boltzmann distribution

$$N_n = N\frac{(2J_n+1)}{Q_{rot}}\exp[-E_n/kT] \tag{12.23}$$

where the rotational partition function Q_{rot} is generally given by

$$Q_{rot} = \sum_{J=0}^{\infty}(2J_n+1)\exp[-E_n/kT] \tag{12.24}$$

In the case of large kT/hcB, the summation can be replaced by integration and one obtains

$$Q_{rot} = kT / hcB$$

where B is the rotational constant for a particular species, the value of which changes in electronic transitions.

Substituting Eqs. (12.22) and (12.23) into Eq.(12.21), the rotational line intensity is then given by

$$I_r = \eta_c A_{nm} h\nu_{nm} N \frac{(2J_n + 1)}{Q_{rot}} \exp[-E_n / kT] \tag{12.25}$$

The calculation of temperature, T, would require measurement of the absolute intensity, I, and the number of molecules, N. The use of more than one line avoids these absolute measurements. The measurement of the intensity ratio of two or more spectral lines and knowledge of the Einstein coefficients for spontaneous emissions will yield a value of the temperature from

$$\frac{I_a}{I_b} = \frac{A^a_{nm}}{A^b_{nm}} \frac{(2J^a_n + 1)}{(2J^b_n + 1)} \frac{\nu^a_{nm}}{\nu^b_{nm}} \exp\left[-\frac{E^a_n - E^b_n}{kT} \right] \tag{12.26}$$

or

$$T = \frac{E^a_n - E^b_n}{k \ln\left(\frac{I_a}{I_b} \frac{A^a_{nm}}{A^b_{nm}} \frac{(2J^a_n + 1)}{(2J^b_n + 1)} \frac{\nu^a_{nm}}{\nu^b_{nm}} \right)} \tag{12.27}$$

Here, $k = 1.3806 \times 10^{-23}$ J/K is the Boltzmann constant. Eq. (12.27) forms the basis of gas temperature measurement by means of rotational line intensities. For known spectra, the energy levels E can be obtained from published tabulations. Established values for the Einstein coefficients are also available for many spectra. It is seen from Eq. (12.27) that the larger the difference between E^a_n and E^b_n, the more sensitive the intensity ratio is to a change in T, while from a practical standpoint the lines should not be too different either in intensity or in wavelength for accurate temperature measurements.

The species that has been most studied for rotational temperature measurements is OH at 3064 Å, corresponding to the (0, 0) vibrational band of the $A^2\Sigma - X^2\Pi$ transition. The rotational energies of OH in the first excited electronic and ground vibrational level are listed in

Table 12.1 [Dieke and Crosswhite, 1962]. Table 12.2 gives the values of Einstein coefficients for spontaneous emissions from OH [Chidsey and Crosley, 1979]. Before using the values of E_n [cm^{-1}] in Table 12.1, they need to be multiplied by 1.986×10^{-23} to convert into joules.

Table 12.1 Rotational energies of OH at the ground vibrational level (v = 1) of the $^2\Sigma$ state

K	F_1 [cm^{-1}]	F_2 [cm^{-1}]
0	32440.16	32440.5
1	32474.62	32474.3
2	32542.56	32541.98
3	32644.22	32643.45
4	32779.49	32778.49
5	32948.31	32947.05
6	33150.14	33148.73
7	33384.97	33383.26
8	33652.29	33650.38
9	33951.8	33949.67
10	34282.99	34280.64
11	34645.53	34642.92
12	35038.61	35035.86
13	35462.01	35459.02
14	35914.82	35911.59
15	36396.66	36393.24
16	36906.50	36902.9
17	37443.91	37440.15
18	38007.9	38003.93
19	38597.79	38593.62
20	39212.69	39208.99

In additional to OH radicals, CH emission bands at 4315 and 3900 Å and the Swan bands of C2 were also used in flame temperature measurements [Gaydon and Wolfhard, 1970; Gaydon, 1974].

The main sources of error in this method are self-absorption and the background intensity. To accurately measure the intensity, conditions must be chosen for which the signal intensity greatly exceeds that of the background. This implies limiting measurements to sufficiently strong lines and to gases sufficiently free of solid particles. The measurement of relative spectral line intensities may also be confused by the tendency of the gas to reabsorb in the wavelength of the line, hence the gaseous layers must be sufficiently thin.

Table 12.2 Einstein A coefficients for the (0, 0) band of OH, normalized to $P_1(1) = 1000$

K	O_{12}	P_1	P_2	Q_{12}	P_{12}	Q_1	Q_2	R_{21}	Q_{21}	R_1	R_2	S_{21}
1	0	1000	0	140	559	477	561	210	667	93	140	26
2	130	671	461	131	310	599	578	182	307	162	199	31
3	86	575	447	107	218	663	634	144	187	212	239	29
4	61	529	446	85	160	701	678	113	127	247	269	25
5	44	501	444	67	120	724	708	89	91	272	290	21
6	33	481	440	53	93	736	726	71	68	290	305	17
7	25	464	434	43	73	740	735	58	53	302	316	14
8	20	449	427	35	58	739	739	47	42	309	322	12
9	16	435	420	29	47	734	737	40	34	314	326	10
10	13	421	410	24	39	726	731	33	27	315	327	9
11	10	408	401	20	33	715	722	28	23	315	327	7
12	9	395	390	17	27	701	711	24	19	313	324	6
13	7	382	380	14	23	686	697	21	16	309	320	5
14	6	369	368	12	20	668	682	18	14	304	315	5
15	5	355	357	10	17	650	664	16	12	297	309	4
16	4	342	344	9	15	623	645	14	10	290	301	4
17	4	328	332	8	13	607	624	12	9	281	293	3
18	3	314	319	7	12	585	603	11	8	272	284	3
19	3	300	305	6	10	561	580	10	7	261	274	2
20	2	285	292	5	9	536	556	9	6	251	263	2

12.4.1.2 Implementation and Application of Rotational Temperature of OH to an SI Engine

Fig. 12.11 shows the experimental setup used by Bach and his co-workers [1996] for measurement of the burned gas temperature in a production spark ignition engine. The flame radiation was transmitted through an optical fiber mounted in the spark plug and dispersed by a spectrometer. The spectrum of the dispersed light was then detected by an intensified CCD camera which was mounted at the exit slit of the spectrometer. In order to protect the optical fiber, a sapphire rod of 0.8 mm length was mounted at the tip of the spark plug probe. The field of view or the probe volume was defined by the dimension of the gap between the sapphire rod and the optical fiber. The rotational lines were identified from the recorded spectrum averaged over several cycles at the same crank angle based on the published results [Learner, 1962; Dieke and Crosswhite, 1962]. The flame temperature was determined from the five lines, $R_2(3)$, $R_2(4)$, $Q_1(14)$, $Q_1(15)$, and $Q_2(12)$, of the $A^2\Sigma$ - $X^2\Pi$ (0,0) band of the OH radical at 306.36 nm. From these values, six line intensity ratios, $R_2(3)/Q_2(12)$, $R_2(3)/Q_1(14)$, $R_2(3)/Q_1(15)$, $R_2(4)/Q_2(12)$, $R_2(4)/Q_1(14)$, and $R_2(3)/Q_1(15)$, were used to calculate a mean value of the resulting six temperatures. Burned gas temperatures during combustion were measured under different loads at the spark plug and a position further away from the spark plug. The NO concentrations were then calculated from the measured temperature data using the extended Zeldovich mechanism.

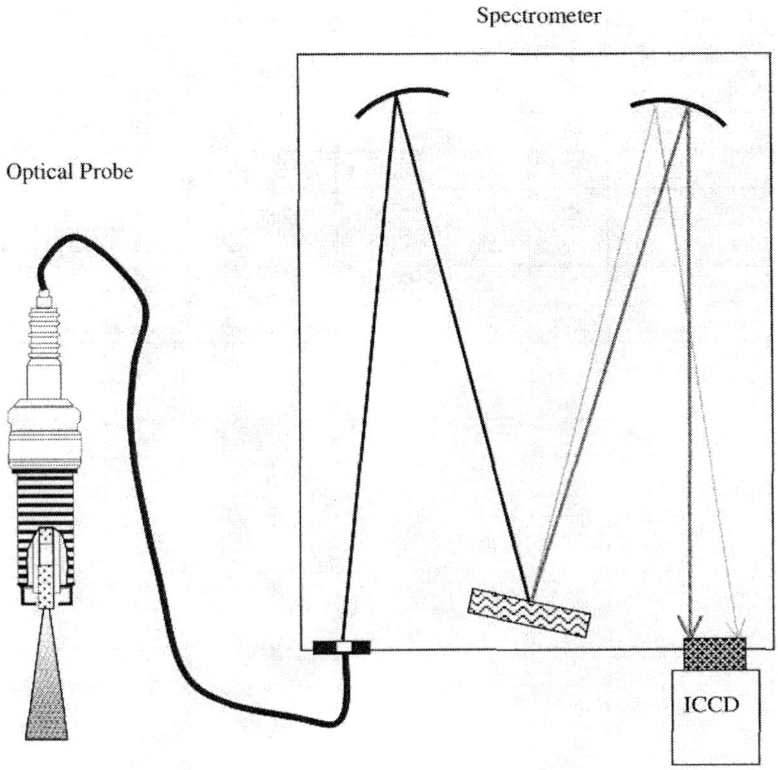

Fig. 12.11 Experimental setup for OH rotational temperature measurements.

The deviation of mean temperatures was rather large (about ±20%) during the initial combustion period (20 °CA after ignition) and decreased to less than ±10% 40 °CA after ignition. It was also found that the combustion temperatures were slightly higher than expected during the initial period of combustion. This is likely caused by the departure from thermodynamic equilibrium of OH radicals in the excited energy state in the reaction zone and the interference of chemiluminescence [Gaydon, p. 266, 1974]

12.4.2 Line Absorption Thermometry

Observations of spectra are not limited to those of emitted radiation. Many spectral lines can be reversed and made to appear as absorption lines. When the radiation from a continuous source, such as that of a blackbody, is passed through a gas volume, spectra of the constituent atoms or molecules appear as absorption spectra. The loss of light intensity due to energy absorbed may be written

$$\Delta I_{nm} = I_0 N_m B_{mn} h\nu_{nm} L / c \qquad (12.28)$$

where N_m is the population of the species at the lower energy level, and is given by the Boltzmann distribution (Eq. [12.23]). B_{mn} is the Einstein coefficient for absorption, and L is the optical path length. Eq. (12.28) forms the basis of absorption thermometry. Single-line or two-line absorption can be used to infer the gas temperature. The experimental setup is similar to that for the monochromatic absorption-emission technique described previously.

Chen et al. [1954] applied the iodine absorption technique to the measurement of end-gas temperatures in a CFR SI engine. The cylinder charge was seeded with iodine vapor by dissolving 10.43 g of iodine per gallon of fuel. The absorption spectral lines for the diatomic iodine molecule at 5230 Å, 5750 Å, and 6050 Å were recorded in order to measure unburned gas temperatures. These bands correspond to the vibrational transitions of the iodine molecule. A set of calibration curves was provided by heating an iodine-filled quartz absorption cell. When plotted against crank angle, the temperature curves showed a second peak during the late stage of compression stroke. Subsequent measurements carried out by Burrows [1961] using both iodine-absorption and infrared monochromatic detection system shows no evidence of the anomalous behavior. The difference between the earlier results by Chen and later ones by Burrows was likely caused by the departure from local thermal equilibrium of the iodine molecules.

12.5 Temperature Measurement by Spontaneous Raman Scattering (SRS)

12.5.1 Principles of Raman Thermometry

12.5.1.1 Raman Spectrum Synthesis

The theoretical treatment of spontaneous Raman scattering (SRS) in Chapter 5 provides the fundamental formulation upon which SRS temperature measurements are based. Species concentration measurements discussed in Chapter 7 derive from the strength of the Raman scattering, whereas temperature measurements derive from the shape of the Raman spectrum.

In Chapter 5, the vibrational Raman scattering signal integrated over the vibrational band spectrum was derived. However, Raman thermometry requires a knowledge of the shape of the Raman spectrum, as well as its intensity, as a function of temperature. The process of calculating a spectrum is known as spectral synthesis. This process involves the construction of a spectrum by examining the spectral position of each transition, the line strength of that particular transition, and the number of molecules in the initial quantum state. Then, convolutions are performed over the spectrum to account for the excitation laser line width, the detection bandwidth, and shape of the measuring instrument. The shape of the Raman spectrum is a convolution of the spectral power and the slit function of the monochromator.

Here, we concentrate on the Raman spectrum of diatomic molecules. The individual line strength of the vibrational-rotational Raman scattering spectrum for diatomic molecules is given by [Long, 1977]

$$I_{rot,vib} = \frac{\pi^2}{\varepsilon_0^2} \left(\nu_{ex} \pm \nu_{rot,vib}\right)^4 N_{ini} \phi\left(\alpha'^2, \gamma'^2, {}^i\rho_s\right) I_o \tag{12.29}$$

where ${}^i\rho_s$ indicates the polarization state, e.g., ${}^{\|}\rho_{\perp}$ shows that the incident light is plane-polarized parrallel ($\|$) to the scattering plane and the light scattering component perpendicular($\perp$) to the scattering plane is detected. The scattering plane is defined in Fig. 5.3. N_{ini} is the number of molecules in the initial vibrational-rotational state, given by

$$N_{J,ini} = \frac{N_{v,ini}}{Q_{rot}} g_{ini}\,(2J+1)\exp\left(-B_v hcJ(J+1)/kT\right) \tag{12.30}$$

where $Q_{rot} = kT/hcB_v$. g_{ini} is the statistical weight of the initial vibrational-rotational state, and for N_2 molecules, it has a value of 3 and 6 for odd and even numbered rotational energy levels, respectively. $N_{v,ini}$ is the number of molecules in the initial vibrational state, as prescribed by the Boltzmann distribution:

$$N_{v,ini} = N\exp\left(-hcv\nu_{vib}/kT\right)\ \left[1-\exp\left(-hc\nu_{vib}/kT\right)\right] \tag{12.31}$$

The grouping hc/k has a value of 1.44 K deg cm^{-1}.

Values of the function $f(\alpha'^2, \gamma'^2, {}^i\rho_s)$ for calculating the individual transition line strength from Eq. (12.29) are dependent on the polarization state of incident light and scattering light, as well as the change in vibrational-rotational states [Long, 1977],

Q branch ($\Delta v = \pm 1,\ \Delta J = 0$):

$$\phi\left(\alpha'^2, \gamma'^2, {}^{\perp}\rho_{\perp}\right) = k(v)\left[(\alpha')^2 + \frac{4}{45} b_{J,J}(\gamma')^2\right] \tag{12.32a}$$

or

$$\phi\left(\alpha'^2, \gamma'^2, {}^{\perp}\rho_{\|}\right) = k(v)\left[\frac{1}{15} b_{J,J}(\gamma')^2\right] \tag{12.32b}$$

O and S branch ($\Delta v = \pm 1,\ \Delta J = \pm 2$):

$$\phi\left(\alpha'^2, \gamma'^2, {}^{\perp}\rho_{\perp}\right) = k(v)\left[\frac{4}{45} b_{J\pm 2,J} (\gamma')^2\right] \tag{12.33a}$$

$$\phi\left(\alpha'^2, \gamma'^2, {}^{\perp}\rho_{\parallel}\right) = k(v)\left[\frac{1}{45} b_{J\pm 2,J} (\gamma')^2\right] \tag{12.33b}$$

where k(v) is dependent on whether it is a Stokes and anti-Stokes line

$$k_s(v) = \frac{h}{4\pi\omega_e}(v+1) \quad \text{Stokes Line}$$

$$k_{as}(v) = \frac{h}{4\pi\omega_e} v \qquad \text{Anti-Stokes Line}$$

where v is the vibrational number of the initial energy level. In the case of N_2, $\omega_e = \nu / c =$ 2360 cm^{-1}.

The Placzek-Teller coefficients $b_{J',J'}$ are given by

$$b_{J+2,J} = \frac{3(J+1)(J+2)}{2(2J+1)(2J+3)}$$

$$b_{J-2,J} = \frac{3J(J-1)}{2(2J+1)(2J-1)}$$

$$b_{J,J} = \frac{J(J+1)}{(2J-1)(2J+3)}$$

Note from Eqs. (12.29) to (12.33) that the O- and S-branch transitions arise solely from the anisotropy invariant γ' of the derived polarizability tensor, while the Q branch transitions arise from both the mean α' and the anisotropy invariant γ' of the derived polarizability tensor. For linear molecules, the value of α' greatly exceeds that of the anisotropy invariant γ'. This means that in vibrational-rotational Raman scattering, Q-branch transitions are much stronger than the O- and S-branch transitions. Furthermore, the O and S branches are spectrally quite spread out and can be neglected in the limited spectral region about the Q-branch. The anisotropy contribution can also be ignored due to its weak contribution. Within a given vibrational band, then, the individual Q-branch vibrational-rotational line strengths will mainly scale with the initial rotational state population.

Eqs. (12.29)–(12.33) also show that the relative strength of the various vibrational bands will scale as (v + 1) for Stokes, v for anti-Stokes, in addition to weighting by the relative populations. The effect of relative populations is that the anti-Stokes bands are much weaker than the Stokes bands due to less populated upper vibrational energy states.

Here we will use N_2 as the example due to its importance in combustion thermometry. The spectra of other diatomic molecules are similar. Because of their relative strong signal strength and hence, wider use in Raman thermometry, we will consider the fundamental Q-branches. The fundamental Q-branch vibrational-rotational line locations are given by (Eq. [5.4])

$$\nu = E_v / hc = \nu_{ex} \pm \left[\omega_e - 2\omega_e x_e (v+1) - \alpha_e J(J+1)\right] = \nu_{ex} \pm \nu_{vib} \qquad (12.34)$$

The plus and minus signs represent the anti-Stokes and Stokes vibrational band, respectively. From Eq. (12.34), it is apparent that without significant vibrational-rotational interaction, i.e., $\alpha_e = 0$, all Q-branch transitions would reside upon one another. In N_2, with $\alpha_e = 0.017$, a given vibrational band is shaded toward smaller Raman shift with increasing J. The effect of anharmonicity, $\omega_e\chi_e$, is to shift and thus separate the various vibrational sequences from each other. Without this, all of the vibrational bandheads would coincide with each other as well. These features are illustrated in the calculated Stokes vibrational-rotational Raman spectrum as a function of temperature, as shown in Fig. 12.12. The shifting of the vibrational-rotational state populations with temperature is reflected in the increasing appearance of wings in the vibrational-rotational Raman spectrum.

12.5.1.2 Raman Thermometry

Fig. 12.13 illustrates the four common methods of determining temperature from N_2 vibrational Raman scattering as given by Lapp [1974] and Drake et al. [1982]. All of the techniques rely on the increased population in the upper N_2 vibrational levels with rising temperature. In the band area method (Fig. 12.13[a]), the temperature is determined from the relative intensities of the upper and ground level vibrational bands within the Stokes Q branch. For the contour fit method (Fig. 12.13[b]), the temperature is determined by theoretical replication of the relative Stokes spectral intensity profile. In the band peak intensity method (Fig. 12.13[c]), the temperature is obtained by using the intensity ratio for two spectral bands at the peaks of the contour.

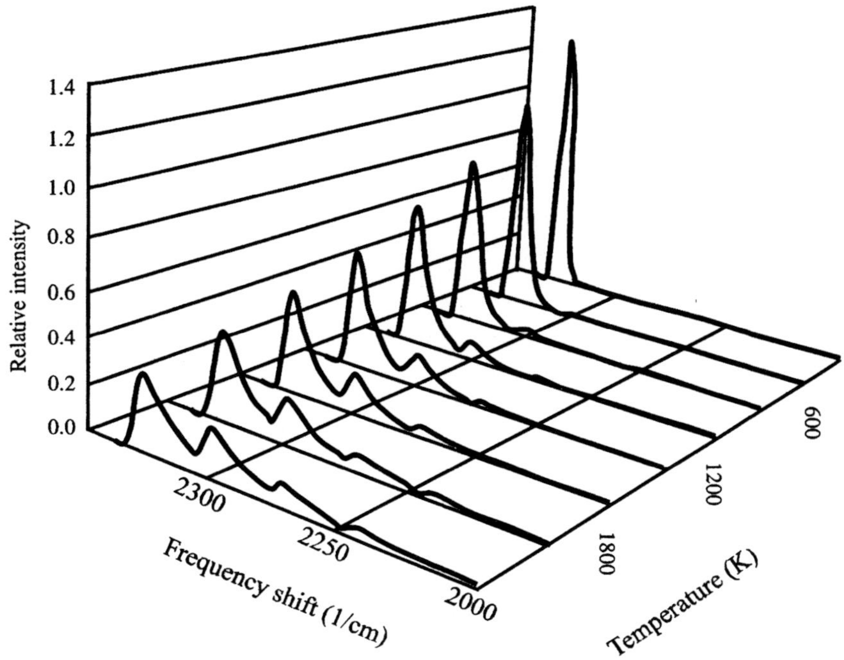

Fig. 12.12 N_2 vibrational Q-branch Raman spectrum as a function of temperature at constant concentration.

The Stokes/anti-Stokes method (Fig. 12.13[d]) involves a wider spectral bandpass and hence provides the maximum collected signal. The ratio of Stokes/anti-Stokes fundamental vibrational Raman scattering for N_2 is related to temperature by

$$P_s / P_{as} = \frac{(\nu_o - \nu_{vib})^4}{(\nu_o + \nu_{vib})^4} \exp(-hc\nu_{vib} / kT) \frac{f^s(T)}{f^{as}(T)} \tag{12.35}$$

The Raman shift, ν_{vib} , corresponds to the energy difference between the ground and first excited vibrational levels. The exponential term in Eq. (12.35) indicates that, the higher the gas temperature is, the smaller the ratio becomes as the anti-Stokes signal increases. The bandwidth factors, $f^i(T)$, are included to account for the fraction of the total Raman scattering actually "seen" by the detector, as discussed in Chapter 7. In Fig. 12.14, the anti-Stokes/ Stokes bandwidth factor ratio in N_2 is displayed for laser excitation at 532 nm and various interference filter bandwidths.

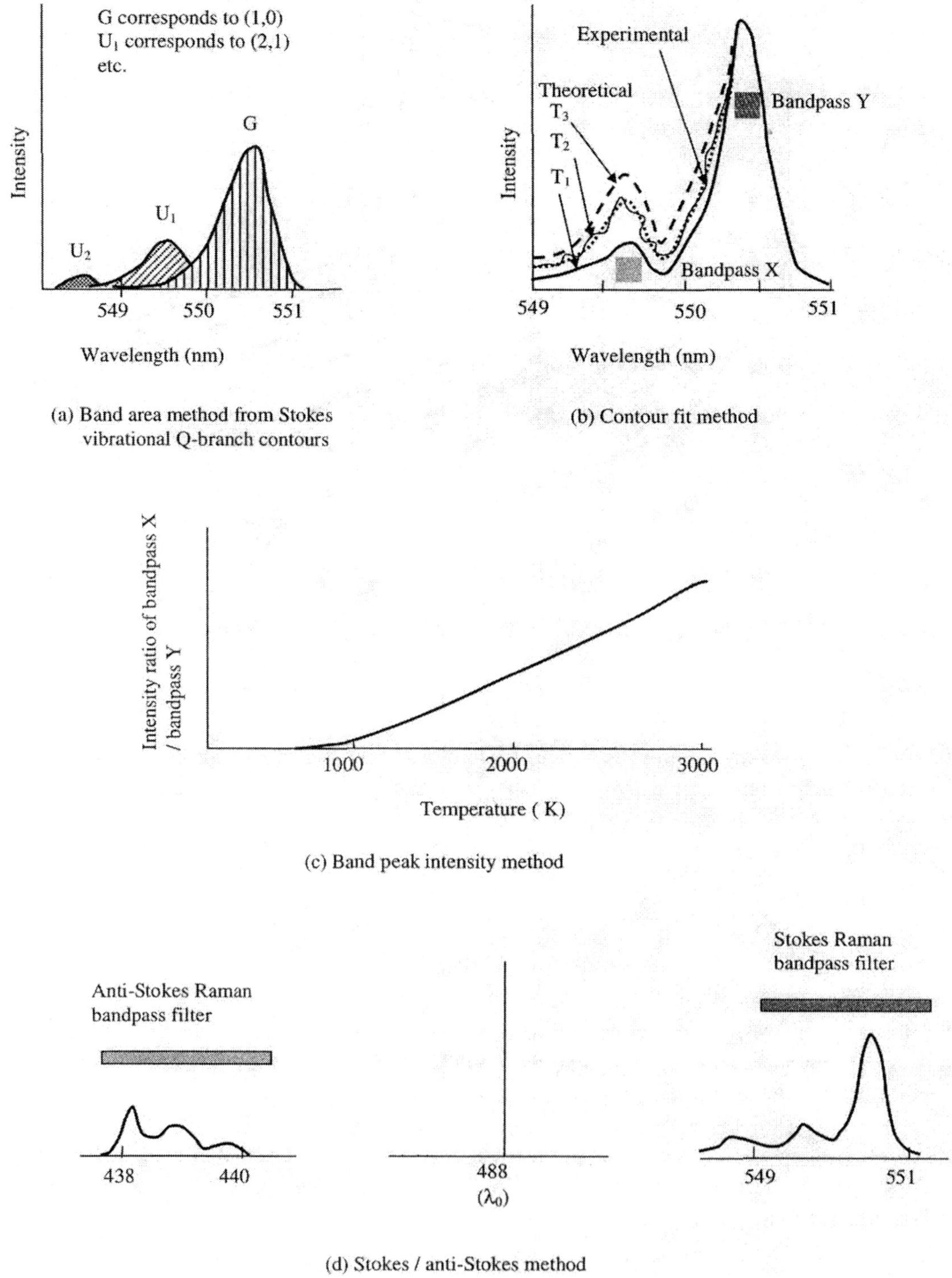

Fig. 12.13 Schematic representation of Raman thermometry for N_2 vibrational Raman scattering, adapted from Drake et al. [1982].

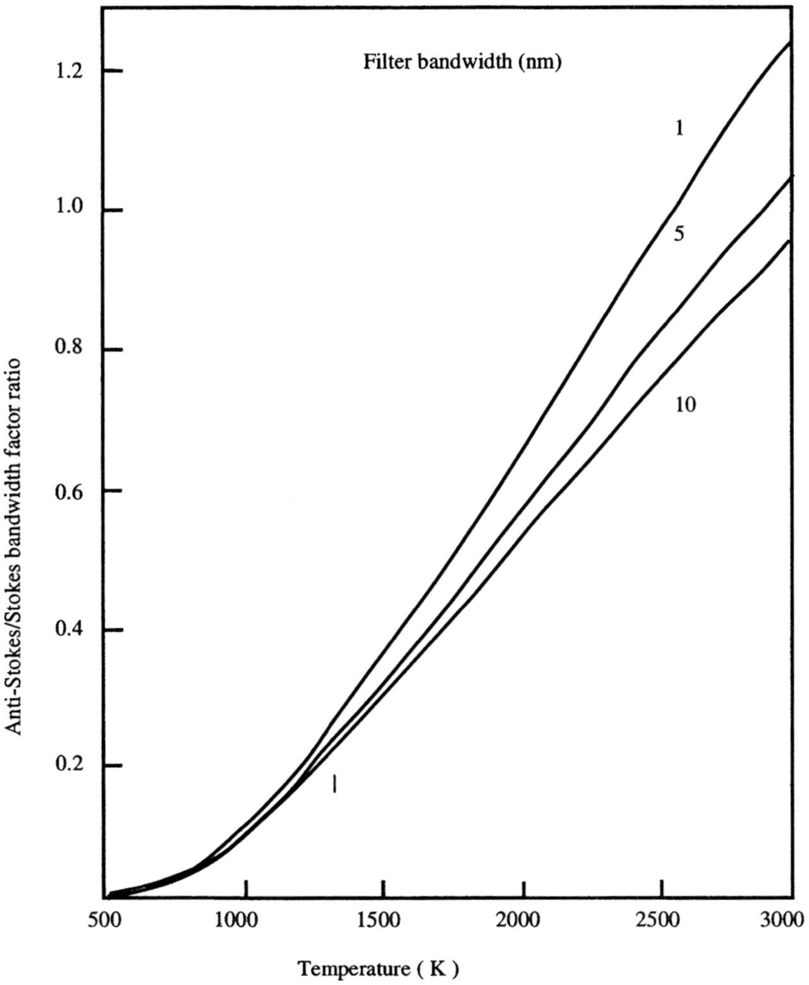

Fig. 12.14 Temperature dependence of the anti-Stokes/Stokes bandwidth factor in N_2 for different filter bandwidths.

The Stokes/anti-Stokes method for N_2 cannot be used at temperatures below about 800 K because of an insufficient number of photons in the anti-Stokes signal [Drake et al., 1982]. The other three methods provide reasonably accurate measurements down to 300 K, since these techniques employ only the Stokes signal.

Pure rotational SRS is not widely employed for combustion diagnostics. The Raman shifts are small, i.e., less than a few hundred wave numbers, and in a multicomponent mixture, spectral interference and overlap make interpretation difficult. Discrimination against the incident laser light is also difficult to achieve because of the small spectral separations. However, the pure rotational Raman scattering cross section for a single transition is generally an order of magnitude stronger than that for an entire vibrational branch. Advances in both atomic/molecular and thin-film filters are helping to mitigate this problem. However, pure rotational Raman scattering will likely remain limited to simple mixtures or to situations where one constituent is dominant.

12.5.2 Implementation and Application of Raman Thermometry to IC Engines

The four methods of Raman thermometry presented in the previous section can be realized through the use of a high-power pulsed laser and employment of sensitive detectors, as shown in Fig. 7.4. The implementation of the Raman scattering technique discussed in Section 7.3 equally applies to Raman thermometry. Other than the Stokes/anti-Stokes method, all other techniques require the use of a spectrometer as the dispersing part of the detection system due to the high spectral resolution required. Narrow bandpass filters can be used to separate Stokes and anti-Stokes bands effectively.

The first temporally and spatially resolved single-shot temperature measurement in a combustion engine was reported by Smith [1980] using the Stokes/anti-Stokes method. The laser beam from an Nd:YAG laser was focused onto a small scattering volume ($0.25 \times 0.5 \times 1.25$ mm) by a tilted 150 mm focal length lens. The scattered light was then collected through the transparent cylinder head and focused onto a spectrometer. Two photomultipliers mounted at the exit of the spectrometer detected the Stokes Raman signal at 607 ± 1 nm and anti-Stokes vibrational Raman signal at 474 ± 1 nm, respectively. The temperature sensitivity of the technique was maximized by choosing a very narrow bandpass filter (1 nm) so as to increase the temperature sensitivity of the system (Fig. 12.14).

A gated integrator was used to reject the flame radiation. The relationship between the temperature and measured anti-Stokes/Stokes signal ratio was first calibrated with a radiation-corrected thermocouple in the post flame region of a premixed flame. The single-shot temperature uncertainty was estimated to be 125 K. When data were averaged over 350 measurements the precision of measurements was estimated less than 10 K. Burned gas temperature history was obtained for four equivalence ratios. The results confirmed that slightly rich-mixture ($\phi = 0.9$) resulted in the highest combustion temperature, as expected. Some burned gas temperature probability distributions were also reported.

Prior to late 1980s, most Raman scattering experiments were carried out using the experimental setup, shown in Fig. 7.4(a), where the Raman spectrum was recorded by a photomultiplier and a scanning spectrometer. This limited the application of band area and contour fit methods to laminar flows or averaged temperature measurements, because of the amount of time required to obtain a complete Raman spectrum. One such example was that by Zur Loye and Santavicca (1983), who attempted to measure in-cylinder burned gas temperature by comparing measured and theoretical nitrogen vibrational Q-branch Raman spectra. They employed a monochromator and a photomultiplier to record the Stokes Raman spectrum of N_2 from an Nd:YAG laser

(50 mJ) at 532 nm. The spatial resolution along the beam was 3.3 mm, which was determined by the spectrometer slit height and the magnification of the collection optics. The measured nitrogen spectra were recorded by scanning the monochromator with a 4.4Å slit width while the laser fired at the same crank angle for every third engine cycle. The scans took about 40 minutes, corresponding to 7500 laser pulses in each firing and non-firing spectrum. The resulting temperature was therefore ensemble-averaged over many hundreds of cycles. Because of the long period required to perform scans, data obtained were affected by window fouling during the test.

To determine the temperature history of the end-gas in the Sandia optical engine, Green and Smith [1984] and Smith et al. [1984] resorted to the contour fit approach. A pulsed Nd:YAG laser excitation at 532.0 nm was used to generate Raman scattering from nitrogen. The vibrational Q-branch spectrum was obtained by signal integration over many engine cycles. These spectra were then contour-fitted with a theoretical model to determine the temperature. Only the fundamental band spectra were recorded because the higher vibrational levels were sparsely populated at these end-gas temperatures. The temperatures obtained were estimated to be accurate to within ±50–70 K. Their results showed higher end-gas temperatures with knocking cycles than non-knocking cycles. However, the accuracy and ensemble-averaged nature of these temperature measurements is not ideal for knocking combustion studies because of cyclic variations in the end-gas temperature.

Within the last decade, significant advances have been made in laser technology and multi-channel detectors. The availability of powerful UV excimer lasers can significantly enhance Raman signal levels. The use of an ICCD-based imaging spectrograph system allows multiple Raman spectra to be recorded, from which the temperature distribution along the laser beam may be determined by one of the Raman methods shown in Fig. 12.13.

12.6 Temperature Measurement by Laser Rayleigh Scattering (LRS)

12.6.1 Principle of Rayleigh Thermometry

As discussed in Chapter 7, Rayleigh scattering is proportional to the total number of scattering molecules. If gases are assumed to be ideal, Rayleigh scattering intensity is proportional to the pressure and the mixture scattering cross section, and is inversely affected by the gas temperature, as given by, (Eq. [5.31]),

$$I = I_0 \frac{pA_0}{RT} \sum_{i=1}^{j} \chi_i \sigma_i \tag{12.36}$$

If the Rayleigh signal from the same gas mixture at a reference temperature T_o and pressure p_o is I_{cal}, then the temperature, T, can be obtained from Eq. (5.32),

$$\frac{T}{T_o} = \frac{\sum_{i=1}^{j} \chi_i \sigma_i}{\sum_{i=1}^{j} \chi_{i,o} \sigma_{i_{i,o}}} \frac{p}{p_o} \frac{I_{cal}}{I} \tag{12.37}$$

Calibration is usually effected at room temperature. Normalization with respect to the instantaneous laser signal is employed to reduce uncertainties caused by variations in laser power. As indicated by Eq. (12.37), an important attribute of Rayleigh thermometry is the inherently large dynamic range of the method. The simplicity and relatively large scattering cross sections of Rayleigh scattering permit two-dimensional temperature measurements.

In Eq. (12.37), the gas pressure can be measured accurately by a pressure transducer. The main difficulty for the accurate interpretation of temperature lies in the precise determination of the scattering cross section of the mixture. In a homogeneous non-reacting flow of a known mixture composition, the scattering cross section of the mixture remains constant. The gas temperature can be determined accurately using this equation. However, in a reacting flow field, since both the mixture number density and the gas composition vary simultaneously, the application of the LRS technique for quantitative temperature measurement becomes difficult.

12.6.2 Implementation and Application of Rayleigh Thermometry to IC Engines

The experimental setup for Rayleigh thermometry is identical to that used for density measurement, as shown in Fig. 7.1. A vertically polarized laser output should be used as the excitation source to maximize the Rayleigh signal. Since Rayleigh thermometry relies on the absolute level of signal, it is necessary to normalize the scattered signal to account for variations in laser output by monitoring the laser output on a pulse-to-pulse basis. The biggest problem with Rayleigh thermometry in an IC engine is the interference of spurious scattered light from air-borne particles and cylinder wall, which significantly reduce the signal-to-noise ratio and the dynamic range of temperature measurements. This limits Rayleigh thermometry to very clean environment. Procedures for minimizing background noise as discussed in Chapter 7 should be followed.

At the University of Heidelberg [Orth, et al. 1994; Bräumer, et al.1995; Schulz, et al. 1996], LRS was used to obtain two-dimensional temperature distributions in a transparent square piston SI engine. A tunable KrF excimer laser was used because of its high energy output and short wavelength so as to boost the Rayleigh signal, which increases with $1/\lambda^4$. Scattered light was imaged by a gated ICCD camera at right angle. Narrow bandpass filters were used to prevent fluorescent light from reaching the detector.

The temperature at each point (x, y) in the laser sheet was calculated from the measured Rayleigh intensity, I

$$\frac{T}{T_0} = \frac{\sum_{i=1}^{j} \chi_i \sigma_i}{\sum_{i=1}^{j} \chi_{i0} \sigma_{io}} \frac{p}{p_0} \frac{I_{cal}}{I} \quad (12.38)$$

The calibration signal was obtained from the fuel/air mixture in a motored engine at the crank angle at which the measurement was to be made. This eliminated the difference in scattering cross section between air and the fuel/air mixture which would have occurred if the calibration had been done with air only. The calibration temperature, T_o, was calculated from the polytropic compression relationship and engine geometry. The pressure ratio was evaluated from a pressure transducer.

Additional measures were taken in order to obtain quantitative temperature measurements. Since LRS is an elastic scattering process, the spurious scattered light from surfaces and air-borne particles cannot be directly removed by an optical filter. To overcome this problem, images were taken of helium in the engine first. Because the Rayleigh cross section of helium is almost negligible, the measured light was assumed to be from the cylinder walls. The subtraction of the helium image (background image) from the measured Rayleigh images therefore minimized the effect of spurious light. Background correction was also used to exclude offsets of the digitizers and camera noise. All of the images were subject to normalization to a reference image to correct for laser intensity distribution and non-uniform response of the camera. The reference image was obtained from a homogeneous medium, such as air, for normalization of all other images.

Other variables that had to be considered were the scattering cross sections of different mixtures. For the unburned mixture, accurate temperature measurements can be achieved if the scattering cross section of the mixture is known and the mixture is homogeneously distributed in the cylinder. Otherwise, Rayleigh thermometry will suffer errors associated with temporal and spatial variations of scattering cross section due to mixture inhomogeneity. In a reacting flow, the scattering cross section of burned gas may be very different from that of the unburned mixture. To overcome this difficulty, Orth et al. [1994] used methane in their experiments. Because the scattering cross section of methane and air mixture varied less than 2% before and after the combustion, accurate two-dimensional temperature measurement were obtained. When the burned gas temperature distribution in the engine was measured for propane fuel, the uncertainty due to variation in the mixture scattering cross section was estimated to be 15% [Braumer, et al. 1995; Schulz, et al. 1996]. Apparently, LRS is not suitable for measuring burned gas temperatures if higher hydrocarbons or gasoline fuel are used, due to the larger variations in scattering cross sections between burned and unburned mixtures.

12.7 Temperature Measurement by CARS

12.7.1 Principle of CARS Thermometry

Compared with laser Rayleigh scattering (LRS), spontaneous Raman scattering (SRS), and laser induced fluorescence (LIF), CARS (coherent anti-Stokes Raman scattering) entails both complicated theoretical treatment and complex and expensive optical setup. Its use in IC engines is very much limited to specialists in laser diagnostics. A detailed description of the CARS technique is beyond the scope of the book. Such information can be found in the book by Eckbreth [1996] and the review by Goss [1993].

The fundamental basis of the CARS technique is outlined in Fig. 12.15. In this method, a medium is irradiated by two superimposed waves, one of a fixed frequency ω_1 and the other of tunable frequency ω_2. They interact in a nonlinear fashion through the 3rd order susceptibility $\chi^{(3)}$ of the medium to generate a third wave at the frequency $\omega_3 = 2\omega_1 - \omega_2$, which is the CARS signal. ω_1 is termed the pump laser and ω_2 the Stokes beam. In the CARS process, adjustment of the frequency difference $(\omega_1 - \omega_2)$ to a particular Raman transition allows CARS signals from different Raman vibrational and rotational lines to be monitored. Because of the nonlinear wave-mixing process involved, the CARS beam at frequency ω_3 is laser-like. The laser-like CARS beam is directional and allows high collection efficiency, thereby leading to a much greater signal-to-noise ratio than SRS. The frequency difference $(\omega_1 - \omega_2)$ can be adjusted by scanning the frequency of the Stokes beam, or by using a broadband tunable laser. The use of a broadband laser generates the entire CARS spectrum by accessing simultaneously all of the Raman resonance in a given spectral band.

The temperature sensitivity of the CARS technique is the result of variation in rotational-vibrational populations of the Raman transitions of the species being probed, which is reflected in the 3rd order susceptibility that governs the CARS process. The susceptibility can be expressed in abbreviated form as

$$\chi^{(3)} = \chi_{nr}^{(3)} + \chi_r^{(3)}$$

where $\chi_r^{(3)}$ is the Raman resonant contribution and $\chi_{nr}^{(3)}$ the non-resonant contribution. The non-resonant susceptibility is proportional to the total number density of the gases in the sampling volume. The Raman resonant contribution $\chi_r^{(3)}$ is a complex quantity arising from Raman allowed transitions of molecules, and is proportional to the population difference, $(N_a - N_b)$, between the levels involved in the wave mixing process, as shown in Fig. 12.15. The population difference is affected by both temperature and species concentration. If the species concentration is known, the gas temperature can be found by recording the CARS signal beam spectrum and fitting it to the theoretical one.

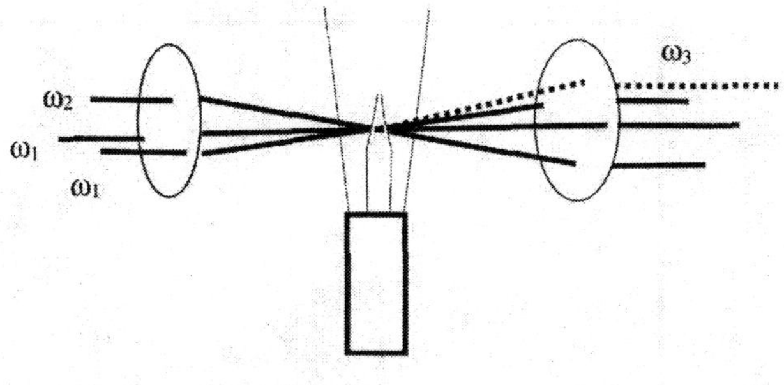

(a) Cross-beam arrangement of CARS technique

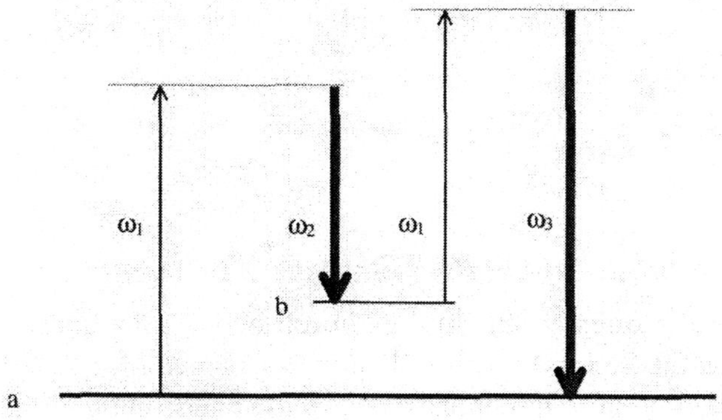

(b) Energy diagram of CARS process

Fig. 12.15 Illustration of CARS principle.

CARS thermometry is generally performed when a major species is in sufficient abundance, so that non-resonant background effects are relatively unimportant. Nitrogen is the dominant constituent of the in-cylinder charge in an IC engine and is present in large concentrations. Nitrogen vibrational CARS has therefore been mostly used for thermometry. In addition, N_2 has the advantage that the spectroscopic and line-width parameters needed for accurate measurements are now relatively well known. CARS spectra for species in high concentration are qualitatively similar to Raman spectra, and tend to mimic the vibrational-rotational state population distributions. Fig. 12.16 shows the Q-branch spectra of N_2 for different temperatures. At low temperatures, the N_2 spectrum is dominated by a strong band from v = 1 to v = 0, whose width increases with temperature because of the increased population of high rotational levels. As the temperature approaches 1000 K, the hot-band transition (v = 2 to v = 1) becomes populated. The band-shape of the N_2 spectrum is therefore a good indicator of temperature. In practice, temperatures are derived by comparing a measured spectrum with a calculated spectrum (Fig. 12.16) using a least squares fitting procedure. Since the computation of CARS spectra is complex, sophisticated computer programs are required.

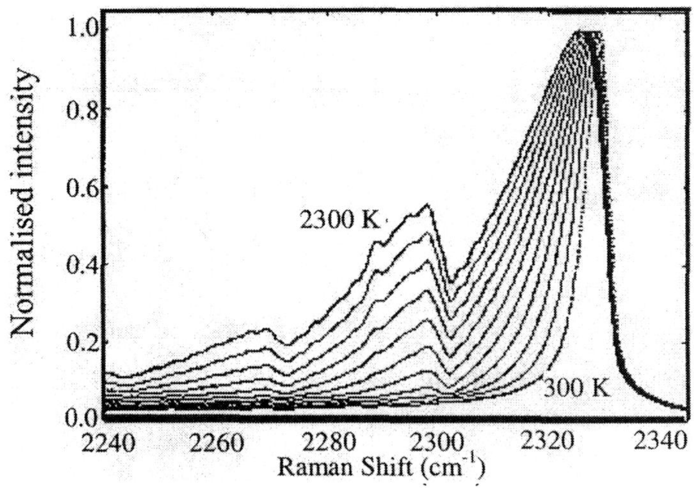

Fig. 12.16 CARS N_2 spectrum as a function of temperature.

12.7.2 Implementation of CARS Nitrogen Thermometry

In additional to the frequency matching required between the pump and Stokes beams, the pump beam ω_1 and the Stokes beam ω_2 should be arranged in such a way that the three beams are phase matched in order to maximize CARS signal generation. Phase matching means that,

$$\bar{k}_3 = 2\bar{k}_1 - \bar{k}_2$$

where k_i is the wave vector at a frequency ω_i. The wave vector is oriented along the beam propagation direction. The absolute magnitude of this vector is,

$$|k_i| = (n_i \omega_i / c)$$

where n_i is the refractive index of the medium at frequency ω_1, and c, the speed of light. Phase matching dictates how the three beams must be oriented. Since the refractive index of gases is essentially invariant with frequency, conservation of momentum in the generation of the CARS signal at $\omega_3 = 2\omega_1 - \omega_2$ indicates that, when only two input beams are used, they should be collinear (Fig. 12.17[a]). If three input beams are used, cross-beam phase matching in gases can be used (Fig.12.17[b] and [c]).

Practical implementation of the CARS technique requires a trade-off between signal intensity and desired spatial resolution. In applications where optical access is not limited, the cross-beam scheme is used because of its ability to isolate the CARS signal from the stronger laser beams as well as high spatial resolution. However, this scheme is optically more complex and results in lower signal intensities because of the smaller interaction volume. The collinear CARS scheme is advantageous for IC applications where optical access is limited. Due to its cubic relationship to total input laser power $\left(P_{CARS} \propto |\chi^{(3)}| P_1^2 P_2\right)$, the CARS signal generated at any point is strongly dependent on the flux density at that point. In CARS experiments, a region of high flux is produced by focusing the laser beams into the sample; this effectively defines an interaction volume in the form of a cylinder whose dimensions are a function of the focal length of the lens used. Thus, it is possible to obtain good spatial resolution using collinear beam arrangement if

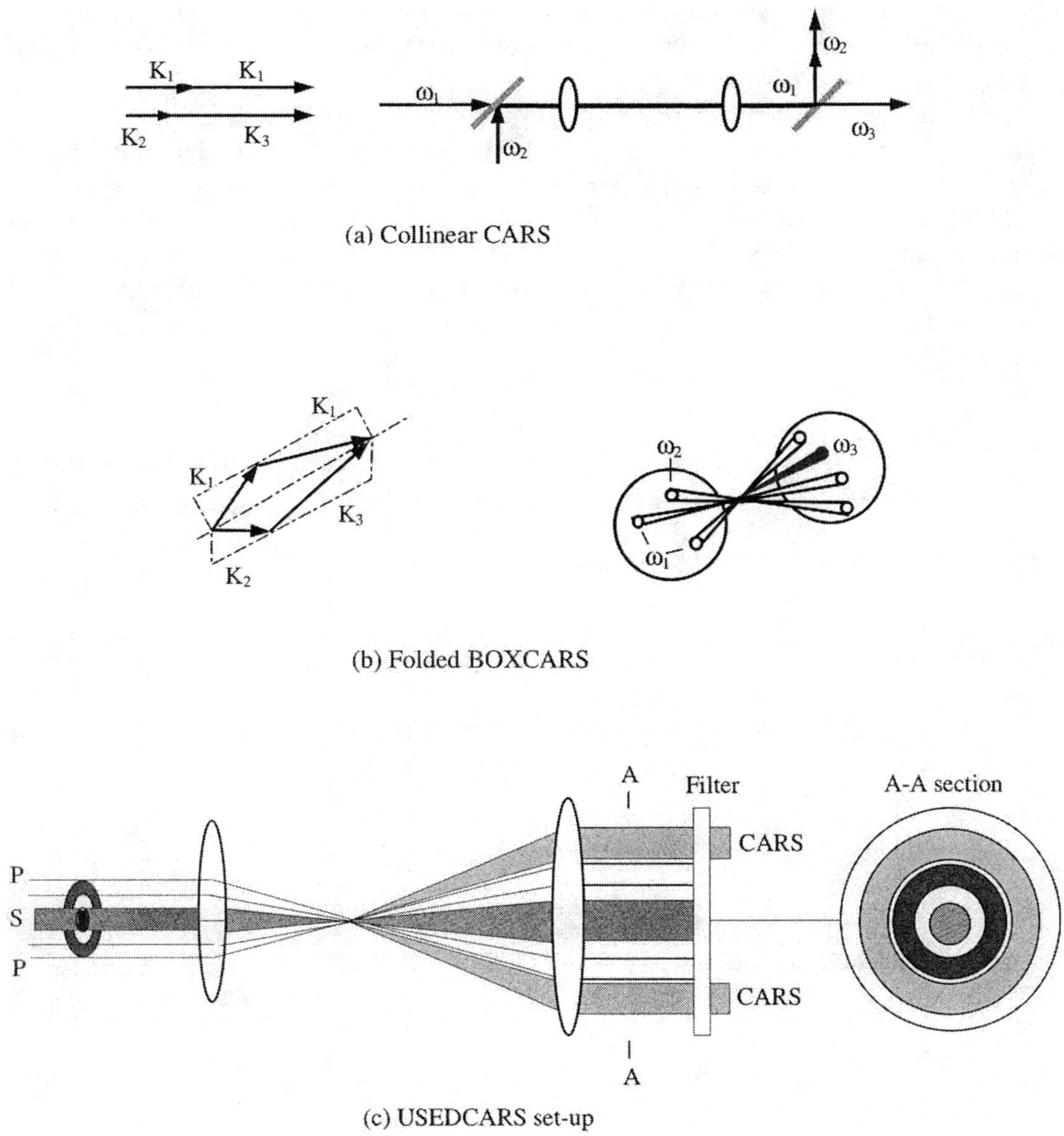

Fig. 12.17 Examples of phase-matching arrangements for the CARS technique.

tight focusing is achieved with a short focus lens. In addition, the collinear CARS scheme is also preferred when there are large refractive index gradients in the probed medium, which would cause serious beam steering problem in cross-beam schemes.

Fig. 12.18 shows the experimental setup of a broad-band collinear CARS setup. The pump laser is typically a pulsed narrow-bandwidth frequency-doubled Nd:YAG laser. A broadband dye laser is used to allow the entire Q-branch of N_2 to be obtained in a single shot, thereby allowing single-shot temperature measurements to be obtained from the band-shape of the CARS signal. The 532 nm output of the Nd:YAG laser serves as a pump for both the CARS process and the broadband dye laser.

Part of the Nd:YAG laser output (~30%) is used as the pump beam, while the rest of it pumps the dye laser to produce a broadband Stokes beam red-shifted from the pump beam. The generated CARS signal beam is also spectrally broad, and has a blue shift with respect to the pump beam, hence it is termed anti-Stokes. The CARS signal beam is filtered and dispersed by a monochromator onto a one-dimensional diode or CCD array to generate the CARS spectrum. Experience has shown that a spectrometer with a resolution of 1 cm^{-1} or better is required, which can be achieved by a 1 or 1.5 meter spectrometer with a 2400 lines/mm grating.

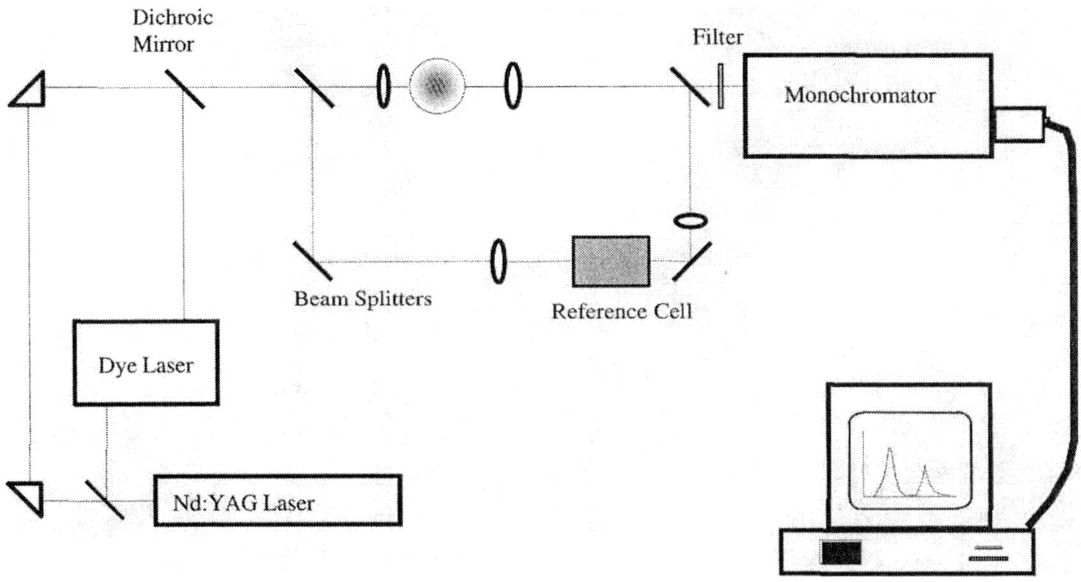

Fig. 12.18 Schematic of a collinear CARS setup.

Raw CARS spectra should first be pre-processed to remove any offset which may arise from the multichannel detector or from scattered light. In addition, the shape of the nitrogen CARS spectrum should be normalized to the CARS spectrum of a broadband dye or Stokes laser. A CARS spectrum for the latter is easily measured by using a non-resonant medium such as inert gases, carbon dioxide, or propane in the reference cell. Because of the high non-resonant susceptibilities of hydrocarbons in the fuel, variations in local air/fuel ratio can lead to changes in inferred temperature. Thus, for an accurate temperature measurement, the measured spectrum should be fitted simultaneously for temperature and nitrogen concentration.

Calibration of a CARS system is typically accomplished using a high-temperature furnace. Excellent accuracy (on the order of 1%) for ambient pressure CARS thermometry is possible [Goss, 1993], but less under high-pressure conditions.

12.7.3 CARS Measurements in IC Engines

The first application of CARS to temperature measurements in a spark ignition engine was reported by Stenhouse and his co-workers [1979]. The collinear scanned method was employed, in which the Stokes beam from a narrow-band dye laser output was aligned collinear with the pump beam from an Nd:YAG laser and scanned over the complete N_2 Q-branch spectrum. The CARS signal was generated in the focal volume of the beams and emerged collinear with them. The scanning took place over a few thousand engine cycles (several minutes). Both unburned and burned gas temperatures in a Ricardo E6 engine were measured. Because the temperature measurements were derived from either the width of the nitrogen Q-branch or the ratio of the first hot band to the fundamental band instead of fitting the complete spectrum, their results should be treated as semi-quantitative. Because of the long ensemble-averaging time involved, temperature readings were affected by unstable combustion during the measurement period.

To overcome the difficulty associated with the scanning method, Klick et al. [1981] employed a broadband dye laser as the Stokes beam. Multi-channel detection of the CARS signal spectrum was realized by an SIT camera coupled to a spectrometer. The CARS signal was generated using USEDCARS (unstable-resonator spatially enhanced detection), as shown in Fig. 12.17(c). The annular cross section of the Nd:YAG pump beam, which was produced by means of an unstable resonator laser cavity, permitted non-collinear phase matching, thus providing spatial resolution even in the longitudinal direction. The CARS signal was emitted in an annular ring on the periphery of the pump beam annulus. A short-wavelength pass filter was used to block the Stokes and pump beams, but allowed the anti-Stokes beam to pass. Both single-pulse and ensemble-averaged temperature measurements were reported in a spark ignition engine with a compression ratio of 3:1.

Pressure has a large effect on the spectral shape of a CARS signal spectrum. With the exception of hydrogen, all the major species possess closely spaced and partially overlapped Raman transitions that are predominantly collision broadened at atmospheric pressure. As the pressure increases, a phenomenon known as collisional narrowing sets in, which causes a major alteration in spectral behavior [Hall et al. 1980]. In the above-mentioned studies, the pressure effect was not properly accounted for. To minimize the pressure effect on temperature determination, Kajiyama, et al. [1982] adopted a data analysis scheme for high-pressure spectra, which extracted temperature from the 20% peak height width of the N_2 Q-branch spectra, a parameter that was found to be affected only by temperature, but not by pressure. Gas temperature measurements across the cylinder bowl of a diesel engine were achieved using the data reduction scheme by means of a collinear CARS system.

When the broadband CARS technique is employed, the CARS signal spectrum must to be normalized to the broad-band Stokes beam via a reference cell filled with a known gas, typically inert, at a fixed temperature and pressure in order to correct the laser fluctuation and laser spectral distribution. Marie and Cottereau [1987] corrected the laser fluctuation and spectral distribution for each measurement by dividing, channel-by-channel, the CARS signal by the non-resonant CARS signal from a reference cell. The temperature was derived by fitting the complete CARS spectrum to the theoretical one, including the collisional narrowing at high pressures. The collinear CARS scheme was used in their experiment, which resulted in a resolution of 10 mm in length. This work can be considered as the first quantitative temperature measurements in internal combustion engines by CARS thermometry. Single-shot as well as averaged gas temperature measurements during the compression and expansion strokes were reported. The end-gas temperature was also measured during knocking cycles. The uncertainty of single-shot measurements was largely due to the weak signals. The accuracy of averaged temperatures was considered within ±50°C.

To measure the burned gas temperature near the wall in a spark ignition engine, Lucht and Maris [1987] employed a cross-beam broadband CARS system. The spatial resolution was about 2 mm × 50 μm. Typically, CARS spectra were averaged over several engine cycles. The normalization of the CARS spectrum was realized by dividing a pre-recorded non-resonant spectrum from oxygen to correct for the spectral profile of the dye laser and the non-uniformity of pixel response of the detector. From the measured gas temperature, a power law expression for the temperature profile in the boundary layer was obtained. The thickening of the boundary layer during the piston expansion stroke was observed.

While the previous studies concentrated on the development of CARS thermometry, more recent CARS temperature measurements focused on its applications. Applications of the CARS technique have been dominated by end-gas temperature measurements. The collinear CARS technique based on the Q-branch N_2 vibration spectrum was used in most studies [e.g., Kalghatgi et al. 1995; Nakada, et al. 1993; Nakano, et al. 1995].

Recently, Bood et al. [1997] reported a dual-broadband rotational CARS process and a BOXCARS configuration for the measurement of end-gas temperature in an SI engine. In their setup, one narrow-band laser beam and two broadband laser beams were focused onto an intersection point that defined the probe volume (1 mm × 40 μm). One of the red dye laser beams (Stokes beams) was superimposed on the green Nd:YAG laser beam (pump beam) using a dichroic mirror. The superimposed laser beams and the single red beam were parallel to each other, separated vertically by 14 mm, before they were focused with an f = 300 mm lens. Raman transitions between rotational levels in the same vibrational state were used. The rotational CARS technique was found to give higher accuracy than vibrational CARS for nitrogen temperature measurements up to 870 K, although the signal strength was less than with vibrational CARS. Using the dual broadband rotational CARS technique, the end-gas temperatures in a spark ignition engine were measured and compared favorably with calculated temperatures.

12.8 Two-Dimensional Temperature Measurement by PLIF

12.8.1 Introduction to Fluorescence Thermometry

Because of their low scattering cross sections, Raman and Rayleigh thermometry is limited to major species of large concentration. Clean measurement conditions are also required, particularly for Rayleigh thermometry, in order to minimize the interference of spurious scattered light. Compared with Raman methods, only fluorescence offers the intensity and spectral selectivity for thermometry in moderately dirty conditions, where interference can arise owing to blackbody radiation, flame emission, or laser induced incandescence. Because of its large scattering cross section, fluorescence can be extended to two-dimensional thermometry.

The in-cylinder high-pressure environment imposes significant difficulty on the application of fluorescence thermometry to IC engines. However, its potential for quantitative two-dimensional temperature measurements has recently attracted a lot of attention and it is currently under development in several laboratories.

There are three major approaches to fluorescence thermometry. They may be generally grouped into excitation or fluorescence scans, monochromatic, and two-line approaches. For excitation scans the spectral bandwidth of the detector is fixed, while the excitation wavelength (laser output) is varied. For fluorescence scans, the excitation frequency is fixed, and the spectral distribution of the emissive signal is recorded by a scanning monochromator or a multi-channel detector. Because of the time required to perform an excitation/fluorescence scan, they are most often employed for steady flame applications. Excitation and fluorescence scans require that collisional quenching Q_{21} does not vary with excitation or fluorescence wavelength, respectively. Therefore, these methods are not suitable for engine applications where collisional quenching is high and often difficult to predict. In the following sections, monochromatic fluorescence thermometry will be introduced. The emphasis, however, will be on two-line fluorescence thermometry, which has the potential to provide quantitative two-dimensional temperature measurements.

12.8.2 Monochromatic Fluorescence Thermometry

12.8.2.1 Principle of Monochromatic Fluorescence Thermometry

The most intuitive monochromatic approach relies on the inherent temperature dependence of the fluorescence signal. Here we will consider linear fluorescence because saturated fluorescence is particularly difficult to achieve for two-dimensional applications. The expression for a fluorescence signal in the linear fluorescence regime has been derived in Chapter 5 and can be rewritten as

$$P_{flu} = \eta_c \Omega V_c f_1(T) N_s B_{12} \frac{A_{21}}{A_{21} + Q_{21}} I_v \tag{12.39}$$

where N_s has replaced grouping $(\chi_m N)$ in Eq. (5.61) and represents the number of fluorescing molecules. At thermal equilibrium, the Boltzmann fraction $f_1(T)$, which is the fraction of the population at the initial vibrational rotational level of the ground electronic state, is given by

$$f_1(T) = \frac{g_J(2J+1)}{Q_{rot} Q_{vib}} \exp(-hcE(v,J)/kT) \tag{12.40}$$

where the rotational and vibrational partition functions

$$Q_{rot} = kT/hcB_v$$

$$Q_{vib} = \left[1 - \exp(-hc\nu_{vib}/kT)\right]$$

and

$$E(v,J) = E_v + E_J = hc\nu_{vib} + B_v J(J+1)hc$$

is the energy (cm^{-1}) of the initial vibrational-rotational level.

If an excited state is chosen such that it is predissociative, the predissociation rate Q_{pre} should be included,

$$P_{flu} = \eta_c \Omega V_c f_1(T) N_s I_v B_{12} \frac{A_{21}}{A_{21} + Q_{21} + Q_{pre}} \tag{12.41}$$

Therefore, temperature measurements by LIF are possible by interrogating individual vibrational rotational states through the Boltzmann fractions in Eqs. (12.40) and (12.41).

Equations (12.40) and (12.41) show that the influence of temperature occurs through the population of the lower vibrational-rotational level and the rate coefficients. If the fluorescence signal from a single transition is measured over a narrow bandwidth, the gas temperature can be derived from Eq. (12.40) with known values of the rate coefficients. The values of A_{21}, B_{12}, and Q_{pre} depend only on the nature of the excited molecule and can be determined accurately, whereas the quenching rate Q_{21} depends on gas composition, density, and temperature and is difficult to evaluate in most cases.

Other approaches based on monochromatic excitation include thermally assisted fluorescence (THAF) thermometry and dual fluorescence detection [Laurendeau, 1988].

12.8.2.2 Implementation and Application of Monochromatic Fluorescence Thermometry

As stated in the previous section, the major difficulty in the method is the accurate determination of the quenching rate Q_{21}. This difficulty can be removed, however, if the loss from the excited level involved is dominated by predissociative rate Q_{pre}. Roller et al. [1995] found that the energy transfer processes of excited oxygen molecules were dominated by predissociative loss, and hence oxygen would be suitable as the fluorescence molecule for monochromatic fluorescence thermometry.

Fig. 12.19 shows the experimental setup for monochromatic fluorescence thermometry used by Roller et al. [1995]. The output of a tunable ArF excimer laser was tuned to the P(17) line of the (10, 2) band of oxygen at 193.424 nm. The resulting fluorescence image from the (10, 5) band emission (203–208 nm) was recorded by an ICCD camera. In order to maintain the optimum excitation efficiency throughout each experiment, the laser output was tuned to obtain the maximum fluorescence signal from oxygen in a Bunsen flame by directing 10% of the laser output to the flame.

In order to derive temperature from the fluorescence signal (Eq. [12.40]), the Einstein coefficients A and B were calculated, and the predissociation rate Q_{pre} and quenching rate Q_{21} were measured experimentally in separate experiments. The system response factor h_c was determined by setting the measured temperature at 77 °BTDC during compression to that calculated by the polytropic compression relation.

Temperature measurements were carried out between 77 and 40 °BTDC during the compression stroke in a diesel engine. Comparison of measured and calculated polytropic compression temperatures showed good agreement. The systematic error was estimated to be less than 10% mainly due to uncertainties of Q_{21} and Q_{pre} at higher pressures.

The advantage of the relatively simple setup for the monochromatic approach is, however, compromised in several respects. First, the application of the monochromatic approach requires that the concentrations of species not vary spatially and temporally as Eq. (12.41) indicates. This requirement was fulfilled because the measurements were performed after the intake

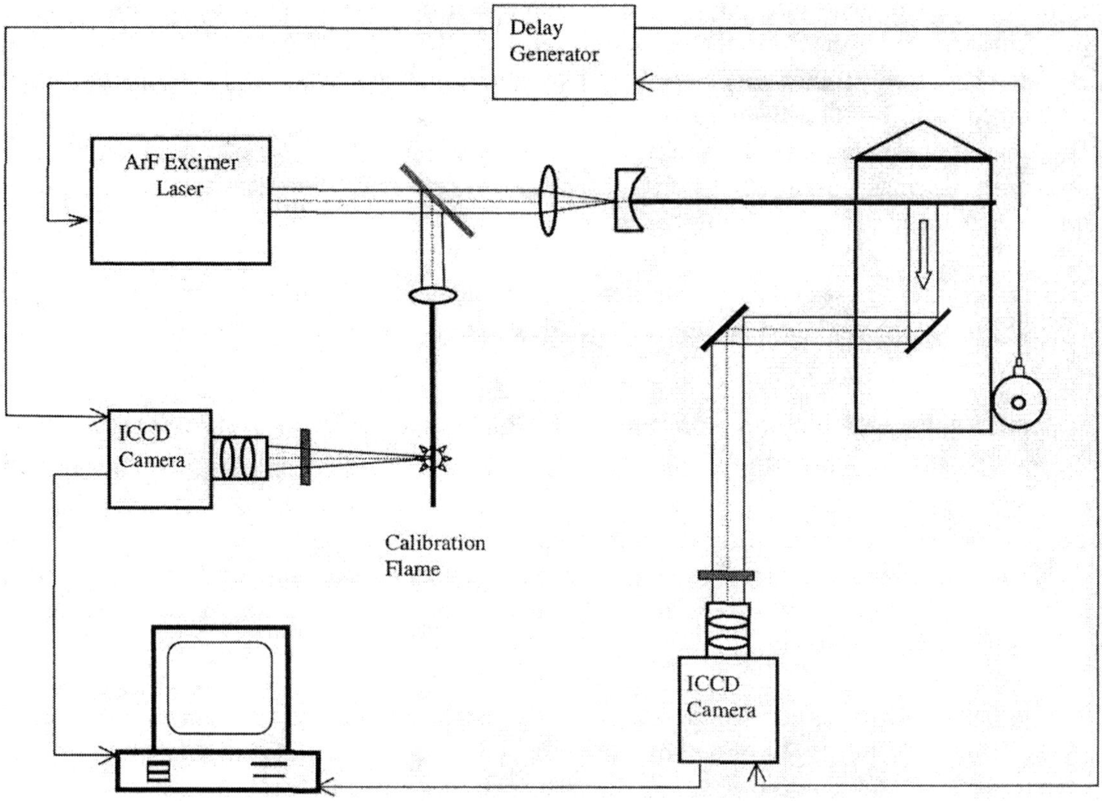

Fig. 12.19 Schematic of the experimental setup for temperature measurements with monochromatic LIF thermometry.

valve closed and there was no spatial variation of air in the motored diesel engine operation. The stratification of charge, due to the presence of burned gas and EGR or fuel, will invalidate the two-dimensional application of the monochromatic approach. Secondly, the accuracy of the monochromatic approach depends on the uncertainty of the quenching coefficient, which is often difficult to determine. In the work reported by Roller et al., measurements were carried out within a pressure range of 3–8 bar, where the predissociation was strong. As the pressure goes up the quenching process will dominate, and hence larger error will occur. In addition, like any absolute measurement technique, the monochromatic approach suffers from errors due to the change in laser output, transmission of windows, and absorption of the gas medium, etc. Finally, a laser beam with such a short wavelength will experience severe attenuation when fuel is present in the combustion chamber.

12.8.3 Two-Line Fluorescence Thermometry

12.8.3.1 Principle of Two-Line Fluorescence Thermometry

The most promising fluorescence thermometry suitable for IC engine applications is probably the two-line method, where a pair of excitation wavelengths is used to generate two broadband fluorescence signals. Two-line fluorescence thermometry can be applied to both atomic and molecular systems.

12.8.3.1.1 Two-Line Molecular Fluorescence Thermometry

The principal feature of two-line molecular fluorescence thermometry (TLMF) is indicated in Fig. 12.20[a]. The basic idea is to measure the relative populations of two states and calculate the temperature via Boltzmann expressions. Two rotational levels in the v = 0 level of the ground electronic state are used to infer the rotational temperature. The upper level could be a single rotational level or, more likely, a vibrational band within an excited electronic state. The excitation is performed sequentially: Level 3 is pumped from level 2 and the broadband fluorescence from level 3 to the ground electronic state monitored, then level 3 is pumped from level 1 and the same broadband fluorescence from level 3 measured. For turbulent combustion applications, both excitations are performed within some characteristic time scale during which the medium is assumed to be unaltered. The dependence on quenching is avoided by exciting lower levels 1 and 2 to the same upper level. According to McKenzie and Gross [1981] and Anderson et al. [1982], if a molecule has closely spaced rotational levels, the broadband fluorescence to the ground electronic state should not depend on the particular rotational level in the excited electronic state. This is due to the fast rotational relaxation and the weak dependence of the quenching rate on rotational levels.

Following Eq. (5.52), the rate equation for level 3 via optical absorption from state 1 is given by

$$\frac{dN_3}{dt} = N_1 B_{13} I_{13} - N_3 \left(B_{31} I_{13} + A + Q \right) \tag{12.42}$$

where B_{13} and B_{31} are the Einstein coefficients for stimulated absorption and emission, A is the Einstein coefficient for spontaneous emission, I_v ($W\ cm^{-2}\ Hz^{-1}$) is the laser spectral intensity, and Q is the non-radiative loss, mainly the quenching loss. The values of A and Q are the sums of those for individual transition lines from the excited state to the ground electronic state.

Similarly, the rate equation for pumping level 3 from level 2 is given by

$$\frac{dN_3}{dt} = N_2 B_{23} I_{23} - N_3 \left(B_{32} I_{23} + A + Q \right) \tag{12.43}$$

The fluorescence from level 3 following the excitations from level 1 and 2 may be written, respectively, assuming steady state, as

$$P_{13} = \eta_{13} N_3 A (h\nu / c) = \eta_{31} \frac{N_1 B_{13} I_{13} A}{(B_{31} I_{13} + A + Q)} (h\nu / c) \tag{12.44}$$

$$P_{23} = \eta_{23} N_3 A (h\nu / c) = \eta_{32} \frac{N_2 B_{23} I_{23} A}{(B_{32} I_{23} + A + Q)} (h\nu / c) \tag{12.45}$$

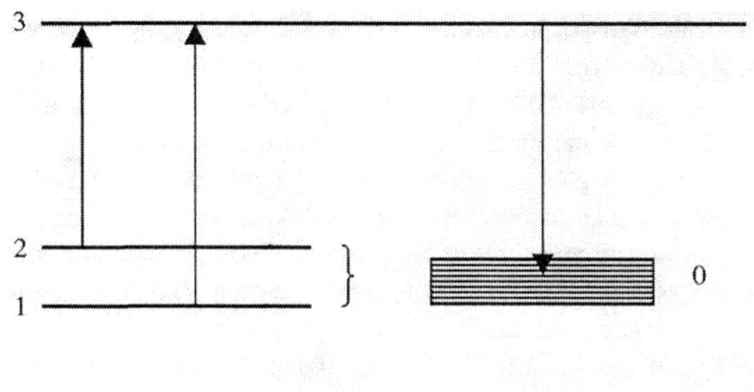

(a) Energy diagram of a molecular system

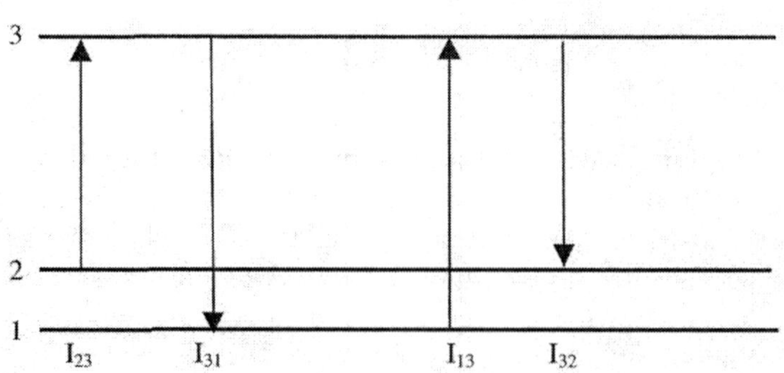

(b) Energy diagram of an atomic system

Fig. 12.20 Two-line fluorescence thermometry.

In the linear fluorescence regime, $B_{ij}I_v$ is much less than (A + Q) and can be ignored. Using the Boltzmann fraction, Eq. (12.40), for N_1 and N_2,

$$N_2/N_1 = (g_2/g_1)\left(\frac{2J_2+1}{2J_1+1}\right)\exp(-hc\Delta E/kT) \tag{12.46}$$

and radiative relations [Measures, 1984],

$$A_{ij} = \left(8\pi h/\lambda_{ij}^3\right)B_{ij} \qquad B_{ij}/B_{ji} = g_j/g_i$$

Hence, from Eqs. (12.44) and (12.45), the fluorescence ratio is obtained,

$$\frac{P_{13}}{P_{23}} = \frac{\eta_{13}}{\eta_{23}}\frac{I_{13}}{I_{23}}\frac{B_{13}}{B_{23}}\frac{(2J_1+1)}{(2J_2+1)}\exp(hc\Delta E/kT) \tag{12.47}$$

Equation (12.47) provides the direct relationship between the ratio of fluorescence signals and the desired temperature. By calibrating at a known temperature, the expression also gives a measured temperature that is independent of any systematic effects on the fluorescence signals.

12.8.3.1.2 Two-Line Atomic Fluorescence Thermometry

For an atom, a low-lying excited electronic state (level 2) and an unpopulated electronic state (level 3) are employed in two-line atomic fluorescence (TLAF) thermometry (Fig. 12.20[b]). Detection of the fluorescence signal occurs by monitoring emission to that lower level not excited by the laser (A_{31} and A_{32}). For the atomic case, a similar analysis for narrowband detection gives

$$\frac{P_{13}}{P_{23}} = \frac{\eta_{13}}{\eta_{23}}\frac{I_{13}}{I_{23}}\frac{B_{13}}{B_{23}}\left(\frac{v_{13}}{v_{23}}\right)^4 \exp(hc\Delta E/kT) \tag{12.48}$$

TLAF thermometry can be achieved at much lower laser power than TLMF thermometry because atomic species have much higher oscillator strengths. However, measurements at lower temperatures ($T < 500$ K) are not feasible using atomic seeds because of their low vapor pressure. There is also difficulty associated with the seeding. In contrast to TLMF thermometry, the TLAF technique is not independent of fluorescence trapping (absorption of fluorescence signal) since the two detection wavelengths are not the same. Because the populations in states 0 and 1 are normally very different at a given T, the relative amount of trapping will differ for P_{13} and P_{23}, and cannot usually be accounted for in a turbulent environment. Note that both Eq. (12.48) and (12.47) are valid under the assumption that no laser absorption takes place in the probed environment and no fluorescence trapping occurs in the case of TLAF.

12.8.3.2 Implementation and Application of Planar Fluorescence Thermometry to IC Engines

12.8.3.2.1 Instrumentation for Two-Line Fluorescence Thermometry

Fig. 12.21 shows the experimental setup for two-line fluorescence thermometry. For temporally resolved measurements, sequential excitation with a very shot time delay is required. The time delay between the two lasers should be short enough to free the flow and long enough to separate the two fluorescence signals. The two laser beams can be obtained in different ways. They can be produced from two separate laser systems as shown in Fig. 12.21, or two dye lasers sharing the same pumping laser [Gross and McKenzie, 1983], or a broadband laser whose bandwidth is wide enough to simultaneously excite two transitions [Andresen et al., 1988]. For field measurements, two gated ICCD cameras are needed to record in sequence the two fluorescence images of the same spectral characteristics. A beam splitter directs half of the fluorescence signal into each camera. Each camera should be synchronized with the corresponding laser excitation. For quantitative measurements, the fluorescence image recorded on each camera should be normalized to the laser output to take into account the pulse-to-pulse variation and difference between the two lasers.

12.8.3.2.2 Gas Temperature Measurement by TLMF Thermometry

The selection of molecular species and energy levels involved in excitation and fluorescence emission is critical to the successful application of TLMF thermometry. Most original examples

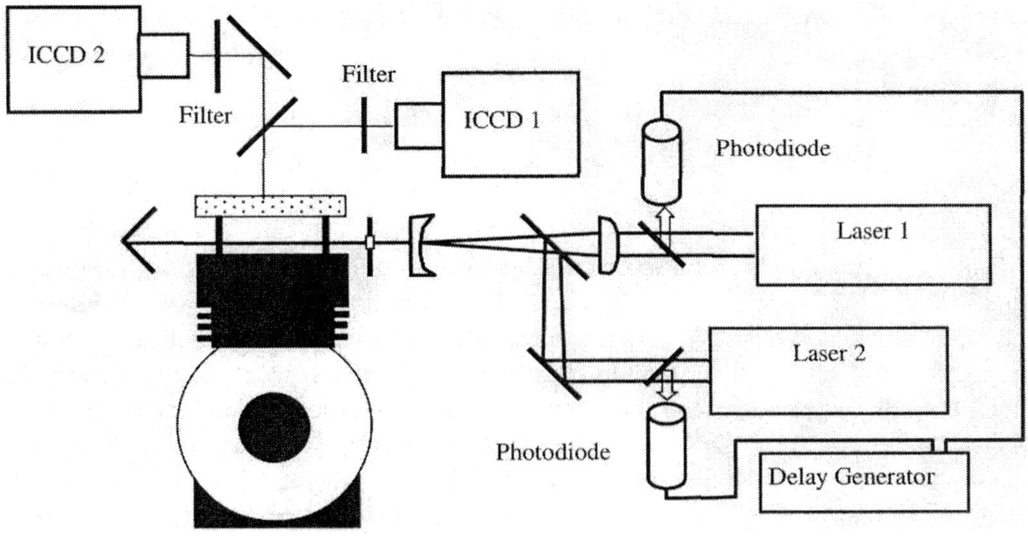

Fig. 12.21 Experimental setup for two-line fluorescence thermometry.

of TLMF employed the OH radical. Kohse-Höinghaus [1994] discussed several two-line excitation schemes that have been applied to flame temperature measurements. OH fluorescence thermometry, however, is limited to OH-abundant regions, and hence cannot be used for temperature measurements in any regions other than high temperature burned gas region.

Besides the OH radical, several other molecules have been utilised as fluorescing species for temperature measurements with two-line fluorescence thermometry. NO and O_2 have been tipped as most promising temperature indicators for TLMF measurements. Arnold et al. [1993] demonstrated that it was possible to obtain two-dimensional temperature distribution during the compression stroke in a diesel engine by exciting R(19) (10,2) and R(29) (11,2) lines of oxygen using a tuneable ArF excimer laser operated at 193 nm.

The most severe limitation of TLMF thermometry is its low signal-to-noise ratio, particularly at higher pressures. The shorter wavelengths required for excitation of NO and O_2 aggravate the problem due to severe laser beam attenuation in the presence of fuels in the combustion chamber. The solution is therefore to find a molecule that has a large fluorescence yield when excited with longer wavelengths and an excitation scheme that possesses good temperature sensitivity. One such example has recently been reported by Einecke et al. [1998a; 1998b].

The temperature indicator used by Einecke et al. was 3-pentanone, which is widely used as the marker for mapping fuel distributions in PLIF. Fig. 12.22 shows the energy states of 3-pentanone involved in the laser induced fluorescence process. Following pumping from the ground electronic state S_0 to the first excited singlet state S^1, the molecule can be deactivated by:

(a) fluorescence emission to S_0,
(b) non-radiative loss to S_0,
(c) inter-system crossing to the triplet state T_1,
(d) chemical decomposition.

The last process does not occur for temperatures below 573 K [Grossmann et al., 1996].

Fig. 12.23 shows the absorption spectrum of 3-pentanone at 298 and 573 K [Grossmann et al., 1996]. According to the Boltzmann distribution, the population of vibronic levels in the ground electronic state is concentrated at lower levels. As the temperature increases, higher levels become more populated. Therefore, the absorption peak at 275 nm shifts to longer wavelengths with increasing temperature (Fig. 12.23). If two lasers are employed to excite 3-pentanone molecules at 250 nm and 300 nm, respectively, according to Fig. 12.23, the fluorescence signal due to 250 nm excitation will fall at higher temperature and the fluorescence signal from the longer excitation wavelength will be greater at elevated temperature. The ratio of these two signals can then be used to obtain temperature.

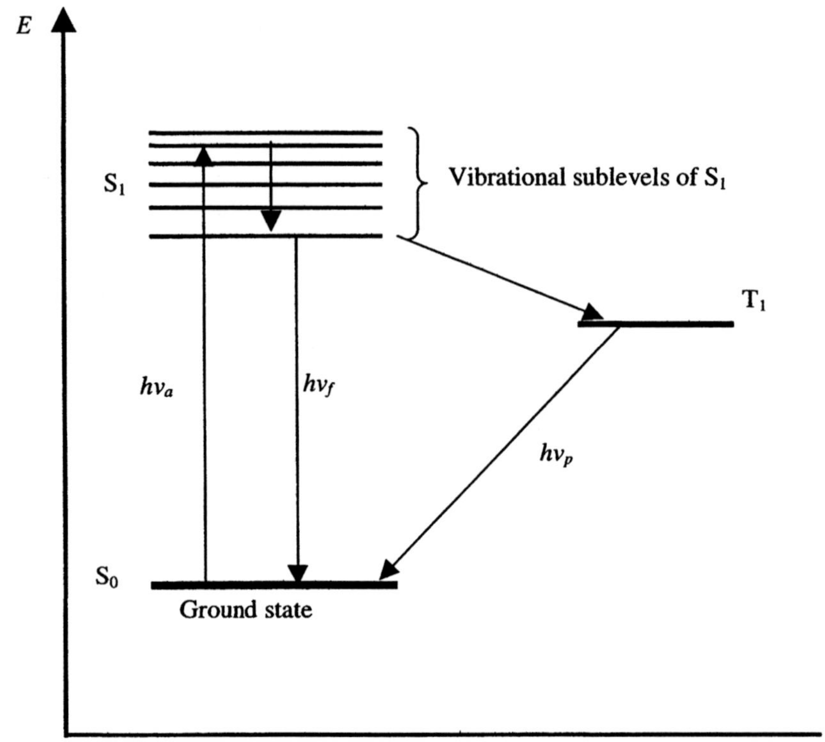

Fig. 12.22 Schematic energy diagram and mechanisms of fluorescence and phosphorescence in 3-pentanone

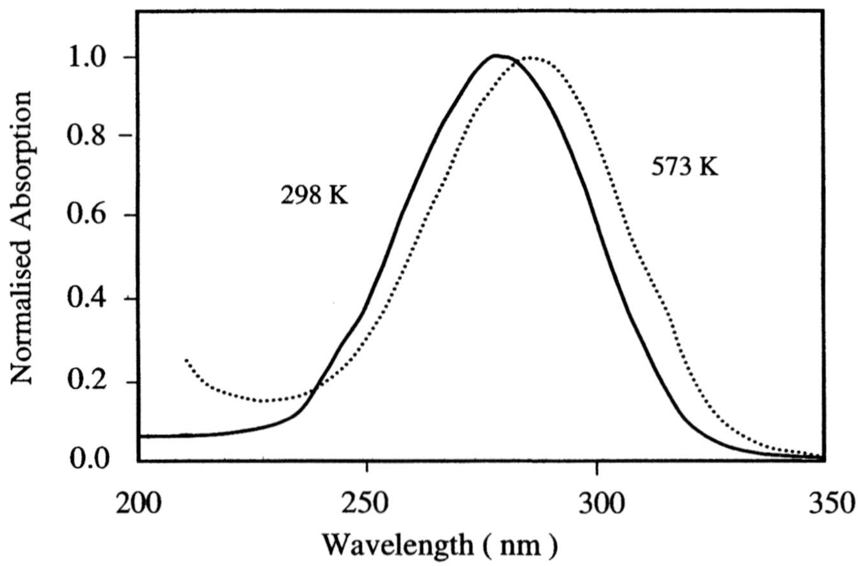

Fig. 12.23 Absorption spectrum of 3-pentanone at two different temperatures and atmospheric pressure, adapted from Grossmann et al. [1996].

Because of overlapping vibrational rotational bands in the electronic absorption spectra, single vibronic excitations of most polyatomic molecules, such as 3-pentanone, are not realizable. In terms of fluorescence thermometry, this means that broadband excitations can be used in the case of 3-pentanone. However, simplification of the broadband excitation is compromised by the increased overhead to obtain accurate temperature measurements. The mathematical expression, Eq. (12.47), was derived on the basis of excitation of two single transition lines to vibronic levels closely distributed in the excited electronic state. In the case of 3-pentanone, overlapping vibrational rotational bands in the absorption spectra means that multiple vibronic levels are excited simultaneously from the ground electronic state to many different vibronic levels. The fluorescence signal in the linear regime, Eq. (12.41), can be rewritten as

$$P_{\lambda,flu} = \sigma_\lambda(\xi, p, T)\eta\Omega V_c N_s I_v \varphi_\lambda(\xi, p, T) \tag{12.49}$$

where the first term σ represents the molecular absorption cross section and is related to the values of the Einstein coefficient B_{12} in Eq. (12.41), the last term φ is the fluorescence quantum yield corresponding to $A/(A + Q)$ in Eq. (12.41). Both σ and φ are affected by mixture composition, ξ, pressure, p, and temperature, T.

By taking the ratio of two fluorescence signals, most parameters in Eq. (12.49) cancel out,

$$\frac{P_{1,flu}}{P_{2,flu}} = \frac{\sigma_1(\xi, p, T) I_v^1 \varphi_1(\xi, p, T)}{\sigma_2(\xi, p, T) I_v^2 \varphi_2(\xi, p, T)} \tag{12.50}$$

The effect of quenching loss, which is the main factor in the fluorescence quantum yield, is largely removed. The remaining effect of pressure is related to the inter-level relaxation between different vibrational levels and is relatively weak. Grossmann et al. [1996] found that the fluorescence intensity increased substantially before it leveled off at pressures above 5 bar for excitation at 248 nm. The pressure effect became less pronounced as the excitation wavelength increased. Experiments in a high-temperature and pressure cell showed that a linear relationship between the ratio of two-line fluorescence intensity and gas temperature could be obtained at pressures above 5 bar (Fig. 12.24) [Einecke et al., 1998b].

Based on these experimental observations by Grossmann et al., Einecke et al. [1998b] set out to develop a two-line fluorescence technique for unburned gas temperature measurements in an SI engine. Isooctane doped with 10% 3-pentanone by volume was admitted into a single-cylinder two-stroke engine with transparent cylinder head. The optical setup was similar to that shown in Fig. 12.21. Excitations at 248 nm and 308 nm were achieved by two excimer lasers, and fluorescence was detected by two separate gated ICCD cameras. In order to prevent cross-talk of the signals, the KrF (248 nm) and XeCl (308 nm) lasers were fired with a fixed time delay of 150 ns with exposure times of 70 ns for each camera. Laser output

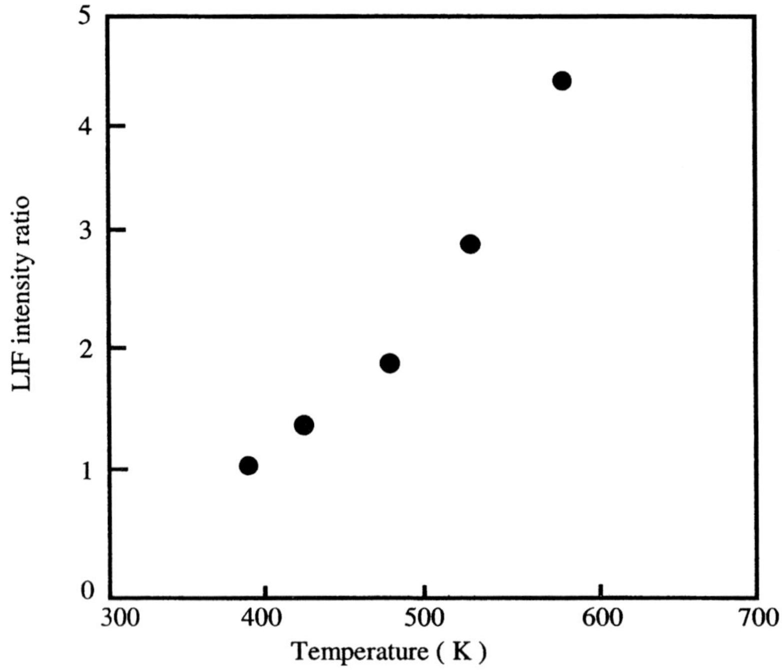

Fig. 12.24 Two-line LIF intensity ratio vs. temperature after excitation at 248 nm and 308 nm, adapted from Einecke et al. [1998].

energies were monitored on a shot-to-shot basis. Two-dimensional temperatures over 140 single shots at four different crank angles in the compression stroke were obtained from a calibration curve obtained in a calibration cell. In order to avoid the pressure effect at lower pressures, measurements were limited to the later compression stage.

12.8.3.2.3 Burned Gas Temperature Measurement by TLAF Thermometry

The most suitable atomic additive for TLAF thermometry is indium, because the gap between its two lower energy levels is such that high sensitivity is available for temperatures between 500 and 2500 K. Excitation takes place at 410 and 451 nm, which corresponds to transitions from the ground 5p $^2P_{1/2}$ state to the excited 6s $^2S_{1/2}$ state, and from the metastable 5p $^2P_{3/2}$ state to the same excited state, respectively [Haraguchi et al., 1977]. These excitation wavelengths can be readily obtained from dye lasers.

Following the previous work by Dec and Keller (1986) on TLAF thermometry in flames, Kaminski et al. (1998) demonstrated temporally resolved two-dimensional TLAF measurements in an SI engine. Indium was added to the intake charge by dissolving $InCl_3$ salt in isopropanol. The resulting solution with isooctane was sprayed into the intake manifold. The concentration of indium was selected based on the trade-off between adequate signal-to-noise ratio and fluorescence trapping. Temporal resolution was obtained by sequential excitation using two separate dye lasers with a time delay of 500 ns. The resulting fluorescence was split by a dichroic mirror into the 410 and 451 nm beams required for detection of indium. Each beam was directed to a gated ICCD camera, and its laser power was monitored by a PIN photodiode. The fluorescence and laser signals were analyzed on a shot-by-shot basis with on-line division of each fluorescence signal by the relevant laser power signal. Laser sheet profile measurements were obtained using PLIF of biacetyl, and ensemble-averaged sheet corrections were carried out to correct for variation in the beam profiles. After background subtraction, and a laser power and beam profile normalization, each individual image pair was divided on a pixel by pixel basis so that a two-dimensional temperature distribution could be obtained. The measured temperature range covered 800–2800 K. A precision of 14% on single-shot temperature distributions was claimed, which could be significantly improved if the laser beam profile was obtained on a shot-by-shot basis.

Since indium atom is only formed in the presence of a flame via electron ion recombination or molecular dissociation, the technique is limited to measuring the burned gas temperature.

12.9 Summary

Early in-cylinder gas temperature measurements were carried out entirely using line-of-sight techniques. Significant contributions were made by these temperature measurements in understanding normal and abnormal combustion processes in IC engines. The advent of laser-based techniques has made it possible to make temporally and spatially resolved temperature measurements in internal combustion engines. The temperature measurement techniques discussed in this chapter are summarized in Table 12.3.

The monochromatic absorption-radiation method is the most appropriate radiation thermometry method for engine applications, because it can provide continuous temperature readings without knowing the emissivity of gas. The velocity-of-sound technique is characterized by its compact design and can be readily adapted to production engines. It is particularly suited to unburned gas temperature measurements. CARS thermometry is the most accurate technique for both burned and unburned gas temperature measurements. However, it is limited to single-point and single-shot measurements. For engine applications, two-dimensional temperature imaging techniques are particularly important because of the inhomogeneous temperature distributions inside the combustion chamber. Such temporally resolved planar thermometry favors fluorescence methods because of their intrinsically high signal-to-noise ratios. Amongst various approaches to fluorescence thermometry, two-line fluorescence thermometry techniques have demonstrated their potential for quantitative two-dimensional temperature measurements in an IC engine.

Table 12.3 Summary of techniques for in-cylinder gas temperature measurements

Technique (category)	Marker	Spatial resolution Temporal resolution	Other Comments	Applications and references
Chopped method (radiation thermometry)	CO_2, H_2O	Line-of-sight Continuous, cycle resolved	Most useful and practical radiation thermometry	End gas temperature [Burrows et al., 1961] Burned gas temperature [Takagi et al., 1990]
Null method (radiation thermometry)	CO_2, H_2O	Line-of-sight Single, cycle averaged	Tedious though more accurate than the chopped method	Unburned gas temperature [Burrows et al., 1961]
Line reversal (radiation thermometry)	Sodium	Line-of-sight Single, cycle averaged	Oldest method, unsuitable for IC engines	Burned gas temperature [Rassweiler et al., 1935]
Two-color method (radiation thermometry)	CO_2, H_2O	Line-of-sight Continuous, cycle resolved	Accurate emissivity ratio required	End gas temperature [Agnew, 1960]
Velocity of sound (speed of sound)	Gas mixture	Line-of-sight Continuous, cycle resolved	Compact design, Less accurate for burned gas	End gas temperature [Livengood, 1954] Unburned and burned gas, [Bauer et al., 1997]
Rotational temperature (spectroscopic technique)	OH	Line-of-sight Continuous, cycle resolved	Departure from equilibrium at high temperatures	Burned gas temperature [Bach et al. 1996]
Line absorption (spectroscopic technique)	Iodine	Line-of-sight Continuous, cycle resolved	Seeding required, anomalous results	End gas temperature [Chen et al., 1954]
Raman thermometry (laser diagnostics)	N_2	Single point, Single or cycle averaged	Weak signal, Potential quantitative multi-point measurements	Burned gas temperature [Smith, 1980; Zur Loye & Santavicca, 1983] End-gas temperature, [Green & Smith, 1984]
Rayleigh thermometry (laser diagnostics)	Gas mixture	Two-dimensional Single or cycle averaged	Extremely clean environment, Unsuitable for burned gas	Unburned and burned gas temperatures [Orth, 1994; Braumer, 1995; Schulz et al., 1996]
CARS thermometry (laser diagnostics)	N_2	Single point Single or cycle averaged	The most accurate method, But limited to a single point	Unburned and burned gas temperatures (See reference on CARS.)
Monochromatic fluorescence thermometry (laser diagnostics)	O_2	Two-dimensional Single or cycle averaged	Suffer from quenching, Beam absorption by fuel	Compressed air in a diesel engine [Roller et al, 1995]
Two-line fluorescence thermometry (laser diagnostics)	3-Pentanone Indium	Two-dimensional Single or cycle averaged	Quantitative 2-D temperature measurements, Complicated equipment	Unburned gas temperature [Einecke et al., 1998 a and b] Burned gas temperature [Kaminski et al., 1998]

References

Agnew, W.G. (1960), "End Gas Temperature Measurement by a Two-Wavelength Infrared Radiation Measurement," *SAE Trans.*, Vol. 68, pp. 495–513.

Anderson, W.R., Decker, L.J., and Kotlar, A.J., (1982), "Temperature Profile of a Stoichiometric CH_4/N_2O Flame from Laser Excited Fluorescence Measurements on OH," *Combustion and Flame*, Vol. 48, pp. 179–190.

Andresen, P., Bath, P.A., Groger, W., Lulf, H.W., Meijer, G., and ter Meulen J.J., (1988), "Laser-Induced Fluorescence with Tunable Excimer Lasers as a Possible Method for Instantaneous Temperature Field Measurements at High Pressures: Checks with an Atmospheric Flame," *App. Opt.*, Vol. 27, pp. 365–378.

Arnold, A., Brauerle, A., Buschmann, A., Decker, M., Dinkelacker, F., Heitzmann, T., Orth, A., Schafer, M., Sick, V., and Wolfrum, J. (1993), "2D-Diagnostics in Industrial Devices," *Ber. Bunsenges, Phys. Chem.*, Vol. 97, No. 12, 1650–1661.

Bauer, W., Tam, C., Heywood, J.B., and Ziegler, C., (1997), "Fast Gas Temperature Measurement by Velocity of Sound for IC Engine Applications," SAE Paper No. 972826.

Bach, M., Reissing, J., and Spicher, U. (1996), "Temperature Measurement and NO Determination in SI Engines Using Optical Fibre Sensors," SAE Paper No. 961922.

Bood, J., Bengtsson, P., Mauss, F., Burgdorf, K., and Denbratt, I., (1997), "Knock in Spark Ignition Engines: End-Gas Temperature Measurements Using Rotational CARS and Detailed Kinetic Calculations of the Autoignition Process," SAE Paper No. 971669.

Bräumer, A., Sick, V., Wolfrum, J., Drewes, V., Zahn, M., and Maly, R. (1995), "Quantitative Two-Dimensional Measurements of Nitric Oxide and Temperature Distributions in a Transparent Square Piston SI Engine," SAE Paper No. 952462.

Burrows, M.C., Shimizu, S., Meyers, P.S., and Uyehara, O.A. (1961). "The Measurement of Unburnt Gas Temperatures in an Engine by an Infrared Radiation Pyrometer," *SAE Transaction*, Vol. 69, pp. 514–528.

Chen, S.K., Beck, N.J., Uyehara, O.A., and Myers, P.S. (1954), "Compression and End-Gas Temperatures from Iodine Absorption Spectra," SAE Trans. Vol. 62, 503–513.

Chidsey, I.L. and Crosley, D.R. (1979), "Calculated Rotational Transition Probabilities for the A-X System of OH," *J. Quant. Spectrosc. Radiat. Transfer.*, Vol. 23, 187–199.

Dec, J.E. and Keller, J.O., (1986), " High-Speed Thermometry Using Two-Line Atomic Fluorescence," *Proc. 21st Symposium on Combustion*, The Combustion Institute, 1737–1745.

Drake, M.C. Lapp, M., and Penney, C.M., (1982), *Temperature: Its Measurement and Control in Science and Industry*, Vol. 5, Ed. by J.F. Schooley, American Institute of Physics, New York.

Dieke, H.H. and Crosswhite, H.M. (1962), "The Ultraviolet Bands of OH," *J. Quant. Spectrosc. Radiat. Transfer*, 2, 97–199.

Eckbreth, A.C. (1996), *Laser Diagnostics for Combustion Temperature and Species*, 2nd Edition, Gordon and Breach Publishers, Amsterdam.

Einecke, S., Schulz, C., and Sick, V. (1998a), "Dual-Wavelength Temperature Measurements in an IC Engine Using 3-Pentanone Laser Induced Fluorescence," in *Laser Applications to Chemical and Environmental Analysis*, Vol. 3 of OSA Technical Digest Series.

Einecke, S., Schulz, C., Sick, V., Dreizler, A., Shiessi, R., and Maas, U. (1998b), "Two-Dimensional Temperature Measurements in an SI Engine Using Two-Line Tracer LIF," SAE Paper No. 982468.

Ely, J.F. and Huber, M.L. (1990), *SUPERTRAPP—Version 1.04: NIST Thermophysical Properties of Hydrocarbon Mixtures*, Thermophysics Division, NIST.

Gaydon, A.G. (1970), *The Spectroscopy of Flames*, 2nd Ed., Chapman and Hall Ltd., London.

Gaydon, A.G. and Wolfhard, H.G. (1974), *Flames, Their structure, Radiation and Temperature*, 3rd Ed. Chapman and Hall Ltd., London.

Gluckstein, M.E. and Walcutt, C. (1961), "End-Gas Temperature-Pressure Histories and Their Relation to Knock," *SAE Trans*., Vol. 69, 529–553.

Gordon, S. and McBride, B.J. (1971), "Computer Program for the Calculation of Complex Chemical Equilibrium Composition," NASA Publication SP-273.

Goss, L.P. (1993), "CARS Instrumentation for Combustion Applications," pp. 251–322, *Instrumentation for Flows with Combustion*, Edited by A.M.K.P. Taylor, Academic Press, London.

Green, R.M. and Smith, J.R. (1984), "An Experimental Study of Engine Knock," Sandia National Laboratory Report, SAND84-8630.

Gross, K.P. and McKenzie, R.L. (1983), "Single-Pulse Gas Thermometry at Low Temperatures Using Two-Photon Laser-Induced Fluorescence in NO-N_2 Mixtures," *Opt. Lett.*, Vol. 8, pp. 368–370.

Grossmann, F., Monkhouse, P.B., Ridder, M., Sick, V., and Wolfrum, J. (1996), "Temperature and Pressure Dependence of the Laser Induced Fluorescence of Gas-Phase Acetone and 3-Pentanone," *Appl. Phys. B*, Vol. 62, pp. 249–253.

Hall, R.J., Verdieck, J.F., and Eckbreth, A.C. (1980), "Pressure-Induced Narrowing of the CARS Spectrum of N_2," *Opt. Commun*. Vol. 35, 69–75.

Haraguchi, H., Smith, B., Weeks, S., Johnson, D.J., and Winefordner, J.D., (1977), "Measurement of Small Volume Flame Temperatures by the Two-Line Atomic Fluorescence Method," *Appl. Spectrosc.*, Vol. 31, 156–163.

Heywood, J.B, (1988), *Internal Combustion Engine Fundamentals*, McGraw-Hill.

Hottel, H. and Broughton, F. (1932), "Determination of True Temperature and Total Radiation from Luminous Gas Flames," *Ind. and Eng. Chem.*, 4-2, pp. 166–174.

Hottel, H.C. and Egbert, R.B. (1942), "Radiant Heat Transmission from Water Vapour," *AIChE Trans*, Vol. 38, 531–565.

Kajiyama, K., et al. (1982), "N_2 CARS Thermometry in Diesel Engine," *SAE Trans.* Vol. 91, 3243–3251, SAE Paper No. 821036.

Kalghatgi, G.T., Snowdon, P., and McDonald, C.R. (1995), "Studies of Knock in a Spark Ignition Engine with CARS Temperature Measurements and Using Different Fuels," SAE Paper No. 950690.

Kaminski, C.F., Engstrom, J., and Alden, M., (1998), "Quasi-Instantaneous Two-Dimensional Temperature Measurements in a Spark Ignition Engine Using 2-Line Atomic Fluorescence," Paper presented at 26th Symposium on Combustion, the Combustion Institute.

Kleppe, J.A. (1989), *Engineering Applications of Acoustics*, Artech House.

Klick, D., Marko, K.A., and Rimai, L. (1981), "Broadband Single-Pulse CARS Spectra in a Fired Internal Combustion Engine," *Appl. Opt.*, Vol. 20, pp. 1178–1181.

Kohse-Hoinghaus, K. (1994), "Laser Techniques for the Quantitative Detection of Reactive Intermediates in Combustion Systems," Progress in Energy and Combustion Science, Vol. 20, pp. 203, 279.

Lapp, M. (1974), *Laser Raman Gas Diagnostics*, Ed. by Lapp, M. and Penny, C.M., p. 107, Plenum, New York.

Laurendeau, N.M., (1988), "Temperature Measurements by Light-Scattering Methods," *Prog. Energy and Combustion Science*, Vol. 14, No. 2, pp.147–170.

Learner, R.C.M. (1962), "The Influence of Vibrational-Rotational Interaction on Intensities in the Electronic Spectra of Diatomic Molecules: I. The Hydroxyl Radical," *Proc. Roy. Soc. A*, Vol. 269, pp. 311–326.

Livengood, J.C., Rona, T.P., and Baruch, J.J., (1954), "Ultrasonic Temperature Measurement in Internal Combustion Engine Chamber," *J. Acoust. Soc. Am.*, Vol. 26 No. 5, pp. 924–830.

Livengood, J.C., Taylor, C.F., and Wu, P.C. (1958), "Measurement of Gas Temperature in an Engine by the Velocity of Sound Method," *SAE Trans*. Vol. 66, pp. 683–695.

Long, D.A. (1977), *Raman Spectroscopy*, McGraw-Hill, New York.

Lucht, R. and Maris, M. (1987), "CARS Measurements of Temperature Profiles Near a Wall in an Internal Combustion Engine," SAE Paper No. 870459.

Marie, J. and Cottereau, M. (1987). "Single-Shot Temperature Measurements by CARS in an IC Engine for Normal and Knocking Conditions," SAE Paper No. 870458.

McKenzie, R.L. and Gross, K. . (1981), "Two-Photon Excitation of Nitric Oxide Fluorescence as a Temperature Indicator in Unsteady Gas Dynamic Processes," *Applied Optics*, Vol. 20, pp. 2153–2165.

Measures, R.M. (1984), *Laser Remote Sensing*, John Wiley.

Millar, G.M., Winans, J.G., and Uyehara, O.A. (1953), "A Fast, Electro-Optical, Hot-Gas Pyrometer," *J. Optical Society of America*, Vol. 43, pp. 609–617.

Nakada, T., Itoh, T., and Takagi, Y. (1993), "Application of CARS to Development of High Compression Ratio Spark Ignition Engine," SAE Paper No. 932644.

Nakano, M., et al. (1995). "Predictions of the Knock Onset and the Effects of Heat Release Pattern and Unburned Gas Temperature on Torque at Knock Limit in SI Engines," SAE Paper No. 952408.

Olikara, C. and Borman, G.L., (1975), "A Computer Program for Calculating Properties of Equilibrium Combustion Products with Some Applications to IC Engines," SAE Paper No. 750468.

Orth, A. et al. (1994), "Simultaneous 2D Single-Shot Imaging of OH Concentrations and Temperature Fields in an SI Engine Simulator," *Proc. of 25th Symposium (International) on Combustion*, The Combustion Institute, pp. 143–150.

Rassweiler, G.M. and Withrow, L. (1935), "Flame Temperatures Vary with Knock and Combustion Chamber Position," *SAE Journal*, Vol. 36, pp. 125–133.

Roller, A., Arnold, A., Decker, M., Sick, V., Wolfrum, J., Hentschel, W., and Schindler, K.P. (1995), "Non-Intrusive Temperature Measurements During the Compression Phase of a DI Diesel Engine," SAE Paper No. 952461.

Ryu, H. and Asanuma, T. (1984), "Combustion Analysis with Gas Temperature Diagrams Measured in a Prechamber Spark Ignition Engine," *Proc. 20th Symposium (International) on Combustion,* The Combustion Institute.

Sakurauchi, Y., Ryu, H., Iijima, T., and Asanuma, T., (1987), "Combustion Gas Temperature in a Prechamber Spark Ignition Engine Measured by Infrared Pyrometry," SAE Paper No. 870457.

Schultz, C., Sick, V., Wolfrum, J., Drewes, V., Zahn, M., and Maly, R. (1996), "Quantitative 2D Single Shot Imaging of Nitric Oxide Concentrations and Temperature in a Transparent SI Engine," *Proc. of 26th Symposium (International) on Combustion*, The Combustion Institute.

Siegel, R. and Howell, J.R. (1993), *Thermal Radiation Heat Transfer*, 3rd Ed. Hemisphere, New York, 1993.

Smith J.R. (1980), "Temperature and Density Measurements in an Engine by Pulsed Raman Spectroscopy," SAE Paper No. 800137, 1980.

Smith, J. R. et al. (1984), "An Experimental and Modelling Study of Engine Knock," *Proc. of 20th Symposium (International) on Combustion,* The Combustion Institute, pp. 91–100.

Stenhouse, I.A., Williams, D.R., Cole, J.B., and Swords, M.D. (1979), "CARS Measurements in an Internal Combustion Engines," *Applied Optics*, Vol. 18, 3819–3825.

Tien, C. (1968), "Thermal Radiation Properties of Gases," *Adv. Heat Transfer*, Vol. 5, pp. 254–321.

Takagi, H., Ohno, T., and Asanuma, T. (1990), "Temperature Measurements of Combustion Gas in a Spark Ignition Engine by Infrared Monochromatic Pyrometry," SAE Paper No. 900483.

Withrow, L. and Rassweiler, G.M. (1932), "Emission Spectra of Engine Flames," *Ind. Eng. Chem.*, Vol. 24, pp. 528–538.

Zur Loye, A. and Santavicca, D.A. (1983), "Temperature and Concentration Measurements in an Internal Combustion Engine Using Laser Raman Spectroscopy," AIAA 18th Thermophysics Conf. Paper AIAA-83-1551.

Chapter 13

In-Cylinder Measurements of Combustion Products

Outline

13.1 Introduction

Prior to combustion, the in-cylinder mixture consists of fresh fuel and air mixture and burned gas, either left over from the previous cycle (residual gas) or externally recirculated for emission control (exhaust gas recirculation, EGR). The composition of the in-cylinder charge and the distribution of each species have significant effects on the subsequent combustion and pollutant formation processes. Techniques for measuring in-cylinder fuel concentration have been discussed in Chapters 7 and 8. Techniques for measuring burned gas concentration and composition will be considered introduced in this chapter.

While combustion photography enables researchers to see what happens inside an engine, spectroscopic techniques have helped engineers to understand the chemical reactions involved in the combustion process and pollutant formation mechanisms. The principles and applications of these spectroscopic techniques are presented in Section 13.3.

Knocking combustion is one of the important limiting factors in SI engines. Numerous studies on knocking combustion were carried out by both spectroscopic and in-cylinder sampling techniques. The results of these studies are discussed in Section 13.4. Finally, two-dimensional techniques for in-cylinder species measurements are introduced and their applications to IC engines are presented.

13.2 Measurement of In-Cylinder Burned Gas

13.2.1 Introduction

The presence of burned gas in the cylinder during the compression stroke comprises two sources: the residual gases and recirculated exhaust gas. The residual gases consist of the charge left over from the previous cycle. They are a consequence of: the clearance volume at TDC; the relative pressures in the inlet manifold, cylinder, and exhaust manifold; and the valve timing. Higher levels of residual burned gases will occur when there is: a larger clearance volume, a wider valve overlap, or a lower inlet manifold pressure.

The residual gas fraction in an SI engine significantly influences the engine behavior, especially at low load or idle conditions. As the residual gas concentration increases, the development of the flame kernel becomes more susceptible to the in-cylinder flows, which is believed to be one of the causes of high cyclic variability at low load. Thus, there is a need to characterize the residual gas fraction as a function of the engine operating parameters.

In addition, a fraction of exhaust gases is recycled back to the cylinder in order to suppress the formation of NO_x in both SI and diesel engines. The exhaust gas recirculation (EGR) level can be defined volumetrically on a molar basis, or gravimetrically. The volumetric definition is simpler, and the EGR level is defined as the percentage reduction in volume flow rate of air at a fixed operating point:

$$\%EGR = \frac{\dot{V}_o - \dot{V}_e}{\dot{V}_o} \times 100 \tag{13.1}$$

where $\dot{V}_o$ = volume flow rate of air with no EGR
$\dot{V}_e$ = volume flow rate of air with EGR

This volumetric definition of EGR is often used in diesel engines. With a diesel engine, the fixed operating point will correspond to the same speed, and either the same fueling level or the same torque.

In SI engines, the EGR level is usually deduced, by comparing the concentration of a particular species in the exhaust, $(CO_2)_e$, with its concentration in the inlet manifold, $(CO_2)_m$. It is usual to measure carbon dioxide, because it is present in the most significant quantities, and it can be measured accurately. There are two definitions of EGR on a molar basis:

$$\%EGR = \frac{(CO_2)_m}{(CO_2)_e} \times 100 \tag{13.2}$$

or

$$\%EGR = \frac{(CO_2)_m}{(CO_2)_e - (CO_2)_m} \times 100 \tag{13.3}$$

Sometimes a gravimetric definition of EGR can also be used. Hence, it is important to state which method of measuring EGR has been used, and how it has been defined.

Thus, the externally recirculated burned gas (EGR) in the cylinder can be calculated by either the reduction in the air flow or measurements of carbon dioxide. However, the concentration of residual gas cannot be readily determined. Furthermore, burned gases are sometimes recirculated internally into the combustion chamber (internal EGR), by altering valve times or exhaust throttling. Special techniques are therefore needed for the measurement of residual gas and/or internal EGR. Both physical probes and optical techniques have been developed for in-cylinder burned gas measurements. The use of a physical probe involves the withdrawal of a representative sample of combustion products at a given instant in the engine cycle. The sample obtained may then be routinely analyzed by one of the standard exhaust gas analyzers. The physical probes used include the fast FID and in-cylinder sampling techniques introduced in Chapter 3 and 4, respectively.

13.2.2 Residual Gas Concentration Measurement by a Fast FID

Because the cylinder contents during the compression stroke of an SI engine comprise both air/fuel mixture and combustion residuals, the HC concentration in the cylinder contents will be lower than the HC concentration in the fresh air-fuel mixture being supplied to the engine. In other words, the fresh air-fuel mixture being supplied to the engine is diluted by the combustion residuals remaining in the engine cylinder at the end of the exhaust stroke. As a consequence of the dilution, one should expect the signal plateau level to be lower than the level corresponding to the fresh air fuel mixture supplied to the engine. How much lower the signal for the cylinder contents would be dependent on the amount of combustion residuals trapped in the engine cylinder. Based on this fact, Galliot et al. [1990] developed a technique that uses the difference between the two signals to obtain an estimate of the amount of combustion residuals trapped in the engine cylinder. The technique is relatively simple, and it was also applied by Ladommatos [1992] (see Table 3.3, sets 5 and 7 for a summary of fast FID configurations used). A brief description of the technique follows.

The sampling probe was mounted in the ground shell of a spark plug. A continuous flow of sample at a constant mass flow rate was withdrawn from the cylinder and analyzed in a flame ionization detector, as described in Chapter 3. The HC concentration in the engine cylinder was first measured using the plateau of the FID signal obtained while the engine cylinder was being motored (i.e., spark disabled) at steady state and supplied with air-fuel mixture. The engine cylinder was then fired continuously as normal at steady state and the HC concentration in the cylinder was measured again from the corresponding signal plateau. Fig. 13.1 shows the FID signals for these two cases. The top line was the hydrocarbon concentration obtained from the motored engine and the lower one represented the hydrocarbon concentration reading from the firing operation. Galliot et al. [1990] has shown that the residual molar fraction (number of moles of residuals/total number of moles of cylinder contents) can be calculated from: (difference between motored and fired HC concentrations fired)/(motored HC concentration).

It should be noted that the residual gas concentration obtained is the mole fraction of the residual gas, x_{mole}, in the cylinder which is related to the mass fraction of the residual gas, x_{mass}, by

$$x_{mole} = \frac{x_{mass}}{1 - \dfrac{\phi(1 - x_{mass})}{23.85 + 2\phi}} \tag{13.4}$$

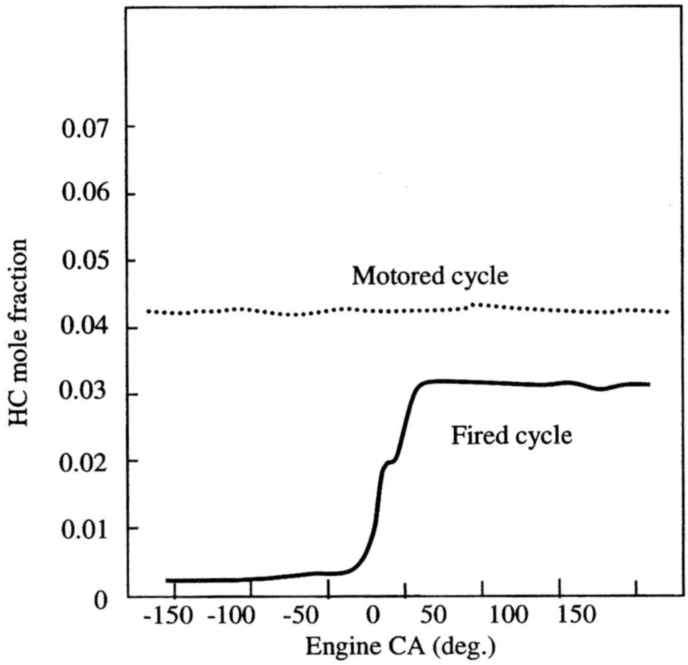

Fig. 13.1 Idealized FID signals from motored and fired cycles.

where ϕ is the fuel/air equivalence ratio. For lean to stoichiometric conditions, the difference between x_{mole} and x_{mass} is less than 4% [Galliot et al.,1990].

Meaningful interpretation of the result was only possible after the FID signal had reached a plateau value, and there was a finite time for the FID signal to reach a plateau. At a low engine speed and high load, a definite plateau region in the FID signal existed, from which the HC concentration or the residual gas fraction was determined. At low load and high engine speed, the FID signal fell sharply due to the arrival of the flame before a plateau could be reached. In order to obtain a good signal (one with a definite plateau region), Galliot et al. used skip-firing every 10 cycles. The FID signal of the skip-fired cycle was used to determine the HC concentration under low load and high-speed conditions.

This technique allows the residual gas fraction to be measured easily for each engine cycle. For example, by measuring the fraction for 300 individual cycles, Galliot et al. calculated the standard deviation of the fraction over the 300 cycles as about 1.5; results from Ladommatos show a variation of similar order. Also, both Galliot et al. and Ladommatos measured the decaying level of the residual fraction for successive engine cycles, after the spark plug was abruptly disabled. Galliot et al. found that about 10 successive unfired cycles were required to expel the combustion residuals from the engine cylinder at 900 rpm and low load (starting from a fraction of 25% just before spark disablement), while Ladommatos found that in his engine around 6 engine cycles were required at 1200 rpm and low load (starting from a fraction of 22%).

13.2.3 Combustion Product Analysis by In-Cylinder Sampling Techniques

The major advantage of in-cylinder sampling over other techniques is its capability of multi-species detection. There are two types of in-cylinder sampling techniques: a high-speed intermittent sampling valve, and the total cylinder dumping method. Their principles and operation have been discussed in detail in Chapter 4. A summary of sampling valve operating parameters and their applications are given in Table 4.1 and Table 4.3, respectively. Here we consider some applications of sampling valves for multi-species detection in IC engines.

Matsui et al. [1979] sampled CO_2 at the spark plug gap of a spark ignition engine, using their needle-type sampling valve incorporated in a spark plug. The sample was diluted by means of helium and then fed to a purpose-built CO_2 analyzer. The CO_2 analyzer had a reduced measuring cell capacity and, hence, fast response, which allowed cycle-resolved measurements of CO_2 concentration.

Aoyagi et al. [1980] sampled NO and NO_2 from the combustion chamber of a DI diesel engine. The line conveying the sample from the valve was heated to 120°C, as was the glass vessel in which the sample was collected. Prior to each sample collection, the sample lines and the glass vessel were evacuated, and then the sample was allowed to fill the glass vessel to a pressure of 1.2 bar. A chemiluminescence analyzer was then used to identify the concentrations of NO and NO_2. For other gaseous species, Aoyagi et al. diverted sample gas from the valve into an evacuated syringe, maintained at 200°C. Gas chromatography was then used to measure the concentration of N_2, O_2, Ar, CO_2, CO, and CH_4 in the sample. Similar techniques were also employed by Barbella et al. [1989].

Using NDIR, FID, and chemiluminescence exhaust gas analyzers, Bäcker et al. [1997] obtained the concentrations of various gaseous species in a sample taken from a DI diesel engine. The sample was diluted with nitrogen gas. Dilution with nitrogen increased the flow rate of the diluted sample and this allowed exhaust gas analyzers to be used to analyze the sample.

While the use of a high-speed intermittent sampling valve enables local gas near the valve tip to be analyzed, the large spatial variation in NO distribution in a diesel engine requires the use of the total cylinder dumpling technique to obtain a cycle resolved total NO concentration. The total cylinder sampling technique was used exclusively by two research groups at the Universities of Wisconsin [Voiculescu and Borman,1978] and Minnesota [Pipho et al., 1991].

Voiculescu and Borman [1978], who pioneered the total cylinder dumping technique, measured total in-cylinder NO concentrations as a function of crank angle in a DI diesel engine. The results represented NO concentrations spatially averaged over 75% of the cylinder mass at a given crank angle. The experimental data showed that the NO histories had a rapid rise time of 17 to 25 °CA. The maximum in-cylinder NO concentration could exceed the exhaust concentrations by as much as 14%.

While emissions of NO_2 are generally lower than NO emissions, the higher toxicity of NO_2 (about five times that of NO), makes the control of NO_2 important. In order to understand its formation, measurements of NO_2 were carried out in the combustion chamber of a DI diesel engine by Pipho et al. [1991] using the total cylinder sampling technique. Samples obtained by total cylinder sampling showed NO_2 /NO ratios of 25 to 50%, much higher than the 1 % to 3% detected in the exhaust. This was ascribed to the conversion of NO to NO_2 due to the sudden dilution of samples during the sampling process.

13.2.4 Burned Gas Visualization by PLIF

13.2.4.1 Introduction

A fast FID indirectly measures the residual gas content from the difference in the HC content of sampled gas from motored and fired cycles. The limited time resolution of a fast FID system makes it very difficult to obtain time-resolved measurements. While in-cylinder sampling has proved to be an effective means of obtaining simultaneous measurements of several species, it lacks spatial or temporal resolution, or both. At an engine speed of 1000 rpm, a sampling period of 1 ms corresponds to 8°CA, which is a rather long period in some applications. In addition, such a relatively long sampling time aggravates the spatial resolution due to in-cylinder flow. The effect of in-cylinder flow on the spatial resolution of a sampling valve has been discussed in detail in Chapter 4.

In order to provide spatial and temporal resolved measurements, PLIF and broadband SRS have been developed and applied to IC engines. Due to its signal strength, the LIF technique has been used for two-dimensional visualization of the burned gas distribution. Two approaches have been employed in the visualization of burned gas distribution by PLIF. One is to directly excite one of the species in the burned gases by a UV laser source and detect its fluorescence signal. Because most species in the burned gases posses electronic resonant frequency in the low UV regime, it is often difficult to directly apply the LIF technique to them. An alternative approach has therefore been developed, where the burned gas fraction is measured indirectly from the LIF signal of a fluorescing dopant.

13.2.4.2 Direct Imaging of Water Vapor by the Two-Photon LIF Technique

In-cylinder burned gas measurements using LIF require the presence of fluorescing molecules in the burned gases. Since LIF relies on the absorption of laser light at the electronic resonance frequency of the molecule to generate the subsequent spontaneous emission (fluorescence), the molecule must have a resonant frequency that is accessible by a laser. Water vapor is present in the cylinder as one of the burned gases in significant quantity. However, its resonance lines are in the low UV region. The LIF technique discussed so far cannot be applied to the detection of water vapor. Instead, two-photon LIF is used to excite water vapor.

Fig. 13.2 illustrates schematically the difference between the single and two-photon LIF processes. For LIF measurements discussed in the previous chapters, the molecule under study absorbs one photon. The energy $E_{21} = h\nu_{21}$ of this photon of frequency ν_{21} matches the energy difference between levels 1 and 2. The excited species in level 2 may emit fluorescence of frequency ν_f. In the two-photon scheme, the molecule simultaneously absorbs two photons. In this case, the sum of the energies of both photons corresponds to the energy difference between levels 1 and 2, i.e., $E_{21} = h(\nu_{2v} + \nu_{v1})$. In the two-photon process, one can imagine that the first photon excites the molecule to a virtual level and then the second photon lifts the molecule further to the energy level 2. Both frequencies, ν_{2v} and ν_{v1}, may be the same, so that a single laser may be used, or they may be different. Similar to one-photon LIF, the excited state emits fluorescence which may be detected as a measure of the species concentration.

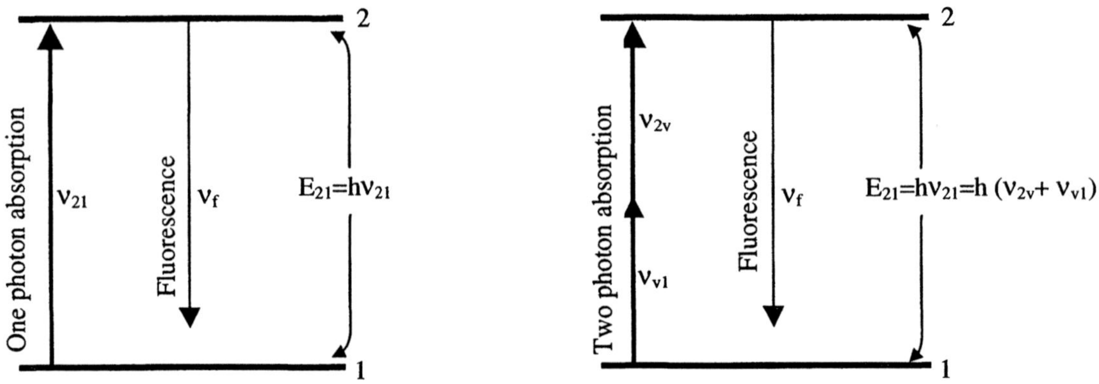

Fig. 13.2 Schematic of the energy level diagrams of one-photon (left) and two-photon (right) LIF processes.

In the two-photon fluorescence process, the Einstein coefficient for stimulated absorption, B_{12} in Eq. (5.61), would be replaced by $(c/h\nu)\alpha_{12}I_o$ [Eckbreth, 1996]. Hence, the two-photon fluorescence signal can be expressed as

$$P_{flu} \propto \chi_m N P_o^2 \alpha_{12} \frac{A_{21}}{A_{21} + Q_{21}} \tag{13.5}$$

where α_{12} is the two-photon absorption cross section from state 1 to state 2. Since the absorption cross section is very small, the two-photon transitions are difficult to saturate. Eq. (13.5) shows that the fluorescence signal will be a quadratic function of laser power. The fluorescence signal will be reduced by quenching.

Fig. 13.3 shows the energy-level diagram for water vapor, including the transitions relevant to combustion diagnostics. If wavelengths near 125 nm are used for excitation from the ground state X^1A_1, the third excited state C^1B_1 can be reached. Because there is no excitation source at 125 nm, instead the transition can be achieved by the two-photon process. The excited water molecule can then be returned to the first excited state by emitting light in the range of 400–500 nm. Fig. 13.4 shows the excitation and emission spectra of water vapor. The lines in Fig. 13.4(a) correspond to one or more rotational lines and are numbered according to the notation by Meijer et al. [1986]. The emission spectrum (Fig. 13.4[b]) exhibits a broad, structureless spectrum that extends from 380 nm to almost 600 nm, with a maximum at approximately 425 nm. The emission spectrum (Fig. 13.4[b]) is independent of the rotational absorption lines as shown in Fig. 13.4(a).

Neij and Aldén [1994] found that water molecules could be excited through a two-photon excitation process at 248 nm with a tunable excimer laser. They estimated that the detection limit at 300 K was 1% water in air. The LIF signal intensity obtained from peak 4 in Fig. 13.4(a) was found to be roughly 20 times stronger than the Raman signal under atmospheric

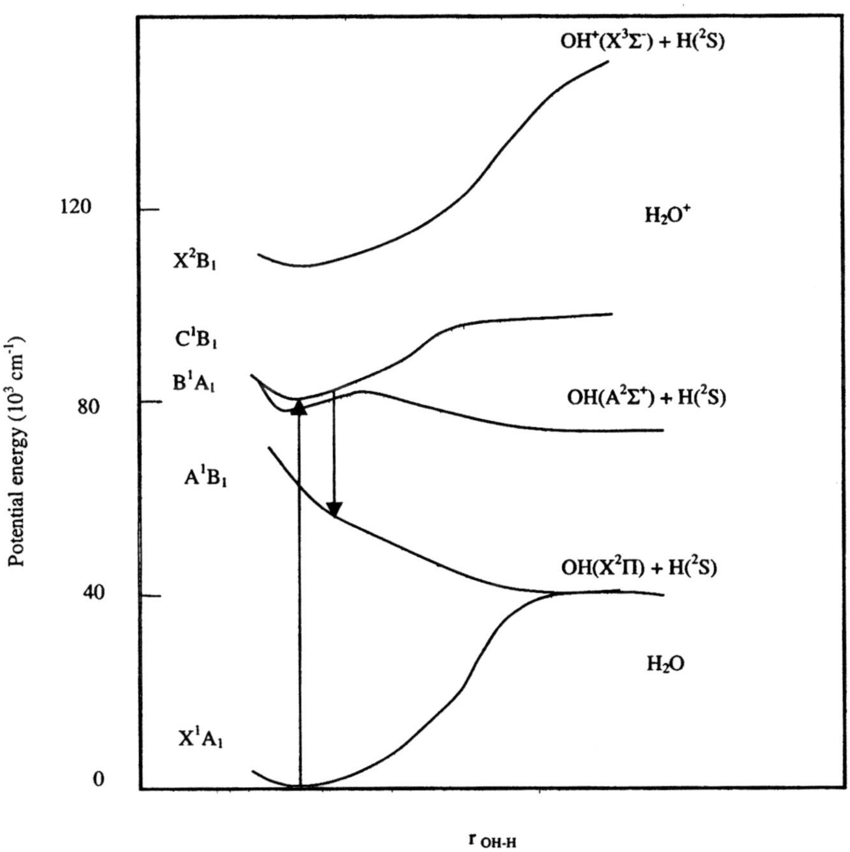

Fig. 13.3 Potential energy level diagram of water vapor.

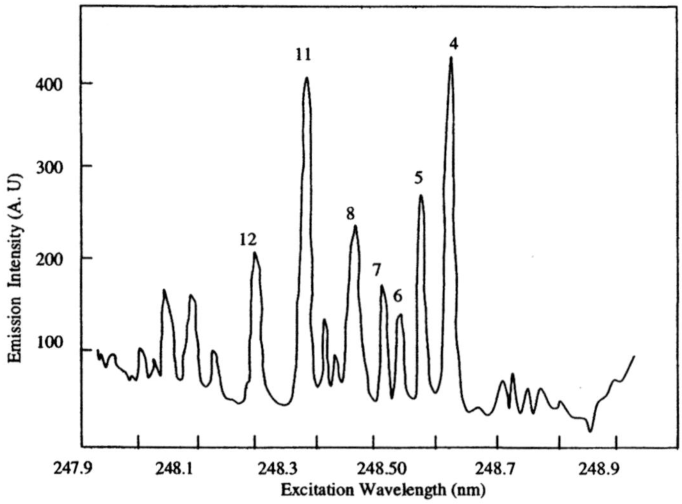

(a) Two-photon water-vapour excitation spectrum

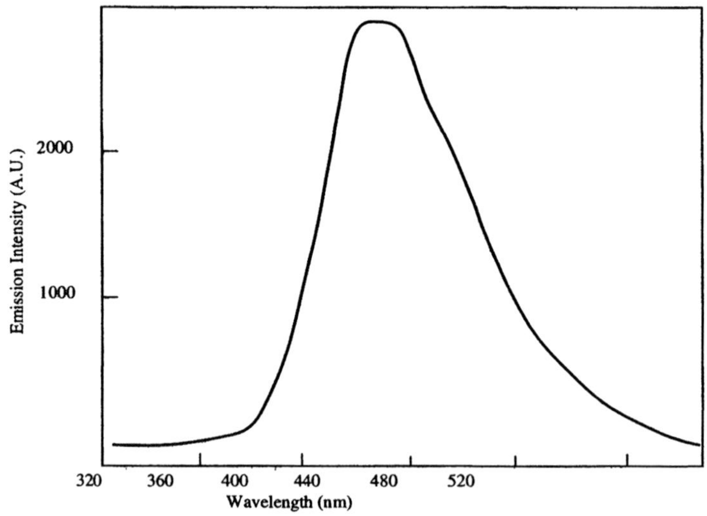

(b) Emission spectrum from water vapour

Fig. 13.4 Two-photon water vapor excitation and emission spectra.

conditions. The fluorescence signal was, however, reduced by about two-thirds as the partial vapor pressure of water was increased from 32 mbar to 320 mbar.

Following the exploratory work by Neij and Aldén, Johansson et al. [1995] applied the two-photon PLIF technique to visualize the water vapor distribution in an SI engine. The experimental setup for their two-photon LIF measurements was very similar to that used for PLIF imaging of fuels, discussed in Chapter 7. The laser was tuned over the strong peak in the excitation spectrum at 248.285 nm (peak 11 in Fig. 13.4[a]) or 248.523 nm (peak 4 in Fig. 13.4[a]). Since the two-photon LIF signal was much weaker than the one-photon LIF intensity, this necessitated the use of a narrow and thin laser sheet in order to get a sufficient fluorescence signal. The output from the tunable KrF excimer laser was formed into a narrow sheet of 7 × 0.1 mm. The fluorescence image was captured by an ICCD camera. Because of the second-order dependence of the two-photon LIF signal on laser intensity, laser power output was corrected on a shot-by-short basis.

The water vapor distribution close to the spark plug was measured in a 1.6 liter single cylinder SI engine. In order to avoid interference from LIF of fuel and inhomogeneities in the mixture, the engine was operated with natural gas. From the LIF images of water vapor, Johansson et al. found that the amount of water was fairly independent of engine load, indicating a larger amount of residual gas under lower load conditions. Simultaneous measurements of cylinder pressures revealed that the cycle-to-cycle variations in early combustion was strongly influenced by the cycle-to-cycle variations in water content (residual gas) at low loads, which diminished at full load conditions.

13.2.4.3 Nitric Oxide as a Residual Gas Marker

There are a few species in the combustion gas that could potentially be used as the residual gas marker. Water vapor accounts for about 10% of combustion gas. This is translated to about 1% during the compression stroke if 10% residual gas is assumed. Since water content in the normal intake air is of the same order, it will be difficult to correlate quantitatively the residual concentration with the measurement of water vapor concentration unless dry air is used. Although CO and H_2 detection by the two-photon LIF technique has been demonstrated in a flame [Seitzman, et al., 1987; Lempert, et al., 1991], the lower concentration and the weaker two-photon processes of these components renders the LIF signal less than the detection limit. Based on these considerations, Berckmüller et al. [1997] adopted NO as the residual gas marker and used PLIF to detect its presence in the cylinder.

The experimental detail is given in Table 13.1 and a detailed description of PLIF of NO will be given later. NO was excited by an Nd:YAG pumped dye laser (3 mJ) using the $A^2\Sigma^+ \Leftrightarrow X^2\Pi$ (0, 0) band at 226 nm. Measurements within the 50 mm wide laser sheet were taken in the intake stroke of a single-cylinder Honda V-TEC engine. Since NO concentration was diluted by the fresh charge by a factor of 10, each PLIF NO image was obtained by integrating 100 shots from successive cycles on the CCD chip in order to increase the signal-to-noise ratio. Ten of these accumulated images were then ensemble-averaged to produce the final PLIF image of NO distribution. The averaged image showed that there was a 10% higher residual concentration on the inlet side than the exhaust side.

In order to study cycle-by-cycle variations in residual gas concentration, Berckmüller et al. carried out cycle-resolved measurements using a narrower laser sheet (10 mm). From 58 individual images, they estimated that the cycle-by-cycle variation was about 17%.

Table 13.1 Summary of NO PLIF measurements reported in the literature

Laser source	Excitation Wavelength	Fluorescence Wavelength	Application	Comment	Reference
Tunable ArF excimer laser	193 nm $D^2\Sigma^+ \Leftrightarrow X^2\Pi(0, 1)$ $Q_1(32.5)$, $R_1(26.5)$		NO formation in an SI engine with isooctane.	Strong absorption of laser and NO fluorescence signal.	Andresen et al. [1990]
Tunable ArF excimer laser 40 mJ	193 nm $D^2\Sigma^+ \Leftrightarrow X^2\Pi(0, 1)$	208 ± 10 nm $D^2\Sigma^+ \Leftrightarrow X^2\Pi(0, 3)$	NO formation in an IDI diesel engine with n-heptane and diesel.	Strong absorption of laser and LIF signal.	Brugman et al. [1993]
ArF excimer laser (227 mJ)	193 nm $D^2\Sigma^+ \Leftrightarrow X^2\Pi(0, 1)$	210–285 nm $A^2\Sigma^+ \Leftrightarrow X^2\Pi$ (0, 0 to 0, 4)	NO formation in a single-cycle SI engine with propane/air mixture.	O_2 LIF interference, NO LIF measurements limited to low pressure.	Tanaka et al. [1997]
KrF excimer laser and Raman cell, 10mJ	225 nm $A^2\Sigma^+ \Leftrightarrow X^2\Pi(0, 0)$ $R_1(21)$	248 ± 5 nm $A^2\Sigma^+ \Leftrightarrow X^2\Pi(0, 2)$	NO formation in a square piston SI engine with propane/air mixture.	Absorption of laser beam, quantitative NO images of an accuracy of 30%.	Bräumer et al. [1995]
Nd:YAG pumped dye laser, 2.8 mJ	226 nm $A^2\Sigma^+ \Leftrightarrow X^2\Pi(0, 0)$	248 nm $A^2\Sigma^+ \Leftrightarrow X^2\Pi(0, 2)$	NO formation in a specially designed DI diesel engine.	Strong absorption of fluorescence signal, O_2 enrichment to enhance NO signal.	Alatas et al. [1993]
Nd:YAG pumped dye laser, 3 mJ	225.967 nm $A^2\Sigma^+ \Leftrightarrow X^2\Pi(0, 0)$ Q_2 (22.5)	248 ± 10 nm $A^2\Sigma^+ \Leftrightarrow X^2\Pi(0, 2)$	Marker for residual gas concentration imaging in an SI engine.	Ensemble-averaged to achieve good S/N ratio.	Berckmüller et al. [1997]
Nd:YAG pumped OPO, 12 mJ	226.035 nm $A^2\Sigma^+ \Leftrightarrow X^2\Pi(0, 0)$ $Q_1 + P_{21}(14)$, $Q_1 + P_{21}(14)$	237–276 nm $A^2\Sigma^+ \Leftrightarrow X^2\Pi$ (0, 1) (0, 2) (0, 3) (0, 4)	NO formation in a single-cylinder large DI diesel engine.	Single-shot NO image and quantitative total NO concentration.	Dec and Canaan [1998]
Tunable KrF excimer laser, 50 mJ	247.95 nm $A^2\Sigma^+ \Leftrightarrow X^2\Pi(0, 2)$ $P_{12}(8.5)$ $P_{12}(9.5)$ $P_{12}(10.5)$	230 ± 10 nm, $A^2\Sigma^+ \Leftrightarrow X^2\Pi$ (0, 0) (0, 1)	NO formation in an SI engine fueled with isooctane.	Less laser beam absorption, High-temperature and NO application.	Knapp et al. [1996]
Tunable KrF excimer laser, 50 mJ	247.95 nm $A^2\Sigma^+ \Leftrightarrow X^2\Pi(0, 2)$ $P_{12}(8.5)$ $P_{12}(9.5)$ $P_{12}(10.5)$	272 nm $A^2\Sigma^+ \Leftrightarrow X^2\Pi$ (0, 4)	NO formation in a production SI engine fueled with isooctane and gasoline.	Low absorption and radiation trap, high-temperature and NO application.	Knapp et al. [1997a]
Tunable KrF excimer laser, 45 mJ	248 nm $A^2\Sigma^+ \Leftrightarrow X^2\Pi(0, 2)$ O_{12}	230 ± 10 nm $A^2\Sigma^+ \Leftrightarrow X^2\Pi$ (0, 0) (0, 1)	NO formation in a square piston SI engine with propane/air mixture.	Quantitative NO results of an accuracy of 30%, high-temperature and NO application.	Schulz et al. [1996]
Tunable KrF excimer laser, 45 mJ	248 nm $A^2\Sigma^+ \Leftrightarrow X^2\Pi(0, 2)$ O_{12}	230 ± 10 nm $A^2\Sigma^+ \Leftrightarrow X^2\Pi$ (0, 0) (0, 1)	NO formation in a single-cylinder SI engine with isooctane.	Quantitative NO results of an accuracy of 30%, high-temperature and NO application.	Hildenbrand et al. [1998]

Because of the weak LIF signal, interference from scattered laser light off of spark plug and inlet valves was observed. Since measurements were taken before the start of combustion, absorption of the laser light was found to be insignificant. There are, however, several sources of uncertainties that must be considered when the NO LIF signal is interpreted. The NO fluorescence will be quenched more heavily in a region with more residual gas due to the large quenching cross sections of CO_2 and H_2O [Drake and Ratcliffe, 1993]. Since the NO concentration in the residual gas is much more sensitive to the combustion temperature of the previous cycle than other major species in the residual gas, the cycle-by-cycle variation in the residual gas concentration may be exaggerated by the measurements of NO concentration.

13.2.4.4 Visualization of Burned Gas Distribution by the Negative PLIF Technique

In order to study the effects of burned gas distribution on combustion and emission in an SI engine, Deschamps and Baritaud [1996] devised a means of obtaining burned gas fraction from normal LIF measurements.

Principle of the Negative PLIF Technique

The principle of the negative PLIF technique is to visualize the unburned mixture fraction and then deduce the burned gas fraction. The intake charge is seeded with a fluorescing dopant which can be readily excited by a UV laser source. The fluorescing dopant is burned during the combustion process so that no fluorescence will be detected in the burned gases. During the compression process, the in-cylinder mixture consists of the fresh intake charge seeded with the fluorescing dopant and residual burned gases without the fluorescing dopant. As a consequence of dilution by non-fluorescing burned gases, the LIF intensity from the seeded charge is reduced. A reference image is recording during a cycle with no residual gas, but with the same level of seeding. The difference in LIF signals between the reference image and the unburned mixture image obtained during a cycle with residual gas, then, represents the burned gases. The above process can be represented by

$$rf(x,y) = 1 - \frac{I_u(x,y) - I_b(x,y)}{I_{ref}(x,y) - I_b(x,y)} \tag{13.6}$$

where $rf(x, y)$ = residual gas fraction on a volumetric basis at point (x, y)
$I_u(x, y)$ = the LIF intensity of unburned mixture at point (x, y)
$I_b(x, y)$ = the averaged background image at point (x, y)
$I_{ref}(x, y)$ = the LIF intensity of the averaged reference image at point (x, y)

The accuracy of this method necessitates a good reference image obtained from a homogeneous distribution of the seeding. In their seeding system, Deschamps and Baritaud introduced 1.5% biacetyl into air far upstream of the intake valve to ensure a homogeneous distribution of biacetyl in the intake air by means of a venturi system. Since their seeding system worked on the venturi principle, the quantity of biacetyl in the intake charge depended on the prĕssures both in the seeding vessel and the intake manifold. The pressures in the intake and the seeding vessel were monitored so that they were the same for reference images and images of unburned gases with residual gases.

Reference images are obtained before and after cycles with combustion (residuals) for a normally operated engine. Then, the averaged reference image can be used for normalization in Eq. (13.6). Alternatively, the reference and unburned mixture images can be obtained successively under skip-firing operating conditions. Depending on the operating conditions, residual gases may not be completely scavenged in the cycle immediately following the firing cycle. Hence, each reference image should be taken proceeding the firing cycle with sufficient number of cycles skipping firing.

Effect of Port Fuel Injection

Up to now, the intake charge has been assumed to comprise the premixed air-fuel mixture with homogeneously seeded fluorescing dopant. This is true when fluorescence dopant is introduced after fuel injection. In a port-injection SI engine, fuel is injected very closely or directly to the back of the intake valves. The seeded air will be diluted by the vaporized fuel in the cylinder. Measurements of the mean burned gas distribution require knowledge of the volumetric fuel fraction as well as the air fraction because

$$1 = (V_a' + V_f' + V_{rf}')/V_{cyl} \tag{13.7}$$

where V'_a = the partial volume of air in the cylinder
V'_f = the partial volume of vaporized fuel in the cylinder
V'_{rf} = the partial volume of residual gases in the cylinder
V_{cyl} = the volume of the cylinder at the time of measurement

Since the LIF signal is proportional to the mole fraction of fluorescing dopant, images of fuel distribution can be obtained by seeding the fuel with the same level of dopant as for the reference image. The LIF signal in the fuel image is given by

$$I_f \propto \frac{V_f' \cdot \chi_s}{V_{cyl}} \tag{13.8}$$

where χ_s is the percentage of seeding in the volume.

Similarly, the LIF signal of fluorescing dopant in air is given by

$$I_a \propto \frac{V_a' \cdot \chi_s}{V_{cyl}} \tag{13.9}$$

where the seeding concentration χ_s is assumed to be the same as that in fuel.

The LIF signal in the reference image is proportional to the percentage of seeding, i.e.,

$$I_{ref} \propto \chi_s \tag{13.10}$$

Substituting Eqs. (13.8–13.10) into Eq. (13.6), one obtains for the burned gas fraction

$$rf(x,y) = 1 - \frac{I_a(x,y)}{I_{ref}(x,y)} - \frac{I_f(x,y)}{I_{ref}(x,y)} \tag{13.11}$$

or

$$rf(x,y) = 1 - I_a^n - I_f^n \tag{13.12}$$

where the averaged reference image, and fuel and air images have been corrected for the background noise.

Calibration of the Negative PLIF Technique

Since the averaged intensity of air and fuel images represent directly the volumetric fractions of air and fuel, respectively, as shown by Eqs. (13.8) and (13.9), the ratio of these two averaged intensities should correlated with the relative air/fuel ratio (lambda value, λ) measured in the exhaust. Assuming a perfect gas mixture, the correlation is given by

$$\bar{I}_f = \bar{I}_a \frac{M_a}{M_f} \frac{1}{\lambda} \left(\frac{F}{A}\right)_s \tag{13.13}$$

where M_a = the molecular weight of air
M_f = the molecular weight of fuel
$(F/A)_s$ = the stoichiometric fuel/air ratio

$\bar{I}_a = \left\langle \frac{I_a(x,y)}{I_{ref}(x,y)} \right\rangle$, mean air fraction of the image

$$\bar{I}_f = \left\langle \frac{I_f(x,y)}{I_{ref}(x,y)} \right\rangle, \text{ mean fuel fraction of the image}$$

Using this correlation, Eq. (13.11) can be written as a function of the averaged intensity of air image and the lambda value measured by a lambda sensor or an exhaust gas analyzer,

$$rf(x,y) = 1 - \bar{I}_a \frac{I_a^n}{\bar{I}_a} - \frac{M_a}{M_f} \frac{1}{\lambda} \left(\frac{F}{A} \right)_s \bar{I}_a \frac{I_f^n}{\bar{I}_f} \tag{13.14}$$

or

$$rf(x,y) = 1 - \bar{I}_a \bar{I}'_a - \frac{M_a}{M_f} \frac{1}{\lambda} \left(\frac{F}{A} \right)_s \bar{I}_a \bar{I}'_f \tag{13.15}$$

Equations (13.14) and (13.15) provide a better measure of the residual gas fraction from the LIF signals and the measured air/fuel ratio.

Effect of External Exhaust Gas Recirculation

So far it has been assumed that the burned gas fraction in the in-cylinder mixture is replaced by the dopant seeded air in the reference image, so that the division of the two images gives the mean air fractions. This assumption is valid if there is no externally recirculated EGR. As external EGR is introduced in a spark ignition engine, the total in-cylinder charge will be increased due to wider opening throttle. In addition, the use of external EGR leads to a higher pressure in the intake system, which results in a lower seeding level for a seeding system based on the venturi principle. To overcome this problem, the air flow rate for the reference images has to be adjusted so that the volumetric efficiency, $\eta_{vol,\,ref}$, is the same as the air volumetric efficiency, $\eta_{vol,\,a}$, for the real image given by

$$\eta_{vol,a} = \eta_{vol} \frac{1}{1 - \%EGR} \tag{13.16}$$

where η_{vol} is the total volumetric efficiency for the intake charge (air + EGR).

Application to a Stratified EGR Engine

Using the negative LIF technique, Deschamps and Baritaud [1996] obtained images of residual gas distributions for different fuels (propane and isooctane), fuel distributions, and intake pressures. They found that in-cylinder residual gas distributions were affected by all the engine parameters investigated and concluded that they were difficult to control because of the complexity of interactions between various parameters. When the negative LIF technique was applied in an SI engine with external EGR, the air flow for the reference image was adjusted according to Eq. (13.16) in order to take into account the dilution effect of external EGR on the PLIF technique. Based on PLIF images of exhaust gas distribution in the combustion chamber, they were able to identify and optimize key engine parameters for stratified EGR operation.

The negative LIF technique has proved to be an effective means for visualization of the in-cylinder burned gas distribution. Like other LIF techniques, it is quite difficult to quantify the accuracy of this technique. It is not possible to differentiate residual gas from externally recirculated exhaust gas. Because the technique relies on normalization to an averaged reference image, it cannot be used to evaluate a single-cycle burned gas distribution. In fact, it is assumed that there is no cyclic variation between the reference condition and the real case, which may not be true in many cases.

13.2.5 Quantitative Measurements of Burned Gas Composition and Concentration by Broadband SRS

In the previous sections, we described how PLIF was adopted to obtain two-dimensional images of burned gas distributions in SI engines. These techniques, however, are limited to qualitative visualization of burned gas distributions, due to fluorescence quenching by O_2 or other species. Because either one species (water vapor) or one fluorescing dopant is used in LIF measurements of burned gases, LIF techniques cannot be used to determine the burned gas composition.

In order to obtain insight into the effects of mixture composition on combustion performance and pollutant formation, and to gather quantitative species concentration data for validation of the prediction of computational models, quantitative measurements of the burned gas composition as well as the concentrations of the gases are needed. Such quantitative information can be obtained using Raman scattering.

Coherent anti-Stokes Raman scattering (CARS) was used by Lebel and Cottereau [1992] to measure CO_2 concentration and temperature simultaneously in an IC engine. Due to its complexity and cost, particularly its single-point and single-species measurement capability, CARS has found limited use, except in gas temperature measurements in IC engines. The availability of new powerful laser sources in the UV region, coupled with recent developments in optical multi-channel detectors, has led to renewed interest in spontaneous Raman scattering (SRS). With broadband detection, the SRS technique permits the simultaneous measurement of all major species in burned gases with excellent accuracy, including H_2O, CO_2, CO, N_2, O_2, and fuel, dispensing the need for a sophisticated CARS system. Multi-point measurements can also be obtained along a portion of the probing laser beam by means of an imaging spectrograph and a 2-D detector array.

The principle of the SRS technique has been described in Chapter 5. The implementation of broadband SRS has been discussed in detail in Chapter 7. Here, we shall concentrate on the applications of broadbard SRS to the measurement of burned gas composition and concentration in SI engines.

SRS was used extensively by a research group at Laser Laboratorium of Göttingen to measure the residual gas concentration in SI engines [Grünefeld et al., 1994a; 1994b and 1995; Knapp et al., 1997b]. The optical setup used can be found in Table 7.2, and has been described in detail in Chapter 7. Initial experiments were carried out in a cylinder with an extended piston, as shown in Fig. 7.7 [Grünefeld et al., 1994b]. The residual water content was measured from the ratio of water vapor and nitrogen signals. The simultaneous detection of the air/fuel ratio and water content allowed the effect of equivalence ratio and residual gas concentration on the cycle-by-cycle variation to be separated.

Using the back-scattering setup shown in Fig. 7.8, Grünefeld et al. [1995] was able to measure the residual gas concentration (water content) in a production engine with only slight modification. By operating the CCD camera in the fast readout mode, Grünefeld et al. were able to measure the residual gas composition for the first 100 consecutive cycles at a fixed crank angle during the cold start and warm-up periods. Their results showed disproportionately high residual water content during cold-start, which was attributed to the condensation of water vapor on the cold cylinder walls. From the simultaneously measured air/fuel ratio and residual water content, they concluded that dilution of the charge by residual gases, particularly water vapor, was the main reason for the fuel enrichment needed during cold-start.

More recently, Knapp et al. [1997b] extended the technique to 1-D spatially resolved cycle-averaged SRS measurements, as well as time-resolved spatially-averaged ($0.5 \times 10 \times 14$ mm^3) measurements in a production engine. Their spatially averaged results did not exhibit any correlation between the residual gas content and the IMEP value of the engine. This was probably a consequence of the much smaller sampling volume (0.1% of the charge volume) used in the SRS experiments. However, their cycle-averaged results did confirm that there was an increased amount of residual gases with decreasing load or speed. The spatially resolved SRS measurements at 30°CA BTDC revealed that there were significant differences in residual gas distributions at different loads. Under the low-load condition (32.5 kPa manifold pressure, 1500 rpm, $\lambda = 0.9$), a gradient of about 40% in the residual gas distribution was observed across the 40 mm path length. This gradient disappeared when the intake manifold pressure was increased to 63 kPa.

In the above experiments, the accuracy of results was confirmed to be better than 6%. Calibration measurements conducted by Knapp et al. [1997b] also showed that the ratio of water vapor signal to nitrogen signal varied little with temperature up to 1100 K so that the temperature effect could be ignored.

Quantitative measurements of water vapor and CO_2 were reported by Miles and Dilligan [1996] using broadband SRS. Their experimental setup is given in Table 7.2. A significant advancement to the state-of-the-art in SRS measurements was made by Miles and Hinze [1998] and Hinze and Miles [1999]. A detailed discussion of their SRS systems and results have been given in Chapter 7. They carried out the first complete characterization of in-cylinder composition fields in terms of mean concentrations, fluctuations, and scalar length scales at light-load and idle engine operation. These measurements included single-shot, quantitative, multi-species, multi-point measurements in both a pancake-chamber of simplified cylinder head geometry and a modern four-valve pent-roof combustion chamber.

13.3 Spectroscopic Studies of Combustion Species in IC Engines

13.3.1 Types of Spectra from Combustion Species in IC Engines

Prior to the advent of laser diagnostics, combustion intermediates and products were detected principally by emission and absorption spectroscopy. Emission and absorption of combustion species are characterized by band spectra in the visible and UV regions. Free radicals such as CH, OH, C_2, CN, and NH are responsible for the strongest band spectra in the visible or near ultraviolet region. Table 13.2 shows prominent spectral peaks of certain species in the visible and near UV region. For the more permanent constituents of combustion gases, such as O_2, N_2, H_2O, and CO, the electronic transitions correspond to band systems far down in the UV. However, H_2O and CO_2 can be detected through their vibrational-rotational spectra in the infrared region (Table 13.2). It should be noted that the wavelengths given in Table 13.2 are peaks of each emission spectrum, which often spread out due to different vibrational/rotational levels involved in the transition process, as well as various line broadening mechanisms. In addition to the combustion generated species, emission and absorption spectroscopy has been used to investigate the pre-flame reactions that are responsible for knocking combustion in SI engines and autoignition processes in CI engines. One of the most important species is formaldehyde (CH_2O), which has strong emission bands in the visible region as given in Table 13.2.

Table 13.2 Strong emission peaks of major combustion species in IC engines

CH (nm)	OH (nm)	C_2 (nm)	CH_2O (nm)	CO_2 (μm)	H_2O (μm)
314	281–283	470–474	368	2.69–2.77	1.13–1.87
387–389	302–309	516	384	4.25–4.30	2.66–2.74
431		558–563	395		3.17
			412–457		6.27

The continuous spectrum is the other type of spectrum that has been employed to detect in-cylinder species and combustion process in an engine. In a diesel engine, continuous spectra are associated with solid soot particles. The intensity distribution on this type of continuum is related to the blackbody radiation from hot soot particles with increasing emissivity at shorter wavelengths. Blackbody radiation from soot particles has been used to measure diesel combustion temperature and soot concentration by means of the two-color technique, as already discussed in Chapter 10. Continuous spectra may also be caused by gas-phase processes such as dissociation of molecules, ionization of atoms, or recombination processes. Continua of particular importance in engine combustion are those due to

$$CO + O = CO_2 + h\nu$$

and

$$NO + O = NO_2 + h\nu$$

where $h\nu$ represents the emission of light associated with each reaction, which is referred to as chemiluminescence. During the above reactions, the colliding molecules/atoms in their ground electronic states form a new molecule in an excited electronic state, and an electronic transition to its ground state then takes place during the brief time of the collision. The energy is disposed of as light emission, $h\nu$, and the new molecule will be stabilized. The CO + O continuum is mainly responsible for the blue color of flames containing carbon monoxide. The NO + O continuum is responsible for the pale yellowish-green emission from flames containing oxides of nitrogen.

13.3.2 Principle of Emission and Absorption Spectroscopy

In the following analysis, the emission and absorption spectroscopy of band spectra from radicals and combustion products will be discussed briefly. A more complete description of the spectroscopic study of combustion is given by Gaydon [1970].

The intensity of a spectral line in emission is determined by the population of the upper level of the transition, N_n, and the Einstein coefficient for spontaneous emission, A_{nm},

$$I_{nm} = \eta_c A_{nm} N_n h\nu_{nm} \tag{13.17}$$

Since the population of the jth level, N_j, is related to the total number density of a particular species, N, by the Boltzmann's formula,

$$N_j = \frac{N}{Q(T)} g_j \exp\left(-E_j / kT\right) \tag{13.18}$$

Eq. (13.17) can be rewritten as

$$I_{nm} = \eta_c N A_{nm} \exp[-E_n / kT] \tag{13.19}$$

Equation (13.19) states that the emission signal is proportional to the total number density of the emitting species. Emission spectroscopy can only be used as a means of detecting the presence of an emitting species in most engine combustion studies due to the lack of temperature information.

For the same transition, the loss of irradiance due to energy absorbed in an absorbing medium of thickness L, and an incident light intensity $I_0(\lambda)$ is given by

$$I(\lambda) = I_0(\lambda) \exp\left(-\int_0^L K_{abs}(\lambda) dx\right) \tag{13.20}$$

If the absorbing medium is homogeneous, so that the absorption coefficient K_{abs} is independent of the position x, the optical depth of the medium $\int_0^L K_{abs}\, dx$ becomes $K_{abs}L$, and we have

$$I(\lambda) = I_0(\lambda) \exp(-K_{abs}(\lambda) L) \tag{13.21}$$

from which the absorption coefficient can be experimentally determined by

$$K_{abs}(\lambda) = \frac{1}{L} \ln\left(\frac{I_0(\lambda)}{I(\lambda)}\right) \tag{13.22}$$

which is related to the number density of the absorbing species at the ground state, N_1, by

$$\int_{\lambda 1}^{\lambda 2} K_{abs}(\lambda) d\lambda = \frac{h\nu_{21}}{c} N_1 B_{12} \tag{13.23}$$

The integration of the aborption coefficient in Eq. (13.23) is necessary because the absorption from an absorbing species at a particular spectral line is always spread over a finite frequency in practice by various broadening mechanisms (as shown in Fig. 13.5[a]). The experimentally determined absorption coefficient has a profile as shown in Fig. 13.5(b).

Equation (13.23) forms the basis of absorption spectroscopy, where the number density of the absorbing species is determined from the total area under the absorption coefficient profile as

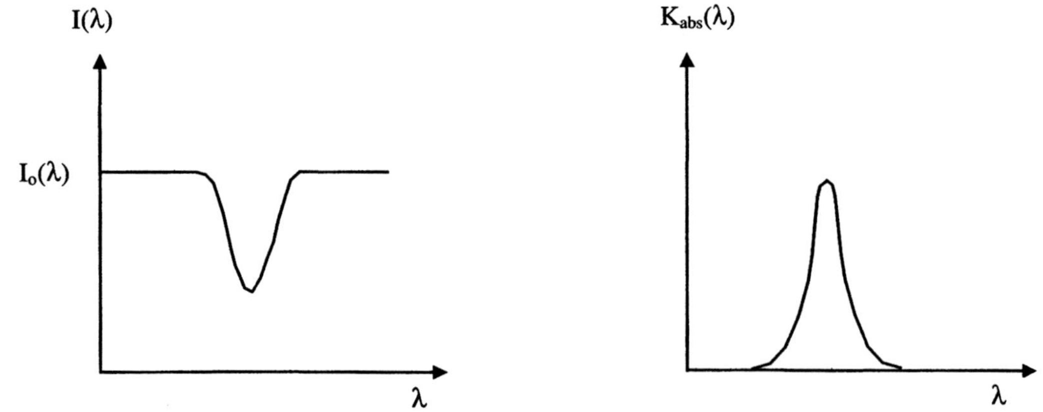

(a) absorption line profile (b) absorption coefficient profile

Fig. 13.5 Absorption intensity and absorption coefficient profiles.

a function of wavelength. This is done after the absorption coefficient at a particular wavelength has been found from the ratio of the incident light and attenuated light from Eq. (13.22).

13.3.3 Implementation of Emission and Absorption Spectroscopy

Emission or absorption of these band spectra can be easily detected and measured using spectrographic equipment. Fig. 13.6 shows schematically the necessary elements of a spectroscopic system for emission (Fig. 13.6[a]) and for absorption (Fig. 13.6[b]). In emission, these band spectra are recorded directly by the spectrographic equipment. The spectrographic equipment essentially consists of a spectrograph and a recording medium. Prisms were first employed as the dispersing element in early spectroscopic instruments, but the achievable dispersion was not very high or uniform. Thus, almost all present day spectrometers employ gratings. The Czerny-Turner configuration, as shown in Fig. 13.7, is the most commonly used spectrometer setup. If an exit slit is used, as shown in Fig. 13.7, a single wavelength emission will be detected at the exit slit. This setup is thus referred to as a monochromator, i.e., a single color instrument. A photomultiplier tube is often used as the photodetector at the exit slit due to its extremely high sensitivity. By removing the exit slit, the monochromator can be converted into a spectrograph, where a spectrum of wavelengths is formed at the exit focal plane of the spectrometer. The photographic method was exclusively used in very early spectroscopic studies as the recording medium. This has been replaced with multi-channel detectors (e.g., diode arrays or a CCD camera), due to their simplicity, linear response, and high sensitivity when an image intensifier is employed.

In absorption, a background source giving a bright continuous spectrum over the relevant wavelength range is required. Early work was mostly done with tungsten filament lamps for the visible region. The brightness temperature of the background source must exceed the flame temperature or, more precisely, the effective excitation temperature of the band system

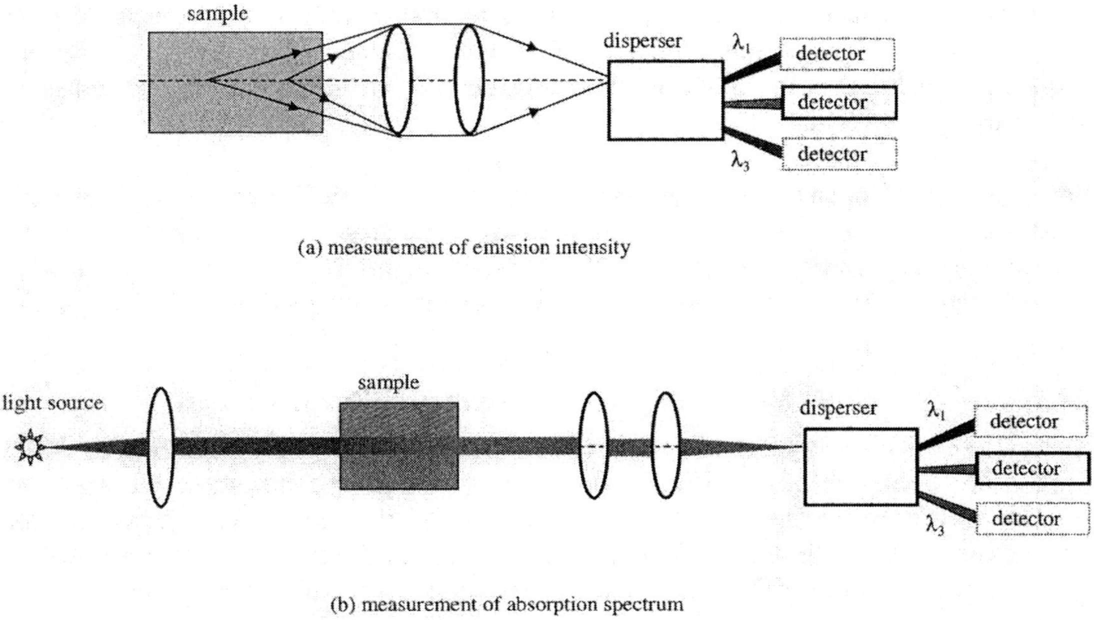

Fig. 13.6 Schematic of emission and absorption systems.

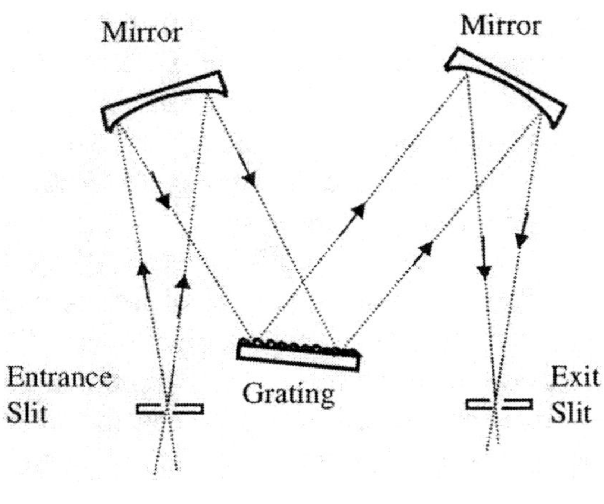

Fig. 13.7 Schematic of a Czerny-Turner monochromator.

in combustion. Tungsten filament lamps are not adequate for absorption spectra of engine combustion because they cannot be used above a brightness temperature of about 3000 K, which is comparable to that of engine combustion. One of the most favored background sources is the high-pressure xenon arc lamp. The hottest part of the arc may have a brightness temperature exceeding 5000 K, especially at shorter wavelengths. Its emission spectrum covers both visible and near UV regions. Dye lasers produce outputs over a wide range of wavelengths by choosing a suitable dye and frequency-doubling crystals for the UV region. Their extremely bright beam output and short pulse duration are ideal for the study of transient phenomena such as combustion.

Emission and absorption techniques provide measurements averaged over a line-of-sight. Both techniques have been almost exclusively used to detect the presence of certain species in the cylinder with a few attempts to quantify their concentrations. The absorption technique is preferred when a more sensitive and quantitative measurement is required.

13.3.4 Spectroscopic Studies of Flame Fronts and Burned Gases

Among the earlier investigations are those of Clark and co-workers [Clark and Henne, 1927]. Spectroscopic research was carried out with respect to knocking combustion. Clark [1928] investigated the emission spectra and attempted to correlate them and knocking tendencies for thirteen commercial fuels, but no clear correlation was found. Thee [1929] summarized the major observations derived from these studies. They found a continuous spectrum extending from 220 to 450 nm under knocking conditions, while the spectrum extended only to approximately 350 nm with normal combustion. In addition, ionized nitrogen near the spark plug and water were positively identified, and tentative identifications were made of hydroxyl radicals and carbon monoxide.

During the 1930s, Withrow and Rassweiler employed spectroscopic techniques to study the emission/absorption spectrum during an entire engine cycle. Direct photographs of combustion revealed a strong reillumination of the gases during the pressure rise accompanying the burning of the last part of the charge, which was referred to as afterglow. To understand the nature of afterglow, they used both emission [Withrow and Rassweiler, 1931; Rassweiler and Withrow, 1932] and absorption [Withrow and Rassweiler, 1933] spectroscopic measurements. Fig. 13.8 shows the optical system employed for their absorption measurements. Absorption through either the burned gas behind the flame front or the unburned end gas ahead of the flame front could be measured. They found that afterglow came principally from CO_2 and OH, not from the CH and C_2 radicals that were major contributors of flame front radiation. Thus, it appeared that afterglow was due not to continued burning of hydrocarbons, but to the delayed combustion of carbon monoxide to carbon dioxide. This was also noticed when Steele [1931; 1935] measured the infrared emission at 4.4 μm and 2.8 μm, and found that the radiation continued to increase for some 20 °CA after the passage of the flame front.

The spectroscopic studies of Withrow and Rassweiler [1931] and Rassweiler and Withrow [1932] included both normal combustion and knocking combustion in SI engines. The flame front under normal conditions showed the band spectra due to CH, C_2, and OH. For knocking combustion, the spectrum sometimes showed a strong continuum in the red due to blackbody radiation from soot formed by cracking of the lubricating oil at the high temperature of knocking combustion. They also noticed that the bands of C_2 and CH were less strong from the knocking

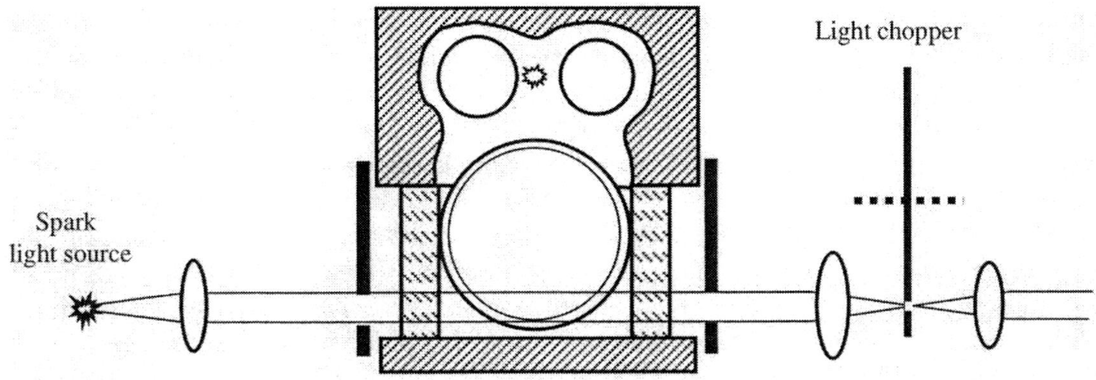

Fig. 13.8 Absorption spectroscopy setup for measuring species in burned gas and end-gas region [Rassweiler and Withrow, 1932].

combustion zone. Later studies indicated that the reduction in C_2 and CH emissions was due to an alteration in the basic kinetics of combustion involved. In normal combustion, the flame front propagates by the radical diffusion processes and involves the burning of C_2 and CH. But in the knocking combustion zone, combustion is dominated by thermal autoignition due to the higher compression ratio, and spectra tend to be more similar to the radiation spectrum observed just prior to ignition in the high-temperature regime of the autoignition process.

Lyn [1962] studied the emission characteristics of diesel combustion. He found that the emission was dominated by a continuum extended into the infrared region. Even the strong CO_2 bands at 2.8 and 4.4 μm produced only slight changes in the continuum. Their studies showed three burning phases during the diesel combustion. The first was characterized by a weak bluish appearance due to the combustion of the premixed fuel and air prepared in the ignition delay period. This was followed by the main highly luminous phase associated with diffusion combustion, where the emissivity was found to be quite high and the intensity distribution in the continuum to be close to that of a blackbody. The less well-defined final phase was also luminous with a continuum spectrum caused by the delayed combustion of carbon monoxide and residual soot particles.

13.3.5 Spectroscopic Measurements of NO Concentration in an SI Engine

Until the late 1960s, most of the research on nitric oxide (NO) production had involved correlation of NO concentrations measured in exhaust gases with engine operating parameters, such as the air/fuel ratio, engine speed, and spark advance [Wimmer and McReynolds, 1962; Hazen and Holiday, 1962; Alperstein and Bradow, 1966]. In these experiments, the measured NO levels were found to be nearer the equilibrium concentration corresponding to the peak cylinder temperature and pressure, than to the equilibrium concentration at exhaust conditions.

In order to account for this non-equilibrium NO concentration in the exhaust, Newhall and Starkman [1967] developed a spectroscopic system suitable for the study of transient NO concentrations during the engine expansion process. The NO concentration was inferred from its infrared radiation at 5.3 μm based on Eq. (13.17). The spectral absorption/emission coefficients were obtained from Malkmus [1963] and Breeze and Ferriso [1964]. Because emission at 5.3 μm arises from both NO and CO_2 spectral bands, the CO_2 concentration had to be known in order to determine NO concentration from the measured emission intensity. In their experiments, Newhall and Starkman determined the CO_2 concentration by chemical equilibrium calculations based on the measured in-cylinder pressure, calculated temperature, and measured air/fuel ratio. Emissions from CO and H_2O in the 5.3 μm spectral region were considered insignificant by Newhall and Starkman.

The experiment was conducted in a single-cylinder CFR engine. The measurement volume was located in the middle of the two diametrically opposite positioned spark plugs. Emission was detected perpendicularly to the common diameter of the spark plugs so that the temperature gradient in the measurement volume was reduced. The measurement volume was confined to a cone of 1.7° half angle by means of a tube placed between the observing window and focusing optics. The infrared radiation was detected by a monochromator and an infrared detector. The time-resolved output from the detector was then recorded by an oscilloscope and averaged over 50–100 cycles. Their results confirmed the theoretical prediction that nitric oxide, once formed in approximate equilibrium quantities in the combustion process, was frozen at a level well above the equilibrium value during the early part of expansion stroke. But their time-resolved measurements gave NO concentrations considerably in excess of equilibrium compositions at the peak pressure and temperature, as predicted by their theoretical studies. The overestimated levels of NO concentrations could have been caused by errors inherent in the inferred temperature and CO_2 concentrations.

A significant contribution toward the understanding of the chemical kinetics of NO formation and decomposition was made by Lavoie et al. [1970a and b]. A time-resolved spectroscopic technique was developed for the measurement of the in-cylinder NO concentration by means of the emitted light intensities from the NO + O and CO + O recombination continua.

Light emission intensities were measured at wavelengths characteristic of the recombination continua: $CO + O \rightarrow CO_2 + h\nu$, $NO + O \rightarrow NO_2 + h\nu$, and the band transition of OH ($A^2\Sigma - X^2\Pi$)(0, 0) and were related to the species concentrations [CO], [O],[OH], and [NO] by

$$I_{CO+O} = K_1(T)[CO][O] \tag{13.24}$$

$$I_{NO+O} = K_2(T)[NO][O] \tag{13.25}$$

$$I_{OH} = K_3(T)[OH] \tag{13.26}$$

where $K_i(T)$ were known functions of temperatures. The concentrations of CO, O, and OH were assumed to be at equilibrium determined by the measured gas pressure and the burned gas temperature calculated from a multi-burned gas zone engine simulation model.

Because of the extremely sensitive nature of NO concentrations to temperature, an accurate temperature reading was required in order to determine the value of $K_2(T)$. OH radiation and CO + O radiation were employed to give two estimates of temperature, T_{OH} and T_{CO}, from Eq. (13.24) and Eq. (13.26), respectively. The averaged value of these two temperatures was then used to calculate the value of $K_2(T)$. I_{OH} and I_{CO+O} were measured at 0.31 μm and at 0.38 μm, respectively. The K factors were obtained from the literature [Fontijn, et al., 1964; Hartunian et al., 1966; Myers and Bartle, 1967; Learner 1962]. The temperatures measured agreed well with the predicted values, and showed a similar trend to those measured previously by Rassweiler and Withrow [1935]. The accuracy of temperature readings was ±125°C.

Fig. 13.9 shows the spectral shapes of major radiation components in the visible range [Lavoie, 1970]. Except for some identifiable discrete line radiation, the spectrum consists of the CO + O and NO + O recombination continua. Although the NO + O recombination radiation was the main contributor toward the green-yellow spectral region of the burned gas radiation, there were also other sources of radiation in the same spectral region. At the wavelength (0.61 μm) where NO + O emission was detected, there was also a contribution from the CO + O recombination continuum and thermal radiation from NO_2. The CO + O contribution was determined for a rich mixture, which produced a negligible amount of radiation from the NO + O continuum. Lavoie found that the CO + O contribution at 0.61 μm was 45% of that at 0.38 μm. Based on the published data [Fontijn et al., 1964; Levitt, 1965] and his own calculation, Lavoie found that, apart from the CO + O continuum, 80% of the radiation at 0.61 μm was due to chemiluminescence of the NO + O reaction. Therefore, the net chemiluminescence radiation of the NO + O reaction was determined by subtracting the contribution of the CO +O continuum from the total radiation at 0. 61 μm and then multiplying the result by 0.8 to correct for the thermal radiation of NO_2.

The major source of absolute error was the uncertainty in the rate constants, k, and this introduced an absolute error of perhaps plus or minus a factor of 2. Other sources of errors were window fouling and reading inaccuracies, and they were considered to be less than ±60%.

The experiments were conducted on a single-cylinder, L-head CFR engine. A number of small quartz windows allowed light emission measurements to be made at different positions along the flame path. The light emitted from a small volume of gas was detected by photomultiplier/interference filter pairs at 0.31 μm, 0.38 μm, and 0.61 μm. Measurements made by Lavoie at two different windows, one closer to the spark plug than the other, showed that the NO concentration was higher near the spark plug. They also found that there was a negligible amount of NO produced at the flame front under the engine operating conditions tested. The measured values of NO concentrations exhibited non-equilibrium behavior and compared favorably with the predicted values using the extended Zeldovich mechanism.

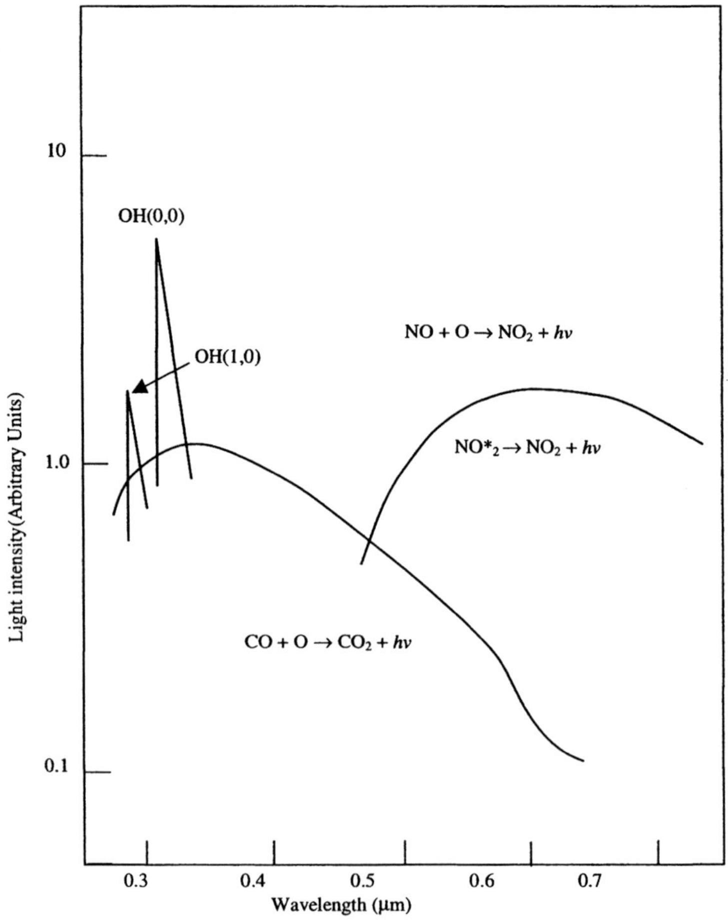

Fig. 13.9 Spectral shapes of major radiation components, adapted from Lavoie [1970].

13.4 Autoignition Studies in SI Engines

The maximum compression ratio that can be used in SI engines is limited by the presence of knocking combustion. Substantial spectroscopic studies have been carried out in the past, in order to understand the chemical kinetics involved in the autoignition process that leads to knocking combustion.

13.4.1 Spectroscopic Studies of the Autoignition Process

From the absorption spectra of end gases in a firing engine, Withrow and Rassweiler [1933] and Rassweiler and Withrow [1933] found absorption lines due to formaldehyde appeared prior to the arrival of the flame and were more intense under knocking conditions. This work established that pre-combustion reactions did occur and that they were oxidative in nature, in that formaldehyde was observed. Measurements of emission spectra of a motored engine by Downs et al. [1953] confirmed that excited formaldehyde was the source of the cool flame radiation.

Spectroscopic measurements in the infrared, visible, and UV regions were carried out by Pipenberg and Pahnke [1957] during the oxidation of n-heptane in a motored engine. The experimental setup permitted both absorption and emission measurements at discrete times throughout the engine cycle. In addition to the carbonyl group and the various C–C and C–H groups from the fuel, the spectral characteristics related to the formyl radical, formaldehyde, hydrogen, water, carbon monoxide, and carbon dioxide were observed.

A significant step forward was taken by Abata et al. [1978] when they made measurements of OH radical concentration in the end-gas of a firing engine fueled with ethane. This work was the first quantitative OH measurement made in the end-gas of a firing engine. The absorption spectroscopy technique was employed using a broadband flash lamp pumped dye laser. The dye laser output was tuned to the $^2\Sigma^+ - {}^2\Pi$ electronic transition of OH with a central wavelength of 308 nm and a bandwidth of 1.6 nm. The absorption spectrum was recorded by a spectrometer on a photographic film, from which the OH radical concentration was measured. The measurement was taken in a CFR engine with six spark plugs so that the end-gas was located in the center of the combustion chamber. Prior to the flame arrival, the concentration of OH radicals was found to be approximately 1×10^{-9} mol/cm^3. Due to the difficult nature of the measurement technique and the challenges presented in applying it to a firing engine, the results were not extensive. However, the experiments did prove that such measurements were possible.

In an SI engine, combustion is initiated by a spark plug and continues as a propagating premixed flame. However, Noguchi et al. [1979] found that another combustion process took over at low loads in a two-stroke SI engine. This new combustion process was initiated by multiple ignition sites due to compression of a premixed air/fuel mixture before the spark discharge took place. The new combustion process was found to be very stable and produced extremely low NO emission in the exhaust. In order to understand this new combustion process, Noguchi et al. set up a spectroscopic system to investigate the initiation and development of this new combustion process. Emission was collected through a sapphire window in the cylinder wall of an opposed piston two-stroke SI engine, as shown in Fig. 13.10. The transmitted light through a slit was equally divided into each of the branched optical passages. The emission from radicals of OH (306.4 nm), CH (431.5 nm), C_2 (473.7 and 516.5 nm), and H (578 nm) was detected separately by a pair of a photomultiplier tube and a narrow bandpass filter via an optical fiber, as shown in Fig. 13.10. The emission from CHO, HO_2 and O radicals was observed together at 400 nm. The emission from lithium and sodium, small quantities of which were added to the fuel, was used to measure flame temperature above 1000°C. The detection of the infrared radiation at 4.3 μm from CO_2 gas was employed to determine temperatures below 1000°C. Because the optical fiber absorbed both in the infrared and UV regions, a hollow brass pipe was placed between the optical fibers to transmit the emissions from CO_2 [4.3 μm] and OH radicals [306.4 nm].

The time-resolved light emission results showed that, in the case of conventional spark ignition combustion, all of the radicals were observed almost at the same time; in the case of the new combustion process, CHO, HO_2, and O radicals were first detected, followed by CH, C_2, and

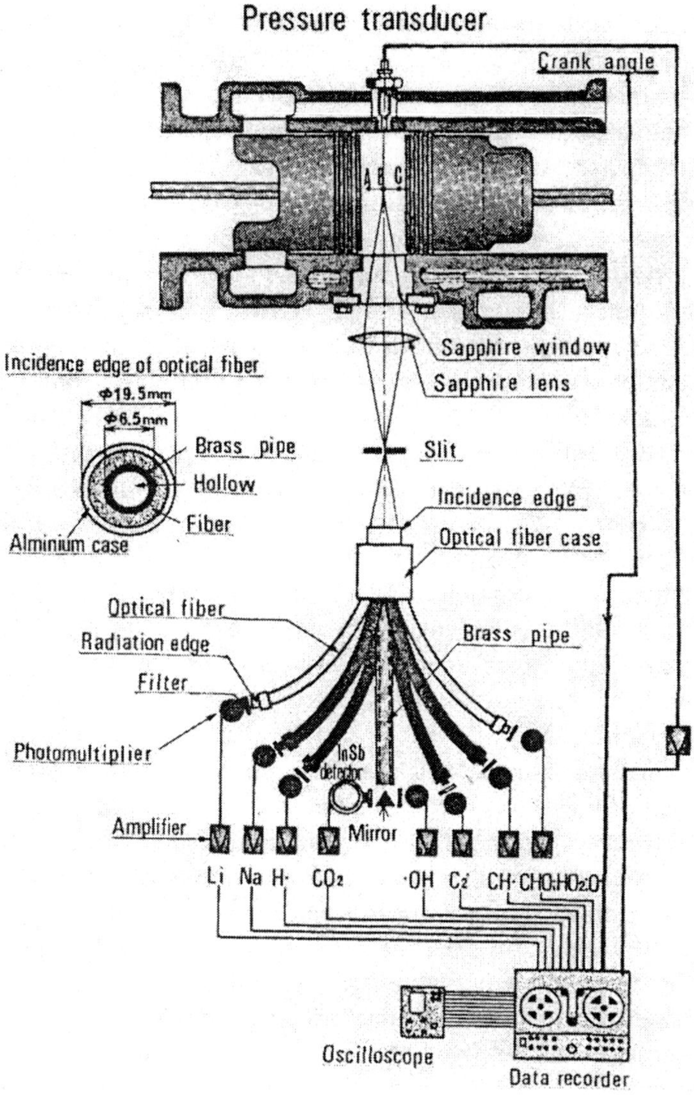

Fig. 13.10 Experimental setup for detecting radicals in premixed compression ignition combustion, Noguchi et al. [1979].

H radicals, and finally, the OH radical. Such an observation seemed to confirm that the new combustion process was initiated by the autoignition of premixed mixture due to compression.

Spectroscopic studies of this new combustion process were also reported by Iida [1994], who investigated emissions from OH, CH, and C_2 radicals from a two-stroke SI engine running with gasoline and methanol. During the premixed compression ignition combustion process, OH radical emission was observed before the start of heat release when the CH radical emission first appeared. This was different from the normal spark ignited combustion, where

OH and CH radicals were detected at the same time in the flame front. In addition, OH emission was observed at the same crank angle for both gasoline and methanol fuels, but no emission from CH radicals was detected with methanol.

The most recent spectroscopic research on pre-flame reactions in an SI engine was performed by Shoji et al. [1994; 1996]. In this work, the pre-flame reactions of n-heptane and isooctane were studied using emission and absorption spectroscopy. Formaldehyde was detected by emission at 395.2 nm and the OH radical was measured simultaneously by absorption from a xenon lamp at 306.4 nm. Shoji et al. concluded that formaldehyde and OH radicals were produced at the same time in the end-gas region. This is inconsistent with the findings discussed previously, where formaldehyde was found to form earlier than OH radicals. The discrepancy may be ascribed to the fact that absorption measurements are more sensitive than emission measurements. Hence, smaller amount of OH would have been detected by absorption, while emission due to formaldehyde would not have been detected until a larger amount had been formed in the end-gas region.

13.4.2 End-Gas Sampling and Analysis

The fast in-cylinder sampling valve has contributed significantly toward the understanding of end-gas reactions leading to knocking combustion in the SI engine. In fact, the earlier use of sampling valves was largely motivated by the desire for such understanding. The first published work involving end-gas sampling appeared in a paper by Egerton et al. [1935]. Using a mechanically driven poppet sampling valve, Egerton et al. were able to sample the end-gas with a sampling duration varying from 2 degrees (near TDC) to 10 degrees (near BDC) at 600 rpm, depending on the pressure in the cylinder. From the sampled gas, they claimed to show that a hydroperoxide existed close to TDC in sufficient quantities to cause knock.

Extensive end-gas sampling experiments were carried out by Downs et al. [1951] for the combustion of mixed isooctane isomers, methane, benzene, and a mixture of benzene and isooctane. A three degree crank angle sampling period was achieved at an engine speed of 1500 rpm by means of an electromagnetically operated needle valve. In order to sample only the end-gas, an ionization probe was used to detect the flame arrival in the vicinity of the sampling probe. Chemical analysis revealed that the aldehydes comprised predominantly formaldehyde and was an order of magnitude greater than the peroxides. While the aldehydes rose monotonically to their peak, the peroxides showed a two-stage curve with an inflection point occurring near TDC. In addition, the concentrations of aldehydes and peroxides were studied and found to decrease with more knock-resistant fuels.

The use of a motored engine for the study of knock in an SI engine was vindicated by Davis et al. [1955]. They studied the production of carbonyl species in a CFR engine under both fired and motored conditions by sampling gases around TDC. Based on experimental results obtained from the three fuels tested (n-heptane, n-pentane, and isooctane), they stated that the motored engine was a convenient and valid research tool for pre-combustion reactions studies.

One of the most complete and comprehensive measurements of chemical species produced in the end-gas of a fired engine using a fast-acting sampling valve was reported by Alperstein and Bradow [1965, 1966]. Their study involved time-resolved sampling of end-gas in a CFR engine at 5 degree crank angle intervals from 20 degrees BTDC to 20 degrees ATDC using an electromagnetically controlled needle valve. They also found that isobutylene in DIB (di-isobutylene) had a significant inhibiting effect on isooctane oxidation by consuming active chain carriers as a radical sink. Their results from the isooctane experiments are still being used as an excellent test case for the development of current detailed autoignition models of isooctane.

Most fast-acting sampling valves can extract one sample from each engine cycle due to the time period required to activate the valve again. Luck et al. [1973] described a novel high-speed sampling system which was capable of extracting multiple end-gas samples in a single cycle. The sample diluted with nitrogen at the tip of a sampling tube was continuously passed through a sampling valve for 60 or 120 degrees crank angle. A slide valve activated by a spring-loaded timing pin and a timing disc determined the dwell angle of the sampling. After passing through the sampling valve, the diluted sample was directed into 48 traps equi-spatially mounted on a rotating disc. By adjusting the speed of rotation of the disc, relative to that of the engine, the sample in each trap could be made to represent either 5 or 20 degrees crank angles. The advantage of the valve in following individual cycles was illustrated by samples from two similar cycles but with different knocking intensities. For the two cycles studied, very different profiles of fuel and intermediates were obtained from the multiple samples in each cycle. While the system was capable of sampling at every 5 degrees at 800 rpm within a single cycle, the sampling valve suffered from small quantities of gases collected and changing sampling volumes as the cylinder pressure changed. Therefore, detection limits were rather high, and even the relative concentration of each gas from sample to sample was sometimes questionable. The other difficulty associated with this setup was the in-cylinder pressure-dependent time lags due to the presence of a sampling tube upstream of the sampling valve.

More recently, end-gas sampling in a firing engine fueled with n-butane has been reported and results were compared with a detailed chemical kinetic scheme [Green et al., 1987].

13.5 Visualization of Combustion Species by PLIF

13.5.1 Introduction

The hydroxyl (OH) radical has been the most studied combustion species in combustion and flames using LIF. This is because of its importance as a highly reactive chemical species in combustion, its relative abundance during combustion, and the availability of a convenient UV laser system for its excitation. LIF measurements of OH radical have been applied to a variety of combustion systems since 1970s. Application of such measurements to IC engines was not carried out until the late 1980s because of the practical difficulties of in-cylinder measurements and quenching properties of LIF.

Detection of NO by LIF in a combustion environment is considerably more difficult than detection of OH because NO has a much smaller absorption cross section and requires UV light which is prone to absorption by other combustion products and fuels.

Formaldehyde (CH_2O) is known to be an intermediate generated during the early stage of the autoignition process. As autoignition proceeds into the high-temperature regime, formaldehyde is rapidly consumed in the degenerate branching reactions that precipitate high-temperature ignition [Benson, 1981]. Formaldehyde is therefore a good natural tracer of autoignition, because it is not only consumed in flames (with one exception, the flame of methanol, [Gaydon, 1970]) but also in the later stages of the autoignition process. Its abundance in the end-gas region and the high sensitivity of its LIF process can therefore be employed to visualize the occurrence of autoignition sites.

13.5.2 Experimental Considerations

The experimental setup for PLIF measurement of these species is shown in Fig. 13.11. The optical setup and detection systems are very similar to those for PLIF fuel concentration measurements except that a calibration flame is often included for spectral calibration purpose.

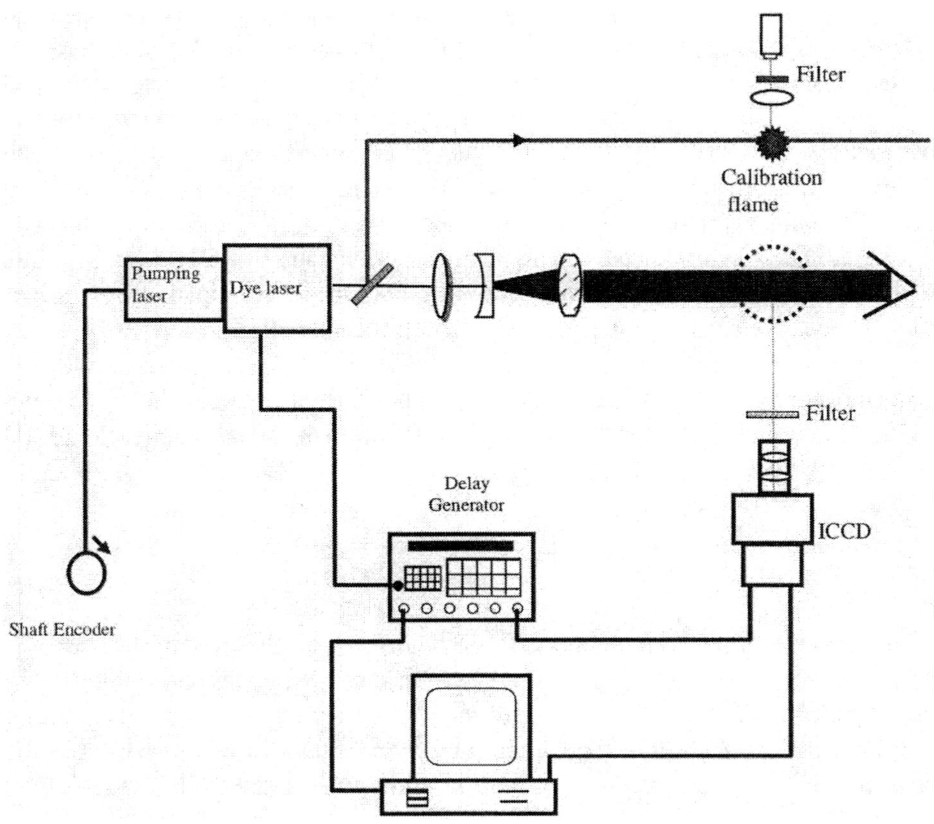

Fig. 13.11 Optical setup for PLIF measurement of in-cylinder species.

As for any PLIF measurements in IC engines, a high-power laser is required to compensate for the reduced LIF signal level due to quenching at high pressure. The most common excitation source for LIF measurement of these species is an Nd:YAG pumped dye laser. The pump beam from the Nd:YAG laser is set at either the second (532 nm) or third (355.5 nm) harmonic for dye pumping. The pump beam frequency employed depends on the dye being pumped, which, in turn, depends on the final frequency required. The dye laser is typically pumped transversely. In order to obtain the desired output frequency in the UV region, the dye laser beam is frequency-doubled or sum-mixed in crystals in a frequency conversion unit. The tuning of the output frequency is achieved by rotating a grating in the laser system. Some dye lasers can also be pumped by an XeCl excimer laser at 308 nm. For OH and NO, tunable excimer lasers can also be used for direct fluorescence excitation, as to be discussed below.

When LIF from fuel or dopant is detected, as demonstrated in Chapter 7, broadband excitation and detection are employed because the detailed vibrational-rotational spectrum cannot be resolved for large polyatomic molecules. The concentration of combustion species is much smaller than that of fluorescing dopant, and hence LIF signals are weaker. However, the absorption process will be greatly enhanced if the laser output can be tuned to a specific spectral line between the two electronic states. Simple polyatomic molecules, especially diatomic molecules, have well-separated vibrational-rotational energy levels within each electronic energy state. Hence, excitation and emission of specific spectral lines are employed in LIF measurements of OH and NO. Although there are four atoms in a formaldehyde (CH_2O) molecule, the vibrational-rotational lines of formaldehyde are sufficiently separated and can be excited individually. The tunable laser output can be accurately tuned to a particular spectral line of the species by performing excitation scans in a calibration source. The excitation scan is carried out by tuning the laser across a spectral region near the absorption line of the probed species, and the maximum fluorescence signal can then be used to identify the strongest absorption line. The calibration source can be the reaction zone (for OH) or post combustion zone (for NO) in a small laboratory burner. Fine-tuning of the laser frequency to the absorption line of formaldehyde can easily be performed in a calibration cell.

There are a number of particular issues associated with quantitative LIF measurements of combustion species. These can be understood by looking the LIF signal in the linear regime (Eq. [5.61]),

$$I_{flu} = C f_1(T) \chi_m N I_v B_{12} \frac{A_{21}}{A_{21} + Q_{21}}$$

which is valid if the laser spectral intensity, I_v, does not saturate the excitation transition. This is usually the case, particularly in IC engine applications where pressure is high. The laser energy that is to be transferred to the probe volume should remain constant, which requires the monitoring of individual laser pulse energies and correction for the change in window transmission. The laser beam may be attenuated by absorption on its way to the probe volume by fuel or combustion products.

The selection of spectral lines for excitation has large effects on the detection and interpretation of LIF signals. The excitation wavelength should be chosen such that the laser is not absorbed by other species in the cylinder. When OH radicals are excited at near UV wavelengths around 300 nm, beam attenuation is less of a problem. However, when shorter wavelengths are employed for OH or NO excitation and detection, absorption by fuel and combustion products may become significant. Additional difficulty arises from the LIF of other species, particularly hot oxygen (O_2) when the excitation of LIF is obtained with shorter wavelengths. The LIF signal depends on the absorption cross section of the absorbing species at the excitation wavelength. Excitation of spectral lines that have large absorption cross sections will increase the fluorescence yield, ($A_{12}/[A_{12} + Q]$).

Pressure broadening and shifting of the rotational lines significantly changes the species absorption spectrum and, hence, the overlap between the laser and the species absorption profile. Di Rosa and Hanson [1994] suggested that this effect may be taken into account by adding an overlap correction coefficient, $g_\lambda(P)$, to Eq. (5.61),

$$I_{flu} = Cf_1(T)\chi_m N g_\lambda(P) I_v B_{12} \frac{A_{21}}{A_{21} + Q_{21}} \qquad (13.27)$$

The value of $g_\lambda(P)$ can be determined experimentally [Hildenbrand et al., 1998] or accounted for by performing in situ calibration, which corrects for effects such as quenching, system response, etc.

In Eq. (13.27), the grouping ($f_1(T)\chi_m N$) represents the number density of fluorescing species at the lower rotational state N_J, from which molecules are excited to a higher energy state. It is a function of temperature as described by the Boltzmann redistribution,

$$N_J = \frac{N}{Q_{rot}} g_I (2J+1) e^{-BJ(J+1)hc/kT} \qquad (13.28)$$

where the rotational partition function, Q_{rot} = kT/hcB, and g_I represents the nuclear spin degeneracy. The temperature dependence of the lower state population therefore must be corrected for quantitative measurements. Alternatively, the lower state should be chosen such that its population is insensitive to temperature variations.

After excitation, the excited species will either fluoresce or lose its energy without radiation due to collisions with other molecules, i.e., be quenched. Because combustion species are normally measured during the combustion period when the pressure and temperature are high, collisional quenching seriously reduces their fluorescence yield. The quenching rate is proportional to pressure and approximately inversely proportional to the square root of the temperature. Since the gas pressure is virtually uniform throughout the combustion chamber and the temperature gradient is small in the burned gas region, the quenching rate may be considered to be constant throughout the burned gas region. The quantification of quenching rate, however, is particularly difficult since the quenching rate is a strong function of temperature, pressure, and the nature of the colliding species.

The pressure, temperature, and quenching effects described so far, together with system calibration factor C, may be accounted for by performing in situ calibration, where known amounts of NO are seeded into the intake charge, and the fluorescence intensity is measured to generate a correlation between the LIF signal and NO concentration at a particular condition [Hildenbrand et al, 1998]. Alternatively, a detailed spectroscopic code [Paul et al. 1993; 1994] has been used by Dec and Canaan [1998] to make these corrections to the total NO concentration. Calibration of the LIF signal is more difficult to implement for OH radicals , the concentration of which cannot be accurately known or determined.

Having survived the quenching process, the fluorescence signal from different transition lines can be detected. Resonant detection, in which the LIF signal is measured at the excitation wavelength, should be avoided because it suffers from interference of Rayleigh scattering and spurious scattering from dust or walls. In addition, radiation trapping should be considered. Radiation trapping takes place when the spontaneous emission (fluorescence) coincides with an absorption line of the same species at different energy states or that of different species. For example, if OH is excited via the (2, 0) $A^2\Sigma$–$X^2\Pi$ transition, the emission from the (2, 1) line could be absorbed by OH radicals at the first vibrational energy state of the ground electronic state ($X^2\Pi$) if their population is large. Measurement geometry may be used to advantage when permissible, e.g., performing measurements near the outer combustion region to minimize the optical path and distance through the medium that the signal travels. Radiation trapping can be avoided by proper selection of excitation and detection schemes as well.

Absorption by partially oxidized fuels and PAHs poses severe interference problems for the detection of fluorescence signals in the UV region, e.g., the detection of the NO fluorescence signal.

The detection of an LIF signal requires the separation of the signal light from other sources of radiation at different wavelengths, often at intensity levels well above that of the signal. Since the OH LIF signal lies in the near UV region, a bandpass interference filter can be used to reject spurious scattered light and flame luminescence. Whereas the fluorescence signal of NO lies in the low UV region, dielectric coated mirrors are preferred because they can be optimized for narrow-band reflections with a very high reflectance (>0.995). A UV filter setup consisting of four dielectric coated mirrors centered around 230 nm, as shown in Fig. 13.12, has been used by Schulz et al. [1996] to isolate NO fluorescence from other fluorescence.

The use of a gated intensified CCD camera is necessary because of the weak nature of OH and NO LIF signals. When the laser and the detector are synchronized by a shaft encoder, the short gating period of the intensifier will minimize the interference from ambient light and flame luminescence.

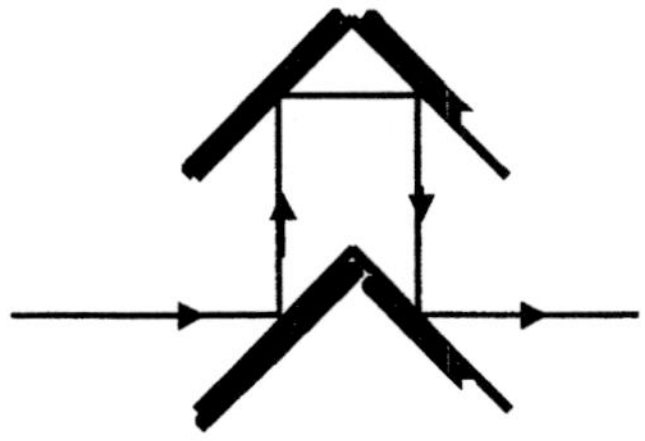

Fig. 13.12 A UV bandpass filter made up of four dielectric coated mirrors.

In addition, the removal of systematic errors present in the imaging system should be carried out, as discussed in Chapter 5, before the final results can be obtained.

13.5.3 Visualization of OH by PLIF

13.5.3.1 Laser Induced Fluorescence of OH Radicals

In the case of LIF measurements of OH, two electronic states are involved: the ground electronic state, $X^2\Pi$, and the first excited electronic state, $A^2\Sigma$, as shown in Fig. 5.1. Three vibrational bands of the $A^2\Sigma$–$X^2\Pi$ transition have been used in OH LIF in IC engines. Table 13.3 summarizes the experimental PLIF setups for OH measurements in IC engines. The most common setup employs an Nd:YAG pumped dye laser to excite the OH radical around 284 nm, which corresponds to the $Q_1(9)$ and $Q_2(8)$ lines of the (1, 0) vibrational band of the $A^2\Sigma$–$X^2\Pi$ transition, because of their high fluorescence yields and low sensitivity to temperature variations [Dieke and Crosswhite, 1962].

13.5.3.2 Visualization of OH in SI Engines

The first PLIF measurement of OH radicals in IC engines was reported by Felton et al. [1988]. Measurements were carried out in a single-cylinder two-stroke SI engine at 600 rpm with propane and skip firing. An Nd:YAG pumped dye laser operating at 284.26 nm was employed to excite the $Q_2(9)$ line. This rotational level ($k'' = 9$) was chosen because its population fraction remained constant to within 3% from 1800 K to 3000 K. A 9 mJ laser pulse was delivered into the combustion chamber. The OH LIF signal from the (1,1) vibrational band was detected using a bandpass filter of 313 ± 5 nm. The detection of the (1, 1) vibrational band of the OH radical reduced the effect of quenching rate on LIF signals in comparison with the detection of LIF from the (0, 0) vibrational band. The detection of (1, 1) fluorescence also reduced the effect of radiation trapping since the population of vibrationally excited OH radicals ($v'' = 1$) responsible for radiation trapping was much smaller than that of the OH radicals at the ground vibrational level ($v'' = 0$). Two-dimensional OH measurements during combustion were obtained for different ratios. Felton concluded that the resolution of their imaging system had to be improved in order to match the quality and precision of the Mie scattering images taken from TiO_2 seeded flames for the observation of flame propagation in an SI engine. A

Table 13.3 Summary of the PLIF setups for OH measurement in IC engines

Laser	Excitation Wavelength	Detection Wavelength	Applications	Comment	Reference
Nd:YAG pumped dye laser, 3mJ	283.92 nm (1, 0) $Q_2(8)$, $Q_1(9)$	310.9 ± 11 nm (1, 1) (0, 0)	Diesel combustion in a RCM fueled with low sooting fuel	Strong signal	Kosaka et al. [1996]
Nd:YAG pumped dye laser, 24 mJ	284.01 nm (1, 0) $Q_2(8)$, $Q_1(9)$	312 ±16 nm (1, 1) (0, 0)	Single-cylinder DI diesel engine with low sooting fuel	Strong signal	Dec and Coy [1996]
Nd:YAG pumped dye laser	283.92 nm (1, 0) $Q_2(8)$, $Q_1(9)$	310 nm (0, 0)	Single-cylinder SI engine fueled with gasoline	Strong signal	Tanaka and Tabata [1994]
Nd:YAG pumped dye laser	283.553 nm (1, 0) $Q_2(8)$, $Q_1(9)$	305–340 nm (0, 0)	DI diesel engine with isooctane and n-tetradecane in O_2 enriched air	Strong signal	Nakagawa et al. [1997]
Nd:YAG pumped dye laser, 10mJ	281 nm (1, 0)	312 ± 10 nm (1, 1) (0, 0)	SI engine with transparent head fueled with propane	Strong signal	Schipperijn et al. [1988]
Nd:YAG pumped dye laser, 9 mJ	284.26 nm (1, 0) $Q_2(9)$	313 ± 5 nm (1, 1)	2-stroke SI engine with transparent head with propane	Reduced radiation trapping	Felton et al. [1988]
XeCl excimer laser, 250 mJ	308.2 nm (0, 0) $Q_1(3)$	310 ± 10 nm (1, 1) (0, 0)	Square piston SI engine with methane	Suffer from spurious scattering	Suntz et al. [1988]
XeCl excimer laser, 20 mJ	308.2 nm (0, 0) $Q_1(3)$	310 ± 10 nm (1, 1) (0, 0)	Side valve L-head SI engine with propane or isooctane	Suffer from spurious scattering	Serpenguzel et al. [1993]
Tunable KrF excimer laser, 250 mJ	248.45 nm (3, 0) $P_2(8)$	297 ± 10 nm (3, 2)	Four cylinder SI engine with isooctane or gasoline	Insensitive to quenching (LIPF), weak signal	Andresen et al. [1990] Vannobel et al. [1993]

similar optical system was also employed by Schipperijn et al. [1988] to obtain two-dimensional OH distributions in the Sandia optical research engine. In both cases, gaseous fuels were burned in order to simplify their experiments, and the fluorescence signal was detected through the transparent cylinder head, which permitted a greater amount of fluorescence to be collected.

Using a similar setup, Tanaka and Tabata [1994] visualized the OH distribution in a single cylinder four-stroke SI engine running on gasoline. Optical access was achieved through an extended piston and transparent cylinder liner. LIF signals of OH were detected through a bandpass filter centered at 310 nm. By simultaneous detection of OH and flame radiation via two ICCD cameras, they found that OH radicals were present both in the flame front and burned gas zone, with more in the latter region. Because combustion was intensified with increasing swirl, the OH concentration increased to a peak value, when the uHC emission in the exhaust reached a minimum.

Resonant detection of the vibrational band (0, 0) was used in some earlier experiments. The $Q_1(3)$ rotational line of the (0, 0) vibrational band was excited by an XeCl excimer laser at 308.2 nm [Suntz et al., 1988]. Resonant fluorescence at the same wavelength was detected because other fluorescence bands were too weak due to collisional quenching at high pressures. Resonant detection resulted in severe interference of elastically scattered light from the walls and dust particles. Nevertheless, OH images from propane combustion in a low-speed (500 rpm) and low compression ratio (4.5) square piston engine were obtained and were found to be comparable in quality to 2-D Mie scattering images of TiO_2 particles for identifying the flame front.

By employing resonant detection of OH, Serpenguzel et al. [1993] were able to obtain simultaneous measurements of unburned gas and flame front distribution from a single image. The unburned gas was marked by Rayleigh scattering superimposed by bright Mie scattering spots from fuel droplets. Burned gas was defined by the stronger LIF signal of OH. By means of manual inspection, they identified the reaction zone as the dark region between the unburned and burned gas zones. The reaction zone thickness was then measured. In this work, the OH LIF signal was found to be stronger than Rayleigh scattering at a peak cylinder pressure of 10 bar. However, in a more realistic engine environment where the peak cylinder pressure is much higher, Rayleigh scattering may become comparable to or stronger than the LIF signal of OH.

13.5.3.3 Visualisation of OH in Diesel Engines

Visualization of OH in a diesel engine is considerably more difficult. There are two main sources of interference in diesel combustion. One is the elastically scattered laser light from soot particles and liquid fuel droplets. The other is the broadband emission from soot particles, LII of soot particles, and fluorescence from PAHs. Since the LIF signal of OH is collected at a wavelength longer than the laser output, the elastically scattered laser light can be rejected from OH images with a long-pass filter or bandpass filter. The broadband emission, however, is more difficult to remove. The fluorescence from PAHs is typically very broadband, extending from UV into visible. When excited by a laser beam at 284 nm, the majority of PAH emission is more red-shifted than the OH fluorescence of 308–320 nm [Berlman, 1971]. Both high-temperature soot radiation and LII signals are characterized by blackbody radiation. They drop off significantly toward the UV region, being relatively weaker below 320 nm. A short-pass filter can be used to suppress both PAH and soot related radiation. A combination of long-pass and short-pass filters, or a bandpass filter (308–320 nm), will be effective in minimizing both elastically scattered light and broadband radiation.

Dec and Coy [1996] found that a combination of a 312 ± 16 nm bandpass filter and a 358 nm short-pass filter gave the best compromise between OH signal strength and low PAH and LII interference. Although the elastic scattering from liquid fuel droplets could be virtually eliminated by adding an additional long-pass filter at 305 nm, it was not excluded from OH images. The elastic scattering was found to be spatially separated from the OH LIF signal and was used to identify the location of OH distribution with respect to the liquid fuel.

Fig. 13.13 shows a temporal sequence of OH images of low-sooting diesel combustion in a single-cylinder DI diesel engine obtained by Dec and Coy [1996]. The laser sheet entered the images from the right along the axis of the fuel jet, as shown in Fig. 13.14, and images were taken through the piston-crown window with a set of a 312 ± 16 nm bandpass filter and a 358 nm short-pass filter. Fig. 13.13 shows that the first image is dominated by the strong elastically scattered laser from the liquid fuel droplets. OH concentrations in the premixed burn phase were too low to be detected, as the equivalence ratio of this premixed burn phase was found to be in the range of 2 to 4, which was too rich to generate much OH radicals. The first indication of OH is seen at points a and b in the second image taken at 5.5 deg after the start of injection (ASI). Five CA degrees later, OH extended around a large fraction of the periphery of the premixed fuel vapor/air region in the leading part of the combusting jet. The rest of sequence shows that OH was present around the periphery of the combusting jet. Dec and Coy considered that these OH radicals were a result of diffusion combustion. The attenuation of the laser sheet was experienced after 7.0° ASI due to absorption and/or scattering by OH, PAH, and soot. Severe attenuation prevented OH images after 9.5° ASI from being taken.

Kosaka et al. [1996] studied the structure of an unsteady diesel flame inside a rapid compression machine using PLIF of OH and elastic scattering by soot particles. The details of their optical setup is given in Table. 13.3. A fuel mixture consisting of 90% paraffin reference fuel and 10% polypropylene-glycol was used to reduce the optical attenuation caused by the dense soot cloud. Simultaneous images of OH and soot distributions were obtained by a single intensified CCD camera via an image-doubling prism and two bandpass filters for OH (311 ± 11 nm) and soot (283 ± 10 nm) detection, respectively. Their results showed that OH formed earlier than soot and was separated from soot in the leading portion of the diesel flame.

13.5.3.4 Visualization of OH by LIPF

The PLIF measurements of OH radicals discussed so far suffer from severe quenching. During their experiment, Serpenguzel et al. [1993] reported that the LIF signal during engine combustion was 15 times lower than that from an atmospheric flame because of the rapid increase in collisional quenching and transition line width at high pressure. Corrections for quenching are extremely difficult, and hence OH measurements discussed so far were limited to qualitative images.

To overcome the difficulty associated with quenching, Andresen et al. [1988] proposed the use of a new LIF approach to the measurement of OH radicals. It is termed laser induced pre-dissociative fluorescence (LIPF) and its principle of operation is as follows.

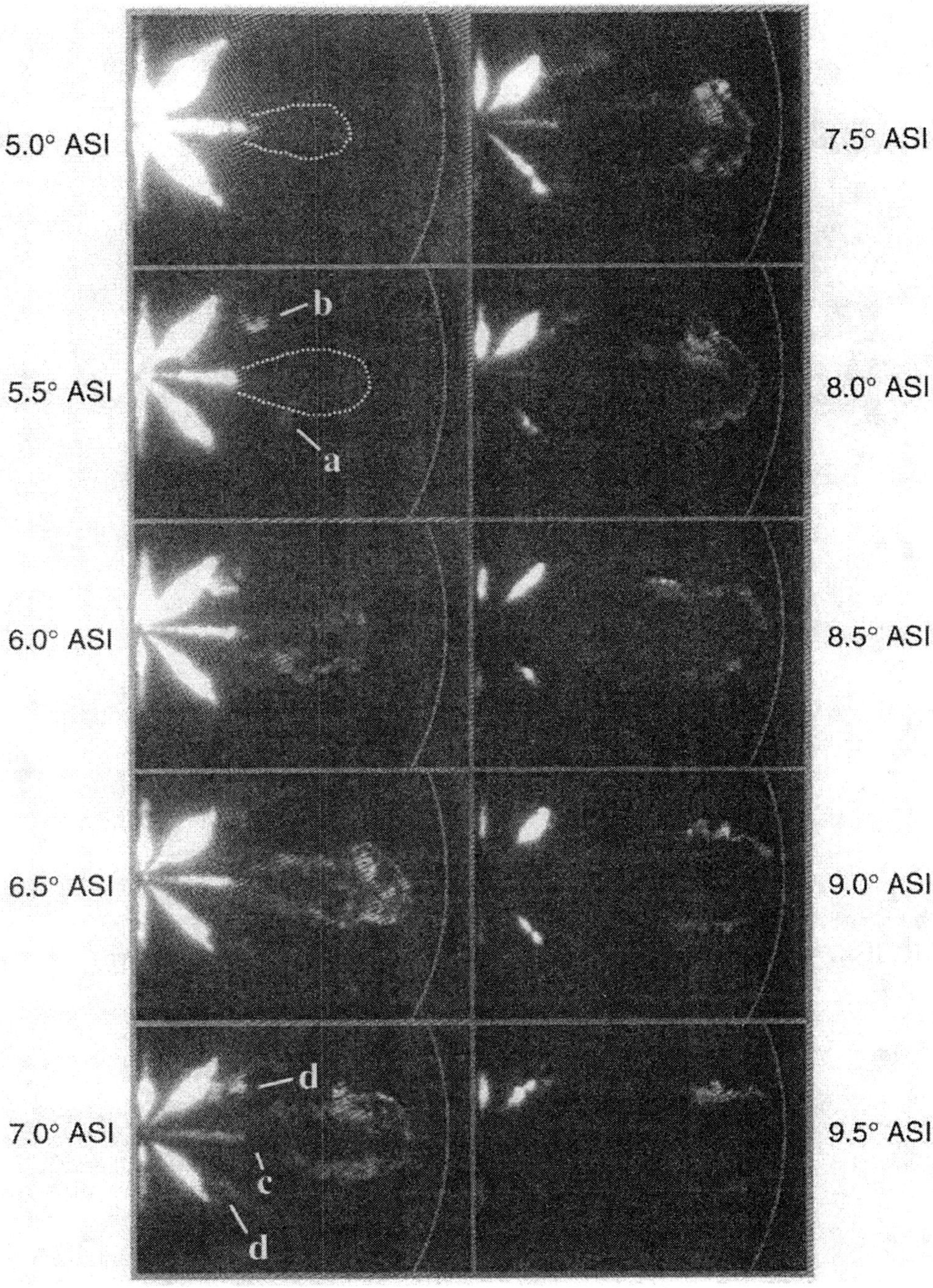

Fig. 13.13 Temporal sequence of OH PLIF images, Dec and Coy [1996].

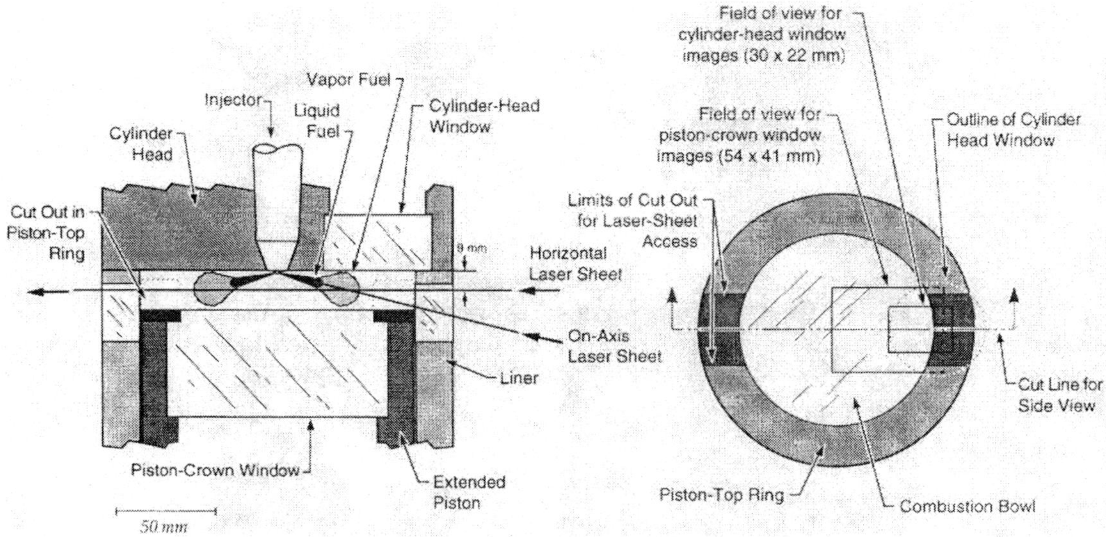

Fig. 13.14 Geometry and optical configuration of the combustion chamber, Dec and Coy [1996].

Recall the expression for the LIF signal in the linear regime (Eq. 5.61),

$$I_{flu} = C f_1(T) \chi_m N I_v B_{12} \frac{A_{21}}{A_{21} + Q_{21}}$$

which has been derived on the assumption that the excited states are not pre-dissociative, i.e., $Q_{pre} = 0$. If an excited state is chosen such that it is pre-dissociative, the equation becomes

$$I_{flu} = C f_1(T) \chi_m N I_v B_{12} \frac{A_{21}}{A_{21} + Q_{21} + Q_{pre}} \tag{13.29}$$

Hence, the fluorescence signal is affected by pre-dissociative loss as well as quenching loss. If the excited state selected for fluorescence is highly pre-dissociative, i.e., $Q_{pre} >> Q_{21}$, then Eq. (13.29) can be simplified to

$$I_{flu} = C f_1(T) \chi_m N I_v B_{12} \frac{A_{21}}{Q_{pre}} \tag{13.30}$$

Note from Eq. (13.30) that the fluorescence intensity is not affected by the quenching rate Q_{21}, but is related to the values of the Einstein coefficients A_{21}, B_{12}, and Q_{pre}. The values of A_{21}, B_{12}, and Q_{pre} depend only on the nature of the excited molecule, while the quenching rate Q_{21} depends on gas composition, density, and temperature. This means that quantitative interpretation can be realized once calibration of the system has been done at a particular condition.

For the excited state to be dominated by pre-dissociation loss, it should be so short-lived that there is no time for collisions to take place. The short-lived molecule at the excited state would lead to a reduced fluorescence signal. Andresen et al. [1990] estimated that the fluorescence yield for LIPF $A_{21}/Q_{pre} \approx 10^{-4}$ and for LIF $A_{21}/Q_{21} \approx 10^{-3}$. Fortunately, the LIPF line of OH coincides with the spectral region of a tunable KrF laser, which has a spectral brightness of 2–3 orders higher than that from a frequency-doubled dye laser used in normal LIF measurements. The lack of quenching also improves the fluorescence yield of LIPF measurements.

Gray and Farrow [1991] measured the pre-dissociative rate constant for OH, and found it was on the order of 10^{10} sec^{-1} and varied with the upper rotational state in the (3, 0) band. Since the quenching rate is about 10^9 sec^{-1} at atmospheric pressure, it will approach the pre-dissociative rate constant at higher pressures. In engine applications, the quenching rate cannot be simply ignored if quantitative measurements are desirable. In addition, this approach as a general strategy is limited due to the absence of a pre-dissociative state with an acceptable lifetime.

LIPF has been applied to an SI engine simulator [Andresen et al., 1988] and a modified four-cylinder production engine [Andresen et al., 1990; Vannobel et al.,1993]. A tunable KrF excimer laser was used to excite the $P_2(8)$ line (248.456 nm) of the (3, 0) vibrational band of the $A^2\Sigma$–$X^2\Pi$ transition because of its high fluorescence yield. Because of pre-dissociation of the $v'' = 3$ level, quenching effect was reduced. The population fraction of OH radicals in $K'' = 8$ was calculated to be constant within 10% in the temperature range between 1300 and 3000 K. The fluorescence of the (3, 2) vibrational band was detected with a bandpass filter at 297 ± 6 nm. The (3, 2) vibrational band was selected because it was found to be least affected by quenching. The use of the 2nd vibrational energy level in the ground electronic state resulted in reduced radiation trapping. Combined with PLIF images of fuel, they were able to obtain simultaneous images of OH and fuel distributions in one of the cylinders of a production four-stroke engine through the piston-crown window.

The application of LIPF of OH images to a diesel engine was also demonstrated by Arnold et al. [1992]. They succeeded in obtaining OH images during the initial period of combustion of n-heptane until 23° ATDC, after which the signal-to-noise ratio deteriorated rapidly. Since the cylinder pressure was close to 50 bar at the time of measurement, the quenching effect was considerably high and contributed to large uncertainty in the resulting OH images.

13.5.4 Visualization of NO by PLIF

13.5.4.1 Laser Induced Fluorescence of NO

The terminology and salient features of NO spectroscopy are similar to those of OH spectroscopy presented in Chapter 5. Five electronic states are normally involved in the band spectra of NO in the UV and visible regions: the ground electronic state, $X^2\Pi$, and the first four excited electronic states, $A^2\Sigma^+$, $B^2\Pi$, $C^2\Pi$, and $D^2\Sigma^+$. The corresponding four bands of NO in the UV and visible regions are known as β, γ, δ, and ε systems, depending upon the transitions of electronic levels [Pearse and Gaydon, 1965]. They are: $B^2\Pi \Leftrightarrow X^2\Pi(\beta)$, $A^2\Sigma^+ \Leftrightarrow X^2\Pi(\gamma)$, $C^2\Pi \Leftrightarrow X^2\Pi(\delta)$, $D^2\Sigma^+ \Leftrightarrow X^2\Pi(\varepsilon)$. The δ and ε systems are stronger, but lie at shorter wavelengths. The spectral lines are labeled similarly to the notation used for the OH radical.

General experimental considerations regarding combustion species measurements by PLIF have been discussed above. However, there are a number of issues that are pertinent to the PLIF measurements of NO.

The selection of spectral lines for excitation is particularly important for the LIF measurement of NO because of the shorter wavelength required. The detection of NO in IC engines by LIF has been realized at three excitation wavelengths: 193 nm, 226 nm, and 248 nm.

In the first excitation scheme, an ArF excimer laser (193 nm) is used and its output corresponds to the spectral lines of the $D^2\Sigma^+ \Leftrightarrow X^2\Pi(0, 1)$ band. After excitation to the $D^2\Sigma^+$ state, some NO molecules will jump to the slightly lower level $C^2\Pi(v' = 0)$ due to collisional quenching. The resulting fluorescence displays two clearly distinguishable sequences: $C^2\Pi(v' = 0) \leftrightarrow X^2\Pi(v'' = 1\text{–}6)$ and $D^2\Sigma^+(v' = 0) \Leftrightarrow X^2\Pi(v'' = 2\text{–}6)$. The strongest fluorescence originates from the primary excited level. Therefore, the $D^2\Sigma^+ \Leftrightarrow X^2\Pi(0, 3)$ transition at 208 nm should be employed for PLIF measurements. A bandpass interference filter consisting of four dielectric coated mirrors, as shown in Fig. 13.12, can be used to isolate the NO LIF signal with a maximum transmission of 80% at 208 ± 10 nm.

The most popular approach for PLIF measurements of NO is the excitation of the rotational-vibrational transitions of the $A^2\Sigma^+ \Leftrightarrow X^2\Pi(0, 0)$ band near 226 nm. The excitation source is usually an Nd:YAG laser pumped tunable dye laser. Excitation can also be achieved by the first anti-Stokes output of an H_2 Raman cell excited by a tunable ArF excimer laser [Bräumer et al., 1995] or a tripled Nd:YAG laser pumped optical parametric oscillator (OPO) [Dec and Canaan, 1998]. The Nd:YAG pumped OPO provides 12–15 mJ/pulse at 226 nm, which is about four times the pulse energy output of a typical Nd:YAG pumped dye laser. The subsequent fluorescence from the $A^2\Sigma^+ \Leftrightarrow X^2\Pi(0, 0)$ band at 237 nm and/or the (0, 2) band at 248 nm is detected for NO measurements.

The other excitation scheme using the direct output of a tunable KrF excimer laser is achieved by exciting the NO $A^2\Sigma^+ \Leftrightarrow X^2\Pi(0, 2)$ band at 248 nm.

Commercial gasoline absorbs at all three wavelengths. However, it has been found that isooctane, a test fuel often used in optical studies, does not absorb laser beam at 248 nm. In addition, the longer excitation wavelength is less attenuated by partial combustion products. Spectrally resolved broadband transmission measurements revealed strong absorption in the spectral region from 193 to 243 nm [Knapp et al., 1996]. At 226 nm, the transmission during the combustion stroke was about 40% in contrast to more than 80% at 248 nm. Therefore, the advantage of NO excitation at 248 nm is the possibility of NO measurements during combustion without severe attenuation when a liquid fuel is used. However, probing a vibrationally excited state ($v'' = 2$) results in a lower sensitivity and a signal that is strongly dependent on temperature, because the number density of vibrationally excited NO molecules is a strong function of temperature. The NO excitation and detection at 248 nm is therefore particularly suitable for high temperature and high NO concentrations.

The absorption problem responsible for laser beam attenuation is also present for fluorescence light at shorter wavelengths because of intermediates formed during combustion, including partially oxidized hydrocarbons and PAHs. Additionally, fluorescence trapping by transitions with large absorption rates may occur. This has been observed by Knapp et al. [1997a] when fluorescence at 226 nm ($\gamma(0, 0)$) and 237 nm ($\gamma(0,1)$) was detected after laser excitation at 248 nm ($\gamma(0,2)$). To avoid absorption and trapping, NO emission at 272 nm (0,4) was detected instead.

Because the hot-band spectrum of the $B^3\Sigma_u$–$X^3\Sigma_g$ Schumann-Runge system of O_2 [Krupenie, 1972] coincides with that of NO, the fluorescence from O_2 must be accounted for in order to obtain quantitative LIF measurements of NO. The term hot-band here refers to the spectra that appear only at elevated temperatures owing to the Boltzmann redistribution into the upper vibrational levels of the ground electronic state. This may be important for certain combustion situations and excitation-detection schemes. For example, it is difficult to separate oxygen fluorescence from NO fluorescence when an ArF laser is employed [Sick and Wolfrum, 1993].

13.5.4.2 Visualization of NO Distribution Using an ArF Excimer Laser

Excitation at 193 nm by a tunable ArF excimer laser was used in earlier investigations of the PLIF measurements of NO in an SI engine by Andresen et al. [1990], in a DI diesel engine by Arnold et al. [1992], and in an IDI diesel engine by Brugman et al. [1993]. Table 13.1 summarizes the experimental setups used in these experiments. In general, these measurements suffered from severe laser absorption in the presence of combustion. Strong attenuation of the laser beam at the earlier stages of combustion limited their NO measurements to the later stage of combustion. Quantitative measurements of NO using this excitation scheme was further complicated by the difficulty of spectral interference from oxygen fluorescence excited at 193 nm [Sick and Wolfrum, 1993].

The same excitation scheme was also employed by Tanaka et al. [1997] in the measurement of NO distributions in a constant-volume combustion bomb and in an SI engine. In order to avoid absorption of the laser radiation by the large hydrocarbon molecules generated during gasoline combustion, the engine was operated in a single compression and expansion stroke cycle. Propane was burned in the combustion chamber in order to minimize laser attenuation and fluorescence absorption. Two laser sheets were introduced into the combustion chamber through two opposite windows to enhance the fluorescence signal. The fluorescence signal was detected in the range 210–285 nm. The time histories of qualitative NO distributions in the engine were measured for different air/fuel ratios up to 25% mass fraction burned at a pressure lower than 10 bar in order to avoid quenching. From the recorded NO PLIF images, they observed that the LIF intensity was higher near the flame front than in the post-flame region. Maximum fluorescence intensity was found with a slightly lean mixture, consistent with that predicted by the extended Zeldovich NO formation mechanism.

13.5.4.3 Visualization of NO Distribution by Laser Excitation Near 226 nm

This excitation and detection scheme was employed by Alatas et al. [1993] to obtain NO images in a specially designed DI diesel engine using an Nd:YAG pumped dye laser. The fluorescence from the $A^2\Sigma^+ \Leftrightarrow X^2\Pi(0, 2)$ band at 248 nm was detected for NO measurements. The experimental setup is given in Table 13.1. A low sooting fuel mixture of 50% isooctane and 50% n-tetradecane was used to reduce the soot production. NO images could not be obtained before 40° CAATDC due to absorption of the laser beam by fuel and partial combustion products. In order to increase the NO concentration and reduce the attenuation of the UV beam, oxygen was added to the intake air. With oxygen enrichment, PLIF imaging of NO was realized at almost all of the crank angles following ignition.

Using a similar setup, Nakagawa et al. [1997] measured the NO distribution in a larger-bore single-cylinder diesel engine with a single-hole injector. To obtain sufficient signal, they boosted oxygen concentrations to 35% in the intake charge, and images were ensemble-averaged over five engine cycles. Simultaneous measurements of NO and OH showed that NO was distributed slightly outside the flame luminescence zone as compared with the region of OH distribution, and reached a maximum value in the later stage of the combustion process.

The first attempt to obtain quantitative two-dimensional measurements of NO was carried out by Braümer et al. [1995]. The $A^2\Sigma^+ \Leftrightarrow X^2\Pi(0, 0)$ band was employed for NO excitation at 225 nm by means of a KrF excimer laser and an H_2 Raman cell. Measurements were carried out in a square piston SI engine with propane in order to avoid absorption by combustion products of higher hydrocarbon fuels. In order to obtain the absolute concentration of NO from LIF measurements, the fluorescence signals were calibrated for known amounts of NO during the motored engine operation. The temperature effect on the ground state population was minimized by the selection of the $R_1(21)$ spectral line, which had a small temperature dependence in the temperature range of NO formation. The quenching rates of NO were assumed to be constant based on results obtained from the study of a premixed methane/air flame at 1 bar. The effect of pressure on the LIF signals was the only factor that was taken into account by a calibration factor. An accuracy of 30% was claimed for the absolute NO concentration measurements.

NO formation in a large DI diesel engine was investigated by Dec and Canaan [1998] using PLIF. An Nd:YAG pumped narrow-line OPO was employed due to its higher energy output (12–15 mJ) at 226 nm compared with that from a frequency-doubled tunable dye laser (4 mJ). The wavelength at 226.035 was selected because it was at the peak of one of the strongest NO fluorescence excitation lines, but it was located near the middle of the weakest oxygen absorption region [DiRosa et al., 1996] so as to minimize oxygen interference. The fluorescence emission from 237–276 nm, corresponding to (0, 1 to 0, 4) vibrational bands, was detected to maximize the fluorescence signal while it allowed rejection of elastically scattered light, PAH fluorescence, oxygen interference, and LII. The combined higher laser energy and the selected detection scheme allowed single-shot NO images to be obtained under realistic diesel-engine operating conditions without oxygen enrichment of the intake air. In addition to 2-D images of NO distribution, total NO formation was examined by integrating the NO PLIF signal over a large probe volume ($15 \times 4.5 \times 140$ mm^3). These total NO results were then corrected for pressure, temperature, and surrounding-species effects on collision broadening and line-shift, Doppler broadening, Boltzmann distribution, and quenching rate using a detailed spectroscopic code developed by Paul et al. [1993; 1994].

Fig. 13.15 shows a sequences of flame images with no optical filters on the left and NO PLIF images on the right. These images were obtained through a piston-crown window with the field of view shown in Fig. 13.14. The tip of the injector is evident at the left of the no-filter images, and the black curve at the right of the images marks the edge of the combustion bowl. Before examining these images, it is important to note that the no-filter images are dominated by natural soot luminosity.

According to the image sequence shown in Fig. 13.15, there is no evidence of NO in the first two images although premixed combustion has taken place (the signal seen in these two images was confirmed to be the fluorescence of fuel). Then at 355°, a small region of NO signal becomes visible. In the following images, the NO signal region becomes larger and brighter as the jet becomes larger and reaches out to the bowl wall, as the combustion reaction progresses. These images demonstrate that NO is not produced by the initial fuel-rich premixed combustion because this combustion is too fuel-rich to produce significant NO. The first appearance of NO begins around the periphery of the spray jet after diffusion combustion has started. The location and timing of NO formation shown in these images indicate that NO is produced on the lean side of the diffusion flame and in the hot post-combustion gases. These results also indicate that the temperature in these regions is the critical parameter for the formation of NO in DI diesel engines.

13.5.4.4 Quantitative Measurements of NO Distribution Using a Tunable KrF Excimer Laser

In order to overcome absorption problems with short wavelength excitation, the γ band of NO transitions were investigated by Knapp et al. [1996]. They found that when the output of a KrF excimer laser was tuned to 247.94 nm, three rotational lines $P_{12}(9.5)$, $P_{12}(8.5)$, and $P_{12}(10.5)$ of the NO $\gamma(0, 2)$ band could be excited simultaneously. In order to avoid the

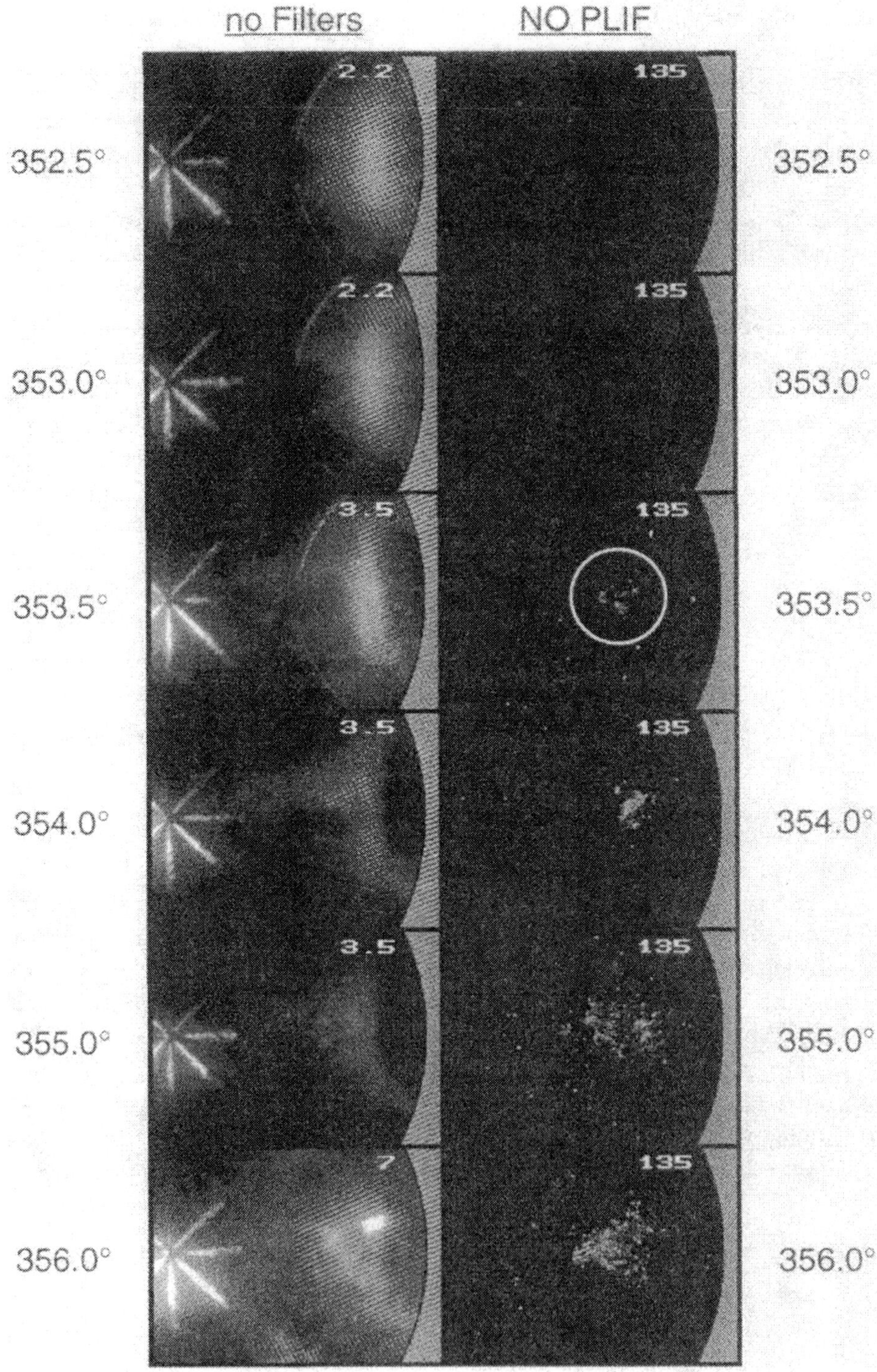

Fig. 13.15(a) Part 1—Temporal sequences of luminous flame and NO images, Dec and Canaan [1998].

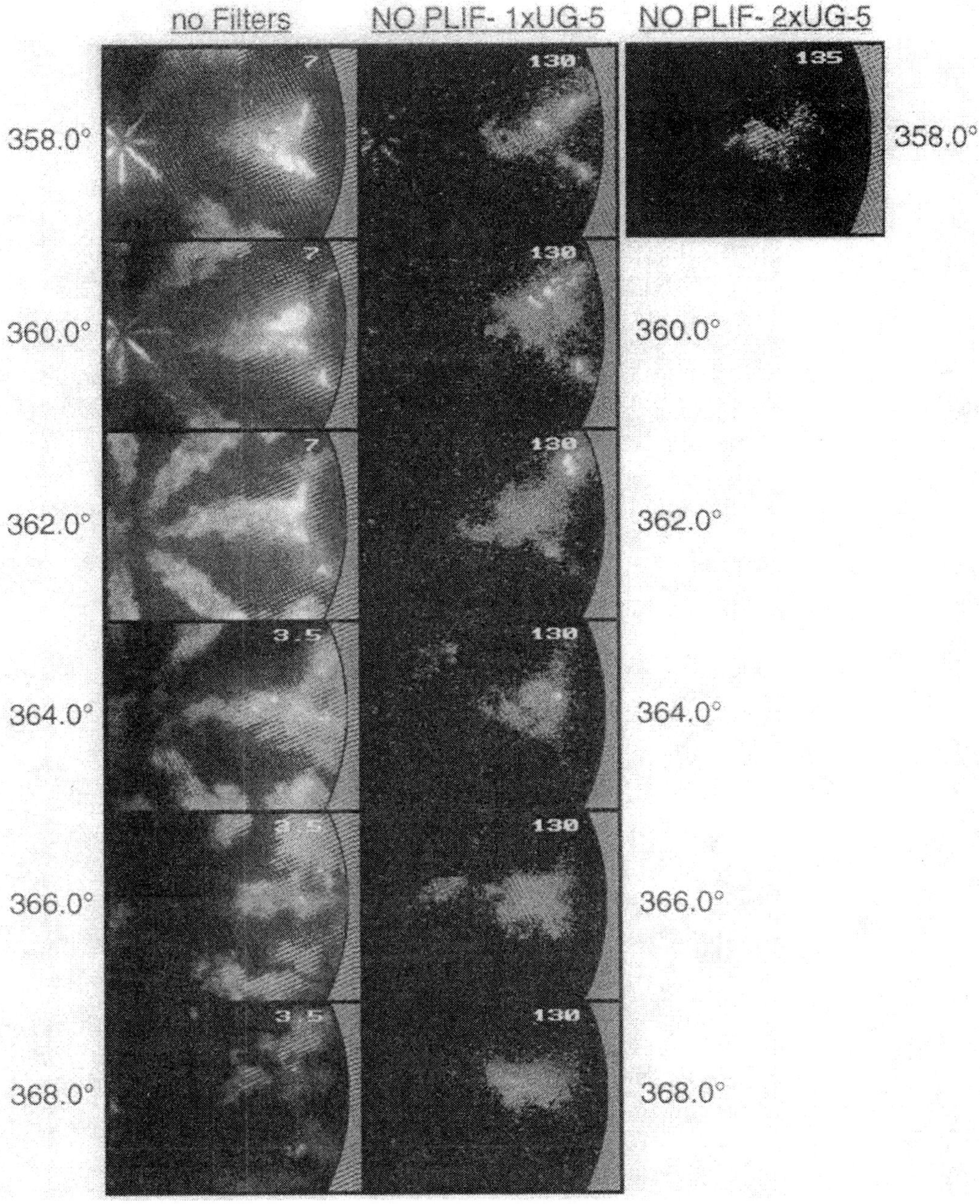

Fig. 13.15(b) Part 2—Temporal sequences of luminous flame and NO images, Dec and Canaan [1998].

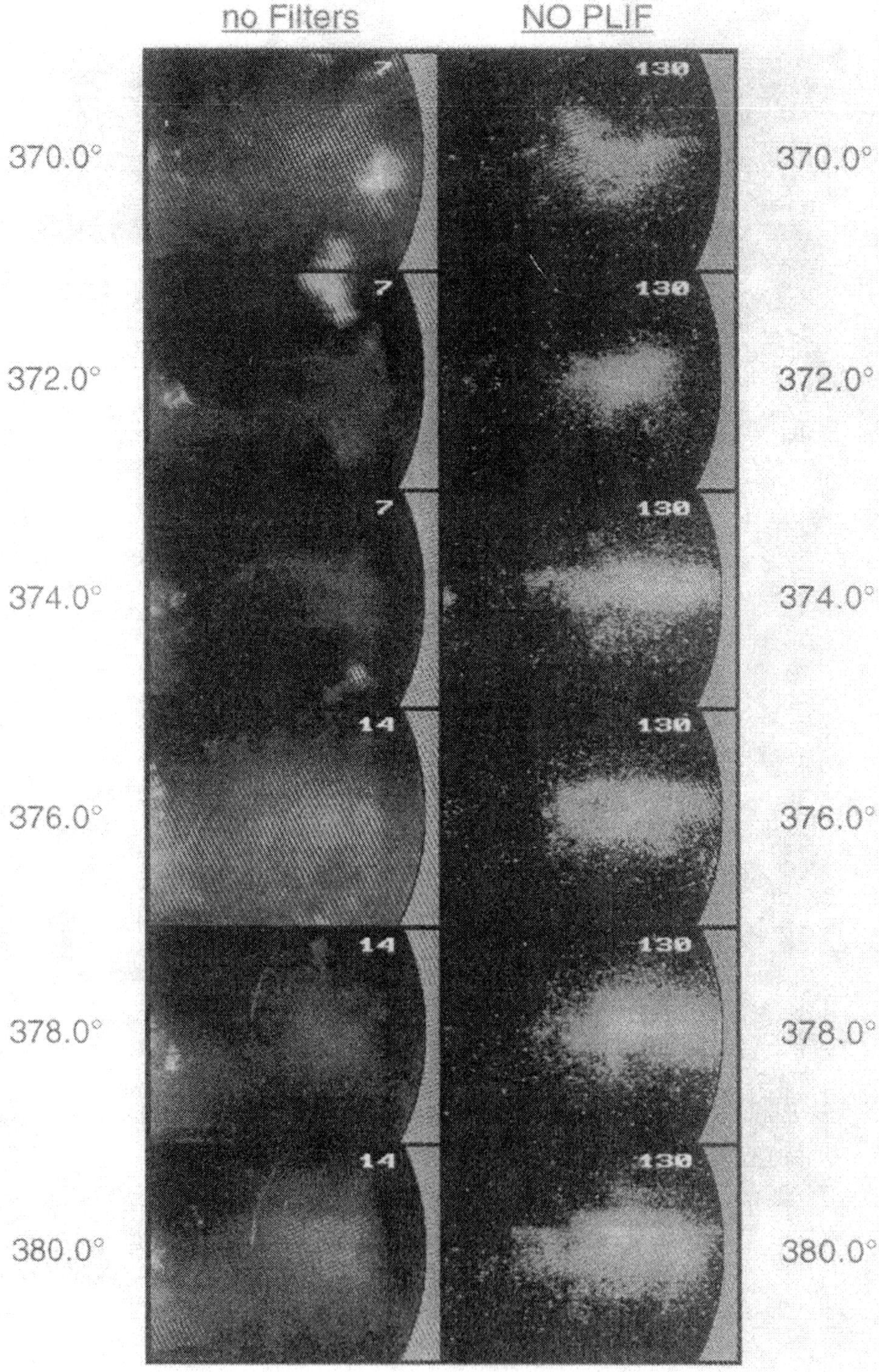

Fig. 13.15(c) Part 3—Temporal sequences of luminous flame and NO images, Dec and Canaan [1998].

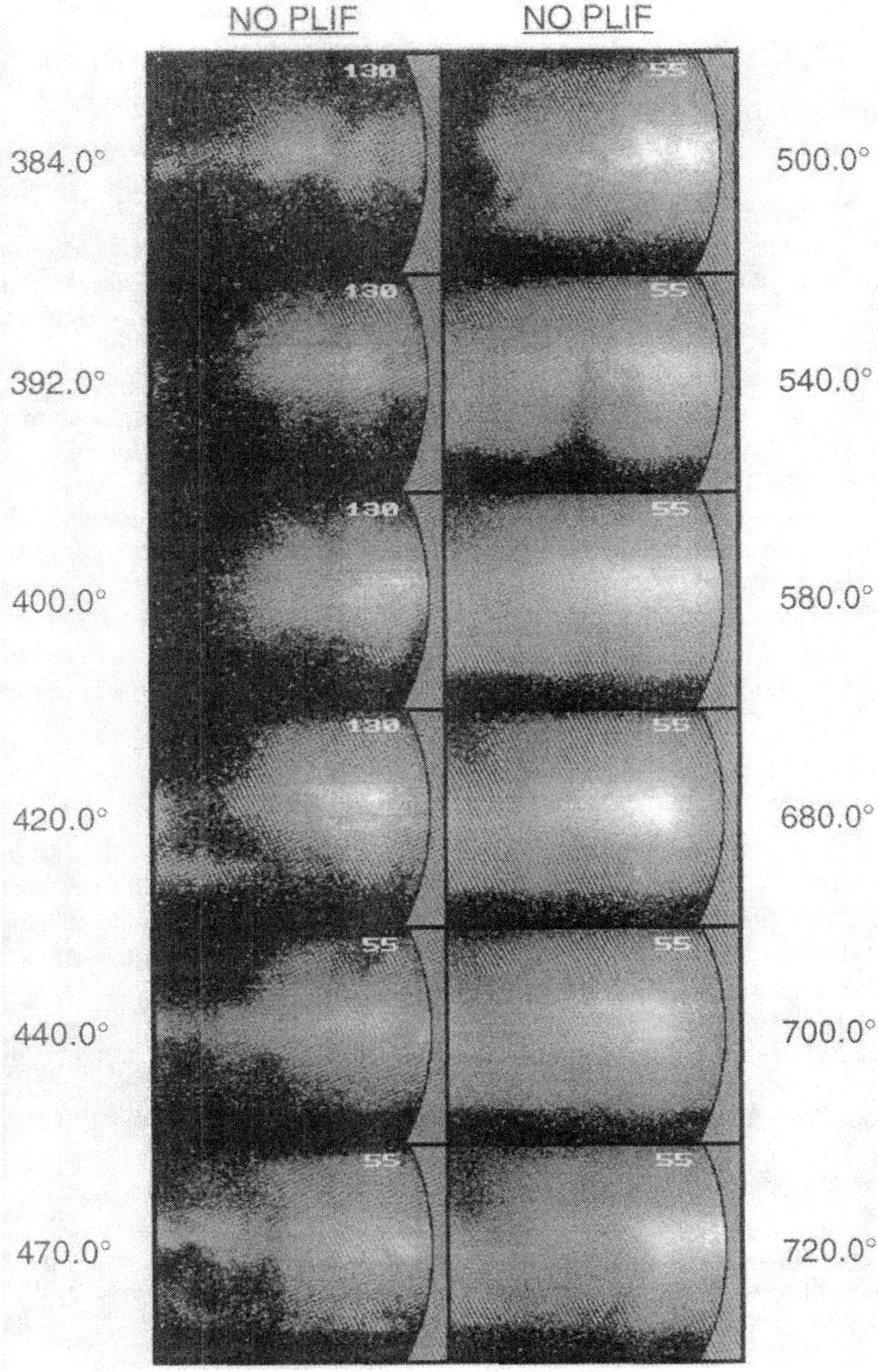

Fig. 13.15(d) Part 4—Temporal sequences of luminous flame and NO images, Dec and Canaan [1998].

interference of O_2 fluorescence at longer wavelengths, the blue-shifted (shorter-wavelength) fluorescence at 230 nm was detected through a UV filter.

Following the initial application to a square transparent piston engine by Schulz et al. [1996], the same excitation and detection scheme was used by Hildenbrand et al. [1998] to obtain quantitative NO measurements in a single-cylinder engine with a stoichiometric isooctane/air mixture. The pressure, temperature, and quenching effects were considered and in situ calibration performed in order to obtain quantitative NO measurements. The peak NO concentration of 3.7×10^{17} cm^{-3} at a pressure 9.5 bar was measured in the middle of the combustion chamber at a spark timing of 30° BTDC. They estimated that the accuracy of their NO measurements was about 30%. The fluorescence trapping was calculated to be 5%.

In a separate study, Knapp et al. [1997a] showed that the fluorescence signals at the shorter wavelength 226 nm was strongly trapped, while the 237 nm fluorescence was not significantly affected until the NO concentration reached 2000 ppm. Therefore, the NO emission at 272 nm was measured. It was found that NO detection at 272 nm was only affected by H_2O Raman scattering that was less than 2% of the NO LIF signal. The measurement accuracy was estimated to be about 10%. 1-D spatially resolved NO measurements revealed that there was a spatial gradient of NO in the burned gas region with maximum NO concentration near the spark plug. A large cycle-to-cycle variation in NO concentration was noticed which was considered to be a consequence of the small probe volume used (<0.1% of total gas volume) and the spatial inhomogeneities of the NO distribution.

13.5.5 Visualization of Formaldehyde by PLIF

Traditionally, high-speed shadowgraph/Schlieren techniques have been used to visualize the auto-ignition sites in a knocking combustion engine. However, they rely on the change in density integrated over the optical path and hence lacks spatial resolution. Laser induced fluorescence, on the other hand, is characterized by its high sensitivity and its species-specific nature, as well as its ability to produce a spatially resolved two-dimensional image of the fluorescing species.

Experimental and theoretical studies discussed in previous sections show that autoignition is initiated at one or more sites by exothermic reactions. Chemical kinetic studies indicate that a temperature difference of 10 K [Dagaut et al., 1994] could influence the start of the autoignition process. However, it is not possible yet to measure temperature fluctuations as small as 10 K or less at temperatures of 700 K or higher.

On the other hand, formaldehyde (CH_2O) has long been associated with the autoignition process in the end-gas region. It was first found in engines prior to knocking combustion by Rassweiler and Withrow [1933]. Since then, numerous studies have confirmed that formaldehyde is formed during the first stage of the autoignition process, and is responsible for the weak blue light emission of cool flames. As autoignition proceeds into the higher-temperature region, formaldehyde is consumed rapidly in the second stage of the autoignition process [Benson, 1981]. The presence of formaldehyde is therefore a good marker of the autoignition process in the end-gas region.

The principle and experimental setup for PLIF measurements of formaldehyde are similar to those for OH and NO. Because of the greater number of atoms in a CH_2O molecule, there are many more vibrational energy levels in each electronic energy state of formaldehyde. As a result, an electronic transition is accompanied by a large number of vibrational-rotational changes spread over a range of frequencies. Fig. 13.16 shows the fluorescence excitation spectrum of formaldehyde [Miller and Lee, 1975]. The fluorescence excitation spectrum has a peak at 353.2 nm corresponding to the 4_0^1 line of the $\tilde{A}^1A_2 \leftarrow X^1A_1$ band. The emission spectrum slightly shifted to the red, starts at about 360 nm. Most formaldehyde fluorescence is found in the range of 390 to 460 nm. Although the frequency tripled output (355 nm) of an Nd:YAG laser may be used for the excitation of the $\tilde{A}^1A_2$ state of formaldehyde, it induces weak absorption by formaldehyde and hence lower fluorescence intensity. A tunable dye laser is preferred because it can be easily tuned to 353.2 nm for the maximum fluorescence signal.

Formaldehyde distribution in a two-stroke engine was measured by Bäuerle et al. [1994] using PLIF. The excitation at 353.2 nm was achieved by a tunable dye laser with BMQ dye (24 mJ). The fluorescence signal from a 50 mm wide laser sheet was detected by an ICCD camera through the transparent cylinder head. A bandpass filter (400–450 nm) was placed in front of the camera to eliminate spurious laser scattering. PLIF CH_2O images of the complete

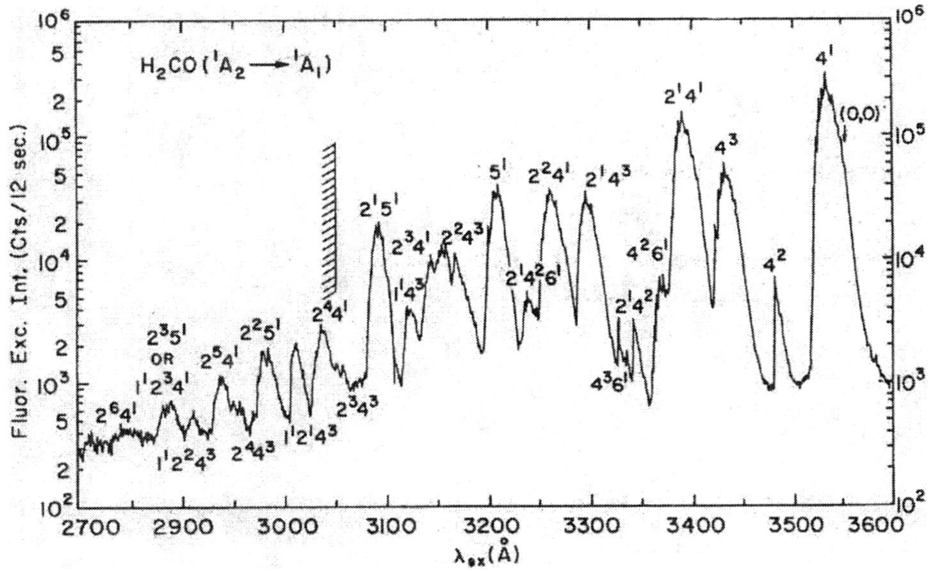

Fig. 13.16 Fluorescence excitation spectra of CH_2O, adapted from Miller and Lee [1975].

cylinder charge were obtained from different knocking cycles. This experimental setup was later extended to record a pair of fluorescence images of CH_2O separated by 50 μs so that the temporal progress of the autoignition process could be tracked [Bäuerle et al., 1996]. The pair of fluorescence images was obtained by double-pulse excitation (2.6 mJ and 1.3 mJ, respectively) of the same dye laser sequentially pumped by two excimer lasers.

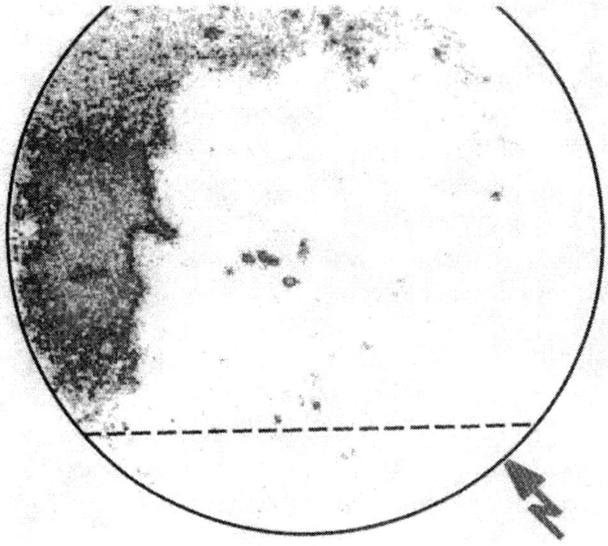

(a) First frame of 2D double pulse LIF of formaldedhyde (6.6° CA ATDC)

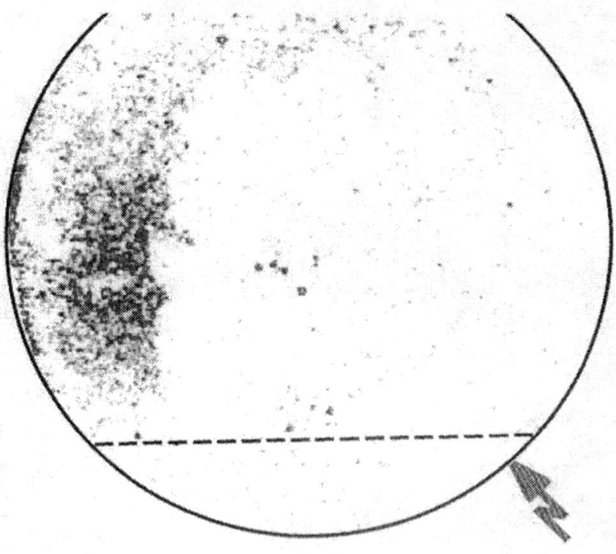

(b) Second frame of 2D double pulse LIF of formaldedhyde (+0.06° CA)

Fig. 13.17 Two-dimensional images of formaldehyde in a knocking cycle.

Fig. 13.17 presents a pair of fluorescence images of formaldehyde taken during a violently knocking cycle. In the left parts of the frames, the roughly crescent-shaped, colored areas represent the end-gas region where the fluorescence of CH_2O was detected. The spark plug is visible in the lower right corner of the frame. The right-hand edges of the fluorescent regions represent the leading edges of the flame front. The blank region to the right-hand side of the flame front is the burned gas region, where no fluorescence of formaldehyde was detectable. Compared with images taken from non-knocking cycles, the regular flame front shows an unwrinkled contour. The first frame, which was taken at the onset of autoignition as detected from the in-cylinder pressure, shows that there is a blank spot right in the highest intensity region and several small spots along the cylinder wall. Bäuerle et al. described these blank spots as hot spots as formaldehyde formed in the low temperature regime was consumed rapidly in the high temperature regime of the auto-ignition process. These so-called hot spots become more evident in the second frame, which was taken 10 μs (0.06° CA) later. From these two frames, Bäuerle et al. calculated the hot spots expansion speed in this cycle. The hot spots near the wall expanded at a speed of 350–500 $m \cdot s^{-1}$ and the enclosed one at a speed up to 750 $m \cdot s^{-1}$. Furthermore, the regular flame front was driven backward at a speed of 30–180 $m \cdot s^{-1}$, which is rather moderate compared with that measured by other researchers using high-speed Schlieren photography.

13.6 Summary

In this chapter techniques for in-cylinder combustion species have been reviewed and discussed.

Because of the importance of the in-cylinder burned gas fraction (residual gas + EGR) to combustion stability and pollutant formation, several techniques have been developed to measure its concentration and distribution inside the cylinder of a combustion engine. A fast FID or a high-speed intermittent sampling valve can be easily adapted to a production engine with limited spatial and temporal resolution. The use of a sampling valve permits simultaneous multi-species detection and hence burned gas composition, as well as concentration measurement. The PLIF technique provides a useful means of visualizing burned gas distribution through one of the combustion species (H_2O or NO) or a fluorescing dopant in the fresh charge. SRS is the only technique that is capable of quantitative simultaneous multi-species and multiple point measurements of in-cylinder burned gases with high temporal and spatial resolution.

Absorption and, particularly, emission spectroscopy was extensively used in early studies of combustion and pollutant formation in IC engines before the advent of laser diagnostics. It has greatly enhanced our knowledge of the autoignition process leading to knocking combustion as well as normal combustion processes. Its application has been made easier with the introduction of fiber optics and modern spectroscopic equipment, and will continue to be a useful tool for IC engine combustion research.

Laser induced fluorescence has proved to be an extremely useful means of imaging combustion species inside an IC engine, although, in most cases, PLIF measurements remain qualitative as a consequence of the signal quenching and other effects. The development in laser technology and modern imaging equipment and advances in spectral physics involved have made it possible to achieve quantitative two-dimensional images of NO, an important pollutant from IC engines.

References

Abata, D.L., Myers, P.S., and Uyehara, O.A. (1978), "Spectroscopic Investigation of Hydroxyl Radical Formation in the End Gases of a Spark Ignited Engine Utilising a Dye Laser," SAE Paper No. 780970.

Alatas, B., Pinson, J.A., Litzinger, T.A., and Santavicca, D.A. (1993), "A Study on the Formation Processes of Soot and NO in a DI Diesel Engine," SAE Paper No. 930973.

Alperstein, M. and Bradow, R.L. (1965), "Combustion Gas Sampling Valve," *The Review of Scientific Instruments,* Vol. 36, No. 7, pp. 1028–1031.

Alperstein, M. and Bradow, R.L. (1966), "Investigations into the Composition of End Gases from Otto Cycle Engines," SAE Paper No. 660410.

Alperstein, M. and Bradow, R.L. (1966), "Exhaust Emissions Related to Engine Combustion Reactions," SAE Paper No. 660781.

Andresen, P., Bath, A., Groger, W., Lulf, H.W., Meijer, G., and ter Meulen, J.J. (1988), "Laser-Induced Fluorescence with Tunable Excimer Lasers as a Possible Method for Instantaneous Temperature Field Measurements at High Pressures: Checks with an Atmospheric Flame," *Appl. Opt.*, Vol. 27, pp. 365–378.

Andresen, P., Meijer G., Schluter, H., Voges, H., Koch, A., Hentschel, W., Oppermann, W., and Rothe, E. (1990), "Fluorescence Imaging Inside an Internal Combustion Engine Using Tunable Excimer Lasers," *Applied Optics*, Vol. 29, No. 16, pp. 2392–2404.

Aoyagi, Y., Kamimoto, T., Matsui, Y., Matsuoka, S. (1980), "A Gas Sampling Study on the Formation Processes of Soot and NO in a DI Diesel Engine," SAE Paper No. 800254.

Arnold, A., Dinkelacker, F., Heitzmann, T., Monkhouse, P., Schafer, M., Sick, V., and Wolfrum, J., (1992), "DI Diesel Engine Combustion Visualized by Combined Laser Techniques," *Proc. 24th Symposium (International) on Combustion*, The Combustion Institute, pp. 1605–1612.

Arnold, A., Brauerle, A., Buschmann, A., Decker, M., Dinkelacker, F., Heitzmann, T., Orth, A., Schafer, M., Sick, V., and Wolfrum, J. (1993), "2D-Diagnostics in Industrial Devices," *Ber. Bunsenges. Phys. Chem.*, Vol. 97, No. 12, pp. 1650–1661, 1993.

Bäcker, H., Hild, O., and Lepperhoff, G. (1997), "Investigation on Particle Size Distribution in the Cylinder of a Passenger Car DI Diesel Engine Using a Fast Gas Sampling Valve," SAE Paper No. 970875.

Barbella, R., Betoli, C., Ciajolo, A., D'Anna, A., Masi, S. (1989), "In-Cylinder Sampling of a High Molecular Weight Hydrocarbons from a DI Light-Duty Diesel Engine," SAE Paper No. 890437.

Bäuerle, B., Hoffmann, F., Behrendt, F., and Warnatz, J. (1994), "Detection of Hot Spots in the End Gas of an Internal Combustion Engine Using Two-Dimensional LIF of Formaldehyde," *Proc. 25th Symposium (International) on Combustion*, The Combustion Institute, pp. 135–141.

Bäuerle, B., Warnatz, J., and Behrendt, F. (1996), "Time-Resolved Investigation of Hot Spots in the End Gas of an SI Engine by Means of 2D Double-Pulse LIF of Formaldehyde," *Proc. 26th Symposium (international) on Combustion*, The Combustion Institute, 1996.

Benson, S.W. (1981), "The Kinetics and Thermochemistry of Chemical Oxidation with Application to Combustion and Flames," *Progr. Energy Combust. Sci.*, Vol. 7, pp. 125–134.

Berckmüller, M., Tait, N., and Greenhalgh, D. (1997), "The Influence of Local Fuel Concentration on Cyclic Variability of a Lean Burn Stratified-Charge Engine," SAE Paper No. 970826.

Berlman, I.B. (1971), *Handbook of Fluorescence Spectra of Aromatic Molecules*, Academic Press.

Bräumer, A., Sick, V., Wolfrum, J., Drewes, V., Zahn, M., and Maly, R. (1995), "Quantitative Two-Dimensional Measurements of Nitric Oxide and Temperature Distributions in a Transparent Square Piston SI Engine," SAE Paper No. 952462, 1995.

Breeze, J.C. and Ferriso, C.C. (1964), "Integrated Intensity Measurements of the 5.3 micron Fundamental and 2.7 micron Overtone Bands of NO between 1400 K and 2400 K," *Journal of Chemical Physics*, Vol. 41, p. 3420.

Brugman, M., Klein-Douwel, R., Huigen, G., Walwijk, E., and Meulen, J. (1993), "Laser Induced Fluorescence Imaging of NO in an n-Heptane and Diesel Fuel Driven Diesel Engine," *App. Physics B*, Vol. 57, pp. 405–410.

Clark, G.L. and Henne, A.L. (1927), "Ultraviolet Spectroscopy of Engine-Fuel Flames," SAE Trans., Vol. 22, No. 1, pp. 15–23.

Clark, G.L. (1928), "Spectroscopic Study of Fuels and Analysis of Detonation Theories," *SAE Trans.*, Vol. 23, No. 2, pp. 351–357.

Dagaut, P., Reuillion, M., and Cathonnet, M. (1994), "Experimental Study of the Oxidation of n-Heptane in a Jet Stirred Reactor from Low to High Temperature and Pressures up to 40 atm," *Combust. Sci. and Tech.*, Vol. 95, pp. 233–260.

Davis, W.C., Smith, M.L., Malberg, E.W., and Bobbitt, J.A. (1955), "Comparison of Intermediate-Combustion Products Formed in Engine with and without Ignition," SAE Paper No. 550262.

Dec. J. and Coy, E.B. (1996), "OH Radical Imaging in a DI Diesel Engine and the Structure of the Early Diffusion Flame," SAE Paper No. 960831.

Dec. J. and Canaan, R. (1998), "PLIF Imaging of NO Formation in a DI Diesel Engine," SAE Paper No. 980147.

Deschamps, B. and Baritaud, T. (1996), "Visualisation of Gasoline and Exhaust Gases Distribution in a 4-Valve SI Engine: Effects of Stratification on Combustion and Pollution," SAE Paper No. 961928.

Dieke, G.H. and Crosswhite, A.M. (1962), "The Ultraviolet Band of OH," *J. Quan. Spectrosc. Radiat. Transfer*, Vol. 2, 97–199.

Di Rosa, M.D. and Hanson, R.K. (1994), "Collisional Broadening and Shifting of NO g(0,0) Absorption Lines by O_2 and H_2O at High Temperatures," *J. Quant. Spectrosc. Radiat. Transfer*, Vol. 52, pp. 515–529.

Di Rosa, M., Klavuhn, K., and Hanson, R. (1996), "LIF Spectroscopy of NO and O_2 in high-pressure flames," *Comb. Science and Technology*, Vol. 118, pp. 257–283.

Downs, A.D., Walsh, A.D., and Wheeler, R.W. (1951), "A Study of the Reactions that Lead to Knock in the Spark-Ignition Engine," Phil. Tans. Roy. Soc. London, Vol. 243, No. A870, pp. 463–524.

Downs, D., Street, J.C., and Wheeler, R.W. (1953), "Cool-Flames in a Motored Engine," *Fuel*, Vol. 32, p. 279.

Drake, M.C. and Ratcliffe, J.W. (1993), "High Temperature Quenching Cross Sections for Nitric Oxide Laser Induced Fluorescence," *J. Chem. Phys.*, Vol. 98(5), pp. 3850–3865.

Eckbreth, A.C. (1996), *Laser Diagnostics for Combustion Temperature and Species*, 2nd Edition, Gordon and Breach Publishers, Amsterdam.

Egerton, A., Smith, F.L., and Ubbelohde, A.R. (1935), "Estimation of the Combustion Products from the Cylinder of the Petrol Engine and Its Relation to Knock," *Phil. Tans. Roy. Soc. London*, Vol. 234, No. 744, pp. 433–521.

Felton, P., Mantzaras, J., Bomse, D., and Woodin, R. (1988), "Initial Two-Dimensional Laser Induced Fluorescence Measurements of OH Radicals in an Internal Combustion Engine," SAE Paper No. 881633.

Fontijn, A., Meyer, C.F., and Schiff, H.I. (1964), "Absolute Quantum Yield Measurements of the NO-O Reaction and Its Use as a Standard for Chemiluminescent Reactions," J. Chem. Phys., Vol. 40, p. 64.

Galliot, F., Cheng, W., Cheng. C., Sztenderowicz, M., Heywood, J.B., and Collings, N., (1990), "In-Cylinder Measurements of Residual Gas Concentration in a Spark Ignition Engine," SAE Paper No. 900485.

Gaydon, A.G. (1970), *The Spectroscopy of Flames*, 2nd Ed., Chapman and Hall Ltd.

Gray, J.A. and Farrow, R.L. (1991), "Predissociation Lifetimes of OH $A^2\Sigma^+$ ($v' = 3$) Obtained from Optical-Optical Double-Resonance Linewidth Measurements," *J. Chem. Phys.*, Vol. 95, pp. 7054–7060.

Green, R.M., Cernansky, N.P., Pitz, W.J., and Westbrook, C.K.(1987), "The Role of Low Temperature Chemistry in the Autoignition of n-Butane," SAE Paper No. 872108.

Grünefeld, G., Beushausen, V., Andresen, P., and Hentschel, W. (1994a), "Spatially Resolved Raman Scattering for Multispecies and Temperature Analysis in Technically Applied Combustion Systems: Spray Flame And Four Cylinder In Line Engine," *Appl. Phys. B.*, Vol. 58, pp. 333–342.

Grünefeld, G., Beushausen, G., Andresen, P., and Hentschel, W. (1994b), "A Major Origin Of Cyclic Energy Conversion Variations in SI Engines: Cycle by Cycle Variations of the Equivalence Ratio and Residual Gas of the Initial Charge," SAE Paper No. 941880.

Grünefeld, G., Knapp, M., Beushausen, G., Andresen, P., Hentschel, W., and Manz, W. (1995), "In Cylinder Measurements and Analysis on Fundamental Cold Start and Warm Up Phenomena of SI Engines," SAE Paper No. 952394.

Hartunian, R.A., Thompson, W.P., and Hewitt, E.W. (1966), "Glow-Discharge Shock Tube for Studying Chemiluminescent, Surface-Catalytic, and Gas Phase Reaction Rates: Temperature Dependence of NO-O and CO-O Chemiluminescence," *J Chem. Phys.*, Vol. 44, p. 1765.

Hazen, D. F. and Holiday, G. (1962), "The Effects of Engine Operating and Design Variables on exhaust Emissions," SAE Paper No. 486 C.

Hildenbrand, F., Schulz, C., Sick, V., Josefsson, G., Magnusson, I., Andeersson, O., and Alden, M. (1998), "Laser Spectroscopic Investigation of Flow Fields and NO Formation in a Realistic SI Engine," SAE Paper No. 980148.

Hinze, P. and Miles, P. (1999), "Quantitative Measurements of Residual and Fresh Charge Mixing in a Modern SI Engine Using Spontaneous Raman Scattering," SAE Paper No. 1999-01-1106.

Iida, N. (1994), "Combustion Analysis of Methanol Fueled Active Thermo Atmosphere (ATAC) Combustion Engine using a Spectroscopic Observation," SAE Paper No. 940684.

Johansson, B., Neij, H., Greger, J., and Aldún, M.(1995), "Residual Gas Visualisation with Laser Induced Fluorescence," SAE Paper No. 952463.

Knapp, M., Luczak, A., Schluter H., Beushausen, V., Hentschel, W., and Andresen, P. (1996), "Crank-Angle-Resolved Laser Induced Fluorescence Imaging of NO in a Spark-Ignition Engine at 248 nm and Correlation to Flame Front Propagation and Pressure Release," *Applied Optics*, Vol. 35, No. 21, pp. 4009–4017.

Knapp, M., Luczak, A., Beushausen, V., Hentschel, W., Manz, P., and Andresen, P. (1997a), "Quantitative In-Cylinder NO LIF Measurements with a KrF Excimer Laser Applied to a Mass-Production SI Engine Fuelled with Isooctane and Regular Gasoline," SAE Paper No. 970824.

Knapp, M., Beushausen, V., Hentschel, W., and Manz, P., Grünefeld, G. and Andresen, P. (1997b), "In-Cylinder Mixture Formation Analysis with Spontaneous Raman Scattering Applied to a Mass-Production SI Engine," SAE Paper No. 970827.

Kosaka, H., Nishigaki, T., and Komimoto, T. (1996), "Simultaneous 2-D Imaging of OH Radicals and Soot in a Diesel Flame by Laser Sheet Technique," SAE Paper No. 960834.

Krupenie, P.H. (1972), "The Spectrum of Molecular Oxygen," *J. Phys. Chem. Ref. Data,* Vol. 1, pp. 423–534.

Ladommatos, N. (1992), "Cyclically Resolved Measurements of Hydrocarbons in the Cylinders of Internal-Combustion Engines by Means of a Fast Flame-Ionisation Detector," *J. Inst. Energy*, Vol LXV, pp. 94–101.

Lavoie, G., Heywood, J., and Keck, J. (1970a), "Experimental and Theoretical Study of Nitric Oxide Formation in Internal Combustion Engines," *Combustion Science and Technology*, Vol. 1, pp. 313–326.

Lavoie, G. (1970b), "Spectroscopic Measurements of Nitric Oxide in Spark Ignition Engines," *Combustion and Flame*, Vol. 15, pp. 97–108.

Learner, R.C. (1962), "The Influence of Vibration-Rotation Interaction on Intensities in the Electronic Spectra of Diatomic Molecules," *Proc. Roy. Soc. (London), Ser. A.*, Vol. 269, pp. 311–326.

Lebel, M. and Cottereau, J.M. (1992), "Study of the Effect of the Residual Gas Fraction on Combustion in a SI Engine Using Simultaneous CARS Measurements of Temperature and CO_2 Concentration," SAE Paper No. 922388.

Lempert, W., Diskin G., Kumar, V., Glesk, I., and Miles, R. (1991), "Two-Dimensional Imaging of Molecular Hydrogen in a H_2-Air Diffusion flame using Two-Photon Laser Induced Fluorescence," *Opt. Lett.*, Vol. 16.

Levitt, B.P. (1965), "Temperature and Wavelength Dependence of the Thermal Emission and O-NO Recombination Spectra of NO_2," *J. Chem. Phys.*, Vol. 42, p. 1038.

Luck, C.J., Burgess, A.R., Desty, D.H., Whitehead, D.M., and Pratley, G., (1973), "A Study of the Combustion of n-Heptane in an Engine Using a Novel High-Speed Sampling Technique," *Proc. of 14th Symposium (International) on Combustion*, The Combustion Institute.

Lyn, W.T. (1962) "Study of Burning Rate and Nature of Combustion in diesel Engines," Proc. of 9th Symposium (International) on Combustion, pp. 1069–1082.

Malkmus, W. (1963), "Infrared Emissivity of Carbon Dioxide (4.3 micron band)," *J. of Optical Society of America*, Vol. 53, p. 951.

Matsui, K., Tanaka, T., and Ohigashi, S. (1979), "Measurement of Local Mixture Strength at Spark Gap of SI Engines," SAE Paper No. 790483.

Meijer, G., ter Meulen, J.J., Andresen, P., and Bath, A. (1986), "Sensitive Quantum State Selective Detection of H_2O and D_2O by (2+1)-Resonance Enhanced Multiphoton Ionisation," *J. Chem. Phys.*, Vol. 85, pp. 6914–6922.

Miles, P. and Dilligan, M. (1996), "Quantitative In-Cylinder Fluid Composition Measurements Using Broadband Spontaneous Raman Scattering," SAE Paper No. 960828.

Miles, P. and Hinze, P. (1998), "Characterisation of the Mixing of Fresh Charge with Combustion Residuals Using Laser Raman Scattering with Broadband Detection," SAE Paper No. 981428.

Miller, R.G. and Lee, K.C. (1975), "Single Vibronic Level Photochemistry of Formaldehyde: Radiative and Non-Radiative Transitions," *Chemical Physics Letters*, Vol. 33, No. 1, pp. 104–107.

Myers, B.I. and Bartle, E.R. (1967), "Shock-Tube Study of the Radiative Processes in Systems Containing Atomic Oxygen and Carbon Monoxide at High Temperature," *J. Chem Phys.*, Vol. 47, p. 1783.

Nakagawa, H., Endo, H., Deguchi, Y., Noda, M., Oikawa, H., and Shimada, H. (1997), "NO Measurement in Diesel Spray Flame Using Laser Induced Fluorescence," SAE Paper No. 970874.

Newhall, H.K. and Starkman, E.S. (1967), "Direct Spectroscopic Determination of Nitric Oxide in Reciprocating Engine Cylinders," SAE Paper No. 670122.

Neij, H. and Aldén, M. (1994), "Application of Two-Photon Laser Induced Fluorescence for Visualisation of Water Vapour in Combustion Environments," *Applied Optics*, Vol. 33, No. 27, pp. 6514–6523.

Noguchi, M., Tanaka, Y., Tanaka, T., and Takeuchi, Y. (1979), "A Study on Gasoline Engine Combustion by Observation of Intermediate Reactive Products During Combustion," SAE Paper No. 790840.

Orth, A., Sick, V., Wolfrum, J., Maly, R., and Zahn, M. (1994), "Simultaneous 2D Single-Shot Imaging of OH Concentrations and Temperature Fields in an SI Engine Simulator," *Proc. of 25th Symposium (International) on Combustion*, The Combustion Institute, pp. 142–150.

Paul, P.H., Gray, J.A., Durant Jr., J.L., and Thoman, J.W., Jr. (1993), "A Model for Temperature-Dependent Collisional Quenching of NO $A^2\Sigma^+$," *Appl. Phys. B*, Vol. 57, pp. 249–259.

Paul, P.H., Gray, J.A., Durant J.L., Jr., and Thoman, J.W., Jr. (1994), "Collisional Quenching Corrections for Laser Induced Fluorescence Measurements of NO $A^2\Sigma^+$," *AIAA Journal*, Vol. 32, pp. 1670–1675.

Pearse, R.W.B. and Gaydon, A.G. (1965), *The Identification of Molecular Spectra*, 3rd Edition, Chapman and Hall, London.

Pipenberg, K.J. and Pahnke, A.J., "Spectrometric Investigations of n-Heptane Preflame Reactions in a Motored Engine," *Ind. Engng Chem.*, Vol. 49, p. 2067, 1957.

Pipho, M.J., Kittelson, D.B., and Zarling, D.D. (1991), "NO_2 formation in a diesel engine," SAE Paper No. 910231.

Rassweiler, G.M. and Withrow, L. (1932), "Emission Spectral of Engine Flames," *Ind. Eng. Chem.*, Vol. 24, pp. 528–538.

Rassweiler, G.M. and Withrow, L. (1933), "Spectrographic Detection of Formaldehyde in an Engine Prior to Knock," *Ind. Eng. Chem.*, Vol. 25, pp. 1359–1366.

Rassweiler, G.M. and Withrow, L. (1935), "Flame Temperatures Vary with Knock and Combustion Chamber Position," *SAE Trans.*, Vol. 36, pp. 125–133.

Schipperijn, F., Nagasaka, R., Sawyer, R., and Green, R. (1988), "Imaging of Engine Flow and Combustion Processes," SAE Paper No. 881631.

Schultz, C., Sick, V., Wolfrum, J., Drewes, V., Zahn, M., and Maly, R. (1996), "Quantitative 2D Single-Shot Imaging and Mathematical Modeling of NO Concentrations and Temperatures in a Transparent SI Engine," *Proc. of 26th Symposium (International) on Combustion*, The Combustion Institute.

Seitzman, J.M., Haumann, J., and Hanson, R.K. (1987), "Quantitative Two-Photon LIF Imaging of Carbon Monoxide in Combustion Gases," *Applied Optics*, Vol. 26, No. 14, pp. 2892–2899.

Serpenguzel, A., Hahn, R., and Acker, W. (1993), "Single Pulse Planar Laser Induced Fluorescence Imaging of Hydroxyl Radicals in a Spark Ignition Engine," SAE Paper No. 932701.

Shoji, H., Tosaka, Y., Yoshida, K. and Saima, A. (1994), "Radical Behaviour in Preflame Reactions Under Knocking Operating in a Spark Ignition Engine," SAE Paper No. 942061.

Shoji, H., Shimizu, T., Nishizawa, T., Yoshida, K., and Saima, A. (1996), "Spectroscopic Measurement of Radical Behaviour Under Knocking Operation," SAE Paper No. 962104.

Sick, V. and Wolfrum, J. (1993), "Absolute Laser-Spectroscopic Measurements of NO Concentrations to Study the Formation of NO in Laminar Counterflow Diffusion Flames," Joint Meeting of the British/German Sections of the Combustion Institute, Cambridge, U.K.

Steele, S., 1931, *Nature*, Vol. 128, p. 185.

Steele, S., 1935, *Nature*, Vol. 135, p. 268.

Suntz, R., Becker, H., Monkhouse, P., and Wolfrum, J. (1988), "Two-Dimensional Visualisation of the Flame Front in an Internal Combustion Engine by Laser Induced Fluorescence of OH Radicals," *Appl. Phys.*, B47, pp. 287–293.

Tanaka, T. and Tabata, M. (1994), "Planar Measurements of OH Radicals in an SI Engine Based on Laser Induced Fluorescence," SAE Paper No. 940477.

Tanaka, T., Fujimoto, M., and Tabata, M. (1997), "Planar Measurements of NO in a SI Engine Based on Laser Induced Fluorescence," SAE Paper No. 970877, 1997.

Thee, W.C., (1929), J. *SAE*, Vol. 25, 388.

Vannobel, F., Arnold, A., Buschmann, A., Sick. V., Wolfrum, J., Cousyn, B., and Decker, M. (1993), "Simultaneous Imaging of Fuel and Hydroxyl Radicals in an In-Line Four Cylinder SI Engine," SAE Paper No. 932696.

Voiculescu, I.A. and Borman, G.L. (1978), "An Experimental Study of Diesel Engine Cylinder-Averaged NOx Histories," SAE Paper No. 780–228.

Wimmer, D.B. and McReynolds, L.A. (1962), "Nitrogen Oxides and Engine Combustion," *SAE Transactions*, Vol. 70.

Withrow, L. and Rassweiler, G.M. (1931), "Spectroscopic Studies of Engine Combustion," *Ind. Eng. Chem.*, Vol. 23, pp. 769–776.

Withrow, L. and Rassweiler, G.M. (1933), "Absorption Spectra of Gaseous Charges in a Gasoline Engine," *Ind. Eng. Chem.*, Vol. 25, pp. 923–931.

Withrow, L. and Rassweiler, G.M. (1934), "Formaldehyde Formation in Preflame Reactions in an Engine," *Ind. Eng. Chem.*, Vol. 26, pp. 1256–1262.

Appendix A

Experimental Error Analysis

Outline

A.1 Introduction

It is a fact of life that when we repeat measurements in the laboratory of what we may believe to be a fixed quantity, we often get several answers. For example, consider an experiment that involves dropping a coin vertically to the floor from a height of 1.2 m, and measuring the free-fall time by means of a digital stopwatch. When this experiment was repeated 84 times, a list of free-fall times was obtained as shown in Table 1. It can be seen from this table that there was considerable variation in the free-fall times, which ranged from 0.38 to 0.60 s. The average free-fall time for the 84 tests was 0.496 s.

A number of important questions arise from this experiment, including: What causes the test-to-test variation in the coin free-fall time? How close is the average time of 0.496 s to the true free-fall time? What is the true free-fall time? Does the number of repeated tests affect the accuracy of the average time? Is the accuracy of the average time affected by the large range of measured free-fall times (0.39 to 0.60 s)?

Here, experimental error will be defined as the difference between a measured value and the true value of a parameter. The true value of a parameter is defined as the average value obtained when an experiment is repeated an infinite number of times.

The following sections discuss the causes of experimental error, how these errors can be quantified, and how they can be reduced.

A.2 Types of Experimental Error

A.2.1 Systematic Errors

Much of the skill and experience of an experimenter is often utilized in eliminating systematic errors. An example of a systematic error is illustrated by the coin free-fall experiment discussed above. The average free-fall time for the 84 repeated tests shown in Table 1 is 0.496 s. Suppose now that another experimenter is also asked to measure the free-fall time of the same coin over the same vertical distance of 1.2 m. However, through carelessness, the second experimenter marks the height above the floor at 1.1 rather than 1.2 m. Not surprisingly, he finds that his average free-fall time is somewhat shorter at 0.476 s. Clearly, this second average value suffered a systematic error of –0.02 s, which affected all tests results.

Consider now another example of systematic error. Suppose that an electronic digital thermometer is used to measure the melting temperature of pure ice in water, and the reading is 1°C. The thermometer is then used to measure the boiling temperature of pure water, and the reading is 101°C. Clearly, this thermometer is suffering from a systematic offset error of +1°C. That is, the thermometer reading is always too high by 1°C throughout the range 0 to 100°C.

Another type of systematic error is illustrated by the following example. A different thermometer than the one discussed above is immersed in a mixture of melting pure ice and water, and the reading is correctly shown as 0°C. The thermometer is then used to measure the temperature of boiling pure water, and the reading is erroneously shown as 101°C. This thermometer is suffering from a systematic **gain** error of +1 per 100°C. That is, the thermometer reading is progressively higher between 0 and 100°C: At 0°C the thermometer reads correctly; at 50°C it reads 0.5°C too high; and at 100°C it reads 1°C too high.

Systematic errors can be a serious problem in experimental work, because an experimenter can perform an experiment without being aware that the errors exist. A systematic error cannot be eliminated by repeating the experiment a very large number of times and averaging the results. In the case of the electronic thermometer with the +1°C offset error, this error could go undetected for a considerable period of time, unless the experimenter frequently checked the thermometer calibration. In the case of the coin free-fall experiment, the error of −0.1 m in the height above the floor could go undetected, unless the experimenter re-traced and meticulously checked all of the steps in his experiment, or he calculated the theoretical free-fall time. In most experimental work, the true value of the parameter being measured will not be known. Therefore, conscious effort is needed in order to detect and eliminate systematic errors.

A.2.1.1 The Effect of Systematic Errors on "True" Value

In the preceding section, the true value of a parameter was defined as the average value obtained when an experiment is repeated an infinite number of times. This is a "statistical" definition of the true free-fall time. However, this definition is significantly defective when systematic errors are present, which cannot be eliminated by repeating the experiment a large number of times. Nevertheless, experimenters often have to resort to this statistical definition because the true value of a parameter is not normally known.

A.2.2 Random Errors

Random errors arise from unpredictable test-to-test variability in the measurements. If the same variable is measured repeatedly several times, random errors will cause a spread or scatter in the measurements. In the example of the coin free-fall experiment above, the measured free-fall time varies significantly and unpredictably from test to test. Reasons for this variability might include: the stopwatch not being started precisely when the coin is released; likewise, the stopwatch not being stopped precisely when the coin reaches the floor; small variations in the height from which the coin is released. The stopwatch used for this experiment displayed digitally the time elapsed down to 1/100 s, and this introduced an error of up to −0.01 s, for the following reason. The stopwatch was designed to increment the time displayed from, say, 0.48 s to 0.49 s when an additional whole period of 0.01 s had elapsed. The display was not incremented if only a part of this 0.01 s period had elapsed (e.g., 0.006 s), giving rise to a maximum error or underestimate of the coin free-fall time of −0.01 s.

In general, random errors are more easily identified and dealt with than systematic errors. Although random errors cannot usually be eliminated altogether, they can be corrected for, by making repeated measurements of the same variable, followed by statistical analysis of the measurements.

A.2.2.1 Frequency Distribution

Repeated measurements of a variable, such as the coin free-fall time, usually show that the measurements are distributed evenly around the average value of these measurements. Figure A.1 shows, for example, the frequency distribution for the 84 repeated measurements of the coin free-fall time. Each bar in the figure corresponds to a range of free-fall times. For example, the second bar corresponds to the free-fall time range of 0.401 to 0.420 s. The height of the bar corresponds to the 5 measurements of the free-fall time that happened to be within this range of 0.401 to 0.420 s.

The average free-fall time for the 84 repeated tests was found to be 0.496 s. It can be seen from Figure 1 that free-fall times close to this average occurred more frequently. For example, 12 tests had free-fall times within the range 0.461 to 0.480 s, while only 2 tests had free-fall times within the range 0.381 to 0.400 s. In general, test results that are subject to random error tend to give a bell-shaped distribution, such as the one shown in Fig. A.1. Usually, the bell shape becomes more well formed and smooth when the number of repeated tests is increased. This is illustrated in Fig. A.2, which shows the frequency distributions for progressively increasing number of repeated coin free-fall tests. The shape of the distribution for the first 24 tests is irregular. However, as the number of repeated tests is progressively increased from 24 to 48, and finally to 84, the corresponding distribution shapes become increasingly more regular. Ultimately, if the coin free-fall experiment was repeated a very large (infinite) number of times, then the frequency distribution would become bell-shaped and very regular, as illustrated by the broken-line curve in Fig. A.2. Often, the actual distribution, obtained from a finite number of tests, is approximated by a theoretical distribution such as the one shown by the broken-line curve in Fig. A.2. If the coin free-fall test was repeated a very large number of times, then the actual frequency distribution would, indeed, become very similar to the theoretical curve shown by the broken line. The question is, why would one wish to approximate an actual frequency distribution with a theoretical distribution? In practice, this has major advantages, because statistical analysis can be applied readily to the theoretical distribution. Over the years, a well-established theory of experimental errors has been developed which is founded on such theoretical distributions. There are several theoretical distributions which can be used to approximate the real distributions obtained from actual test data. These include the Gaussian or normal distribution (shown in Fig. A.2), the binomial, and the Poisson distribution. Often, in practice, the normal (Gaussian) distribution is used. For more details on distributions and the theory of errors, the reader can consult specialized texts on experimental methods and measurements, such as Barford [1985] and Kirkup [1994].

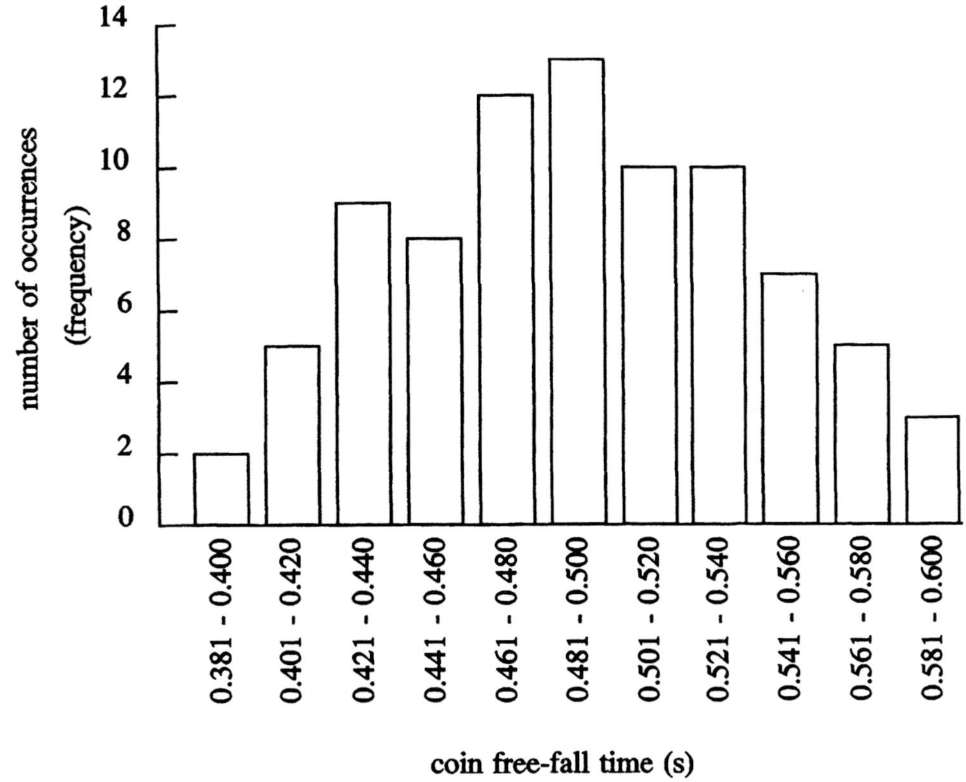

Fig. A.1 Distribution of coin free-fall times for 84 repeated tests.

A.3 Mean Value

The average value of a group of measurements is called the mean value. Suppose that five measurements of the same variable are made, x_1, x_2, x_3, x_4, and x_5. The mean value, x_{mean} is given by:

$$x_{mean} = \frac{x_1 + x_2 + x_3 + x_4 + x_5}{5}$$

or,

$$x_{mean} = \frac{\sum_{i=1}^{i=n} x_i}{n} \tag{A.1}$$

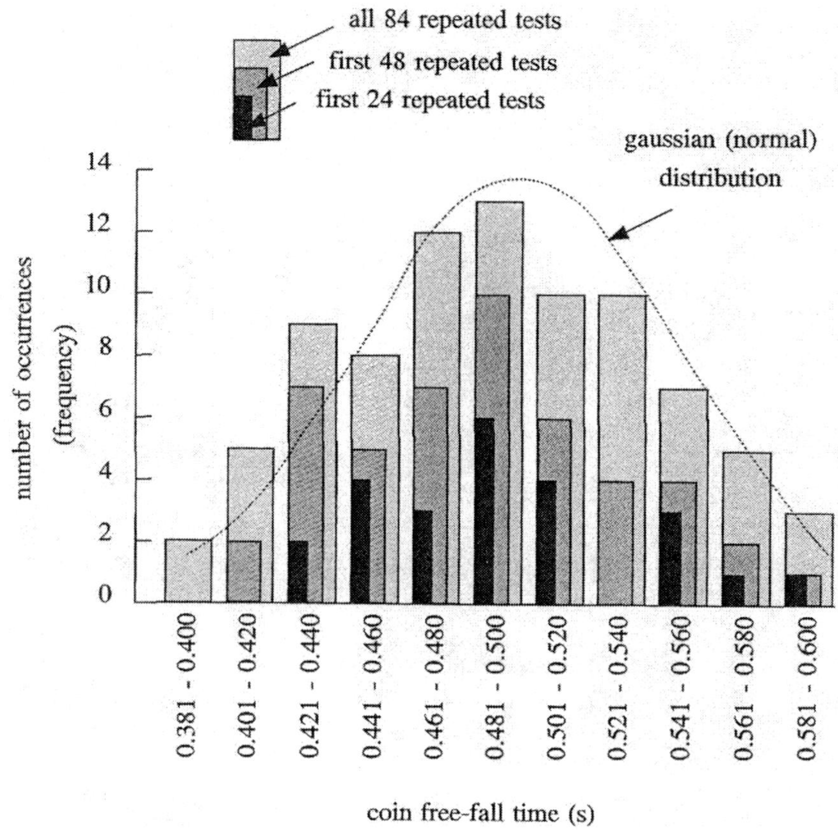

Fig. A.2 Effects of number of repeated tests on distribution.

where x_{mean} = mean value
i = ith measurement (e.g., 1st or 5th repeated measurement)
n = total number of repeated measurements made

In the case of the coin free-fall experiment, the mean value of the 84 repeated tests is 0.496 s. It is intuitively obvious that the bigger the number of repeated tests used, the more "representative" will be the calculated mean value. In other words, the mean value calculated from a large number of repeated tests will be closer to the true value. For example, in the case of the coin free-fall experiment, the very first test showed a free-fall time of 0.56 s; after 6 repeated tests, the mean free-fall time was 0.518 s, after 12 tests it was 0.508 s, after 48 tests it was 0.492 s, and after 84 tests it was 0.496 s. The last mean value (0.496 s) is likely to be closer to the true value of the free-fall time than either the first measurement (0.56 s) or the mean value of the first 6 tests (0.518 s). The following section will examine the accuracy of estimated mean values.

A.4 Estimation of Errors

A.4.1 Error in the Mean Value

In the coin free-fall experiment, the mean value of the free-fall time was measured as 0.496 s after 84 repeated tests. An important question now is, how close is this value to the true free-fall time? This question can be answered by calculating the error in the mean value of 0.496 s. The measured mean value can then be quoted as $0.496 \pm S_m$s, where S_m denotes the error in the mean value. This means that 0.496 s is, most probably, not further than $\pm S_m$s from the true value of the free-fall time (the reader is reminded that true value is defined as the average free-fall time from an infinite number of coin free-fall tests).

The probable error in the mean value, S_m, can be calculate from the following expression:

$$S_m = \frac{s}{\sqrt{n-1}} \tag{A.2}$$

where

$$s = \sqrt{\frac{\sum_{i=1}^{i=n} (x_i - x_{mean})^2}{n-1}}$$

This error is usually called the standard error of the mean. It will be interesting to calculate the standard error for mean coin free-fall time. Using n = 84, S_m is calculated to be 0.0055 s. Therefore, the mean value of 0.496 s is subject to an error (or uncertainty) of ±0.0055 s. Hence, the true value of the free-fall time will, most probably, be within the range 0.496 ± 0.0055 s, or within the range 0.4905 to 0.5015 s. The above expression for S_m can be used to estimate the standard error of the mean value when the coin free-fall experiment is repeated for fewer than 84 times. For example:

For the first 6 tests:	mean value = 0.518 ± 0.0229 s	(0.518 ± 4.4%)
For the first 12 tests:	mean value = 0.508 ± 0.0157 s	(0.508 ± 3.1%)
For the first 48 tests:	mean value = 0.492 ± 0.0066 s	(0.492 ± 1.3%)
For the first 84 tests:	mean value = 0.496 ± 0.0055 s	(0.496 ± 1.1%)

It is clear from the above list of results that the error in the mean value becomes smaller as the number of repeated tests increases. This is consistent with our intuition and experience. We can state our conclusion from the above results in a number of different ways: We are more confident that 0.496 s is closes to the true free-fall time than 0.518 s is; or, simply, we feel that 0.496 s is a more accurate estimate of the free-fall time than 0.518 s is.

In the above expression for S_m, the parameter s is the sample standard deviation. The following section will consider the nature of this parameter.

A.4.2 Variability in Measurements

Consider again the results of the coin free-fall experiment shown in Fig.A.1 and listed in Table A.1. It can be seen that the free-fall times for individual tests vary considerably. For example, the free-fall time for the 8th test was 0.60 s, which is 21% greater than the mean value of 0.496 s. In the case of the 77th test, the free-fall time was 0.38 s, which is 23% smaller than the mean value. It so happens that the coin free-fall experiment suffered from a large test-to-test variability. In other words, for a significant number of the 84 repeated tests, the free-fall time was substantially different from the mean value of 0.496 s. In general, the variability for some laboratory experiments may be large, while for others it may be small. Some types of experiment have no variability at all. A rather trivial example of this is the counting of the coins in someone's pocket. If we repeat this experiment several times, it is unlikely that there will be any variability in the result at all—unless a counting or recording error is made!

Table A.1 Flight time for coin to fall vertically 1.2 meters

Test No.	1	2	3	4	5	6	7	8	9	10	11	12
Free-fall Time (s)	**0.56**	**0.49**	**0.57**	**0.48**	**0.56**	**0.45**	**0.49**	**0.60**	**0.45**	**0.48**	**0.45**	**0.51**
Test No.	13	14	15	16	17	18	19	20	21	22	23	24
Free-fall Time (s)	**0.49**	**0.51**	**0.48**	**0.50**	**0.51**	**0.45**	**0.51**	**0.56**	**0.49**	**0.49**	**0.43**	**0.44**
Test No.	25	26	27	28	29	30	32	32	33	34	35	36
Free-fall Time (s)	**0.53**	**0.43**	**0.58**	**0.49**	**0.52**	**0.44**	**0.49**	**0.50**	**0.43**	**0.51**	**0.50**	**0.42**
Test No.	37	38	39	40	41	42	43	44	45	46	47	48
Free-fall Time (s)	**0.44**	**0.42**	**0.54**	**0.54**	**0.44**	**0.47**	**0.46**	**0.47**	**0.54**	**0.47**	**0.55**	**0.48**
Test No.	49	50	51	52	53	54	55	56	57	58	59	60
Free-fall Time (s)	**0.54**	**0.46**	**0.51**	**0.49**	**0.44**	**0.42**	**0.46**	**0.48**	**0.42**	**0.52**	**0.57**	**0.51**
Test No.	61	62	63	64	65	66	67	68	69	70	71	72
Free-fall Time (s)	**0.53**	**0.57**	**0.39**	**0.42**	**0.54**	**0.48**	**0.56**	**0.54**	**0.57**	**0.52**	**0.54**	**0.49**
Test No.	73	74	75	76	77	78	79	80	81	82	83	84
Free-fall Time (s)	**0.55**	**0.56**	**0.60**	**0.48**	**0.38**	**0.50**	**0.47**	**0.43**	**0.48**	**0.59**	**0.46**	**0.54**

A.4.2.1 Standard Deviation

In general, it is helpful to have a way of expressing the variability in a set of test results. If the number of test results is very large, then this is achieved by using a parameter called standard deviation, which is defined as follows:

$$\sigma = \sqrt{\frac{(x_1 - x_{mean})^2 + (x_2 - x_{mean})^2 + ... + (x_n - x_{mean})^2}{n}}$$

$$\text{i.e.,} \qquad \sigma = \sqrt{\frac{\sum_{i=1}^{i=n} (x_i - x_{mean})^2}{n}} \qquad \text{(A.3)}$$

Often, the number of times an experiment can be repeated is quite small due to, for example, cost and time constraints. When the number of repeated tests is small, a more appropriate indication of variability is the adjusted root mean square deviation, given by:

$$s = \sqrt{\frac{\sum_{i=1}^{i=n} (x_i - x_{mean})^2}{n - 1}} \qquad \text{(A.4)}$$

Evidently, as the number of repeated tests, n, becomes large, s becomes equal to σ. For example, in the case of the 84 coin free-fall tests:

$$\sigma = \pm 0.0497 \text{ and } s = \pm 0.0500$$

The value of s is very nearly the same as that of σ. Often, σ is called the population standard deviation while s is called the sample standard deviation. Here, the term sample standard deviation will be used for s, in preference to the term adjusted root mean square deviation, simply because it is the term most commonly used. The sample standard deviation is a measure of the scatter or the spread in repeated test results. A high sample standard deviation indicates that the results show a large spread. When the standard deviation is large, the bell-shaped frequency distribution shown in Fig. A.1 has a wide base compared with its height. Conversely, when the sample standard deviation is small, the distribution is narrow, making its shape appear to be tall and narrow. Unlike the standard error of the mean (S_m), the sample standard deviation for an experiment is not greatly affected by the number of repeated tests. For example, in the coin free-fall experiment, the sample standard deviation was:

For the first 6 tests:	s = ±0.0512
For the first 12 tests:	s = ±0.0522
For the first 48 tests:	s = ±0.0453
For the first 84 tests:	s = ±0.0500

This result is consistent with what we might expect intuitively. That is, if repeated test results show large spread or scatter, then repeating the test many more times does not usually reduce the spread. The increased number of tests merely allows the spread to be established more accurately, much in the same way that the mean becomes more accurate when the number of repeated tests is large.

One way of assessing whether the sample standard deviation is large or not, is to compare it with the value of the mean. This is done using the coefficient of variation, as follows:

$$\text{coefficient of variation} = \frac{s}{x_{mean}} \times 100\,\% \qquad \text{(A.5)}$$

A large value of the coefficient of variation will obviously imply a large test-to-test variability. In the case of the coin free-fall experiment, the mean for the 84 tests is 0.496 s and the sample standard deviation is ±0.05 s. The value of the coefficient of variation is 10.1%.

Let us now consider a more informative way of utilizing the sample standard deviation. Consider, again, the coin free-fall experiment. The sample standard deviation for the 84 tests is ±0.05 s and the mean is 0.496 s. Let us assume, for a moment, that the actual frequency distribution shown in Fig. A.1 can be approximated by a theoretical normal (Gaussian) distribution, as shown in Fig. A.2. Statistical analysis of the normal distribution shows that, in theory, we can expect:

- 68% of the 84 free-fall times to be within $x_{mean} \pm s$; that is, within 0.496 ± 0.05 s (0.446 to 0.546 s)
- 95% of the 84 free-fall times to be within $x_{mean} \pm 2s$; that is, within 0.496 ± 0.1 s (0.396 to 0.596 s)
- 99.7% of the 84 free-fall times to be within $x_{mean} \pm 3s$; that is, within 0.496 ± 0.15 s (0.346 to 0.646 s)

We can check whether these theoretically expected percentages agree with the actual ones. Using the results listed in Table 1, we can see that 53 of the 84 free-fall times (i.e., 63%) lie within the range 0.446 to 0.546 s; we can also see that 81 of the 84 times (i.e., 96%) lie within the range 0.396 to 0.596 s; finally, we can see that all 84 free-fall times (i.e., 100%) lie within the range 0.346 to 0.646 s. These results are quite similar to the theoretically expected percentages listed above, because the actual and normal distributions shown in figure two are similar.

A.4.2.2 Error in an Individual Measurement

The sample standard deviation can provide a useful indication of the error likely to be involved in making an individual observation. Consider, again, the 84 coin free-fall tests. The mean free-fall time is 0.496 s and the standard deviation is 0.05 s. If we now repeat the experiment for an 85th time, there is a 68% chance that the free-fall time we get will be within 0.496 ± 0.05 s. We can say the same thing in a number of different ways, as follows:

- There is 68% chance that the time obtained will not be further than ±0.05 s from the mean value of the 84 tests.
- There is 68% chance that the error in the 85th test time will not be greater than ±0.05 s.
- We are 68% confident that the error will not be greater than ±0.05 s; similarly, we can be 95 % confident that the error in the 85th test time will not be greater than ±0.1 s (2 × 0.05 s); finally, we can be very nearly 100% confident that the error will not be greater than ±0.15 s (3 × 0.05 s).

A.5 Indicating Experimental Errors

Usually, it is advisable for an experimenter to repeat tests, and then calculate the sample standard deviation as well as the standard error of the mean. Similarly, when the display of an instrument is fluctuating, the experimenter should read the display several times and then calculate the sample standard deviation and the standard error of the mean. The resulting values should be quoted in the text of a report or shown graphically when the results are plotted. Fig. A.3 shows how error limits can be used to indicate experimental errors on bar charts and graphs. Fig. A.3(a) and Fig. A.3(b) show two alternative error limits. In Fig. A.3(a), the error limits are for one sample standard deviation (±s) either side of the mean value. In Fig. A.3(b), the error limits correspond to the standard error of the mean ($\pm S_m$). The error limits in Fig. A.3(a) inform the reader about the test-to-test variability or the extend of fluctuation in an instrument reading. On the other hand, the limits in Fig. A.3(b) inform the reader of the likely error involved in the mean value. Of the two types of error limits, the second is often the more appropriate one. Fig. A.3(c) shows a graph with error bars to indicate the two error limits at each data point.

A.5.1 Elimination of Outliers

Data points that lie far off a trend or a cluster of repeated measurements are called "outliers." An example of an outlier is shown in Fig. A.3(d). This figure shows seven repeated measurements. While six of the points lie close to each other in a fairly compact cluster, the seventh point lies distinctly outside this cluster.

Sometimes, an outlier may be due to an intermittent fault in the instrument, or due to a mistake made in reading or recording the value of a measurement. More rarely, an outlier may be caused by an unexpected change in the conditions of the experiment, which may be of special interest to the experimenter. An outlier should be eliminated only if it is known than the instrument was faulty or the observer made a mistake.

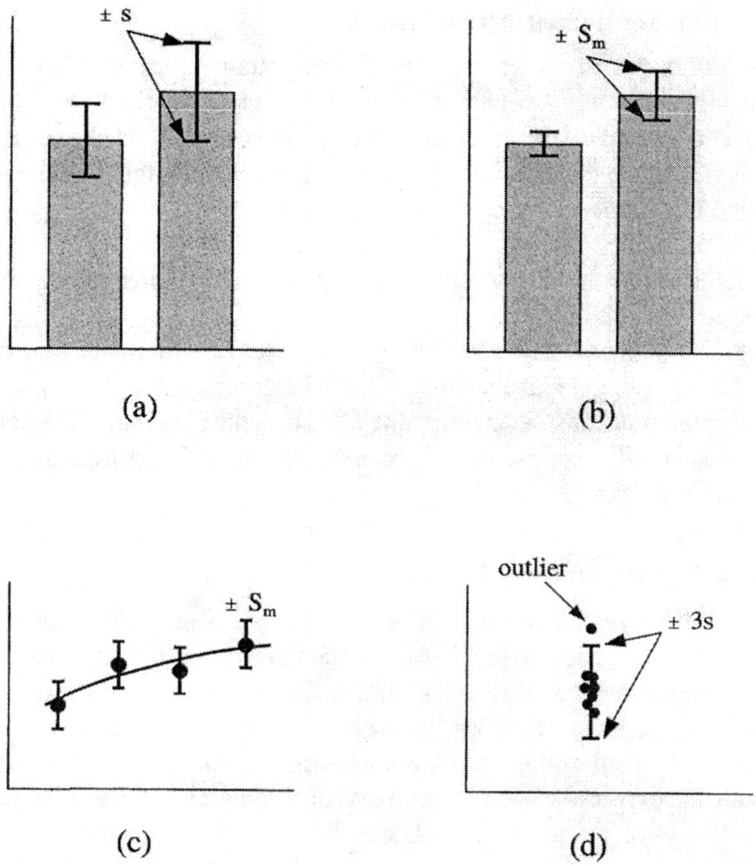

Fig. A.3 Ways of indicating experimental errors.

A more rigorous way of eliminating an outlier is to calculate the mean value (x_{mean}) of a set of data obtained from repeated tests, as well as the sample standard deviation(s). Then, any individual reading that is outside the range $x_{mean} \pm 3s$ would be a candidate for elimination. This outlier could be eliminated if there are good reasons to believe that an instrument fault occurred, or a mistake was made in observing or recording the data, or control of the experiment was temporarily lost when an outlying reading was taken. The logic behind the $x_{mean} \pm 3s$ criterion is as follows. In a group of data, obtained from repeated tests, very nearly all of the values (99.7%) should lie within $x_{mean} \pm 3s$, provided that the data are normally distributed about the mean.

A.6 Combination of Errors

In many engine tests, variables such as engine speed, brake torque, exhaust gas temperature, fuel consumption rate, and air flow rate are measured. These variables are then combined in order to calculate other parameters, such as engine power output and specific fuel consumption.

This section will deal with the estimation of the errors in the calculated parameters.

The problem at hand can be illustrated by the following simple example. Suppose we wish to know the perimeter of a room. This is calculated from the expression:

$$P = W_m + X_m + Y_m + Z_m \tag{A.6}$$

where P is the perimeter of the room and W_m, X_m, Y_m, and Z_m are the measured mean lengths of the four room walls. In this case, the estimation of the room perimeter requires four separate sets of measurements. Each set requires measuring, repeatedly, the length of a wall (e.g., W_1, W_2, W_3, etc., for the first wall). From these repeated measurements, the mean length for each wall can be calculated (e.g., W_m, X_m, etc.) as well as the standard error of each mean length (e.g., $S_{m,W}$ in the case of the first wall, $S_{m,X}$ for the second wall, etc.). The estimated perimeter must also be subject to error (say, $S_{m,P}$). An important question is, how can the errors for the four mean lengths ($S_{m,W}$, $S_{m,X}$, $S_{m,W}$, and $S_{m,X}$) be used to obtain the error ($S_{m,P}$) in the estimated value of the room perimeter? The answer to this question is given in the following section.

A.6.1 Error in a Sum

Let,

X_m	=	mean value of parameter X calculated from repeated readings $X_1, X_2, X_3, \ldots$ etc.
Y_m	=	mean value of parameter Y calculated from repeated readings $Y_1, Y_2, Y_3, \ldots$ etc.
Z	=	estimated parameter using the mean values, X_m, Y_m, ... etc.
$S_{m,X}$	=	standard error of the mean X_m
$S_{m,Y}$	=	standard error of the mean Y_m
$S_{m,Z}$	=	standard error of estimated parameter Z

If $Z = X_m + Y_m + \ldots$

then $$S_{m,Z} = \sqrt{S_{m,X}^2 + S_{m,Y}^2 + \ldots} \tag{A.7}$$

It should be noted that the above expression for $S_{m,Z}$ remains the same when the sum has a constant term (say, k) as for example, $Z = k + X_m + Y_m + \ldots$

Likewise, the above expression for the estimation of $S_{m,Z}$ also remains the same when one or more of the terms are negative, as for example, $Z = X_m - Y_m + \ldots$

A.6.1.1 Example

A gasket is to be made of cork stuck onto a metal backing. It is necessary to predict the overall thickness of the gasket. The thickness of the cork sheet is repeatedly measured, and from these measurements the mean thickness and its standard error are calculated as 4 ± 0.02 mm. These measurements are then repeated for the metal backing and the mean thickness and its standard error are calculated as 0.2 ± 0.01 mm. The overall thickness of the gasket can now be predicted (ignoring the effects of the glue):

Estimated mean value of the overall thickness = 4 + 0.2 = 4.2 mm
Estimated standard error of overall mean thickness = $[0.02^2 + 0.01^2]^{0.5}$ = 0.022 mm

Therefore, the gasket is predicted to have a thickness of 4.2 ± 0.022 mm.

Consider a second example involving two temperature measurements. The first, 95 ± 0.5°C, is the coolant temperature at the exit of an engine; the second, 79 ± 0.4°C, is the corresponding coolant temperature at the entry to the engine. The temperature difference is:

$16 \pm [0.5^2 + 0.4^2]^{0.5} = 16 \pm 0.64°C$

A.6.2 Error in a Linear Combination

If $\quad Z = aX_m + bY_m + \ldots$

$$\text{then} \qquad S_{m,Z} = \sqrt{a^2 S_{m,X}^2 + b^2 S_{m,Y}^2 + \ldots}$$

It should be noted that the above expression for $S_{m,Z}$ remains the same when a constant term is involved (e.g., $Z = k + aX_m + bY_m + \ldots$). The expression for $S_{m,Z}$ also remains the same (i.e., the error terms are still added) when one term is negative (e.g., $Z = k + aX_m - bY_m + \ldots$).

A.6.2.1 Example

A rectangular room has length, L = 3 ± 0.005 m, and width, W = 2 ± 0.004 m. Calculate the perimeter of the room.

The mean value of the perimeter is, P = 2L + 2W = 10 m.
The standard error of the mean perimeter is, $[2^2 \times 0.005^2 + 2^2 \times 0.004^2]^{0.5} = 0.013$ m.

Therefore, the estimated value for the room perimeter is 10 ± 0.013 m.

A.6.3 Error in Scaling a Parameter

If $Z = aX_m$

$$\text{then} \quad S_{m,Z} = \sqrt{a^2 S_{m,X}^2} = aS_{m,X} \tag{A.8}$$

A.6.4 Error in a Product

If $Z = X_m Y_m \cdots$

$$\text{then} \quad \frac{S_{m,Z}}{Z} = \sqrt{\frac{S_{m,X}^2}{X_m^2} + \frac{S_{m,Y}^2}{Y_m^2} + \ldots} \tag{A.9}$$

A.6.4.1 Example

A rectangular room has length, $L = 3 \pm 0.005$ m, and width, $W = 2 \pm 0.004$ m. Calculate the floor area of the room.

The mean value of the area is, $A = L \times W = 6\ m^2$.
The **relative** standard error of the mean area is, $[0.005^2/3^2 + 0.004^2/2^2]^{0.5} = 0.0026$. Thus, the standard error of the area is 0.26% of the mean value of 6 m^2 or 0.016 m^2.

Therefore, the floor area of this room is estimated to be 6 m ± 0.26 % or 6 ± 0.016 m^2.

A.6.5 Error in the Sum of Powers

If $Z = X_m^u + Y_m^v + \ldots$

$$\text{then} \quad S_{m,Z} = \sqrt{\left[uX_m^{(u-1)}\right]^2 S_{m,X}^2 + \left[vY_m^{(v-1)}\right]^2 S_{m,Y}^2 + \ldots} \tag{A.10}$$

A square room has side, $L = 3 \pm 0.005$ m. Calculate the floor area of the room. The above expressions for sum of powers can be used with only the first term of the expressions taken into account, that is, $A = L^2$.

The estimated floor area is, $A = L^2 = 9\ m^2$.
The standard error of the estimated area is, $[(2 \times 3^1)^2\ 0.005^2]^{0.5} = 0.03\ m^2$.

Therefore, the floor area is estimated to be 9 ± 0.03 m^2.

A.6.6 Error in the Product of Powers

If $Z = X_m^u Y_m^v \cdots$

$$\text{then} \quad \frac{S_{m,Z}}{Z} = \sqrt{u^2 \frac{S_{m,X}^2}{X_m^2} + v^2 \frac{S_{m,Y}^2}{Y_m^2} + \ldots} \tag{A.11}$$

As another example, consider the estimation of the specific fuel consumption (sfc) for an internal combustion engine. In general,

$$sfc = \frac{\dot{m}_f}{\dot{W}} = \frac{\rho_f \dot{V}_f}{\dot{W}}$$

where $\dot{m}_f$ = fuel mass flow rate; $\dot{W}$= engine brake power.

Assuming that the various parameters have been measured as:

fuel density = 0.81 ± 0.015 g/cm^3
fuel volumetric rate = 8521 ± 82 cm^3/hr
brake power = 20 ± 0.3 kW

Rewriting the above expression for sfc as,

$$sfc = \rho_f^1 \dot{V}_f^1 \dot{W}^{-1}$$

the estimated value of the sfc is 345.1 g/kWhr.

Applying the error expression for the product of powers,

$$S_{m,sfc} = 345.1\sqrt{1^2 \frac{0.015^2}{0.81^2} + 1^2 \frac{82^2}{8521^2} + (-1)^2 \frac{0.3^2}{20^2}} = 8.89 \text{ g/kWhr}$$

Therefore, the sfc for this engine is estimated as 345.1 ± 8.89 g/kWhr or 345.1 ± 2.6%.

It is worth noting here that the error in sfc, $S_{m,sfc}$, is dominated by the largest percentage error in its individual component parts. Here, the individual percentage errors in density, volumetric flow rate, and brake power are of similar magnitude (1.9%, 1%, and 1.5% respectively). Had one of the three errors been much greater than the other two, then this error would dominate $S_{m,sfc}$. The experimenter would then be better off concentrating on reducing this dominant error rather than all three.

A.6.7 Error in a Logarithmic Function

If $Z = \log X_m$

$$\text{then} \qquad S_{m,Z} = \frac{S_{m,X}}{X_m} \tag{A.12}$$

A.6.8 Error in a General Function of Many Variables

The combination of errors can be generalized for any function of one or more variables by means of the following expression (Barford, 1985):

If $Z = f\left(X_m, Y_m, \ldots\right.$

$$\text{then} \qquad S_{m,Z} = \sqrt{\left[f'(X_m)\right]^2\left[S_{m,X}\right]^2 + \left[f'(Y_m)\right]^2\left[S_{m,Y}\right]^2 + \ldots} \tag{A.13}$$

where

$$f'(X_m) = \frac{\partial f(X_m, Y_m)}{\partial X_m} = \frac{\partial Z}{\partial X_m}$$

$$f'(Y_m) = \frac{\partial f(X_m, Y_m)}{\partial Y_m} = \frac{\partial Z}{\partial Y_m}$$

i.e., $f'(X_m)$ is the derivative of $f(X_m,Y_m)$ evaluated at X_m. Similarly, $f'(Y_m)$ is the derivative of $f(X_m,Y_m)$ evaluated at Y_m.

It should be noted that the variables X_m, Y_m, ..., in the function $Z = f(X_m, Y_m, \ldots)$ should be independent of one another.

A.6.8.1 Examples

This general expression can be used to obtain the particular cases listed in the preceding sections. As a first example, consider the case of a product:

$$Z = kX_m Y_m$$

where, k is constant. The partial derivatives for the above expression are:

$$f'(X_m) = \frac{\partial Z}{\partial X_m} = kY_m$$

$$f'(Y_m) = \frac{\partial Z}{\partial Y_m} = kX_m$$

substituting the partial derivatives in the expression,

$$S_{m,Z} = \sqrt{[f'(X_m)]^2 [S_{m,X}]^2 + [f'(Y_m)]^2 [S_{m,Y}]^2 + \ldots}$$

gives:

$$S_{m,Z} = \sqrt{[kY_m]^2 [S_{m,X}]^2 + [kX_m]^2 [S_{m,Y}]^2 + \ldots}$$

Dividing both sides by Z,

$$\frac{S_{m,Z}}{Z} = \sqrt{\frac{[kY_m]^2 [S_{m,X}]^2}{[kX_m Y_m]^2} + \frac{[kX_m]^2 [S_{m,Y}]^2}{[kX_m Y_m]^2} + \ldots}$$

Simplifying, gives the final result:

$$\frac{S_{m,Z}}{Z}=\sqrt{\frac{S_{m,X}^2}{X_m^2}+\frac{S_{m,Y}^2}{Y_m^2}+\ldots}$$

As a second example, consider the case of the product of powers:

$$Z=k$$

where, k is constant. The partial derivatives for the above expression are:

$$f'(X_m)=\frac{\partial Z}{\partial X_m}=kuX_m^{u-1}Y_m^v$$

$$f'(Y_m)=\frac{\partial Z}{\partial Y_m}=kX_m^u vY_m^{v-1}$$

Substituting the partial derivatives in the expression,

$$S_{m,Z}=\sqrt{[f'(X_m)]^2[S_{m,X}]^2+[f'(Y_m)]^2[S_{m,Y}]^2+\ldots}$$

gives:

$$S_{m,Z}=\sqrt{\left[kuX_m^{u-1}Y_m^v\right]^2\left[S_{m,X}\right]^2+\left[kX_m^u vY_m^{v-1}\right]^2\left[S_{m,Y}\right]^2+\ldots}$$

Dividing both sides by Z,

$$\frac{S_{m,Z}}{Z}=\sqrt{\frac{\left[kuX_m^{u-1}Y_m^v\right]^2\left[S_{m,X}\right]^2}{\left[kX_m^u Y_m^v\right]^2}+\frac{\left[kX_m^u vY_m^{v-1}\right]^2\left[S_{m,Y}\right]^2}{\left[kX_m^u Y_m^v\right]^2}+\ldots}$$

Simplifying gives the final result:

$$\frac{S_{m,Z}}{Z}=\sqrt{u^2\frac{S_{m,X}^2}{X_m^2}+v^2\frac{S_{m,Y}^2}{Y_m^2}+\ldots}$$

As a third example, consider the following expression:

$$Z = X_m Y_m - P_m Q_m$$

It is not necessary to use the generalized formula in this case, since we can rely on the expressions given in previous sections.

The first term is a product, and the combined standard error for the first term is:

$$S_{m,XY} = X_m Y_m \sqrt{\frac{S_{m,X}^2}{X_m^2} + \frac{S_{m,Y}^2}{Y_m^2}}$$

Similarly, the second term is a product, and the combined error of the second term can be obtained in a similar way.

$$S_{m,PQ} = P_m Q_m \sqrt{\frac{S_{m,P}^2}{P_m^2} + \frac{S_{m,Q}^2}{Q_m^2}}$$

Having obtained the standard errors for the first and second terms, these two standard errors can be combined using the expression for the error of a difference given in a previous section.

$$S_{m,Z} = \sqrt{S_{m,XY}^2 + S_{m,PQ}^2}$$

$$S_{m,Z} = \sqrt{Y_m^2 S_{m,X}^2 + X_m^2 S_{m,Y}^2 + Q_m^2 S_{m,P}^2 + P_m^2 S_{m,Q}^2}$$

We can confirm this result by using the generalized formula. The four partial derivatives are:

$$f'(X_m) = \frac{\partial Z}{\partial X_m} = Y_m \qquad f'(Y_m) = \frac{\partial Z}{\partial Y_m} = X_m$$

$$f'(P_m) = \frac{\partial Z}{\partial P_m} = Q_m \qquad f'(Q_m) = \frac{\partial Z}{\partial Q_m} = P_m$$

When inserted into the generalized formula for $S_{m,Z}$ the result is:

$$S_{m,Z} = \sqrt{Y_m^2 S_{m,X}^2 + X_m^2 S_{m,Y}^2 + Q_m^2 S_{m,P}^2 + P_m^2 S_{m,Q}^2}$$

which is the same expression obtained above by using the expressions for products and differences.

References

Barford, N.C. (1985), *Experimental Measurements: Precision, Error, and Truth*, John Wiley and Sons, 2nd Edition, Chichester.

Kirkup, L. (1994), *Experimental Methods: An Introduction to the Analysis and Presentation of Data*, John Wiley and Sons, Brisbane.

Appendix B

Gas Flow through a Sampling Valve

Outline

For most sampling applications from the engine cylinder, the flow through the valve is likely to be turbulent, compressible, close to choking or choking, and approximately one-dimensional. A departure from these conditions might occur when the valve lift is very small or the cylinder pressure is very low, when laminar flow may prevail. Then, in principle, the flow ceases to be one-dimensional. The following two sections deal with the prediction of the mass sample flow rate through the valve. First, in section one, one-dimensional compressible flow is considered. Then, in section two, laminar flow is considered.

B.1 Turbulent Choked Flow

B.1.1 Compressible Flow Analysis

The mass flow rate through a sampling valve when the flow is one-dimensional, steady, adiabatic, and compressible can be predicted using compressible flow analysis through a restriction, without fluid frictional effects. Fluid friction is then taken into account by means of an empirically determined coefficient of discharge, C_D. The analysis is as follows:

Let,

C_D = discharge coefficient accounting for fluid frictional and inertial effects (C_D = actual mass flow/ideal mass flow)
A_R = geometric flow area at the flow restriction
p_c = upstream stagnation pressure (cylinder gas pressure)
T_c = upstream stagnation temperature (cylinder gas pressure)
p_T = static pressure at the restriction; p_T is assumed to be equal to pressure at the sampling valve outlet

R = specific gas constant
γ = ratio of specific heat capacities

then,

$$\dot{m} = \frac{C_D A_R p_c}{(RT_c)^{1/2}} \left(\frac{p_T}{p_c}\right)^{1/\gamma} \left[\frac{2\gamma}{\gamma-1}\left[1-\left(\frac{p_T}{p_c}\right)^{(\gamma-1)/\gamma}\right]\right]^{1/2}$$

The fluid flow at the restriction will become choked when the pressure ratio across the restriction, p_c/p_T, exceeds a critical value; that is, when:

$$\frac{P_c}{P_T} \geq \left(\frac{\gamma+1}{2}\right)^{\gamma/(\gamma-1)}$$

P_T is taken as the pressure at the sampling valve outlet. For example, if the flow is adiabatic, for combustion gases $\gamma \approx 1.35$ and $[(\gamma+1)/2]^{(\gamma/(\gamma-1)} = 1.86$. Therefore, choking will occur when the ratio of the cylinder gas pressure to the sample valve outlet pressure is greater than 1.86. In practice the valve outlet pressure is likely to be close to the atmospheric value. Therefore, during most of the engine compression and expansion strokes, the ratio of cylinder gas to atmospheric pressure will exceed this critical value of 1.86, and the flow through the sampling valve will be choked. With choked flow, the mass flow rate can then be predicted from the following expression which is obtained by substituting $p_c/p_T = [(\gamma+1)/2]^{(\gamma/(\gamma-1)}$ in the above equation:

$$\dot{m} = \frac{C_D A_R p_c}{(RT_c)^{1/2}} \gamma^{1/2} \left(\frac{2}{\gamma+1}\right)^{[\gamma+1]/[2(\gamma-1)]}$$

One problem in applying this expression is that the value of C_D depends on the restriction geometry, the flow Reynolds number at the restriction, the Mach number, and the gas properties. The value of C_D usually must be determined experimentally. Often a constant value of $C_D = 0.6$ is used, but the value could be lower than this for small valve lifts and can vary considerably at low Reynolds numbers.

The flow area at the flow restriction, A_R, will depend on the type of sampling valve used and on the valve lift. In the case of poppet valve designs, the flow restriction is likely to be the annular flow passage formed between the poppet and its seat. Referring to Fig. 4.11 (poppet valve), and assuming that c << a and l << a (this is normally the case in practice), the flow area will be:

$$A_R = \pi a l \cos\psi$$

where l = valve lift
a = poppet seat diameter
ψ = valve seat angle
c = valve seat length in the flow direction

In the case of needle valve designs (see Fig. 4.11), the flow restriction is likely to be either the sample hole itself or the annular flow passage formed between the valve head and its seat. Another possibility is that at small valve lifts, the flow restriction is at the seat and at large valve lifts, the flow restriction shifts to the sample hole. If the flow restriction is at the seat, the flow area, A_R, can be calculated from the above expression using the sample hole diameter in place of the poppet diameter, a. Otherwise, the cross-sectional area of the sample hole will be the appropriate value for A_R.

The above expression gives the mass flow rate of the sample through the valve when the valve is open continuously. Consider now intermittent valve operation. For the sake of simplicity, assume that the valve opens instantaneously, remains open for a brief period, t_s, and then closes instantaneously. Also assume that the valve opens once every engine cycle. The mass of the gas sample withdrawn from the combustion chamber can be estimated approximately from:

$$\Delta m = \dot{m} t_s$$

where t_s is the period that the sampling valve remains open.

The volume of the gas sample withdrawn from the combustion chamber per engine cycle can also be calculated. Often, it is more useful to be able to estimate the sample volume per cycle, ΔV_o, at atmospheric conditions (i.e., at atmospheric temperature, T_o, and pressure, p_o):

$$\Delta V_o = \frac{\dot{m}}{\rho_o} t_s = \dot{m} \frac{R T_o}{p_o} t_s$$

where ΔV_o = sample volume per cycle at atmospheric conditions
T_o = atmospheric temperature
p_o = atmospheric pressure
ρ_o = sample gas density at atmospheric conditions
R = specific gas constant for sample (~287 J/kgK)

Also,

$$t_s = \frac{\theta_s}{360} \frac{1}{(N/60)}$$

where θ_s = engine crank angle during which valve is open
N = engine speed in rpm

Hence,

$$\Delta V_o = \dot{m}\frac{RT_o}{p_o}\frac{\theta_s}{6N}$$

The quasi-steady, time-averaged volume flow rate, V_o, out of the valve is:

$$V_o = \Delta V_o \frac{N}{60n} = \dot{m}\frac{RT_o}{P_o}\frac{\theta_s}{360n}$$

where V_o = quasi-steady state time-averaged volume flow rate out of the valve at atmospheric p_o and T_o (m^3/s)
N = number of revolutions per engine cycle (four-stroke, n = 2)

The above analysis is most likely to overestimate the sample mass per engine cycle, as well as the sample flow rate out of the valve. This is because the above relations are for steady flow, while the flow through the valve will be unsteady. When the sample valve opens, the flow will take some time to be established due to the fluid inertial and other transient effects, and the flow will not be fully developed. Also, the sample valve cannot open and close instantaneously. Nevertheless, the above relations can provide a useful indication of the sample volume for valve design purposes. However, estimation of V_o is useful to select the right valve geometry so that an adequate sample flow rate can be matched to the requirement of the analytical equipment to be used for gas analysis.

B1.2 Example (Choked Flow)

The above expressions were used in order to estimate the sample volume per engine cycle, both at cylinder and atmospheric conditions (pressure and temperature).

Fig. B.1 shows the estimated sample volume when the flow is choked. The sample volume is shown at both cylinder and atmospheric conditions, for a range of cylinder gas pressure/temperature combinations. The results are for an inward-opening needle sampling valve, with the following data:

Sample hole diameter = 2 mm
Seat angle = 45°
Valve lift = 0.1 mm
Valve flow discharge coefficient = C_D, 0.6

$\gamma = 1.35$
Sampling duration per engine cycle = 1 ms
Engine speed = 2000 rpm (four-stroke engine)

First, consider the three curves shown in Fig. B.1 for the sample gas volume at atmospheric conditions. When the cylinder gas pressure is 30 bar, the sample gas volume is around 1.5 cm^3 at atmospheric pressure and temperature. When the cylinder gas pressure is doubled to 60 bar, the sample volume also doubles to around 3 cm^3 at atmospheric conditions. Clearly, at the higher cylinder gas pressure, the sample mass flow rate through the open valve will be higher, and a larger mass of gas and volume will be collected per engine cycle at atmospheric conditions. Although not shown in the figure, the quasi-steady state volume flow rate out of the sample valve is around 25 cm^3/s with a cylinder pressure of 30 bar, and this roughly doubles to around 50 cm^3/s when the cylinder pressure is doubled to 60 bar. On the other hand, the effect of cylinder gas temperature on the sample volume is fairly small.

Now, consider the single curve shown in Fig. B.1 for the sample gas volume at cylinder conditions. The volume of the sample obtained per engine cycle is around 0.3 cm^3 and it is, of course, much smaller than the volume it occupies at atmospheric conditions. It can also be seen from Fig. B.1 that the sample volume at cylinder conditions is independent of the cylinder gas pressure. Also noteworthy is the fact that the sample volume is not very sensitive to cylinder gas temperature. Therefore, Fig. B.1 shows that the sample volume at cylinder gas conditions is almost independent of the cylinder gas pressure and temperature. This is an interesting result, and arises as follows.

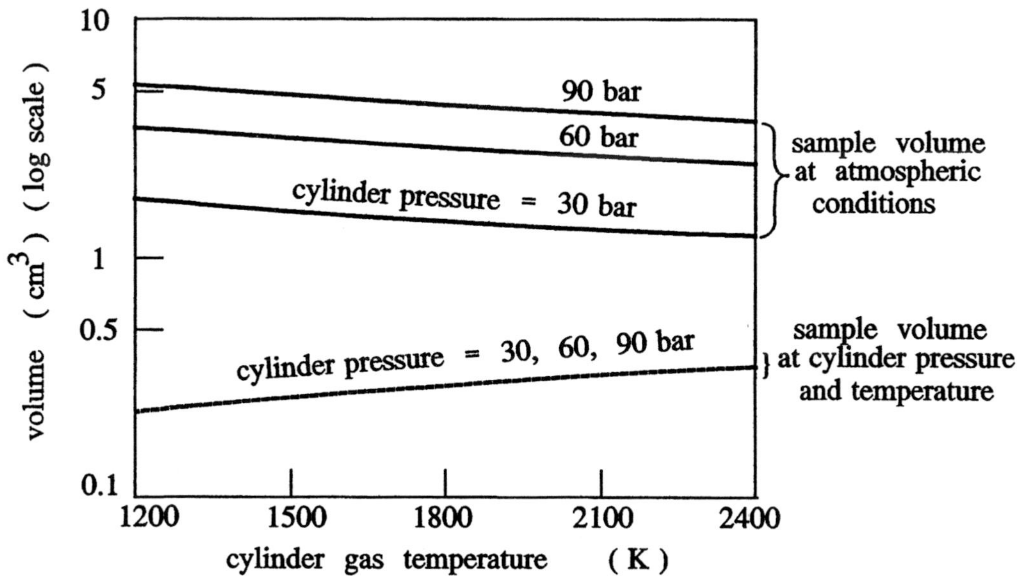

Fig. B.1 Volume of sample per valve opening.

The sample volume per engine cycle at cylinder temperature and pressure is given by,

$$\Delta V_c = \frac{\Delta m}{\rho_c} = \Delta m \frac{RT_c}{p_c}$$

where, ΔV_c, and ρ_c are the sample volume and density at cylinder gas temperature, T_c, and pressure, p_c. Now, the mass of sample obtained per engine cycle is given by,

$$\Delta m = \dot{m} t_s$$

Therefore,

$$\Delta V_c = \dot{m} t_s \frac{RT_c}{p_c}$$

From the preceding section, when flow through a restriction is choked, the mass flow rate through the valve is given by:

$$\dot{m} = \left[\frac{C_D A_R P_c}{(RT_c)^{1/2}} \gamma^{1/2} \left(\frac{2}{\gamma + 1} \right)^{(\gamma+1)/2(\gamma-1)} \right]$$

Therefore,

$$\Delta V_c = \left[\frac{C_D A_R P_c}{(RT_c)^{1/2}} \gamma^{1/2} \left(\frac{2}{\gamma + 1} \right)^{(\gamma+1)/2(\gamma-1)} \right] ts \frac{RT_c}{P_c}$$

Now, it should be remembered that Fig. B.1 is drawn for a given constant valve geometry and valve opening (i.e., constant A_R), and given values of valve flow discharge coefficient, C_D, ratio of specific heat capacities, γ, and sampling duration per engine cycle, t_s. Simplifying,

$$\Delta V_c \propto T_c^{1/2}$$

Therefore, for a given sampling period and a given valve opening, the volume occupied by the sample inside the engine cylinder will not vary greatly over a considerable range of cylinder gas pressure and temperature.

B.2 Laminar Flow (Poiseuille Flow)

In most sampling applications from the engine cylinder, the mass flow rate through a sampling valve can be estimated using the one-dimensional compressible flow expressions of Section B.1, with an appropriate value of C_D. However, at very low valve lifts or when the cylinder gas pressure is low, the flow through the sampling valve seat may be laminar. Although the flow is still compressible, it is, in principle, no longer one-dimensional. In other words, flow quantities such as velocity, pressure, density, and so on, are not constant over the cross-sectional area of the flow annulus (in addition to varying in the direction of the flow). In these circumstances, laminar flow analysis would be more appropriate. In practice, however, a flow regime within the valve seat may never become established fully. This is due, partly, to the fact that the flow is highly unsteady, as noted at the end of Section B.1 above. In addition, the sample valve seat is relatively short. With a valve seat length of, say, 0.5 mm and a maximum valve lift of 0.1 mm, the valve length-to-lift ratio is only 5. For the flow to be fully developed, the ratio will have to be orders of magnitude greater than this value.

If the flow through the valve became fully developed, laminar, and compressible, the mass flow rate will be given by the Poiseuille flow equation [Massey, 1970], modified so that it is applicable to the sampling valve geometry:

$$\dot{m}_{Pois} = \rho \frac{Bh^3}{12\mu}\left(\frac{dp}{dx}\right)$$

where ρ = gas density
h = valve opening normal to the flow ($h = l \cos\psi$; see Fig. 4.11)
B = seat perimeter ($= \pi a$)
dp/dx = pressure gradient in the direction of flow

This expression includes, implicitly, fluid friction (introduced by means of the fluid dynamic viscosity, μ). The above expression applies to both compressible and incompressible flow, and assumes fully developed flow.

Integration of the above expression presents considerable problems in the case of a sampling valve, because the density and the seat perimeter are likely to be varying in the direction of flow. For the few cases when the pressure drop across the sampling valve is small enough for the flow to be considered approximately incompressible, the above expression can be written as:

$$\dot{m}_{Pois} = \bar{\rho} \frac{\bar{B}h^3}{12\mu}\left(\frac{p_c - p_e}{c}\right)$$

where average values are used for the density and the perimeter, between the inlet and outlet of the valve seat (alternatively, B can be expressed as a function of x prior to integration); $(p_c - p_e)/c$ is the average pressure gradient over the valve seat length, c; and the dynamic viscosity, μ, is assumed constant along the valve seat.

Integration is more problematic when, as is usually the case, the pressure drop across the valve seat is large, making the flow through the seat compressible. A number of strategies can then be adopted in order to allow integration of the mass flow rate equation. The density can be assumed, crudely, to be constant with respect to the distance along the valve seat, x, as for incompressible flow; the constant value of the density can be taken as the average value between the cylinder and the exit from the valve seat. Alternatively, the density can be assumed to vary, say, linearly with distance along the valve seat, between the density value in the cylinder and that at the exit from the valve seat. Yet another alternative is the following: Noting that for an ideal gas, $\rho = p/(RT)$, the expression for mass flow rate can be written as:

$$\dot{m}_{Pois} = \frac{p}{RT}\frac{Bh^3}{12\mu}\left(\frac{dp}{dx}\right)$$

T is then taken as having a constant value which is equal to the average between the temperature in the cylinder and that at the exit from the valve seat. The above expression can then be integrated to give:

$$\dot{m}_{Pois} = \frac{\bar{B}h^3}{24\mu R\bar{T}}\left(\frac{p_c^2 - p_e^2}{c}\right)$$

There is some evidence in the work of LoRusso et al. [1980] that for very low valve lifts, Poiseuille flow may be present in the seat. LoRusso et al. measured the flow of high-pressure nitrogen through a poppet-type sampling valve as a function of poppet lift. The nitrogen was at room temperature and the valve was opened intermittently at a frequency of 50 Hz. The poppet lift was measured directly at the front of the valve in order to exclude the elongation of the valve stem. At low valve poppet lifts between 0.003 and 0.02 mm, the sample mass per valve opening increased proportionately to h^3, which is characteristic of Poiseuille flow. At higher valve lifts, beyond 0.02 mm, the sample mass per valve opening was directly proportional to valve lift, which is characteristic of choked flow, as described by the compressible flow equations presented in Section B.1.

The maximum valve opening of 0.02 mm, at which LoRusso et al. found that the flow changes from Poiseuille to choked flow, is significantly lower than that which might be used for most sampling applications (see Tables 4.1 and 4.3). As mentioned above, LoRusso et al. [1980] selected a low valve lift so as to extract tiny gas samples from the wall quench layer, which had thickness of ~0.1 mm.

B.3 Comparison of One-Dimensional Choked and Laminar Flows

The mass flow rate through a hypothetical inward-opening needle sampling valve has been calculated using both the one-dimensional choked flow theory of Section B.1 and the laminar flow analysis of Section B.2. The calculations have been carried out at progressively increasing valve lift, so that the flow changes from laminar to turbulent flow conditions. The sampling valve geometry and gas conditions chosen are listed below:

Valve sample hole diameter = 2 mm
Seat angle = 45°
Seat length along the direction of flow = 0.6 mm
Valve flow discharge coefficient, $C_D = 0.6$
Ratio of specific heat capacities, $\gamma = 1.35$
Engine cylinder gas pressure = 30 bar
Exit pressure from the valve seat = 1 bar
Engine cylinder gas temperature = 1200 K
Exit temperature from the valve seat
(assuming isentropic flow through seat) = 497 K

Results shown in Table B.1 were evaluated as follows: The Reynolds numbers for both types of flow were based on the hydraulic diameter of the flow annulus; the Reynolds number for laminar flow was based on average fluid properties between inlet and outlet of seat; the Reynolds number for one-dimensional choked flow was based on critical gas properties (critical p and T and sonic velocity); the valve flow area and valve seat perimeter were calculated at the mean diameter between the inlet and outlet from the seat. It should also be noted that in Table B.1, the sonic velocities for one-dimensional choked flow and laminar flow are different, since for one-dimensional flow the sonic velocity is calculated from the critical temperature while for laminar flow it is calculated from the average temperature across the valve seat.

Now, consider the results of Table B.1. Both sets of calculations (i.e., for one-dimensional choked flow and for laminar flow) show that at a valve lift of 0.013 mm, the flow Reynolds number just exceeds 2000, which will be taken as denoting the transition from laminar to turbulent flow. Interestingly, the mass flow rates predicted by both sets of calculations at 0.013 mm valve lift are very similar: 0.334 and 0.332 g/s. For this sampling valve and operating conditions, it would appear that up to about 0.013 mm lift, the mass flow rate should be predicted using laminar flow analysis. At greater valve lifts, one-dimensional choked flow analysis would be more appropriate.

To a certain extent, the agreement in Table B.1 between the predicted mass flow rates for one-dimensional and laminar flows at the transition from laminar to turbulent flow is fortuitous. This is because the predicted mass rate is dependent on the valve seat length in the case of laminar flow, but not in the case of one-dimensional choked flow. The valve seat length chosen for the above example was 0.6 mm, and a different choice of seat length would have resulted in a somewhat lesser agreement.

Table B.1 Comparison of mass flow rates calculated using turbulent choked and laminar flow analysis

	One-dimensional choked flow			Laminar flow		
Valve lift mm	Flow velocity[(a)] m/s	Reynolds number	Mass rate g/s	Flow velocity m/s	Reynolds number	Mass rate g/s
0.006				159	277	0.033
0.007				216	439	0.052
0.008				282	656	0.077
0.009				357	934	0.110
0.01				441	1281	0.151
0.011	629	1734	0.283	534	1705	0.201
0.012	629	1892	0.308	573[(a)]	1998	0.261
0.013	629	2050	0.334	573[(a)]	2164	0.332
0.014	629	2207	0.360	573[(a)]	2331	0.415
0.015	629	2365	0.385			
0.016	629	2522	0.411			
0.017	629	2680	0.437			
0.018	629	2838	0.462			

Note: [(a)] Flow is choking

References

LoRusso, J.A., Lavoie, G.A., and Kaiser, E.W. (1980), "An Electro-Hydraulic Gas Sampling Valve with Application to Hydrocarbon Emissions Studies," SAE Paper No. 800045.

Massey, B.S. (1970), *Mechanics of Fluids*, Van Nostrand Rheinhold, 2nd Edition, London.

Appendix C

Heat Transfer in the Valve

Outline

C.1 Introduction

When gas flows through the valve seat, its temperature falls due to expansion. If the expansion of the flow were adiabatic (no heat transfer to the seat surfaces), the temperature of the gas would fall from T_c in the engine cylinder to T_T in the valve seat. When flow is choked, T_T at the choking restriction can be calculated from:

$$T_T = \frac{T_c}{\left(\frac{\gamma+1}{2}\right)}$$

For example, with a value of $\gamma = 1.35$ and a cylinder gas temperature of $T_c = 1800$ K, the adiabatic gas temperature due to expansion in the seat will be 1532 K at the choking restriction.

In practice, there will be heat transfer from the gas to the valve seat, and the actual value of temperature can be expected to be lower than that for adiabatic expansion.

As explained above, during most of the compression and expansion strokes of the engine piston, the flow through the seat will be choked. The flow velocity in the seat (i.e., the flow restriction) will then reach the sonic value, calculated from:

$$v_S = \sqrt{\gamma R T_T}$$

Using the value of $T_T = 1532$ K, calculated above, $v_s = 770$ m/s. With such high velocity levels in the valve seat and a valve seat length of only around 0.6 mm, an element of the gas sample would spend only a few microseconds in contact with the seat surfaces. In reality, this period is likely to be somewhat longer because the sample element will take some time to accelerate through the seat when the valve opens. Nevertheless, the actual period that an element of sample takes to pass through the seat would still be very short. In practice, most of the charge cooling takes place after the sample leaves the valve seat; that is, while the sample flows through the annular passage surrounding the valve stem. The following two sections will examine, first, heat transfer during the sample passage through the annular space surrounding the valve stem and, secondly, while the sample flows through the valve seat.

C.2 Heat Transfer in Annulus Surrounding Valve Stem

After leaving the valve seat region, the sample flow enters the annular passage surrounding the valve stem, formed between diameters d and b shown in Fig. 4.11. Heat transfer to the walls of this annulus causes the sample to cool down. An indication of the fall in the sample temperature can be obtained by means of the following unsteady heat transfer analysis. The analysis considers an annular element of the sample, as it flows through the passage, and predicts the temperature of this element. Although the analysis is a very approximate one, it nevertheless provides a useful indication of the cooling effects on this sample element.

Fig. C.1(a) shows the annular sample element. The element travels along the annular passage surrounding the valve stem. The element loses energy as heat to wall and stem surfaces. A number of assumptions will be made: The temperature throughout the element is uniform; there is no heat or mass transfer axially between adjacent sample elements; changes in the kinetic energy of the element are neglected; the sample specific heat capacity and the convection surface heat transfer coefficient are assumed to be independent of temperature; the mass flow rate through the valve has reached steady state.

Let,

M_s = mass of sample element
c_{ps} = specific heat capacity of sample element
T_s = bulk temperature of sample element
t = time
h_s = convection, surface heat transfer coefficient
A_s = sample surface area in contact with stem and outer wall
T_w = temperature of stem and outer wall surfaces
ρ_s = density of sample element
p_s = pressure in sample element
T_{si} = initial sample element temperature
L = axial length of sample element
β_a = $[4h_sRT_w]/[c_{ps}p_s(d - b)]$
β_s = $[2h_sRT_w]/[c_{ps}p_s l \cos\psi]$

Nu_s = $(h_s d_h)/k_s$ (Nusselt number)
Re_s = $(\rho_s u_s d_h)/\mu_s$ (Reynolds number)
Pr_s = $(c_{ps}\mu_s)/k_s$ (Prandtl number) ≈ 0.7
b = inner diameter of annular sample element
d = outer diameter of annular sample element
d_h = $[4\pi(d^2 - b^2)]/[4\pi(d + b)] = d - b$ (hydraulic diameter of annular flow passage)
k_s = thermal conductivity of annular sample element
μ_s = dynamic viscosity of annular sample element evaluated at T_s
μ_w = dynamic viscosity of annular sample element evaluated at T_w
u_s = sample flow velocity in annular passage
m_s = sample mass flow rate in the annular passage
T_{mean} = average temperature = $(T_w + T_s)/2$
S = length of annular flow passage
l = valve lift

The energy balance for the element can be written as:

$$-M_s C_{ps}\frac{dT_s}{dt} = h_s A_s (T_s - T_w)$$

$$-\rho_s\left[\frac{\pi}{4}\left(d^2 - b^2\right)L\right]c_{ps}\frac{dT_s}{dt} = h_s\left[\pi(d+b)L\right](T_s - T_w)$$

which simplifies to,

$$-\frac{dT_s}{T_s(T_s - T_w)} = \frac{4h_s R}{c_{ps}\rho_s(d-b)}dt$$

Integrating with the initial condition: at $t = 0$, $T_s = T_{si}$

$$\ln\left[\left(\frac{T_s - T_w}{T_{si} - T_w}\right)\left(\frac{T_{si}}{T_s}\right)\right] = -\beta_a t \quad \text{where} \quad \beta_a = \frac{4h_s R T_w}{c_{ps}\rho_s(d-b)}$$

Isolating, T_s,

$$T_s = \frac{T_w}{1 - \left(1 - \frac{T_w}{T_{si}}\right)e^{-\beta_a t}}$$

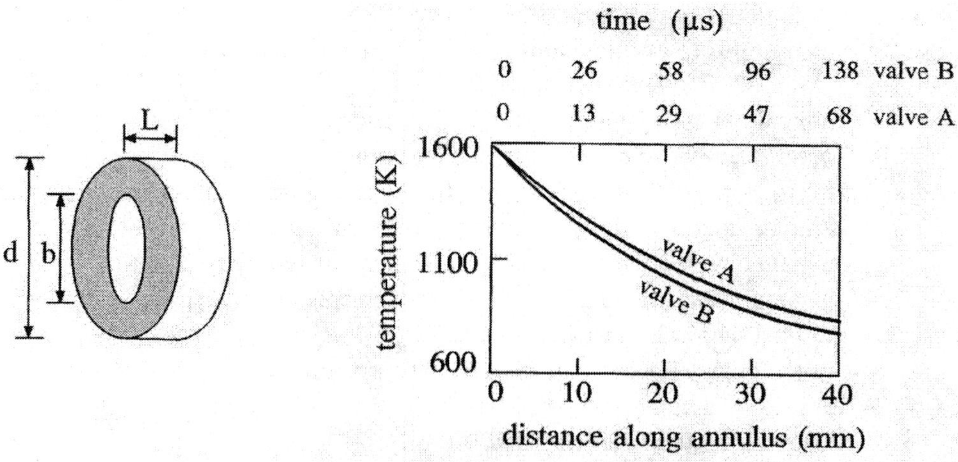

(a) heat transfer for annular sample element

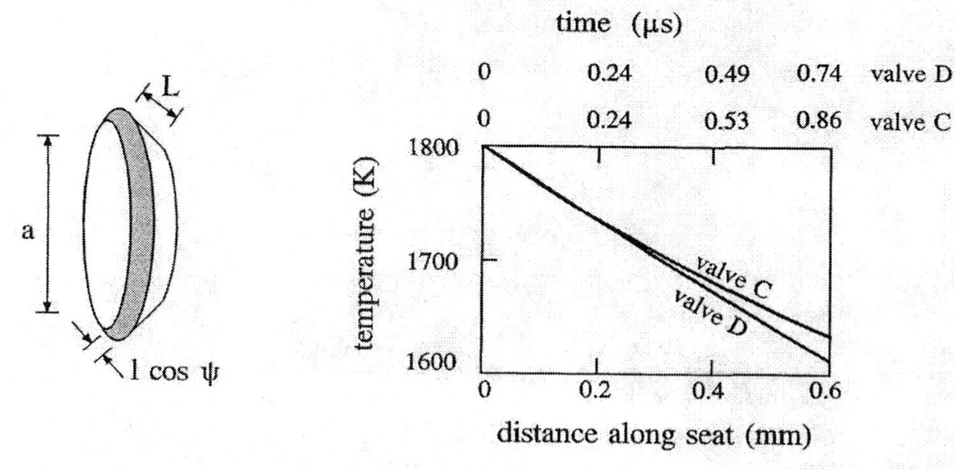

(b) heat transfer for annular sample element in valve seat

Fig. C.1 Heat transfer in the valve.

The above expression gives the temperature, T_s, of the sample element as a function of time, t. A rough estimate of the value of the heat transfer coefficient, h_s, can be obtained by considering the steady flow of sample through the annular passage. For example, a suitable semi-empirical correlation for turbulent flow is [Incropera and DeWitt, 1985]:

$$Nu_s = 0.027 Re_s^{0.8} Pr_s^{0.33} \left(\frac{\mu_s}{\mu_w} \right)^{0.14}$$

and for laminar flow (including entry effects and large variations in sample properties) [Incropera and DeWitt, 1985]

$$Nu_s = 1.86 Re_s^{0.33} Pr_s^{0.33} \left(\frac{d_h}{S} \right)^{0.33} \left(\frac{\mu_s}{\mu_w} \right)^{0.14}$$

The Nusselt and the Reynolds numbers are both based on the hydraulic diameter, d_h, of the annular flow passage (4 × flow area/total wetted perimeter). Also, the fluid properties in Nu_s, Re_s, and Pr_s are evaluated at the mean temperature, T_{mean} (except μ_w and μ_s).

It should be noted, in passing, that the above correlations for Nu_s are intended to be used for steady flow. In the case of the sampling valve, the sample flow is highly unsteady, and this tends to inhibit the establishment of a thermal boundary layer on the internal surfaces of the annular flow passage. In reality, therefore, the actual heat transfer coefficient may be somewhat greater than the value predicted from the above correlations for Nu_s.

The distance along the annulus, x, that the sample travels in a given period of time, t, can be calculated from, $x = u_s t$, that is:

$$x = \frac{4 m_s R T_s}{\pi \left(d^2 - b^2 \right) p_s} t$$

C.3 Heat Transfer in the Valve Seat

The analysis above can be repeated, with appropriate modification, for the valve seat. Consider the valve seat region. Fig. C.1(b) shows the geometry of the sample element as it travels through the valve seat. Letting,

p_m = average pressure in valve seat = $(p_c + p_e)/2$
p_c = cylinder gas pressure
p_e = sample pressure at valve seat exit
c = length of valve seat in the direction of flow

Again, the energy balance for this element can be written as before:

$$-M_s C_{ps} \frac{dT_s}{dt} = h_s A_s (T_s - T_w)$$

$$-\rho_s (L\pi a l \cos\psi) C_{ps} \frac{dT_s}{dt} h_s (2\pi a L)(T_s - T_w) \quad (a \gg L)$$

which simplifies to,

$$-\frac{dT_s}{T_s (T_s - T_w)} = \frac{2h_s R}{c_{ps} p_s l \cos\psi} dt$$

The pressure in element, p_s, will obviously fall as the element passes through the seat, and this pressure must be expressed as a function of time before the above expression can be integrated. To simplify the analysis, a fixed average value of this pressure in the seat region, $p_m = (p_c + p_e)/2$, will be taken as a first approximation. With the sample pressure fixed, the above expression can be integrated with the initial condition, at $t = 0$, $T_s = T_{si}$. Integrating and then isolating the sample temperature, T_s:

$$T_s = \frac{T_w}{1 - \left(1 - \frac{T_w}{T_{si}}\right) e^{-\beta_s t}} \quad \text{where } \beta_s = \frac{2h_s R T_w}{c_{ps} p_m l \cos\psi}$$

Again, h_s can be estimated from the correlations for Nu_s given in the preceding section, with d_h/S replaced by d_h/c, where d_h for the seat is equal to $2l\cos\psi$. The distance along the seat, x, that the sample travels in a given period of time, t, can again be calculated, roughly, from, $x = u_s t$, that is:

$$x = \frac{m_s R T_s}{(\pi a l \cos\psi) p_m} t \quad (a >> L)$$

C.4 Examples of Heat Transfer in the Valve

C.4.1 Example of Heat Transfer in the Stem Annulus

The following example illustrates the way in which an annular element of the sample, such as the one shown in Fig. C.1(a), cools down during its journey through the annular passage surrounding the valve stem. A reasonably good approximation is to take the value of the Prandtl number, Pr_s, as equal to 0.7 and independent of temperature; also, the values of the sample properties μ_s, k_s, and c_{ps} in Nu_s and Re_s were assumed to be constant (i.e., independent of temperature); ρ_s was allowed to vary with temperature. Two annular passage geometries were chosen as examples, named valve A and B, respectively. Valve A has a larger passage and 2.5 times greater mass flow rate through it than valve B. The various parameters for these two passages were:

Valve A	Valve B	
Temperature of stem and outer wall surfaces, T_w, K	500	500
Pressure in sample element (assuming no pressure recovery after seat), p_s, bar	1.013	1.013
Initial sample element temperature upon entering annular passage, T_{si}, K	1600	1600
Outer diameter of annular sample element, d, mm	6.0	5.0
Inner diameter of annular sample element, b, mm	4.5	3.5
Sample mass flow rate in the annular passage, m_s, kg/s	0.0025	0.001
Specific heat capacity of bulk sample, c_{ps}, J/(kgK)	1200	1200
Thermal conductivity of sample at T_s, k_s, W/(mK)	0.08	0.08
Thermal conductivity of sample at T_w, k_w, W/(mK)	0.04	0 04
Dynamic viscosity of sample at T_s, μ_s, kg/(ms)	4.0×10^{-5}	4.0×10^{-5}
Dynamic viscosity of sample at T_w, μ_w, kg/(ms)	2.7×10^{-5}	2.7×10^{-5}

Fig. C.1(a) shows the calculated fall in the temperature of the sample annular element as it travels through the passage surrounding the stem of valves A and B. It can be seen that the flow passage is quite effective in cooling the element. The Reynolds number in the passage of valve A is around 12,000 and in that of valve B is about 6,000. In the case of both valve annular geometries, the sample element temperature falls from 1600 to about 800 K within the first 40 mm or so of the annular passage length. What is also quite interesting is that this distance is covered in quite a short period of about 0.07 ms in the case of valve A and 0.14 ms in the case of valve B.

C.4.2 Example of Heat Transfer in the Valve Seat

A similar analysis was carried out for the valve seat region of two distinctly different valve geometries, named valve C and valve D. Valve C is of the needle-type and has a seat inlet diameter of 2 mm (equal to the sample hole diameter) and a sample gas mass flow rate of 0.0035 kg/s. Valve D has a large seat inlet diameter of 8 mm and a correspondingly large sample gas mass flow rate of 0.0135 kg/s. The data used for these valves are summarized below:

	Valve C	Valve D
Temperature of seat surfaces, T_w, K	750	750
Pressure in sample element (mean between pressure in the cylinder an at the choked seat exit), bar	46	46
Initial sample element temperature upon entering seat passage, T_{si}, K	1800	1800
Diameter at point where gas enters the valve seat, a, mm	2.0	8.0
Seat length, c, mm	0.6	0.6
Seat angle, $\psi°$,	45	45

	Valve C	Valve D
Valve lift, l, mm	0.1	0.1
Sample mass flow rate through valve seat, m_s, kg/s	0.0035	0.0135
Specific heat capacity of sample, c_{ps}, J/(kgK)	240	1240
Thermal conductivity of sample at T_s, k_s, W/(mK)	0.098	0.098
Thermal conductivity of sample at T_w, k_w, W/(mK)	0.04	0.04
Dynamic viscosity of sample at T_s, μ_s, kg/(ms)	5.8×10^{-5}	5.8×10^{-5}
Dynamic viscosity of sample at T_w, μ_w, kg/(ms)	2.7×10^{-5}	2.7×10^{-5}

Fig. C.1(b) shows the calculated decrease in the sample element temperature as it passes through the seat of the two valves. It can be seen that the element temperature decreases from 1800 K at the seat entrance to almost 1600 K at the exit of the two valves. This decrease is entirely due to heat transfer. The fall in temperature due to the expansion of the flow is not included: Expansion alone without heat transfer (adiabatic) would involve a temperature reduction from 1800 to 1532 K at the choking restriction in the valve. Fig. C.1(b) shows that the element is estimated to take a remarkably short period of time to travel through the seat (at steady state flow), about 0.009 ms in the case of needle valve C and about 0.008 ms in the case of poppet valve D. The flow Reynolds number in the seat of both valves is high at around 35,000 in both cases.

C.5 Quenching of Chemical Reactions in the Valve Stem Annulus and Seat

The above calculations indicate that a gas sample entering the valve from the engine cylinder at about 1800 K will be cooled down rapidly (quenched) to less than 800 K within, say, 0.1 ms or so. An important question is whether during this short period of time the chemistry of the sample traveling through the valve changes significantly. If it does, then the sample composition at the exit of the valve will not be representative of the gas composition in the engine cylinder.

LoRusso et al. [1983] quote a characteristic chemical time scale for hydrocarbon oxidation of 0.020 ms at 1500 K. In general, with gas temperatures above about 1250 K the characteristic time for HC oxidation is less than 1 ms, while at temperatures below 1000 K, the characteristic time is greater than 50 ms [Heywood, 1988]. Therefore, depending on the sampling timing and sampling location in the engine cylinder, some hydrocarbon oxidation might take place after the sample enters the valve. Similarly, formation of CO is relatively fast and some changes in CO composition might take place in the valve. On the other hand, formation of NO_x and CO_2 are relatively slow, and quenching within the sampling valve of these species to arrest chemical reactions may be more effective. For example, oxidation to CO_2 of 3000 ppm of CO in combustion products at 1100 K would require a residence time of around 50 ms. Likewise, the formation time scale for NO_x in post flame gases is greater than 1 ms [Heywood, 1988]. In the case of soot, the oxidation period is relatively long. For example, the oxidation period of a 0.030 μm soot particle can be estimated using the Nagle and Strickland-Constable semi-empirical correlation [Heywood, 1988]. At 1800 K, the estimated burn-up period is about 3 ms and at 1300 K about 135 ms. It is therefore unlikely that significant soot oxidation will take place within the sampling valve, except when sampling cylinder gases over 2000 K.

C.6 Valve Tip Cooling Arrangements

When the sample valve protrudes into the combustion chamber, cooling of the valve tip may be required. Cooling may be necessary to preserve the integrity of the valve tip, and, also, in order to ensure that the chemical reactions in the gas sample are arrested as soon as possible after the sample enters the valve. In practice, cooling of the valve tip can be accomplished with either a liquid or gaseous coolant.

Bennethum et al. [1975] measured the temperature at the uncooled tip of an inward-opening needle valve installed in a diesel engine operating at 1600 rpm (see Tables 4.3 and 4.4). This was done by replacing the valve in the engine by a probe having the same geometry as the valve tip and embedding a thermocouple in the tip. The tip had a diameter of 5.7 mm and was made of titanium. With 1 mm penetration, the tip temperatures measured varied between 350°C at idling to around 650°C at full load. With 8.6 mm tip penetration, the corresponding temperatures were remarkably high, up to 1000°C.

The valve tip temperature will depend significantly on the detailed design of the tip and the tip installation on the engine. For example, greater tip wall thickness and greater valve stem diameter increases heat transfer by conduction away from the tip, and helps keep the tip temperature lower. Good contact between the valve tip region and the engine cylinder surfaces also helps to enhance heat dissipation away from the tip. If the tip diameter can be large, then it is easier to design the tip with good heat transfer in mind. However, other design considerations will usually dictate a small tip diameter. Ladommatos et al. [1996] and Zhao et al. [1996] successfully used uncooled sampling valves in DI diesel engines at speeds ranging from 1000 to 2000 rpm and up to full load.

A number of investigators used tip cooling for sampling valves operated in gasoline and diesel engines. Duggal et al. [1978] employed tip cooling by circulating water close to the valve seat, as shown in Fig. C.2(a). The valve was installed in the swirl chamber of a high-speed IDI diesel engine. Cooling was required because of high heat transfer rate to the tip, induced by the high flow velocities and large tip penetrations into the swirl chamber. Garner et al. [1952] also used water to cool the tip of a sampling valve installed in the pre-chamber of a diesel engine. Cooling water-jacket arrangements were also used by Egerton et al. [1935] and by Rifkin and Walcutt [1956] for sampling valves installed in gasoline engines.

A number of tip cooling arrangements considered by Yates [1988] are shown in Fig. C.2. Fig. C.2(b) shows the lower part of the valve enclosed in a close-fitting tube. Before the tube was fixed in place, a two-start spiral groove was cut around the outside of the lower part of the valve to carry cooling water. Fig. C.2(c) shows water channels formed by four flats machined onto the outside of the lower part of the valve. The channels are enclosed within a close-fitting tube. This latter method provided effective cooling and was simple to manufacture [Yates, 1988]. Fig. C.2(d) shows the valve head being cooled by water circulated through the hollow valve stem.

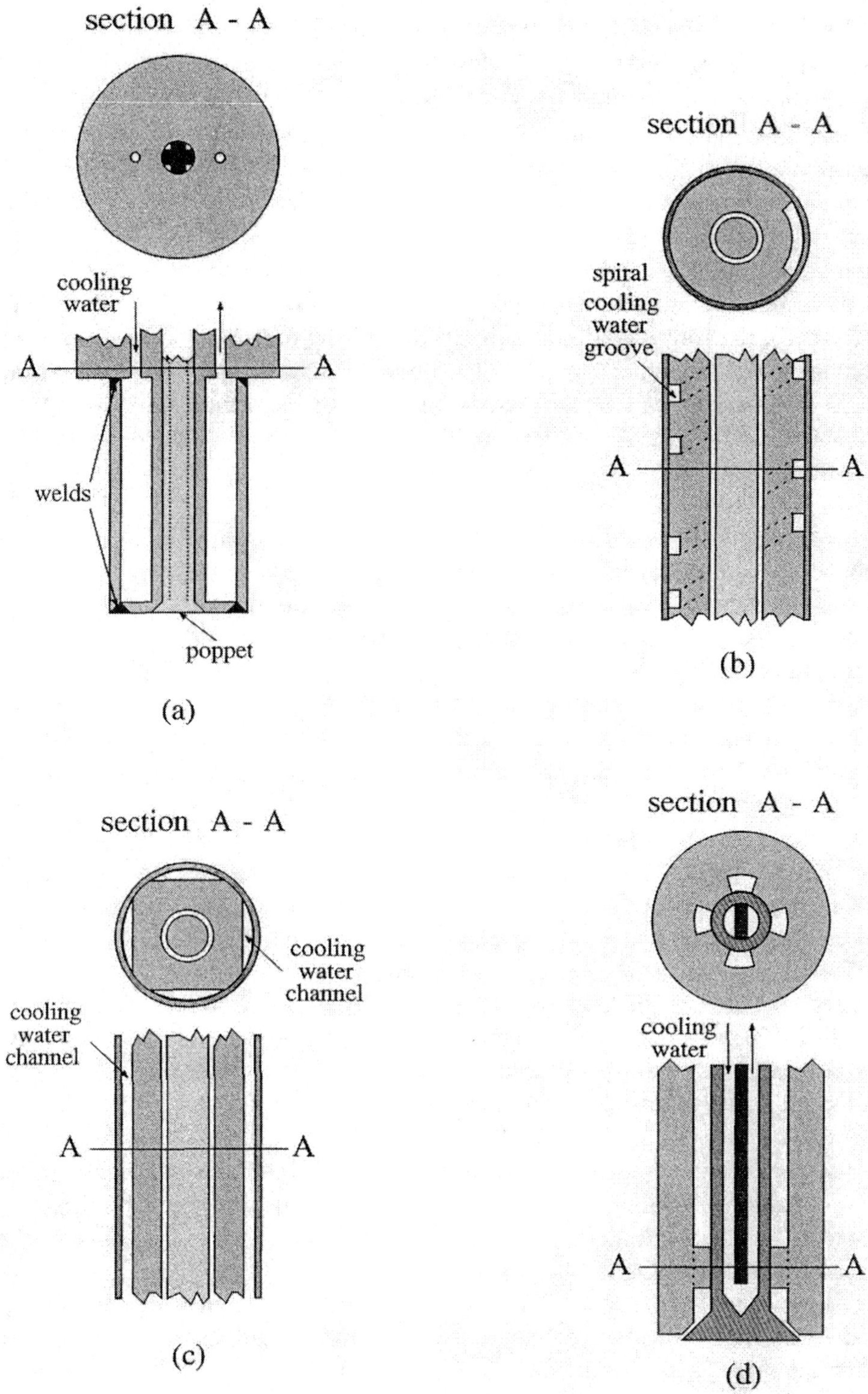

Fig. C.2 Cooling arrangements for the valve tip region.

References

Bennethum, J.E., Mattavi, J.N., and Toepel, R.R. (1975), "Diesel Combustion Chamber Sampling—Hardware, Procedures, and Data Interpretation," SAE Paper No. 750849.

Duggal, V.K., Priede, T., and Khan, I.M. (1978), "A Study of Pollutant Formation Within the Combustion Space of a Diesel Engine," SAE Paper No. 780227.

Egerton, A., Smith, F.L., and Ubbelohde, A.R. (1935), "Estimation of the Combustion Products from the Cylinder of the Petrol Engine and Its Relation to Knock," *Phil. Trans. Roy. Soc. London*, Vol. 234, No. 744, pp. 433–521.

Garner, F.H., Malpas, W.E., Morton, F., Reid, W.D., and Wright, E.P. (1952), "Pre-Flame Reactions in Diesel Engines (Part II)," *Inst. Petroleum*, 38, No. 341, pp. 312–343.

Heywood, J.B. (1988), *Internal Combustion Engine Fundamentals*, McGraw-Hill, New York.

Incropera, F.P., and DeWitt, D.P. (1985), *Fundamentals of Heat and Mass Transfer*, John Wiley, 2nd Ed., New York.

Ladommatos, N., Balian, R., Horrocks, R., and Cooper, L.(1996), "The Effect of Exhaust Gas Recirculation on Soot Formation in a High-Speed Direct-Injection Diesel Engine," SAE Paper No. 960841.

LoRusso, J.A., Kaiser, E.W., and Lavoie, G.A. (1983), "In-Cylinder Measurements of Wall Layer Hydrocarbons in a Spark-Ignition Engine," *Combustion Science and Technology*, Vol. 33, pp. 75–112.

Rifkin, E.B. and Walcutt, C. (1956), "Decomposition of Tetra-Ethyl Lead in an Engine," *Industr. and Eng. Chemistry*, Vol. 48, No. 9, pp. 1532–1539.

Yates, D.A. (1988), "The Compact High-Speed Gas Sampling Valve for an Internal Combustion Engine," IMechE Conference, *Experimental Methods in Engine Research and Development*, London.

Zhao, H., Lowry, G., and Ladommatos, N. (1996), "Time-Resolved Measurements and Analysis of In-Cylinder Gases and Particulates in Compression-Ignition Engines," SAE Paper No. 961168.

Appendix D

Stress Considerations for Sampling Valves

Outline

Almost all of the sample valve designs reported in the literature employed either a poppet- or a needle-type valve head, as shown in Fig. 4.11. Of these two types of valve head, the needle-type and associated seat area are easier to design so as to withstand the stresses imposed during valve operation. However, a more careful consideration of the stresses on the head is necessary in the case of the poppet head. The following section will consider the stresses on the poppet-type of valve head in detail.

D.1 Maximum Closing Impact Force

Fig. 4.11 shows a simplified dynamic model of a poppet valve. m_e is the equivalent mass of the valve stem and other associated moving parts. l_s is the valve stem length which is subject to elongation. This is, of course, an idealized model of reality. In practical valve designs, the mass of the moving parts will be distributed, unevenly, along the stem of the valve. As a first approximation, the equivalent mass can be taken to be the mass of the moving parts of the valve. Also, l_s can be taken to be the actual distance between the poppet seat and the point at which the spring acts on the stem, as shown in Fig. 4.11.

When the poppet impinges on the seat, the valve stem will be elongated to a maximum length due to two forces [LoRusso et al, 1980]:

(i) A variable dynamic force, F_d, arising from the impact of the poppet onto its seat and the subsequent deceleration of the moving parts of the valve
(ii) A force on the valve stem, F_s, arising from the pre-compression of the spring, which keeps the poppet seated

The dynamic impact force, F_d, is short-lived and ceases to exist after the moving parts associated with the valve stem have come to rest. F_d increases sharply upon impact, from zero to a maximum value when the stem attains its maximum elongation. However, the static force, F_s, continues to be exerted by the pre-compressed spring after the poppet has settled on its seat. As will be seen below, F_s is practically constant because the elongation of the valve stem is much smaller than the length by which the spring is pre-compressed.

The following analysis will consider, first, the spring force F_s and, secondly, the dynamic force F_d.

Let,

A	=	cross-sectional area of the valve stem
b	=	diameter of the valve stem
E	=	Young's modulus
F_d	=	variable dynamic force in the x direction due to deceleration of the valve moving parts associated with the stem
$F_{d,max}$	=	maximum value of F_d
F_s	=	static force straining the valve stem due to spring pre-compression
$F_{s,max}$	=	maximum value of F_s
$F_{t,max}$	=	maximum total force straining the valve stem
	=	$F_{d,max} + F_{s,max}$
k	=	spring stiffness
l_s	=	length of valve stem capable of elongation
m_e	=	effective mass of valve moving parts associated with the stem
$U_{d,max}$	=	maximum strain energy along x direction due to dynamic force F_d straining the valve stem
V_i	=	velocity of valve moving parts when poppet impinges on its seat
x_d	=	elongation of the stem due to dynamic force F_d
$x_{d,max}$	=	maximum elongation of the valve stem due to dynamic force $F_{d,max}$
$x_{s,max}$	=	maximum elongation of the valve stem due to spring force $F_{s,max}$
$x_{t,max}$	=	maximum elongation of the valve stem due to the combined dynamic and spring forces
y_o	=	length by which spring is pre-compressed
l_{max}	=	maximum valve opening (lift)
σ	=	stress
ε	=	strain

Consider, first, the spring force, F_s. This can be calculated easily from the spring stiffness, k, and the length, y_o, by which the spring is pre-compressed when the poppet is seated. That is:

$$F_s = ky_o$$

F_s is practically constant because the length by which the spring is compressed does not change significantly during impact. In other words, the elongation of the valve stem is very small compared with the length, y_o, by which the spring is pre-compressed. Therefore, denoting the maximum spring force by $F_{s,max}$,

$$F_{s,max} = F_s = ky_o$$

Consider now the variable dynamic force, F_d. An expression for this force can be obtained by employing an energy method [Ryder, 1973].

Assumptions:

- The valve stem behaves elastically and obeys Hooke's Law.
- There is no energy dissipation due to impact between the poppet and its seat.
- All available energy is transformed to normal elastic energy of the valve stem without dissipation.

At the moment of impact, the moving parts of the valve possess a certain amount of kinetic energy. After the poppet impinges on its seat, the valve stem elongates elastically due to the inertial force associated with the deceleration of the valve moving parts. It is assumed that all of this kinetic energy is converted and temporarily stored as strain energy as a result of the momentary elongation of the valve stem. The maximum strain energy stored, $U_{d,max}$, can be calculated from the energy balance as follows:

$$U_{d,max} = \frac{1}{2} m_e V_i^2$$

However, the maximum strain energy,

$$U_{d,max} = \text{work done by a force } F_d \text{ when straining the stem}$$

Therefore,

$$U_{d,max} = \int_0^{x_{d,max}} F_d \, dx_d$$

However, F_d is a function of x_d, and before the integration can be carried out an expression for F_d in terms of x_d is needed. This expression can be obtained from Hooke's Law ($\sigma = E\varepsilon$)

$$\frac{F_d}{A} = E\frac{x_d}{l_s}$$

Hence,

$$U_{d,max} = \frac{AE}{l_s}\int_0^{x_{d,max}} x_d dx_d$$

Integrating,

$$U_{d,max} = \frac{1}{2}\frac{AE}{l_s}x_{d,max}^2 = \frac{1}{2}m_e V_i^2$$

Since,

$$x_{d,max} = \frac{F_{d,max} l_s}{AE}$$

Hence,

$$F_{d,max}^2 = m_e V_i^2 \frac{\pi b^2}{4}\frac{E}{l_s}$$

The maximum straining force on the valve stem can therefore be written as:

$$F_{t,max} = F_{d,max} + F_{s,max}$$

$$F_{t,max} = \sqrt{m_e V_i^2 \frac{\pi b^2}{4}\frac{E}{l_s}} + ky_o$$

In practice, the impact velocity, V_i, may not be known, but can easily be estimated from the energy balance:

kinetic energy at impact = energy stored in valve spring at maximum valve lift

= work done in opening valve to maximum lift

$$\frac{1}{2} m_e V_i^2 = k y_o l_{max}$$

where l_{max} is the maximum valve lift; it is also assumed that $l_{max} << y_o$. Substituting for V_i^2 in the above expression for $F_{t,max}$:

$$F_{t,max} = \sqrt{\frac{1}{2} k y_o l_{max} \pi b^2 \frac{E}{l_s}} + k y_o$$

D.2 Maximum Elongation of Valve Stem

As discussed above, the maximum straining force, $F_{t,max}$, elongates the valve stem to a maximum elongation, $x_{t,max}$. This maximum elongation of the stem is given by:

$$x_{t,max} = x_{s,max} + x_{d,max}$$

where $x_{s,max}$ is the maximum elongation of the valve stem due to the spring force $F_{s,max}$, and $x_{d,max}$ is the maximum elongation of the valve stem due to the dynamic force $F_{d,max}$. The maximum elongation, $x_{t,max}$, can be calculated as follows:

$$x_{t,max} = F_{t,max} \frac{l_s}{\left(\frac{\pi b^2}{4}\right) E}$$

Example:

k = 6000 N/m

y_o = 0.013 m

l_{max} = 0.00012 m

b = 0.005 m

E = 1.5×10^{11} Pa

l_s = 0.2 m

Therefore,

$$\begin{aligned} F_{t,max} &= F_{d,max} + F_{s,max} \\ &= 525 + 78\ \text{N} \\ &= 603\ \text{N} \\ x_{t,max} &= x_{d,max} + x_{s,max} \\ &= 0.036 + 0.005 \\ &= 0.041\ \text{mm} \end{aligned}$$

It is worth noting that in this example, the contributions of the spring force $F_{s,max}$ toward the maximum force $F_{t,max}$ and maximum elongation $x_{t,max}$ are small but not insignificant.

D.3 Shear Stress in Valve Tip

D.3.1 Linearly Varying Valve Tip Load

The shear stress distribution in the valve seat region can be estimated using a method outlined in LoRusso et al. [1980]. Fig. D.1 shows a loading diagram for the valve tip region. Of interest is the shear stress distribution in the tip region over a distance denoted by d in this figure (Fig. D.1). The shear stress in this region must not, of course, exceed the 0.2% yield strength of the material of which the poppet is to be made.

For the calculation of the shear stress distribution, the valve tip is represented by a tapered (non-prismatic) beam (see Fig. D.1) having:

length = d
width = πa (valve tip circumference)
height = varying linearly between $d(\tan\psi + \tan\theta)$ and zero

Let,

θ = stress relieving angle
x = horizontal coordinate
y = vertical coordinate
c' = projected seat length ($c' = c\cos\psi$)
$P(x)$ = distributed load per unit length

The valve tip is in contact with its seat over a seat length equal to c (see Figs 4.11 and D.1). The above model becomes a better representation of reality when $c \ll a/2$, which will be the case in most sampling valve designs.

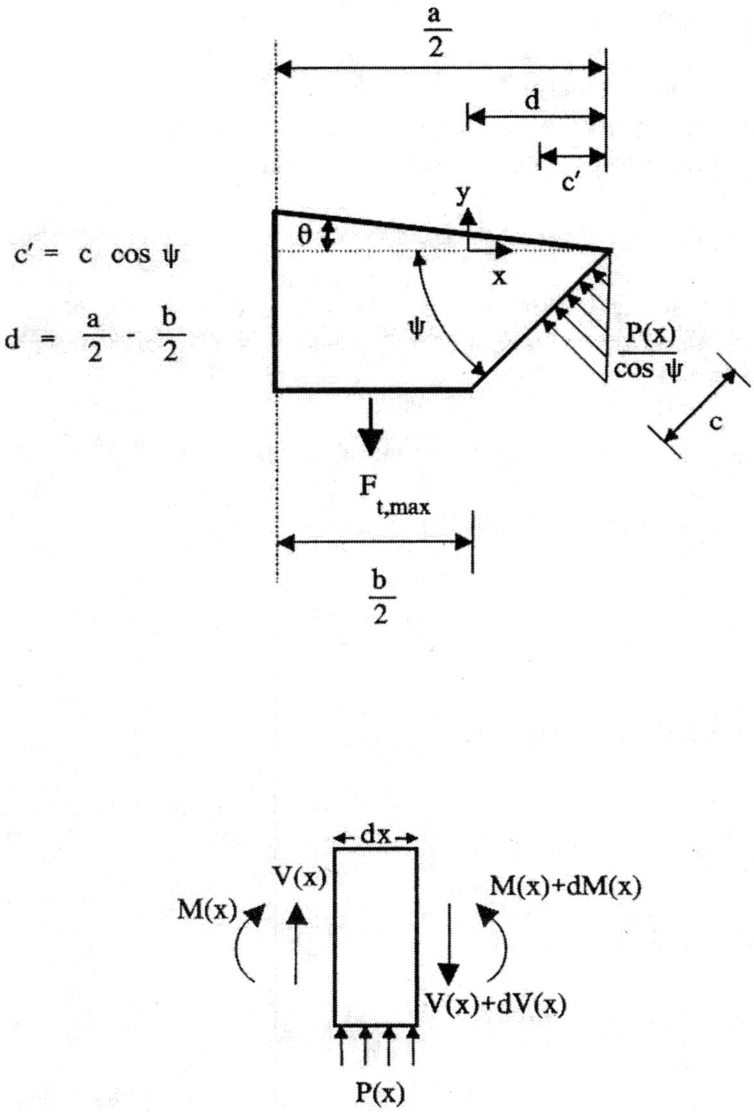

Fig. D.1 Poppet model used for shear stress analysis.

In general, the shear stress in a plane xy of a tapered beam will vary along the beam length. The shear stress, $\tau_{xy}(x)$, for such a beam is given by (Gere, 1985):

$$\tau_{xy}(x) = \frac{V(x)Q(x)}{I(x)} + \frac{M(x)h(x)}{4I(x)}\left(1 - \frac{Q(x)h(x)}{I(x)}\right)\frac{dh(x)}{dx}$$

where h(x) = beam height
I(x) = second moment of area
M(x) = bending moment
Q(x) = first moment of the beam cross-sectional area
V(x) = shear force per unit width of the beam

The magnitude of all of the above parameters varies along the length of the beam.

Consider, again, the valve tip shown in Fig. D.1. As mentioned previously, the tip is being modeled as a beam, having a constant width equal to πa. At a distance x along the beam length, the height of the beam is,

$$h(x) = (d - x)(\tan\psi + \tan\theta)$$

The first moment of area of the beam is,

$$Q(x) = \frac{\pi a [h(x)]^2}{8}$$

The second moment of area of the beam is,

$$I(x) = \frac{\pi a [h(x)]^3}{12}$$

The first derivative of the beam height is,

$$\frac{dh(x)}{dx} = -(\tan\psi + \tan\theta)$$

The only parameters left for evaluation now are V(x) and Q(x). Expressions for these two parameters are obtained from the equilibrium equations. The bottom part of Fig. D.1 shows the forces acting on a small element of the beam. For equilibrium of vertical forces:

$$\frac{dV(x)}{dx} = P(x)$$

where P(x) is a distributed load acting on the beam in the upward direction. The distributed load may not be uniformly distributed, so it is expressed as a function of x. The sign convention adopted is: P(x) is positive when acting upward.

For equilibrium of couples:

$$\frac{dM(x)}{dx} = V(x)$$

where counter-clockwise moments are considered positive.

The poppet head closes against the seat and this generates a reaction force acting on the poppet. Thus, the section of the valve tip $d - c' \leq x \leq d$ is assumed to be loaded in the vertical direction by a distributed load. This distributed load may be uniformly distributed along the valve tip, or it may by non-uniformly distributed. Consider first the case of a non-uniformly distributed load. Fig. D.1 shows the distributed load to be varying linearly between $P(x) = 2F_{t,max}/c'$ at $x = d - c'$, and $P(x) = 0$ at $x = d$. The justification for assuming a linearly increasing load distribution is as follows. In practice, the tip of the valve poppet (i.e., outer rim of the poppet) will be deflected upward due to contact forces between the valve tip and its seat. If both the valve tip and the seat are made from hard materials, the deflection will cause contact at the outer edge of the valve seat to be lost. In this case, the reaction force from the valve seat can be expected to increase inward along the valve seat. Here, it is assumed that this variation in the reaction forces is linear, as illustrated in Fig. D.1.

The linear variation in the distributed load P(x) can be expressed by the following equation:

$$P(x) = \frac{2F_{t,max}}{(c')^2}(-x + d)$$

Of course, if this distributed load is summed up (integrated) between $x = d - c'$ and $x = d$, the resulting total upward load is then equal to the total downward force on the valve stem, $F_{t,max}$.

It is now possible to integrate the equilibrium equations. Mathematically, it is more convenient to consider first the outer valve portion:

Section A $\quad$ ***d-c′ ≤ x ≤ d***

$$\frac{dV(x)}{dx} = \frac{2F_{t,max}}{(c')^2}(-x + d)$$

Integrating,

$$V(x) = \frac{2F_{t,max}}{(c')^2}\left(-\frac{x^2}{2} + xd\right) + C_{A1}$$

where C_{A1} is the constant of integration. The constant C_{A1} is evaluated from the boundary condition:

at $\quad x = d \quad V(x) = 0$

This gives,

$$C_{A1} = -\frac{F_{t,max}}{(c')^2} d^2$$

Therefore,

$$V(x) = \frac{F_{t,max}}{(c')^2}\left[2\left(-\frac{x^2}{2}\right) - d^2\right]$$

The second equilibrium equation can now be integrated to obtain an expression for M(x). Substituting for V(x) in the second equilibrium equation:

$$\frac{dM(x)}{dx} = \frac{2F_{t,max}}{(c')^2}\left(\frac{-x^2}{2} + xd\right) - \frac{F_{t,max}}{(c')^2} d^2$$

Integrating,

$$M(x) = \frac{2F_{t,max}}{(c')^2}\left(-\frac{x^3}{2} + d\frac{x^2}{2}\right) - \frac{F_{t,max}}{(c')^2} d^2 x + C_{A2}$$

The constant C_{A2} is evaluated from the boundary condition

at $\quad x = d \quad M(x) = 0$

This gives,

$$C_{A2} = \frac{1}{3}\frac{F_{t,max}}{(c')^2} d^3$$

Therefore,

$$M(x)\frac{F_{t,max}}{(c')^2}\left[2\left(-\frac{x^3}{6}+d\frac{x^2}{2}\right)-d^2x+\frac{d^3}{3}\right]$$

Section B $\quad 0 \leq x \leq d\text{-}c'$

Now, $$P(x) = 0$$

Hence,

$$\frac{dV(x)}{dx}=0$$

Integrating, $$V(x) = C_{B1}$$

where C_{B1} is the constant of integration. The constant C_{B1} is evaluated by matching V(x) at the interface between Zone A and Zone B. That is, the values of V(x) for the two zones at $x = d - c'$ must be equal.

This gives,

$$C_{B1}=-\frac{F_{t,max}}{(c')^2}(c')^2$$

Therefore,

$$V(x)=-\frac{F_{t,max}}{(c')^2}(c')^2$$

Substituting for V(x) in the second equilibrium equation

$$\frac{dM(x)}{dx}=-\frac{F_{t,max}}{(c')^2}(c')^2$$

Integrating to obtain an expression for M(x),

$$M(x) = -\frac{F_{t,max}}{(c')^2}(c')^2 x + C_{B2}$$

where C_{B2} is the constant of integration. The constant C_{B2} is evaluated by matching M(x) at the interface between Zone A and Zone B. That is, the values of M(x) for the two zones at $x = d - c'$ must be equal.

This gives,

$$C_{B2} = \frac{F_{t,max}}{(c')^2}(c')^2 d - \frac{2}{3}\frac{F_{t,max}}{(c')^2}(c')^3$$

Therefore,

$$M(x) = \frac{F_{t,max}}{(c')^2}\left[-(c')^2 x + (c')^2 d - \frac{2}{3}(c')^3\right]$$

The equations for the shear stress can now be evaluated for Zones A and B using the above expressions above for V(x), M(x), Q(x), I(x), h(x), and dh(x)/dx

D.3.2 Uniformly Distributed Valve Tip Load

The analysis in the preceding section dealt with the case of a distributed load which increased linearly along the valve tip. This section will consider an alternative load distribution: the case of a uniformly distributed load along the tip (not illustrated in Fig. D.1). In many sampling-valve designs, the valve poppet is made from a hard material such as high-strength steel, while the valve seat is made of a softer material such as copper or high-purity gold. With such a combination of a hard poppet and a soft seat, the load along the valve tip is likely to be fairly uniformly distributed. The analysis below is similar to that presented in the preceding section. However, this time, the distributed load along the valve tip is assumed to be uniform.

Section A $\quad d-c' \le x \le d$

The uniformly distributed load P(x) can now be expressed by the following equation:

$$P(x) = \frac{F_{t,max}}{c'} = \frac{F_{t,max}}{(c')^2}c'$$

Hence,

$$\frac{dV(x)}{dx} = \frac{F_{t,max}}{(c')^2} c'$$

Integrating,

$$V(x) = \frac{F_{t,max}}{(c')^2} c'x + C_{A1}$$

where C_{A1} is the constant of integration. The constant C_{A1} is evaluated from the boundary condition:

at $\quad x = d \quad V(x) = 0$

This gives,

$$C_{A1} = -\frac{F_{t,max}}{(c')^2} c'd$$

Therefore,

$$V(x) = \frac{F_{t,max}}{(c')^2} c'(x - d)$$

The second equilibrium equation can now be integrated to obtain an expression for M(x). Substituting for V(x) in the second equilibrium equation and integrating:

$$M(x) = \frac{F_{t,max}}{(c')^2} c' \left(\frac{x^2}{2} - dx \right) + C_{A2}$$

The constant C_{A2} is evaluated from the boundary condition

at $\quad x = d \quad M(x) = 0$

This gives,

$$C_{A2} = \frac{F_{t,max}}{(c')^2} c' \frac{d^2}{2}$$

Therefore,

$$M(x) = \frac{F_{t,max}}{(c')^2} c' \left(\frac{x^2}{2} - dx + \frac{d^2}{2} \right)$$

Section B $\quad 0 \le x \le d\text{-}c'$

Now, $$P(x) = 0$$

Hence,

$$\frac{dV(x)}{dx} = 0$$

Integrating, $$V(x) = C_{B1}$$

where C_{B1} is the constant of integration. The constant C_{B1} is evaluated by matching V(x) at the interface between Zone A and Zone B. That is, the values of V(x) for the two zones at x = d-c′ must be equal.

This gives,

$$C_{B1} = -\frac{F_{max}}{(c')^2} (c')^2$$

Therefore,

$$V(x) = -\frac{F_{t,max}}{(c')^2} (c')^2$$

Substituting for V(x) in the second equilibrium equation and integrating to obtain an expression for M(x),

$$M(x) = -\frac{F_{t,max}}{(c')^2}(c')^2 x + C_{B2}$$

where C_{B2} is the constant of integration. The constant C_{B2} is evaluated by matching M(x) at the interface between Zone A and Zone B. That is, the values of M(x) for the two zones at $x = d - c'$ must be equal.

This gives,

$$C_{B2} = \frac{F_{t,max}}{(c')^2}(c')^2\left(d - \frac{c'}{2}\right)$$

Therefore,

$$M(x) = \frac{F_{t,max}}{(c')^2}(c')^2\left(-x + d - \frac{c'}{2}\right)$$

The equations for the shear stress can now be evaluated for Zones A and B using the above expressions for V(x), M(x), Q(x), I(x), h(x), and dh(x)/dx

D.3.2.1 Examples

The shear stress along the poppet tip of a sampling-valve was evaluated. The results of the calculations are discussed below. The geometry of the valve tip chosen as an example is summarized below, and also illustrated in the top diagram of Fig. 4.D.2:

Valve poppet diameter,	a	=	8 mm
Valve stem diameter,	b	=	6 mm
Valve seat length,	c	=	0.5 mm
Valve seat angle,	ψ	=	45°
Stress relieve angle,	θ	=	0°
Valve closing force,	$F_{t,max}$	=	600 N

D.3.2.2 Linearly Varying Load Distribution

Consider, first, the case of a linearly varying load distribution. The shear stress distribution is shown in the bottom diagram of Fig. D.2. The calculated shear stress (τ_{xy}) distribution along the valve tip is plotted at the top of this figure. It can be seen from the figure that the maximum shear

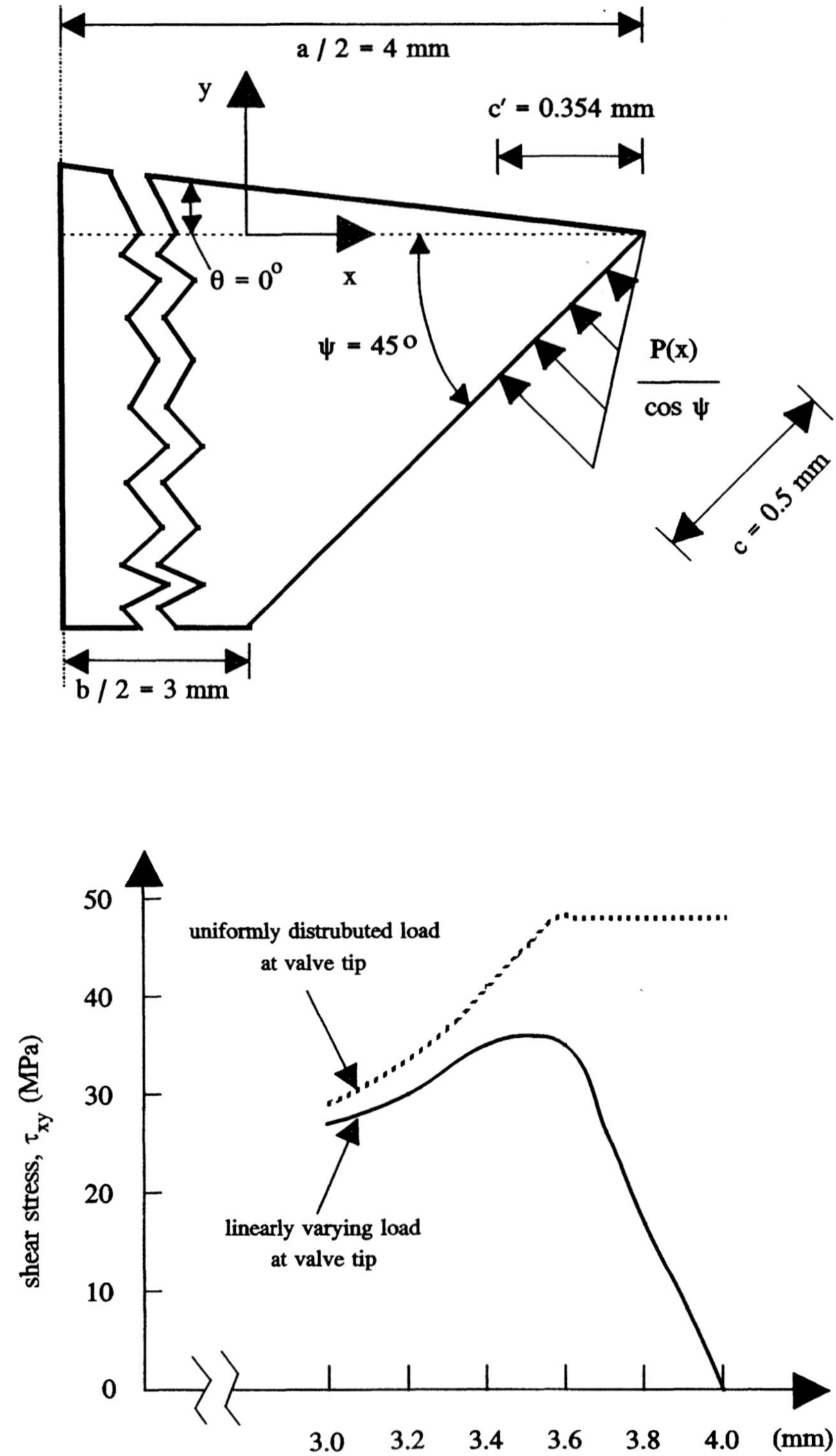

Fig. D.2 Poppet used as example for shear stress analysis.

stress is about 36 MPa, and it occurs at 3.5 mm from the center-line of the poppet, that is, at the inner edge of the valve seat, where the distributed load reaches its peak level.

The calculated maximum shear stress of 36 MPa is much smaller than, for example, the 0.2% proof yield strength of 724 MPa for high carbon AISI 1335 hardened steel.

D.3.2.3 Uniform Load Distribution

Consider now the case of the uniformly distributed load. The calculated shear stress (τ_{xy}) distribution along the valve tip is also plotted at the bottom of Fig. D.2. It can be seen from this figure that the maximum shear stress is now 33% higher, at about 48 MPa. The maximum shear stress of 48 MPa occurs along the entire portion of the valve tip that is in contact with the seat (i.e., approximately along 3.6 mm < x < 4 mm). It is interesting to note that the shear stress is constant for this portion of the valve tip. This is coincidental and it is, briefly, due to the following. Although the shear force is increasing inward from the outer edge of the poppet, so is the depth of the valve tip; as a result, the shear stress remains constant.

D.3.2.4 Effect of Valve Geometry and Closing Force on Maximum Shear Stress

Consider, again, the case of a linearly varying load distribution as shown in Fig .D.2. The shear stress (τ_{xy}) distribution along the valve tip was calculated for a number of valve geometries, and maximum values of the shear stress are listed in Table D.1.

Consider, first, the effect on maximum shear stress of increasing the valve seat length. It can be seen from the table that the maximum shear stress decreases proportionately. For example, doubling the valve seat length from 0.3 to 0.6 mm halves the maximum shear stress from 60 to 30 MPa. This is an interesting result, with a number of other implications. For example, increasing the seat length would improve the sealing of the valve against leakage of high-pressure gases in the combustion chamber.

Consider now the effect of increasing the seat angle. It can be seen from the table that the maximum shear stress decreases. For example, doubling the seat angle from 30 to 60° nearly halves the maximum shear stress from 51 to 29 MPa. However, increasing the valve seat angle reduces the valve gas flow area for a given valve lift. On the other hand, increasing the seat angle would improve the sealing of the valve against leakage of the high-pressure combustion chamber, by raising the sealing stress at the seat.

Next, consider the effect of increasing the stress relief angle, θ. The effect on the maximum shear stress is quite small. For example, tripling this angle from 5 to 15° reduced the maximum shear stress by about 15% from 33 to 28 MPa.

Finally, increasing the closing force, F_{max}, increases the maximum shear stress in direct proportion. For example, doubling the force from 500 to 1000 N increases the maximum shear stress from 30 to 60 MPa.

Table D.1 The effect of valve geometry and closing force on maximum shear stress of valve poppet

Outer diam.	Inner diam.	Seat length	Seat angle	Stress relieve angle	Maximum valve closing force	Location of maximum shear stress	Maximum shear stress
a (mm)	b (mm)	c (mm)	ψ (°)	θ (°)	F_{max} (N)	(Fig. 11) (mm)	τ_{xy} (MPa)
8	6	**0.3**	45	0	600	3.70	60
8	6	**0.4**	45	0	600	3.60	45
8	6	**0.5**	45	0	600	3.50	36
8	6	**0.6**	45	0	600	3.40	30
8	6	**0.7**	45	0	600	3.30	26
8	6	0.5	**20**	0	600	3.36	75
8	6	0.5	**30**	0	600	3.40	51
8	6	0.5	**40**	0	600	3.47	40
8	6	0.5	**45**	0	600	3.50	36
8	6	0.5	**50**	0	600	3.55	33
8	6	0.5	**60**	0	600	3.64	29
8	6	0.5	45	**0**	600	3.50	36
8	6	0.5	45	**5**	600	3.50	33
8	6	0.5	45	**10**	600	3.50	31
8	6	0.5	45	**15**	600	3.50	28
8	6	0.5	45	**20**	600	3.50	26
8	6	0.5	45	0	**400**	3.50	24
8	6	0.5	45	0	**500**	3.50	30
8	6	0.5	45	0	**600**	3.50	36
8	6	0.5	45	0	**700**	3.50	42
8	6	0.5	45	0	**800**	3.50	48
8	6	0.5	45	0	**900**	3.50	54
8	6	0.5	45	0	**1000**	3.50	60

Note: 0.2% proof yield strength for high carbon AISI 1335 hardened steel is 724 MPa.

References

LoRusso, J.A., Lavoie, G.A, and Kaiser, E.W. (1980), "An Electro-Hydraulic Gas Sampling Valve with Application to Hydrocarbon Emissions Studies," SAE Paper No. 800045.

Ryder, G.H. (1973), *Strength of Materials*, 3rd Ed., MacMillan, London.

Subject Index

Author Index

About the Authors

Dr. Hua Zhao graduated from Tianjin University, China with a BEng in Laser Technology in 1984 and received his PhD in Combustion Measurements from Leeds University. He was a College Research Fellow at Cambridge University for three years. After that he worked for one year at Imperial College of Science, Technology, and Medicine in London, and then joined the Department of Mechanical Engineering at Brunel University, where he is currently a Reader on internal combustion engines. Together with Prof. Nicos Ladommatos, Dr. Zhao is responsible for the development of engines research at Brunel, with particular emphasis on the development and application of laser diagnostics to internal combustion engines and combustion flows.

Prof. Nicos Ladommatos received a BSc in Mechanical Engineering from Queen Mary and Westfield College, London University in 1975 and his MSc and PhD from Imperial College, where he also worked for a number of years as a research assistant. Since 1985, he has been responsible for the development of engines research at Brunel. Prof. Ladommatos is the author of over 100 technical papers. He is a Fellow of the Institute of Mechanical Engineers. Currently he is the Dean of the Faculty of Technology and the head of the Department of Mechanical Engineering at Brunel University.